Advantages of Web

Gain maximum with the minimum use of time and efforts through these online contents

Image Bank—a practical approach for enhancing the subject knowledge

- The image bank consists of a variety of colourful line diagrams and images to help learn in a fun and interesting way.

- Nearly 450 simple diagrams contained in the book are incorporated, which will help to memorize the subject with ease and reproduce in the examinations.

- These diagrams are realistic to the extent that make the journey of discovering anatomy and physiology interesting as well as informative.

- These images can be downloaded which are to be used only for teaching purposes such as powerpoint presentations or other teaching aids.

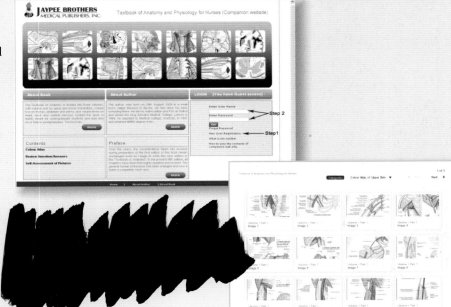

Free Access to Two Renowned Titles—avoid cramming and increase your knowledge

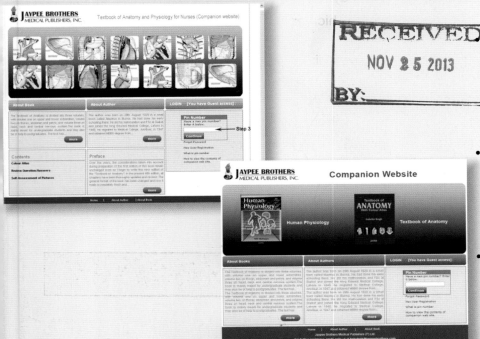

The priority of these books is on picture memory rather than a verbal one, and on understanding the facts rather than cramming, which helps in gaining in-depth knowledge of subjects.

- A very effective mode of learning anatomy and physiology through easy-to-remember illustrations, which helps in gaining in-depth knowledge of subjects.

- Help students in fast-tracking their preparation efforts by providing them textual matter which is easy-to-comprehend and reproduce during examinations.

Textbook of
Anatomy and Physiology
for Nurses

Textbook of
Anatomy and Physiology
for Nurses

Third Edition

PR ASHALATHA

BSc MBBS MS (Anatomy)
Associate Professor
Department of Anatomy
Government Medical College
Calicut, Kerala, India

Co-author
G DEEPA

MBBS MD (Physiology)
Assistant Professor
Department of Physiology
Government Medical College
Thiruvananthapuram,
Kerala, India

JAYPEE BROTHERS MEDICAL PUBLISHERS (P) LTD

New Delhi • Panama City • London

Jaypee Brothers Medical Publishers (P) Ltd.

Headquarter

Jaypee Brothers Medical Publishers (P) Ltd
4838/24, Ansari Road, Daryaganj
New Delhi 110 002, India
Phone: +91-11-43574357
Fax: +91-11-43574314
Email: jaypee@jaypeebrothers.com

Overseas Offices

J.P. Medical Ltd.,
83 Victoria Street London
SW1H 0HW (UK)
Phone: +44-2031708910
Fax: +02-03-0086180
Email: info@jpmedpub.com

Jaypee-Highlights Medical Publishers Inc.
City of Knowledge, Bld. 237, Clayton
Panama City, Panama
Phone: 507-317-0160
Fax: +50-73-010499
Email: cservice@jphmedical.com

Website: www.jaypeebrothers.com
Website: www.jaypeedigital.com

Inquiries for bulk sales may be solicited at: jaypee@jaypeebrothers.com

This book has been published in good faith that the contents provided by the author(s) contained herein are original, and is intended for educational purposes only. While every effort is made to ensure a accuracy of information, the publisher and the author(s) specifically disclaim any damage, liability, or loss incurred, directly or indirectly, from the use or application of any of the contents of this work. If not specifically stated, all figures and tables are courtesy of the authors(s). Where appropriate, the readers should consult with a specialist or contact the manufacturer of the drug or device.

Publisher: Jitendar P Vij
Publishing Director: Tarun Duneja
Editor: Richa Saxena
Cover Design: Seema Dogra

Textbook of Anatomy and Physiology for Nurses

First Edition: 2006
Reprint: 2007
Second Edition: 2010
Third Edition: **2011**

ISBN 978-93-5025-423-3

Printed at Replika Press Pvt. Ltd.

I dedicate this work to the memory of my
beloved parents Sri P S Ramachandra Shenoi
and Smt Valsala R Shenoi who were a guiding force in my life

Foreword

It is indeed with a great sense of pleasure and privilege that I give this Foreword to the *Textbook of Anatomy and Physiology for Nurses* by Dr PR Ashalatha. The Author is a dedicated teacher in the subject of Anatomy for Medical, Nursing and Paramedical students of Calicut Medical College for several years. The BSc nursing students have always appreciated the efforts of this great teacher and her ability to make Anatomy and Physiology easily comprehensible and interesting. There was a long-felt need for a comprehensive textbook which is tailored to the needs of nursing students. The book has been largely directed to the broad and specific learning needs of the Nursing Students. Simplicity and clarity have been emphasized. The students can easily assimilate the logical sequence in which the subjects have been presented not only for them to understand the same but also to perform well in the examinations.

The presentation style and the diagrammatic presentations with neatly labeled diagrams which is short and sweet that makes the subject comprehensive and revision easy. The approach utilized in dealing with the subject of Anatomy and Physiology would be appreciated by other teachers as well. I have no doubt that it is a boon to the nursing students to possess this book. It can also be used for advanced learners and teachers for quick reference in Anatomy and Physiology.

My professional colleagues and I congratulate Dr PR Ashalatha on her great effort.

M. Sham Bhat

Prof (Mrs) Mary Sham Bhat
MSc (N) MPhil (N)
HOD and Professor in Medical–Surgical Nursing
Mangalore, Karnataka, India

Preface to the Third Edition

I feel proud to bring out the third edition of *Textbook of Anatomy and Physiology for Nurses*, with several changes.

I have included new photographs of bones and Histology slides. I thank Dr S Arivuselvan, Professor and Head, Department of Anatomy, Calicut Medical College for permitting me to take photographs of bones. I also thank Dr VK Girijamony, Professor and Head, Dr Lincy Davis, Assistant Professor, Department of Anatomy, Thrissur Medical College for providing me with photographs of Histology slides.

I have added a note on development and congenital anomalies of various organs for the benefit of BSc MLT students, who also follow this textbook.

Dr Deepa has done a wonderful job in thoroughly revising several chapters in Physiology. Five new chapters namely *The Cell, Transport Across the Cell Membrane and the Resting Membrane Potential, The Nerve Physiology, The Muscle Physiology and Immunity* have been added.

Nearly 600 diagrams and 122 new photographs will make the study of Anatomy and Physiology easy and interesting.

I thank all my students and colleagues who have helped me by giving valuable suggestions for the improvement of this book.

Valuable comments and healthy criticisms are most welcome.

PR Ashalatha
ashalathapr@rediffmail.com

Preface to the First Edition

Learning Anatomy has always been a Himalayan task for the medical and paramedical students. This is partly due to the lack of a simple and complete book on the subject.

The need of a simple and "easy-to-learn" book in the field of Anatomy and Physiology was constantly pointed out to me by my students and colleagues. In fact, the seeds of this idea were sown in me by Professor Mary Shyam Bhat, presently Vice Principal of Yenapoya Nursing College, Mangalore.

Rapid advances in scientific and technological fields have led to much more being expected from a modern nurse or a paramedic. This necessitates a rigorous and advanced training and understanding of modern scientific equipment and the theories behind it, demanding thorough knowledge, and expertise, which is in no way inferior to that of a doctor. This requires a sound understanding of the basic concepts in Anatomy and Physiology.

A humble attempt is being made to fulfil the long-felt need of a complete textbook in the field of Anatomy and Physiology, mainly for nursing and paramedical students and as a foundation for the medical undergraduates too. Details are included in an organized, methodical way, in a simple and easily understandable style. All the relevant aspects of a topic like gross anatomy, histology and applied anatomy are given systematically, so that the students need not have to hunt for material in different books. There are nearly 450 simple diagrams, which will help in memorizing the subject with ease and reproduce it in the examinations.

This book can be used as a reference for the paramedical students and nurses as it contains in places more details than may be needed for the purpose of passing the examination. A few model question papers are also included.

It is the constant encouragement of my students and the staff of Medical, Nursing and Pharmacy Colleges, Calicut, which has prompted me to embark on this difficult task. This work has been done in a very short time in the midst of heavy schedule of the department and it is very likely that there may be certain omissions and discrepancies in the text. I will be too pleased and thankful to invite valuable comments and suggestions to improve the work in future editions.

I am greatly indebted to Dr Deepa, Postgraduate Student in Physiology, Calicut Medical College, for the immense help rendered in the preparation of the manuscript, diagrams and charts in Physiology.

I also keep on record the secretarial assistance given by Mr Sajeesh and Mrs Vijitha of Calicut Medical College Alumni Association, in typing this manuscript. I thank Dr Shirly Vasu and her colleagues of the Department of Forensic Medicine, Medical college, Calicut, for permitting me to take photographs of various organs.

My particular thanks are due to the printers and publishers M/s Jaypee Brothers Medical Publishers (P) Ltd, New Delhi in bringing out this work in a short duration of time. This work would not have been possible, but for the unstinted co-operation and encouragement of my family members, especially my children Ashwin and Anjali. My husband Dr Jayakumar, Consultant Surgeon, National Hospital, Calicut has taken a keen interest at every stage in the preparation of this book and has given invaluable assistance in editing the text.

I hope the students will enjoy reading this book and not only benefit in mastering the subject but also achieve success in their academic goal.

PR Ashalatha

Contents

PART 2: PHYSIOLOGY

CP1: Sphenoid bone – Anterior view

CP2A: Ethmoid bone-inferior view

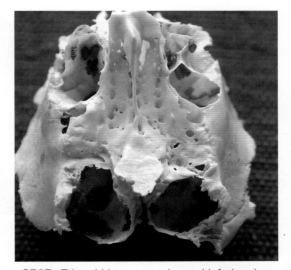

CP2B: Ethmoid bone-superior and inferior view

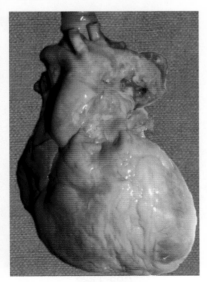

CP3: Heart

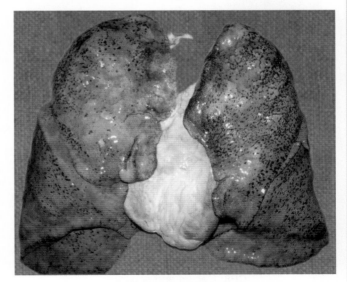

CP4: Heart and Lungs

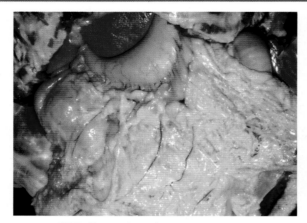

CP5: Stomach and greater omentum

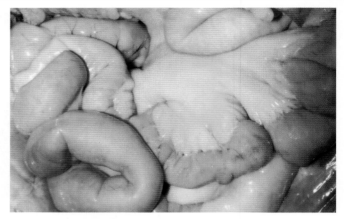

CP6: Small intestine and mesentery

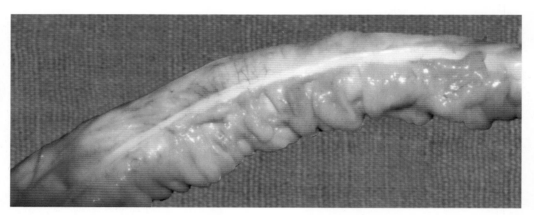

CP7: Large intestine

CP8: Cecum and appendix

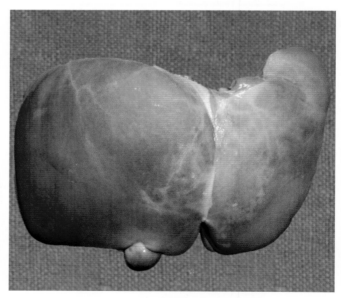

CP9: Liver

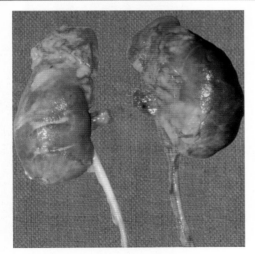

CP10: Kidneys, suprarenals and ureters

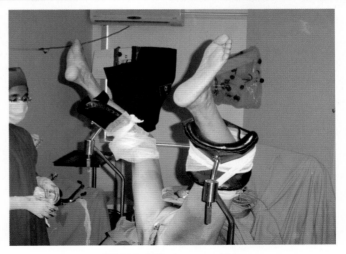

CP11A: Lithotomy position

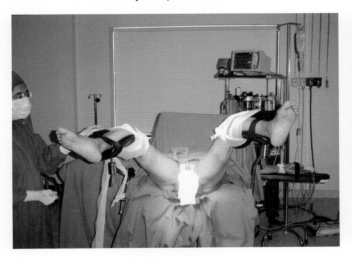

CP11B: Lithotomy position

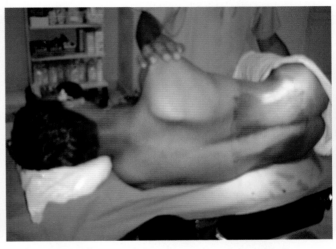

CP12: Patient positioned for spinal anesthesia

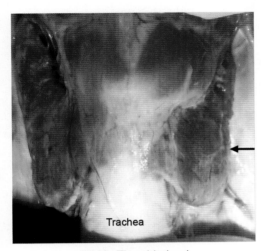

CP13: Thyroid gland

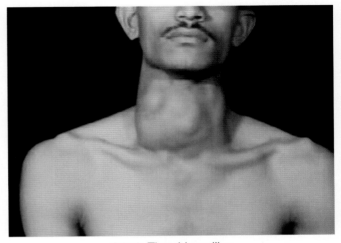

CP14: Thyroid swelling

PART 1

ANATOMY

Section A

General Anatomy

- ❖ Introduction to Anatomy
- ❖ Primary Tissues
- ❖ Special Connective Tissues
- ❖ Nervous Tissue
- ❖ Muscle Tissue
- ❖ Lymphoid Tissue
- ❖ Skin and its Appendages

1

Introduction to Anatomy

INTRODUCTION

Anatomy is the study of structure and function of the body. Aristotle (384-322 BC) was the first person to use the term "anatome", a Greek word meaning "cutting up or taking apart". The Latin word "dissecare" has a similar meaning.

Anatomy is one of the oldest basic medical sciences; it was first studied formally in Egypt. Human Anatomy was taught in Greece by Hippocrates (460-377 BC) who is regarded as the "Father of Medicine". He has written several books on Anatomy.

MEDICAL AND ANATOMICAL TERMINOLOGY

Although students entering the new world of Medicine are familiar with the common terms for many parts and regions of the body (e.g. heart, brain, liver, lung), they should learn to use the internationally adopted nomenclature, the Nomina Anatomica.

Anatomical terminology is important because it introduces the student to a large part of Medical Terminology. Since most terms are derived from Latin and Greek, medical language can be difficult at first,

but as the student learns the origin of medical terms, the words make sense.

Example: Levator palpebrae superioris muscle

Levator	= one which elevates	⎫	The muscle
Palpebrae	= eyelid	⎬	which elevates
Superioris	= superior or upper	⎭	the upper eyelid

Clear communication is fundamental in Clinical Medicine. To describe the body clearly and to indicate the position of its parts and organs relative to each other, Anatomists and Clinicians use the same descriptive terms of position and direction.

The Anatomical Position (Fig. 1.1)

All descriptions in Human Anatomy and Clinical Medicine are expressed in relation to "Anatomical Position".

A person in the anatomical position is standing erect (or lying supine) with the head, eyes and toes directed forward, the upper limbs by the sides with the palms facing anteriorly. The student must always visualize the anatomical position in his "mind's eye" when describing patients lying on their backs, sides or fronts. Always describe the body as if it were in the anatomical position.

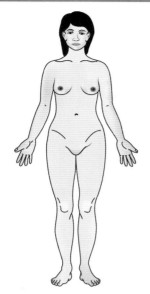

Fig. 1.1: Anatomical position

The Anatomical Planes

Anatomical descriptions are also based on four imaginary planes that pass through the body in the anatomical position. They are as follows:
1. Median
2. Sagittal
3. Coronal
4. Horizontal

The Median Plane (Fig. 1.2)

This is the imaginary vertical plane passing longitudinally through the body from front to back, dividing it into right and left halves.

The Sagittal Planes

These are parallel to the median plane. They are named after the sagittal suture of the skull (Fig. 1.3). The sagittal plane that passes through the median plane can be called the midsagittal plane; those passing parallel to the midsagittal plane and away from the median plane may be called the parasagittal planes.

The Coronal Planes

These are imaginary vertical planes passing through the body at right angles to the median plane, dividing it into anterior (front) and posterior (back) portions. These planes are named after the coronal suture of the skull, which is in a coronal plane (Fig. 1.3).

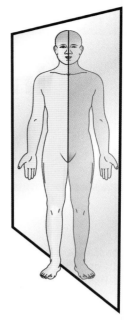

Fig. 1.2: Median plane

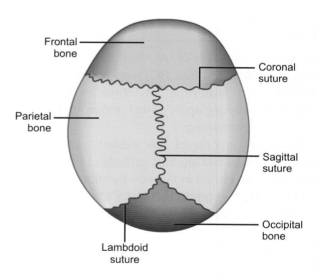

Fig. 1.3: Skull—viewed from above, showing sagittal and coronal sutures

The Horizontal Planes

These are imaginary planes passing through the body at right angles to both the median and coronal planes (They are parallel to the "horizon"). A horizontal plane divides the body into superior (upper) and inferior (lower) parts. A horizontal plane is also referred to as the transverse plane (Fig. 1.4).

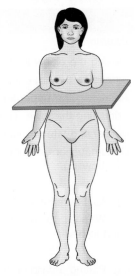

Fig. 1.4: Horizontal or transverse plane

■ TERMS

Terms of Relationship (Table 1.1)

Various terms (adjectives) are used to describe the relationship of parts of the body in the Anatomical Position.

Terms of Movement (Fig. 1.5, Table 1.2)

Various terms are used to describe the different movements of the limbs and other parts of the body. Movements take place at joints where two or more bones meet or articulate with one another.

The Meaning of Terms

Most of the anatomical terms are derived from Greek and Latin. Some of them are translated to English (e.g. musculus = muscle). Many anatomical terms indicate the shape, size, location and function or resemblance of a structure to something.

Examples

1. *According to shape*
   a. Deltoid – delta or triangular
   b. Sphenoid – wedge-shaped
   c. Styloid – pillar-shaped
   d. Uvula – grape-like
   e. Pisiform – Pea-shaped
2. *According to the number of heads of origin*
   a) Biceps – 2 heads
   b. Triceps – 3 heads
   c. Quadriceps – 4 heads

No.	Term	Meaning	Example
		Table 1.1: Commonly used terms of relationship and comparison	
1.	Superior (cranial)	Nearer to the head	The lung is superior to the diaphragm
2.	Inferior (caudal)	Nearer to the feet (tail)	The stomach is inferior to the heart
3.	Anterior(ventral)	Nearer to the front	Cornea is anterior to the lens
4.	Posterior(dorsal)	Nearer to the back	Lens is posterior to the cornea
5.	Medial	Nearer to the median plane	Heart is medial to the lung
6.	Lateral	Away from the median plane	Kidney is lateral to the vertebral column
7.	Proximal	Nearer to the trunk or point of origin	The knee is proximal to the ankle
8.	Distal	Farther from the trunk or away from the origin	The wrist is distal to the elbow
9.	Superficial	Nearer to the surface	Muscles of the thigh are superficial to the bone femur
10.	Deep	Farther from the surface	The femur is deep to the muscles of thigh
11.	External (outer)	Towards the exterior	The sclera is the external coat of the eyeball
12.	Internal (inner)	Towards or in the interior	Retina is internal to the sclera and choroid
13.	Central	Nearer or toward the center	Brain is a part of the central nervous system
14.	Peripheral	Farther or away from the center	The spinal nerves are part of the peripheral nervous system
15.	Parietal	Pertaining to the external wall of a body cavity	Parietal peritoneum lines the abdominal wall
16.	Visceral	Pertaining to the covering of an organ	Visceral pleura covers the external surface of lung

Table 1.2: Commonly used terms of movement

No.	Term	Meaning	Example
1.	Flexion	Bending or decreasing the angle between body parts	Flexing the elbow joint
2.	Extension	Straightening or increasing the angle between body parts	Extending the knee joint
3.	Abduction	Moving away from the median plane	Abducting the upper limb
4.	Adduction	Moving toward the median plane	Adducting the lower limb
5.	Rotation	Moving around the long axis	Medial and lateral rotation of UL
6.	Circumduction	Circular movement combining flexion, extension, abduction and adduction	Circumduction of upper limb e.g. bowling
7.	Eversion	Moving the sole of the foot away from the median plane	
8.	Inversion	Moving the sole of the foot toward the median plane	e.g. As if to remove the thorn
9.	Supination	Rotating the forearm and hand laterally, palm faces anteriorly. Radius lies parallel to ulna	e.g. When a person extends a hand to beg
10.	Pronation	Rotating the forearm and hand medially so that palm faces posteriorly. Radius crosses ulna diagonally	e.g. Patting a child on the head
11.	Protrusion	Moving anteriorly	Sticking the chin out
12.	Retrusion or retraction	Moving posteriorly	Tucking the chin
13.	Elevation	To lift	Elevation of eyeball to look upwards
14.	Depression	To lower	Depression of eyeball to look at the feet

3. *According to function*
   For example,
   a. Depressor anguli oris – Muscle which depresses the angle of mouth
   b. Tensor tympani – Muscle which tenses the tympanic membrane
4. *According to size*
   a. Gluteus maximus – Largest among the gluteus muscles
   b. Gluteus minimus – Smallest among the gluteus muscles
5. *According to length*
   a. Abductor pollicis longus – Long abductor of thumb
   b. Abductor pollicis brevis – Short abductor of thumb
6. *According to consistency*
   Pancreas – Pan = throughout; Kreas = Flesh; fleshy throughout
   Dura mater – Dura = tough; Mater = mother; Tough mother
7. *According to location*
   a. Biceps brachii – Biceps muscle of arm
   b. Biceps femoris – Biceps muscle of thigh
   c. Triceps suri – 3 muscles of calf

8. *According to sites of attachment*
   For example,
   a. Sternocleidomastoid muscle – attached to sternum, clavicle and mastoid
   b. Omohyoid; omos = shoulder
      Omohyoid – from scapula (shoulder blade) to hyoid.
   Some of the commonly used anatomical and clinical abbreviations are given in Table 1.3.

APPROACHES IN STUDYING ANATOMY

The three main approaches are as follows:
1. Regional Anatomy
2. Systemic Anatomy
3. Clinical Anatomy

Regional Anatomy or Topographical Anatomy

It is the study of the body by regions such as head, neck, thorax, abdomen and limbs.

Systemic Anatomy (Table 1.4)

It is the study of the body systems, e.g. digestive system, cardiovascular system, nervous system.

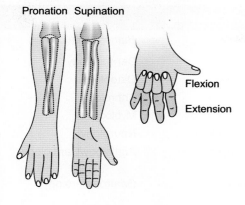

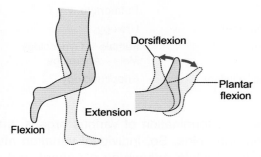

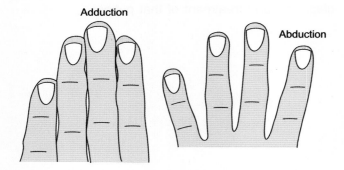

Fig. 1.5: Movements

Table 1.3: Commonly used anatomical and clinical abbreviations	
a, aa	Artery, arteries
ANS	Autonomic nervous system
A-V	Atrioventricular
C_1-C_7 (C_8)	First to seventh cervical vertebrae or 1st to 8th spinal nerves
Ca	Cancer, carcinoma
CAD	Coronary artery disease
CAT or CT	Computerized axial tomography
CN	Cranial nerve
CNS	Central nervous system
CSF	Cerebrospinal fluid
ECG	Electrocardiogram
EEG	Electroencephalogram/graphy
G; GK	Greek
GI / GIT	Gastrointestinal/gastrointestinal tract
IP	Interphalangeal
IV	Interventricular or intervertebral
IV	Intravenous
IVC	Inferior vena cava
IVF	*In vitro* fertilization
Jt	Joint
L_1–L_5	1st to 5th Lumbar Nerves/vertebrae
LA	Left atrium
LICS	Left Intercostal space
Lig	Ligament
LP	Lumbar puncture
LV	Left ventricle
m, mm	Muscle, muscles
MCP	Metacarpophalangeal
MI	Myocardial Infarction
MRI	Magnetic resonance imaging
MTP	Medical Termination of Pregnancy
MV	Mitral valve
n, nn	Nerve, nerves
PA	Posteroanterior
PNS	Peripheral nervous system, paranasal sinuses
RA	Right atrium
RV	Right ventricle
S_1-S_5	First to fifth sacral vertebrae/nerves
SA	Sinuatrial/sinoatrial
SVC	Superior vena cava
T_1–T_{12}	1st to 12th thoracic vertebrae/nerves
TIA	Transient ischemic attack
TMJ	Temporomandibular joint
V, VV	Vein/veins

Clinical Anatomy

Correlation of anatomy with clinical signs and symptoms to arrive at a diagnosis.

Gross Anatomy and Histology

Gross Anatomy

It is the examination of body structures that can be seen without a microscope.

Histology

Microscopic study of a tissue.

No.	System	Organ/Organs studied	Branch of study
	Table 1.4: Systems and their branches of study		
1.	Integumentary system	Skin	Dermatology
2.	The skeletal system	Bones and cartilages	Osteology
3.	The articular system	Joints and ligaments	Arthrology
4.	The muscular system	Muscles	Myology
5.	The nervous system	Central and peripheral nervous system	Neurology
6.	The circulatory system/ cardiovascular system	Heart, blood vessels and lymphatics	Angiology/Cardiology
7.	The digestive system	Digestive tract and glands assisting digestion	Gastroenterology
8.	The respiratory system	Air passages and lungs	Pulmonology
9.	The urinary system	Kidneys, ureters, bladder and urethra	Urology
10.	The reproductive or genital system	Genital organs—male or female	Female – Gynecology Male – Andrology
11.	The endocrine system	Ductless glands, e.g. Thyroid, pituitary	Endocrinology

Anatomical Variations

Individuals differ in physical appearance. Similarly variations are seen in the size, shape, weight, origin, course and termination of various organs, arteries, nerves and veins. So, individual variation must be considered while examining a patient and in the diagnosis and treatment of that patient.

2 Primary Tissues

- Introduction
- Definition
- Classification
- Epithelium
  - Definition
  - Meaning
  - Development
  - Endothelium
- Functions of Epithelium
- Classification of Epithelium
- Unilaminar or Simple
- Compound or Multilaminar Epithelia
- Glands
- The Connective Tissue
  - Components of Connective Tissue
  - Types of Connective Tissue

■ INTRODUCTION

The body is composed of only 3 basic elements, i.e. cells, intercellular substances and body fluids. The cells are derived from the 3 cellular layers (Ectoderm, endoderm and mesoderm) of the embryo. These cells continue to divide and gradually specialize structurally and functionally.

■ DEFINITION

A primary or basic tissue may be defined as a collection of cells and associated intercellular materials specialized for a particular function or functions. In turn, organs are formed from these tissues, and usually, all four basic tissue types are present in a single organ.

■ CLASSIFICATION

There are 4 primary tissues in the body. They are:
1. Epithelium
2. Connective tissue
3. Muscle, and
4. Nervous tissue.
   According to structure and function they are classified into subdivisions (Table 2.1).

■ EPITHELIUM

Epithelium is one of the 4 primary tissues of our body.

Definition

It is a layer or layers of cells covering body surfaces and all the body cavities opening onto it.

Meaning

Epi = above; Thelos=nipple.
Epithelium = covering of nipple
This term originally referred to the cellular covering of the nipple.

Development

The cells of the epithelium are derived from the 3 germ layers of the embryo, viz. ectoderm, endoderm and mesoderm.

Ectodermal cells give rise to epidermis, glandular tissue of breast, cornea, junctional zones of buccal cavity and anal canal.

Endodermal cells form the epithelial lining of alimentary canal and its glands, most of the respiratory tract and distal urogenital tract.

Table 2.1: Classification of tissues

1. **Epithelium**
   Covering external body surface or internal surface
   1. *Simple epithelium* (Unilaminar epithelium)
      a. Squamous, e.g. alveoli of lungs
      b. Cuboidal, e.g. collecting tubule of kidney
      c. Columnar:
         i. Ciliated, e.g. uterine tube
         ii. Non-ciliated, e.g. gallbladder
      d. Myoepithelial cells – In glands
      e. Sensory epithelium → olfactory
         Gustatory, Vestibulo-Cochlear
   2. *Pseudostratified columnar:*
      a. Ciliated, e.g. trachea
      b. Non-ciliated, e.g. male urethra
   3. *Stratified epithelium* (compound or multilaminar)
      a. *Stratified squamous:* Keratinized, e.g. thick skin (palm)
         Non-keratinized, e.g. cornea
      b. Stratified cuboidal: e.g.Sweat gland ducts
      c. Stratified columnar: e.g.Male urethra
      d. Transitional epithelium: (urothelium)–ureter, urinary bladder
   4. *Multicellular glands:*
      a. *Exocrine:* simple – e.g. gastric, sweat
         compound – e.g. salivary, pancreas
      b. *Endocrine* – e.g. pituitary, thyroid
2. **Muscle**
   1. **Smooth (involuntary)** – GIT, Blood vessels
   2. **Striated (voluntary)** – Skeletal muscles
   3. **Cardiac (striated, involuntary)** – Heart
3. **Connective Tissue**
   1. **General**
      a. *Loose:*
         i. Mesenchyme – in embryo / fetus
         ii. Mucoid – Wharton's jelly of umbilical cord
         iii. Areolar – loose packing tissue of almost all organs
         iv. Adipose – Subcutaneous tissue
         v. Reticular – Bone marrow, lymph node
      b. *Dense:*
         i. Irregular – Dermis, capsule of organs
         ii. Regular – Tendons
   2. **Special**
      a. Cartilage:
         i. Hyaline – Tracheal rings
         ii. Fibrous – Intervertebral disc
         iii. Elastic – Epiglottis
      b. Bone:
         i. Compact – shaft of long bone
         ii. Cancellous – center of long bone
      c. Hemopoietic:
         i. Myeloid – bone marrow
         ii. Lymphoid – lymphoid – spleen, lymph node
      d. Blood
      e. Lymph
4. **Nervous Tissue**
   1. Neurons
   2. Neuroglia

Table 2.2: Classification of epithelium

Unilaminar or simple	Multilaminar or compound or (Stratified)
— Simple squamous	— Stratified squamous Keratinizing
— Simple cuboidal	
— Simple columnar – ciliated – non-ciliated	— Stratified squamous non-keratinizing
— Pseudostratified	
— Myoepitheliocytes	— Transitional or Urothelium
— Sensory epithelium	— Stratified cuboidal and columnar

Mesoderm gives rise to the lining of internal cavities (pleura, peritoneum and pericardium) and a part of the urogenital tract.

Endothelium

It is the lining of blood vessels and lymphatics.

Functions of Epithelium

1. Form selective barriers.
2. Protection from dehydration, chemical and mechanical damage
3. Secretion
4. Sensory function, e.g. Pain, touch, temperature, etc.
5. Absorption.

Classification of Epithelium

See Table 2.2.

Unilaminar or Simple

Simple Squamous Epithelium (Pavement epithelium)

Simple squamous epithelium (Pavement epithelium) is composed of a single layer of flattened interlocking, polygonal cells or squames (resembling the scales of fish). Nucleus usually bulges into the overlying space (Fig. 2.1).

Sites of occurrence Cells lining the alveoli, Bowman's capsule of kidney.

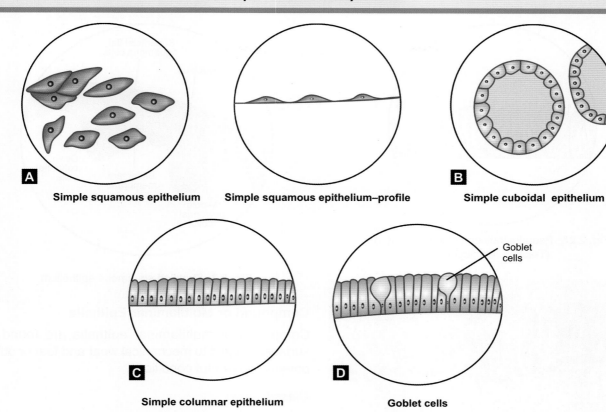

Figs 2.1A to D: Simple epithelia

Functions

1. Because it is so thin, the squamous epithelium allows rapid diffusion of gases and water.
2. Active transport of molecules.

Cuboidal and Columnar Epithelia

Cuboidal and columnar epithelia are found at sites where there is a high metabolic activity. These types consist of single, regular rows of cuboidal or columnar epithelial cells.

Cuboidal The cells are square in vertical section. Commonly, microvilli are present on their free surfaces, providing a large absorptive area.

Sites of occurrence Proximal and distal convoluted tubules, thyroid follicles.

Columnar The cells are taller than their diameter. Commonly microvilli are present. The cells may be ciliated or non-ciliated.

Sites of occurrence:

1. Most of the respiratory tract. Here the cells are ciliated.
2. Uterine tubes – ciliated columnar cells
3. Gastrointestinal tract – (major part)

Some columnar cells are *glandular,* their apices are filled with mucus, giving a characteristic appearance. They are called *Goblet cells (wine-glass),* e.g. Larges intestine—epithelium shows numerous goblet cells (Fig. 2.1D).

Pseudostratified Epithelium

Pseudostratified epithelium (Figs 2.2A and B) is a simple columnar epithelium in which nuclei lie at different levels. Not all cells extend through the whole thickness of the epithelium. Some cells constitute a basal layer. They are able to replace damaged mature cells.

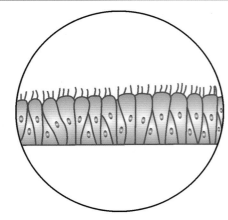

Fig. 2.2A: Pseudostratified columnar ciliated (Trachea, epididymis)

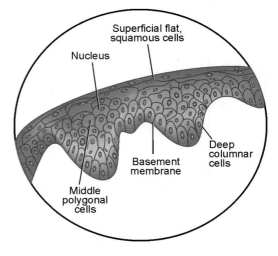

Fig. 2.3: Stratified squamous epithelium

Compound or Multilaminar Epithelia

Compound or multilaminar epithelia are found at surfaces subject to mechanical wear and tear or other potentially harmful conditions.

Classification

Compound epithelium
1. Stratified squamous
   a. Keratinizing
   b. Non-keratinizing
2. Stratified cuboidal
3. Stratified columnar
4. Transitional (urothelium).

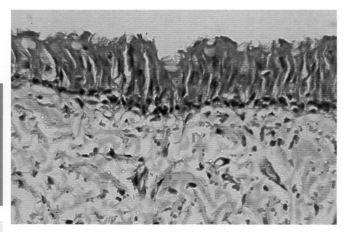

Fig. 2.2B: Pseudostratified epithelium

Sites
1. Much of the ciliated lining of respiratory tract is of pseudostratified type.
2. Sensory epithelium of olfactory area
3. Parts of male urethra.

Sensory Epithelia

Sensory epithelia are seen in olfactory (smell), gustatory (taste) and vestibulocochlear (hearing and balance) receptor systems. They are highly specialized cells.

Myoepitheliocytes or Basket Cells

Myoepitheliocytes or basket cells are specialized simple epithelial cells; they are star-shaped, containing actin and myosin filaments. They surround glands and ducts of the glands, squeezing out their contents.

For example, mammary, salivary and sweat glands.

Stratified Squamous Epithelium (Fig. 2.3)

1. Keratinizing
2. Non-keratinizing

These are multilayered epithelia in which there is a constant formation, maturation and loss of cells. The cells are formed at the innermost or basal layer; gradually move superficially and eventually shed from the surface.

The basal cells are columnar; intermediate cells are cuboidal or polygonal; superficial cells are flattened.

The 2 major types of stratified squamous epithelia are a) keratinizing or keratinized and b) non-keratinizing or non-keratinized.

Keratinized Epithelium Found at sites which are subject to drying, mechanical stresses and high levels of abrasions.

Sites

• Skin, mucocutaneous junctions of lips, nostrils, distal anal canal, outer surface of tympanic membrane, parts of gums, hard palate, etc.

The superficial cells of the epithelium are flattened; they synthesize a protein, keratohyaline. Most superficial cells lose their nuclei; the cells are dead and flattened (squames). The keratin filaments form a coat on the cells surface.

This arrangement makes this type of epithelium an excellent barrier against different types of injury and water loss.

Non-keratinized Epithelium It is present at surfaces subject to stress, but protected from drying; includes the buccal cavity, oropharynx, esophagus, part of the anal canal, vagina, distal part of uterine cervix, cornea, conjunctiva and inner surfaces of eyelids.

There are 6-10 layers of cells. Basal cells are columnar. Surface cells are flattened, resembling simple squamous cells. In between, the cells are polygonal. Except in cornea, the underlying connective tissue is raised into ridges and folds that appear like finger-like processes.

Stratified Cuboidal

Stratified cuboidal epithelium is seen in the ducts of sweat glands.

Stratified Columnar

Stratified columnar epithelium is rare. May be seen in some parts of male urethra.

Transitional Epithelium or Urothelium (Figs 2.4A and B)

Transitional epithelium or urothelium is so called, because originally it was believed to represent a transition between the stratified squamous and stratified columnar types. This type of epithelium exclusively lines the urinary tract (from renal pelvis, ureter, bladder and part of the urethra). It is exposed to internal pressure and capacity. Its appearance varies with the degree of distension.

There are 4-6 layers of cells. Basal layer of cells is cuboidal or columnar; intermediate layers are polyhedral; superficial layers are cuboidal when relaxed, and squamous , when distended. Cells of the superficial layers are often binucleate. Their cell membranes

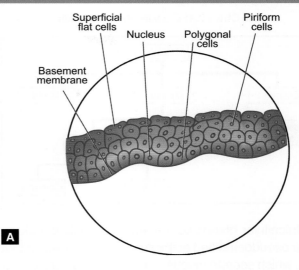

A

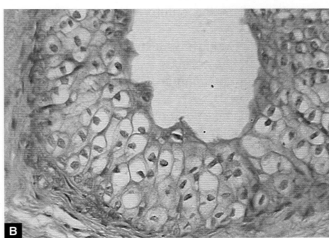

B

Figs 2.4A and B: Transitional epithelium

are protected by a glycoprotein – lipid complex. The adjacent cells are tightly held together. So, the urothelium forms an effective barrier, preventing urine from passing into the epithelium or the underlying tissues (The superficial cells are sometimes known as "umbrella cells").

Glands

In addition to protection and absorption, many cells of the epithelium secrete materials. Such cells, present singly or in groups are called glands.

Classification of Glands (Flow Chart 2.1)

Endocrine glands discharge their secretions directly into the blood stream, i.e. they are ductless glands. Exocrine glands discharge their secretions via a duct.

Flow Chart 2.1: Classification of glands

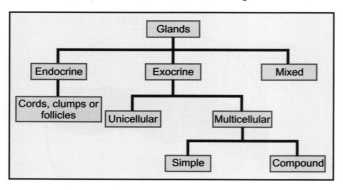

Unicellular glands lie among other cells of columnar or pseudostratified epithelium. For example, Goblet cells, which secrete mucus. There are some unicellular endocrine cells – the entero-endocrine cells in the mucosa of GIT.

The simple glands have an unbranched duct which may or may not be coiled.

The compound glands have a branching duct system.

The secreting part of a multicellular gland may be tubular or alveolar (acinar) having a flask-shaped secretory unit.

Depending on the type of secretions, the exocrine glands may be mucus secreting or serous secreting or mixed type.

Endocrine Glands

Two types of arrangement of cells are seen. In some glands, the cells are arranged in cords or clumps.

For example Pituitary gland. There are large numbers of capillaries or sinusoids which come in contact with these cells, so that, their secretions are discharged directly into the blood stream.

In some endocrine glands, as in thyroid gland, the cells form a rounded vesicle or follicle having a cavity. The secretions are stored in this cavity; when required, they are released into the blood stream.

There are some glands having both exocrine and endocrine functions, e.g. pancreas.

■ THE CONNECTIVE TISSUE

The connective tissue (CT), one of the four primary tissues of the body, connects various structures of our body with each other. It is also called the supporting

tissue or communicative tissue. It is derived from the mesodermal layer of the embryo.

Classification (Flow Chart 2.2)

Flow Chart 2.2: Classification of connective tissue

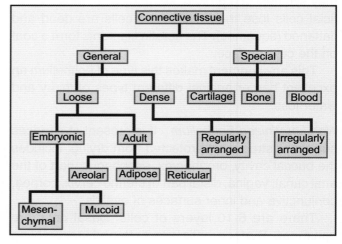

Components of Connective Tissue

Any CT is made up of 3 basic elements. They are:
1. An amorphous ground substance or intercellular-substance (It can be compared to cement)
2. Fibers (Can be compared to metal or iron rods)
3. Cells (can be compared to bricks or granite)

These 3 elements are bathed in tissue fluid or extracellular fluids.

Intercellular Substances

Intercellular substances are non-living, in which the cells live. They provide strength and support to the tissues. They act as a medium for diffusion of tissue fluid between blood capillaries and cells.

Properties
a. Transparent
b. Colorless
c. Homogeneous.

Composition: Composed of glycosaminoglycans and glycoproteins.

Types of Glycosaminoglycans

The different types of glycosaminoglycans are:
1. Hyaluronic acid
2. Chondroitin sulfate

3. Dermatan sulfate
4. Keratan sulfate
5. Heparan sulfate.

Hyaluronic acid
• Largest glycosaminoglycan
• Present in nearly all types of connective tissues.

Important sites of occurrence—Wharton's jelly, synovial fluid and vitreous body of eye.

Chondroitin sulfate is abundant in cartilage, bone and intervertebral disc.

Dermatan sulfate is abundant in skin, tendons and valves of heart.

Keratan sulfate—There are 2 types. Type I and II. Type I is exclusively present in cornea. Type II is a component of cartilage and intervertebral disc.

Heparan sulfate is found in relation to cell surfaces.
  Glycoproteins are compound molecules of proteins and polysaccharides.

Fibers of Connective Tissue (Figs 2.5A to C)

Function: To provide tensile strength and support for the tissues. There are 3 types of fibers: (i) Collagenous, (ii) reticular, and (iii) elastic.
  All are complex proteins, insoluble in water or neutral solvents.

(i) Collagen Fibers ("One which yields jelly or glue")

Characteristic features
• Found in all connective tissues
• Made up of a polypeptide, collagen
• Extremely tough.
• In the fresh state, white in color; so, they are also called "white fibers".
• Fibers are transparent, wavy, soft and flexible.
• On boiling, collagen fibers are denatured, become soft and become "gelatin".

Types of collagen fibers—mainly type I, II, III, IV, and V. See Table 2.3 for the distribution of different types of collagen fibers.

Reticular Fibers

Reticular fibers are very fine collagen fibers, arranged to form a net-like supporting framework or "reticulum".

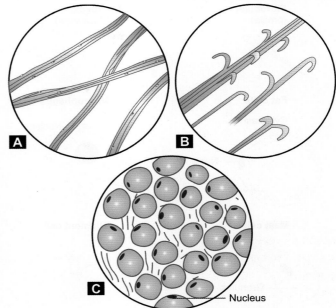

Figs 2.5A to C: Connective tissue
    A. White fibrous tissue
    B. Yellow elastic tissue
    C. Adipose tissue

Nucleus

Table 2.3: Distribution of collagen fibers	
Type	*Sites of occurrence*
I	90% of all collagen in the body Abundant in bones, tendons, dermis of skin, Teeth and practically all other CT
II	Mainly cartilage
III	Blood vessels, uterus, GIT, skin
IV	Basal laminae
V	Blood vessels, fetal membranes

Sites of occurrence Occur as fine network around muscle fibers, nerve fibers and fat cells, in the fine partitions of lung and lymphoid tissues.

Elastic Fibers

Elastic fibers are long, thin, highly refractile cylindrical threads or flat ribbons. Tissues rich in elastic fibers appear yellow in the fresh state. So they are also called yellow elastic fibers. Composed of elastin.

Sites of occurrence Walls of major blood vessels (e.g. aorta), elastic cartilage (pinna, epiglottis).

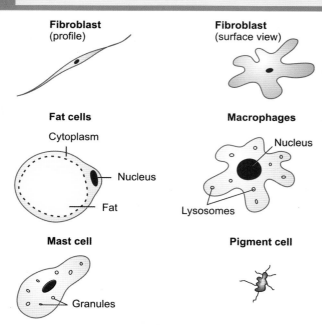

Fig. 2.6: Cells of connective tissue

Cells of Connective Tissue (Fig. 2.6)

1. Fibroblasts }
2. Macrophages } most numerous
3. Fat cells
4. Undifferentiated mesenchymal cells
5. Pigment cells
6. Blood leukocytes
7. Mast cells

Fibroblasts (Fig. 2.6)

They occur in large numbers in the connective tissue. They are responsible for the production of fibers and ground substances. They are large, flat, branching cells, (when viewed from the surface) or spindle shaped when viewed from a side (profile).

Macrophages (Fig. 2.6)

They are also called histiocytes. Seen in highly vascular areas. Two types—fixed or wandering (free) macrophages. These are the 2 functional phases of the cells of same origin. When stimulated, the fixed cells become free cells. Example of fixed macrophages and Kupffer cells of liver.

Wandering macrophages are found in blood. Macrophages are irregularly shaped, capable of free amoeboid movement and capable of phagocytosis.

Functions

1. Macrophages are important agents of defense. They act as scavengers. Several macrophages fuse together to form a multinucleated, foreign body giant cell to attack a large "foreign body".
2. Contribute to immunological reaction of the body.
3. Secrete important substances, enzymes such as lysozymes, elastase and collagenase.

Mast Cells (Fig. 2.6)

These cells occur in groups in relation to blood vessels. They are irregularly oval, with cytoplasmic granules. These granules secrete heparin, histamine and serotonin. These chemical mediators promote allergic reactions (Immediate hypersensitivity reaction).

Undifferentiated Mesenchymal Cells

Undifferentiated mesenchymal cells are developed from mesoderm. They are pleuripotent cells (capable of developing into different types of cells, when required). They are seen around blood vessels.

Fat Cells (Fig. 2.6)

Fat cells are large cells seen in groups. When seen in large groups, the tissue is called an adipose tissue. Each cell has a large glistening droplet of fat, surrounded by a thin rim of cytoplasm. Nucleus is pushed to one side and it is flattened.

Blood Leukocytes

Although leukocytes are transported by the blood stream, they perform their functions outside the blood vessels. That is why they are encountered within connective tissue. The two most frequently seen cells are lymphocytes and eosinophils.

Lymphocytes are the smallest of the free cells of connective tissue. They possess a spherical, dark staining nucleus that occupies most of the cell, surrounded by a thin rim of cytoplasm. Two types of lymphocytes are seen in connective tissue; one population with brief life span; the other, living for months or years. Functionally there are 2 types—
a. *T-lymphocytes*—They are long-lived, concerned with cell mediated immunity.

b. *B-lymphocytes* are short-lived cells, which, when stimulated by an antigen, are capable of active division and differentiation into plasma cells that synthesize antibodies against the antigen.

Eosinophils also migrate from the blood stream into the connective tissue. They are abundantly seen in the connective tissue of lactating breast, respiratory and gastrointestinal tracts. Nucleus is bilobed or kidney-shaped. Cytoplasm contains spherical granules.

Eosinophils accumulate in blood and tissues in allergic and inflammatory conditions resulting from parasitic diseases.

Pigment Cells (Fig. 2.6)

Pigment cells are seen in the skin, the choroid coat of the eye, pia mater, etc. The main pigment found in these cells is melanin.

Functions
1. Protection of deeper tissues from ultraviolet radiation (e.g. skin)
2. Prevention of light from reaching deeper tissues, e.g. choroid of eye.

Types of Connective Tissue

Depending on the proportion of cells, ground substance and fibers, different types of connective tissues are formed. The loose connective tissue has few fibers and more cells. The dense connective tissue has abundant, densely packed fibers.

Adipose Tissue

Adipose tissue is made up of fat cells, reticular fibers and large number of blood vessels.

Distribution
1. In the subcutaneous tissue all over the body, except eyelids, scrotum and penis
2. Hollow spaces like orbit, axilla (arm pit)
3. Bone marrow cavities of long bones.
4. Around abdominal organs, e.g. perinephric fat
5. Peritoneal folds, mesentery.

Functions
1. Store house of energy
2. Mechanical support for organs. It also acts as a shock absorber.
3. Insulation.

Reticular Tissue

Reticular tissue forms a framework for lymphoid organs, bone marrow and liver.

Dense irregular connective tissue is found in fasciae, periosteum, dermis of skin and capsule of organs.

Dense regular connective tissue is found in tendons and ligaments. Fibers are dense and parallel.

Section A

General Anatomy

3 Special Connective Tissues

INTRODUCTION

Special connective tissues are:
1. Cartilage
2. Bone
3. Hemopoietic tissue
4. Blood
5. Lymph

(For description of 3, 4 and 5, refer Part II – Physiology).

CARTILAGE

Cartilage is also known as "gristle". It is a modified connective tissue. It forms a part of our skeleton. In lower vertebrates, the skeleton is mainly made up of cartilage. In man, during fetal life, cartilage is more widely distributed in the skeleton.

Properties

Cartilage is firm in consistency. It is derived from mesenchyme. Newly formed cartilage grows by multiplication of cells. But, further growth takes place only by addition of new cartilage over the surface of the existing cartilage. Once destroyed by injury or disease, regeneration of cartilage is very limited. The defects are usually filled by fibrous tissue.

Replacement of cartilage by bone is called ossification. The cartilaginous parts of the fetal skeleton are gradually ossified.

As the age advances, calcium salts get deposited in some cartilages making them hard and brittle. This is known as calcification.

Cartilage is a relatively avascular tissue. It derives its nutrition from adjacent tissues mainly by diffusion or through small vessels passing through cartilage canals.

Components

Cartilage has 3 components:
1. Cells
   a. Chondroblasts—young cells
   b. Chondrocytes—mature cells
2. Ground substance
3. Fibers.

Cells

The mature cells of cartilage are called chondrocytes. These cells lie in small spaces called lacunae. The younger cells, chondroblasts are smaller and divide mitotically.

Ground Substance

The ground substance is made up of complex molecules containing proteins, carbohydrates, minerals and water. The carbohydrates are chemically glycosaminoglycans. They include chondroitin sulfate and hyaluronic acid.

Fibers

Fibers are predominantly Type II collagen fibers.
There are 3 types of cartilage:
1. Hyaline cartilage
2. White fibrocartilage
3. Yellow elastic cartilage.

Hyaline Cartilage (hyalos = glass) (Figs 3.1A and B)

It is called "hyaline" because of its glass-like transparency.

The free surface of hyaline cartilage is covered by a fibrous membrane called perichondrium (the articular cartilage, covering the bony ends, is an exception). The chondrocytes are large; they occupy the center of a mass of hyaline cartilage. The cells are usually present in groups of 2,4,8, etc. These groups, called "cell nests", are formed by division of a single parent cell. The daughter cells are prevented from moving apart by the dense matrix surrounding them.

The newly formed matrix, which stains dark blue, surrounding the lacunae is called lacunar capsule or territorial matrix. The pale-staining matrix separating the lacunae is called interstitial matrix.

A cell nest surrounded by its lacunar capsule is called a chondron.

Just beneath the perichondrium, there are numerous elongated fibroblasts.

Note: Calcification occurs only in the hyaline cartilage.

Distribution of hyaline cartilage

1. *Costal cartilages:* These are bars of hyaline cartilage, at the ventral ends of ribs connecting them to sternum or other ribs.
2. *Articular cartilage:* The articular surfaces of most synovial joints are lined by hyaline cartilage. The bony ends become smooth, so that, friction is reduced. The articular cartilage also acts as a shock absorber.

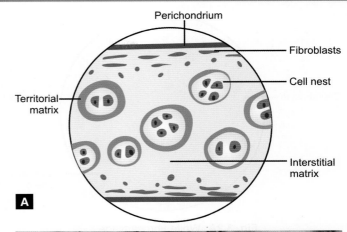

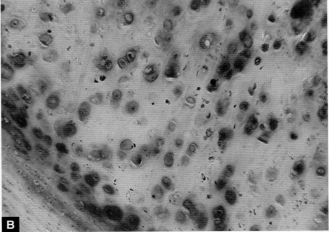

Figs 3.1A and B: Hyaline cartilage

Peculiarity: Even though, it is a hyaline cartilage, its free surface is not covered by perichondrium. It is kept moist by the synovial fluid, which provides nutrition for the cartilage.

3. Cartilages of larynx, e.g. thyroid and cricoid cartilages
4. Tracheal rings
5. Part of nasal septum (septal cartilage)
6. Parts of long bones (Epiphyseal plate – essential for bone growth) in children.

Fibrocartilage (White Fibrocartilage) (Figs 3.2A and B)

Macroscopically it looks like dense fibrous tissue. Microscopically, it is found to be made up of large bundles of Type I collagen fibers. The cells are frequently arranged in rows. There is no perichondrium. The fibers of the cartilage merge with fibers of surrounding connective tissue. This type of cartilage has great tensile strength and elasticity.

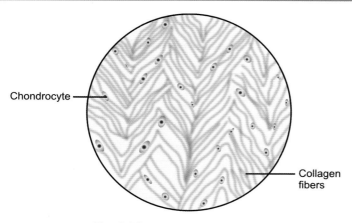

Fig. 3.2A: White fibrocartilage

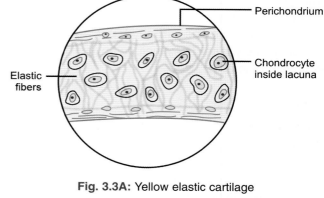

Fig. 3.3A: Yellow elastic cartilage

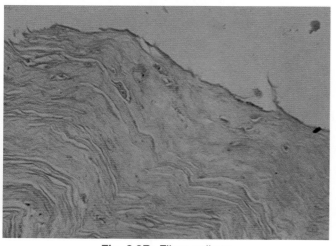

Fig. 3.2B: Fibrocartilage

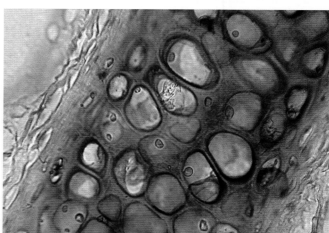

Fig. 3.3B: Elastic cartilage

Distribution

1. In secondary cartilaginous joints or symphysis, e.g. Symphysis pubis.
2. Glenoidal and acetabular labrum—A rim of fibrocartilage, which increases the concavity of articular surfaces. (Refer shoulder joint and hip joint)
3. Menisci of knee joint (C-shaped fibrocartilage plates of knee joint—Refer: knee joint)
4. Intervertebral discs.

Elastic Cartilage (Yellow Elastic Cartilage) (Figs 3.3A and B)

In fresh section, it appears yellow due to the presence of yellow elastic fibers.

The free surface is covered by perichondrium. Large chondrocytes are seen within their lacunae.

Elastic cartilage is flexible and readily recovers its shape after being deformed.

Distribution: In general, we can say that, elastic cartilage is found in sites where sound is produced (larynx) and sound is received (ear).

For example epiglottis (larynx), pinna, parts of external auditory meatus, and Eustachian tube.

For comparison of properties of 3 types of cartilages, see Table 3.1.

▋ BONE

Bone is a specialized connective tissue. It is derived from mesoderm.

Elements of Bone Tissue

1. Cells
2. Ground substance
3. Fibers.

Sl. No	Properties	Hyaline	Elastic	White fibrocartilage
	Table 3.1: Comparison of 3 types of cartilages			
1.	Perichondrium	Present	Present	Absent
2.	Cells	Occur in groups—called cell nests, inside lacunae	Large; single cells in lacunae	Rows of small cells in between fibers
3.	Fibers	Type II collagen fibers	Elastic fibers	Type I collagen fibers
4.	Calcification	+	–	–

Cells

a. Young cells—osteoblasts
b. Mature cells—osteocytes
c. Bone-removing cells—osteoclasts.

Ground Substance

Ground substance consists of proteins, carbohydrates, water and mineral salts (organic and inorganic). The inorganic mineral salts are predominantly calcium and phosphorus salts. They form 65% of the dry weight of the bone. Ninety-seven per cent of total calcium in the body is located in the bone.

Gross Structure (Fig. 3.4)

In a longitudinal section, the shaft of a long bone shows a cavity called the marrow cavity. The wall of this tubular cavity is made up of hard, dense material called *compact* bone.

The marrow cavity does not extend into the ends of long bones. These ends are filled with a network of tiny rods of plates of bones with numerous spaces in between, resembling a sponge. This kind of bone is called spongy bone or cancellous bone (cancel = cavity).

The articular ends of a long bone are covered with hyaline cartilage. The entire outer surface (excepting the area covered by articular cartilage) is covered by a fibrous membrane called periosteum. The wall of the marrow cavity is lined with endosteum.

The marrow cavity is filled with a highly vascular tissue called bone marrow.

Structure (Fig. 3.5)

Bone is made up of layers or lamellae. Each lamellus is a thin plate of bone consisting of collagen fibers and mineral salts deposited in a gelatinous ground substance.

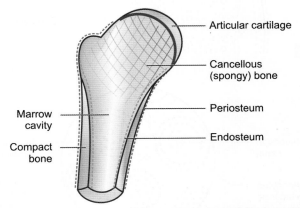

Fig. 3.4: Gross structure of long bone—longitudinal section

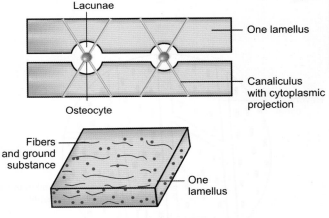

Fig. 3.5: Lamellae

Between adjoining lamellae, there are flattened spaces called lacunae, which contain osteocytes. The osteocytes communicate with each other with cytoplasmic processes, passing through minute canaliculi.

Microscopic Structure of Compact Bone (Figs 3.6A and B)

In a *transverse section* of decalcified compact bone, most of the lamellae are seen in the form of concentric rings, which surround a narrow Haversian canal. This canal transmits blood vessels, nerves and lymphatics.

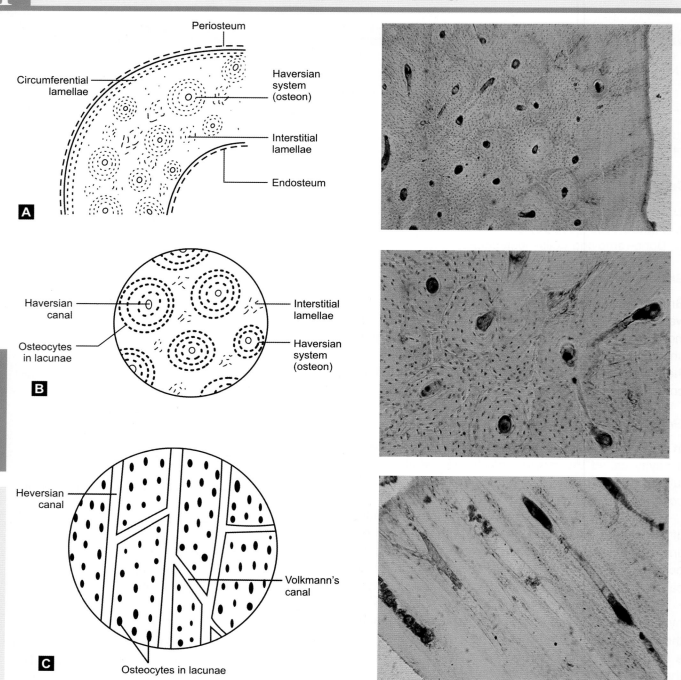

Figs 3.6A to C: Microscopic structure of compact bone (A) TS of compact bone, (B) Compact Bone TS LP, (C) Compact Bone LS

One Haversian canal and the lamellae surrounding it, constitute a Haversian system or an osteon. Compact bone is made up of several osteons.

In the angular intervals between adjoining osteons, there are interstitial lamellae; near the surface, deep to the periosteum, there are circumferential lamellae.

In a *longitudinal section* (LS) (Fig. 3.6C) the Haversian canals are found to run along the length of the bone. These canals branch and communicate with each other. They also communicate with marrow cavity and external surface of bone through minute canals called Volkmann's canals.

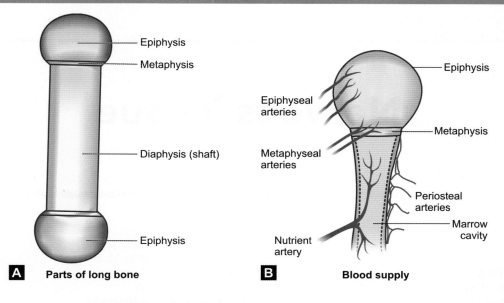

A   **Parts of long bone**   **B**   **Blood supply**

Figs 3.7A and B: Parts of long bone and its blood supply

The periosteum is a highly vascular membrane which covers the entire outer surface except the area covered by articular cartilage. It provides a medium through which tendons, ligaments and muscles are attached to the bone. It also helps in the nutrition, development and repair of bone.

Parts of a Long Bone and its Blood Supply (Figs 3.7A and B)

Bone is a highly vascular tissue. The blood supply is provided by 4 sets of vessels.
1. Nutrient artery—Pierces the shaft (diaphysis) near its middle and enters the marrow cavity through the nutrient foramen.
2. The ends (epiphysis) are supplied by epiphyseal arteries
3. Metaphyseal arteries—near the ends of shaft (diaphysis)
4. Periosteal vessels.
   All these vessels form a rich plexus. They enter the Haversian canal and Volkmann's canal.

Functions of Bone

1. Support or framework of body
2. A rich mineral store
3. Hemopoietic function.

Applied Anatomy

1. Fracture: A fracture may be a complete break in the continuity of a bone or it may be an incomplete break or crack.
   In children, bones do not contain much calcium salts. So bones are more elastic, resulting in partial fracture, described as green-stick fracture.
2. Epiphyseal injury in children will interfere with growth
3. Rickets—In calcium and vitamin D deficiency
4. Tumors
5. Bone marrow transplantation: Bone marrow is directly harvested from the red marrow in the crest of ilium, under anesthesia. The marrow is transplanted to the recipient just like blood transfusion, through intravenous route. Bone marrow transplantation is done in severe cases of diseases of bone marrow.
6. Bone marrow aspiration: To diagnose diseases involving bone marrow, e.g. leukemia, a sample is obtained from sternum or iliac crest under local/general anesthesia.

Section A

General Anatomy

4 Nervous Tissue

INTRODUCTION

The nervous tissue is made up of highly specialized tissue that has the property of conducting impulses from one part of the body to another.

The nervous tissue is composed of:
1. Neurons
2. Neuroglia.

NEURONS

Neurons (Fig. 4.1) are the functional units of nervous system. Each neuron consists of a cell body (soma or perikaryon) and a variable number of processes. Neurons vary in size, shape and other features.

The Cell Body

Like a typical cell, it consists of:
a. Cytoplasm
b. A cell membrane
c. Large central nucleus with a prominent nucleolus— usually described as an "owl's eye" appearance
d. Numerous mitochondria
e. Golgi complex
f. Nissl substance or Nissl granules—these are intensely staining rough endoplasmic reticulum
g. Neurofibrils—a network of fibrils permeating the cytoplasm.

Processes of Neuron

The processes of neuron are called neurites. They are of two types (a) Dendrites (b) Axons.

Dendrites

Dendrites are short, branching processes. They terminate near the cell body. They have numerous spine-like processes. Nerve impulses travel towards the cell body in a dendrite.

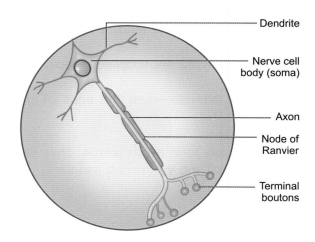

Dendrite

Nerve cell body (soma)

Axon

Node of Ranvier

Terminal boutons

Fig. 4.1: Neuron

Axon

Axon extends for a considerable distance from the cell body. Nerve impulses travel away from the cell body in the axon. The axon may or may not be myelinated. The myelin sheath is provided by Schwann cells in the peripheral nervous system and by the oligodendrocytes in the CNS.

A myelinated axon is related to a large number of Schwann cells over its length. Each Schwann cell provides a short segment of myelin. At the junction of any two segments there is a short gap in the myelin sheath called nodes of Ranvier (Fig. 4.1).

Unmyelinated axons, devoid of myelin sheath, just invaginate into the cytoplasm of Schwann cells (Fig. 4.2).

Classification of Neurons

Neurons vary in (a) size and shape of cell body; and (b) length and manner of branching processes. Accordingly, they are classified as:

a. *Multipolar neurons (Fig. 4.3):* Most common type. There is an axon and several dendrites.
b. *Bipolar neurons (Fig. 4.3):* The neuron has two processes; one afferent and one efferent, e.g. Bipolar neurons of retina, olfactory neurons.
c. *Unipolar or pseudounipolar neurons (Fig. 4.3):* The neuron has a single process, which immediately divides into two; one of these branches acts as a dendrite; the other one acts as an axon, e.g. Neurons of dorsal root ganglion.

Synapses

Synapses are sites of junctions between neurons.
There are different types of synapses.
1. Axodendritic (Axon of one neuron → Dendrite of another neuron)
2. Axosomatic (Axon of one neuron → cell body of another neuron)
3. Axoaxonal (Axon of one neuron → axon of another neuron)
4. Dendrodendritic (Dendrite of one neuron → Dendrite of another neuron).

Neurotransmitters

Neurotransmitters are chemicals released by neurons to help in the transmission of impulses at the synapse, e.g. Adrenaline, Acetylcholine.

Grey and White Matter

Sections through brain and spinal cord show whitish and darker greyish areas—they are called white matter and grey matter respectively.

Microscopic examination of grey matter shows the cell bodies of neurons and the beginning or termination of processes. The white matter contains predominantly myelinated fibers. It is the reflection of light by myelin that gives this region its whitish appearance.

Neuroglia and blood vessels are present both in white and grey matter.

Arrangement of Grey and White Matter

In a transverse section of spinal cord, the grey matter is seen to be situated inner to white matter (Fig. 4.4).

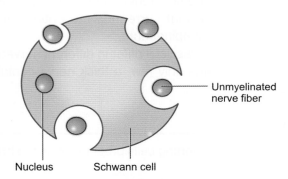

Fig. 4.2: Relationship of unmyelinated axons to Schwann cell

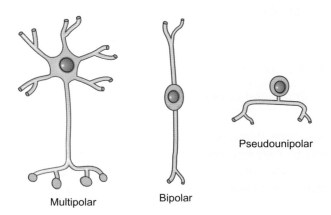

Fig. 4.3: Different types of neurons

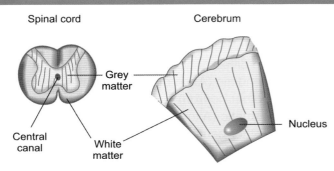

Fig. 4.4: Arrangement of grey and white matter in spinal cord and cerebrum

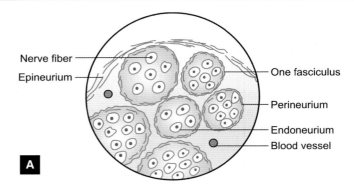

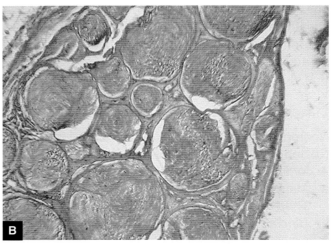

Figs 4.5A and B: Structure of peripheral nerve

In the cerebrum, grey matter is external to white matter. Situated deep in the white matter, there can be collections of grey matter. They are called nuclei (Fig. 4.4).

"Nuclei" These are groups of nerve cell bodies anywhere in the CNS "GANGLION". [Collections of nerve cell bodies outside the CNS are called "ganglia" (Singular-Ganglion)].

"TRACT" It is an aggregation or bundle of axons connecting one mass of grey matter to another mass of grey matter, e.g. Corticospinal tract connects cerebral cortex to spinal cord. Larger collections are also called funiculi, fasciculi or lemnisci.

Peripheral nerves: These are aggregations of processes of neurons outside the CNS.

They are of 2 types:
1. Efferent or motor
2. Afferent or sensory.

Efferent or motor nerve

Afferent or sensory nerve

Structure (Histology) of peripheral nerve (Figs 4.5A and B): Peripheral nerves are collections of nerve fibers (processes of neurons) outside the CNS. Each nerve fiber has a central core formed of the axon. This is

called the axis cylinder. It is surrounded by a myelin sheath. Each nerve fiber is surrounded by a layer of connective tissue called the endoneurium. It holds the adjoining fibers together and helps to form bundles or fasciculi.

Each fasciculus is surrounded by a thicker layer of connective tissue called the perineurium.

The fasciculi are held together by a dense layer of connective tissue that surrounds the entire nerve, which is called the epineurium.

The number of fasciculi differ in different nerves. A thin nerve contains only few; a thick nerve contains several fasciculi.

NEUROGLIA

Neuroglia are supporting cells present within the brain and spinal cord.

Classification

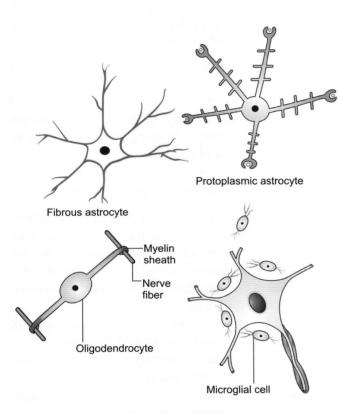

Neuroglia

Macroglia   Microglia

Astrocytes   Oligodendrocytes

Fibrous   Protoplasmic

Fibrous astrocyte

Protoplasmic astrocyte

Myelin sheath

Nerve fiber

Oligodendrocyte

Microglial cell

Fig. 4.6: Neuroglia

Macroglia

Macroglia are large glial cells. They are ectodermal in origin.

Astrocytes (Fig. 4.6): These are star-shaped cells. The protoplasmic astrocytes are seen mainly in the grey matter, whereas the fibrous astrocytes are seen mainly in the white matter.

Oligodendrocytes: These are pear-shaped cells with few processes; they form myelin sheath in the CNS. When one Schwann cell ensheaths one segment of one axon, an oligodendrocyte ensheaths several axons. Oligodendrocytes have no role in regeneration whereas, Schwann cells are actively involved in regeneration.

Microglia

Microglia are the smallest neuroglial cells; they are mesodermal in origin. The cell body is flattened; they have short processes. They are frequently associated with capillaries; they are capable of migration, proliferation and phagocytosis.

Functions of Neuroglia

1. They provide mechanical support for neurons.
2. They are good insulators; they prevent spread of impulses.
3. Phagocytosis.
4. Repair—by a process called gliosis.
5. Help in the growth of brain.
6. Help in the normal function of neurons.
7. Provide a blood-brain barrier.

> **Applied Anatomy**
>
> Tumors—Gliomas, astrocytomas, etc. can arise from neuroglia.

Section A

General Anatomy

5

Muscle Tissue

■ INTRODUCTION

Muscle tissue is composed of cells that are specialized to shorten in length by contraction. This contraction results in movement. It is in this way that practically all movements within the body or movements of the body in relation to the environment are produced.

Muscle tissue is made up of cells called myocytes or muscle fibers.

Classification

There are 3 types of muscles:
1. Skeletal muscle
2. Cardiac muscle
3. Smooth muscle.

■ SKELETAL MUSCLE

Skeletal muscle is present in the limbs and body wall. Because of its close relationship to the bony skeleton, this variety is called skeletal muscle. They form about 40% of the total body weight.

Specialized Functions of Skeletal Muscles

1. *Contraction:* It is the ability to shorten and thicken
2. *Elasticity:* The skeletal muscle can be stretched by a weight, when the weight is removed, muscle returns to its normal length.

3. Maintenance of posture: The skeletal muscles can counteract the force of gravity, thereby helping in the maintenance of posture.

Structure of Skeletal Muscle (Figs 5.1 and 5.2)

Under light microscope, the cells (muscle fibers) show prominent transverse striations. So, the skeletal muscle is also known as striated muscle. A skeletal muscle is composed of numerous elongated cells (muscle fibers). These fibers are arranged in bundles called fasciculi (Fig. 5.2).

Each fiber is surrounded by a loose areolar connective tissue sheath called endomysium.

Each fasciculus is covered by a stronger connective tissue sheath called perimysium.

The whole muscle is enveloped by a tough connective tissue sheath, the epimysium.

Muscle Fibers

Muscle fibers are elongated cells, ranging between 10-100 μ in diameter. Length varies according to size and shape of the muscle. Each fiber is surrounded by a cell membrane called sarcolemma, just beneath which lie many flattened nuclei. The cytoplasm is called sarcoplasm which is rich in mitochondria and bundles of fine longitudinal threads called myofibrils.

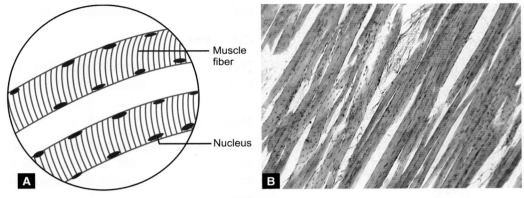

Figs 5.1A and B: Skeletal muscle fibers

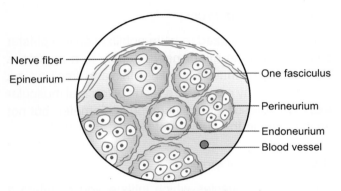

Fig. 5.2: Structure of skeletal muscle

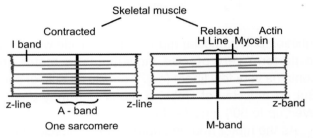

Fig. 5.3: Skeletal muscle fiber—under electron microscope

Myofibrils are the contractile units of muscle fibers. Under electron microscope, each myofibril is composed of still smaller units, the thick and thin myofilaments. The thick myofilaments are composed of a protein called myosin and thin filaments are composed of a complex protein called actin.

Group of thick myofibrils lie partially overlapping groups of thin myofibrils (Fig. 5.3) giving rise to alternate dark and light bands. This arrangement produces striations in a skeletal muscle. The darker bands are called A-bands (anisotropic) and lighter bands as I-bands (isotropic). The lighter bands are again bisected by a thin dark line called Z-line. The middle of A-band is bisected by a less dense line called as H-line. (For details, see Part II – Physiology).

Sarcomere

The part of one myofibril between the two Z-lines is called a sarcomere and it is considered as a unit of a striated muscle.

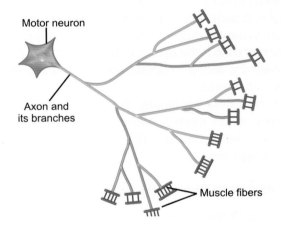

Fig. 5.4: A motor unit

Innervation of Skeletal Muscle

Each skeletal muscle is innervated by one or more nerves containing both sensory and motor nerve fibers. After entering the muscle, the nerve divides into branches to supply the muscle fibers.

A motor neuron and the muscle fibers innervated by it is called a motor unit (Fig. 5.4). The number of muscle fibers in a motor unit varies according to the

Section A

General Anatomy

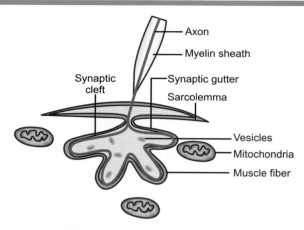

Fig. 5.5: Neuromuscular junction

function of a particular muscle. In a muscle which has to perform very precise movements, a motor unit contains very few fibers (e.g. extrinsic muscle of eye). But, in muscles of limbs, a motor unit may contain even up to 500 muscle fibers.

As the motor neuron approaches a muscle fiber, it loses its myelin sheath and branches to form a complex of nerve terminals, the motor end plate (Fig. 5.5). The motor end plate lies in an invagination of sarcolemma called synaptic gutter. The space surrounding the motor end plate is the synaptic cleft. The area of contact between the motor endplate and the muscle fiber is known as the neuromuscular junction.

In the nerve terminals, there are several small vesicles, which store a neurotransmitter acetyl choline. The skeletal muscle fibers contract in response to nervous stimuli.

Attachment of Skeletal Muscle

At the extremities of muscles, the endomysium, perimysium and epimysium fuse to form a strong fibrous non-elastic cord called tendon; it attaches the muscle to a bone.

Sometimes, the tendon forms a broad, flat expansion called an aponeurosis (e.g. aponeurosis of muscles of anterior abdominal wall).

Most skeletal muscles are attached to bone; some are attached to cartilage, ligaments or adjacent soft tissues.

Common Terms Associated with Muscles

Origin and Insertion

The fixed point of muscle attachment is called origin; movable point of attachment is called insertion.

Muscle Tone

A muscle is always partially contracted even when it appears to be at rest. So it is ready for immediate action. This state of partial contraction is called muscle tone.

Atrophy and Hypertrophy

If a muscle is not used (e.g. immobilization in a plaster cast or when the motor nerve supply to a muscle is destroyed), the muscle becomes thin and small in size.

Hypertrophy occurs as a result of forceful muscular exercise. It results in an increase in diameter; but not an increase in the number of fibers.

Agonists and Antagonists

Agonists are muscles which initiate and maintain a movement (also known as prime movers).

Antagonists are muscles which oppose a movement or reverse it.

For example, Biceps brachii muscle is an agonist; it flexes the elbow joint.

The triceps muscle is its antagonist; it extends the elbow joint.

■ CARDIAC MUSCLE (FIGS 5.6A and B)

It is present exclusively in the heart and in the beginning of large vessels arising from it. It is involuntary, striated; it has inherent rhythmic contractility. The muscle fibers are supplied by autonomic nervous system.

Adjacent muscle fibers are separated by the modification of cell membrane called intercalated discs. The nucleus of each muscle fiber is centrally placed. The cells are mainly parallel with numerous anastomoses and branching giving the false impression of a syncitial network.

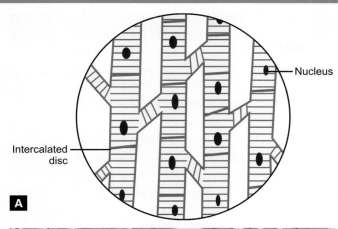

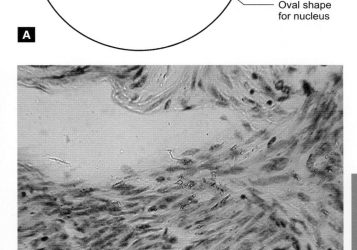

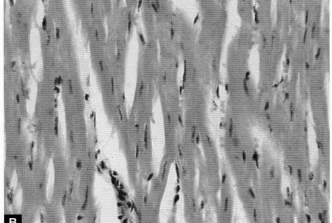

Figs 5.6A and B: Cardiac muscle

Fig 5.7A and B: Smooth muscle fibers

SMOOTH MUSCLE

Smooth muscle is present in relation to the walls of hollow viscera like stomach, intestine, etc. and in the walls of blood vessels. As the muscle fibers do not show transverse striations, it is called smooth muscle, non-striated, unstriped on plain muscle. As a rule, contraction of smooth muscle is not under our will or voluntary control. So, smooth muscle is also called involuntary muscle. It is innervated by autonomic nervous system.

The cells or myocytes are long, spindle-shaped, having a broad central part and narrow tapering ends (Figs 5.7A and B). Nucleus is oval or elongated, lies in the central part of the cell. The length of smooth muscle fibers are highly variable (15 μm to 500 μm). Under

light microscope, the cells show indistinct longitudinal striation; but there are no transverse striations.

Distribution of Smooth Muscle

1. In the walls of hollow viscera, e.g. stomach, intestines, urinary bladder (detrusor muscle), uterus.
2. In the walls of tubular structures, e.g. blood vessels, bronchi, ureters, deferent ducts, uterine tubes, ducts of several glands.
3. Muscles of iris—sphincter and dilator pupillae
4. Other regions —skin (arrector pili), prostate gland, Müller's muscle of upper eyelid, Dartos muscle of scrotum, sphincters like pyloric sphincter, sphincter of Oddi, etc.

For a comparison of the 3 types of muscles, see Table 5.1.

Section A

General Anatomy

Table 5.1: Comparison of skeletal, cardiac and smooth muscle			
Characteristics	*Skeletal muscle*	*Cardiac muscle*	*Smooth muscle*
Location	Flesh of limbs and trunk	Myocardium	Hollow viscera, Blood vessels, iris
Shape and arrangement of fibers	Elongated, parallel fibers	Parallel fibers with branching and anastomosis	Elongated, fusiform
Transverse striations	Present	Present	Absent
Nucleus	Multinucleate; peripheral	Single nucleus, central	Single nucleus, central
Sarcoplasmic reticulum	Complex; very well-developed	Less well-developed	Poorly developed
Capillary plexus	Rich	Extremely rich	Not obvious
Nerve supply	Somatic	ANS	ANS
Contraction	Powerful; voluntary	Powerful, constant, involuntary	Slow, involuntary, rhythmic

6 Lymphoid Tissue

INTRODUCTION

When blood reaches the capillaries, part of its fluid content passes into the surrounding tissues as tissue fluid. Most of this fluid re-enters the capillaries at their venous ends.

A part of this fluid, returns to the circulation through a separate system of vessels called lymphatic vessels or lymphatics.

The lymphatics from different parts of the body ultimately end in large vessels like the thoracic duct, which drains lymph into the blood stream. The thoracic duct ends at the junction of left subclavian and left internal jugular veins.

Scattered along the course of the lymphatics, there are numerous lymph nodes. They act as filters, removing bacteria from lymph.

Collections of lymphoid tissue are also found at various other sites. They are:

1. Spleen.
2. Thymus
3. Tonsils.

Collection of lymphoid tissue are found in small intestine (Peyer's patches) and appendix. Trachea and bronchi also show the presence of lymphoid tissue.

LYMPH NODES

They are small bean-shaped structures, usually present in groups along the course of lymphatic vessels.

Structure (Figs 6.1A to C)

The lymph node is bean-shaped. Its concave surface is called the hilum, through which blood vessels enter and leave the lymph node. Several lymph vessels enter the node on its convex surface.

The lymph node is covered by a connective tissue capsule. A number of septa extend into the node from the capsule, dividing the node into a number of lobules.

Lymph node has an outer *cortex* and inner *medulla*.

The Cortex

The cortex is made up of densely packed lymphocytes and stains darkly. There are several rounded masses

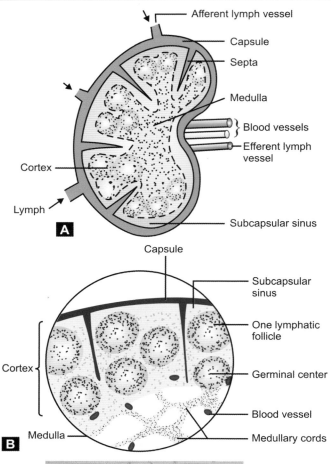

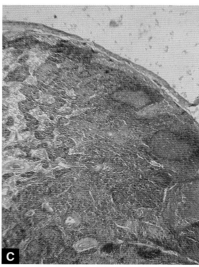

Figs 6.1A to C: Structure of lymph node

of lymphocytes, called lymphatic follicles or lymphatic nodules. Each nodule has a paler germinal center, surrounded by dark staining, densely packed lymphocytes.

The Medulla

In this zone, the lymphocytes are arranged in the form of branching and anastomosing cords.

The remaining space within the node is filled by a network of reticular fibers.

Cells of Lymph Node

The cells are predominantly lymphocytes. Both B-lymphocytes and T-lymphocytes are present in lymph nodes.

The lymphatic nodules of the cortex are mainly composed of B-lymphocytes. Their pale germinal centers are made up of lymphoblasts.

The medulla has both B-lymphocytes and T-lymphocytes.

Other Cells

Macrophages and fibroblasts.

Circulation of Lymph through Lymph Node

There are endothelium-lined spaces called sinuses, within the lymph node. They are: (1) The subcapsular sinus – just beneath the capsule; (2) cortical sinuses– in the cortex, and (3) Medullary sinuses, in the medulla.

Afferent lymph vessels reach the convex, outer surface of the lymph node. The lymph reaches the subcapsular sinus, from where the lymph goes to cortical sinuses, then medullary sinuses. Here they join to form one or two efferent vessels which come out of the hilum.

When lymph flows through the lymph node, it comes into contact with the macrophages which remove bacteria and other particulate matter.

Functions of Lymph Nodes

1. Production of lymphocytes
2. Phagocytosis of "foreign bodies"
3. Production of antibodies (by B-lymphocytes).

Applied Anatomy

1. Inflammation of lymph nodes is called lymphadenitis. Infection of any part of the body leads to enlargement and inflammation of the lymph nodes draining the area.
2. Metastasis of cancer cells from primary site to lymph nodes can occur.
   There are 3 sets of lymph nodes which are clinically important.
   a. The cervical lymph nodes
   b. The axillary lymph nodes
   c. Inguinal lymph nodes

The Cervical Lymph Nodes (Figs 6.2A and B)

The entire lymph from the head and neck drains ultimately into the deep cervical nodes either directly or through superficial nodes (peripheral nodes).

The superficial cervical lymph nodes are submental, submandibular, buccal, preauricular, post-auricular and occipital.

The deep cervical lymph nodes form a vertical chain situated along the entire length of internal jugular vein. They are grouped into upper deep cervical and lower deep cervical nodes. *Jugulodigastric node* is a member of the upper deep cervical nodes. It lies between the angle of the mandible and the anterior border of sternocleidomastoid muscle. It is the main node draining the *palatine tonsil.*

The *jugulo-omohyoid* node is a member of the lower deep cervical nodes. It lies under cover of the sternocleidomastoid muscle (above the intermediate tendon of omohyoid muscle). It is the main lymph node of the tongue.

The *supraclavicular nodes* also belong to the lower deep cervical group of lymph nodes. In addition to the structures in the neck, the mammary glands also drain lymph into these nodes. The left supraclavicular nodes are clinically important. They are called the *"Sentinel Nodes"*. The thoracic duct, draining lymph from both halves of the body below the diaphragm and left half of the body above the diaphragm ends at the junction of left subclavian vein and left internal jugular vein. Due to the proximity of the left supraclavicular nodes to the thoracic duct, cancer cells coming from the drainage area of the thoracic duct may reach the left supraclavicular nodes. For example, the only symptom in a patient with cancer of the stomach may be enlarged, left supraclavicular nodes (Troisier's sign).

Efferents from the deep cervical lymph nodes join to form jugular trunks. The left jugular trunk opens into the thoracic duct; the right jugular trunk opens at the junction of right subclavian vein and right internal jugular vein.

The Axillary Lymph Nodes

These lymph nodes, about 20-30 in number, drain the lymphatics from:
1. The upper limb
2. Most of the lymph from the mammary gland

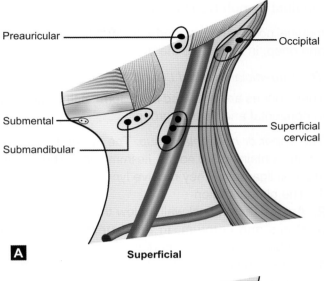

A   **Superficial**

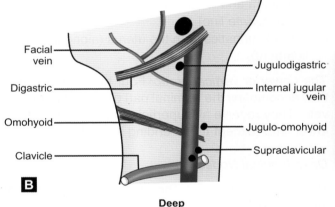

B

Deep

Figs 6.2A and B: Cervical lymph nodes

3. Cutaneous lymph from the trunk, above the level of umbilicus and below the level of clavicle.

They are divided into 5 groups – anterior, posterior, lateral, central and apical. The apical nodes, situated at the apex of axilla, medial to the axillary vein receive lymph from the other 4 groups, lymphatics of upper limb accompanying the cephalic vein and upper part of mammary gland.

Efferents from the apical group form the subclavian trunk which may end at the junction of subclavian vein and internal jugular vein or into the thoracic duct.

Lymphatics from one mammary gland drains into the axillary nodes of same side and also, to the axillary nodes of the opposite side, so, in suspected cases of Ca breast, both axillae should be examined, for enlarged axillary nodes.

Section A

General Anatomy

Inguinal Lymph Nodes (Fig. 6.3)

Inguinal nodes are arranged in two groups – superficial and deep, in relation to the deep fascia of thigh.

The Superficial Inguinal Lymph Nodes

These nodes are arranged in the form of the letter "T", in horizontal and vertical groups.

The upper or horizontal group: The upper, horizontal group, containing 5-6 nodes, lie immediately below the inguinal ligament. They receive lymph from:
1. The gluteal region
2. Anterior abdominal wall, below the umbilicus
3. Penis and scrotum, in male (testis drains into lumbar or para-aortic nodes)
4. Vulva and vagina (below the hymen) in female
5. Perineum and lower part of the anal canal (below the pectinate line)
6. A few lymphatics from the uterus (along the round ligament).

The Lower or vertical group: Four to five in number, they accompany the terminal part of great saphenous vein. They receive all superficial lymph vessels from lower limb (except those ending in popliteal nodes).

Deep Inguinal Nodes

They are 1-3 in number. They lie on the medial side of femoral vein. They receive: (1) lymph vessels accompanying the femoral vessels; (2) glans penis or clitoris; (3) a few lymphatics from superficial inguinal nodes.

Applied Anatomy

The *upper group* of lymph nodes may be enlarged in infection or malignant growth extending from the lymphatic territory drained by these nodes, i.e. infra-umbilical part of anterior abdominal wall, gluteal region, perineal region including external genitalia, vagina (below hymen) anal canal (below pectinate line) and the uterus. The *lower groups* are enlarged in the disease of the lower limb.

■ THE SPLEEN (GK: SPLEN, L: LIEN) (FIG. 6.4)

The spleen is the largest single mass of lymphoid tissue in the body. It is red-brown in color, soft and friable.

The spleen is situated in the left hypochondrium opposite the levels of 9th, 10th and 11th ribs, its long axis is parallel to the 10th rib.

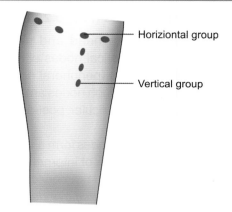

Fig. 6.3: Superficial inguinal lymph nodes

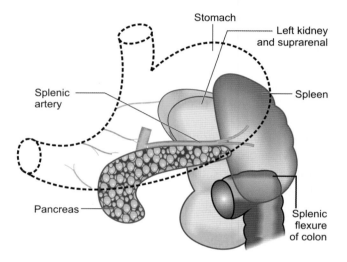

Fig. 6.4: Relation of spleen—visceral surface

It is a wedge-shaped organ. It is about 1 inch thick, 3 inches broad, 5 inches long, 7 ounces in weight and is related to 9th to 11th ribs (Harris' dictum).

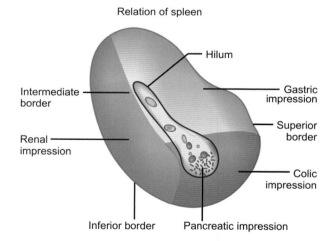

External Features

The spleen has 2 ends, 3 borders and 2 surfaces. The anterior end is expanded. It is directed downwards, forwards and laterally.

The *posterior end* is rounded. It is directed upwards, backwards and medially and rests on the upper pole of left kidney.

The *superior border* shows notches; one of them, called the splenic notch, is prominent, it lies near the anterior end. The *inferior border* is rounded. The *intermediate border,* short and rounded, is directed to the right.

The outer, *diaphragmatic surface* is convex and smooth. The *visceral surface* shows the impressions of adjacent viscera (a) the *gastric impression* produced by the fundus of the stomach, lies between the superior and intermediate borders. It is the most concave and largest visceral impression; (b) the *renal impression*, for the left kidney, is seen between the intermediate and inferior borders; (c) *colic impression* for the splenic flexure of colon; close to the anterior end of spleen; (d) the *pancreatic impression*; for the tail of pancreas, close to the hilum of spleen.

The *hilum* of the spleen is inferomedial to the gastric impression. It transmits the splenic vessels and nerves. It also provides attachment to the gastrosplenic and lienorenal (splenorenal) ligaments.

The gastrosplenic ligament, is a peritoneal fold, extending between hilum of spleen and greater curvature of stomach, containing short gastric vessels, lymphatics and nerves. The lienorenal ligament, is also, a peritoneal fold, extending from the hilum of spleen to the kidney; it contains the tail of the pancreas, the splenic vessels and lymph nodes.

The anterior end of the spleen is supported by a fold of peritoneum extending from the splenic flexure of colon to the diaphragm. This fold is called the *phrenico-colic ligament*. It prevents the vertical descent of spleen in splenomegaly. So, the spleen tends to move towards the right iliac fossa.

The diaphragm separates the diaphragmatic surface of spleen from the left costodiaphragmatic recess, the lung and ribs.

Blood Supply

Arterial Supply

Arterial supply is by the splenic artery, the largest branch of coeliac trunk. It is tortuous; passes through the lienorenal ligament to the hilum, where the artery divides into 4-5 branches.

Venous Drainage

The splenic vein emerges from the hilum of the spleen. It joins with the superior mesenteric vein behind the neck of the pancreas to form the portal vein.

Structure (Histology) (Figs 6.5A and B)

1. Outer serous layer, made of peritoneum.
2. Deep to it, a thick fibrous capsule. The capsule sends numerous trabeculae into the substance of the spleen.

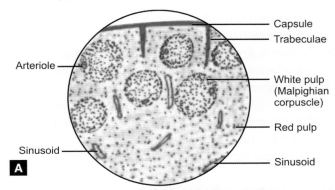

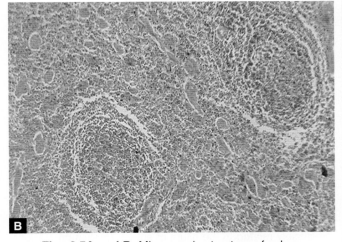

Figs 6.5A and B: Microscopic structure of spleen

3. The *white pulp;* It is made of lymphatic tissue (lymphocytes) which form dense collections around the arterioles. They form cord-like aggregations, and follow the branching pattern of the arterioles. At regular intervals, this collection of lymphoid tissue shows nodular swellings called Malpighian corpuscles or splenic corpuscles. In a section, each corpuscle is seen as a circular mass of dense collection of lymphocytes, with an eccentrically placed arteriole. These are predominantly B-lymphocytes. The remaining white pulp is made up of T-lymphocytes.

4. The *Red pulp:* It fills the rest of the spleen; contains blood, RBC, macrophages and lymphocytes. There are a number of sinusoids, lined by reticulo-endothelial cells, in spleen.

Functions of Spleen

1. Both B-lymphocytes and T-lymphocytes multiply in the spleen and play an important role in immune responses.

2. Spleen is an important component of the reticulo-endothelial system. It contains a large number of macrophages. Their main function is the destruc-tion of RBC that have completed their life span. The macrophages also destroy worn out leukocytes and bacteria.

3. *Hemopoiesis,* during fetal life.

Applied Anatomy

1. Enlargement of spleen is called splenomegaly. A pathologically enlarged spleen extends downwards and medially; the direct downward displacement is prevented by the phrenico-colic ligament. The enlarged spleen can be palpated under the left costal margin. The splenic notch can be felt easily. The spleen becomes just palpable when it has enlarged twice its normal size. Start palpating from right iliac fossa.

2. In obstruction of the branches of splenic artery, splenic infarction results; this causes referred pain to the left shoulder (Kehr's sign).

3. Rupture of spleen is common, in blunt injuries to the abdomen. There will be massive hemorrhage; treatment is splenectomy. While doing splenectomy, care should be taken to avoid injury to the tail of the pancreas which is closely related to the hilum of spleen (maximum number of Islets of Langerhans are situated in the tail of the pancreas).

THE TONSIL (PALATINE TONSIL)

At the entrance into the alimentary tract, there are collections of lymphoid tissue, which are collectively called the "Waldeyer's ring". The following lymphatic collections form the Waldeyer's ring (Fig. 13.10).

1. Posteriorly, nasopharyngeal tonsil
2. Posterolaterally, tubal tonsils
3. Laterally, the palatine tonsils
4. Ventrally, the lingual tonsil.

The tonsils are the first line of defense of the body against bacterial invasion. They differ from a lymph node in having an incomplete capsule and no lymph sinus, so that, the tissue fluid is filtered directly.

The Palatine Tonsils or 'the Tonsils"

They are the largest collections of lymphatic tissue in the Waldeyer's ring.

The tonsils are situated one on each side of the lateral wall of oropharynx, in the *tonsillar fossa.* The tonsillar fossa is bounded anteriorly by the palatoglossal fold and posteriorly by the palatopharyngeal fold.

The *tonsillar bed* is formed by (1) the pharyngo-basilar fascia; (2) palatoglossus muscle and (3) the superior constrictor muscle of pharynx.

The tonsil is almond-shaped. It has 2 poles—upper and lower; 2 surfaces—medial and lateral and 2 borders—anterior and posterior.

Capsule: It covers only the lateral surface of the tonsil. It is formed by the pharyngobasilar fascia. It sends septa into the tonsil, which conducts blood vessels and nerves into it.

Relations: The *medial surface* is covered by mucous membrane of the oral cavity, with stratified squamous epithelium. There are a number of tonsillar crypts (pits) on this surface.

Laterally, the structures in the tonsillar bed and the structures attached to the styloid process are related to the tonsil. There is a deep, *intratonsillar cleft* at the upper pole (also called supratonsillar fossa). It is the remnant of 2nd pharyngeal pouch, from which the tonsil develops.

Blood Supply

Arteries: The tonsillar branch of the facial artery is the main artery of the tonsil.

Tonsillar branches of lingual and ascending palatine arteries supply the mucous membrane, capsule and adjacent muscles.

Venous Drainage

The veins form a plexus; they may end in lingual vein or pharyngeal plexus of veins. A large *paratonsillar vein* descends from the soft palate to the pharyngeal plexus across the upper part of the tonsil. It is usually a source of bleeding in tonsillectomy.

Nerve Supply

Branches from lesser palatine and glossopharyngeal nerves go to the upper and lower poles respectively.

Lymphatic Drainage

Mainly into jugulodigastric lymph node.

Histology (Figs 6.6A and B)

1. The mucous membrane: It is the continuation of mucous membrane of the oral cavity. It is lined by the stratified squamous epithelium. Tonsillar crypts can be seen as cleft-like spaces.
2. The lymphocytes form dense collections and form the main part of the tonsil.
3. Mucous secreting acini are seen in the connective tissue, beneath the epithelium.

Applied Anatomy

1. The tonsils are large in children. The size decreases gradually after puberty.
2. Infection of tonsils—tonsillitis—is common in children. Infection may spread to the middle ear, causing otitis media.
3. Tonsillectomy is the surgical removal of tonsils. The paratonsillar vein may cause bleeding during and after surgery.
4. Intratonsillar cleft is often the site of beginning of inflammation, which can lead to a peritonsillar abscess or quinsy.

■ THE THYMUS

The thymus is a bilobed, roughly pyramidal mass of lymphoid tissue, situated in the superior mediastinum (it may extend into the anterior mediastinum and also into the root of the neck).

At birth the thymus is relatively large; its size increases gradually, till puberty. Then it atrophies gradually, getting infiltrated by fatty and fibrous tissue.

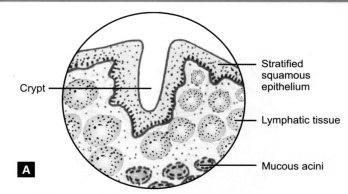

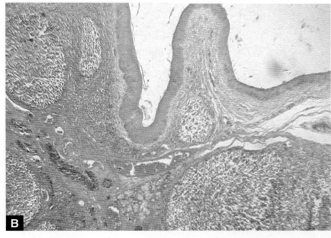

Figs 6.6A and B: Microscopic structure of palatine tonsil

The thymus is situated behind the manubrium sterni, and anterior to the aortic arch and its branches. It is supplied by branches of internal thoracic and inferior thyroid arteries. The veins drain into the corresponding veins.

Histology (Figs 6.7A to C)

1. There is a fibrous *capsule;* a number of septa arise from the capsule and divide the thymus into numerous lobules;
2. Thymus lobule has an outer *cortex* and inner *medulla.*

   In the *cortex,* there are dense collections of lymphocytes, supported by a branching network of reticular cells (large branching cells).

   In the *medulla* the lymphocytes are fewer in number. There may be Hassall's corpuscles, in the medulla. The Hassall's corpuscle has a central, homogeneous material, surrounded by concentric layers of flattened epithelial cells.

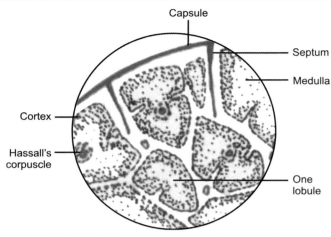

Fig. 6.7A: Structure of thymus diagrammatic

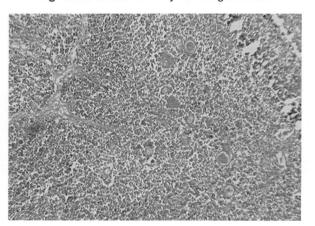

Fig. 6.7B: Under low power

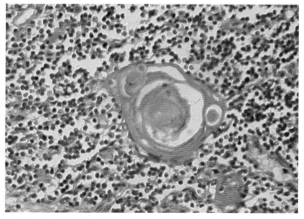

Fig. 6.7C: Hassall's corpuscle (Under high power)

Lymphocytes of Thymus (Thymocytes)

The cortex in each lobule of the thymus is densely packed with lymphocytes. The stem cells formed in the bone marrow travel to the thymus. They come to lie in the superficial part of the cortex. They divide repeatedly and small lymphocytes are formed which move to the deeper layers of cortex and finally reach the medulla; ultimately they leave the thymus by passing into blood vessels and lymphatics.

Functions of Thymus

1. The stem cells coming to the thymus from the bone marrow, mature in the thymus and become immunologically competent, i.e. they react only against proteins foreign to the body. These lymphocytes are thrown into the circulation. They lodge themselves in lymph nodes and spleen.
2. Thymus produces a number of hormones like thymulin, thymopoietin and thymosin.

Applied Anatomy

Enlargement of thymus and its tumor may be associated with an autoimmune disease called myasthenia gravis, in which there is great weakness of skeletal muscles. Removal of thymus may result in improvement, in some cases.

■ THE THORACIC DUCT (FIG. 6.8)

The thoracic duct is a great lymph channel which conveys chyle and greater part of the lymph into the venous system.

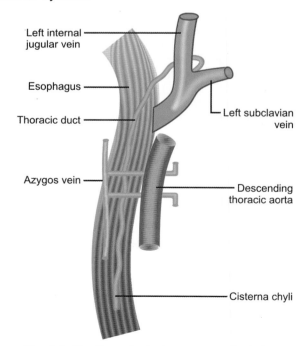

Fig. 6.8: The thoracic duct—course and relations

Areas of Drainage

The thoracic duct drains lymph from (roughly) all parts of the body except the right half of the body above the diaphragm.

It starts from the upper end of the cisterna chyli (an elongated lymph sac, about 5-7 cm, situated in front of L1 and L2 vertebrae). It enters the thorax through the aortic opening of the diaphragm, and ascends in the posterior mediastinum, and then, the superior mediastinum and ends in the root of the neck by joining the junction of left subclavian vein and left internal jugular vein. It is 45 cm long. It has a beaded appearance due to the presence of valves.

Applied Anatomy

Malignant tumor cells pass from an abdominal organ (or any one of the drainage areas), through the thoracic duct into the root of the neck. Some tumour cells enter the venous system; some cells enter the left supraclavicular nodes (by retrograde permeation). The cancer cells form new malignant tumors here. Thus, the left supraclavicular nodes may be enlarged when there is a cancer of the bronchus, stomach or any other abdominal organ. The supraclavicular lymph nodes are therefore referred to as "sentinel lymph nodes", because their enlargement alerts the examiner to the possibility of malignant disease in the thoracic or abdominal organs.

Section A

General Anatomy

7

Skin and its Appendages

■ INTRODUCTION

The skin is an extensive organ which forms the outer covering of the body. It is continuous with mucous membrane lining the body orifices.

The skin has 2 layers (Figs 7.1A to C). (1) A superficial layer – the *epidermis* – made up of stratified squamous epithelium; and (2) a deeper layer, the *dermis* made up of connective tissue.

The junction between these two layers is markedly wavy because of the presence of finger-like extensions from the dermis – the dermal papillae (Fig. 7.1B).

■ THE EPIDERMIS

The epidermis consists of stratified squamous keratinizing epithelium; there are 5 layers. From deep to superficial, they are:
1. Stratum basale
2. Stratum spinosum
3. Stratum granulosum
4. Stratum lucidum
5. Stratum corneum

Stratum Basale

It is the deepest, basal layer of epidermis. It is made up of a single layer of columnar cells that rest on a basal lamina. These cells undergo mitotic divisions and give off cells called keratinocytes, which form the more superficial layers of the epidermis. The basal layer, is therefore, also known as the germinal layer or stratum germinativum.

Stratum Spinosum

Consists of several layers of polygonal keratinocytes. These cells appear to have a number of "spines". That is why this layer is called stratum spinosum.

Stratum Granulosum

Consists of 1-5 layers of flattened cells, containing deeply staining granules in their cytoplasm. The granules consist of a protein called keratohyalin.

Stratum Lucidum (Lucid = Clear)

This layer appears clear, homogeneous, with distinct cell boundaries.

The Stratum Corneum

It is the most superficial layer of the epidermis. The cells or corneocytes are dead; they have lost their nuclei and organelles. These cells are extremely flattened; they are scale-like (squames). These cells

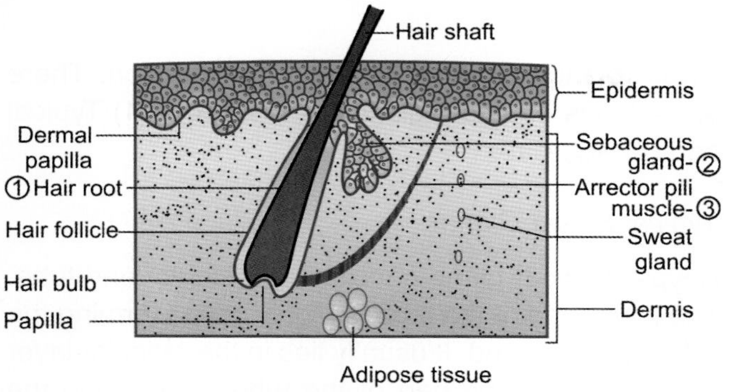

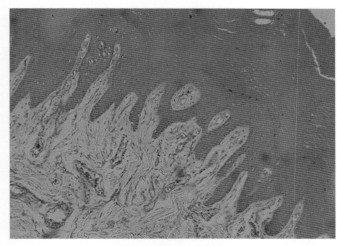

Fig. 7.1A: Basic structure of skin

THE DERMIS

This layer is made up of connective tissue. Just deep to the epidermis, the connective tissue is dense. It is known as the papillary layer. Deep to this layer, the connective tissue fibers form interlacing, network; it is known as the reticular layer.

There are capillary loops, tactile corpuscles, collagen fibers, elastic fibers, adipose tissue and plexuses of nerve fibers in the dermis.

FUNCTIONS OF SKIN

1. The skin provides mechanical protection for underlying tissues.
2. The skin acts as a physical barrier against the entry of microorganisms and chemicals.
3. Prevents loss of water. The major cause of death following burns, is dehydration.
4. *Protection from UV radiations:* The cells of the basal layer and stratum spinosum contain a brown pigment called melanin. This pigment protects the underlying tissues from the harmful effects of ultraviolet light.
5. It forms an important sensory organ.
6. Skin has an important role in temperature regulation; there are 2 mechanisms (1) by controlling the blood flow (2) by the production of sweat.
7. Storage of water and fat.
8. Formation of Vitamin-D.
9. Absorption of some molecules; this is useful in the administration of certain drugs, e.g. nitroglycerine "patch" in angina.

APPENDAGES OF SKIN

They are:
1. Hairs.
2. Nail.
3. Sweat glands.
4. Sebaceous glands.

Hairs

Hairs are present on the skin covering almost all parts of the body. The sites where hairs are absent are palms, soles, sides of the fingers and some parts of the external genitalia.

The length, texture and distribution of hairs are different in males and females.

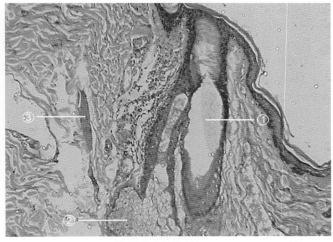

Fig. 7.1B: Thick skin

Fig. 7.1C: Thin skin

contain keratin. They are held together by a layer of lipid, which makes the layer "waterproof".

In palms and soles, where the skin is subject to maximum friction, the stratum corneum is extremely thick (Thick skin see Fig. 7.1B).

Section A

General Anatomy

Parts of a Hair (Fig. 7.1A)

1. Shaft – the visible part
2. Root – the embedded part
3. The bulb – the expanded lower end
4. The hair follicle – surrounding tubular sheath
5. The hair papilla – invaginating layer of dermis.

The hair roots are always attached to the skin obliquely; so, the hair lies flat on the skin surface.

Arrector Pili Muscle (Fig. 7.1C)

These are bands of smooth muscles, attached at one end to the dermis, just below the dermal papilla; and, at the other end, to the hair follicle. A sebaceous gland lies in the angle between the hair follicle and the arrector pili muscle. Contraction of the muscle has 2 effects.

1. The hair follicle becomes vertical from its original oblique position; "the hair stands on end". The skin surface overlying the attachment of the muscle becomes depressed; the surrounding area is raised. So, the skin has the appearance of "goose flesh".
2. The sebaceous gland, lying between the muscle and the hair follicle is squeezed; so, its secretions are squeezed out into the hair follicle.

Sebaceous Glands (Fig. 7.1C)

They are seen in relation to the hair follicles. Each gland consists of a number of alveoli, that are connected to a duct. This duct opens into the hair follicle. As mentioned earlier, the sebaceous gland is situated between the hair follicle and the arrector pili muscle. When the muscle contracts, it squeezes the gland, which facilitates the discharge of its secretions into the hair follicle.

The secretion of sebaceous glands is called "sebum".

Functions of Sebum

1. Its oily nature helps to keep the skin and hair soft
2. It prevents drying of skin
3. It makes the skin resistant to moisture.

Modified Sebaceous Glands

The tarsal glands of eyelids are modified sebaceous glands. They are called Meibomian glands.

Sweat Glands

Sweat glands produce sweat or perspiration. There are 2 types of sweat glands in the body: (1) Typical and (2) Atypical.

Typical Sweat Glands

A typical sweat gland consists of a single long tube. The lower end of the tube is highly coiled forming the "body" of the gland. It usually lies in the reticular layer of the dermis. The part of the tube, connecting the "body" to the skin surface is the "duct". Within the epidermis it has a spiral course. The duct is lined by a single layer of cuboidal epithelium.

Sweat glands are innervated by cholinergic nerves. Secretion of the glands (sweat) has a high water content. Evaporation of sweat causes cooling of the body.

The number and size of the sweat glands vary in different parts of the body. The palm and soles have the largest number of sweat glands.

Atypical Sweat Glands

Atypical sweat glands are found in the following regions.

1. Axilla
2. Nipple and areola
3. Perianal region
4. The glans penis
5. Some parts of female external genitalia.

These glands are larger in size. They show branching. Their ducts open into hair follicles. Their secretions are viscous.

The secretions of atypical glands are odorless; but due to bacterial decomposition, they give off body odors.

Modified Sweat Glands

1. Ceruminous glands of external auditory meatus
2. Ciliary glands of eyelids
3. Mammary gland.

Nails

They are solid plates of modified horny cells. They form a protective covering on the dorsal surface of fingers and toes.

Parts of a Nail

The main part of a nail is called its body. It has a free distal edge. The proximal part of the nail is implanted into a groove on the skin – this is the root of the nail. The tissue on which the nail rests, is the nailbed. It is highly vascular; that is why the nail appears pink in color.

The main function of nails is to provide a firm support for the finger tips. This support, increases the sensitivity of the finger tips and increases their efficiency in carrying out delicate functions.

■ BLOOD SUPPLY OF SKIN

Skin is a highly vascular organ. It derives its arterial blood from a number of plexuses. One plexus of arteries is present over the deep fascia; another plexus, just below the dermis, is called reticular plexus; the papillary plexus lies just below the dermal papilla. Capillary loops arising from this plexus pass into each dermal papilla.

The epidermis has no blood supply. It derives its nutrition entirely by diffusion from the capillary loops of the dermal papillae. There are numerous arteriovenous anastomoses in the skin, which have an important role in temperature regulation.

Cutaneous Receptors or Exteroceptive Receptors (Fig. 7.2)

These receptors, present in the skin, are concerned with touch, pain, temperature and pressure.

1. *Free nerve endings:* When terminals of sensory nerves do not show any specialization of structure, they are called free nerve endings. This type of nerve endings are numerous in relation to hair follicles. They respond mainly to the deformation of hair.
2. *Tactile corpuscles of Meissner:* These are small oval or cylindrical structures seen in relation to dermal papillae in the hand and foot. They are receptors of touch.
3. *Lamellated corpuscles of Pacini:* They are large receptors, found in the subcutaneous tissue of palm, sole and digits. They respond to vibration and pressure.
4. *Tactile menisci or Merkel cell endings:* These are small disc-like structures seen in relation to specialized epithelial cells or Merkel cells present in the stratum spinosum of the epidermis. They are sensitive to pressure.

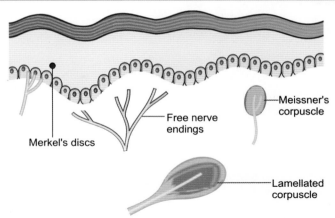

Fig. 7.2: Sensory end organs of skin

Applied Anatomy

1. *Callosity* is a localized, thickened part of skin caused by friction. It is commonly occupational, e.g. on a Gardener's hand.
2. *Infections of skin:* A 'boil' or furuncle is an acute staphylococcal infection of a hair follicle.
3. *Stye* or hordeolum is due to infection of an eyelash follicle.
4. A sebaceous cyst follows obstruction to the mouth of a sebaceous duct. It commonly occurs on the face or scalp.
5. *Malignant disease of skin:*
   a. Basal cell carcinoma or Rodent ulcer is a low grade malignancy occurring in white skinned people. Exposure to sunlight is a predisposing factor. Ninety per cent lesions are found on the face; the commonest site is the medial canthus of the eye.
   b. *Squamous cell carcinoma:* Usually occurs in pre-existing skin lesions, several years after irradiation, prolonged irritation by various chemicals like dyes, tar, etc.
   c. *Malignant melanoma:* It can arise from a pre-existing mole or from a normal skin. It should be suspected if a mole enlarges, itches or bleeds. Malignant melanoma spreads by local extension, by lymphatics or by the blood stream to lungs, liver, brain, bones and skin.
6. *Skin grafting:* Skin grafting becomes necessary when a large area of skin has been lost by injury (e.g. burns) or by disease or by surgical excision (e.g. malignant melanoma). This procedure enhances healing, limits the deformity and disability.
7. Examination of nails is a part of the general examination of a patient. It may give us a clue regarding the diagnosis, e.g. In iron deficiency anemia, the nails become concave and spoon-shaped, a condition called koilonychia. The nails become thin and brittle.

   "Clubbing" of the nails can be associated with congenital heart disease (Fallot's tetralogy) or bronchiectasis.

Section B

Musculoskeletal System

- ❖ Osteology
- ❖ Muscular System
- ❖ Joints

8

Osteology

INTRODUCTION

Osteology is the study of bones. The bony and cartilaginous framework of the body constitutes the skeleton. The human skeleton is internal to muscles– so it is described as an endoskeleton. In lower animals such as insects, the muscles are attached to the inner aspects of rigid material which also offers protection— this type of skeleton is called an exoskeleton. The human skeleton is derived from original endoskeleton, exoskeletal dermal elements (e.g. nail), modified branchial arches and sesamoid bones developed in tendons.

BONES

Functions of Bones

1. Constitute the framework of body, give shape and form to the body
2. Form the central axis of body
3. Support and transmit the weight of the body
4. Give attachment to muscles and ligaments
5. Protect vital organs such as brain, heart and lungs
6. Storage of calcium
7. Formation of blood elements in the bone marrow.

Number of Bones

There are 206 bones.
- Upper limbs – 64
- Lower limbs – 62
- Vertebrae – 33 (C_7 T_{12} L_5 S_5 CO_4)
- Skull – 22
- Ribs – 24
- Sternum – 1

Classification

According to Position

1. Axial skeleton
2. Appendicular skeleton

Axial skeleton—Bones forming the axis of the body, e.g. Ribs, skull, sternum, vertebrae

Appendicular skeleton—Bones forming the skeleton of appendages or limbs.

According to Size and Shape

1. Long bones, e.g. femur, humerus.
2. Short long bones, e.g. metacarpals, metatarsals.
3. Short bones, e.g. tarsal, carpal.
4. Flat bones, e.g. parietal, scapula, sternum.
5. Irregular bones, e.g. vertebrae.

6. Pneumatic bones (bones having large air-filled spaces)—Maxilla, frontal, ethmoid.
7. Sesamoid bones—"seed-like" bones seen in tendons, e.g. patella, pisiform.

Sesamoid bones do not have a periosteum; they do not have Haversian systems and they ossify after birth.

Functions of sesamoid bones: They diminish friction and modify the pull of muscles or change the direction of pull of muscles.

According to Structure

Compact bone—It is the outer cortical part of long bones. Hard in consistency.

Spongy or cancellous—is the inner part of bone. Less hard; spongy in appearance.

Diploic bones have inner and outer layers ("tables") of compact bone with spongy substance in between, which is rich in bone marrow, e.g. most of the cranial bones.

According to Development
i. Membranous, e.g. skull bones
ii. Cartilaginous, e.g. long bones.

Frequently used anatomical terms in Osteology, their meanings and examples are given in Table 8.1.

Sl. No.	Terminology	Meaning	Example
		Table 8.1: Common terms in osteology	
1.	Foramen	Opening or a hole in a bone	Foramen magnum of skull
2.	Canal	A bony tunnel	Mandibular canal
3.	Canaliculus	A very narrow tunnel or canal	Canaliculi for blood vessels
4.	Meatus	A narrow passage	External auditory meatus
5.	Sulcus	A furrow or a groove	Intertubercular sulcus of humerus
6.	Pit or fovea	A small depression	Pit on the head of femur
7.	Fossa	A large depression	Supra and infraspinous fossa of scapula
8.	Facet	A smooth articular area of bone	Superior and inferior facets of vertebra
9.	Linea	An elongated elevation / Line	Linea aspera of femur
10.	Ridge	A rough elevation	Supracondylar ridge of humerus
11.	Crest	A broad ridge	Crest of spine of scapula
12.	Process	A large projection	Coracoid process of scapula
13.	Tubercle	A small round thickening	Conoid tubercle of clavicle
14.	Tuberosity	Large round thickening	Tibial tuberosity
15.	Head	A round, articular area of a bone	Head of femur
16.	Condyle	Smooth, rounded articular surface at the end of a bone	At the lower end of femur – 2 condyles
17.	Epicondyle	A non-articular bony projection above the condyle	At the lower end of femur
18.	Trochlea	A pulley-shaped articular surface	At the lower end of humerus

Contd...

Contd...

Sl. No.	Terminology	Meaning	Example
19.	Trochanter	A large non-articular projection	Greater trochanter of femur
20.	Malleolus	A hammer-shaped bony projection	Medial malleolus of tibia
21.	Spine	A sharp, pointed projection	Spine of vertebrae
22.	Hamulus	A hook-like bony projection	Pterygoid hamulus
23.	Lamina	A thin plate of bone	Laminae of vertebrae
24.	Squama	A scale-like part of a bone	Squamous part of temporal bone
25.	Sinus	A hollow or cavity inside a bone	Air sinuses of maxilla, ethmoid, frontal and sphenoid

APPENDICULAR SKELETON

■ UPPER LIMB

The parts of the upper limb and their bones are shown in Table 8.2; Figures 8.1A and B.

Shoulder Region

Clavicle: (meaning "Little key") (Figs 8.2A and B)

The clavicle is a long bone. It is a part of the pectoral girdle or shoulder girdle, which serves to attach the upper limb to the trunk; (the only point of articulation is at the sternoclavicular joint).

The clavicle supports the shoulder so that the arm can swing clearly away from the trunk. It transmits the weight of the limb to the sternum.

Peculiarities of clavicle
1. It is the only long bone that lies horizontally
2. It is subcutaneous throughout
3. It is the first bone to start ossifying
4. It is the only long bone which ossifies in membrane
5. It has no medullary cavity.

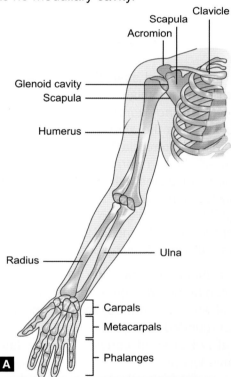

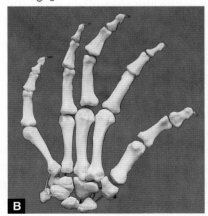

Figs 8.1A and B: Skeleton of hand

Table 8.2: Bones of upper limb	
Parts	*Bones*
1. Shoulder region	Bones of shoulder girdle a. Clavicle b. Scapula
2. Upper arm (arm or brachium)	Humerus
3. Forearm (antebrachium)	a. Radius b. Ulna

4. Hand
   a. Wrist (carpus)
 	
   b. Hand proper (Metacarpus)
   c. Five digits (Thumb 1st Little finger 5th)
	
	a. Carpus, made up of 8 carpal bones
	b. 5 Metacarpal bones
	
	c. Phalanges; 2 in thumb and 3 each in other fingers

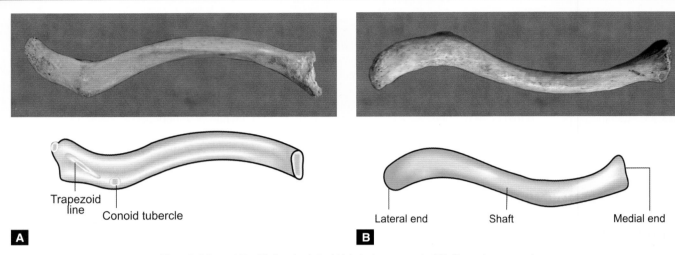

Figs 8.2A and B: Right clavicle (A) Interior aspect, (B) Superior aspect

Parts: (i) A cylindrical shaft (ii) Medial end (sternal) (iii) Lateral end (acromial).

1. The *lateral end* (acromial end) is flattened. It articulates with the acromion process of scapula to form the acromioclavicular joint.
2. The *medial end* (sternal end) is larger, almost rounded, articulates with manubrium sterni to form sternoclavicular joint.
3. The *shaft* is divisible into medial two-thirds and lateral one-third. The medial 2/3 is convex forwards. The lateral 1/3 is concave forwards.

Side determination: The side to which a clavicle belongs can be determined from the following points.

1. Lateral end is flattened; medial end large and almost rounded
2. Medial 2/3 of shaft convex forward; lateral 1/3 concave forward
3. There is a longitudinal groove on the inferior surface.

Muscles attached to clavicle

1. Sternocleidomastoid
2. Pectoralis major
3. Subclavius
4. Deltoid and
5. Trapezius.

At the junction of medial ¾ and lateral ¼ on the inferior surface is a conoid tubercle and a trapezoid line to which the coracoclavicular ligament is attached. This ligament transmits the weight of upper limb to axial skeleton.

Applied Anatomy

The clavicle is commonly fractured by falling on the outstretched hand; the commonest site of fracture is the junction of medial 2/3 and lateral 1/3, which is the weakest point. The lateral fragment is displaced downwards (shoulder droops) by the weight of the limb.

The Scapula (Figs 8.3A to C)

The scapula "shoulder blade", is a large, flat triangular bone. It has 2 surfaces, three borders, three angles and three processes.

Surfaces: (i) Costal surface, and (ii) Dorsal surface.

1. *The costal surface* (which faces the ribs) is concave. It is also called the subscapular fossa. The subscapularis muscle takes origin from this fossa. The serratus anterior muscle is inserted along the medial border of the costal surface.
2. *The dorsal surface:* A prominent spine divides this surface into a smaller, upper supraspinous fossa and a larger infraspinous fossa giving origin to supraspinatus and infraspinatus muscles respectively.

The borders

1. The *superior border*, thin and short, presents a suprascapular notch which transmits suprascapular vessels and nerve.
2. The *lateral border* is thick. At its upper end, it presents the infraglenoid tubercle. This tubercle gives origin to the long head of triceps. Below this, the teres minor and teres major muscles are attached to the lateral border.

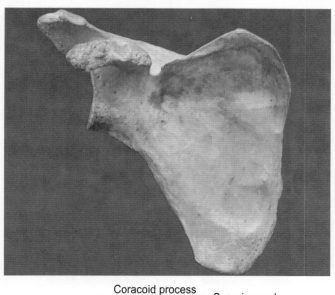

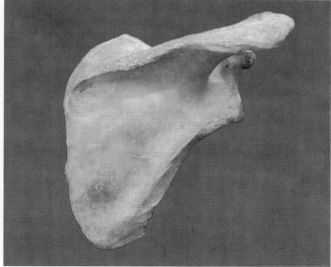

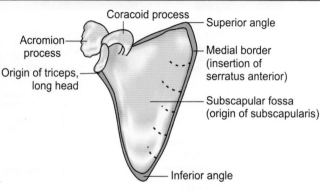

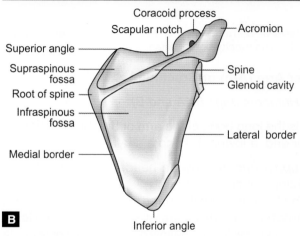

Figs 8.3A and B: Right scapula (A) Costal aspect, (B) Dorsal aspect

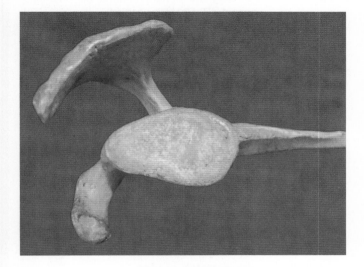

Fig. 8.3C: Glenoid cavity

3. The *medial border* is thin; gives attachment to levator scapulae, rhomboideus minor and rhomboideus major, from above downwards.

The angles

1. The superior angle – covered by trapezius muscle
2. The inferior angle covered by latissimus dorsi
3. The lateral angle or glenoid angle or "head" is broad and bears an articular, concave surface, the glenoid cavity or fossa. The head of the humerus articulates with it to form a "ball and socket" joint – the shoulder joint.

The processes

1. *The spine* – It is a prominent projection on the dorsal surface, dividing it into 2 unequal fossae – supra and infraspinous fossae. It gives attachment to trapezius and deltoid muscles

2. *The acromion* – Projecting forwards, almost at right angles from the lateral end of the spine; it is a visible and subcutaneous bony part.
3. *The coracoid process* is directed forwards; arises from the summit of the scapular head (glenoid angle). It gives attachment to three muscles and 3 ligaments – Pectoralis minor, coracobrachialis and short head of biceps muscles; coracoclavicular, coracoacromial and coracohumeral ligaments.

Side Determination
1. Lateral angle is large and bears the glenoid cavity
2. The dorsal surface is convex with a prominent spine. Costal surface is concave.
3. The thicker, lateral border runs from glenoid cavity to inferior angle.

Applied Anatomy

Paralysis of Serratus anterior muscle causes "winging" of scapula (The medial border "stands out").

Upper Arm (Arm or Brachium)

The Humerus (Figs 8.4A and B)

This is the long bone of the arm or brachium. It has an upper end, a lower end and a shaft.

The upper end: It shows a head, greater and lesser tubercles (tuberosities).
1. The head is smooth, forms about 1/3 of a sphere; directed medially, backwards and upwards. It articulates with the shallow glenoid cavity of scapula to form the shoulder joint.
   The head is separated from the rest of the upper end by a groove called anatomical neck.
2. The greater tubercle is an elevation that forms the lateral part of upper end. Three muscles—supraspinatus, infraspinatus and teres minor (SIT) are inserted here.
3. The lesser tubercle is an elevation on the anterior aspect of upper end. The subscapularis muscle is inserted here.
4. The two tubercles are separated by an inter-tubercular sulcus – it lodges the tendon of long head of biceps.
5. The line separating the upper end of the humerus from the shaft is called the *surgical* neck. The axillary nerve and circumflex humeral vessels are closely related to the surgical neck.

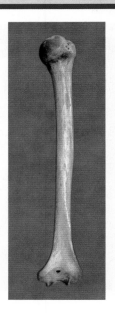

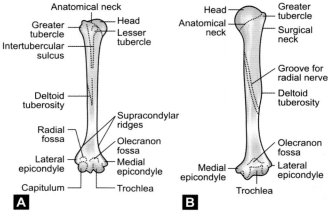

Figs 8.4A and B: Diagrams of right humerus (A) Anterior aspect, (B) Posterior aspect

The shaft is circular in cross-section in the upper part and triangular in cross-section in the lower part. A little above the middle, on the lateral surface is a V-shaped deltoid tuberosity. Behind this tuberosity is a spiral groove or radial groove, which is closely related to the radial nerve and profunda brachii vessels.

Lower end is expanded from side to side, having articular and non-articular parts.
1. The *articular parts* are the capitulum and the trochlea. The capitulum is a rounded projection which articulates with the head of the radius. The trochlea is a pulley-shaped surface; it articulates with the trochlear notch of ulna.

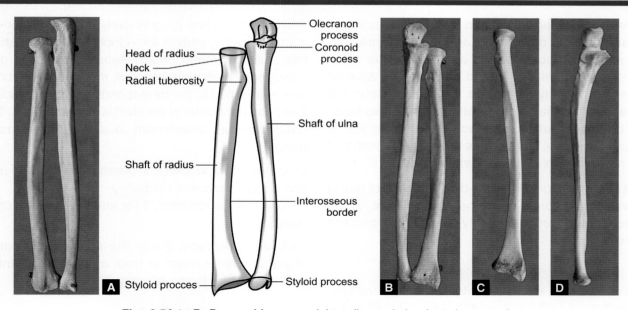

Figs 8.5A to D: Bones of forearm—right radius and ulna (anterior aspect)

2. The non-articular part includes the medial and lateral epicondyles and the medial and lateral supracondylar ridges. The medial epicondyle is a prominent bony projection. It is subcutaneous; posteriorly, it is closely related to the ulnar nerve.

There is a small depression just above the trochlea. It accommodates the coronoid process of ulna when the elbow is flexed. The radial fossa (notch) lies just above the capitulum, lodging the head of radius in full flexion.

The olecranon fossa lies just above the posterior aspect of the trochlea. It accommodates the olecranon process of ulna when the elbow is extended.

Side determination
1. Upper end shows the head; directed medially
2. Lower end is expanded from side to side
3. Medial epicondyle is larger
4. Olecranon fossa is posterior.

Applied Anatomy

1. The humerus is closely related to three nerves, which are liable to injury in fracture or dislocation; the axillary nerve at the surgical neck; the radial nerve at the radial groove; the ulnar nerve behind the medial epicondyle.
2. The common sites of fracture are the surgical neck and the supracondylar region. Supracondylar fracture is common in children. This fracture may injure the brachial artery, leading to a condition called Volkmann's ischemic contracture.

Forearm Bones (Figs 8.5A to D)

Two long bones, radius and ulna form the skeleton of forearm. In supination, they lie parallel to each other, whereas, during pronation, the radius lies diagonally across the ulna. In the anatomical position, radius lies lateral and parallel to ulna.

Radius

Radius is the lateral bone of forearm. It has an upper end, shaft and lower end.

Upper end consists of head, neck and a radial tuberosity.
1. The head presents a disc-like concavity on its upper surface, which articulates with the capitulum of humerus and forms the humeroradial part of elbow joint. The head articulates medially with the radial notch of ulna.
2. The neck of radius is a constricted area just distal to the head. The brachial artery divides into its two terminal branches at this level.
3. The radial tuberosity lies on the medial side of the lower part of neck. The biceps brachii is inserted here.

The shaft: It presents a sharp, medial interosseous border, to which a strong interosseous membrane is attached. It is connected to the corresponding border of ulna.

The rest of the shaft provides attachment to forearm muscles.

Lower end: It is the widest part of the bone. Its lateral surface projects downwards as the styloid process. The medial surface presents a concave ulnar notch for articulation with the head of ulna. The anterior surface is closely related to the radial artery; the pulsations of radial artery is felt against this surface.

The inferior articular surface of the lower end of radius is concave; it articulates laterally with scaphoid and medially with lunate to form the wrist joint.

Side determination: Place the disc-like head of radius above, gentle concavity of the shaft in front, sharp interosseous border medially and styloid process laterally.

Applied Anatomy

Since the radius is a force transmitting bone, it is commonly fractured about 2 cm above its lower end due to a fall on the outstretched hand. This is known as Colle's fracture, in which the distal fragment is displaced upwards, backwards and laterally producing a classical "dinner fork" deformity.

Ulna

Ulna is the medial bone of forearm. It corresponds to the fibula of the lower limb. It resembles the shape of a pipe wrench.

Ulna consists of upper and lower ends and shaft in between.

Upper end: It presents a hook-like massive part with the concavity anteriorly. The upper end has 2 processes—olecranon process and a coronoid process; and 2 articular areas—the trochlear and radial notches.

1. The olecranon process is a beak-like projection overhanging the trochlear notch. It occupies the olecranon fossa of humerus. This subcutaneous bony prominence gives insertion to the triceps brachii muscle.
2. The coronoid process is a shelf-like projection giving insertion to brachialis muscle.

The trochlear notch is a C-shaped concavity between the anterior surface of olecranon process and upper surface of coronoid process. It articulates with the trochlea of humerus to form the humeroulnar part of elbow joint.

Laterally, the upper end of ulna presents a concave radial notch for articulation with the head of radius.

Shaft: It diminishes progressively from above downward. It is gently convex throughout its entire extent. It has a sharp lateral (interosseous) border, to give attachment to the interosseous membrane; which in turn, is attached to the corresponding border of radius. The posterior border of the shaft is subcutaneous. The shaft provides attachment to several muscles of forearm.

Lower end is expanded and presents the head of ulna and a styloid process medially.

The head is rounded; it fits into the ulnar notch of radius.

Side determination: Place the olecranon process above, trochlear notch in front and the sharp interosseous border facing laterally.

Applied Anatomy

Fracture of ulna alone or both forearm bones is not uncommon.

The Skeleton of Hand

The hand consists of carpal bones at the carpus or wrist; metacarpal bones in the palm or hand proper and phalanges in the digits.

Carpal Bones

These are short bones; 8 in number, arranged in 2 rows – proximal and distal. Each row presenting four bones, as follows:

	Distal				
Medial	HAMATE	CAPITATE	TRAPE-ZOID	TRAPEZIUM	Lateral
	PISIFORM	TRIQUETRAL	LUNATE	SCAPHOID	
	Proximal				

When these bones articulate with each other, a concavity or arch is formed. The lateral pillar of this concavity is formed by the scaphoid and trapezium and the medial pillar by the hamate and pisiform. Both pillars are connected by a flexor retinaculum (a band of deep fascia) which acts as a tie beam and converts the concavity into a carpal tunnel. The long flexor tendons of the fingers and the median nerve are transmitted through the carpal tunnel (Fig. 8.6).

The scaphoid and lunate articulate with the distal surface of lower end of radius to form the wrist joint.

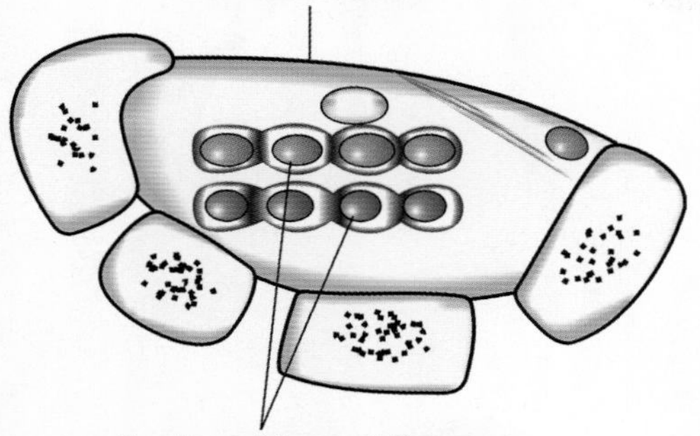

Median nerve in "carpal tunnel"

Flexor retinaculum

Tendons of flexor digitorum superficialis and profundus

Fig. 8.6: Carpal tunnel and contents

The capitate is the largest bone among the carpal bones; pisiform is the smallest, it is considered as a sesamoid bone in the tendon of flexor carpi ulnaris (The corresponding bone is absent in the foot).

Distal row of carpal bones articulate with the bases of 5 metacarpal bones.

Applied Anatomy

1. The median nerve may be compressed in the carpal tunnel leading to carpal tunnel syndrome.
2. A fall on the outstretched hand may lead to fracture of scaphoid. As a result, the blood supply to the bone may be cut off leading to avascular necrosis.

Metacarpal Bones

Metacarpal bones are 5 in number; they are short long bones. They are numbered from lateral to medial side. Each metacarpal bone consists of a head, shaft and base. The head articulates with the proximal phalanx. The head is slightly larger than the base and forms the knuckles of the hand.

The bases of the metacarpals articulate with the distal row of carpal bones.

The first metacarpal is placed on a more anterior plane than the rest of the metacarpals and is rotated medially. This helps in the free movement of first metacarpal bone and the thumb.

The shafts provide attachment to the intrinsic and extrinsic muscles of hand.

Applied Anatomy

A fracture of shaft of metacarpal is not uncommon.

Phalangeal Bones or Phalanges (Singular-phalanx)

There are 14 phalanges; 2 for the thumb and three each for the medial four fingers. They are referred to as proximal, middle and distal phalanges. Each phalanx has a base, shaft and a head.

The base of the proximal phalanx articulates with the head of metacarpal to form the metacarpo-phalangeal joint (MP joint). Each of the bases of middle and distal phalanges present two concave facets to articulate with the adjacent phalanges to form proximal and distal interphalangeal joints (IP joints).

The head of the distal phalanx has a horse-shoe shaped tuberosity on its palmar side and a smooth dorsal surface, covered by finger nails.

The phalanges provide attachment to several intrinsic and extrinsic muscles of hand.

■ THE SKELETON OF LOWER LIMB

The lower limb is specialized for: (i) locomotion, (ii) bearing weight, and (iii) maintenance of balance. It consists of 4 major parts.

1. Hip—containing the hip bone (innominate bone)
2. The thigh—containing the femur or thigh bone
3. The leg—containing the tibia medially and fibula laterally
4. The foot—containing the *tarsus* (bones of posterior and middle parts of foot) and *metatarsus* (bones of anterior part of foot) and phalanges (bones of digits or toes).

The Hip Bone (Figs 8.7A to D)

This is a large, irregular bone. It has three parts— Ilium, ischium and pubis. The three parts are fused together at a cup-shaped hollow called acetabulum. The acetabulum articulates with the head of the femur to form the hip joint.

The two hip bones form the pelvic girdle. The bony pelvis is formed by the two hip bones along with the sacrum and coccyx.

Ilium

Ilium is the upper expanded part of the hip bone. It has an upper end called iliac crest. It has 3 surfaces– Gluteal surface, iliac fossa and sacropelvic surface.

Section B

Musculoskeletal System

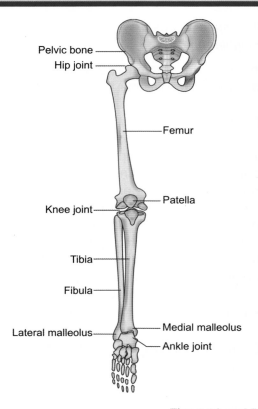

Figs 8.7A and B: Skeleton of lower limb

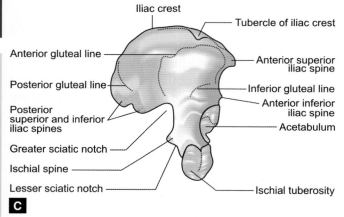

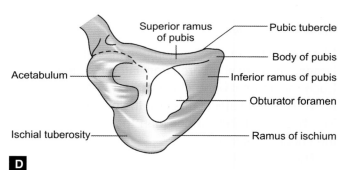

Figs 8.7C and D: Hip bone

Iliac Crest

The iliac crest is the broad convex ridge forming the upper end of ilium. The highest point of iliac crest lies at the level of the interval between L3 and L4 vertebrae. This is an important landmark while doing a lumbar puncture.

1. The anterior end of iliac crest is called anterior superior iliac spine. The lateral end of inguinal ligament is attached to it.
2. The posterior end of iliac crest is called posterior superior iliac spine, which lies at the level of S2 spine.

3. A little distance below the anterior and posterior superior iliac spines are the anterior and posterior inferior iliac spines.

Gluteal surface: The gluteal surface is the outer surface of ilium. It is divided into 4 areas by 3 gluteal lines – the posterior, anterior and inferior gluteal lines. The gluteus maximus, medius and minimus muscles take origin from the gluteal surface.

Iliac fossa: The iliac fossa is a large concave area on the inner surface of ilium from where the iliacus muscle takes origin.

Sacropelvic surface: The sacropelvic surface is an uneven area for articulation with the sacrum.

Ischium

Ischium forms the postero-inferior part of the hip bone. It forms the posterior boundary of obturator foramen (It is a large oval opening which separates ischium and pubis). The ischium has a body and ramus. The upper end of the body forms a part of the acetabulum. The lower end forms the ischial tuberosity, from which the hamstring muscles arise.

The posterior border of ischium is continuous above with the posterior border of ilium. It forms a part of greater sciatic notch, the ischial spine and the lesser sciatic notch.

Conjoined ischiopubic rami: The inferior ramus of pubis unites with the ramus of ischium on the medial side of obturator foramen to form the conjoined rami.

Pubis

It forms the antero-inferior part of the hip bone and part of the acetabulum; it also forms the boundary of obturator foramen.

It has (1) a body anteriorly; (2) a superior ramus superolaterally and (3) an inferior ramus inferolaterally.

Body of pubis: It has a superior border called pubic crest (2) a pubic tubercle at the lateral end of pubic crest, to which the medial end of inguinal ligament is attached. Pubic tubercle is an important bony landmark.

The medial or symphyseal surface of body of pubis articulates with the opposite pubis to form the pubic symphysis.

Applied Anatomy

A fracture of the hip bone can occur in severe accidents; usually associated with rupture of urethra or bladder.

Femur (Figs 8.8A and B)

Femur or thigh bone is the longest and strongest bone of the body. It has an upper end, lower end and shaft.

Side determination: The upper end bears a rounded head, directed medially. The lower end has 2 large condyles. The shaft is convex forwards.

Upper end has a head, neck, greater and lesser trochanters, intertrochanteric line and intertrochanteric crest.

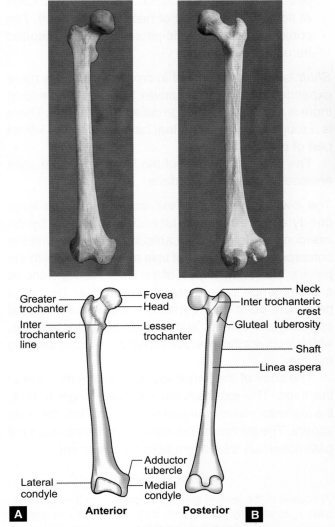

Figs 8.8A and B: Right femur

Section B   Musculoskeletal System

1. The head: Forms more than half a sphere; directed medially; articulates with the acetabulum of hip bone to form the hip joint. A small pit (fovea) is seen just below the center of the head.
2. Neck: Connects the head with the shaft. It makes an angle – the neck-shaft angle – with the shaft (125° in males; less in females). The junction of neck with shaft is marked anteriorly by the intertrochanteric line and posteriorly by the intertrochanteric crest.
3. Greater trochanter: It is a large, quadrangular prominence located at the upper part of the junction of neck with the shaft. The lateral rotators of hip, the gluteus medius and minimus are inserted here.
4. Lesser trochanter: A small conical eminence, directed medially and backwards from the junction of postero-inferior part of neck with the shaft. The conjoined tendon of ilio-psoas muscle is inserted here.

Shaft is almost cylindrical in cross-section, it is more expanded inferiorly. It is convex forwards. Posteriorly, there is a broad, rough ridge called linea aspera. There is a rough area called gluteal tuberosity on the lateral part of posterior surface.

The posterior surface of the lower 1/3 of the shaft encloses the popliteal surface.

The lower end is widely expanded; it has two large condyles, medial and lateral, separated by an intercondylar fossa. They articulate inferiorly with the corresponding condyles of tibia and anteriorly with the patella. The lateral aspect of the lateral condyle shows a prominence called lateral epicondyle. The most prominent point on the medial condyle is the medial epicondyle. Posterior to this is the adductor tubercle, which receives the insertion of part of adductor magnus muscle.

The shaft of the femur lies deep to the muscles of the thigh. The extensor muscles take origin from it; the adductor muscles are inserted mainly into the linea aspera. The popliteus, the medial and lateral heads of gastrocnemius take origin from the lower end.

Applied Anatomy

A fracture of the neck of femur is common over the age of 60 years, especially in females due to thinning of bones (osteoporosis). A spiral fracture of the shaft of femur in young individuals results from accidents.

Skeleton of the Leg

The skeleton of the leg is formed by two long bones – tibia and fibula. The tibia lies medial to fibula. The tibia is more massive; it articulates with the femur to form the knee joint and helps in the transmission of body weight.

Tibia (Figs 8.9A and B)

The tibia has an upper end, lower end and a shaft.

The upper end is expanded, it has two large-medial and lateral-condyles which articulates with the corresponding condyles of the femur. The upper surface of both condyles are articular; they show a concavity. The peripheral part of articular surface is flat upon which a fibrocartilaginous meniscus rests. The intercondylar area, situated between the two condyles gives attachment to several structures (Fig. 8.10). From anterior to posterior, they are:
1. Anterior horn of medial meniscus
2. Anterior cruciate ligament
3. Anterior horn of lateral meniscus
4. Posterior horn of lateral meniscus
5. Posterior horn of medial meniscus; and
6. Posterior cruciate ligament

On the anterior surface of both condyles, there is a tibial tuberosity, which provides attachment for the ligamentum patellae.

The shaft: It is triangular in cross-section. It has three borders and three surfaces.
1. Anterior border or shin is entirely subcutaneous. It extends from the tibial tuberosity and is continuous below with the anterior border of medial malleolus.
2. The medial border is continuous below with the posterior margin of medial malleolus. The lateral or interosseous border is sharp and gives attachment to the interosseous membrane which unites the tibia with the fibula.
3. The medial surface is entirely subcutaneous. Its lower end is related to the great saphenous vein and the saphenous nerve. Its upper part receives the insertions of sartorius, gracilis and semi-tendinosus muscles.
4. The major part of lateral surface provides origin for the tibialis anterior muscle. The lower third is related to the extensor tendons and neurovascular bundle passing from the leg to the dorsum of foot.

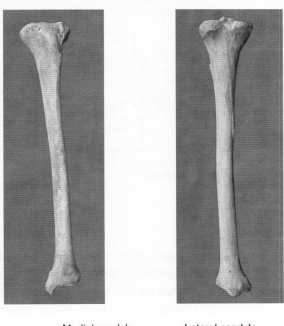

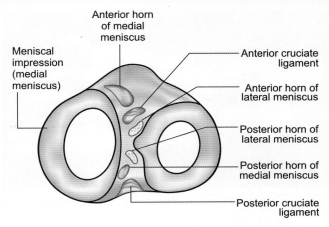

Fig. 8.10: Superior view of upper end of the right tibia

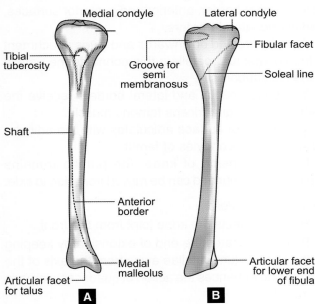

Figs 8.9A and B: Tibia (right) (A) Anterior view, (B) Posterior view

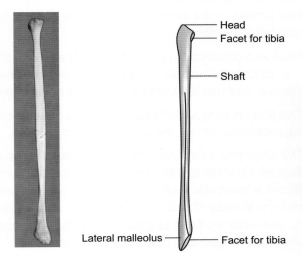

Fig. 8.11: Right fibula

5. The upper third of posterior surface receives the insertion of popliteus muscle. Below this is the soleal line. The area below the soleal line provides attachment to the flexor muscles.

The lower end of the tibia is expanded and projects medially as the medial malleolus. The inferior surface of lower end and medial malleolus articulate with the talus to form a part of the ankle joint.

Side determination: The upper end is larger. Tibial tuberosity faces anteriorly; medial malleolus faces medially.

Applied Anatomy

1. Because the body (shaft) of tibia is unprotected anteromedially throughout its course and is relatively slender at the junction of middle and inferior thirds, it is usually fractured at this site.
2. Because of its extensive subcutaneous surface, the tibia is accessible for obtaining pieces of bone for grafting.
3. The rickets usually affects the tibia at the junction of middle and inferior thirds during infancy and childhood.

Fibula (Latin : Fibula = Pin)

Fibula lies lateral to the tibia and consists of upper and lower ends and an intervening shaft (Fig. 8.11).

Section B

Musculoskeletal System

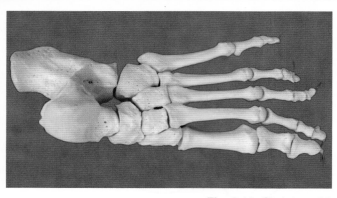

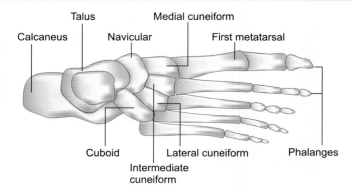

Fig. 8.12: Skeleton of the foot (viewed from above)

Upper end or head is expanded. It shows an oval articular facet, for the lateral condyle of tibia. It provides attachment to the biceps femoris tendon and fibular collateral ligament.

Neck is a constriction at the junction of head with shaft. It is closely related to the common peroneal nerve laterally and medially, with the anterior tibial vessels.

The shaft is long and slender, provides attachment to interosseous membrane and muscles of the leg.

The lower end is expanded to form the lateral malleolus. Its lateral surface is subcutaneous. The medial surface shows a triangular facet for articulation with the lateral surface of body of talus, to form a part of ankle joint. Its posterior surface is grooved by the tendons of peroneus longus and peroneus brevis muscles.

Functions of Fibula
1. It provides attachment for most of the muscles of leg
2. It completes the ankle joint from the lateral side
3. Lateral malleolus acts as a pulley for the peroneal tendons.

Applied Anatomy

1. Fracture of fibula along with tibia is quite common in accidents
2. Forceful eversion of foot can lead to Pott's fracture, in which there is a fracture of lower end of shaft of fibula and medial malleolus.
3. Fibula is a common source of bone for grafting
4. The common peroneal nerve winds superficially around the neck of fibula. The common fibular (peroneal) nerve may be injured during fracture of the neck of fibula; it is susceptible to pressure exerted in the region of the head of the fibula by a tightly applied plaster cast. Injury of the common peroneal nerve results in paralysis of all the dorsiflexor and eversion muscles of the foot resulting in "foot-drop".

Patella

Patella is the largest sesamoid bone, ossified in the tendon of quadriceps femoris. It lies in front of the knee joint; so, it is also known as the "knee-cap".

It is somewhat triangular in outline. It has an apex, base or upper border, anterior and posterior surfaces, medial and lateral borders.

Apex is directed downwards and gives attachment to the ligamentum patellae; this connects the patella to the tibial tuberosity.

The base, medial and lateral borders receive the insertion of the quadriceps femoris muscle.

The posterior surface articulates with the articular surfaces of the condyles of femur.

In full extension of knee, the patella remains free from the femur and can be moved from side to side.

Functions of Patella
1. Patella protects the knee joint from the front
2. Patella facilitates the end of extension by keeping the ligamentum patellae away from the axis of the joint and enhances the action of quadriceps femoris muscle.

Applied Anatomy

Fracture of patella is common, which affects the movement at the knee joint.

Skeleton of the Foot (Fig. 8.12)

The foot is set at right angle to the leg. This permits the sole to have a proper grip on the ground.

The skeleton of the foot is composed, from proximal to distal, of:
1. Tarsal bones
2. Metatarsal bones
3. Phalangeal bones.

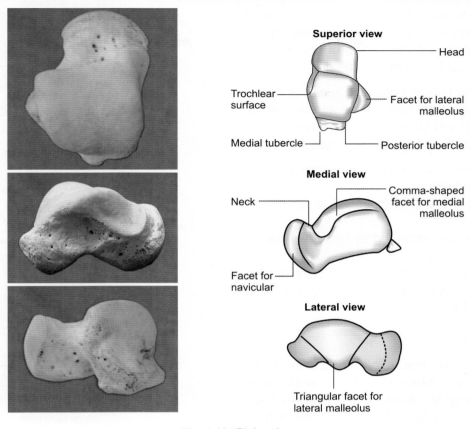

Superior view

Head

Trochlear surface

Facet for lateral malleolus

Medial tubercle

Posterior tubercle

Medial view

Neck

Comma-shaped facet for medial malleolus

Facet for navicular

Lateral view

Triangular facet for lateral malleolus

Fig. 8.13: Right talus

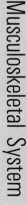

Tarsal Bones

These are short bones, seven in number. They are:
1. Talus
2. Calcaneus or calcaneum
3. Navicular
4. Cuboid
5. Medial cuneiform
6. Lateral cuneiform
7. Intermediate cuneiform.

These bones are arranged in 3 rows. The talus and calcaneus in the proximal row, the navicular in the middle row, the three cuneiforms and cuboid in the distal row.

The talus (Fig. 8.13) forms the connection link between the bones of the leg and foot. It is situated on the upper surface of anterior 2/3 of calcaneus. The *head* of the talus articulates with the navicular in front. The superior surface, medial and lateral surfaces of the *body* of the talus are articular. The superior surface articulates with the inferior surface of lower end of tibia; its medial surface with medial malleolus and lateral surface, with lateral malleolus. The posterior surface of the body is grooved by the tendon of flexor hallucis longus.

No muscles are attached to the talus. It articulates with 4 bones—tibia, fibula, calcaneus and navicular and forms three joints—ankle joint, sub-talar joint and pre-talar joints.

The calcaneus is the largest of tarsal bones, situated below the talus; it extends behind the talus. Anteriorly, it articulates with the cuboid. The posterior surface is large and receives the attachment of tendocalcaneus. The plantar (inferior) surface is rough; its medial tubercle strikes the ground on standing.

Navicular bone: The navicular bone belongs to the middle row of tarsal bones and occupies the medial margin of foot. It articulates with the three cuneiforms anteriorly and posteriorly with the talus. It has got a tuberosity medially, to which greater part of tibialis posterior tendon is attached.

Cuneiforms: The three cuneiforms are wedge-shaped bones. They articulate behind with the navicular; in front, they articulate with the bases of first, second and third metatarsal bones mediolaterally.

Cuboid bone: The cuboid bone occupies the lateral margin of foot. Posteriorly it articulates with calcaneus; anteriorly with 4th and 5th metatarsals. Inferiorly, it is grooved by the tendon of peroneus longus.

Comparison between carpal and tarsal bones
1. Both are short bones.
2. Carpal bones are 8 in number, tarsal bones are 7 in number
3. Carpal bones are arranged in two rows; tarsal bones, in three rows.
4. Pisiform is a sesamoid bone ossified in the tendon of flexor carpi ulnaris; its corresponding bone is absent in the foot.

Metatarsal Bones

Metatarsal bones are five, short long bones, numbered from medial to lateral side. First metatarsal is shortest and strongest; second is the longest. Each metatarsal presents a head in front, a base behind and a shaft between them. The bases articulate with the distal row of tarsal bones; their heads articulate with the bases of proximal phalanges. The heads are generally smaller than the bases (In the metacarpals, head is generally larger than the base).

Phalanges

Phalanges are miniature long bones; 14 in number. There are 3 phalanges each for the lateral four toes and 2 for the big toe. These bones give insertion to intrinsic and extrinsic muscles of the leg and foot.

Arches of Foot (Fig. 8.14)

The human foot has two functions to perform:
1. To support the weight of the body
2. To serve as a lever to propel the body forward during locomotion

Both these functions are carried out efficiently since the skeleton of the foot is composed of a series of short bones. The foot is made pliable and can adapt itself to uneven surfaces.

The skeleton of the foot is arched, both longitudinally and transversely.

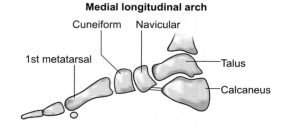

Medial longitudinal arch

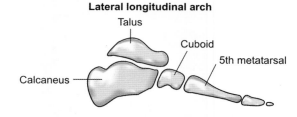

Lateral longitudinal arch

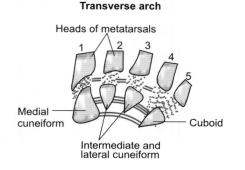

Transverse arch

Fig. 8.14: Arches of foot

An arched foot is a distinctive feature of human beings.

Classification of arches
1. Longitudinal
   (a) Medial, and (b) Lateral
2. Transverse

Longitudinal arches (Fig. 8.14)
1. The *medial longitudinal* arch is higher, and more mobile than the lateral. It is formed by calcaneus, talus, navicular, the three cuneiforms and the first three metatarsals.
2. The lateral longitudinal arch is formed by the calcaneus, cuboid and lateral two metatarsals. This arch is low, has limited mobility and is built to transmit weight to the ground.

Transverse arch is formed by the heads of five metatarsal bones. Each foot forms a "half dome" because only the lateral end of the foot comes in contact with the ground. When the two feet are kept side by side, the transverse arch is completed.

Factors responsible for the maintenance of arches
a. Shape of the bones.
b. Powerful ligaments
c. Muscles and tendons.

Functions of arches
1. The arches of the foot distribute the body weight to the ground, through the heel and heads of metatarsals.
2. The arches act as springs and levers, which help in walking and running
3. They act as shock absorbers during jumping
4. The concavity of the arches protect the nerves and blood vessels.

Applied Anatomy

1. Absence or collapse of arches leads to "flat foot" or pes planus.
2. Exaggeration of arches of foot is called "pes cavus"
3. The commonest deformity of the foot is talipes equinovarus or "club foot". In this condition, the foot is inverted, adducted and plantar flexed.

AXIAL SKELETON

■ SKULL

Introduction

The skeleton of the head is called skull. It consists of the cranium and mandible. The upper part of cranium forming the brain box is termed the calvaria and the remainder of the skull forms the facial skeleton.

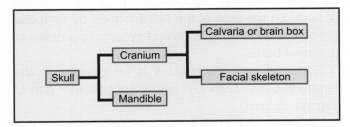

Functions of Skull

1. Protection of brain
2. Protection of special sense organs like ears, eyes, tongue and nose
3. Protection of beginnings of respiratory and digestive tracts.

Bones of Skull

The skull consists of 22 bones, which are named below.

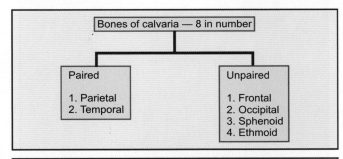

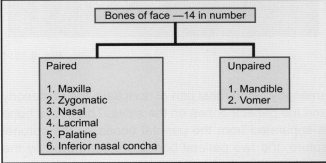

Joints of Skull

With the exception of the temporomandibular joint, which permits free movements, the skull bones are united by fibrous, immovable joints called sutures.

Methods of Study of Skull

1. The skull can be studied as a whole from different views.
   a. Superior view or Norma verticalis
   b. Posterior view or Norma occipitalis
   c. Anterior view or Norma frontalis
   d. Lateral view or Norma lateralis
   e. Inferior view or Norma basalis.
2. Interior of skull can be studied after removing the roof or skull cap. The details of interior of cranial vault and the internal surface of base can be studied. The base is subdivided into anterior, middle and posterior cranial fossae.
3. Study of individual bones.

Exterior of Skull

Norma Verticalis (Fig. 8.15)

When viewed from above, the skull is oval in shape; In this view, the upper part of frontal bone is seen

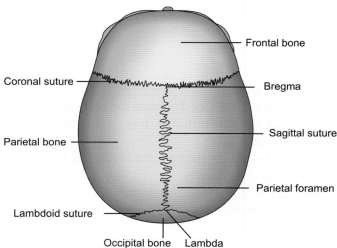

Fig. 8.15: Norma verticalis

anteriorly; uppermost part of occipital bone posteriorly and the parietal bones on the sides. The frontal bone is separated from the parietal bones by the coronal suture; the two parietal bones are separated by the sagittal suture; the lambdoid suture is situated between the parietal bones and occipital bone.

The bregma is the meeting point between the coronal and sagittal sutures. The lambda is the meeting point of sagittal and lambdoid sutures.

The bregma is the site of a diamond-shaped membranous gap called anterior fontanelle in the fetus. It closes 1½ years after birth.

The lambda is the site of posterior fontanelle which closes 2-3 months after birth.

Norma Occipitalis

This view shows the posterior parts of parietal bones, the squamous part of occipital bone and mastoid parts of temporal bones.

The external occipital protuberance is a median prominence. It marks the junction of head and neck. Two lines extend laterally from here—the superior nuchal lines (Figs 8.21A and B).

Norma Frontalis

Norma frontalis is roughly oval in outline (Fig. 8.16).

Bones seen

1. Frontal bone forms the forehead
2. The right and left maxillae form the upper jaw

3. The right and left nasal bones form the bridge of the nose
4. The zygomatic bones form the bony prominence of the cheeks
5. Mandible forms the lower jaw.

The orbital openings are cone-shaped, with a wide, open base in front and narrow apex behind. The norma frontalis also shows the piriform aperture of the nose.

Norma Lateralis (Figs 8.17A to C)

Shows the following bones—frontal, parietal occipital, temporal, sphenoid, zygomatic, maxilla and mandible.

Other features: The zygomatic arch is a horizontal bar of bone on the side of the head formed by temporal process of zygomatic bone and zygomatic process of temporal bone.

External acoustic meatus is an opening in the temporal bone, seen just below the posterior part of zygomatic arch.

The mastoid part of temporal bone is seen just behind the external acoustic (auditory) meatus.

The styloid process is a long, thin projection from the temporal bone. It is directed downward, forwards and slightly medially.

Pterion is the junction of frontal, parietal, greater wing of sphenoid and squamous part of temporal bone. These bones are united by an H-shaped suture. The skull is relatively thin here. A fracture in this region

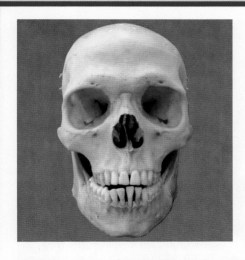

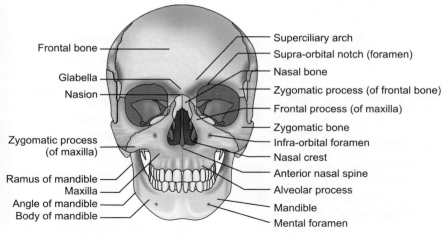

Frontal bone
Glabella
Nasion
Zygomatic process
(of maxilla)
Ramus of mandible
Maxilla
Angle of mandible
Body of mandible

Superciliary arch
Supra-orbital notch (foramen)
Nasal bone
Zygomatic process (of frontal bone)
Frontal process (of maxilla)
Zygomatic bone
Infra-orbital foramen
Nasal crest
Anterior nasal spine
Alveolar process
Mandible
Mental foramen

Fig. 8.16: Norma frontalis

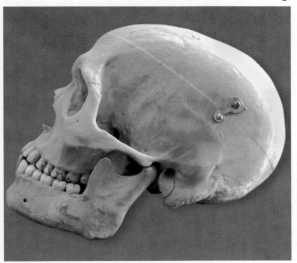

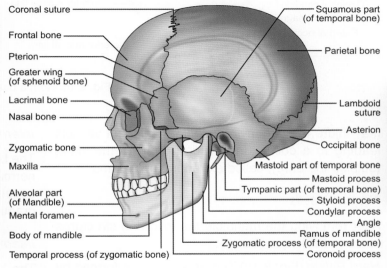

Coronal suture
Frontal bone
Pterion
Greater wing
(of sphenoid bone)
Lacrimal bone
Nasal bone
Zygomatic bone
Maxilla
Alveolar part
(of Mandible)
Mental foramen
Body of mandible
Temporal process (of zygomatic bone)

Squamous part
(of temporal bone)
Parietal bone
Lambdoid
suture
Asterion
Occipital bone
Mastoid part of temporal bone
Mastoid process
Tympanic part (of temporal bone)
Styloid process
Condylar process
Angle
Ramus of mandible
Zygomatic process (of temporal bone)
Coronoid process

Fig. 8.17A: Normal lateralis

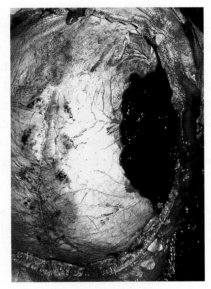

Fig. 8.17B: Extra dural hematoma

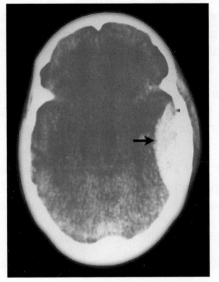

Fig. 8.17C: CT of head showing dural hematoma

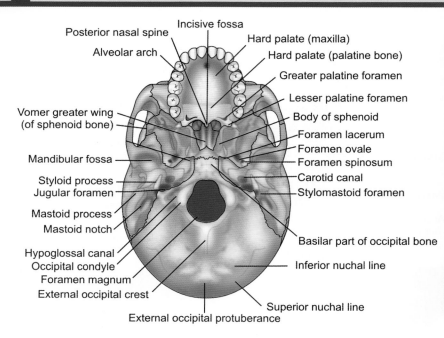

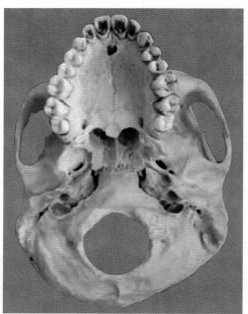

Incisive fossa
Posterior nasal spine
Hard palate (maxilla)
Alveolar arch
Hard palate (palatine bone)
Greater palatine foramen
Lesser palatine foramen
Vomer greater wing
(of sphenoid bone)
Body of sphenoid
Foramen lacerum
Foramen ovale
Mandibular fossa
Foramen spinosum
Styloid process
Carotid canal
Jugular foramen
Stylomastoid foramen
Mastoid process
Mastoid notch
Hypoglossal canal
Basilar part of occipital bone
Occipital condyle
Inferior nuchal line
Foramen magnum
External occipital crest
Superior nuchal line
External occipital protuberance

Fig. 8.18: Norma basalis

may injure the underlying middle meningeal vessels leading to a collection of blood deep to the skull, (outside the dura), called extradural hematoma.

Norma Basalis

After removing the mandible the following structures can be seen:

Hard Palate (Fig. 8.18) The anterior 2/3 is formed by the palatal processes of maxillae; the posterior 1/3 by the horizontal plates of palatine bones. The hard palate shows a cruciform suture between these bones. The greater palatine foramen is seen just behind the palato-maxillary suture.

Middle part of norma basalis extends from the posterior border of the hard palate to an imaginary transverse line passing through the anterior margin of foramen magnum.

Posterior part of norma basalis shows the largest opening in the skull—the foramen magnum. On either side of this foramen are the occipital condyles.

Interior of the Skull

The cranium is lined internally by endocranium.

Internal Surface of Vault

The bones and sutures have been described with the norma verticalis. The bones are grooved by the gyri of cerebral cortex, blood vessels and arachnoid granulations.

Internal Surface of Base of the Skull

The base of the skull presents anterior, middle and posterior cranial fossae.

Anterior cranial fossa
1. *Boundaries*
   • Anteriorly and on the sides—frontal bone
   • Posteriorly—lesser wing of sphenoid
   • Floor—anteriorly by cribriform plate of ethmoid
   • Posteriorly by a part of the body of sphenoid
   • On each side by orbital plate of frontal bone.
2. *Contents*—Frontal lobe of cerebrum, olfactory bulb.

Applied Anatomy

A fracture of anterior cranial fossa may cause bleeding and discharge of CSF through nose (CSF–rhinorrhea). If not diagnosed, this can lead to meningitis.

Middle cranial fossa: It is deeper than the anterior cranial fossa and shaped like a butterfly, being narrow and shallow in the middle and wide and deep on each side, lodging the temporal lobe of cerebral hemisphere.

1. Boundaries
   - Anteriorly—Posterior or lesser wing of sphenoid
   - Posteriorly—superior border of petrous temporal bone and dorsum sellae of sphenoid
   - Laterally—Greater wing of sphenoid, antero-inferior angle of parietal bone.
2. Floor
   - Median plane—body of sphenoid
   - Laterally—greater wings of sphenoid, petrous part of temporal and squamous part of temporal bones.
3. Important features
   a. Optic canal – transmits optic nerve to the orbit.
   b. *The sella turcica* ("Turkish saddle"). The upper surface of body of sphenoid is hollowed out in the form of a Turkish saddle. It has a hypophyseal fossa in the middle, which lodges the pituitary gland.

Applied Anatomy

A fracture of middle cranial fossa leads to:
1. bleeding or discharge of CSF through the ear, nose or mouth
2. 7th and 8th cranial nerves may be damaged if the fracture passes through the internal acoustic meatus.
   The middle cranial fossa is most commonly fractured.

Posterior cranial fossa: Largest and deepest of the three cranial fossae. Contains the hindbrain which consists of the cerebellum behind and the pons and medulla in front.

1. Boundaries:
   - Anteriorly
     — Superior border of petrous temporal bone
     — Dorsum sellae of sphenoid
   - Posteriorly
     — Squamous part of occipital bone
   - On each side
     — Mastoid part of temporal bone
     — Mastoid angle of parietal bone
2. Floor
   — Sloping area of sphenoid and occipital (clivus)
   — Foramen magnum

— Squamous part of occipital, behind.
- Laterally
  — Condylar part of occipital
  — Posterior surface of petrous temporal
  — Mastoid part of temporal.

Individual Skull Bones

The Frontal Bone

The frontal bone (Figs 8.19A and B) is an irregular, flat bone which forms the forehead region. It has large air filled spaces called frontal sinuses which communicate with the nasal cavity via frontonasal canal.

On the anterior aspect of frontal bone are two eminences—the frontal tuberosities. Below them, there are two curved superciliary arches, which are more prominent in males. These arches are joined across the midline by a smooth eminence, the glabella.

Below the superciliary arches are the supraorbital margins which form the upper borders of orbital openings. At the junction of lateral 2/3 and medial 1/3 in the supraorbital margin is a supraorbital notch which transmits supraorbital vessels and nerve.

A portion of frontal bone – the nasal part – projects downwards between the supraorbital margin, which articulates with the nasal bones.

The orbital plate of the frontal bone separates the anterior cranial fossa from the orbit. It supports the orbital surface of the frontal lobe of the brain. The orbital plate covers the ethmoidal air cells also. The posterior margin articulates with the lesser wing of sphenoid.

The following bones articulate with the frontal bone— the parietal, sphenoid, nasal, zygomatic and ethmoid.

The Parietal Bones (Figs 8.20A and B)

Two in number; right and left; joined superiorly by the sagittal suture. They form greater part of the side walls and roof of the cranium. Each bone has 4 sides and four angles.

Anterior border articulates with frontal bone at the coronal suture.

Posterior border articulates with occipital bone at the lambdoid suture; inferiorly with the squamous part of temporal bone.

At birth, there are membrane-bound intervals at the four angles of the parietal bone (fontanelle).

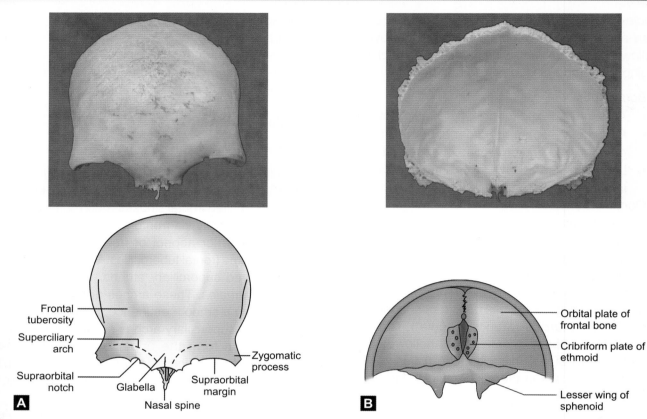

Figs 8.19A and B: (A) Frontal bone (anterior view). (B) Anterior cranial fossa showing parts of frontal, ethmoid and sphenoid bones

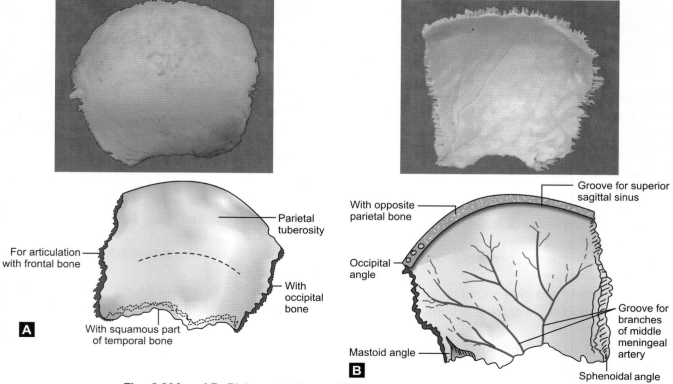

Figs 8.20A and B: Right parietal bone. (A) External surface, (B) Internal surface

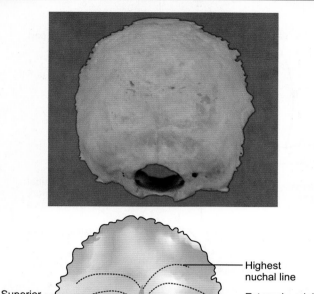

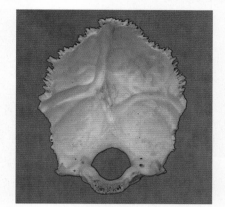

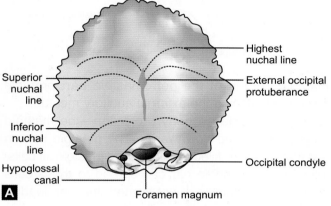

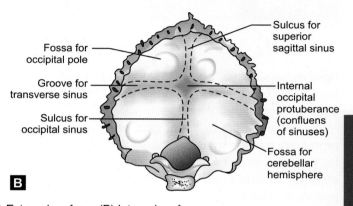

Figs 8.21A and B: Occipital bone. (A) External surface, (B) Internal surface

Internal surface of parietal bone is marked by impressions of gyri of cerebrum and furrows for middle meningeal vessels. The groove for middle meningeal vessels run from the anteroinferior angle towards posteriorly.

A shallow groove for superior sagittal sinus runs along the superior border.

The point of maximum convexity of parietal bone is called parietal tuberosity.

Occipital Bone (Figs 8.21A and B)

Situated at the back and base of the skull, this unpaired bone forms the greater part of the floor of posterior cranial fossa.

Articulates in front with parietal bones at the lambdoid suture, on either side, it articulates with temporal bone.

Most striking feature is the presence of a large foramen—the *foramen magnum. Structures passing through foramen magnum* are.

1. Lower part of medulla, continues as spinal cord
2. The meninges
3. Spinal part of accessory nerve
4. Vertebral arteries
5. Anterior and posterior spinal arteries
6. Apical ligament of dens

Parts of Occipital Bone

1. Behind foramen magnum—the squamous part
2. In front of foramen magnum—basilar part
3. On either side—condylar part.

Squamous part is the expanded, plate-like portion. Its inner side shows 4 fossae; two upper, for the occipital lobes. 2 lower, for the cerebellar hemispheres. At the junction of the four fossae is the internal occipital protuberance. The sulcus for the superior sagittal sinus runs upwards from the internal occipital protuberance; on either side are the sulci for transverse sinuses.

Five dural venous sinuses meet at the internal occipital protuberance—the superior sagittal sinus, 2

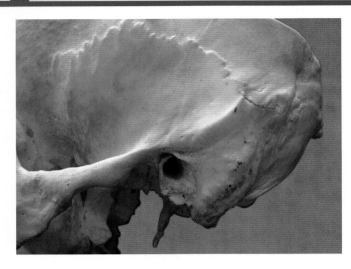

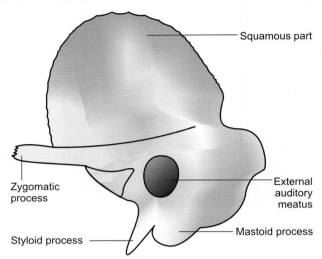

Fig. 8.22: Left temporal bone—external features

transverse sinuses, occipital sinus and straight sinus from front. This junction is known as the *confluence of sinuses* (confluens of sinuses).

External aspect of squamous part shows a bony prominence in the median plane—the external occipital protuberance. Arching on either side, are the superior nuchal lines; below these lines, there are inferior nuchal lines.

Basilar part: It is a thick bar of bone; extends anteriorly to articulate with the sphenoid bone to form the clivus, which supports the pons and basilar artery.

Condylar part: Inferior surface shows two occipital condyles for articulation with the atlas vertebra.

Temporal Bone (Fig. 8.22)

These are paired bones, seen on the lateral side of skull.

Parts It has 4 parts:
1. Squamous part
2. Petromastoid part (Petrous + mastoid)
3. Tympanic part
4. Styloid process

Squamous part is the thin, expanded portion. It articulates with parietal and sphenoid bones. On the outer aspect is the external auditory meatus; it is an opening leading to external auditory canal. A zygomatic process projects anteriorly, which articulates with the temporal process of zygomatic bone to form zygomatic arch.

Mastoid portion of the petromastoid part projects downwards behind the external auditory meatus.

There is a dense bony projection—the mastoid process—directed downwards. The mastoid portion has numerous air-filled cavities called mastoid air cells. Infection from middle ear can spread to these cells leading to mastoiditis. If not treated, this can lead to meningitis and brain abscess.

Muscles attached to mastoid process are sternocleidomastoid and posterior belly of digastric.

Petrous part: (This dense ivory-hard bone is named after Peter, a disciple of Jesus Christ, whose faith in Jesus Christ was as hard as rock !). It projects anteromedially, wedged between sphenoid and occipital bones. It separates the middle and posterior cranial fossae. Posterior surface shows a foramen—the *internal auditory meatus.* It transmits the 7th and 8th cranial nerves and labyrinthine vessels.

Petrous part contains the inner ear—cochlea, vestibule and semicircular canals.

Tympanic part forms the anterior wall, floor and lower part of posterior wall of external auditory meatus.

Styloid process projects from the under surface of temporal bone. It is a long, thin process to which the following muscles and ligaments are attached:
• Muscles
  – Stylohyoid
  – Styloglossus
  – Stylopharyngeus

- Ligaments
  - Stylohyoid
  - Stylomandibular

These structures, with the styloid process, together form the styloid apparatus.

The Sphenoid (Figs 8.23A and B); (See CP1 in Color Plate 1)

(sphene = wedge). This unpaired bone, resembling a bat, is wedged between other skull bones.

Parts
1. Body
2. 2 lesser wings
3. 2 greater wings
4. 2 pterygoid processes.

The body is hollow, contains two large air-filled cavities called sphenoidal air sinuses separated by a bony septum. These sinuses open into the nasal cavity.

Upper surface of the body is shaped like a Turkish saddle. So, it is known as sella turcica. Deepest part of the sella turcica is the hypophyseal fossa, which lodges the pituitary gland. Behind the sella turcica is the dorsum sellae.

Lesser wings: From the anterior part of the body, the lesser wings spread out on either side to form the posterior margin of the anterior cranial fossa.

Greater wings: From the sides of the body, the *greater wings* spread outwards between the frontal and temporal bones. The greater wing shows a well-marked oval foramen—the *foramen ovale*. Structures passing through this foramen are:
1. Mandibular nerve
2. Accessory meningeal artery
3. Lesser petrosal nerve
4. Emissary vein (can be remembered easily— "MALE")

Pterygoid processes The two pterygoid processes—Lateral and medial project downwards from the adjoining parts of the body and greater wings.

Applied Anatomy
1. Greater wing articulates with the squamous part of temporal, frontal and anteroinferior angle of parietal to form "pterion", which is closely related internally to middle meningeal vessels. A fracture at this site may tear the vessels leading to an extradural hematoma.

Contd...

Contd...
2. A tumor of hypophysis cerebri is likely to deepen or erode the hypophyseal fossa which can be detected in an X-ray of the lateral view of skull.
3. Infection from nasal cavity can spread to sphenoidal air sinus, leading to sinusitis.

Optic canal: Lies between the two roots or lesser wing.

Transmits: 1. Optic nerve with meningeal coverings, 2. Ophthalmic artery, with sympathetical fibers.

Superior orbital fissure: It is a triangular passage lying between the greater and lesser wings. It leads from the cranial cavity to orbital cavity.

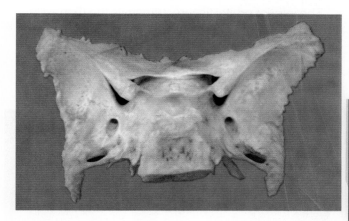

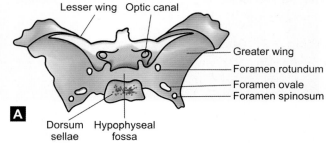

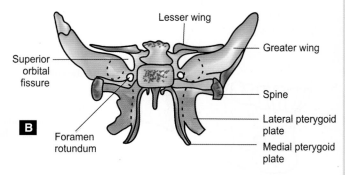

Figs 8.23A and B: Sphenoid. (A) Superior view, (B) Posterior view

Section B

Musculoskeletal System

Structures passing through this fissure are:

1. *From cranial cavity to orbit*
   a. Superior and inferior divisions of oculomotor (Cr. N. III) nerve
   b. Nasociliary nerve
   c. Abducent nerve (Cr. N. VI)
   d. Trochlear nerve (Cr. N. IV)
   e. Frontal nerve
   f. Lacrimal nerve
   g. Orbital branch of middle meningeal artery.
2. *Structures from orbit to cranial cavity*
   a. Superior ophthalmic vein
   b. Inferior ophthalmic vein

The inferior orbital fissure: transmits the (i) maxillary nerve; (ii) infraorbital vessels; (iii) zygomatic nerve.

Foramen rotundum: It is a round foramen just below and behind the medial end of superior orbital fissure, at the antero-medial part of greater wing of sphenoid.

Transmits: Maxillary nerve, a sensory branch of trigeminal nerve (Cr. N. V).

Foramen spinosum lies just behind and lateral to foramen ovale; it transmits the middle meningeal artery and a meningeal branch of mandibular nerve (nervus spinosus).

Some important foramina and fissures of the skull and the structures passing through them are described in Table 8.3.

Mandible (Figs 8.24A and B)

Mandible or lower jaw is the largest, strongest and the only mobile bone of the skull.

Parts (i) Body, and (ii) A pair of rami.

Body is horse-shoe shaped; it has an outer and inner surface; upper and lower border. The right and left halves of mandible meet anteriorly at a faint ridge, the

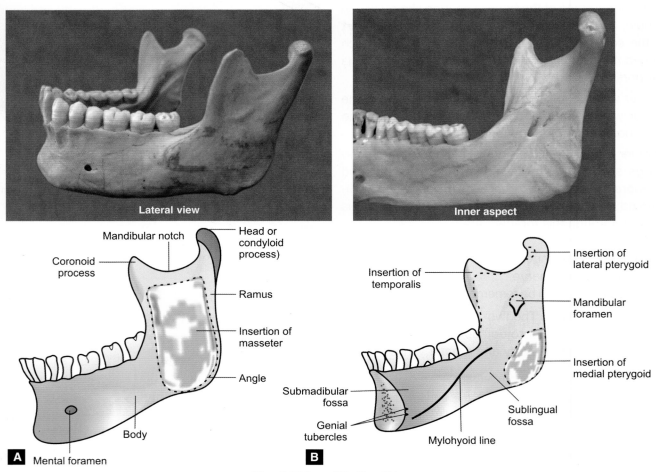

Figs 8.24A and B: Mandible

Table 8.3: Most important foramina and fissures of skull

Foramen magnum	1. Lower part of medulla, continuing inferiorly as spinal cord 2. The meninges 3. Spinal accessory nerves 4. Vertebral arteries 5. Anterior and posterior spinal arteries 6. Apical ligament of dens
Jugular foramen	1. Glossopharyngeal nerve (IX cr. N) 2. Vagus nerve (X cr. N) 3. Accessory nerve (XI cr. N) 4. Sigmoid sinus continues as internal jugular vein 5. Inferior petrosal sinus
Foramen ovale	1. Mandibular nerve 2. Accessory meningeal artery 3. Lessr petrosal nerve 4. Emissary vein
Foramen rotundum	Maxillary nerve
Stylomastoid foramen	1. Facial nerve 2. Stylomastoid vessels
Carotid canal	1. Internal carotid artery 2. Sympathetic plexus
Optic canal	1. Optic nerve, covered by meninges 2. Ophthalmic artery
Internal acoustic meatus	1. Facial nerve 2. Vestibulo cochlear nerve 3. Labyrinthine vessels
Mandibular foramen	1. Inferior alveolar nerve 2. Inferior alveolar vessels
Superior orbital fissure	1. Lacrimal nerve 2. Frontal nerve 3. Trochlear nerve 4. Upper and lower divisions of oculomotor nerve 5. Abducent nerve 6. Superior ophthalmic vein
Inferior orbital fissure	1. Maxillary nerve 2. Infraorbital vessels 3. Zygomatic nerve
Infraorbital foramen	1. Infraorbital nerve 2. Infraorbital vessels

symphysis menti. There is a small opening below the interval between 2 premolars—the mental foramen, transmitting mental nerve and vessels.

The inner surface shows an oblique ridge—the mylohyoid line—which gives attachment to mylohyoid muscle. Below the mylohyoid line there is a shallow depression, the submandibular fossa, to lodge the submandibular gland.

Above the mylohyoid line is the sublingual fossa for the sublingual gland.

The upper border or alveolar process bears sockets for lower set of teeth.

The ramus is quadrilateral in shape; it has 2 surfaces—lateral and medial and 4 borders (upper, lower, anterior and posterior) and 2 processes— coronoid and condyloid processes.

The lateral surface is rough, gives attachment to masseter muscle.

The medial surface shows a mandibular foramen, which transmits the inferior alveolar vessels and nerve. This foramen leads to the mandibular canal.

The upper border of ramus is thin and curved, known as mandibular notch.

The coronoid process is triangular in shape, projecting upwards from the anterosuperior part of ramus. It receives the insertion of temporalis muscle.

The condyloid process is a strong upward projection from the posterosuperior part of ramus. Its upper expanded part is the head, which articulates with the temporal bone to form the temporomandibular joint.

The constriction below the head is the neck. Its anterior surface presents a depression, the pteryoid fovea, where lateral pterygoid muscle is inserted.

To a rough area below and behind the mylohyoid groove, on the medial surface of ramus, the medial pterygoid muscle is attached.

Behind the symphysis menti, there are 4 small tubercles—the genial tubercles—upper two give origin to genioglossus and lower two give origin to geniohyoid muscle.

Applied Anatomy

The mandible is often fractured in accidents. Dislocation at temporomandibular joint is common.

Maxilla (Fig. 8.25)

There are 2 maxillae, forming the upper jaw. Each maxilla has a body and 4 processes—zygomatic, alveolar, frontal and palatine.

The body has a large air-filled cavity, the maxillary sinus, which communicates with the nasal cavity. The anterior surface of the body presents an infraorbital foramen, which transmits infraorbital nerve and vessels. The posterior surface of the body shows a

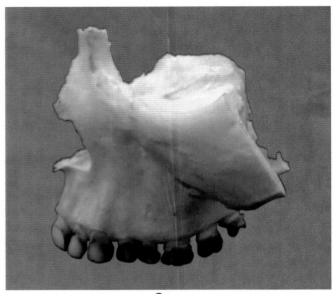
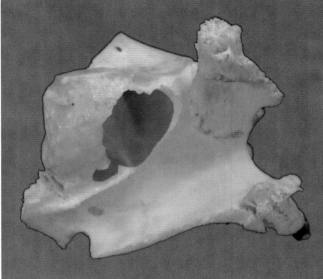

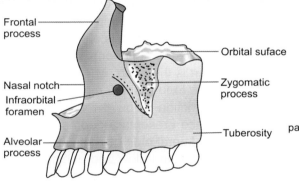

Frontal process

Orbital suface

Nasal notch

Infraorbital foramen

Zygomatic process

Alveolar process

Tuberosity

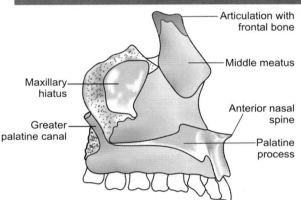

Articulation with frontal bone

Maxillary hiatus

Middle meatus

Greater palatine canal

Anterior nasal spine

Palatine process

Fig. 8.25: Maxilla

bony prominence, the maxillary tuberosity. On the medial surface is a large opening—the maxillary hiatus which leads to maxillary sinus.

Processes of Maxilla

1. The *zygomatic process* is a pyramidal projection from the lateral side of the body, which articulates with zygomatic bone.
2. The alveolar process shows sockets for one half of upper set of teeth (8 in number).
3. The frontal process projects upwards from the body.
4. The palatine process is a horizontal plate-like process which projects medially from the medial surface of maxilla. It forms the anterior 2/3 of hard palate along with the palatine process of the opposite maxilla.

Applied Anatomy

1. Fracture—in accidents
2. Maxillary sinusitis—infection from nose can spread to maxillary sinus.

Ethmoid (See CP2A and B in Color Plate 1)

It is an unpaired bone; light and fragile; takes part in forming:
1. Medial walls of orbit
2. Septum of nose—the perpendicular plate of ethmoid forms a part of the septum
3. Roof of nose
4. Lateral walls of nose.

Parts

a. A horizontal, cribriform plate
b. Perpendicular plate
c. Two lateral masses.

The cribriform (Sieve-like) plate is situated or wedged between the ethmoid notch of frontal bone. It has numerous small openings through which the olfactory nerves reach the olfactory bulb from the nasal cavity. The superior surface of cribriform plate shows a triangular, vertical projection, resembling a cock's comb—it is called crista galli. It gives attachment to the anterior end of falx cerebri (sickle-shaped fold of dura).

The ethmoid bone has numerous air-filled cavities lined with respiratory mucosa—they are called anterior, middle and posterior groups of ethmoid air cells or sinuses. They open into the nasal cavity.

Applied Anatomy

A fracture of the base of skull (anterior cranial fossa) may lead to leak of blood or CSF through nose; this can lead to meningitis. Infections from nose can spread to ethmoid sinus, causing ethmoid sinusitis.

Fetal Skull (Fig. 8.26)

Fetal skull shows the following features:

1. *Presence of fontanelles:* There are some gaps in the bony vault of the skull, which are bridged by membranes formed by the fusion of periosteum and underlying dura mater. There are 6 fontanelles, one in relation to each angle of parietal bones. They are termed anterior (bregmatic), posterior (lambdoid), 1 pair of anterolateral (sphenoidal) and 1 pair of posterolateral (mastoidal). The lateral fontanelles are usually closed at birth. The posterior one closes 3 months after birth; the anterior one, only 1½–2 years after birth.

Applied Anatomy

1. In hydrocephalus, the fontanelles are raised.
2. In severe dehydration, the fontanelles are depressed.

2. Markings for muscles and fasciae are absent in fetal skull.
3. Facial part of skull is small. The maxillae are short owing to the complete absence of alveolar processes and small size of maxillary sinuses. The orbits appear large, but they are shallow.
4. Mandible has 2 halves, united by the fibrous tissue at the symphysis menti.
5. Mastoid process is absent.

A summary of most important foramina of skull and the structures they transmit is given in Table 8.4.

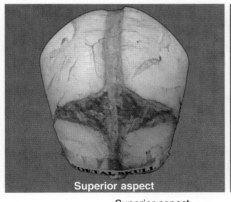
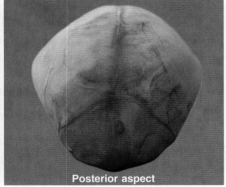
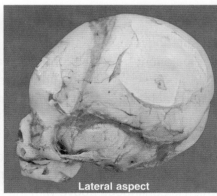

Superior aspect   Posterior aspect   Lateral aspect

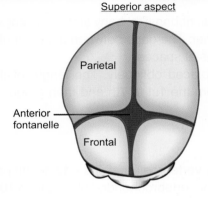
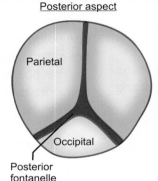
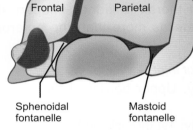

Superior aspect

Parietal

Anterior fontanelle

Frontal

Posterior aspect

Parietal

Occipital

Posterior fontanelle

Lateral aspect

Frontal   Parietal   Occipital

Sphenoidal fontanelle   Mastoid fontanelle

Fig. 8.26: Foetal skull

No.	Structure/Organ	Vertebral level
	Table 8.4: Some important structures and their vertebral levels	
1.	Bifurcation of common carotid artery	C_4
2.	Bifurcation of trachea	T_4
3.	Cricoid cartilage/pharyngo-esophageal junction	C_6
4.	Epiploic foramen	T_{12}
5.	Transpyloric plane	L_1
6.	Celiac trunk (origin)	Disc between T_{12} and L_1
7.	Superior mesenteric artery (origin)	L_1
8.	Inferior mesenteric artery (origin)	L_3
9.	Termination of spinal cord	Lower border of L_1
10.	Termination of abdominal aorta	L_4
11.	Beginning of inferior vena cava	L_5
12.	Portal vein–beginning	L_2
13.	Umbilicus	Disc between L_3 and L_4
14.	Transtubercular plane	L_5

THE SKELETON OF THE THORAX (THORACIC CAGE)

The thoracic cage protects the chief organs of respiration and circulation. It is an osseocartilaginous, elastic cage which is designed to increase or decrease the intrathoracic pressure.

Formation

- Anteriorly   : by sternum
- Posteriorly   : by 12 thoracic vertebrae and Intervertebral discs between them
- On each side   : 12 ribs and their cartilages.

Shape

The thorax resembles a truncated cone, which is narrow above and broad below. The narrow upper end, which is continuous above with the neck, is called the inlet of the thorax.

The Inlet of Thorax

The Inlet of thorax is kidney shaped, its transverse diameter is more than the anteroposterior diameter.

Boundaries of inlet

- Anteriorly: Upper border of the manubrium of sternum
- Posteriorly: 1st thoracic vertebra
- On each side: 1st rib, with its cartilage.

The Outlet (Inferior Aperture)

The outlet (inferior aperture) of thorax is the broad end of thorax which surrounds the upper part of abdominal cavity, but is separated from it by diaphragm.

Boundaries

- Anteriorly: Infrasternal angle between the two costal margins
- Posteriorly: 12th thoracic vertebra on each side.
- Costal margin is formed by the cartilages of 7-10 ribs and the 11th and 12th ribs.

Bones of Thorax

The Ribs (Costae)

There are normally 12 pairs of ribs, forming greater part of the thoracic skeleton (sometimes a cervical or lumbar rib may be present or the 12th rib may be absent).

The ribs are flat, ribbon-like, bony arches, arranged one below the other. The gaps between adjacent ribs are called intercostal spaces.

The ribs are placed obliquely. The length of the ribs increases from the 1st to 7th and then gradually decreases.

Classification

1. a. True ribs or vertebrosternal ribs – 1st to 7th pair
   b. False ribs or vertebrochondral ribs – 8th to 10th pair

c. Floating ribs or vertebral ribs – 11th and 12th rib

2. a. Typical ribs – 3rd to 9th ribs
   b. Atypical ribs – 1,2,10,11,12

The first seven pairs of ribs are connected through their cartilages to the sternum. So, they are called vertebrosternal or true ribs. The cartilages of 8th, 9th and 10th ribs are joined to the next higher cartilage – so, they are called vertebrochondral ribs or false ribs. The anterior ends of the 11th and 12th ribs are free. They are called vertebral ribs, or floating ribs.

Typical Ribs (Figs 8.27A and B)

Each rib has an anterior end, shaft and a posterior end.

The anterior end bears a concave depression for attachment of costal cartilage.

Posterior end shows a head, neck and tubercle. The head has two articular facets; lower and upper. The lower larger facet articulates with the body of the numerically corresponding vertebra, while the upper smaller facet articulates with the next higher vertebra.

The neck lies in front of the transverse process of its own vertebra.

The tubercle is placed at the junction of the neck and shaft.

The shaft is flattened; it has an inner and outer surfaces; upper and lower borders. The shaft is curved

with its convexity outwards 0.5 cm away from the tubercle, the shaft abruptly changes its direction; this is called the angle of the rib.

The inner surface shows a costal groove near the inferior border. It lodges, from above downwards, the intercostal vein, artery and nerve (VAN).

First Rib

The first rib is the shortest, broadest and most curved rib. Its shaft is flattened from above downwards. So, it has an upper and lower surfaces and outer and inner borders.

Features (Fig. 8.28): The anterior end is larger and thicker. It articulates with the 1st costal cartilage. Its small, rounded head articulates with T_1 vertebra.

The upper surface of the shaft shows a scalene tubercle, which separates 2 grooves lying anterior and posterior to it. The anterior groove lodges the subclavian vein; the posterior groove lodges subclavian artery and lower trunk of brachial plexus. The scalenus anterior muscle is attached to the scalene tubercle.

Second Rib

The length is twice that of the first rib. The shaft is sharply curved. The shaft has no twist. Near its middle, the shaft is marked by a large rough tubercle, for attachment of upper part of serratus anterior muscle.

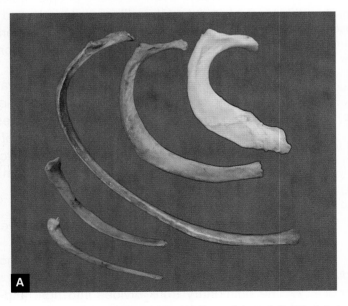

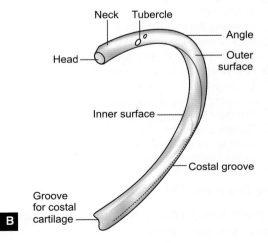

Figs 8.27A and B: (A) From right to left, Rib 1,2, typical, 11 and 12 (B) Typical rib (drawing)

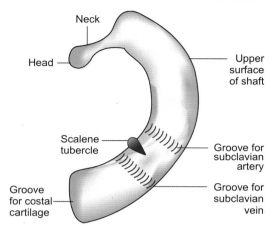

Fig. 8.28: First rib (left)

Tenth Rib

The tenth rib closely resembles a typical rib, but is shorter and has only one articular facet on the head, for the 10th thoracic vertebra.

11th and 12th Ribs

The 11th and 12th ribs are short. They have pointed ends. Necks and tubercles are absent. The angle and costal groove are absent in the 12th rib and poorly marked in the 11th rib.

Applied Anatomy—Ribs

1. *Abnormal number:* Although there are usually 12 pairs of ribs, the number may be increased by the development of a cervical rib or lumbar ribs. The number may be decreased by the failure of development of 12th rib. A cervical rib may compress the subclavian artery and lower trunk of brachial plexus.
2. *Fracture:* Angle of ribs are the weakest points; even though the ribs have the ability to bend under stress, they may be fractured in automobile accidents, especially in adults.
3. *Flail chest* results from multiple rib fractures.
4. *Bone grafting:* Sometimes a piece of rib is used for autogenous bone grafting, e.g. for reconstruction of mandible following excision of a tumor.
5. Rickets cause the ends of the ribs (usually 2nd – 8th) to enlarge. The bead-like bony enlargements are referred to as the "rachitis rosary" or "rickety rosary".

The Sternum (Fig. 8.29)

The sternum is a flat bone, forming the anterior median part of the thoracic skeleton. It resembles a dagger or a short sword in shape.

Parts

The upper part is called the manubrium (handle). The middle part is called the body and the lowest tapering part is called the xiphisternum or xiphoid process (forming the point of the sword).

Length

15-17 cm.

Manubrium

The manubrium is quadrilateral in shape. It is the thickest and strongest part of the sternum. It has an anterior and posterior surface, and 4 borders—superior, inferior and 2 lateral.

The superior border is thick, rounded and concave. It is marked by the suprasternal notch (jugular notch) and by the clavicular notch on each side of it. The clavicular notch articulates with the medial end of clavicle to form the sternoclavicular joint.

The inferior border forms a secondary cartilaginous joint with the body of the sternum. The manubrium makes a slight angle with the body at this junction, called the sternal angle or the angle of Louis.

Sternal Angle or Angle of Louis— Clinical Importance

It is felt as a transverse ridge about 5 cm below the suprasternal notch, even in obese people. It lies at the level of 2nd costal cartilage anteriorly and the intervertebral disc between the 4th and 5th thoracic vertebrae posteriorly. The sternal angle is an important bony landmark for the following reasons:

1. The ribs are counted from this level downwards. The second costal cartilage and rib lie at this level. Counting the ribs and intercostals spaces is important in marking the borders and apex of the heart, the pleura, the upper border of liver, etc.
2. The superior mediastinum is separated from the inferior mediastinum at this level.
3. The ascending aorta ends at this level.
4. The arch of aorta begins and ends at this level.
5. The descending aorta begins at this level.
6. The trachea bifurcates into right and left principal bronchi.
7. The pulmonary trunk divides into right and left pulmonary arteries.

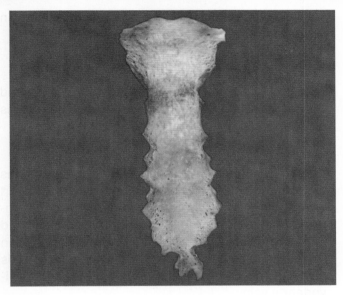

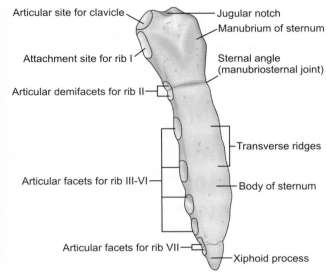

Fig. 8.29: Sternum

8. Upper border of heart lies at this level.
9. Azygos vein crosses the right lung root to end in the superior vena cava.

Body of the Sternum

The body of the sternum is longer, narrower and thinner than the manubrium. The lateral borders form synovial joints with 2nd-6th costal cartilages and upper half of 7th costal cartilage.

The upper end articulates with the manubrium at the sternal angle and lower end articulates with the xiphisternum.

The xiphoid process is the smallest part of the sternum. It varies greatly in size and shape. It may be bifid or may show a foramen. The rectus abdominis and a few slips of diaphragm are attached to the xiphoid process.

Applied Anatomy

1. Sternal puncture: A sample of bone marrow can be obtained from sternum by a procedure called sternal puncture.
2. In "pigeon chest", there is forward projection of sternum; in "funnel chest", the sternum is depressed.
3. To gain access to the mediastinum for surgical operations on heart and great vessels, it is often necessary to divide the sternum in the median plane.
4. Sternal fracture is common following accidents, e.g. in automobile accidents, when the driver's chest is hit against the steering wheel. Sternum is often fractured at the sternal angle.

■ VERTEBRAL COLUMN (Fig. 8.30)

The vertebral column (spine, spinal column or back bone) is the main part of axial skeleton. It consists of 33 bones called vertebrae and the intervertebral discs between them. The vertebral column forms a strong and flexible support for the trunk; it extends from the base of the skull through the neck and trunk.

The vertebral column has important roles in posture, support the body weight, locomotion and protection of spinal cord and nerve roots; it also transmits the body weight to the ground through the lower limbs.

The vertebral column is made up of 33 vertebrae—7 cervical, 12 thoracic, 5 lumbar, 5 sacral and 4 coccygeal; but only 24 of them are movable (7 cervical, 12 thoracic, 5 lumbar). In adults, the 5 sacral vertebrae are fused to form the sacrum and the coccygeal vertebrae are fused to form the coccyx. The abbreviations C,T,L,S and Co are used for the regions of the vertebral column.

The 24 movable vertebrae give the vertebral column considerable flexibility; the intervertebral discs between them also play an important role in movements and in absorbing shocks.

The length of the spine is about 28 inches in males and 24 inches in females. The vertebrae contribute to 4/5 of total length and remaining 1/5 by intervertebral discs.

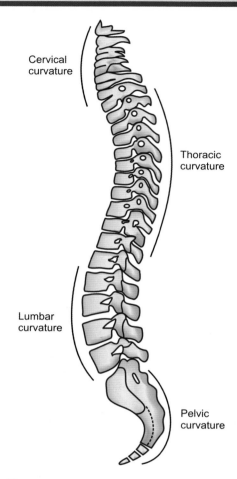

Fig. 8.30: Vertebral column—lateral aspect

Curvatures of Vertebral Column

In the articulated vertebral column and in MRIs, four curvatures are normally visible in the adult; (1) the thoracic and sacral curvatures are concave anteriorly; (2) the cervical and lumbar curvatures are convex anteriorly.

Primary Curvatures

The thoracic and sacral curvatures are called primary curvatures, because they develop during the fetal period.

Secondary Curvatures

The cervical and lumbar curvatures become obvious after birth. So, they are called *secondary curvatures*. The cervical curvature becomes prominent when an infant begins to hold its head erect; the lumbar curvature becomes obvious when a child begins to walk. The lumbar curvature is more prominent in women.

Parts of a Typical Vertebra (Figs 8.31A and B)

A "typical" vertebra (T5 to T8, L1 and L2) is composed of 2 parts, a body and a vertebral arch. The body is anterior and the vertebral arch is posterior.

The Body

The body is the large, heavy, anterior part which has the form of a short cylinder. Its function is to support weight. The bodies of the vertebrae become progressively larger to bear weight. The superior and inferior surfaces are rough and flat. A nutrient foramen is seen on the anterior surface and a larger foramen is seen on the posterior surface for the exit of basi-vertebral vein.

The Vertebral Arch or Neural Arch

The vertebral arch or neural arch encloses the vertebral foramen. It is attached on each side of the body and protects the spinal cord and spinal nerve roots (the neural tissues) from injury.

The arch is formed by two pedicles (L. little feet), which project posteriorly to meet two laminae (L. thin plates). The laminae meet posteriorly to form a spinous process.

Four articular processes and two transverse processes also arise from the vertebral arch.

The space enclosed by the body and arch is the vertebral foramen. The vertebral foramina in an articulated vertebral column form the vertebral canal (spinal canal), which contains:
1. The spinal cord
2. Meninges
3. Nerve roots
4. Blood vessels

Pedicles

The pedicles are short, stout processes attached to the superior part of the body on each side. The superior and inferior surfaces are concave—they are called superior and inferior vertebral notches. When two vertebrae articulate with each other, the vertebral notches of the adjacent vertebrae form a complete, oval, bony ring called intervertebral foramen; the dorsal root ganglion of the spinal nerve lies in this foramen.

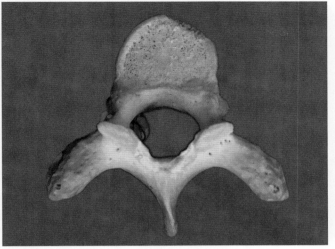
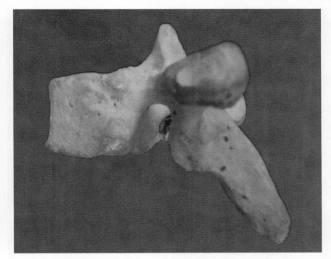

Superior aspect

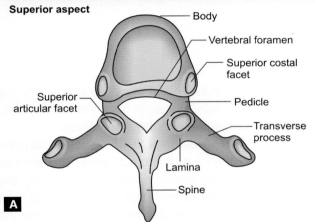

- Body
- Vertebral foramen
- Superior costal facet
- Pedicle
- Transverse process
- Superior articular facet
- Lamina
- Spine

A

Lateral view

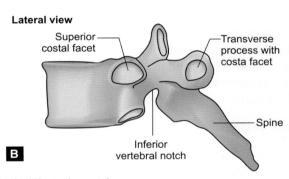

- Superior costal facet
- Transverse process with costa facet
- Spine
- Inferior vertebral notch

B

Figs 8.31A and B: Parts of typical thoracic vertebra

Laminae

The *laminae* are thin plates of bone, forming the roof of the vertebral foramen, where they unite in the median plane; the *spinous process* arises from this junction. The spinous processes or spines project posteroinferiorly. They overlap the vertebrae below. They give attachment to ligaments and muscles.

The *transverse processes* project posterolaterally from the junctions of pedicles and laminae. They provide attachment to the muscles of the back.

The *articular processes* (zygapophyses) also arise from the junction of laminae with pedicles. They are superior and inferior articular processes. The superior articular process of one vertebra articulates with the inferior articular process of the vertebra above it.

The Cervical Vertebrae (Fig. 8.32)

The cervical vertebrae are the smallest in size. They are 7 in number. They form the skeleton of the neck.

Classification

1. Typical cervical vertebrae—C3 to C6
2. Atypical cervical vertebrae—C1, C2 and C7

Typical Cervical Vertebra

1. Typical cervical vertebra can be identified by the following features
   1. Each transverse process has an oval foramen—the *foramen transversarium.* The vertebral artery passes through the foramina transversaria of all cervical vertebrae except that of C7.

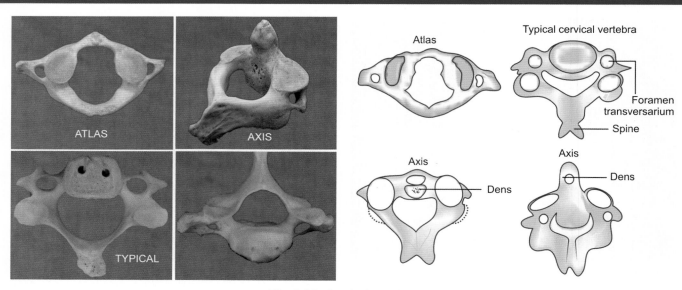

Fig. 8.32: Cervical vertebrae

2. The spines are short and bifid
3. The body is small. Its transverse diameter is greater than the anteroposterior length.
4. The vertebral foramen is large and triangular (this is to accommodate the cervical enlargement of spinal cord).

Atypical Vertebrae

Atlas

The first cervical vertebra (C1) is called the atlas. Because it supports the skull, it was named after Atlas, who, according to Greek mythology, supported the earth on his shoulders.

The atlas has no body or spinous process. The position of the body is occupied by the odontoid process or dens of the axis.

Atlas is a ring of bone consisting of anterior and posterior arches and a lateral mass on each side. The lateral masses have superior and inferior articular facets. The superior facet is concave, kidney-shaped, and articulates with the occipital condyle. The inferior facet is circular and flat. It articulates with the superior facet on axis.

The transverse processes are large and strong. The vertebral artery, after passing through the foramen transversarium, grooves the upper border of posterior arch.

The posterior surface of anterior arch shows an oval facet which articulates with the dens. The dens is retained in position by the transverse ligament of atlas.

Axis

This is the second cervical vertebra (C2). It is the strongest of the cervical vertebrae. It is called "axis" because the atlas carrying the skull rotates on it.

Its distinguishing feature is the blunt, tooth-like dens (odontoid process), which projects superiorly from its body. The dens (Gk.tooth) is held in position by the transverse ligament of the atlas (part of the body of atlas is incorporated into the body of C2; it forms the dens). The dens forms the pivot around which atlas, carrying the head, rotates. On either side of the odontoid process are oval, superior articular facets, which articulate with the atlas.

The laminae are thickest among the cervical vertebrae. The spine is large, bifid and powerful giving attachment to muscles and ligaments. The transverse process has the foramen transversarium for the vertebral artery; it gives attachment to muscles.

C7—called vertebraprominens because it has a long, stout spine.

The Thoracic Vertebrae—12 in Number (Figs 8.33 and 8.31)

The characteristic features:
1. Presence of costal facets on each side of the body for articulation with ribs.
2. Their spinous processes are long, slender and directed downward.

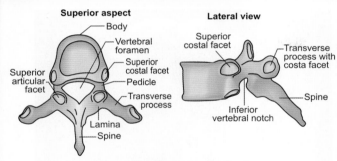

Fig. 8.33: Thoracic vertebra

3. The outline of their bodies, viewed from above, is heart-shaped.
4. Vertebral foramina are circular.

The Lumbar Vertebrae—5 in Number

Characteristic Features (Figs 8.34A and B)

1. Massive bodies
2. Sturdy laminae
3. Absence of costal facets on the sides of the body
4. Bodies are kidney-shaped
5. Vertebral foramina are oval or triangular
6. The spines are quadrangular, horizontal, with thickened posterior and lower borders
7. Transverse processes are thin and long.

The Fifth Lumbar Vertebra

The body of L5 is the largest of all lumbar vertebrae (L5 is the largest of all movable vertebrae). It is characterized by a stout transverse process. L5 is largely responsible for the lumbosacral angle between the lumbar region and sacrum. The body weight is transmitted from L5 to the sacrum.

The Sacrum (L. the Sacred Bone) (Fig. 8.35)

In adults, it is a large, triangular, wedge-shaped bone, formed by the fusion of 5 sacral vertebrae. The triangular shape is due to the rapid decrease in size of the vertebrae from above downwards (This is because, the lower sacral vertebrae have no role in transmission of body weight).

The sacrum transmits the body weight to the pelvic girdle through the sacroiliac joints. Sacrum has two ends – a superior, broad, *base* and an inferior, blunt, *apex,* it has 4 surfaces—anterior (pelvic), posterior (dorsal) and 2 lateral surfaces.

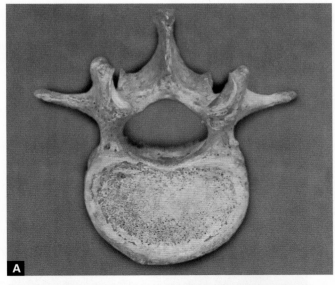

A

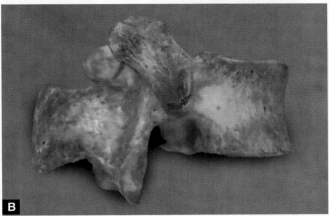

B

Figs 8.34A and B: Lumbar vertebra

The sacrum is divided by two rows of foramina into (a) a median portion traversed by sacral canal; and (b) a pair of lateral masses formed by the fusion of the transverse processes with the costal elements.

Base

The broad *base* is formed by the upper surface of 1st sacral vertebra (S1). It has a large, broad body. Its projecting anterior margin is the sacral promontory.

The pelvic surface is concave and is directed downwards and forwards. There are 4 pairs of pelvic sacral foramina, which communicate with the sacral canal through the intervertebral foramina.

The dorsal surface shows in the midline, the median sacral crest which represents the fused spines of the sacral vertebrae.

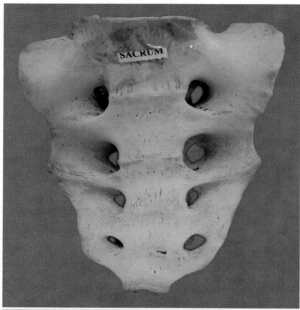

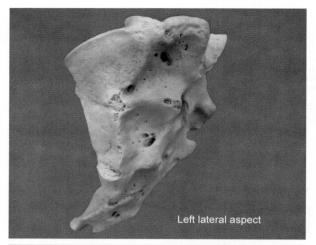

Left lateral aspect

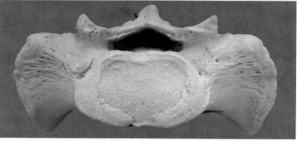

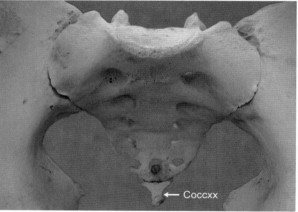

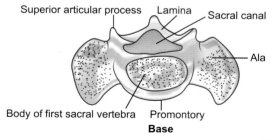

← Coccxx

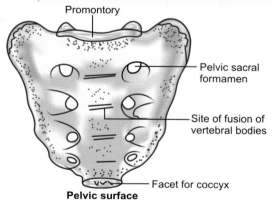

Promontory

Pelvic sacral
formamen

Site of fusion of
vertebral bodies

Facet for coccyx

Pelvic surface

Superior
articular process

Spinous tubercles

Promontory

Auricular surface

Area for ligamentous
attachment

Right lateral aspect

Superior articular process   Lamina   Sacral canal

Ala

Body of first sacral vertebra   Promontory

Base

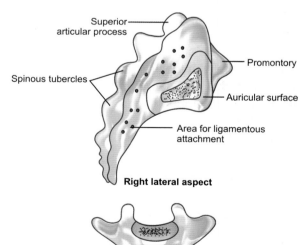

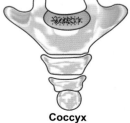

Coccyx

Fig. 8.35: Sacrum

The sacral hiatus is an inverted V-shaped gap in the dorsal wall of sacral canal at the lower end, it is formed by the non-fusion of laminae of S5.

There are 4 pairs of dorsal sacral foramina , which communicate with the sacral canal through the intervertebral foramina.

The Lateral Surface

The broad upper part of this surface has an ear-shaped articular surface called the auricular surface. There is a rough area behind this for attachment of interosseous sacroiliac ligament.

Apex

The apex is formed by the lower surface of body of S5 and articulates with the 1st piece of coccyx.

Sacral Canal

Sacral canal is formed by the contiguous vertebral foramina of the sacral vertebrae.

Contents

a. Cauda equina and filum terminale
b. Spinal meninges—the subarachnoid and subdural spaces end at S2 level (second piece of sacrum)
c. The filum terminale and 5th sacral nerves emerge out through the sacral hiatus.

The features of sacrum are different in males and females (Table 8.5).

Applied Anatomy

A local anesthetic can be injected into the sacral canal through the sacral hiatus. This procedure is called epidural anesthesia or caudal block.

In about 5% of people, L5 vertebra is incorporated into the sacrum, known as sacralization of L5 vertebra. In some individuals, S1 vertebra is separated from the sacrum and fused (partly or completely) with L5 vertebra. This is known as lumbarization of S1 vertebra.

The Coccyx

The coccyx is a small triangular bone made of four rudimentary coccygeal vertebrae fused together (It is also called the "tail bone" because it is the remnant of the tail which human embryos have until the beginning of 8th week). The first coccygeal vertebra (CO1) is largest and broadest. Its articular processes are rudimentary, forming horn-like structures, known as coccygeal cornua (cornua = horns). The last 3 coccygeal vertebrae fuse together, forming a beak-like structure (that is why it is called coccyx (Gk.coccyx = cuckoo).

Coccyx is curved more anteriorly in males than in females. So, the pelvic outlet is wider in females, as it forms a part of the birth canal.

Coccyx has no role in the support of body weight. It gives attachment to part of the gluteus maximus muscle, coccygeus muscle and anococcygeal ligament.

The coccyx articulates with the sacrum at the sacrococcygeal joint which permits movements that are important during childbirth.

A fall directly on the buttocks may result in separation of coccyx from sacrum, producing severe pain while sitting. This condition is called coccydynia (pain in the coccygeal region).

Important identifying features of vertebrae are summarized in Table 8.6.

Intervertebral Discs

They are fibrocartilaginous discs, interposed between the adjacent surfaces of the vertebral bodies from the 2nd cervical to the sacrum. There is no disc between atlas and axis.

The intervertebral discs provide the strongest attachment between the bodies of the vertebrae. The

---

Table 8.5: Sacrum – Sex differences	
Male	*Female*
1. Narrow and long	Shorter and wider
2. Concavity of pelvic surface is less	Concavity is more
3. The auricular surface is large – it lies opposite the upper 3 sacral pieces	Auricular surface shorter, lies opposite the upper 2 ½ pieces
4. The body forms a large portion of the base	Body is smaller
5. *Sacral index = 105	Sacral index = 115

$$*\frac{\text{Breadth of the base}}{\text{Maximum vertical length}} \times 100$$

Table 8.6: Important identifying features of vertebrae

Region	Identifying features
Cervical	Foramina of the transverse processes— foramina transversaria, bifid spines
Thoracic	Facets on the sides of the bodies for articulation with the ribs; heart shaped bodies; circular vertebral foramina
Lumbar	Massive bodies; thick laminae; largest vertebrae; absence of costal facets; short; thick quadrangular spine
Sacrum	Usually the vertebrae are fused to form a triangular, wedge-shaped bone. Four foramina on each side.

discs vary in size and thickness in different parts of the vertebral column. The discs are thinnest in the thoracic region and thickest in the lumbar region. In the cervical and lumbar region, the discs are thicker anteriorly, making them wedge-shaped; this is related to the curvatures in these regions.

Structure

Each disc is composed of an external annulus fibrosus and internal nucleus pulposus (L.pulpa = fleshy). The annulus fibrosus is composed of concentric lamellae of fibrocartilage, which run obliquely from one vertebra to another.

Nucleus pulposus: This central core of the intervertebral disc is highly elastic. It is made of soft gelatinous material with a few, notochordal cells. As age advances, it is invaded by fibrous tissue and its elasticity decreases. The nucleus pulposus represents the remnant of notochord.

Applied Anatomy of Vertebral Column

1. Kyphosis: Refers to the curvature of the vertebral column with the convexity backward. It can be congenital, it can follow tuberculous destruction of the vertebral body (Pott's disease); in old age, due to degeneration of intervertebral discs, senile kyphosis may result.
2. Scoliosis – is an abnormal lateral deviation of the vertebral column.
3. Fractures or fracture dislocations of the vertebral column are caused by forced flexion or due to violent forces transmitted along the column by a fall on the feet from a height. It may be associated with spinal cord injuries.
4. Herniation of nucleus pulposus (IVDP—Intervertebral Disc Prolapse) Due to violence applied to the vertebral column,

Contd...

the nucleus pulposus is displaced. Herniation usually results in a posterolateral direction (This is because, the nucleus is nearer the posterior surface and the posterior longitudinal ligament is present in the midline). Patient experiences severe, acute low back pain. The herniated nucleus pulposus may press upon the adjacent nerve root and the cord giving rise to pressure symptoms.

About 95% of lumbar disc protrusions occur at the L4/L5 or L5/S1 levels. This may irritate the sciatic nerve leading to severe pain during any movement that stretches the sciatic nerve (e.g. flexing the thigh with the leg extended).

5. If the transverse ligament of atlas ruptures, the dens is set free and may be driven into the cervical region of spinal cord causing quadriplegia or into the medulla, causing sudden death.

All neck injuries are potentially serious, because of the possibility of fracture vertebrae and injuring the spinal cord. Injury to the spinal cord in the cervical region, associated with compression or transection of spinal cord, may result in loss of all sensations and voluntary movement inferior to the lesion.

Movements of Vertebral Column

The following movements are possible in different regions of vertebral column:
1. Cervical region: flexion, extension, lateral flexion
2. Thoracic region: flexion, extension, lateral flexion
3. Lumbar region: flexion, extension, free lateral rotation

■ THE BONY PELVIS (Figs 8.36A to C)

The bony pelvis is formed by 4 bones—the two hip bones, the sacrum and coccyx. These bones are united at 4 joints. They are (1) the two sacroiliac joints, (2) the pubic symphysis and (3) the sacrococcygeal joint.

The pelvis is divided into 2 parts by the plane of pelvic inlet or pelvic brim. This plane passes from the sacral promontory to the upper margin of pubic symphysis. The part of the pelvis lying above this plane is the greater pelvis; and the part lying below it, is the lesser pelvis.

The greater pelvis (false pelvis) includes the two iliac fossae. The lesser pelvis (true pelvis) contains the pelvic viscera.

Boundaries

Boundaries of Pelvic Inlet

- Posteriorly—sacral promontory
- Anteriorly—upper margin of pubic symphysis

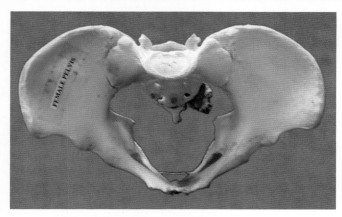

Fig. 8.36A: Female pelvis—Viewed from above

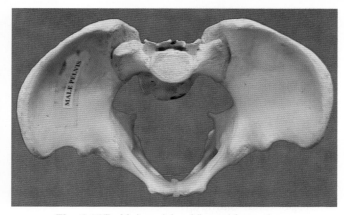

Fig. 8.36B: Male pelvis—Viewed from above

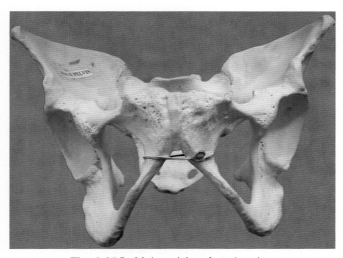

Fig. 8.36C: Male pelvis—Anterior view

- On each side—linea terminalis
- (Linea terminalis includes the anterior margin of ala of sacrum, the arcuate line of ilium, pecten pubis and the pubic crest).

Boundaries of Pelvic Outlet

a. Anteriorly by pubic symphysis
b. Posteriorly by the coccyx
c. On each side, by ischiopubic rami, the ischial tuberosity and the sacrotuberous ligaments.
   Although there are usually clear-cut anatomical differences between male and female pelvis (Table 8.7), the pelvis of any person may have features of the opposite sex. The presence of male

	Table 8.7: Differences between male pelvis and female pelvis	
Features	*Male pelvis*	*Female pelvis*
1. General structure	Thick, heavy, well-marked muscle attachments	Thin, light, poorly marked muscle attachments
2. Greater pelvis	Deep	Shallow
3. Lesser pelvis (True pelvis)	Narrow and deep	Shallow and wide
4. Pelvic inlet	Heart shaped	Oval or rounded
5. Pelvic outlet	Comparatively small	Large
6. Subpubic angle	Narrow (acute angle)	Wide (almost right angle)
7. Obturator foramen	Round	Oval
8. Acetabulum	Large	Small
9. All diameters of the cavity	Lesser	Greater
10. Ischiopubic rami	Prominently everted to give attachment to crura of penis	Not everted

characteristics in a female pelvis may present difficulties during childbirth.

The measurements of the pelvic inlet are important in obstetrics. The anteroposterior diameter of the aperture is the measurement from the midpoint of the superior border of the pubic symphysis to the midpoint of the sacral promontory.

The transverse diameter of the pelvic inlet is measured from the linea terminalis of one side to the linea terminalis of the opposite side. In a typical female pelvis, the transverse diameter is greater than the anteroposterior diameter.

Types

There are four types of pelvis:

1. Anthropoid
2. Platypelloid
3. Android
4. Gynecoid

The 1st two are common in males; third and fourth types are common in females. The gynecoid pelvis is the most spacious type and a woman with this type can have a normal, vaginal delivery.

9 Muscular System

CLASSIFICATION OF MUSCLES

1. Muscles of the head
2. Anterolateral muscles of the neck
3. Muscles of the trunk
4. Muscles of the upper limb
5. Muscles of the lower limb.

MUSCLES OF THE HEAD

a. Craniofacial muscles
b. Muscles of mastication

The ocular and extraocular muscles, the auricular and tympanic muscles, the lingual, the palatal, the pharyngeal and the laryngeal muscles are discussed elsewhere (please read the corresponding topics).

The Craniofacial Muscles

All these muscles are innervated by branches of the facial nerve, because they develop from the mesoderm of the second pharyngeal arch.
These muscles include:
1. The occipitofrontalis muscle
2. The muscles of facial expression.

The Occipitofrontalis or the Epicranius

It is a broad, musculoaponeurotic layer that covers the superior aspect of calvaria from the highest nuchal line posteriorly, to the supraorbital margins anteriorly. The occipitofrontalis has four parts—two occipital bellies (occipitalis) and two frontal bellies (frontalis). They are connected by the epicranial aponeurosis.

The two *frontal bellies* unite along their medial edges. They arise from the subcutaneous tissue over the eyebrows and the root of the nose, (bony origin) and blend with the orbital part of the orbicularis oculi. Posteriorly, the two bellies are inserted into the epicranial aponeurosis.

The occipital bellies arise from the lateral halves of the highest nuchal lines and passing anteriorly, get inserted into the epicranial aponeurosis.

All four parts are supplied by the facial nerve – the occipitalis, by the posterior auricular branches; the frontalis, by the temporal branches.

The muscle raises the eyebrows and produces transverse wrinkles in the skin of the forehead, giving the face a surprised look.

In supranuclear lesions of the facial nerve, only the lower part of the face is paralysed. The frontalis muscle escapes due to bilateral representation in the cerebral cortex.

In infranuclear lesions of the facial nerve, the whole muscles of face, including the frontalis are paralyzed, on the affected side.

The Muscles of Facial Expression (Fig. 9.1)

These are subcutaneous muscles. They bring about different facial expressions. They develop from the mesoderm of the second pharyngeal arch, and are therefore supplied by the facial nerve.

The term "facies" refers to the appearance of the face. Some diseases produce a typical facies, e.g. mask-like facies (expressionless) in Parkinsonism. The muscles of facial expression help to convey mood during communication.

These muscles do not move the facial skeleton. They surround the facial orifices (openings)—the mouth, eyes, nose and ears—and act as sphincters and dilators, i.e. they open and close these openings.

These muscles can be grouped into:
1. Muscles of the forehead – the frontalis, which is a part of the scalp muscle occipitofrontalis.
2. Muscles around the mouth
3. Muscles around the eyelids
4. Muscles around the nose.

Muscles around the mouth: Several muscles change the shape of the mouth and lips during speaking, singing, whistling, etc. The orbicularis oris is the sphincter of the mouth. The dilator muscles radiate outward from the lips. Some of the common facial expressions and the muscles producing them are given in Table 9.1.

The orbicularis oris: This muscle encircles the mouth. It is the sphincter of the mouth. Its fibers are derived from the buccinator and other facial muscles. Its fibers are located within the upper and lower lips.

Actions
1. It closes the mouth
2. It protrudes the lips
3. It compresses the lips against the teeth
4. It has an important role in articulation and chewing.

The buccinator muscle: (Buccinator = Trumpeter) This muscle was given its name because it compresses the cheeks (or buccae).

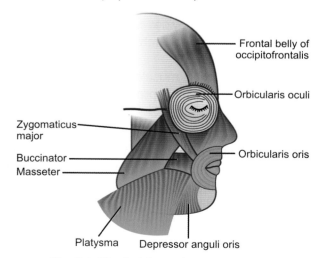

Fig. 9.1: The facial muscles and masseter

Labels: Frontal belly of occipitofrontalis; Orbicularis oculi; Orbicularis oris; Zygomaticus major; Buccinator; Masseter; Platysma; Depressor anguli oris

Table 9.1: Facial expressions and muscles producing them	
Facial expression	*Muscles involved*
1. Smiling and laughing	Zygomaticus major
2. Sadness	Levator labii superioris and zygomaticus minor
3. Grinning	Risorius
4. Doubt	Mentalis
5. Impatience	Depressor labii inferioris
6. Frowning	Corrugator supercilii and procerus
7. Surprise	Frontalis
8. Grief	Depressor anguli oris
9. Anger	Dilator naris
10. Contempt	Zygomaticus minor

It is a thin, flat, rectangular muscle. Its upper fibers arise from the alveolar process of maxilla, opposite the molar teeth. The lower fibers arise from the alveolar process of mandible, opposite the molar teeth. Middle fibers arise from the pterygomandibular raphe.

Medially, its upper fibers go to the upper lip, the lower fibers go to the lower lip. The middle fibers decussate before passing to the lips.

Action: It flattens the cheek against the gum and teeth and prevents accumulation of the food in the vestibule.

Buccinator is also used in whistling and sucking by forcing the cheeks against the teeth.

Muscles around the eyelids (palpebrae) The function of the eyelids is to protect the eye from injury and excess light. They also prevent the cornea from drying. The orbicularis oculi muscle is the sphincter of the eye. The levator palpebrae superioris opens the eye by raising the upper eyelid.

The orbicularis oculi: Its fibers sweep in concentric circles around the orbital margin and eyelids. The orbicularis oculi muscle has 3 parts:

1. A thick *orbital part* which lies on and around the orbital margin. The upper fibers blend with the fibers of frontalis muscle. Their fibers arise from the medial palpebral ligament, which encircle the eye and return to the point of origin. The orbital part of the orbicularis oculi closes the eyelid tightly and protects the eye from bright light and strong wind.
2. A *palpebral part* This part is thin. It is situated in the eyelid. This part closes the eyelid gently during blinking and sleeping.
3. *The lacrimal part* draws the eyelids medially, simultaneously dilating the lacrimal sac. This muscle produces frequent blinking, which facilitates the drainage of lacrimal fluid into the lacrimal sac. Thus it also helps to keep the surface of the cornea always moist.

When all three parts of the orbicularis oculi contract, the eyes are firmly closed.

Nerve supply: Zygomatic branch of facial nerve.

Applied Anatomy

Injury to the facial nerve or some of its branches produces weakness or paralysis of all or some of the facial muscles on the affected side. Paralysis of the facial nerve for no obvious reason is known as *Bell's palsy* (it is probably due to inflammation of facial nerve near the stylomastoid foramen).

Patients with facial paralysis are unable to close their lips and eyelids on the affected side. The cornea becomes dry; they are unable to whistle or to chew effectively. Food and saliva dribbles out of the mouth on the affected side or gets collected in the vestibule, due to the paralysis of buccinator muscle. Transverse wrinkles in the forehead and nasolabial fold of skin are obliterated.

Face is distorted owing to the normal actions of the facial muscles of the opposite side.

As mentioned earlier, in infranuclear lesions of the facial nerve, whole facial muscles are paralysed on the affected side, whereas, in supranuclear lesions (usually a part of hemiplegia) only the lower part of the face is paralysed. The upper part escapes because of bilateral representation in the cerebral cortex.

Masticatory Muscles (Fig. 9.2)

They are also called the muscles of mastication (chewing). They act on the temporomandibular joint. The mandible may be depressed or elevated, it may be protruded or retruded. These movements are controlled mainly by the muscles acting on the temporomandibular joint. The various movements result from the coordinated activity of these muscles bilaterally or unilaterally.

The muscles of mastication (Figs 9.2A to C) are:

1. The masseter
2. The temporalis
3. The medial pterygoid
4. The lateral pterygoid

They develop from the mesoderm of the first pharyngeal arch. They are, therefore supplied by the mandibular nerve.

The Masseter

The Masseter: (Gk : masseter = masticator, chewer). It is a quadrangular muscle, which takes origin from the zygomatic arch. It is inserted into the lateral surface of the ramus of the mandible. This muscle is supplied by the masseteric branch of the mandibular nerve. It elevates and protrudes the mandible and clenches the teeth.

The anteroinferior angle of the masseter muscle is related to the facial artery and its pulsations can be felt here.

The Temporalis Muscle

It is a fan-shaped muscle. It takes *origin* from the temporal fossa and the temporal fascia.

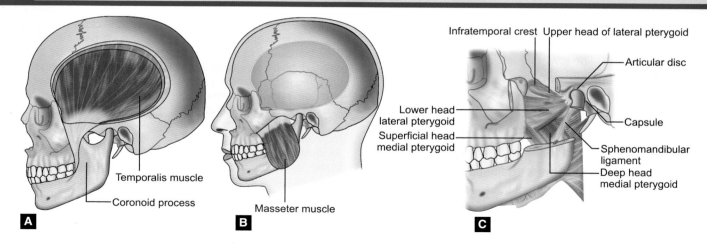

Figs 9.2A to C: (A) Temporalis (B) Masseter (C) Medial and lateral pterygoid

The fibers of the muscles converge inferiorly. They pass deep to the zygomatic arch and are *inserted* into the coronoid process of the mandible and the anterior border of the ramus of the mandible.

Lateral Pterygoid

It is the key muscle of the infratemporal fossa. It takes origin by 2 heads – upper and lower, both arise from the sphenoid bone. The fibers run backwards and laterally and converge to get inserted into the pterygoid fovea on the anterior surface of the neck of the mandible.

Nerve supply: Branch of mandibular nerve.

Actions
1. It depresses the mandible, to open the mouth.
2. When the muscles of the two sides act, they produce side to side movement, as in chewing.

Relations
1. The maxillary artery lies on the lateral pterygoid muscle; then it passes between the two heads of the muscle
2. The pterygoid venous plexus surrounds the muscle
3. The mandibular nerve and its branches lie deep to it.

The Medial Pterygoid

It is a quadrilateral muscle. It has a superficial and deep heads of origin. The superficial head takes origin from the maxillary tuberosity. The deep head takes origin from the medial surface of the lateral pterygoid plate.

Insertion Fibers run downwards, backwards and laterally to be inserted into a rough area on the medial surface of the angle of the mandible, below and behind the mandibular foramen.

Nerve supply: It is supplied by the nerve to medial pterygoid, a branch of the mandibular nerve.

Actions: It elevates and protrudes the mandible, when the muscles of both sides act simultaneously a side-to-side movement of mandible occurs, as in chewing.

ANTEROLATERAL MUSCLES OF THE NECK (FIG. 9.3)

They are classified as:
1. Superficial and lateral cervical muscles
2. Suprahyoid muscles
3. Infrahyoid muscles
4. Anterior vertebral muscles
5. Lateral vertebral muscles.

Superficial and Lateral Cervical Muscles

The superficial and lateral cervical muscles include the platysma, trapezius and the sternocleidomastoid.

Suprahyoid Muscles

The suprahyoid muscles include the digastric, stylohyoid, mylohyoid and geniohyoid.

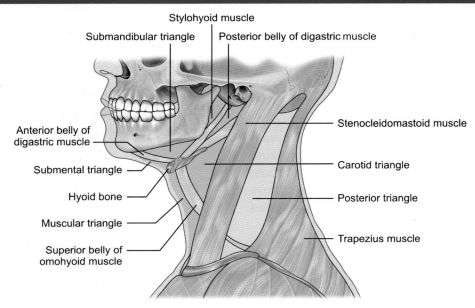

Fig. 9.3: Muscles of the neck

Infrahyoid Muscles

The infrahyoid muscles are sternohyoid, sternothyroid, thyrohyoid and the omohyoid.

The suprahyoid and infrahyoid muscles act together to stabilize the hyoid bone, which serves as a fixed base for the action of muscles of the tongue attached to it. The suprahyoid muscles elevate the hyoid bone, whereas the infrahyoid muscles depress the hyoid.

Anterior Vertebral Muscles

The anterior vertebral muscles include the Longus colli, Rectus capitis anterior and rectus capitis lateralis. They are flexors of the head and neck.

The Lateral Vertebral Muscles

1. The scalenus anterior
2. The scalenus medius
3. The scalenus posterior
   (L. scala = ladder). These muscles extend obliquely like ladders, between the upper two ribs and the cervical transverse processes.

Platysma (Gk – flat plate)

This is a thin, wide subcutaneous muscle, located in the superficial fascia. It takes origin from the fascia covering the pectoralis major and deltoid. The fibers ascend to the face. Some of them are inserted on the mandible. Most of the fibers blend with the muscles of the lower lip and other facial muscles.

Nerve Supply

The cervical branch of facial nerve. This muscle tenses the skin of the neck. This muscle acts with the other muscles to express horror, fright or sadness.

When the cervical branch of the facial nerve is injured, the muscle is paralyzed and the skin overlying it tends to fall away in loose folds.

The Sternocleidomastoid Muscle (Fig. 9.3)

The sternocleidomastoid forms the "key muscle" and prominent landmark of the neck. It stretches obliquely across the neck and divides it into an anterior triangle and a posterior triangle.

Origin

Sternocleidomastoid arises by 2 heads.
1. Sternal head
2. The clavicular head

The *sternal head* arises by a rounded tendon from the upper part of anterior surface of the manubrium sterni.

The *clavicular head* arises from the medial 1/3 of the clavicle.

Insertion

The muscle is inserted into the mastoid process of the temporal bone and the superior nuchal line of the occipital bone.

Relations

It is related superficially to the external jugular vein and the cutaneous branches of the cervical plexus. Situated deep to the muscle are the carotid arteries, the internal jugular vein, vagus nerve, brachial plexus and cervical plexus. The deep cervical lymph nodes are also situated deep to this muscle.

Nerve Supply

The spinal part of the accessory nerve supplies this muscle.

Action

1. Acting alone, it tilts the head towards the shoulder of the same side and rotates the face to the opposite side
2. Acting together from below, the muscles draw the head forwards.

Clinical Relevance

Wry neck or torticollis is due to permanent contracture of the sternocleidomastoid. The head will be tilted to the side of the contracture, but the chin points to the opposite side.

■ MUSCLES OF THE TRUNK

These muscles are classified into:
1. Deep muscles of the back
2. Muscles of the thorax
3. Muscles of the abdomen
4. Muscles of the pelvis
5. Muscles of the perineum.

Deep Muscles of the Back

These muscles are a complex group extending from the pelvis to the skull. They include extensors and rotators of the head, neck and the spine. Collectively, these muscles control the vertebral column.

Muscles of the Thorax

1. Intercostal muscles
2. The diaphragm.

Intercostal Muscles (Fig. 9.4)

These muscles occupy the intercostal spaces and connect the adjacent ribs. They are arranged in 3 layers. They are:
1. The external intercostals
2. The internal intercostals
3. The innermost intercostals.

The external intercostal muscle
Total number = 11 pairs

Extent: In each intercostal space, this muscle extends from the level of the tubercle of the rib posteriorly, to the level of the costal cartilage anteriorly, where it is replaced by the anterior intercostal membrane.

Each muscle stretches from the lower border of the rib above to the upper border of the rib below. The fibers of this muscle are directed downwards, forwards and medially.

The internal intercostal muscles
Total number = 11 pairs

In each intercostal space, this muscle extends from the lateral margin of the sternum to the level of the angle of the rib, where it is replaced by the posterior intercostal membrane.

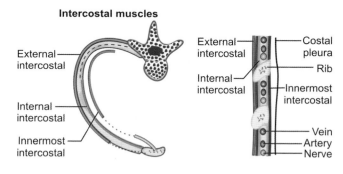

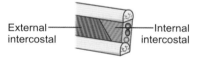

Fig. 9.4: Arrangement of muscles in an intercostal space

Superiorly, it is attached to the costal groove of the upper rib, and inferiorly, to the upper border of the rib below.

The fibers are directed upwards, forwards and medially. Its fibers are arranged at right angles to those of the external intercostal muscle.

The innermost intercostal muscles or the intercostalis intimus Their number is variable; usually there are 11 pairs.

This muscle lies deep to the internal intercostal. These two muscles are separated by the intercostal nerves and vessels. It occupies the middle 2/4 of each intercostal space. The direction of fibers is same as that of the internal intercostals.

The intercostal muscles are elevators of ribs and help in respiration.

Nerve supply by intercostal nerves.

The Diaphragm (Fig. 9.5)

The diaphragm is a large, dome-shaped, fibromuscular partition separating the thoracic cavity from the abdominal cavity. It is an important muscle of respiration. When it contracts, it increases the vertical extent of the thoracic cavity by partially flattening its dome and displacing the abdominal contents downwards.

The fibrous central part of the diaphragm is called the central tendon. It is slightly depressed by the heart,

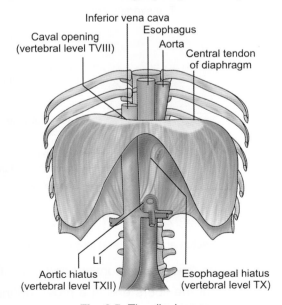

Fig. 9.5: The diaphragm

so that, on either side, there are 2 domes – the right and left domes.

The right dome, supported by the liver lies at a slightly higher level than the left.

Attachments: The diaphragm takes origin from three sites:
1. The sternal origin
2. The costal origin
3. The vertebral origin.

The sternal part – two small slips pass backwards from the xiphoid process.

The costal part – arises by wide slips from the lower 6 costal cartilages.

The vertebral part – arises by the crura and by the arcuate ligaments.

Crura: These thick fleshy bundles are attached on each side of the aorta to the anterior surfaces of the upper 2 lumbar vertebral bodies on the left side and upper 3 lumbar vertebral bodies on the right side and the intervertebral discs between them. Some fibers from the right crus deviate to the left side and surround the esophageal hiatus.

All the muscle fibers of the diaphragm converge on the strong central tendon.

Relations: Superiorly (1) the diaphragmatic surfaces of the lungs and pleurae; (2) the heart covered by the pericardium.

Inferiorly (1) the liver; (2) the fundus of the stomach and (3) the spleen.

Foramina in the Diaphragm
There are 3 major foramina—
1. *The foramen for the inferior vena cava* This foramen is at the level of T8 vertebra, about 2-3 cm to the right of the median plane.
2. *The esophageal opening* The esophagus passes through an oval opening at the level of the T10 vertebra. This opening transmits the esophagus, gastric nerves, esophageal branches of left gastric vessels and some lymphatics.
3. *The aortic hiatus:* It is the lowest and most posterior of the large openings. It is at the level of T12 vertebra. This opening transmits the aorta, the thoracic duct and sometimes the azygos and hemiazygos veins.

In addition to these structures, a number of smaller structures pass from the thorax to the abdomen through or around the diaphragm. They are:
1. The phrenic nerves
2. The superior epigastric vessels – between the sternal and costal origins
3. Lower 5 intercostal nerves
4. Sympathetic trunks
5. The splanchnic nerves
6. Minute veins.

Nerve supply: Motor supply by **phrenic nerve** (C3,C4,C5) and sensory supply, by intercostal nerves. Damage to the phrenic nerve leads to paralysis of the diaphragm on the affected side.

Action: When diaphragm contracts, it is flattened. This facilitates inspiration. When it relaxes, it resumes its dome-shaped appearance, which facilitates expiration.

Sometimes abdominal contents herniate through openings in the diaphragm. They are called diaphragmatic hernia.

Because of the common nerve root origins of phrenic and supraclavicular nerves in the neck, diaphragmatic pain is frequently referred to the shoulder, on the affected side.

Muscles of the Abdomen (Figs 9.6A to C)

Muscles of the Anterior Abdominal Wall

Consists of:
1. 4 large flat muscular sheets that form the abdominal wall—They are the external oblique, Internal oblique, transversus abdominis and rectus abdominis
2. 2 small muscles – The cremaster and pyramidalis.

The external oblique, internal oblique and the transversus abdominis muscles end in extensive aponeuroses that reach the midline. Here, the aponeuroses of the right and left sides decussate to form a median band called the linea alba. The rectus abdominis runs vertically on either side of the linea alba.

External oblique muscle of abdomen
Origin: This muscle arises by 8 fleshy slips from the lower 8 ribs; (these slips interdigitate with serratus anterior and latissimus dorsi).

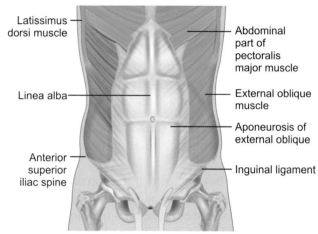
Fig. 9.6A: External oblique

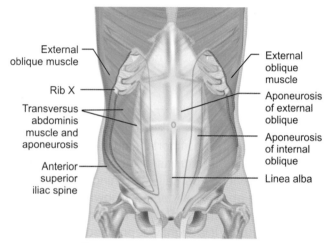
Fig. 9.6B: Transversus abdominis

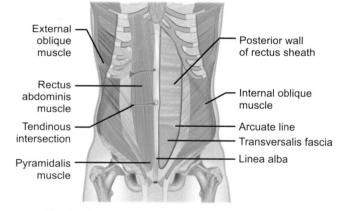
Fig. 9.6C: Rectus abdominis and internal oblique

The fibers run downwards, forward and medially.

Insertion: Most of the fibers end in a broad aponeurosis, through which it is inserted into the xiphoid process, linea alba, the pubic symphysis and the pubic crest. The lower fibers of the muscle are inserted into the outer lip of the iliac crest.

The external oblique muscle is the largest and most superficial of the 3 flat muscles of the anterior abdominal wall.

Between the anterior superior iliac spine and the pubic tubercle, the aponeurosis of the external oblique has a free lower border, which is rolled on itself, forming the inguinal ligament.

Just above the pubic crest, the aponeurosis of the external oblique presents an opening called the superficial inguinal ring.

Internal oblique muscle of the abdomen

Origin: This flat muscle arises from the lateral two-thirds of the inguinal ligament, the anterior 2/3 of iliac crest and the thoracolumbar fascia. Its fibers run upwards, forwards and medially at right angles to the fibers of external oblique.

Insertion: Uppermost fibers are inserted into the lower 3-4 ribs and their cartilages. The greater part of the muscle ends in an aponeurosis, through which it is inserted into the lower ribs, their costal cartilages, the xiphoid process, the linea alba, the pubic crest and pecten pubis.

Transversus abdominis This muscle takes origin from the lateral 1/3 of inguinal ligament, anterior 2/3 of iliac crest and the inner surfaces of the lower 6 costal cartilages. The fibers run horizontally.

Most of the fibers end in a broad aponeurosis which is inserted into the xiphoid process, linea alba, pubic crest and pecten pubis.

The rectus abdominis (Rectus = straight) This muscle takes origin from the pubic crest and pubic symphysis. The fibers run vertically upwards and inserts into 5th, 6th and 7th costal cartilages.

Rectus abdominis is a long strap-like muscle that extends along the whole length of the front of the abdomen. The paired recti are separated in the midline by the linea alba. The muscle fibers of the rectus are interrupted by 3 fibrous bands or tendinous inter-sections.

The lateral border of the rectus abdominis is seen as a curved groove called linea semilunaris, on the surface of the anterior abdominal wall.

All these 4 muscles are supplied by the lower 6-7 intercostal nerves.

The rectus sheath (Figs 9.7A to D) The rectus abdominis is enclosed between the aponeuroses of the external oblique, internal oblique and transversus abdominis, which forms the *rectus sheath.*

The rectus sheath has an *anterior wall* and a *posterior wall.*

The anterior wall is complete, covering the muscle from end to end. It is firmly adherent to the tendinous intersections.

The posterior layer is deficient superiorly and inferiorly.

At the lateral margin of the rectus abdominis the aponeurosis of the internal oblique splits into two layers—an anterior layer and a posterior layer. The anterior layer joins with the aponeurosis of external oblique to form the anterior wall of the rectus sheath.

The posterior layer of aponeurosis of the internal oblique joins with the aponeurosis of the transversus abdominis to form the posterior wall of the rectus sheath.

The fibers of the anterior and posterior walls of the rectus sheath interlace in the anterior median plane to form the linea alba.

Superior to the costal margin, the posterior wall of the rectus sheath is deficient because the transversus abdominis passes internal to the costal cartilages and the internal oblique is attached to the costal margin, so the rectus abdominis lies directly on the thoracic wall.

The posterior wall in the lower one-fourth of the rectus sheath is also deficient because all the aponeurosis fuse to form a single layer which pass anterior to the rectus muscle to form its anterior wall. The inferior limit of the posterior wall of the rectus sheath is marked by a crescentic border called the arcuate line. This line is situated approximately midway between the umbilicus and the pubic crest.

Contents of the rectus sheath

1. The rectus abdominis muscle.
2. The superior and inferior epigastric vessels.
3. Terminal parts of lower 5 intercostal vessels, nerves
4. Subcostal vessels and nerves.
5. The pyramidalis muscle lies in front of the lower part of the rectus abdominis.

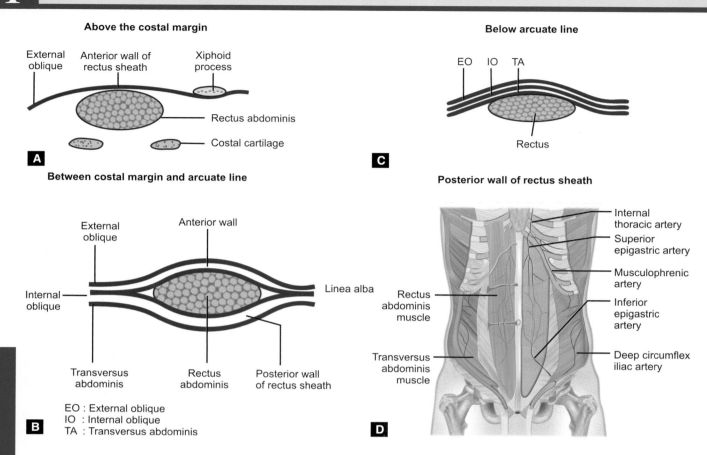

Above the costal margin

External oblique | Anterior wall of rectus sheath | Xiphoid process

Rectus abdominis

Costal cartilage

A

Between costal margin and arcuate line

External oblique

Anterior wall

Internal oblique

Linea alba

Transversus abdominis | Rectus abdominis | Posterior wall of rectus sheath

EO : External oblique
IO : Internal oblique
TA : Transversus abdominis

B

Below arcuate line

EO   IO   TA

Rectus

C

Posterior wall of rectus sheath

Internal thoracic artery
Superior epigastric artery
Musculophrenic artery
Inferior epigastric artery
Deep circumflex iliac artery

Rectus abdominis muscle

Transversus abdominis muscle

D

Figs 9.7A to D: Rectus sheath

Applied Anatomy

1. The linea alba, formed by the interlacing of the fibers of the three aponeuroses, is about 1 cm broad superiorly and narrow, inferiorly. In obese people and in multiparous women, the linea alba may become stretched and weak. The condition is known as divarication of recti. An epigastric hernia may occur in such patients.
2. Supraumbilical incision through the linea alba have several advantages – it provides a bloodless, safe area (because this line transmits only small vessels and nerves to the skin), but a hernia may develop later.
3. Little communication occurs between nerves from the lateral border of the rectus abdominis muscle to the anterior midline. So, a transverse incision through the rectus abdominis provides good access to the abdominal cavity and cause least damage to the nerves.
4. Paramedian incisions are made vertically in a sagittal plane about 2 cm from the median plane. They are made through the anterior layer of the rectus sheath. The muscle is retracted laterally. This avoids injury to its nerve supply. Posterior layer of rectus sheath and peritoneum are then incised to enter the peritoneal cavity.

Actions of abdominal muscles:

These muscles have 4 main actions:

1. Movement of the trunk – mainly flexion
2. Compression of the viscera – this increases the intra-abdominal pressure and produces the force required for defecation, micturition and parturition (childbirth)
3. Protection for the abdominal viscera
4. In respiration—when the diaphragm contracts during inspiration, its domes flatten and descend, increasing the vertical dimension of the thorax. To make room for the abdominal viscera, the anterior abdominal wall expands by the relaxation of the muscles. When the diaphragm relaxes during expiration, the abdominal wall passively sinks in.

Muscles of the Posterior Abdominal Wall

They are:
1. Psoas major
2. Psoas minor
3. Iliacus
4. Quadratus lumborum

The psoas major muscle arises from the transverse processes and the bodies of all lumbar vertebrae and the intervertebral discs between them. The muscle passes behind the inguinal ligament and in front of the hip joint to enter the thigh. It ends in a tendon, which receives the fibers of the iliacus and together, inserted into the lesser trochanter of femur (Fig. 9.27).

Important relations: In the *abdomen,* it is related to the kidney, ureter, renal and gonadal vessels, pancreas, the inferior vena cava, sympathetic trunk, the aorta, aortic lymph nodes, terminal ileum, appendix and external iliac vessels; it is also related to the peritoneum. In the *thigh,* it is related to the femoral artery, capsule of the hip joint, femoral nerve and iliacus muscle.

The *lumbar plexus* lies embedded in the posterior part of the substance of the muscle. Its branches either pierce the muscle or emerge along its medial or lateral borders.

Nerve supply: Branches from L_2 and L_3 spinal nerves.

Action: It is a powerful flexor of the hip joint, along with the iliacus. It flexes the trunk also.

Applied Anatomy

As the psoas major muscle has extensive, clinically important relations to the kidneys, ureters, appendix and nerves, when any of these structures is diseased, movements of this muscle produces pain. Cancer of the pancreas may invade the muscle and nerves of the posterior abdominal wall, producing excruciating pain. Psoas spasm, due to inflamed appendix, forces the patient to flex the hip. Extension of the joint causes abdominal pain. An abscess resulting from tuberculosis of the lumbar vertebra tends to spread into the psoas sheath (fascia enclosing the psoas major) and produces a psoas abscess. The pus from the abscess passes deep to the inguinal ligament and points (surfaces) in the femoral triangle.

Muscles of the Pelvis (Fig. 9.8)

The pelvic muscles include two groups:
1. The levator ani and coccygeus
2. The piriformis and obturator internus

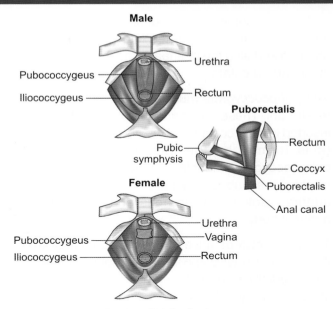

Fig. 9.8: Pelvic diaphragm

The two levator ani and the two coccygeus muscles form a funnel-shaped pelvic diaphragm. This diaphragm separates the pelvic cavity from the perineum. The musculofascial pelvic diaphragm supports the abdominopelvic viscera. The rectum, urethra and vagina in females, penetrate the pelvic diaphragm to reach the exterior.

The Levator Ani Muscles

These are the largest and most important muscles in the pelvic floor. Posterior to them are the coccygeus muscles, forming the small posterior part of the pelvic floor.

The thin, broad levator ani muscles of the two sides unite to form a hammock like (sling-like) sheet of muscle which extends from pubis to coccyx and also from one lateral pelvic wall to the other.

For purposes of description, the levator ani is divided into 3 parts (1) puborectalis; (2) pubococcygeus; and (3) iliococcygeus.

The puborectalis arises from the pubis, passes posteriorly where it unites with the muscle fibers of the opposite side to form a U-shaped sling around the anorectal junction. This sling maintains the anorectal flexure.

Some fibers of the puborectalis sweep around the vagina and get inserted into the perineal body. These fibers are called "pubovaginalis".

In males, some fibers of puborectalis sweep around the prostate and get inserted into the capsule of the prostate and also the perineal body. These fibers form the levator prostate.

The pubococcygeus muscle is the main part of the levator ani. These fibers arise from the pubis and are inserted into the coccyx and the anococcygeal ligament (it is formed by the interlacing fibers of the pubococcygeus muscles of the two sides). In the female, the pubococcygeus muscles encircle the urethra, vagina and the anus and merge with the perineal body.

The *iliococcygeus* arises from the tendinous arch of the obturatory fascia and the ischial spine. These fibers pass medially and posteriorly, attach to the coccyx and the anococcygeal ligament.

Nerve supply: Levator ani is supplied by the perineal branches of S_3 and S_4.

Actions:
1. The pelvic diaphragm supports the pelvic viscera
2. Acting together, the levator ani muscles raise the pelvic floor, thereby helping the muscles of the anterior abdominal wall in compressing the abdominal and pelvic contents. This action is important in coughing, micturition, in lifting a heavy object, etc.
3. The puborectalis muscle prevents the passage of fecal matter from the rectum into the anal canal when defecation is not desired.
4. During parturition (childbirth), the levator ani support the fetal head; meanwhile, the cervix dilates for the passage of the baby.

Applied Anatomy

The pubococcygeus part of the levator ani is likely to get damaged during childbirth. To ease delivery and to prevent damage to the pubococcygeus, an *episiotomy* (Fig. 9.9) may be performed. This incision, passes through the skin, vaginal wall and the bulbospongiosus muscle. Weakness of the pubovaginalis part of levator ani, as a result of stretching and laceration during childbirth, may lead to urinary stress incontinence. It is characterized by dribbling of urine whenever intra-abdominal pressure increases (e.g. During coughing, sneezing, etc).

The Piriformis

These are pear-shaped muscles, which occupy a key position in the gluteal region. The piriformis arises

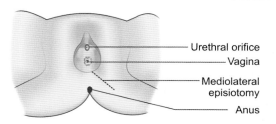

Urethral orifice
Vagina
Mediolateral episiotomy
Anus

Fig. 9.9: Episiotomy

within the pelvis from the pelvic surface of middle 3 pieces of sacrum and upper margin of greater sciatic notch. Its tendon is inserted into the greater trochanter of femur. It rotates the thigh laterally.

Relations

The following structures emerge from the *upper border of piriformis,* through the greater sciatic foramen – the superior gluteal vessels and nerves. The following structures emerge from the *lower border* of piriformis, through the great sciatic foramen.
1. The sciatic nerve
2. Posterior femoral cutaneous nerve
3. The inferior gluteal vessels and nerve
4. Pudendal nerve
5. Internal pudendal vessels
6. Nerve to obturator internus.
    Medial to the piriformis is the sacral plexus and internal iliac vessels. The piriformis forms a muscular bed for the sacral plexus.

MUSCLES OF THE SHOULDER GIRDLE AND UPPER LIMB

These muscles can be classified as follows:
1. Muscles attaching the scapula to the trunk
    — Rhomboideus major and Rhomboideus minor
    — Trapezius
    — Serratus anterior
2. Muscles attaching the humerus to the scapula
    — Supraspinatus
    — Infraspinatus
    — Subscapularis
    — Deltoid
3. Muscles attaching the humerus to the chest wall
    — Pectoralis major
    — Pectoralis minor
    — Latissimus dorsi

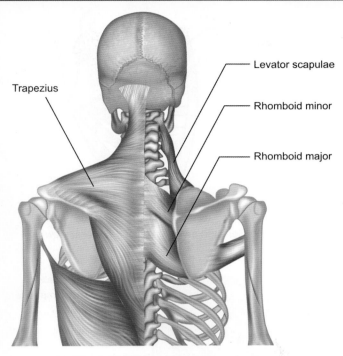

Fig. 9.10: Rhomboid major and minor

4. Muscles of the arm
5. Muscles of the forearm
6. Muscles of the hand and fingers.

**The Rhomboideus Muscles
or Rhomboid Muscles (Fig. 9.10)**

The rhomboid major and minor muscles lie deep to the trapezius muscle. They are called "rhomboid" because they form an oblique parallelogram. The rhomboid major is about 2 times wider than the rhomboid minor. These muscles pass inferolaterally from the vertebrae to the scapulae.

Rhomboid Major

Origin: From the spinous processes of T2 to T5.
Insertion: To the medial border of scapula, from the inferior angle to the root of the spine of the scapula.

Rhomboid Minor

Origin: From the spinous processes of C7 and T1

Insertion: Medial border of scapula, at the apex of scapular spine.

Nerve supply: Both rhomboids are supplied by the dorsal scapular nerve (C5).

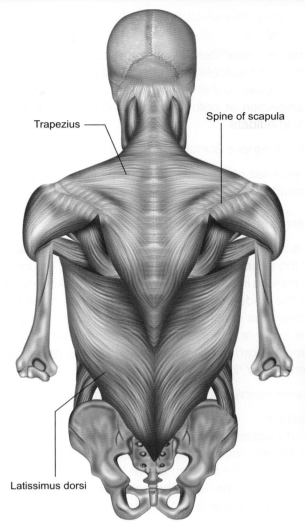

Fig. 9.11: Trapezius muscles

Actions: They retract the scapula and rotate it to depress the glenoid cavity.

Trapezius (Fig. 9.11)

This large, flat, triangular muscle covers the posterior aspect of the neck and superior half of the trunk. It is called trapezius, because the muscles of the two sides form a "trapezion" (Gk – irregular, four-sided figure).

Origin: Its fibers take origin from the superior nuchal line of the occipital bone, the ligamentum nuchae and spines of all the thoracic vertebrae.

Insertion
• Upper fibers—to lateral 1/3 of clavicle
• Middle fibers—to upper lip of the crest of the spine of scapula

Section B

Musculoskeletal System

- Lower fibers—to the middle of the spine of the scapula

Actions: Upper fibers elevate the scapula, middle fibers retract the scapula. The upper and lower fibers rotate the scapula so that the glenoid cavity faces upwards. Contraction of both muscles produces shrugging of the shoulders.

Nerve supply: Spinal part of accessory nerve

Weakness of these muscles result in drooping of the shoulders.

Serratus Anterior (Fig. 9.12)

It is a broad sheet of muscle; it forms the medial wall of the axilla. It arises from the upper 8 ribs (Its lower 4 digitations interdigitate with the external oblique muscle of the abdomen). The digitations pass backwards around the chest wall and reach the costal surface of the scapula. Here, the muscle is inserted along the medial border *(The fleshy digitations resemble a serratus or "saw")*.

Nerve supply Long thoracic nerve (C5, C6, C7)

Actions
1. It protracts the scapula
2. Rotates the scapula forwards and upwards, during elevation of the arm above the head.
   Injury to the long thoracic nerve paralyses the muscle, leading to "winging of the scapula".

Supraspinatus and Infraspinatus (Figs 9.13A and B)

The supraspinatus and infraspinatus take origin from the supraspinous and infraspinous fossae of the scapula respectively. They pass laterally as tendons, which fuse with the fibrous capsule of the shoulder joint and finally get inserted into the greater tubercle (tuberosity) of the humerus. Both are supplied by the suprascapular nerve (C_5, C_6), which arises at the Erb's point of the brachial plexus.

Actions

1. The supraspinatus initiates abduction of shoulder joint to the extent of first 15°.
2. These tendons maintain the stability of the shoulder joint.
3. Infraspinatus is a lateral rotator of shoulder joint.

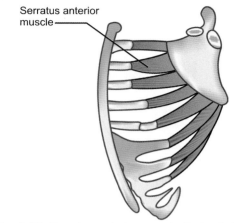

Fig. 9.12: Serratus anterior muscle (Lateral view)

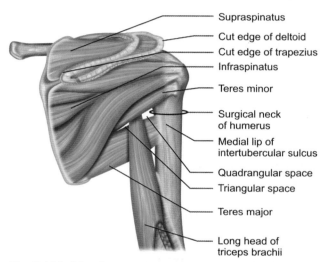

Fig. 9.13A: Muscles attaching the humerus to the scapula

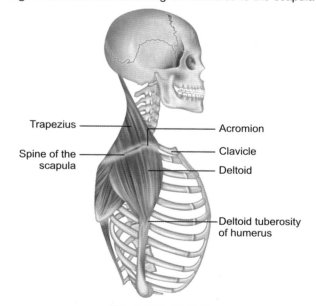

Fig. 9.13B: Deltoid

Deltoid (Fig. 9.13B)

It is a powerful, multipennate muscle. It is shaped like an inverted delta (∇). It forms the rounded contour of the shoulder because it overlaps, the upper end of the humerus.

Origin

1. Anterior border and upper surface of lateral 1/3 of clavicle
2. Lateral border of acromion
3. Crest of the spine of the scapula.

Insertion

All fibers converge to the deltoid tuberosity— a V-shaped impression on the lateral surface of the shaft of the humerus.

Nerve Supply

The axillary nerve supplies the deltoid muscle.

Action

1. Acromial fibers—abduction up to 90°
2. Anterior fibers—flexion, adduction and medial rotation of shoulder joint
3. Posterior fibers—extension and lateral rotation of arm.

Structures Under Cover of the Deltoid

a. Bones: Coracoid process, lesser and greater tuberosities, upper part of shaft and surgical neck of humerus.
b. Muscles: Both heads of biceps, long head of triceps, tendons attached to coracoid process and the tuberosities of humerus.
c. Vessels and nerves: Axillary nerve and circumflex vessels
d. Joint: Shoulder joint and its ligaments.

Applied Anatomy

1. Fracture at the surgical neck of humerus can severely damage the axillary nerve. This leads to paralysis and atrophy of the deltoid muscle. The rounded contour of the shoulder is lost. This gives the shoulder a flattened appearance. Abduction of the arm against resistance is not possible.
2. Deltoid is a common site for giving intramuscular injection. Injection should be given in the lower half of the muscle to avoid injury to the axillary nerve.

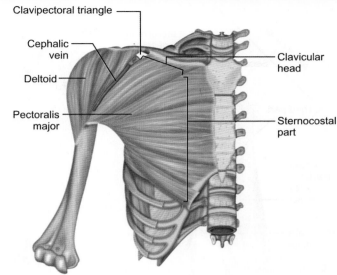

Fig. 9.14: Pectoralis major

The Pectoralis Major (Fig. 9.14)

This large, fan-shaped muscle covers the superior part of the thorax. Along with the pectoralis minor, it is associated with movements of the pectoral girdle and the upper limb; they form the anterior wall of the axilla.

Origin

1. Medial third of the front of the clavicle
2. Anterior surface of the sternum
3. Upper six costal cartilages
4. The aponeurosis of external oblique muscle.

Insertion

To the lateral lip of intertubercular sulcus of humerus.

Nerve Supply

Medial and lateral pectoral nerves.

Actions

Adduction and medial rotation of shoulder joint.

The Pectoralis Minor (Fig. 9.15)

This small triangular muscle lies under cover of the pectoralis major. The pectoralis minor is the landmark of the axilla. Along with the coracoid process, it forms an arch, deep to which pass the vessels and nerves of the upper limb. The axillary artery is subdivided into

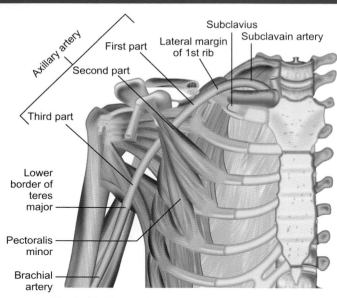

Fig. 9.15: Pectoralis minor and axillary artery

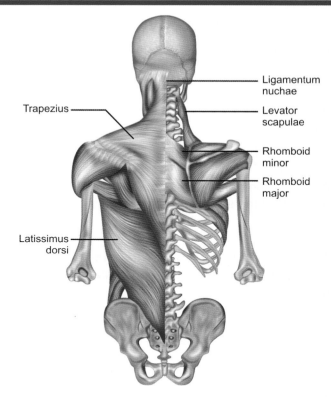

Fig. 9.16: Latissimus dorsi

3 parts by the pectoralis minor. The second part of the artery, surrounded by the 3 cords of the brachial plexus, lies behind the muscle.

Origin

From 3rd to 5th ribs.

Insertion

Coracoid process of scapula.

Nerve Supply

Medial and lateral pectoral nerves.

Action

1. Protraction of scapula
2. It depresses the shoulder along with the rhomboids
   A strong sheet of fascia, called clavipectoral fascia extends between the pectoralis minor and the clavicle. This fascia is pierced by the cephalic vein, lymphatics from the breast to the axillary nodes, acromio-thoracic vessels and lateral pectoral nerve.

Latissimus Dorsi (Fig. 9.16)

The name of this muscle is derived from Latin, meaning "widest of the back". This is a large, wide, fan-shaped muscle, which covers the inferior half of the back-from T6 level to the iliac crest. It extends from the trunk to the humerus and acts on the shoulder joint.

Origin

- Spines of lower six thoracic, lumbar and sacral vertebrae
- Thoracolumbar fascia
  Iliac crest and lower 3-4 ribs.

Insertion

Intertubercular sulcus of humerus, between the insertions of two "majors", pectoralis major and teres major ("Lady between two majors").

Nerve Supply

Thoracodorsal nerve, a branch of posterior cord of brachial plexus.

Actions

1. Extension and medial rotation at the shoulder joint
2. Adduction
3. When the arms are raised above the head, it assists in elevating the trunk during climbing.

Triangles

Two triangles are associated with the latissimus dorsi.

Lumbar triangle: It is bounded by the iliac crest, latissimus dorsi and the external oblique muscle of the abdomen. This is a potentially weak area, through which herniation can occur.

Triangle of auscultation: It is bounded by the upper border of latissimus dorsi, lateral border of trapezius and medial border of the scapula. This area lies close to the chest wall. Deep to this triangle are: (i) the cardiac orifice of the stomach, on the left side; (ii) the apex of lower lobe of both lungs. In esophageal obstruction, the splash of swallowed liquids could be heard, with a stethoscope.

Muscles of the Arm

The arm or brachium is enveloped by a sleeve of deep fascia which projects into the interior as medial and lateral intermuscular septa, which are attached to the medial and lateral supracondylar ridges of the humerus respectively. These septa divide the arm into anterior and posterior compartments.

The Anterior Compartment

Contents are as follows.
1. *Muscles*
   - Biceps brachii
   - Brachialis
   - Coracobrachialis
2. *Artery:* The brachial artery
   Nerves: Musculocutaneous nerve, median, ulnar and part of the radial nerve.

Muscles of the anterior compartment (Figs 9.17A and B)
1. *Biceps brachii:* It is an elongated fusiform muscle which arises by 2 heads.
   a. *Long head* arises within the fibrous capsule of the shoulder joint from the supraglenoid tubercle of the scapula.
   b. *The short head* arises from the tip of the coracoid process.
      Both heads expand into fusiform bellies, which fuse about 7 cm above the elbow joint, where, a flat tendon is formed. The tendon passes through the cubital fossa taking a twisted course and it is inserted into the radial tuberosity.
      A fibrous expansion called bicipital aponeurosis arises from the medial border of

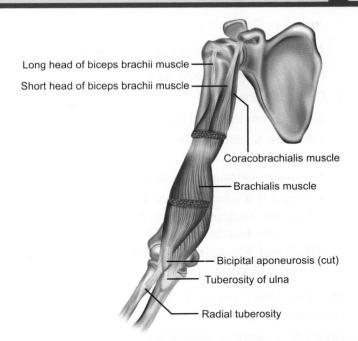

Fig. 9.17A: Muscles of the arm

Long head of biceps brachii muscle
Short head of biceps brachii muscle
Coracobrachialis muscle
Brachialis muscle
Bicipital aponeurosis (cut)
Tuberosity of ulna
Radial tuberosity

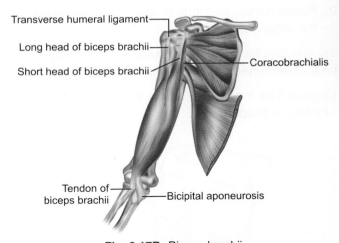

Fig. 9.17B: Biceps brachii

Transverse humeral ligament
Long head of biceps brachii
Short head of biceps brachii
Coracobrachialis
Tendon of biceps brachii
Bicipital aponeurosis

the tendon, which is attached to the subcutaneous posterior border of the ulna.

Nerve supply: The musculocutaneous nerve.
Actions
a. It is a powerful supinator of the forearm
b. It is a flexor of the elbow

2. Brachialis
   Origin:
   a. From the lower half of the shaft of the humerus
   b. Lower part of the spiral groove.

Insertion: Coronoid process of ulna

Nerve supply: Musculocutaneous nerve

Action: Flexion of elbow joint.

3. Coracobrachialis

Origin: From the tip of the coracoid process

Insertion: Medial border of middle of the shaft of humerus

Nerve supply: Musculocutaneous nerve

Action: It is a weak flexor of shoulder joint (Note: The musculocutaneous nerve supplies the 3 muscles—biceps, brachialis and coracobrachialis (BBC)—of the anterior compartment or flexor compartment).

Muscles of the Posterior Compartment (Extensor Compartment)

Contents of posterior compartment:
1. Triceps brachii is the main content
2. Radial nerve, profunda brachii vessels and part of the ulnar nerve.

Triceps Brachii:(Fig. 9.18) It arises by 3 heads—long head, lateral head and a medial head.

Origins: The long head arises from the infraglenoid tubercle of scapula.

The lateral head arises from an oblique ridge, just above the spiral groove (radial groove) of humerus.

The medial head arises from the shaft of humerus, below the radial groove.

Insertion: Triceps is inserted into the posterior part of upper surface of the *olecranon process of ulna.*

Nerve supply: Radial nerve supplies all the three heads of the triceps.

Actions:
1. Extension of elbow
2. The long head supports the humeral head from below (during abduction).

Muscles of the Forearm

The forearm is divided into an anterior or flexor compartment and a posterior or extensor compartment. The flexor muscles are more massive than the extensors, because they are antigravity muscles.

Muscles of anterior (flexor) compartment: There are 8 muscles in this compartment, 5 superficial and 3 deep *The superficial muscles (Fig. 9.19) are* (from lateral to medial)
1. Pronator teres
2. Flexor carpi radialis

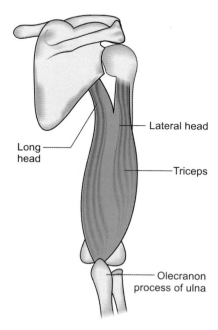

Fig. 9.18: Triceps brachii

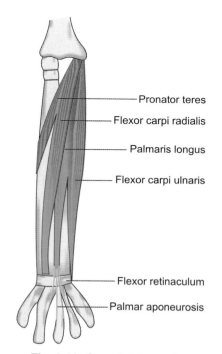

Fig. 9.19: Superficial muscles

3. Palmaris longus
4. Flexor carpi ulnaris
5. Flexor digitorum superficialis or sublimis
   The deep muscles are:
   a. Flexor digitorum profundus
   b. Flexor pollicis longus
   c. Pronator quadratus.

Note:
1. All the superficial flexors of the forearm have a common origin from the front of the medial epicondyle of the humerus. This is called the common flexor origin.
2. The median nerve and its anterior interosseous branch supply all the flexor muscles of the forearm *except* flexor carpi ulnaris and medial part of the flexor digitorum profundus, which are supplied by the ulnar nerve.

Pronator teres

Origins: It arises by humeral and ulnar heads. The *humeral head* arises from the medial epicondyle of humerus.

The *ulnar head* arises from the coronoid process of the ulna.

The *median nerve* lies between the two heads.

Insertion: The two heads join and pass laterally, forming the median boundary of the cubital fossa. The muscle is inserted on the middle of the lateral surface of the shaft of radius.

Action:

1. Pronation of the forearm
2. It is a weak flexor of the elbow joint.

Nerve supply: Median nerve.

Applied Anatomy

The pronator syndrome: This "nerve entrapment syndrome" is caused by the compression of the median nerve between the two heads of the pronator teres muscle. It may be due to trauma or hypertrophy of the muscle.

Flexor digitorum superficialis (Fig. 9.20)
Origins
1. Humeroulnar head
2. Radial head
   The humeroulnar head arises from the medial epicondyle of the humerus, the ulnar collateral ligament

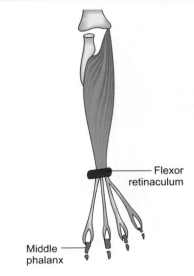

Fig. 9.20: Flexor digitorum superficialis

and the coronoid process of ulna. The radial head arises from the whole length of anterior oblique line of radius.

A fibrous arch connects the two heads, transmitting the median nerve and ulnar artery deep to it. The median nerve lies closely adherent to the deep surface of the flexor digitorum superficialis.

Above the wrist, the muscle forms 4 tendons, which are arranged in two strata – the tendons for the middle and ring fingers are superficial, whereas the tendons for the index and little finger are deep.

The four tendons pass through the carpal tunnel into the palm and diverge and go to the medial 4 digits. At the base of the proximal phalanx, each tendon splits into two, to form a sling, to allow the passage of the tendon of the flexor digitorum profundus. The split tendinous slips are attached to the sides of the middle phalanx of medial four digits.

The flexor digitorum profundus: (Fig. 9.21) It is the bulkiest muscle of the forearm.
Origin
1. Anterior, medial and posterior aspect of upper ¾ of the shaft of ulna
2. Interosseous membrane
   The muscle ends below into 4 tendons for the medial four fingers. These tendons pass through the carpal tunnel.

In the palm, these tendons provide origins of four lumbrical muscles. In the fingers, each tendon passes

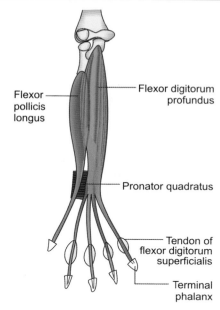

Flexor pollicis longus

Flexor digitorum profundus

Pronator quadratus

Tendon of flexor digitorum superficialis

Terminal phalanx

Fig. 9.21: Flexor digitorum profundus

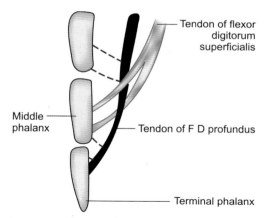

Tendon of flexor digitorum superficialis

Middle phalanx

Tendon of F D profundus

Terminal phalanx

Fig. 9.22: Insertion of flexor digitorum superficialis and profundus

through the split tendon of flexor digitorum superficialis and finally, inserted into the palmar surface of the terminal phalanx of medial 4 fingers (Fig. 9.22).

Nerve supply: Medial part of the muscle is supplied by the ulnar nerve; the lateral part is supplied by the anterior interosseous branch of the median nerve.

Actions: The flexor digitorum profundus flexes the terminal phalanges, metacarpophalangeal joints and the carpal joints. The flexor digitorum superficialis flexes the middle phalanges, the metacarpophalangeal and carpal joints.

The pronator quadratus: It is a quadrilateral muscle which extends across the lower part of both bones of the forearm, in front of the interosseous membrane. It is the principal pronator of the forearm. The deep fibers of the muscle prevent separation of the lower parts of radius and ulna.

Note: The anterior interosseous branch of the median nerve supplies
1. Flexor digitorum profundus—lateral part
2. Flexor pollicis longus
3. Pronator quadratus

Posterior (Extensor) Compartment of Forearm

The extensor muscles of the forearm are arranged in 2 groups—superficial and deep.

All the extensor muscles of the forearm (except brachioradialis, extensor carpi radialis longus and anconeus) have a common extensor origin from the lateral epicondyle of the humerus and are supplied by the posterior interosseous branch of the radial nerve.

The brachioradialis, the extensor carpi radialis longus and the anconeus are supplied directly by the trunk of the radial nerve.

Superficial extensor muscles There are 7 muscles which are arranged in 2 groups:
a. Lateral group consists of brachioradialis, extensor carpi radialis longus and extensor carpi radialis brevis.
b. Posterior group includes from lateral to medial side—the extensor digitorum, extensor digiti minimi, extensor carpi ulnaris and anconeus.

Deep extensor muscles
1. Abductor pollicis longus
2. Extensor pollicis brevis
3. Extensor pollicis longus
4. Supinator.

Brachioradialis

Origin: Upper 2/3 of lateral supracondylar ridge of humerus

Insertion: To the base of styloid process of radius

Nerve supply: Supplied by radial nerve

Action: It is a flexor of elbow joint. It acts best in mid-prone position. It is not a pronator or supinator of the forearm.

Proximal part of the brachioradialis forms the lateral boundary of the cubital fossa.

Supinator: It wraps the upper 1/3 of radius. It has a superficial part and a deep part. The posterior interosseous branch of radial nerve passes between the two parts.

The superficial fibers arise from the lateral epicondyle, radial collateral ligament and the annular ligament.

The deep fibers take origin from the supinator crest of ulna.

Insertion: Inserted into upper 1/3 of lateral surface of radius

Nerve supply: Posterior interosseous nerve

Action: Supination

Intrinsic Muscles of the Hand

These short muscles of the hand are designed for precise movements (e.g. writing, playing musical instruments). Therefore, their motor nerves possess small motor units. The area for hand movements in the motor cortex of brain is extensive.

The intrinsic muscles consist of:
1. Thenar muscles and adductor pollicis
2. Hypothenar muscles and palmaris brevis
3. Lumbrical muscles
4. Palmar and dorsal interossei.

Thenar Muscles

There are 3 short muscles which produce the thenar eminence. They are (1) Abductor pollicis brevis; (2) Flexor pollicis brevis; and (3) Opponens pollicis (Pollex = Thumb). The abductor pollicis brevis abducts the thumb and rotates it medially. The flexor pollicis brevis flexes the proximal phalanx of the thumb. The opponens, lying deep to the other two muscles, flexes and medially rorates the first metacarpal bone and helps in the opposition of thumb, so that the thumb comes in contact with the pulps of semiflexed fingers.

Nerve supply: They are supplied by the recurrent branch of the median nerve.

Adductor Pollicis

It lies deep to the long flexor tendons in the palm. It arises by two heads—a transverse head and an

oblique head (The transverse head arising from the 3rd metacarpal; the oblique head, from the bases of 2nd and 3rd metacarpal). The muscle is inserted into the ulnar side of the proximal phalanx of the thumb.

Nerve supply: From the deep branch of ulnar nerve.
Action: Adduction of thumb.

Hypothenar Muscles

There are 3 short muscles which form the hypothenar eminence. They are abductor and flexor digiti minimi and opponens digiti minimi.

Nerve supply: They are supplied by the deep branch of the ulnar nerve.

Actions: The 3 hypothenar muscles help to deepen the cup of the palm for proper grip of a large object.

The Palmaris Brevis

The palmaris brevis is a subcutaneous skeletal muscle which extends across the proximal part of the hypothenar eminence. It protects the underlying ulnar vessels and nerve. It is innervated by superficial branch of ulnar nerve. It helps to improve palmar grip.

Lumbrical Muscles (Fig. 9.23)

These four small muscles resemble earthworms in shape. That is why they are called lumbricals. They are numbered from lateral to medial side.

They take origin from the four tendons of the flexor digitorum profundus. Distally, their tendons form "distal wing tendons", which join the dorsal digital expansion.

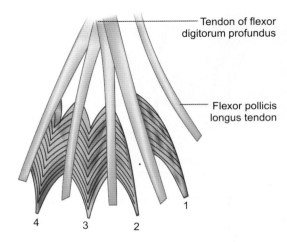

Tendon of flexor digitorum profundus

Flexor pollicis longus tendon

Fig. 9.23: Lumbricals

1st and 2nd lumbricals are unipennate. They are innervated by the median nerve.

3rd and 4th lumbricals are bipennate. They are innervated by the deep branch of the ulnar nerve.

Action: They flex the digits at the metacarpophalangeal joints and extend at the interphalangeal joints.

INTEROSSEOUS MUSCLES

There are 4 palmar interossei and 4 dorsal interossei. They are numbered from lateral to medial side. Palmar interossei are unipennate. They are adductors of fingers (Palmar Adduct—PAD). The dorsal interossei are bipennate muscles. They abduct the fingers (Dorsal Abduct—DAB). Both the palmar and dorsal interossei are innervated by the deep branch of ulnar nerve.

Note: All the intrinsic muscles of the hand are supplied by the *deep branch of the ulnar nerve* except the 3 thenar muscles and the first two lumbricals which are supplied by the median nerve. The ulnar nerve is called the musician's nerve because it innervates the muscles essential for precise movements, e.g. playing a piano.

Flexor Retinaculum of the Hand (Fig. 9.24), (also see Fig. 8.6)

The palmar deep fascia is thickened to form a strong fibrous band, called the flexor retinaculum. It converts the anterior concave surface of the wrist (carpus) into an osseofibrous "carpal tunnel". This tunnel contains the long flexor tendons of the digits and the median nerve.

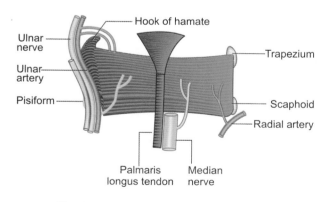

Ulnar nerve
Ulnar artery
Pisiform
Hook of hamate
Trapezium
Scaphoid
Radial artery
Palmaris longus tendon
Median nerve

Fig. 9.24: Flexor retinaculum of the hand

Attachments

Medially: Pisiform and hook of the hamate

Laterally: Tubercle of scaphoid and trapezium

Relations

Structures passing superficial to the retinaculum are:
1. Palmaris longus tendon
2. Ulnar nerve
3. Ulnar vessels
4. Palmar cutaneous branches of median and ulnar nerve.

Structures passing deep to the retinaculum are:
1. Median nerve
2. 4 tendons of flexor digitorum superficialis
3. 4 tendons of flexor digitorum profundus
4. Tendon of flexor pollicis longus.

Applied Anatomy

The medial nerve may be compressed in the carpal tunnel, leading to carpal tunnel syndrome, characterized by—
1. Weakness and wasting of thenar muscles, with the loss of power of opposition
2. Loss of cutaneous sensation of the palmar surface involving lateral three and half digits.

MUSCLES OF THE LOWER LIMB

These muscles can be grouped under the following headings:
a. Muscles of the gluteal region
b. Muscles of the thigh
c. Muscles of the leg
d. Muscles of the foot

Muscles of the Gluteal Region (Figs 9.25A and B)

The buttock or gluteal region is unique in man due to adoption of upright posture and bipedal mode of locomotion. The gluteal prominence is mainly due to the massive gluteus maximus and abundant subcutaneous fat.

Extent of Gluteal Region

• Superiorly—Iliac crest
• Inferiorly—Gluteal fold (a transverse skin crease)

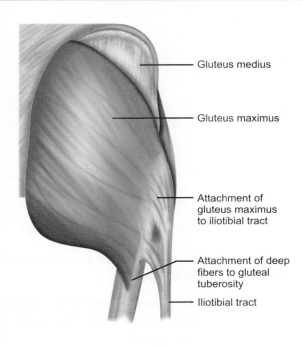

Fig. 9.25A: Gluteus maximus

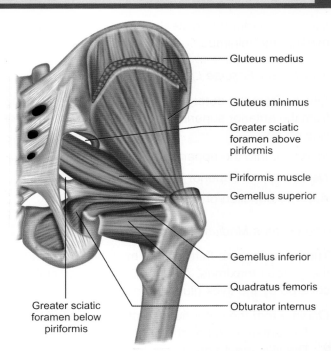

Fig. 9.25B: Gluteus minimus and piriformis

- Anteriorly—An imaginary vertical line extending downward from the anterior superior iliac spine to the greater trochanter.
- Posteriorly—The sacral spines.

The muscles of gluteal region are arranged in 3 layers. The *superficial layer* consists of the gluteus maximus and tensor fasciae latae; the intermediate layer consists of the gluteus medius and the short lateral rotators of the hip joint; the deep layer consists of the gluteus minimus.

The gluteus maximus: It is the largest and most superficial muscle of the gluteal region.

Origins

a. From the gluteal surface of ilium, behind the posterior gluteal line.
b. From the gluteal aponeurosis covering the gluteus medius
c. From the lower part of sacrum and coccyx
d. From the sacrotuberous ligament.

Insertions

a. One fourth of the muscle is inserted into the gluteal tuberosity of femur

b. Three fourth of the muscle, ends in an aponeurosis and is inserted into the iliotibial tract, the linea aspera and lateral supracondylar line of femur.

Nerve supply: The inferior gluteal nerve.

Actions

1. It is an extensor of the hip joint
2. It is a powerful abductor and lateral rotator of the hip joint

Relations

Several important structures are situated deep to the gluteus maximus.

1. Bones and ligaments: Such as ilium, sacrum, coccyx, ischial tuberosity, sacrotuberous ligament
2. Muscles: Gluteus medius, minimus and short lateral rotators of the hip joint.
3. Nerves: The sciatic nerve, superior and inferior gluteal nerves, the pudendal nerve
4. Vessels: Superior and inferior gluteal vessels, the internal pudendal vessels.

Note: Intramuscular injections are often given in the gluteal region. To avoid injury to the sciatic nerve, other vessels and nerves, the injection is given in anterosuperior quadrant of gluteal region, i.e. in the gluteus medius and minimus.

The Tensor Fasciae Latae (Figs 9.27A and B)

Origin: From the outer lip of the iliac crest (extending from the anterior superior iliac spine to the tubercle of iliac crest).

Insertion: Into the upper part of iliotibial tract.

Nerve supply: Superior gluteal nerve.

Action: Abduction of hip joint.

The Gluteus Medius

The posterior one-third of this muscle is covered by the gluteus maximus. Anterior 2/3 is superficial. It is covered by the thick gluteal aponeurosis.

Origin: From the outer surface of ilium, limited by:
1. The anterior and posterior gluteal lines and
2. The iliac crest above

Insertion: Lateral surface of greater trochanter

Gluteus Minimus

It lies beneath the medius

Origin: From the gluteal surface of ilium, between the anterior and inferior gluteal lines.

Insertion: The greater trochanter

Nerve supply: Both the gluteus medius and minimus are supplied by the superior gluteal nerve.

Actions of gluteus medius and minimus
1. Abduction of hip joint
2. Acting from below, both these muscles prevent the unsupported side of the pelvis from sagging downwards during locomotion. So, when a patient stands on the affected limb with paralysis of the medius and minimus muscles, the pelvis sinks on the unsupported side. This is known as Trendelenburg's sign. The person walks with a lurching gait.

Muscles of the Thigh (Fig. 9.26)

Subdivisions of the Thigh

The deep fascia of the thigh (fascia lata) sends three intermuscular septa—lateral, medial and posterior,

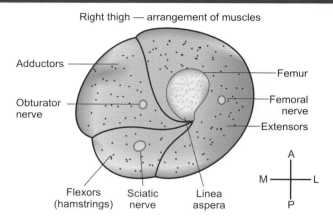

Right thigh — arrangement of muscles

Fig. 9.26: Transverse section through right thigh

which converge to the linea aspera and subdivide the thigh muscles into three groups.

The extensor or anterior compartment It lies between the lateral and medial septa; these muscles are supplied by the femoral nerve.

The adductor or medial compartment It lies between the medial and posterior septa. The adductor muscles are supplied by the obturator nerve.

The flexor or posterior compartment It lies between the posterior and lateral septa. The flexor muscles are supplied by the tibial component of the sciatic nerve.

The Extensor Muscles

1. Quadriceps femoris (Figs 9.27 A and B)
2. Sartorius

The quadriceps femoris: It is a powerful extensor of the knee joint. The quadriceps femoris consists of 4 parts—the rectus femoris, vastus medialis, vastus lateralis and the vastus intermedius (Figs 9.27 A and B). The rectus femoris arises from the ilium and the three vasti arise from the shaft of the femur.

The quadriceps femoris is the biggest muscle in the body. It covers the front and sides of the femur.

Insertion: It is inserted into the tibial tuberosity by a strong, round tendon. The patella (largest sesamoid bone in the body) develops in this tendon. The part of the tendon distal to the patella, is known as the ligamentum patellae.

Nerve supply: Femoral nerve

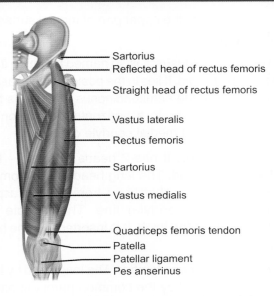

Fig. 9.27A: Muscles of the front of the thigh

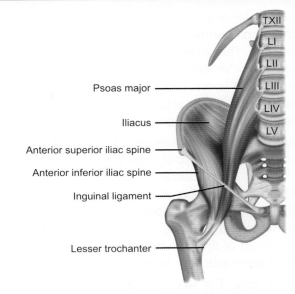

Fig. 9.27B: Psoas major and Iliacus

Action
1. Extension of the knee
2. The rectus femoris flexes the hip along with the iliopsoas

The sartorius (Sartor = tailor) It is the longest muscle in the body. It is a long, narrow, ribbon-like muscle, which extends from the anterior superior iliac spine to the upper part of the medial surface of the tibial shaft.

Nerve supply: Femoral nerve

Actions
1. Adduction and lateral rotation of thigh
2. Flexion of knee joint (all these actions are involved in assuming the position in which tailors sit and work).

The Adductor Muscles (Fig. 9.28)

The muscles of the adductor compartment (medial compartment) are:
1. Adductor longus
2. Adductor brevis
3. Adductor magnus
4. Pectineus
5. Gracilis
6. The obturator externus
   All these muscles are supplied by the obturator nerve.

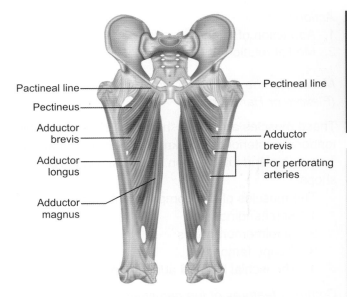

Fig. 9.28: Adductors of thigh

Adductor longus is a triangular muscle, forming the medial part of the floor of the femoral triangle.

Origin: From the front of the body of the pubis.

Insertion: To the linea aspera of femur.

Adductor brevis arises from the body of the pubis and the ischiopubic ramus. It is inserted along a line extending from the lesser trochanter to the upper part of linea aspera.

The adductor magnus: It is the largest muscle of the adductor compartment. It has an adductor part and a flexor part. This extensive, triangular muscle takes origin from the ischiopubic ramus and the ischial tuberosity. Fibers from the ischiopubic ramus represents the adductor part of the muscle; fibers from the ischial tuberosity represent the flexor part.

Insertion: The fibers are inserted into the gluteal tuberosity, the linea aspera, medial supracondylar line and the adductor tubercle.

Along the insertion of the muscle to the femoral shaft, there are 5 osseoaponeurotic openings. Upper 4 openings transmit the perforating branches and terminal part of the profunda femoris artery. The fifth opening gives passage to the femoral vessels.

Nerve supply: The adductor part is supplied by the obturator nerve; the flexor part is supplied by the tibial division of the sciatic nerve.

Actions
1. Adduction of thigh
2. Medial rotation of hip.

Muscles of the Posterior Compartment (Flexors or Hamstrings)

These muscles are called "hamstrings" because their tendons posterior to the knee are used to hang up 'hams' (hip and thigh region of animals) in the butcher shops.

The muscles of this compartment are:
1. Semitendinosus
2. Semimembranosus
3. Biceps femoris
4. The ischial head of adductor magnus.

Common features of the hamstrings
1. They take origin from the ischial tuberosity
2. They are inserted into one of the bones of the leg
3. They are supplied by the tibial part of the sciatic nerve
4. They are flexors of the knee and extensors of the hip.

The semitendinosus: It is called so, because it has a long tendon of insertion. The muscle lies superficially, on the semimembranosus, in the posteromedial aspect of the thigh. It takes origin from the ischial tuberosity

and is inserted into the upper part of medial surface of the shaft of tibia.

The semimembranosus: It is so named because it has a flat tendon of origin. It lies in the posteromedial aspect of thigh, deep to the semitendinosus. It extends from the ischial tuberosity, to a horizontal groove on the posterior surface of medial condyle of tibia.

The biceps femoris: It has 2 heads of origin – long head and short head. The long head arises from the ischial tuberosity. The short head, from the linea aspera and lateral supracondylar line. The muscle lies posterolaterally in the thigh. It is inserted into the head of the fibula.

Nerve supply of biceps femoris: The long head by tibial part and short head by the common peroneal part of the sciatic nerve.

Muscles of the Leg (Fig. 9.29)

The deep fascia of leg sends an anterior and a posterior intermuscular septa, which are attached to the anterior and posterior borders of fibula. They divide the leg into three compartments – Anterior, lateral and posterior.

Muscles of the Anterior Compartment of Leg (Fig. 9.30)

The muscles of the anterior compartment of the leg are:
1. Tibialis anterior
2. Extensor hallucis longus
3. Extensor digitorum longus
4. Peroneus tertius.

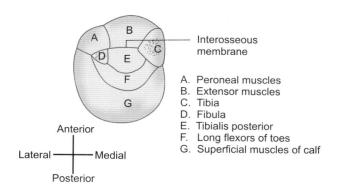

Interosseous membrane

A. Peroneal muscles
B. Extensor muscles
C. Tibia
D. Fibula
E. Tibialis posterior
F. Long flexors of toes
G. Superficial muscles of calf

Anterior

Lateral —— Medial

Posterior

Fig. 9.29: Osteofascial compartments of left leg

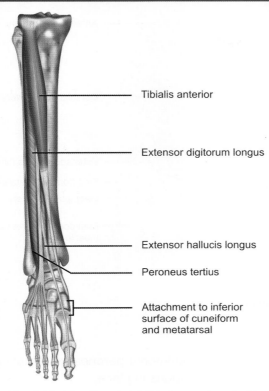

Tibialis anterior

Extensor digitorum longus

Extensor hallucis longus

Peroneus tertius

Attachment to inferior surface of cuneiform and metatarsal

Fig. 9.30: Muscles of anterior compartment of leg

Note:
1. All muscles of the anterior compartment are supplied by the deep peroneal nerve.
2. All muscles of the anterior compartment are dorsiflexors of the foot
3. Paralysis of the muscles of the anterior compartment results in "foot drop"—this is due to the loss of the power of dorsiflexion. As a result, the foot is plantar flexed.

Tibialis anterior: It is a spindle-shaped muscle.

Origin: Upper 2/3 of lateral surface of the shaft of the tibia and the adjacent interosseous membrane.

Insertion: The tendon of tibialis anterior is inserted into the inferomedial side of the base of the first metatarsal and the medial cuneiform bone.

Actions
1. It dorsiflexes the foot at the ankle joint
2. It acts as an invertor of the foot at the sub-talar and mid-tarsal joints
3. It maintains the medial longitudinal arch of the foot.

Extensor Digitorum Longus

Origin: From the upper ¾ of medial surface of the shaft of the fibula, interosseous membrane and lateral tibial condyle.

Insertion: On the dorsum of the foot, the tendon of extensor digitorum longus splits into 4 digital slips for insertion into the lateral four toes.

Action
1. Dorsiflexion of the ankle joint
2. Dorsiflexion of the lateral four toes.

Extensor Hallucis Longus

Origin: Middle 2/4 of medial surface of the shaft of the fibula and adjacent interosseous membrane.

Insertion: Inserted into the base of the terminal phalanx of the big toe.

Action
1. Dorsiflexion of the foot at the ankle
2. Dorsiflexion of the big toe

Peroneus tertius: It may be considered as the fifth tendon of the extensor digitorum longus.

Origin: Lower 1/4 of the medial surface of the shaft of the fibula

Insertion: Into the base of the 5th metatarsal bone

Action:
1. Dorsiflexor of the ankle joint
2. Weak evertor of the foot

Retinacula (Figs 9.31A and B)

On the front of the ankle, the deep fascia is thickened to form bands called retinacula, which retain the tendons in position—these are the *superior* and *inferior extensor* retinacula.

The superior extensor retinaculum is a broad band which extends from the lower part of the anterior borders of the tibia and fibula.

The inferior extensor retinaculum is a Y-shaped band of deep fascia. The stem of the Y is attached laterally to the superior surface of the calcaneum. Medially, the upper band is attached to the medial malleolus and the lower band is attached to the plantar aponeurosis.

Section B

Musculoskeletal System

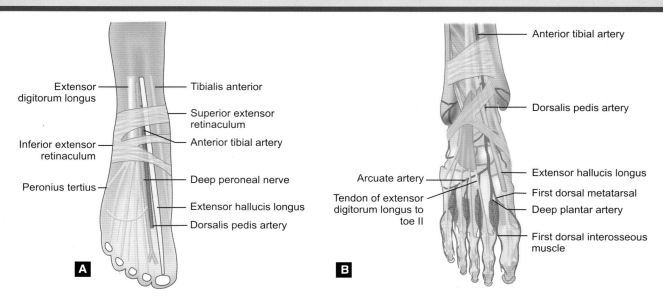

Figs 9.31A and B: Extensor retinacula of leg

Structures passing under cover of the retinacula are from medial to lateral:

1. **T**ibialis anterior
2. Extensor **h**allucis longus
3. **A**nterior tibial **a**rtery
4. Anterior libial **v**eins (venae comitantes), on either side of the artery
5. Deep peroneal **n**erve
6. Extensor **d**igitorum longus
7. **P**eroneus tertius

(The relations can be memorized by following the mnemonic **T**eacher **H**as **A** **V**ery **N**ice **D**ot **P**en. The anterior tibial artery is accompanied by venae comitantes).

Muscles of the Lateral (Peroneal) Compartment (Fig. 9.32)

There are two muscles in this compartment
1. The peroneus longus
2. The peroneus brevis.

Common features
1. They are evertors of the foot
2. They are supplied by the superficial peroneal nerve
3. Along with other muscles present around the ankle joint, the peroneal muscles help to maintain the stability of the joint.

4. The superior and inferior peroneal retinacula hold the peroneal tendons in place.

The peroneus longus

Origin: From the upper 2/3 of the peroneal surface of the fibula.

Insertion: The long tendon of the peroneus longus lies superficial to the tendon of the peroneus brevis. They lie in a groove behind the lateral malleolus. Then they pass deep to the superior peroneal retinaculum. The peroneus longus tendon passes downward and forward. On reaching the lateral side of the cuboid, the tendon passes forward and medially across the sole lying in an osseofibrous tunnel on the plantar surface of the cuboid. Finally, the tendon is inserted into the inferolateral surface of the *base of the first metatarsal bone* and *adjacent medial cuneiform*.

Actions
1. Eversion of the foot
2. It maintains the lateral longitudinal arch of the foot
3. It maintains the transverse arch of the foot.

Peroneus Brevis

Origin: From the lower 2/3 of the peroneal surface of fibula.

Insertion: The tendon grooves the posterior surface of the lateral malleolus, passes deep to the superior and

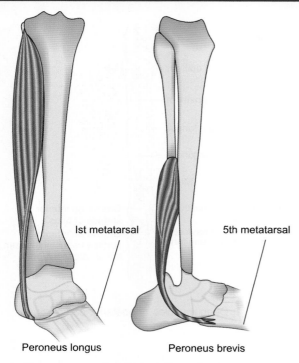

Peroneus longus   Peroneus brevis

Ist metatarsal   5th metatarsal

Fig. 9.32: Muscles of lateral compartment of leg

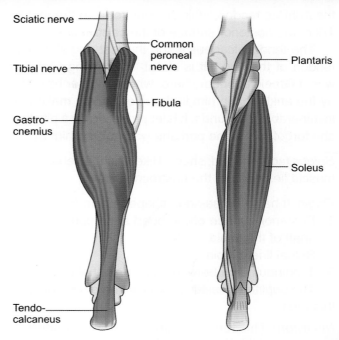

Sciatic nerve

Common peroneal nerve

Tibial nerve

Fibula

Gastro-cnemius

Plantaris

Soleus

Tendo-calcaneus

Fig. 9.33: Superficial muscles of posterior compartment of leg

inferior peroneal retinacula and finally inserted into the base of the 5th metatarsal bone.

Actions
1. It is an evertor of the foot
2. It maintains the lateral longitudinal arch

Muscles of the Posterior (Flexor) Compartment

These muscles can be grouped into two:
1. *Superficial muscles* Gastrocnemius, plantaris and soleus muscles.
2. *Deep muscles* Popliteus, flexor hallucis longus, flexor digitorum longus, tibialis posterior.

Common Features
1. The muscles of the posterior compartment (calf muscles) are strong plantar flexors of the foot at the ankle joint
2. They are supplied by the tibial nerve
3. They are associated with erect posture and bipedal locomotion
4. The muscles of the posterior compartment play an important role in circulation. Contractions of these muscles help in the venous return from the lower limb. *Soleus* is called the "peripheral heart". There

are large valveless veins in its substance. When the muscle contracts, the blood from these veins is pumped out.

Superficial Muscles (Fig. 9.33)

Gastrocnemius: It is a large, powerful muscle. It lies superficial to the soleus. It has two heads, medial and lateral. The two heads of the gastrocnemius and soleus are together known as the triceps surae.

Origins

Medial head arises by a broad, flat tendon from the posterosuperior depression on the medial condyle of femur, the adjoining popliteal surface and the capsule of the knee joint. This head is larger than the lateral head.

Lateral head arises by a broad flat tendon from the lateral surface of the lateral condyle of femur and the capsule of the knee joint. It may contain a sesamoid bone called Fabella.

Insertion: The tendon of this muscle fuses with the tendon of the soleus to form the tendocalcaneus or the Achilles tendon, which is inserted into the middle 1/3 of the posterior surface of the calcaneum.

The tendocalcaneus is the thickest and strongest tendon of the body. It is about 15 cm long (Achilles was a Greek legendary hero, whose mother held him by the ankle to dip him in the river Styx to make him invulnerable to wounds. It later proved to be a pity that she forgot to wet the part she was holding him by).

Soleus (sole shaped; shaped like a fish "soleus"). This muscle lies deep to the gastrocnemius.

Origin: It has a horse-shoe shaped origin from
1. Posterior surface of the head and upper ¼ of the shaft of the fibula
2. Soleal line of tibia
3. Tendinous arch between tibia and fibula
   The popliteal vessels and tibial nerve pass deep to this arch.

Insertion: The tendon blends with the tendon of gastrocnemius to form the tendocalcaneus which is inserted into the middle 1/3 of the posterior surface of the calcaneus.

Plantaris: It is a small fusiform muscle, with a long slendor tendon. It lies between the gastrocnemius and soleus.

Origin: From lower 1/3 of lateral supracondylar ridge of femur and adjoining popliteal surface.

Insertion: The slender belly of the muscle accompanies the lateral head of the gastrocnemius and ends in an elongated tendon, which blends with the tendocalcaneus.
• Plantaris is a vestigial muscle in man.
• Sometimes it is absent.

Actions of triceps surae: Gastrocnemius and soleus, along with plantaris produce *plantar flexion* of the ankle joint. The gastrocnemius and plantaris also produce flexion of the knee joint. "You stroll with the soleus but win the long jump with the gastrocnemius."

Deep Flexor Muscles (Fig. 9.34)

They are, popliteus, flexor hallucis longus, flexor digitorum longus and tibialis posterior.

Popliteus: It is a flat triangular muscle, which forms the floor of the popliteal fossa. Its origin is tendinous and the insertion is fleshy.

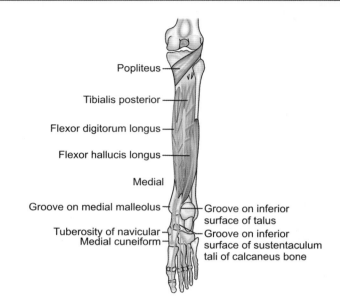

Fig. 9.34: Deep muscles of posterior compartment

Origin: The origin is intracapsular. It arises from:
1. The popliteal groove on the lateral surface of the lateral condyle of the femur
2. Outer margin of the lateral meniscus of the knee joint.

Insertion: It is inserted into the triangular area of the posterior surface of tibia above the soleal line.

Nerve supply: A branch of tibial nerve. This nerve winds round the lower border of the muscle and innervates it from the deep surface.

Actions:
1. It initiates flexion of the knee joint by unlocking the locked knee
2. It pulls the lateral meniscus backward and prevents it from being trapped between the condyles of tibia and femur at the beginning of flexion.

Flexor hallucis longus

Origin: From the lower 2/3 of posterior surface of the shaft of the fibula and interosseous membrane.

Insertion: The tendon grooves the talus and the undersurface of the sustentaculum tali of the calcaneus, and gets inserted into the base of the distal phalanx of the great toe.

Actions
1. It plantarflexes the big toe
2. It plantarflexes the ankle joint
3. It maintains the medial longitudinal arch of the foot.

Flexor digitorum longus
Origin: From posterior surface of tibia below the soleal line.

Insertion: The muscle ends in a tendon, which divides into 4 slips; each one is attached to the plantar surface of the distal phalanx of the digit concerned.

Actions
1. It plantarflexes the lateral four toes
2. It plantarflexes the ankle joint
3. It helps to maintain the medial longitudinal arch of the foot.

Tibialis posterior
It is the deepest muscle of the posterior compartment.

Origin: From the posterior surfaces of fibula and tibia (below the soleal line) and the interosseous membrane.

Insertion: Mainly into the tuberosity of the navicular bone; it also gives tendinous slips to all the tarsal bones except the talus and the 2nd, 3rd and 4th metatarsal bones.

Actions:
1. It is a powerful plantar flexor of the foot
2. It is an invertor of the foot
3. It helps to maintain the longitudinal and transverse arches of the foot.

The flexor retinaculum of the ankle (Fig. 9.35)
It is a thickened band of deep fascia on the medial side of the ankle. It retains the tendons, vessels and nerves in position as they pass from the back of the leg to the sole of the foot.

Attachments
Anteriorly to the posterior border and tip of the medial malleolus.

Posteriorly to the medial tubercle of calcaneum.

Structures passing deep to the retinaculum, from medial to lateral.
1. **T**ibialis posterior tendon
2. Flexor **d**igitorum longus tendon

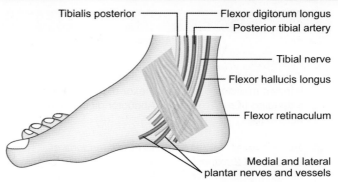

Fig. 9.35: Flexor retinaculum at the ankle

3. Posterior tibial *artery* (with venae comitantes)
4. Tibial nerve
5. Flexor **h**allucis longus (Mnemonic—"**T**om **D**ick **AN** **H**arry")

The tibial nerve divides into medial and lateral plantar nerves deep to the retinaculum. Similarly, the posterior tibial artery also divides into its terminal branches—medial and lateral plantar arteries – deep to the flexor retinaculum.

Muscles of the Foot

They can be grouped into two:
1. The plantar muscles (muscles of the sole)
2. *Intrinsic muscle of the dorsum of the foot—the extensor digitorum brevis.

Plantar Muscles
Muscles of the sole are described in four layers as shown in Table 9.2.

Intrinsic Muscle of the Dorsum of the Foot— The Extensor Digitorum Brevis

Origin: The small, fleshy belly arises from the anterior part of upper surface of the calcaneum.

Insertion: It splits into 4 tendons, into the medial four toes. Its most medial tendon is called extensor hallucis brevis. The tendons are inserted into the dorsal surface of the middle and terminal phalanges.

This muscle lies deep to the tendons of peroneus tertius and the extensor digitorum longus.

---

* Intrinsic = Those contained entirely within the foot.

Table 9.2: The muscles of the sole			
Layer	Muscles	Nerve supply	Action
1. First layer Consists of three intrinsic muscles extending from heel to toes	a. Abductor hallucis b. Flexor digitorum brevis c. Abductor digiti minimi	Medial plantar nerve Medial plantar nerve Lateral plantar nerve	Abducts the great toe Plantar flexion of lateral 4 toes Plantar flexor of little toe

2. Second layer
Consists of 2 extrinsic
tendons, and intrinsic
muscles
	a. Extrinsic tendons of
	  1. Flexor hallucis longus
	  2. Flexor digitorum longus
	b. Intrinsic muscles
	  1. Flexor accessorius
	  2. Four lumbricals
		Lateral plantar nerve
		1st lumbrical – Medial
		plantar, other three
		lumbricals—by
		deep branch of
		lateral plantar nerve
			Permits plantar flexion of lateral four
			toes
			Plantar flexion at MP joint and
			dorsiflexion at the IP joints of lateral
			four toes—This increases the grip

3. Third layer
The 3 muscles of
this layer are confined
to the metatarsal region
	1. Adductor hallucis
	2. Flexor digiti minimi brevis
	3. Flexor hallucis brevis
		1. Deep branch of lateral
		   plantar nerve
		2. Superficial branch of
		   lateral plantar nerve
		3. Medial plantar nerve
			Adducts the big toe towards
			the 2nd toe
			Plantar flexion of little toe
			Plantar flexion of big toe

4. Fourth layer had 7 intrinsic muscles and 2 extrinsic tendons	Intrinsic Muscles a. 4 dorsal interossei b. 3 plantar interossei Extrinsic Tendons a. Tibialis posterior b. Peroneus longus	Abductors of the toes (DAB) Adductors of lateral 3 toes (PAD)	1st and 2nd dorsal interossei—by deep peroneal nerve 3rd and 4th—lateral plantar nerve

Nerve supply: It is supplied by pseudoganglion of the lateral branch of deep peroneal nerve.

Action: It dorsiflexes the medial four toes.

Other Functions of the Intrinsic Muscles of the Foot

1. They act as tie-beams—they prevent the separation of the components of medial and lateral longitudinal arches; hence, they help to maintain these arches.

2. Along with the tarsal bones, they make the foot pliable so that the foot can adapt to uneven surfaces.

3. The lumbricals prevent the toes from buckling under when pulled upon by the flexor digitorum longus.

Note: There are 5 muscles which may be often absent in the body. They are:

1. **P**almaris longus
2. **P**almaris brevis
3. **P**yramidalis
4. **P**soas minor
5. **P**lantaris

Their names can be remembered by a tip, "5 P's".

10   Joints

INTRODUCTION

A joint is formed, where two or more bones come (articulate) together. There may or may not be movement between them.

CLASSIFICATION OF JOINTS (TABLE 10.1)

The joints can be classified according to the tissues that unite the bone ends. There are 3 types of joints.
1. Fibrous joints
2. Cartilaginous joints
3. Synovial joints.

FIBROUS JOINTS (FIG. 10.1)

In this type, the articulating surfaces of bones are connected by fibrous tissue. There are 3 types of fibrous joints:

Sutures or Sutural Joints (Fig. 10.1A)

Sutures occur only in the skull, e.g. coronal suture between the frontal and parietals. No movement is possible (But, in a skull of a newborn infant the skull bones do not make full contact with each other. The sutures form wide areas of fibrous tissue called fontanelles. The anterior fontanelle is the most prominent among the 6 fontanelles).

Syndesmosis (Syndesmos = Ligament) (Fig. 10.1B)

In this type, the bones are united by a sheet of fibrous tissue. It may be a ligament or a fibrous membrane, e.g. an interosseous membrane connects the radius and ulna.

The degree of movement depends on the distance between the bones and the degree of flexibility of the membrane.

Table 10.1: Classification of joints		
Fibrous joints	*Cartilaginous joints*	*Synovial joints*
a. Sutural joints or sutures, e.g. between the skull bones	a. Primary cartilaginous joints, e.g. epiphyseal plate between epiphysis and diaphysis	a. Ball and socket type, e.g. shoulder joint
b. Syndesmosis, e.g. an interosseous membrane, connecting radius and ulna	b. Secondary cartilaginous joints or symphysis, e.g. pubic symphysis	b. Hinge joints, e.g. elbow joint
c. Gomphosis – joint between tooth and its socket		c. Pivot joint, e.g. radioulnar joint, Atlantoaxial joint
		d. Condyloid joint, e.g. wrist joint
		e. Saddle joint, e.g. carpometacarpal joint of thumb
		f. Plane joint, e.g. joint between articular processes of vertebrae

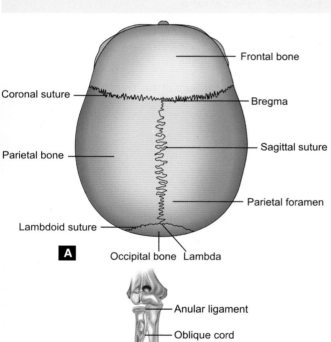

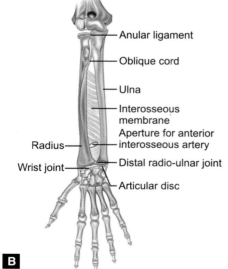

Figs 10.1A and B: Different types of joints:
(A) Sutures, (B) Syndesmosis

The interosseous membrane between the radius and ulna is flexible and wide enough to permit considerable movement during pronation and supination.

Gomphosis

This is a special type of fibrous joint, between a tooth and its socket. The fibrous tissue of the periodontal ligament firmly holds the tooth in its socket. Movement of the tooth is a pathological condition in the adult.

■ CARTILAGINOUS JOINTS

Bones are united either by hyaline cartilage or by fibrocartilage. Depending on this, the cartilaginous joints are classified into primary and secondary cartilaginous joints.

Primary Cartilaginous Joints (Fig. 10.2A) (Hyaline Cartilaginous Joints or Synchondroses)

The bones are united by hyaline cartilage, which permits slight movement during early life. This type of joint is temporary, as in the development of a long bone. During development, the epiphyses (or ends) of the long bone and the body (or diaphysis) of a bone are separated by an epiphyseal cartilaginous plate. This arrangement permits the growth of the bone. When full growth is achieved, the cartilage is converted into bone and the epiphyses fuse with the diaphysis.

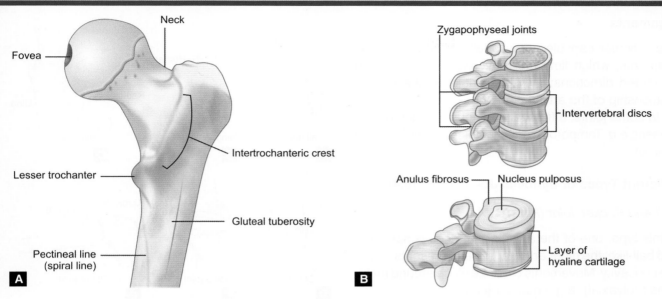

Figs 10.2A and B: Cartilaginous joints: (A) Primary, (B) Secondary

Secondary Cartilaginous Joints (Fibro-cartilaginous Joints or Symphyses) (Fig. 10.2B)

The surfaces of the articulating bones are covered with hyaline cartilage; and the bones are united by strong fibrous tissue or fibrocartilage. These joints usually occur in the midline of the body, e.g. symphysis pubis, joints between vertebral bodies, manubriosternal joint, symphysis menti.

■ SYNOVIAL JOINTS

They are the most common and important joints in the body. They normally provide free movement.

They are called synovial joints because they are lined with a synovial membrane and contain a lubricating fluid called synovial fluid.

Distinguishing features of a synovial joint (Fig. 10.3)
The synovial joints have:

1. A joint cavity
2. An articular cartilage
3. An articular capsule, lined internally by a synovial membrane.

Joint Cavity

The joint cavity is enclosed within a fibrous capsule. The inner surface of the capsule and the non-articulating ends of bones, which are inside the capsule

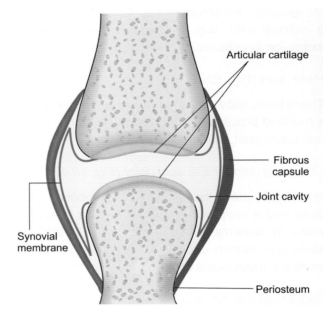

Fig. 10.3: Features of synovial joint

are lined by synovial membrane. The synovial fluid is produced, as well as absorbed by the synovial membrane.

Articular Cartilage

This cartilage is usually hyaline type. It has no nerve supply, blood supply or perichondrium. It is nourished by the synovial fluid covering its free surface.

Ligaments

The articular capsules are usually strengthened by ligaments, which limit the movement of the joint in unwanted directions; they also maintain the normal relationship of the articulating bones.

In some joints, a fibrocartilaginous articular disc is present, e.g. Temporomandibular joint, menisci of knee joint, etc.

Different Types of Synovial Joints (Fig. 10.4)

Ball and Socket Joint (Fig. 10.4A)

In this type, one of the articular surfaces is spherical and ball-like; the other articular surface presents a cup-like concavity. Movements can take place around many axes (polyaxial), e.g. shoulder joint.

Hinge Joint (Fig. 10.4B)

Movements take place in one plane only; It is usually a uniaxial joint, e.g. elbow joint, only flexion and extension is possible.

Pivot Joint (Fig. 10.4C)

These joints allow rotation movement. In these joints, a rounded process of bone rotates within a ring, e.g. radioulnar joint, atlantoaxial joint.

Condyloid Joint or Ellipsoid Joint (Fig. 10.4D)

In this type, one of the articular ends is convex and the other end is reciprocally concave. It is a bi-axial joint where movements can occur in 2 axes. So, adduction, abduction, flexion and extension can occur; but no rotational movements are possible, e.g. wrist joint.

Saddle Joint or Sellar Joint (Fig. 10.4E)

In this variety, the articular surfaces are reciprocally concavoconvex (saddle shaped) and movements can occur in all planes, e.g. the carpometacarpal joint of the thumb.

Plane Joint (Fig. 10.4F)

In this type, the articular surfaces are flat and movements restricted to slight gliding, tilting and rotation, e.g. joints between the articular processes of the thoracic vertebrae, acromio-clavicular joint.

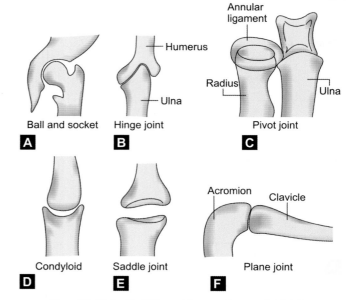

Figs 10.4A to F: Different types of synovial joints

Joints have a rich blood supply and nerve supply. *Hilton's Law* states that the nerves supplying a joint, also supply the muscles moving the joint and the skin covering the attachments of these muscles.

■ JOINTS OF THE UPPER LIMB

Joints of the Shoulder Girdle (Fig. 10.5)

1. Sternoclavicular joint
2. Acromioclavicular joint

The shoulder girdle or the pectoral girdle connects the bones of the upper limb with the axial skeleton. The girdle consists of clavicle and scapula. The clavicle meets the sternum at the sternoclavicular joint; the clavicle unites with the scapula at the acromio-clavicular joint.

The scapula has no direct connection with the axial skeleton. Its glenoid cavity articulates with the head of the humerus to form the shoulder joint.

A brief description of sternoclavicular and acromio-clavicular joints is given below.

The Sternoclavicular Joint

Type: Synovial joint—saddle joint.

Bones forming the joint:
1. Sternal end of the clavicle covered by fibrocartilage
2. Clavicular notch of manubrium sterni
3. Upper surface of first costal cartilage.

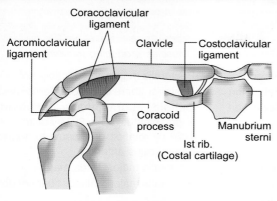

Fig. 10.5: Joints of upper limb: Sternoclavicular and acromioclavicular joints

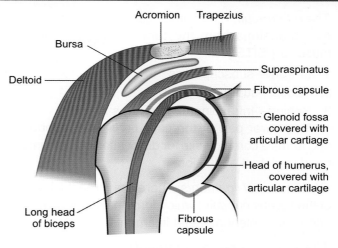

Fig. 10.6: Shoulder joint

Ligaments
1. Outer fibrous capsule
2. Anterior and posterior sternoclavicular ligament
3. Interclavicular ligament
4. Costoclavicular ligament.

An articular disc, made of fibrocartilage intervenes between the sternal notch and clavicle and divides the joint cavity into two.

This joint is so strong that dislocation is very rare; under stress, the clavicle fractures, rather than dislocation from sternum.

The Acromioclavicular Joint

It is a plane, synovial joint.

Bones forming the joint are
1. Lateral end of the clavicle
2. Small articular facet on the acromion process of scapula.

Ligaments
1. Fibrous capsule – a part of the capsule is thickened to form the acromioclavicular ligament.
2. Coracoclavicular ligament—It suspends the scapula from the lateral one-third of clavicle (its conoid tubercle and trapezoid line) and forms a strong bond between them. The weight of the upper limb is transmitted to the axial skeleton through this ligament. A fracture of the clavicle, medial to the attachment of this ligament leads to drooping of the upper limb.

Movements of shoulder girdle are elevation, depression, protraction or forward movement, retraction or backward movement and rotation.

The Shoulder Joint or the Glenohumeral Joint (Fig. 10.6)

It is a multiaxial, ball and socket type of synovial joint.

Bones Forming the Joint
1. The 'ball' is represented by the spherical head of the humerus
2. The 'socket' is formed by the pear-shaped, shallow, glenoid cavity of scapula.

Both the articular surfaces are covered by articular hyaline cartilage. Only one-third of the humeral head comes in contact with the glenoid cavity at any position. The shallow glenoid fossa is deepened by a fibrocartilaginous rim—the glenoidal labrum, which is attached to its peripheral margin.

Ligaments

The fibrous capsule—it forms a loose covering, permitting free movements.

• *Medially* it is attached to the periphery of the glenoid cavity.
• *Laterally* it is attached to the anatomical neck of the humerus.
• *Inferiorly* it extends 1 cm below, to encroach on the surgical neck of humerus.
  — The fibrous capsule is lined internally by synovial membrane.
  — The long head of biceps takes origin from the supraglenoid tubercle. It has an intracapsular origin.

The rotator-cuff: The fibrous capsule is strengthened by expansions from the tendons of the following muscles ("SITS").

1. In front—subscapularis
2. Above—supraspinatus
3. Behind—Infraspinatus and teres major.

The lower part of the capsule is least supported and forms a common site for dislocation of the humeral head in violent abduction.

Glenohumeral ligaments: There are three thickenings of the fibrous capsule, which form the superior, middle and inferior glenohumeral ligaments.

The coracohumeral ligament: Extends from the coracoid process to the anatomical neck.

Transverse humeral ligament: It connects the two lips of the intertubercular sulcus. This ligament holds the long tendon of biceps in position.

Relations of Shoulder Joint

The deltoid muscle covers the joint anteriorly, laterally and posteriorly.

- Superiorly—supraspinatus tendon.
- Inferiorly—The axillary nerve, long head of triceps and posterior circumflex humeral vessels.
- Anteriorly—subscapularis, coracobrachialis and short head of biceps.
- Within the capsule—long head of biceps tendon.

Blood supply Anterior and posterior circumflex humeral vessels.

Nerve supply From axillary nerve.

The movements permitted at the shoulder joint are flexion—extension, adduction—abduction, medial rotation—lateral rotation and circumduction. The muscles producing these movements are given in Table 10.2.

Bursae Around the Joint

There are several bursae around the shoulder joint, containing capillary films of synovial fluid. They reduce friction. Some of them communicate with the joint cavity.

Table 10.2: Movements at the shoulder joint and muscles producing them	
Movements	*Muscles producing movements*
1. Flexion	Pectoralis major, anterior fibers of deltoid
2. Extension	Posterior fibers of deltoid, Teres major
3. Abduction	Deltoid, supraspinatus (initiated by supraspinatus; up to 90° by deltoid; 90 to 180° by rotation of scapula by trapezius and serratus anterior)
4. Adduction	Pectoralis major, Teres major
5. Medial rotation	Pectoralis major, anterior fibers of deltoid
6. Lateral rotation	Posterior fibers of deltoid
7. Circumduction	A combination of all the above movements

Applied Anatomy

1. *Dislocation* of the shoulder joint is common due to laxity of the ligaments and disproportionate articular surfaces. The axillary nerve may be affected in inferior dislocation.
2. *Frozen shoulder* results from tendonitis involving the entire rotator cuff. So, all the shoulder movements are restricted.

The Elbow Joint (Figs 10.7A and B)

It is a hinge joint. It has two parts:
1. Humeroulnar part
2. Humeroradial part.

Humeroulnar Part (Fig. 10.7A)

The humeroulnar part is formed by the articulation between the trochlea of humerus and the trochlear notch of the ulna.

Humeroradial Part (Fig. 10.7B)

The humeroradial part is a ball and socket type of articulation. The 'ball' is represented by the capitulum of the humerus. The disc-like, concave upper surface of the head of the radius forms the socket.

Ligaments:
1. The fibrous capsule, lined internally by the synovial membrane, envelops the joint completely. The capsule is thin in front and behind, to permit flexion and extension.

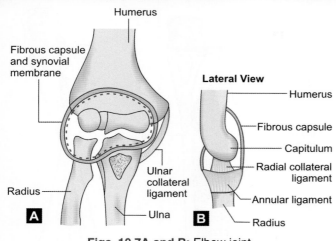

Humerus

Fibrous capsule and synovial membrane

Lateral View

Humerus

Fibrous capsule

Capitulum

Radial collateral ligament

Radius

Ulnar collateral ligament

Annular ligament

Ulna

Radius

A **B**

Figs 10.7A and B: Elbow joint

2. Medial or ulnar collateral ligament—It extends from the medial epicondyle to the medial margin of trochlear notch.
3. The lateral or radial collateral ligament—It extends from the lateral epicondyle to the annular ligament.

Relations of Elbow Joint

- Anteriorly
  — Brachialis
  — Tendon of biceps
  — Median nerve
  — Brachial artery
- Posteriorly
  — Triceps
- Medially
  — Common origin of flexor muscles of the forearm
  — The ulnar nerve
- *Laterally* Common extensor origin, radial nerve and its branches.

Movements The chief movements at the elbow joint are flexion and extension.

Muscles producing flexion are Brachialis, Biceps brachii and Brachioradialis

Muscles producing extension are Triceps and anconeus and this movement is assisted by gravity.

Carrying angle: When the elbow is fully extended and the forearm supinated, the arm and the forearm form an obtuse angle, which is open on the lateral side. This is known as the carrying angle. Because of this,

the ulnar border of the forearm does not come in close contact with the lateral surface of the thigh. This facilitates to carry a heavy object in the hand. The carrying angle disappears when the elbow is fixed and the forearm pronated.

Applied Anatomy

1. Supracondylar fracture of humerus is common in children. It usually occurs due to a fall on the outstretched hand. The brachial artery may be injured, resulting in ischemia of the deep flexors of the forearm, followed by fibrosis and shortening–this phenomenon is known as Volkmann's ischemic contracture.
2. *Dislocation* of elbow joint can occur in adults, often associated with fracture of coronoid process.
3. *Tennis elbow or lateral epicondylitis* is a painful musculoskeletal condition that may follow repetitive forceful pronation-supination movements (this condition is not confined to tennis players).

The Radioulnar Joints

The radius and ulna are united by 3 joints.
1. Superior radioulnar joint
2. Inferior radioulnar joint } synovial (Pivot) joints
3. Middle radioulnar joint—Syndesmosis (fibrous joint).

The Superior Radioulnar Joint (Fig. 10.8A)

It is a pivot joint.

Structures forming the joint
1. The articular circumference of the head of the radius
2. An osseofibrous ring—formed by radial notch of ulna and the annular ligament. 1/5th of this ring is bony and 4/5th, fibrous.

The annular ligament keeps the radial head in position. It is attached to the two ends of the radial notch of ulna. The radial head rotates within this ring.

Middle Radioulnar Joint (Fig. 10.8B)

It connects the shafts of the radius and ulna by syndesmosis, which consists of (1) An oblique cord and (2) An interosseous membrane.

The *oblique cord* passes at right angles to the fibers of the interosseous membrane. It extends from the ulnar tuberosity to the radial tuberosity.

Section B

Musculoskeletal System

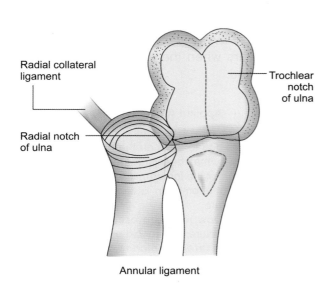

Fig. 10.8A: Radioulnar joints: An osseofibrous ring of superior radioulnar joint

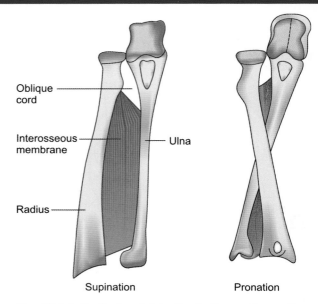

Fig. 10.8B: Radioulnar joints: Supination and pronation

The *interosseous membrane* is a broad fibrous sheet extending between the interosseous borders of radius and ulna. Upper border of the membrane is free and presents a gap between it and the oblique cord for the passage of posterior interosseous vessels. The lower part presents an oval gap for the passage of anterior interosseous vessels to the back of the arm.

The fibers of the membrane slope downward and medially from radius to ulna, except in the lower part, where they are arranged in reverse direction.

The interosseous membrane is stretched in mid-prone position; it is relaxed in extremes of supination and pronation.

Inferior Radioulnar Joint

This is a pivot joint.

Bones forming the joint
1. Articular surface of the head of ulna
2. Concave ulnar notch of radius

They are connected below by an articular disc. Both bones are enclosed in an articular capsule.

The articular disc is a triangular plate of fibrocartilage. It separates the head of the ulna from the wrist joint.

Movements: The radioulnar joints permit the movements of *pronation* and *supination* of the forearm.

Supination: In the anatomical position, the palm is directed forward and the forearm is supinated so that, the radius and the ulna lie side by side and almost parallel to each other.

Pronation: During pronation, the head of the radius spins within the annular ligament around a vertical axis; the head of the radius, still retains its position lateral to the ulna.

But, the lower end of the radius, carrying the hand with it, rotates medially across the lower part of ulna. In this process, the interosseous membrane is spiralised.

In supination, the rotation is reversed. The bones become parallel and the interosseous membrane is despiralized.

The ulna also moves during pronation and supination, eventhough it appears to be stationary. The distal end of the ulna moves backward and laterally in pronation and forward and medially in supination.

The range of pronation and supination, with the fixed elbow, is about 140-150°; when the elbow is extended, the range of movement is increased to almost 360° (This is due to the associated rotation of humeral head at the shoulder joint).

Muscles producing movements
*   *Pronation:*
    — Pronator quadratus
    — Pronator teres
    — Flexor carpi radialis
    — Gravity
*   *Supination:*
    — Biceps brachii (in flexion)
    — Supinator (in extension)

Wrist Joint (Fig. 10.9)

The wrist joint is a bi-axial ellipsoid (condyloid) joint; this joint is also called the radiocarpal joint.

Bones Forming the Joint

1. Distal articular surface of radius and the articular disc of inferior radioulnar joint.
2. Scaphoid, lunate and part of triquetral bones.

Ligaments

1. Capsular ligament, lined by synovial membrane
2. *Radial and ulnar collateral ligaments* they are thickenings of the fibrous capsule. The radial collateral ligament extends from the styloid process of radius to the scaphoid and trapezium. The ulnar collateral ligament extends from the ulnar styloid process to the triquetral and pisiform bones.

Relations

*   In front:
    — The long flexor tendons to the fingers and thumb
    — The median nerve
    — The ulnar nerve and vessels
    — The palmaris longus tendon
    — The radial artery

*   Behind:
    — Extensor tendons to thumb and other fingers

Movements and Muscles Producing them

1. Flexion—Flexor muscles of wrist and fingers
2. Extension—Extensor muscles of wrist and fingers
3. Adduction or ulnar deviation—flexor carpi ulnaris and extensor carpi ulnaris muscles
4. Abduction or radial deviation—abductor pollicis longus
5. Circumduction—it is a combination of flexion adduction, extension and abduction.

Applied Anatomy

1. Colles' fracture results from a fall on the outstretched hand. It involves the distal end of radius. The fracture line is transverse. The lower segment of the radius is displaced upward and backward; it produces a characteristic "dinner fork" deformity.
2. Smith's fracture: It is the reverse of Colles' fracture produced by a fall on the back of the hand.

Carpometacarpal Joint of the Thumb (Fig. 10.10)

The first carpometacarpal joint is a saddle (sellar) joint. It permits great range of movements.

Bones Forming the Joint

1. Distal articular surface of trapezium
2. Base of first metacarpal bone
    The articular ends of both bones are reciprocally convexoconcave.

Ligaments

Capsular ligament lined internally by the synovial membrane.

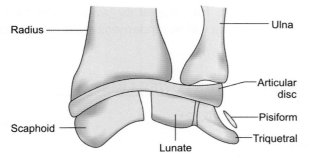

Fig. 10.9: Wrist joint

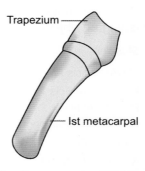

Fig. 10.10: Carpometacarpal joint of thumb

Section B   Musculoskeletal System

Lateral ligament connects lateral surface of trapezium to the lateral side of the base of first metacarpal bone.

The *dorsal and palmar ligaments* are oblique bands extending from the dorsal and palmar surfaces of trapezium to the ulnar side of the base of the 1st metacarpal bone.

Relations

- In front—
    - Abductor pollicis brevis ⎫
    - Flexor pollicis brevis ⎬ The thenar muscles
    - Opponens pollicis ⎭
- Behind—
    - Extensor pollicis longus and brevis
- Laterally—
    - Abductor pollicis longus and extensor pollicis brevis
- Medially—
    - 1st dorsal interosseous muscle and the radial artery

Movements (Fig. 10.11)
and Muscles Producing Them

The first metacarpal bone lies on a more anterior plane than the other metacarpal bones and undergoes medial rotation through 90°, so that its dorsal surface is lateral and palmar surface, medial.

Due to this peculiar anatomical position of the 1st metacarpal, flexion and extension take place parallel to the plane of the palm. Adduction and abduction take place at right angles to the palm.

The combination of flexion, abduction, extension and adduction is known as circumduction.

The opposition is a combination of abduction, flexion and medial rotation at the first carpometacarpal joint, so that the pulp of the thumb can be brought in contact with the tip of any finger (semiflexed). This is essential for grasping an object.

Movement	Muscles
1. Flexion	Flexor pollicis brevis and longus
2. Extension	Extensor pollicis longus, brevis, abductor pollicis longus
3. Abduction	Abductor pollicis longus and brevis
4. Adduction	Adductor pollicis
5. Opposition	It is initiated by the abductors of thumb and maintained by opponens pollicis

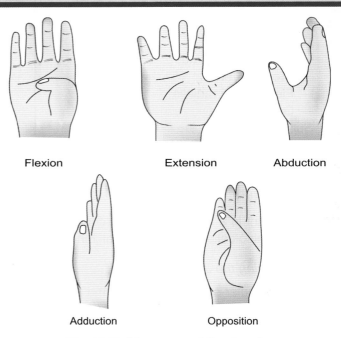

Flexion   Extension   Abduction

Adduction   Opposition

Fig. 10.11: Movements of thumb at first carpometacarpal joint

Because of the complexity of the movement, opposition of the thumb may be affected by nerve injuries in the upper limb, especially the median nerve. If the median nerve is cut in the forearm or wrist, the thumb cannot be opposed. Similarly, an injury to the recurrent branch of median nerve, that supplies the thenar muscles results in the paralysis of these muscles. The thumb, as a result, loses much of its usefulness.

When a person falls on the outstretched hand with the forearm pronated, the mainforce of the fall is transmitted through the wrist (carpus) to the forearm bones, especially, the radius, then to the humerus, scapula and clavicle. So, during such falls, fractures may occur in the wrist, radius, humerus or the clavicle.

Fracture of the scaphoid bone is also a common injury. The bone fractures at its "waist", producing 2 fragments. There will be tenderness in the anatomical snuff box.

JOINTS OF THE LOWER LIMB

Hip Joint (Fig. 10.12)

Type

It is a synovial joint, "ball and socket" type.

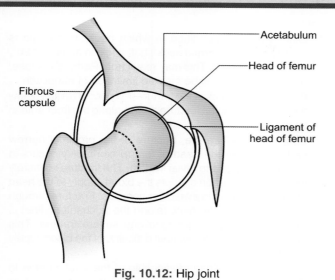

Fig. 10.12: Hip joint

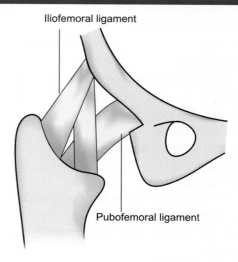

Fig. 10.13: Ligaments of hip joint

Articular Ends

1. The acetabulum of hip bone
2. Head of the femur.

The *acetabulum* is a cup-shaped depression of the hip bone, where ilium, ischium and pubis meet (They are separated by a tri-radiate Y-shaped cartilage, the ossification of which is complete by 20 years). The acetabulum presents a horse-shoe-shaped "lunate surface" internally. Below this lunate surface is a non-articular area filled with a pad of fat. A fibro-cartilaginous ribbon-like structure called labrum acetabulare is attached to the margins of the acetabulum.

The head of the femur is spherical, smooth and covered with articular (hyaline) cartilage. On the upper part of the head is a small pit.

Ligaments

The Capsule

It is a loose, thick fibrous sac enclosing the joint cavity. Its inner surface and the nonarticular surfaces of bone are lined by synovial membrane.

Attachments of fibrous capsule
- *Proximally* – to the margins of acetabulum
- *Distally:* Anteriorly, to the trochanteric line.
  — Posteriorly, 1 cm proximal to the trochanteric crest
  — Above: close to the greater trochanter
  — Below: close to the lesser trochanter.

The capsule is made up of superficial longitudinal fibers and deep circular fibers (zona orbicularis), which encircle the neck of the femur.

Transverse Ligament

It bridges across the acetabular notch, which is a gap in the lower part of acetabular rim.

Iliofemoral Ligament (Ligament of Bigelow) (Fig. 10.13)

It is the strongest ligament in the body. It is inverted 'Y' shaped. Superiorly, it is attached to the anterior inferior iliac spine. Inferiorly, it splits into 2 bands which are attached to the upper and lower parts of the intertrochanteric line of femur. In between these bands, the fibrous capsule is weak.

Pubofemoral Ligament

It extends from the superior ramus of pubis and iliopubic eminence and merges with the fibrous capsule.

Ischiofemoral Ligament

Proximally, it is attached to the ischium below and behind the acetabulum. Distally, it merges with the fibrous capsule.

Ligament of the Head of the Femur

It is triangular in shape. Its apex is attached to the pit on the femoral head. The base is attached to the margins of acetabular notch.

Relations of Hip Joint

- Anteriorly
  — Pectineus
  — Psoas major (*see* Fig. 9.27)
  — Iliacus
  — Rectus femoris
  — Femoral nerve
  — Femoral artery and vein
- Posteriorly
  — The sciatic nerve
  — Lateral rotators of the hip (lying under cover of gluteus maximus)
  — Inferior gluteal vessels and nerve
- Superiorly
  — Gluteus minimus
- Below
  — Pectineus and obturator externus

Arterial supply: From medial and lateral femoral circumflex arteries, superior and inferior gluteal arteries and the obturator artery.

Nerve supply: Mainly the femoral nerve

Unique Features of the Hip Joint

The hip joint is unique because it has a high degree of stability as well as mobility.

The stability of the hip joint is due to:

1. The depth of the acetabulum, which is increased further by a labrum acetabulare.
2. The strength and tension of the ligaments
3. The powerful muscles around it
4. The length and obliquity of the neck of the femur
5. The atmospheric pressure.

The mobility depends upon the neck of the femur, which is narrower than the head.

Movements (Table 10.3): Active movements permitted at the hip joint are flexion—extension, adduction-abduction, medial-lateral rotations, and circum-duction.

Applied Anatomy

1. Congenital dislocation of hip—it is due to the defective development of the upper part of the acetabulum. It is more common in females.
2. Perthes' disease—results from avascular necrosis of the head of the femur, which might have occurred in childhood as a result of trauma.

Contd...

Contd...

3. Coxa vara is a condition in which neck-shaft angle is diminished (The normal neck-shaft angle in adults is 120°, in children is 160°). This may result from Perthes' disease, or rickets. There will be marked limitation of abduction.
4. Fracture of the neck of the femur—May occur (a) just below the head (subcapital fracture) or (b) along the trochanteric line.
   Subcapital fracture (i.e. fracture of the proximal narrow part, just below the head of femur) commonly occurs in elderly people due to trivial trauma. The fracture is entirely intracapsular. It often disrupts the blood supply to the head of femur, resulting in avascular necrosis. Fracture through the distal part of the neck (along the trochanteric line) is more common in adults following severe trauma. This fracture, usually unites without difficulty, as the blood supply is not disrupted.
5. Because of the close relationship of the sciatic nerve to the hip joint, it may be injured (either stretched or compressed) during posterior dislocations or fracture—dislocations of hip joint. This may result in paralysis of the muscles supplied by the sciatic nerve; there will be sensory changes also.
6. In diseases of the hip joint, pain may be referred to the knee because both the hip and knee joints are supplied by the femoral and obturator nerves.

The Knee Joint (Fig. 10.14)

The knee joint is a modified hinge joint. It is the largest and most complex joint of the body. It is called a compound joint because it incorporates two condylar joints between the condyles of the femur and tibia and one saddle joint between the femur and the patella.

Table 10.3: Movements at the hip joint and muscles producing them	
Movements	*Muscles producing movements*
1. Flexion	Ilio-psoas
2. Extension*	Gluteus maximus and hamstring muscles
3. Adduction**	Adductor longus, brevis and non-ischial part of adductor magnus
4. Abduction	Gluteus medius and minimus
5. Medial rotation	Gluteus medius and minimus
6. Lateral rotation	Gluteus maximus and muscles under cover of gluteus maximus (Piriformis, gemelli, obturator internus and quadratus femoris)

* Extension is limited; it is arrested by the tension of the iliofemoral ligament

** Adduction is limited by the apposition of the opposite thigh

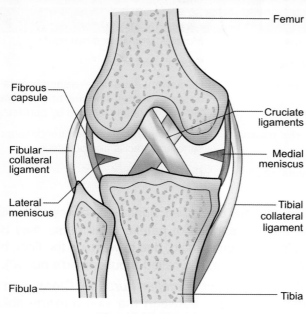

Fig. 10.14: Knee joint

Femur

Fibrous capsule

Cruciate ligaments

Fibular collateral ligament

Medial meniscus

Lateral meniscus

Tibial collateral ligament

Fibula

Tibia

Articular Surfaces

The knee joint is formed by—
1. The condyles of the femur
2. The condyles of the tibia
3. The patella.

The femoral condyles articulate with the tibial condyles below and behind, and with the patella in front.

Ligaments of Knee Joint

1. Capsular ligament with synovial membrane
2. The ligamentum patellae
3. Tibial and fibular collateral ligaments
4. Anterior and posterior cruciate ligaments
5. Oblique popliteal ligament.

The capsular ligament: It envelopes the joint. It is pierced posteriorly by the tendon of the popliteus muscle. The capsule is attached anteriorly to the peripheral margins of the patella and fuses with the ligamentum patellae. Inner side of the fibrous capsule provides attachment to the peripheral margins of medial and lateral menisci.

The synovial membrane lines the inner surface of the fibrous capsule and the portion of the bones within the capsule; but it does not cover the articular surfaces and the menisci.

The ligamentum patellae: It is derived from the tendon of quadriceps femoris and extends from the apex of the patella to the tibial tuberosity.

The tibial collateral ligament: This ligament has 2 parts, superficial and deep. Both parts are attached above to the medial epicondyle of femur. The superficial part extends downward as a flattened band and is attached to the medial condyle and adjacent part of medial border of the shaft of tibia.

The deep part is attached to the periphery of medial meniscus and it blends with the fibrous capsule. In the interval between the superficial and deep parts lie the inferior medial genicular vessels and nerve.

The fibular collateral ligament: It extends as a fibrous cord from the lateral epicondyle of femur to the head of the fibula. The ligament is overlapped superficially by the tendon of biceps femoris.

The deep surface of this ligament is not attached to the fibrous capsule. The gap between this ligament and the fibrous capsule is occupied by the inferior lateral genicular vessels and nerve.

(The tibial collateral ligament is morphologically the distal part of adductor magnus; the fibular collateral ligament is morphologically derived from the primitive origin of peroneus longus muscle).

Cruciate ligaments: There are 2 ligaments which cross like the letter "X", so they are called cruciate ligaments. They are named as anterior and posterior cruciate ligaments according to their tibial attachments.

The anterior cruciate ligament extends from the intercondylar area of tibia to the medial surface of lateral condyle of femur. It binds the tibia and femur together and prevents hyperextension at the knee joint.

The posterior cruciate ligament extends from the posterior part of intercondylar area of tibia to the lateral surface of medial condyle of femur. This ligament binds the tibia and femur and prevents the backward displacement of tibial condyles.

Both cruciate ligaments prevent side to side displacement of femur and tibia.

The oblique popliteal ligament: It is an expansion from the lower end of semimembranosus muscle (from its insertion).

The medial and lateral menisci (see Fig. 8.10) (or semilunar cartilages) The menisci are composed of fibrocartilage. They project from the fibrous capsule as incomplete partitions and occupy a position between the condyles of femur and tibia.

The medial meniscus (see Fig. 8.10): It is semilunar in shape and longer in the anteroposterior direction than the lateral meniscus. It has an anterior and a posterior horn, by which it is attached to the intercondylar area of tibia. Medially, it is attached to the fibrous capsule and the tibial collateral ligament. Its superior surface is concave.

The *lateral meniscus* is smaller. It is attached to the intercondylar area of tibia by anterior and posterior horns. Its upper surface is concave. The lateral margin is attached to the fibrous capsule.

Functions of menisci
1. They increase the concavity of the tibial condyles for better adaptation with femoral condyles.
2. They act as shock absorbers
3. They allow the gliding movements during flexion and extension.

Nutrition of these cartilages Since the peripheral part is attached to the fibrous capsule, it receives nutrition from arteries supplying the capsule. The inner free part of the meniscus is avascular; it derives its nutrition from synovial fluid. The semilunar cartilages can regenerate, provided the cartilage is removed entirely.

Movements at the knee joint (Table 10.4) are extension, flexion, medial rotation and lateral rotation.

Extension or straightening continues till the leg and thigh are in the same vertical line. Hyperextension is prevented by the tension of all ligaments and the tension of antagonistic muscles.

Bursae Around the Knee Joint

A bursa is a flattened sac ("purse"), whose walls are separated by a capillary film of synovial fluid. This fluid acts as a lubricant, enabling its walls to slide over each other without friction.

Bursae are present wherever a muscle or tendon is likely to rub on another muscle, tendon, bone or skin. There will be several bursae around active joints, like the knee joint.

Table 10.4: Movements at the knee joint	
Movements	*Muscles producing movements*
Extension	Quadriceps femoris
Flexion	Hamstring muscles—semimembranosus, semitendinosus and biceps femoris; flexion is initiated by popliteus
Medial rotation	Semimembranosus, semitendinosus, sartorius, gracilis and popliteus
Lateral rotation	Biceps femoris

Subcutaneous prepatellar bursa lies between the skin and anterior surface of patella (This bursa may be swollen in a house-maid, due to washing the floor, by kneeling – hence, it is called "house-maid's bursa").

Subcutaneous infrapatellar bursa or Clergyman's bursa, intervenes between the skin and the tibial tuberosity (It is sometimes swollen in people who pray in kneel-down position with the trunk upright—hence known as "Clergyman's bursa").

There are 4 or 5 bursae medial and lateral to the knee joint.

Arterial supply: Knee joint is highly vascular. It is supplied by genicular branches of popliteal, femoral, anterior and posterior tibial arteries.

Nerve supply: It is richly innervated by branches of femoral, tibial, common peroneal and obturator nerves.

Applied Anatomy

1. The firm attachment of the tibial collateral ligament to the medial meniscus is of great clinical importance, because, injury to the tibial collateral ligament will be often associated with injury to the medial meniscus. It is a common type of injury in football players (when considering soft tissue injuries of the knee, always think of the three C's – the collateral ligaments, cruciate ligaments and cartilages (menisci).
2. Genu valgum is due to the medial condyle and shaft of the femur protruding too far medially.
3. Genu varum or bow-leg presents with outward curving of the lower limb. These conditions may be congenital; they may be associated with rickets.
4. The sprain of anterior cruciate ligament can occur in violent hyperextension of the knee. The posterior cruciate ligament may be injured in the posterior dislocation of the tibia. Complete tear of these ligaments will lead to abnormal anteroposterior movements.

The Ankle Joint (Fig. 10.15)

The ankle joint or the talocrural joint is a uniaxial, hinge joint.

Bones Forming the Joint

1. Above:
   a. The inferior articular surface of the lower end of tibia, with its medial malleolus
   b. The lateral malleolus of fibula
      Articular surfaces of the tibia and fibula form a "tibiofibular mortise" (socket). This socket, receives the talus from below.
2. Below:
   a. The upper surface of the body of talus
   b. The comma-shaped articular facet on the medial side of talus, for medial malleolus
   c. A triangular facet, for the lateral malleolus, on the lateral surface of talus. These three articular areas form a continuous surface known as "trochlea tali".

Ligaments

1. Capsular ligament, with synovial membrane
2. Medially, the deltoid ligament
3. Laterally, anterior and posterior talofibular and calcaneofibular ligaments.

The deltoid or medial ligament (triangular or delta-shaped). It is a strong triangular band, having 2 parts—a superficial part and a deep part.

The superficial part is attached superiorly to the tip of medial malleolus. Inferiorly, it is attached to the following bones, from anterior to posterior—the navicular, the calcaneus and the talus.

The deep part extends from the medial malleolus to the talus, below the comma-shaped facet.

Movements (Fig. 10.16)

Movements permitted at the ankle joint are *dorsiflexion* and *plantar flexion.*

Normally, the leg meets the foot, forming a right angle. The dorsiflexion diminishes the angle between them, so that, the heel strikes the ground and the toes lie above the ground.

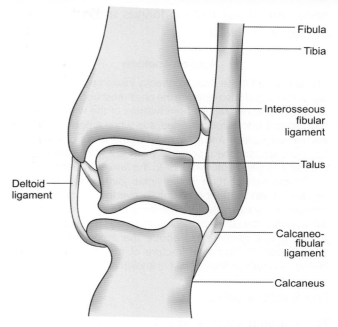

Fig. 10.15: Coronal section through left ankle joint

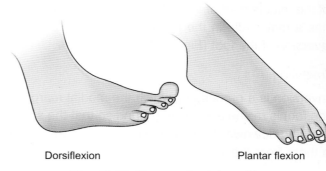

Dorsiflexion   Plantar flexion

Fig. 10.16: Movements at the ankle

The plantar flexion increases the angle between the leg and the foot, so that the person stands on toes.

Muscles Producing Movements

Dorsiflexion is produced by the muscles of the anterior compartment of the leg. They are tibialis anterior, extensor digitorum longus, extensor hallucis longus and peroneus tertius.

Plantar flexion is produced by the muscles of posterior compartment—the gastrocnemius, and soleus are the main flexors; the tibialis posterior, flexor digitorum longus and flexor hallucis longus assist the gastro-cnemius and soleus.

Applied Anatomy

The ankle joint is the most frequently injured major joint in our body. The lateral ligaments are the ones most frequently injured. A "sprained ankle" results from twisting of the weight bearing foot, when it is forcefully inverted, e.g. when a person steps on an uneven surface and falls, most of the fibers of the lateral ligament are stretched; some of them are torn. There will be pain and swelling anteroinferior to the lateral malleolus.

Fracture-dislocation of the ankle joint occurs in severe injuries. When the foot is forcefully everted, there may be a fracture of medial malleolus. The force of impact may push the talus against the lateral malleolus; this may lead to a fracture of fibula, a few centimeters above the ankle joint. Such a combination of horizontal fracture of medial malleolus and oblique fracture of fibular shaft is known as 'Pott's fracture'.

The Subtalar Joint

The subtalar joint is also known as the posterior talocalcanean joint. It is formed between the lower surface of the body of the talus and the upper surface of the middle third of the calcaneus. The subtalar joint has a major role in the movements of *inversion* and *eversion* of the foot (Fig. 10.17).

Inversion

Inversion is a process by which the medial margin of the foot is raised above the ground so that the sole is directed downward and medially (e.g. as if to remove a thorn from the sole).

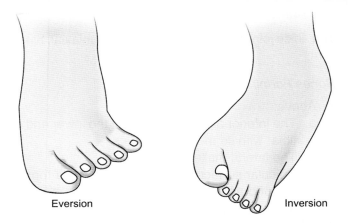

Eversion   Inversion

Fig. 10.17: Movements at the subtalar joint

Eversion

In eversion, the lateral margin of the foot is raised above the ground and the sole faces downward and laterally.

These two movements are best demonstrated when the foot is off the ground.

These movements help us to walk on uneven ground.

Muscles producing inversion
1. Tibialis anterior
2. Tibialis posterior.

Muscles producing eversion:
1. Peroneus longus
2. Peroneus brevis.

Section C

The Organ Systems

- ❖ Cardiovascular System
- ❖ Respiratory System
- ❖ Digestive System
- ❖ Urinary System
- ❖ Reproductive System
- ❖ Nervous System
- ❖ Endocrine System
- ❖ Sense Organs
- ❖ Medical Genetics
- ❖ Appendix

11 Cardiovascular System

◼ INTRODUCTION

Every tissue in the body requires an adequate supply of oxygen, nutrients and hormones. The waste products should be removed from the tissues from time to time. These functions are carried out by the blood. The heart and blood vessels are the mechanism by which a constant circulation is maintained throughout the body. The blood is pumped by the heart into the aorta from which it is distributed to all parts of the body (Fig. 11.1).

◼ THE HEART

It is a hollow muscular organ, which is situated in the middle mediastinum in the thorax. It lies between the two lungs and immediately above the diaphragm. It is situated behind the sternum and adjoining costal cartilages and left ribs (See CP3 in Color Plate 1).

Average weight: Males–300 gm; Females–250 gm

The heart is slightly larger than a clenched fist. The right side of the heart receives deoxygenated blood (low in oxygen, but not without it) from the body and pumps it into the aorta for distribution to the body.

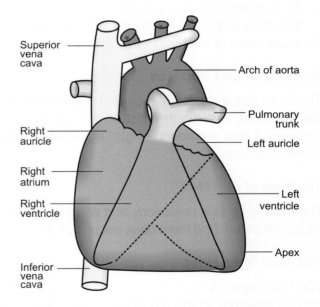

Fig. 11.1: Sternocostal surface of heart

Chambers of Heart (Fig. 11.1)

The heart has 4 chambers – 2 atria (the receiving area) and 2 ventricles (L. little belly), the discharging chambers.

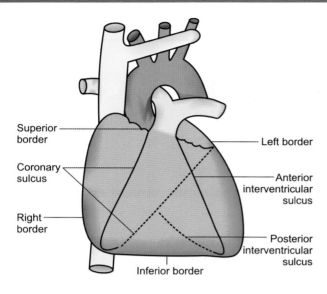

Fig. 11.2: Coronary sulcus and interventricular sulcus

The wall of each chamber consists of 3 layers:
1. An internal layer or endocardium
2. A middle layer or myocardium, composed of cardiac muscle
3. An outer layer or epicardium

The myocardium is the thickest layer and forms the main mass of the heart.

Base, Apex and Surfaces

The heart has a *base,* an *apex* and *3 surfaces*—the sternocostal, the diaphragmatic and pulmonary surfaces; it has *4 borders* – right, left, superior and inferior borders.

The Base of the Heart

The base is located posteriorly and is formed mainly by the left atrium (the heart does not rest on its base).

The Apex of the Heart

The blunt apex is formed by the left ventricle. It is located posterior to the 5th left intercostal space in adults, just medial to the midclavicular line (7-9 cm from the median plane). The "apex beat" (or heart beat) is an impulse imparted by the heart. It is the point of maximal pulsation.

The sternocostal (anterior surface) of the heart is mainly formed by the right ventricle.

Flow chart 11.1A: Systemic circulation

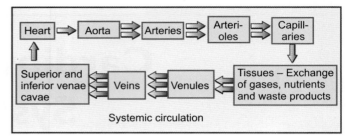

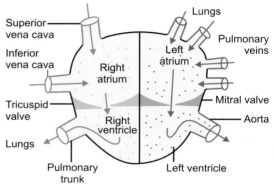

Fig. 11.3: Course of blood through the heart

The diaphragmatic (inferior) surface of the heart is formed by both ventricles, mainly the left one. It is related to the central tendon of the diaphragm. The posterior interventricular groove (Fig. 11.2) divides this surface into right one-third and left two-thirds.

The pulmonary or left surface of the heart is formed mainly by left ventricle, which occupies the cardiac notch of the left lung.

Borders (Fig. 11.2)

The right border is formed by the right atrium, the inferior border is formed mainly by right ventricle and partly the left; the left border is formed mainly by left ventricle and partly by left auricle. The great vessels enter and leave the superior border of the heart. The right and left auricles and the infundibulum are also seen here.

COURSE OF BLOOD THROUGH THE HEART (FIG. 11.3)

Systemic Circulation (Flow Chart 11.1A and B)

Oxygenated blood is pumped by the heart – the left ventricle – into the aorta, from which it is distributed to different parts of the body by arteries. The arteries

Flow chart 11.1B: Systemic circulation

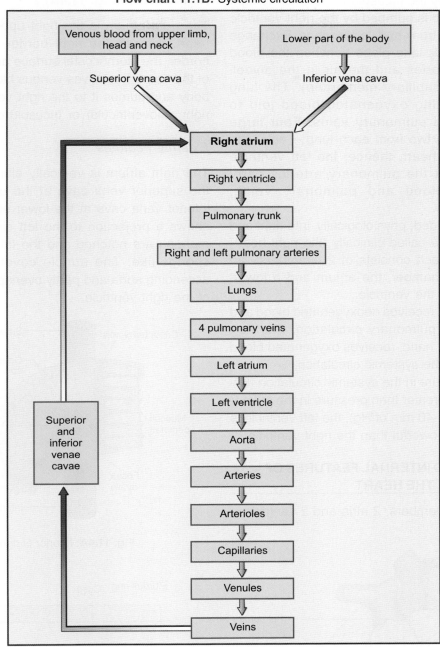

branch and rebranch to form arterioles and capillaries which distribute blood to the tissues. The capillaries have only one layer of flat, endothelial cells; the oxygen, carbon dioxide, nutrients and waste products are exchanged in the capillaries.

The capillaries drain into venules; they unite to form larger veins. Venous blood from the upper part of the body is drained by a large vein, the superior vena cava;

venous blood from the lower part of the body is drained by the inferior vena cava. These two veins carry the deoxygenated blood back to the right atrium of the heart. This completes the *systemic circulation.*

The Pulmonary Circulation (Fig. 11.3)

The deoxygenated blood from the right atrium passes through the right atrioventricular opening, which is guarded by the tricuspid valve, into the right ventricle; this blood is pumped by the right ventricle into the pulmonary trunk and its branches. Exchange of oxygen and CO_2 take place between the blood in the lung capillaries and the air in the alveoli (across alveolar-capillary membrane). The lung capillaries, carrying oxygenated blood join to form venules and pulmonary veins. Four large pulmonary veins – two from each lung – enter the left atrium of the heart, thence, the left ventricle (Please note that the pulmonary arteries carry deoxygenated blood and pulmonary veins, oxygenated blood).

The heart is divided, physiologically, into right and left parts, which are called clinically "the right heart" and "left heart". Each consists of 2 chambers—an upper, receiving chamber, the atrium and a lower, pumping chamber, the ventricle.

The "right heart" receives deoxygenated blood and pumps it into the pulmonary circulation. The "left heart", on the other hand, receives oxygenated blood and pumps it into the systemic circulation.

Since the pressure in the systemic circulation (80-120 mm of Hg) is greater than pressure in the pulmonary circulation (20-40 mm of Hg), the left ventricle is thicker and more powerful than the right ventricle.

EXTERNAL AND INTERNAL FEATURES OF CHAMBERS OF THE HEART

The heart has 4 chambers, 2 atria and 2 ventricles.

Right Atrium (Fig. 11.4A)

The right atrium is the right upper chamber of the heart. It forms the right border, part of the upper border, the sternocostal surface and part of the base of the heart. It receives venous blood from the whole body and pumps it to the right ventricle through the right atrioventricular or tricuspid orifice (opening).

External Features

The right atrium is vertically elongated. It receives the superior vena cava at the upper end and the inferior vena cava at the lower end. The upper end shows a projection to the left side, the auricle. Its margins are notched and the interior is rough and sponge-like. The auricle covers the root of the ascending aorta and partly overlaps the infundibulum of the right ventricle.

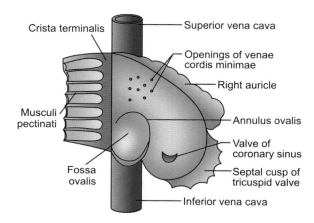

Fig. 11.4A: Interior of right atrium

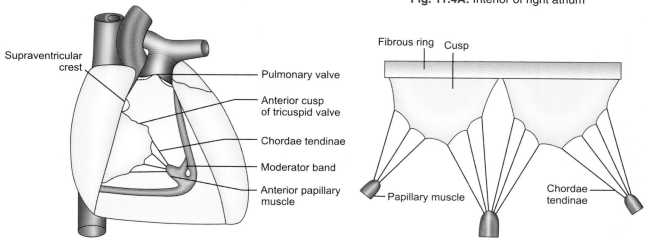

Fig. 11.4B: Interior of right ventricle and AV valve

Extending from the superior vena cava to the inferior vena cava, along the right border of atrium, is a shallow vertical groove - the sulcus terminalis. This groove is produced by an internal muscular ridge called the crista terminalis. The SA node is situated near the upper part of the sulcus terminalis.

The right atrium is separated from the right ventricle by the atrioventricular groove or the coronary sulcus. The right coronary artery and the small cardiac vein lie in this groove.

Veins Draining into the Right Atrium

1. Superior vena cava
2. Inferior vena cava
3. Coronary sinus
4. Anterior cardiac veins
5. Venae cordis minimi (Thebesian veins)

Internal Features

The right atrium has 3 parts:
1. The smooth posterior part or sinus venarum
2. The rough anterior part or pectinate part including the auricle
3. The inter atrial septum.

Smooth Posterior Part or Sinus Venarum

It is called sinus venarum because it develops from the right horn of the sinus venosus.

The superior vena cava opens at its upper end. The inferior vena cava opens at the lower end. This opening is guarded by a rudimentary valve, the Eustachian valve. It has an important function during fetal life, directing the oxygenated blood from the placenta to the left atrium through the foramen ovale.

The coronary sinus, the vein collecting most of the venous blood from the walls of the heart, opens into the right atrium between the opening of inferior vena cava and the right atrioventricular orifice. This opening is guarded by the valve of coronary sinus.

The venae cordis minimi are numerous small veins present in the walls of all the four chambers. They open into the right atrium through small foramina.

The intervenous tubercle is a small projection on the posterior wall of right atrium. During fetal life, it directs the superior vena caval blood to the right ventricle.

Rough Anterior Part or Pectinate Part

Arising from the crista terminalis are a series of ridges which run forwards and downwards towards the atrioventricular orifice. This gives the appearance of the teeth of a comb (Pecten = Comb). The auricle is an extension of the rough anterior part.

Pectinate part and the auricle develop from the primitive atrial chamber.

Interatrial Septum

It separates the right atrium from the left atrium. It develops from the septum primum and septum secundum.

The inter atrial septum shows a shallow saucer-shaped depression in its lower part, the fossa ovalis. It represents the site of the septum primum.

The annulus ovalis or limbus fossa ovalis is the sharp margin of the fossa ovalis. It represents the lower free edge of the septum secundum. It is shaped like an inverted U.

Applied Anatomy

Atrial septal defect: If the foramen ovale does not close after birth (atrial septal defect or ASD), oxygenated blood from the left atrium is shunted to the right atrium through the defect. This overloads the pulmonary system and causes enlargement of the right atrium, right ventricle and the pulmonary trunk.

Right Ventricle (Fig. 11.4B)

The right ventricle forms the inferior border and a large part of the sternocostal surface of the heart.

It is a triangular chamber. It receives deoxygenated blood from the right atrium and pumps it to the lung through the pulmonary trunk and pulmonary arteries.

Interior of Right Ventricle

It has an inflowing part and an outflowing part.

Inflowing Part

It is rough due to the presence of muscular ridges called trabeculae carneae.

Outflowing Part

The outflowing part or infundibulum is smooth. It forms the upper conical part of the right ventricle which gives origin to the pulmonary trunk.

Section C

The Organ Systems

The interior of right ventricle shows two orifices:
1. The right atrioventricular orifice or tricuspid orifice, guarded by the tricuspid valve.
2. The pulmonary orifice guarded by the pulmonary valve.

The interior of the inflowing part shows trabeculae carneae or muscular ridges of 3 types.
1. Ridges or fixed elevations
2. Bridges
3. Papillary muscles or pillars with one end attached to the ventricular wall and the other end connected to the cusps of the tricuspid valve by cordae tendinae.

Papillary Muscles

There are three papillary muscles in the right ventricle-Anterior, posterior and septal. The anterior papillary muscle is the largest. Posterior or inferior papillary muscle is small and irregular. Septal muscle is divided into a number of small projections. Each papillary muscle is attached by chordae tendinae to the contiguous sides of two cusps (Fig. 11.4B).

Septomarginal Trabecula or Moderator Band

It is a muscular ridge extending from the interventricular septum to the base of anterior papillary muscle. It contains the right branch of the AV bundle.

The cavity of the right ventricle is crescentic in section because of the forward bulge of the interventricular septum.

The wall of the right ventricle is thinner than that of the left ventricle (Pressure in the pulmonary trunk is 20-40 mm of Hg whereas the pressure in the aorta is 80-120 mm of Hg. So the left ventricle has to pump against a higher resistance, making its wall thicker.)

Interventricular Septum

It is placed obliquely between the two ventricles. The upper part is thin and membranous and lower part is thick and muscular. The membranous part separates the two ventricles and also, the right atrium from left ventricle. The muscular part separates the two ventricles.

Left Atrium

It forms the left two-thirds of the base of the heart, the greater part of upper border, parts of the sternocostal and left surfaces and part of the left border of the heart.

It is a quadrangular chamber receiving oxygenated blood from the lungs through four pulmonary veins and pumps it to the left ventricle through the left atrioventricular or bicuspid or mitral orifice which is guarded by the mitral valve (bicuspid valve).

The appendage of left atrium, the left auricle projects anteriorly to overlap the infundibulum of right ventricle.

Two pulmonary veins open into the left atrium on each side of the posterior wall. The greater part of interior of the atrium is smooth walled. Musculi pectinati are present only in the auricle. The anterior wall of the atrium is formed by the interatrial septum. A few venae cordis minimi open directly into the left atrium.

Left Ventricle

The left ventricle receives oxygenated blood from the left atrium and pumps it into the aorta.

It forms the apex of the heart, a part of the sternocostal surface, most of the left border and left surface, and the left two-thirds of the diaphragmatic surface.

The interior of left ventricle has two parts
1. The lower rough part with trabeculae carneae
2. The upper smooth part or aortic vestibule gives origin to the ascending aorta.

There are two well-developed papillary muscles, anterior and posterior. Cordae tendinae from both muscles are attached to both cusps of the mitral valve.

The interior of the ventricle shows two orifices:
1. The left atrioventricular or bicuspid or mitral orifice, guarded by the mitral valve.
2. The aortic orifice, guarded by the aortic valve.

The cavity of the left ventricle is circular in shape.

Valves of the Heart

The valves of the heart have two functions to perform:
1. Maintain unidirectional flow of blood.

2. Prevent regurgitation of blood.

There are two pairs of valves in the heart:
1. A pair of atrioventricular valves.
2. A pair of semilunar valves.

Atrioventricular Valves

The right atrioventricular valve is known as the tricuspid valve because it has three cusps. The left atrioventricular valve is otherwise known as the bicuspid or mitral valve because it has two cusps and has a resemblance to a Bishop's mitre.

Parts of Atrioventricular Valves (Fig. 11.4B)

1. A fibrous ring to which the cusps are attached
2. The cusps: Each cusp or leaflet is flat and projects into the ventricular cavity. Each cusp has a free margin and an attached margin; has an atrial and a ventricular surface. The free margins and ventricular surfaces are rough due to the attachment of cordae tendinae. The valves are closed during ventricular systole by apposition.
3. The chordate tendinae: Connect the free margins of cusps to the apices of papillary muscles. They prevent eversion of the free margins of the cusps towards the cavity of the atrium.
4. Papillary muscles : Keep the atrioventricular valves competent, by pulling the cordae tendinae during ventricular systole. The tricuspid valve has three cusps and can admit the tips of three fingers. The mitral or bicuspid valve has two cusps. It admits the tips of two fingers.

Semilunar Valves

The aortic and pulmonary valves are called semilunar valves because their cusps are semilunar in shape. The cusps form small pockets with their mouths directed away from the ventricular cavity. The free margin of each cusp contains a central fibrous nodule from each side of which a thin smooth margin, the lunule extends up to the base of the cusp.

The semilunar valves are closed during ventricular diastole. Opposite the cusps, the vessel walls are slightly dilated to form the aortic and pulmonary sinuses. The coronary arteries arise from two of the three aortic sinuses.

Applied Anatomy

Mitral valve is the most frequently diseased of the heart valves. Rheumatic fever can lead to mitral stenosis (narrowing) or mitral regurgitation (blood from left ventricle flows to left atrium). Similarly, damage of aortic valve can lead to aortic stenosis or regurgitation.

Ventricular septal defect (VSD) is the most common congenital heart disease.

■ BLOOD SUPPLY OF HEART (FIG.11.5)

Arterial Supply

The heart gets its oxygen and nutrients from 2 arteries—the right and left coronary arteries, which are the first branches of aorta.

At the beginning of the ascending aorta, there are 3 dilatations or swellings called aortic sinuses. The right coronary artery arises from the right aortic sinus and the left coronary artery, from the left aortic sinus.

The right and left coronary arteries are called "coronary" because they encircle the base of the ventricles somewhat like a crown.

The Right Coronary Artery (RCA)

It arises from the right aortic sinus and descends in the coronary sulcus between the right auricle and right ventricle. It then passes towards the inferior border of the heart, where it gives off a marginal branch that runs towards the apex. After giving off this branch, the RCA turns to the left and enters the posterior interventricular groove, where it gives off its largest branch, the posterior interventricular branch. This branch supplies both ventricles, runs towards the

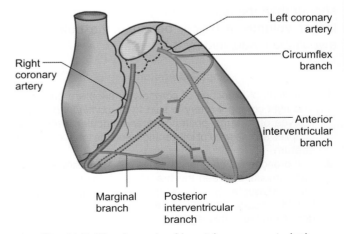

Fig. 11.5: Blood supply of heart (coronary arteries)

apex and anastomoses with the anterior interventricular branch of left coronary artery.

Branches Right coronary artery gives off atrial branches, ventricular branches, nodal artery to SA node and major part of conducting system of heart. Thus, it supplies the right atrium, right ventricle, interventricular septum, SA node, AV node and AV bundles. It also supplies variable parts of left atrium and left ventricle.

The Left Coronary Artery (LCA)

It arises from the left aortic sinus; passes between left auricle and pulmonary trunk, to reach the coronary groove (sulcus). It soon divides into 2 terminal branches (i) The anterior interventricular branch and (ii) the circumflex branch.

The anterior interventricular branch passes along the anterior interventricular groove to the apex of the heart, where, it anastomoses with the posterior interventricular branch of right coronary artery. This artery supplies both ventricles and the interventricular septum.

The circumflex branch is smaller. It follows the coronary groove or coronary sulcus around the left border of the heart to the posterior surface. It terminates to the left of posterior interventricular artery. The circumflex artery supplies the left atrium, left ventricle and the left surface of the heart.

Summary of distribution of left coronary artery. This vessel supplies most of the left ventricle, the left atrium, part of the interventricular septum and also, a part of the right atrium.

The trunk of the left coronary artery is much shorter than the right coronary artery. But it carries a larger volume of blood, because the bulk of the myocardium supplied by it is larger.

Applied Anatomy

1. Variations of branches of coronary arteries are common. The right coronary artery is said to be *dominant* because it gives off a large posterior interventricular artery that supplies a large area of left ventricle and the interventricular septum.
2. The branches of coronary arteries are "end arteries". The anastomosis between the branches of coronary arteries are inadequate. So, if a sudden occlusion of a major branch occurs, the region (myocardium) supplied by it gets infarcted – this is called myocardial infarction or MI

Contd...

Contd...

3. *Ischemic heart disease (IHD)* – Inadequate supply of blood to the myocardium (ischemia) results in substernal or retrosternal pain or discomfort. The most common cause of IHD is coronary insufficiency, resulting from atherosclerosis of coronary arteries (lipid accumulation of the internal walls of blood vessels).
4. *Angina pectoris* (chest pain) is a clinical syndrome characterized by substernal discomfort resulting from myocardial ischemia. Sublingual nitroglycerin dilates the coronary arteries. When the blood flow increases, the pain decreases.
5. *Coronary angiography*: Coronary arteries can be visualized by this method. After introducing a catheter (along femoral or brachial artery) into the opening of coronary artery, a radio-opaque dye is injected. Then X-rays are taken to see the lumen of the coronary artery.
6. *Coronary bypass surgery*: When there is obstruction of a coronary artery, a "bypass graft" surgery is carried out. A segment of a vein (usually, great saphenous vein) is connected to the coronary artery, proximal and distal to the stenosis (Fig. 11.6).

Venous Drainage of the Heart (Fig. 11.7)

The walls of the heart are drained by veins that empty into the coronary sinus and partly by small veins (anterior cardiac veins or venae cordis minimae or Thebesian veins) that open directly into the chambers of the heart, especially the right atrium and right ventricle.

The Veins of the Heart

Coronary Sinus

It is the largest vein of the heart. It is situated in the posterior part of coronary sulcus. It ends by opening

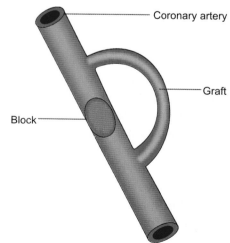

Fig. 11.6: Basis of bypass surgery

Coronary artery

Graft

Block

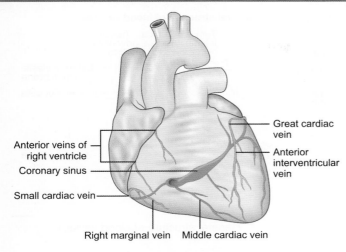

Anterior veins of right ventricle
Coronary sinus
Small cardiac vein
Great cardiac vein
Anterior interventricular vein
Right marginal vein   Middle cardiac vein

Fig. 11.7: Venous drainage of heart

into the right atrium, to the left of opening of inferior vena cava. It receives the following tributaries:

1. *The great cardiac vein:* It accompanies the anterior interventricular artery at first; then it accompanies the left coronary artery and enters the left end of coronary sinus.
2. *The middle cardiac vein:* It accompanies the posterior interventricular artery and joins the middle part of coronary sinus.
3. The small cardiac vein accompanies the right coronary artery and ends in the right end of coronary sinus.
4. The marginal vein accompanies the marginal branch of right coronary artery.

Coverings of the Heart

The heart lies enclosed within a fibroserous sac called the pericardium. It has 2 layers – an outer fibrous layer and an inner serous layers.

Outer fibrous layer It continues *above* with the external layer (tunica adventitia) of great vessels.

Below: It is fused with the central tendon of diaphragm.

Anteriorly it is attached to the posterior surface of sternum by sternopericardial ligaments.

Functions (1) It helps to maintain the general position of the heart; (2) It prevents the overdistension of heart.

Inner serous layer Within the fibrous pericardium is a double-layered serous pericardium – outer parietal

layer and an inner visceral layer or epicardium, with a thin film of fluid in between for lubrication. The potential space between the parietal and visceral layers is called the pericardial cavity.

Nerve supply of pericardium   The nerves are derived from vagus nerve, phrenic nerve and sympathetic trunks.

Applied Anatomy

Pericarditis is the inflammation of pericardium. It causes retrosternal pain and produces pericardial effusion (passage of fluid from pericardial capillaries into the pericardial cavity). Excess fluid in the pericardial cavity can affect the normal function of the heart, by compressing the pulmonary trunk and atria. This leads to a condition called cardiac tamponade.

Inflammation of the pericardium makes its layers rough; the resulting friction is called the friction rub, which can be heard with a stethoscope; it sounds like a "rustle of silk". Tapping or draining excess pericardial fluid to relieve pressure is called pericariocentesis. A wide-bore needle is inserted into the 5th or 6th left intercostal space, near the sternum.

Nerve Supply of Heart

The heart is supplied by autonomic nerve fibers. Parasympathetic fibers are derived from both vagus nerves; sympathetic fibers are derived from sympathetic trunks. Both these fibers form a network called the cardiac plexus.

Stimulation of the sympathetic nerves results in:
1. Increased heart rate (tachycardia)
2. Increased force of contraction
3. Dilatation of coronary arteries, resulting in increased blood flow and availability of more oxygen and nutrients for myocardium.

Stimulation of vagus results in:
1. Slowing of the heart (bradycardia)
2. Reduction in the force of contraction
3. Constriction of coronary arteries.

Conducting System of the Heart

This system consists of specialized cardiac muscle cells, that can initiate impulses and conduct them rapidly through the heart. They co-ordinate the contractions of the 4 chambers of heart. Thus, both atria contract together; and both ventricles contract together. The atrial contraction occurs first. The synchronized contraction of the chambers is essential for the efficient pumping of heart and for the maintenance of systemic and pulmonary circulations.

Parts

1. The sinoatrial or SA node
2. The atrioventricular or AV node
3. The atrioventricular bundle or AV bundle
4. Right and left branches of AV bundle.

SA node is situated in the wall of the right atrium. It is the "natural pacemaker" of the heart, because it initiates the impulses for contraction.

AV node is located in the interatrial septum. Impulses from both atria reach the AV node, which conducts them to the ventricles via the AV bundle.

The AV bundle, also called Purkinje fibers, originates in the AV node; this bundle lies in the interventricular septum. It is the only bridge between the atrial and ventricular myocardium. Within the interventricular septum, the AV bundle divides into right and left limbs or branches. Each branch passes deep to the endocardium into the walls of the ventricles.

Applied Anatomy

The passage of impulses over the heart from the SA node can be amplified and recorded as an ECG (electro-cardiogram). ECG becomes abnormal in many heart problems.

Diseases of the conducting system may result in "complete heart block". Artificial pacemakers are available, which produce electrical impulses that initiate ventricular contractions at a predetermined rate.

Cardiopulmonary resuscitation (CPR)—when the heart stops beating, firm pressure is applied to the chest vertically, over the lower part of sternum. This attempt forces the blood out of the heart into the great vessels.

■ BLOOD VESSELS

General Structure (Histology) of Blood Vessels (Fig. 11.8)

All blood vessels with lumina larger than that of the capillaries exhibit a common pattern of organization. The wall of each vessel contains 3 concentric coats or tunics.

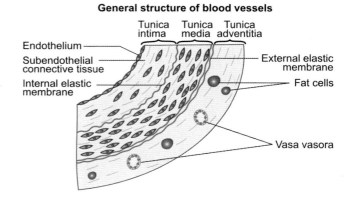

General structure of blood vessels

Fig. 11.8: General structure of blood vessels

1. The Tunica intima – Innermost layer
2. The Tunica media – Middle layer
3. The Tunica adventitia – Outer layer.

The Tunica Intima

It has the following layers:
a. Inner endothelial lining, made up of a single layer of simple squamous epithelium
b. Underlying basal lamina
c. Subendothelial layer of fibroelastic tissue
d. A band of elastic fibers – the internal elastic lamina.

The Tunica Media

The tunica media is made up chiefly of smooth muscle fibers, which are circularly arranged. In addition, the middle layer has elastic fibers and collagen fibers.

The Tunica Adventitia

The tunica adventitia is principally composed of fibro-elastic tissue. Most of the collagen fibers run parallel to the long axis of the vessel. Closest to the media, there may be a concentration of elastic fibers – the external elastic membrane. The tunica adventitia merges with connective tissue of the organ through which the vessel is passing. Vasa vasora (blood vessel supplying the wall of a blood vessel) lie in the tunica adventitia in larger blood vessels. The inner layers of blood vessel get O_2 and nourishment by diffusion. In larger blood vessels, the tunics are too thick to be nourished by diffusion from the lumen.

Veins

Veins are different from arteries in the following aspects:

1. Blood within the veins are under much less pressure; so, veins must accommodate a greater volume of blood.
2. Because of the above reason, veins are generally larger in diameter than their corresponding arteries.
3. Their walls are thinner, mainly due to a reduction in muscular and elastic components.

Structure of a Large Artery (e.g. Aorta) or Elastic Artery (Figs 11.9A and B)

1. It has a large lumen.
2. Tunica intima is lined by endothelium and a subendothelial layer of connective tissue.
3. The internal elastic lamina is not easily distinguishable.
4. Tunica media is very thick; it occupies most of the wall of the vessel. There are 40-60 layers of con-

centrically arranged elastic membranes and few smooth muscle fibers.

5. Tunica adventitia is thin; vasa vasora present.

Function of Large (Elastic) Arteries

They are also called conducting arteries. They absorb some of the pulse beat and make the blood flow continuous.

Structure of Medium Sized (Muscular) Arteries or Distributing Arteries (Figs 11.10A and B)

1. Tunica intima is lined by endothelium. The internal elastic lamina is thrown into refractile, wavy folds (by the contraction of smooth muscle fibers of tunica media).
2. The tunica media is composed of circularly arranged smooth muscle fibers and few elastic fibers.
3. Tunica adventitia is composed of collagen fibers and elastic fibers.

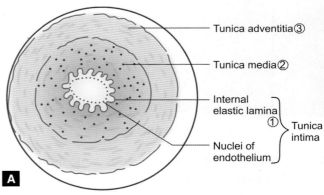

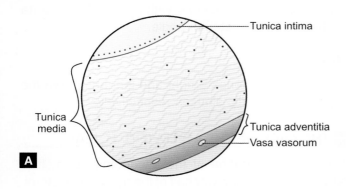

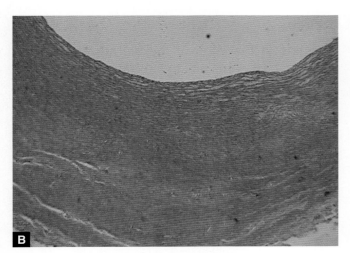

Figs 11.9A and B: Elastic artery (aorta)

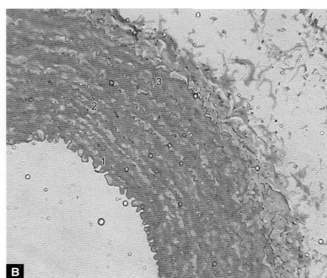

Figs 11.10A and B: Medium sized muscular artery

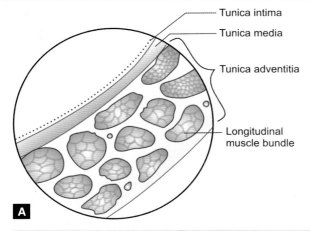

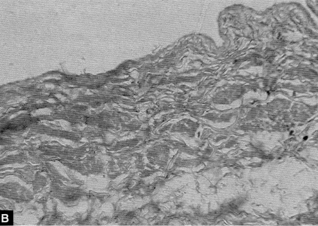

Figs 11.11A and B: Large vein

4. The thickness of tunica media is almost equal to the thickness of the tunica adventitia.

Functions

The medium sized arteries are called muscular arteries owing to large amount of smooth muscle fibers in the tunica media. These vessels regulate the volume of blood to the tissues in response to varying functional demands by vasoconstriction and vasodilatation.

These vessels distribute blood to various organs so, they are also called distributing arteries.

Structure of a Large Vein (e.g. Venae Cavae) (Figs 11.11A and B)

Tunica intima and media are poorly developed. Tunica adventitia is the thickest. It is made up of longitudinally running bundles of smooth muscle fibers and elastic fibers.

■ THE AORTA

It is the largest artery in the body, which carries oxygenated blood from the left ventricle and distributes it to all parts of the body.

Parts

1. The ascending aorta
2. The arch of the aorta
3. The descending aorta:
   a. Thoracic aorta
   b. Abdominal aorta

■ THE ASCENDING AORTA

The ascending aorta arises from the left ventricle. At the root of the aorta, there are three dilatations of the vessel wall called the aortic sinuses. The ascending aorta is about 5 cm long; it is enclosed in the pericardium.

After arising from the heart (at the level of 3rd costal cartilage) it runs upwards, forwards and to the right; it becomes continuous with the arch of the aorta at the level of the sternal angle.

Branches

The coronary arteries – right and left, arise from the right and left aortic sinuses respectively.

■ THE ARCH OF THE AORTA (FIG. 11.12A)

It is the continuation of the ascending aorta. It is situated behind the lower half of the manubrium sterni.

Course

It begins at the level of the sternal angle; runs upwards, backwards and to the left, arching over the root of the left lung. It ends at the lower border of the T4 vertebra (same horizontal plane as that of the sternal angle), by becoming continuous with the descending (thoracic) aorta.

Important Relations (Fig. 11.12B)

1. *Anteriorly and to the left*: (i) Left phrenic and left vagus nerves; cardiac branches of vagus and left sympathetic trunk, and (ii) Left pleura and lung.

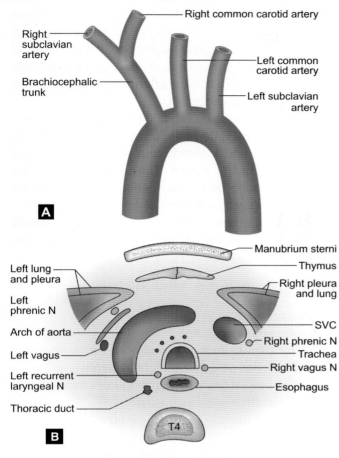

A

B

T4

Figs 11.12A and B: Arch of aorta

2. *Posteriorly and to the right:*
   a. Trachea
   b. Esophagus
   c. Left recurrent laryngeal nerve
   d. Thoracic duct
   e. Vertebral column.
3. *Superiorly:* The three branches of the arch of aorta. From right to left, they are (i) The brachiocephalic trunk (ii) The left common carotid and (iii)The left subclavian arteries.
4. *Inferiorly:*
   a. Bifurcation of pulmonary trunk
   b. Left bronchus
   c. The ligamentum arteriosum
   d. The recurrent laryngeal nerve

The ligamentum arteriosum is the remnant of a short, wide channel connecting the beginning of left pulmonary artery to the arch of aorta. In the fetal life, it conducts most of the blood from the right ventricle to the aorta, because the lungs are not functioning. In the adult, the remnant of ductus arteriosus is a fibrous band called ligamentum arteriosum. The left recurrent laryngeal nerve hooks around it. The duct may remain patent after birth – called patent ductus arteriosus (PDA), causing serious problems.

Applied Anatomy

1. *Aortic knob or aortic knuckle:* In an X-ray of the chest, the arch of aorta can be seen as a projection beyond the left margin of the cardiac shadow; this is called aortic knob or knuckle.
2. Aortic aneurysm is a localized dilatation of aorta
3. Coarctation of the aorta is a localized narrowing of the aorta, opposite to the attachment of the ductus arteriosus.

Branches of the Arch of Aorta

1. The brachiocephalic trunk, it divides into right common carotid and right subclavian arteries
2. The left common carotid artery
3. The left subclavian artery
4. Occasionally, the thyroidea ima artery.

The Brachiocephalic Artery (or Trunk)

The brachiocephalic artery (or trunk) arises behind the manubrium of the sternum from the convexity of the aortic arch. It is the first and largest branch of the aortic arch. It passes upwards, backwards and to the right; after 5 cm, divides into right subclavian and right common carotid arteries.

The Common Carotid Arteries (CCA) (Fig. 11.13)

The right common carotid artery is a branch of the brachiocephalic trunk. The left CCA is a direct branch of the arch of the aorta.

After the origin, their course and distribution is similar. The common carotid artery passes upwards in the carotid sheath along with the internal jugular vein and the vagus nerve. At the level of the upper border of thyroid cartilage, it divides into 2 branches – the internal and external carotid arteries.

The External Carotid Artery (Fig. 11.13)

It is one of the terminal branches of the common carotid artery. It lies anterior to the internal carotid artery. It is the chief artery of supply to structures in the neck and face.

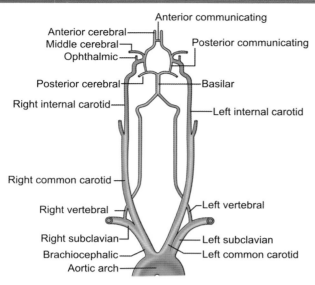

Fig. 11.13: Carotid arteries

The external carotid artery begins in the neck at the upper border of thyroid cartilage (opposite the intervertebral disc between C3 and C4). It runs upwards, backwards and laterally and terminates behind the neck of the mandible by dividing into maxillary and superficial temporal arteries.

Branches (8 branches)

1. Superior thyroid
2. Lingual
3. Posterior auricular
4. Facial
5. Occipital
6. Ascending pharyngeal
7. Superficial temporal
8. Maxillary.

(The branches can be remembered by the following mnemonic "**S**ister **L**ucy's **P**owdered **F**ace **O**ften **A**ttracts **S**illy **M**edicos", first letter of each word stands for the 1st letter of each branch).

The superior thyroid artery supplies the thyroid gland; it anastomoses with the inferior thyroid artery (a branch of thyrocervical trunk, of subclavian artery) It is closely related to the external laryngeal nerve (refer – Thyroid gland – applied anatomy).

The lingual artery Supplies the tongue, submandibular and sublingual salivary glands.

The facial artery It is the chief artery of face. It arises from the external carotid artery just above the level of the tip of greater horn of hyoid bone. It has a cervical part and a facial part. In its cervical course, it passes through the submandibular region, grooving the submandibular gland.

It enters the face by winding round the base of the mandible, at the anteroinferior angle of mandible. It runs upwards and forwards to a point 1.5 cm lateral to the angle of the mouth. Then it ascends by the side of the nose up to the medial angle of eye by anastomosing with a branch of ophthalmic artery.

The facial artery is very tortuous, which prevents it from being stretched during movements of the lips, cheeks and mandible.

Branches

1. Cervical part:
   a. Ascending palatine
   b. Tonsillar branch to palatine tonsil
   c. Branches to submandibular gland
   d. Submental branch to sublingual salivary gland.
2. Facial part:
   a. Inferior labial – to the lower lip
   b. Superior labial – to the upper lip
   c. The lateral nasal – to the nose
   d. Muscular branches – to facial muscles.

The branches of one facial artery anastomoses with the branches of the opposite facial artery. The wounds of the face bleed profusely; but they heal rapidly because of this reason.

Posterior auricular artery It ascends, upwards and backwards, and ascends behind the auricle. It supplies the back of the pinna, the skin over the mastoid process and the back of the scalp.

Ascending pharyngeal artery It supplies the side wall of the pharynx, the tonsil, the medial wall of middle ear and the auditory tube.

The maxillary artery This is the larger terminal branch of the external carotid artery. It begins behind the neck of the mandible, deep to the parotid gland. It enters the infratemporal fossa.

The artery is divided into 3 parts by the lateral pterygoid muscle. It has a wide territory of distribution.

It supplies (i) the external ear, the middle ear and the auditory tube, (ii) the dura mater, (iii) the jaws, (iv) muscles of mastication, (v) the nose, palate and paranasal air sinuses, and (vi) the pharynx.

The middle meningeal artery is an important branch of the maxillary artery, because it is the commonest source of **extradural hemorrhage.**

It arises in the infratemporal fossa, from the first part of maxillary artery. The artery is encircled by the two roots of auriculotemporal nerve. It enters the middle cranial fossa through the foramen spinosum. In the middle cranial fossa it lies between the skull and the dura (extradural). It grooves the squamous part of temporal bone and divides into a frontal and parietal branch.

Applied Anatomy

The frontal branch is anterior and larger, it is closely related to the **pterion** laterally and the motor area of cerebral cortex medially. So, a fracture at the pterion can tear the frontal branch leading to extradural hemorrhage. The hematoma presses on the motor area; it can cause hemiplegia of the opposite side of the body.

The middle meningeal and accessory meningeal arteries, which are branches of the maxillary artery are the main branches of external carotid artery which supply structures *inside* the skull.

The inferior alveolar artery, a branch of first part of maxillary artery, runs along with the inferior alveolar nerve. They enter the mandibular foramen, traverse the mandibular canal and supply the teeth and mandible.

The occipital artery arises from the external carotid, opposite the origin of the facial artery. It supplies the posterior aspect of scalp.

The superficial temporal artery is the smaller terminal branch of external carotid artery. It begins behind the neck of the mandible, deep to the parotid gland. It supplies the scalp, the parotid gland, the facial muscles and the temporalis muscle. Its pulsations can be easily felt where it crosses the root of the zygomatic process (preauricular point).

The Internal Carotid Artery (Figs 11.13 and 11.14)

The internal carotid artery is one of the 2 terminal branches of the common carotid artery. It begins at the level of upper border of thyroid cartilage (opposite the disc between C3 and C4). **This is the principal artery of the brain and eye**.

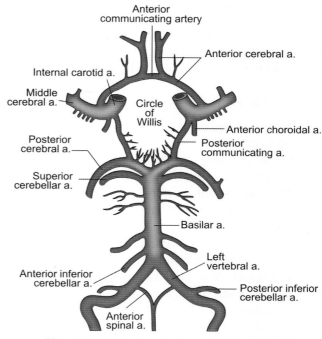

Fig. 11.14: Circulus arteriosus (circle of Willis)

The course of the artery can be described under 4 headings:
1. Cervical part—in the neck
2. Petrous part—in the carotid canal of the petrous part of temporal bone
3. Cavernous part – within the cavernous sinus (paired dural venous sinuses lying on either side of the body of the sphenoid)
4. The cerebral part – related to the base of the brain.

Cervical Part

This part of the internal carotid artery ascends vertically in the neck from its origin to the base of the skull and enters the carotid canal. The cervical part is enclosed in the carotid sheath, along with the internal jugular vein and the vagus nerve.

No branches arise from the cervical part. It shows a dilatation close to its beginning, called the carotid sinus, which acts as a baroreceptor.

The Petrous Part

It runs in the canal of the petrous part of temporal bone. It is surrounded by sympathetic plexus. It is related to the middle ear, auditory tube and cochlea. It sends branches to the middle ear.

The Cavernous Part

The cavernous part lies within the cavernous sinus along with the abducent nerve (VI cranial nerve). Here it gives off branches to the pituitary gland.

The Cerebral Part

This part lies at the base of the brain. It gives off the following branches:

1. Ophthalmic artery—to the structures in the orbit
2. The anterior cerebral artery
3. The middle cerebral artery   Supply the brain
4. The posterior communicating artery

The curvatures of petrous, cavernous and cerebral parts of the internal carotid artery together form an S-shaped curve, described as the **"carotid siphon"**, which can be seen in angiograms.

The anterior and middle cerebral arteries communicate with each other and also with basilar artery (which is the continuation of the vertebral arteries), thereby forming the circle of Willis or circulus arteriosus (Fig. 11.13) which ensures an even distribution of blood to the brain.

Applied Anatomy

In elderly people, the anastomoses between the arteries forming the circle of Willis are often inadequate, when there is a sudden occlusion of a large artery such as the internal carotid artery. This results in a "stroke".

Hemorrhagic stroke follows from rupture of an aneurysm, which results in bleeding into the brain tissue.

Thrombotic stroke results from thrombosis (clotting). It may also result from an embolus ("plug"), which is a floating mass of material from a distant site (such as a blood clot from the heart, gas bubbles, amniotic fluid or fat from fractured long bones).

Aneurysms are localized, abnormal dilatations of arteries, mainly due to the congenital weakness of the wall of an artery. This results in a localized berry-like swelling to develop (berry aneurysm). The most common type of aneurysm is the berry aneurysm arising from the circulus arteriosus. In hypertension, the aneurysm may rupture; characterized by sudden, severe, unbearable headache. Rupture of a berry aneurysm may result in subarachnoid hemorrhage, intracerebral hemorrhage or subdural hemorrhage.

The Subclavian Artery (Fig. 11.15)

It is the main **artery of upper limb**. The right subclavian artery is a branch of the brachiocephalic trunk; the left subclavian artery is a direct branch of the arch of the aorta (Fig. 11.12).

It passes over the first rib and grooves it. Then it passes behind the clavicle to enter the upper part of the axilla. At the outer border of first rib, it continues as the axillary artery.

The scalenus anterior muscle crosses the artery anteriorly and divides it into 3 parts – the first part medial, 2nd part posterior and the 3rd part lateral to the muscle.

Branches

The subclavian artery usually gives off 5 branches:

1. Vertebral artery
2. Internal thoracic artery
3. Thyrocervical trunk – it divides into 3 branches:
   a. Inferior thyroid artery
   b. Suprascapular artery
   c. Transverse cervical artery
4. Costocervical trunk, which divides into 2 branches:
   a. Superior intercostal
   b. Deep cervical
5. Dorsal scapular artery.

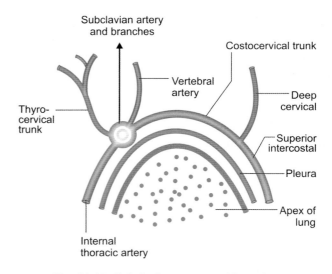

Fig. 11.15: Subclavian artery and branches

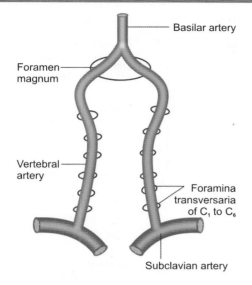

Fig. 11.16: Vertebral arteries

Applied Anatomy

A cervical rib may compress the subclavian artery. The radial pulse decreases or disappears on turning the patient's head upwards and to the affected side after a deep breath (Adson's test). Patient may have pain, which is brought on by use of the arm. The pain is relieved by rest. The hand on the affected side may be colder and paler; when the obstruction is severe, there may be ulceration or gangrene.

The Vertebral Artery (Fig. 11.16)

It is the first and largest branch of the subclavian artery. It is one of the two principal arteries which supply the brain (the other artery is the internal carotid). In addition, it also supplies the spinal cord, meninges, cervical muscles and bones.

It arises from the 1st part of subclavian artery, close to its beginning. After a long course in the neck, it ends in the cranial cavity, supplying the brain.

Parts

The artery has 4 parts:

Part I It extends from the origin of the artery from the subclavian artery to the foramen transversarium of C_6 vertebra.

Part II It runs through the foramina transversaria of the upper 6 cervical vertebrae.

Part III Emerging from the foramen transversarium of atlas, the artery grooves the lateral mass of atlas, then it turns medially and enters, subarachnoid space.

Part IV The last part passes up through the foramen magnum to ascend on the sides of the medulla to end by uniting with the vertebral artery of the opposite side at the lower border of the pons to form the basilar artery.

Branches

In the neck (a) spinal branches, to the spinal cord. They enter through the intervertebral foramina (b) muscular branches to the cervical muscles.

In the cranium
1. Meningeal branches
2. Anterior and posterior spinal arteries
3. Posterior inferior cerebellar artery
4. Branches to medulla

Applied Anatomy

Occlusion or stenosis of the vertebral arteries results in brainstem symptoms such as dizziness and fainting.

The Axillary Artery (Fig. 11.17)

This large vessel begins at the outer border of the first rib. It is the continuation of subclavian artery. The axillary artery ends at the lower border of teres major muscle; where it passes into the arm as the brachial artery.

In the axilla, the artery passes posterior to the **pectoralis minor** muscle, which divides the artery into 3 parts – the first part medial to the pectoralis minor, the second part posterior to it and the 3rd part, lateral to the muscle.

Branches (Table 11.1)

The axillary artery gives one branch from the first part; 2 branches from the 2nd part and 3 branches from the third part.

The subscapular artery is the largest branch of axillary artery. The posterior circumflex humeral artery and the axillary nerve are closely related to the surgical neck of humerus.

Table 11.1: Parts and branches of axillary artery	
Part	*Branches*
Part I	Superior thoracic artery
Part II	a. Thoracoacromial artery
	b. Lateral thoracic artery
Part III	a. Subscapular artery
	b. Anterior circumflex humeral artery
	c. Posterior circumflex humeral artery

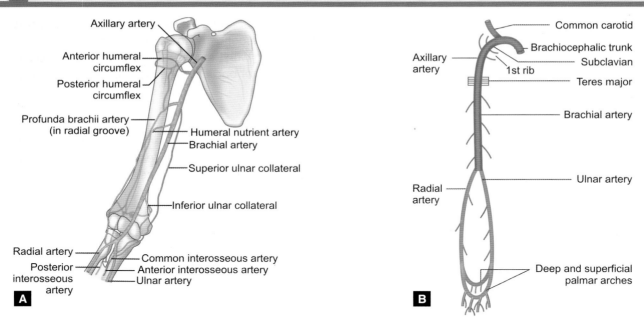

Figs 11.17A and B: Arteries of upper limb

Ligation of the axillary artery, distal to the origin of subscapular artery cuts off the blood supply to the arm.

The Brachial Artery (Figs 11.17A and B)

This is the principal artery of the arm. It begins at the inferior border of teres major as the continuation of axillary artery. In the cubital fossa, it ends opposite the neck of the radius. Under cover of the bicipital aponeurosis, the brachial artery divides into the two terminal branches – the radial and ulnar arteries.

The brachial artery is superficial and palpable throughout its course. It is accompanied by the median nerve in the arm, which crosses anterior to the artery in the middle of the arm.

Branches

1. Muscular branches
2. The profunda brachii artery
3. Nutrient artery to humerus
4. Superior and inferior ulnar collateral arteries.

Applied Anatomy

Arterial blood pressure (BP) is usually recorded by applying the inflatable cuff around the arm (for compressing the brachial artery against the shaft of the humerus). A stethoscope is placed over the artery in the cubital fossa, just medial to the biceps tendon.

Compression of the brachial artery, to control hemorrhage, may be produced anywhere along its course.

Contd...

Contd...

The best place to compress the brachial artery is near the middle of the arm.

Occlusion or laceration (tear) of the brachial artery results in ischemia of deep flexors of forearm due to ischemia. It can produce irreversible damage; the muscle tissue is replaced by fibrous scar tissue; the muscles are shortened. This deformity is known as Volkmann's ischemic contracture. This type of injury usually follows a supracondylar fracture of the humerus.

The Radial Artery (Figs 11.17A and B)

The radial artery begins in the cubital fossa opposite the neck of the radius. It is the smaller of the two terminal branches of brachial artery (the other branch is the ulnar artery).

The course of the radial artery can be represented by a line connecting (i) the midpoint of cubital fossa and (ii) a point just medial to the tip of the styloid process of radius.

In the upper part of the forearm, the radial artery lies under cover of muscles; but in the distal part of forearm, the artery is quite superficial. It lies on the distal end of the radius lateral to the tendon of the flexor carpi radialis muscle, covered only by the skin, superficial and deep fasciae. This is the common site for measuring the pulse rate.

The radial artery, then enters the "anatomical snuff box", bounded by the tendons of abductor pollicis longus and extensor pollicis brevis; the artery lies on the floor of the anatomical snuff box, which is formed by scaphoid and trapezoid. It enters the palm, passing between the two heads of first dorsal interosseous and adductor pollicis muscles.

The radial artery ends by forming the deep palmar arch, with the deep palmar branch of the ulnar artery.

Branches

1. Muscular branches
2. Branches to the elbow joint—they form anastomoses with branches of brachial artery
3. A branch to the superficial palmar arch. This branch anastomoses with the terminal part of the ulnar artery
4. Branches to the wrist joint.

The Ulnar Artery (Figs 11.17A and B)

It begins near the neck of the radius; it is the larger of the two terminal branches of the brachial artery. In the proximal part of the forearm, it accompanies the median nerve; in the distal 2/3 of forearm, the artery lies lateral to the ulnar nerve. It passes superficial to the flexor retinaculum and ends by forming the superficial palmar arch, with a branch of radial artery.

Branches

1. Muscular branches
2. Branches to the elbow joint—they anastomose with branches of the brachial artery
3. Common interosseous artery
4. Carpal branches
5. Superficial palmar arch—it is the continuation of the ulnar artery
6. A branch to deep palmar arch—the arch is completed by the radial artery.

Palmar Arches (Figs 11.17A and B)

There are two palmar arterial arches – the superficial palmar arch and the deep palmar arch.

The Superficial Palmar Arch

This arch is located distal to the deep palmar arch. It is formed, mainly by the ulnar artery. It is convex towards the digits.

Branches This arch gives rise to 3 common palmar digital arteries that anastomose with the palmar metacarpal arteries from the deep arch.

Each common palmar digital artery divides into a pair of proper palmar digital arteries, which run along the sides of the second to fourth digits.

The Deep Palmar Arch

This arterial arch is formed mainly by the radial artery; it is about a finger breadth proximal to the superficial palmar arch.

Branches Three palmar metacarpal arteries arise from this arch. They anastomose with the 3 common digital arteries from the superficial palmar arch.

Applied Anatomy

Because of the rich anastomoses in the hand and digits, bleeding is profuse when there is an injury. In lacerations of the arterial arches, it is useless to ligate (tie) only one of the forearm arteries. To obtain a bloodless field for surgeries of the hand or to arrest bleeding from the palmar arches, it is wise to compress the brachial artery, proximal to the elbow using an inflatable tourniquet.

■ THE DESCENDING AORTA (FIG. 11.18)

The descending aorta, which is the continuation of the arch of the aorta, is divisible into 2 parts:
1. The thoracic part
2. The abdominal part (abdominal aorta).

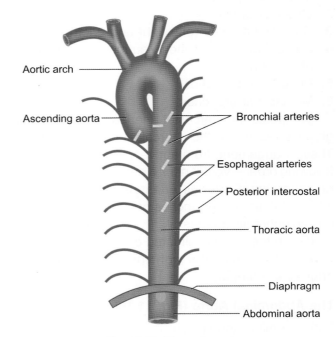

Fig. 11.18: Thoracic aorta

The Thoracic Part

It is the continuation of the arch of the aorta. It lies in the posterior mediastinum.

The descending thoracic aorta begins on the left side of the lower border of T4 vertebra. It descends with a slight inclination to the right and terminates at the lower border of the 12th thoracic vertebra.

Diameter: About 2 cm.

Relations

Anterior relations
1. Root of the left lung
2. Heart, covered by pericardium
3. In the lower part, esophagus
4. The diaphragm.
   Posteriorly – the vertebral column (T5-T12)

On the right side
1. Esophagus (in the upper part)
2. Azygos vein
3. Thoracic duct
4. Right lung and pleura.

On the left side Left lung and pleura.

Branches

1. Nine pairs of posterior intercostals arteries—to 3rd to 11th intercostal spaces
2. One pair of subcostal arteries
3. Bronchial arteries (one right and 2 left)
4. 2 esophageal branches—supply the middle 1/3 of esophagus
5. Pericardial branches
6. Mediastinal branches.

The posterior intercostal arteries run in the costal grooves along with the intercostal nerves and veins. Their arrangement, from above downwards, is in the following order—Vein, Artery, Nerve (VAN). The posterior intercostal arteries anastomose with anterior intercostal arteries (branches of internal thoracic and musculophrenic arteries).

(Since the aorta is situated on the left side of the vertebral column, the right posterior intercostal arteries are longer).

The Abdominal Aorta (Fig. 11.19)

The abdominal aorta begins in the midline at the aortic opening of the diaphragm, opposite the body of T12

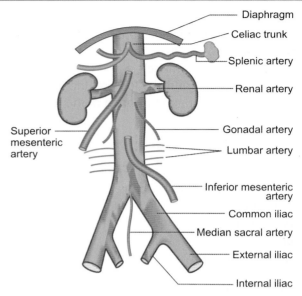

Fig. 11.19: Abdominal aorta—branches

vertebra. It ends in front of the lower part of the body of L4 vertebra, by dividing into its 2 terminal branches, right and left common iliac arteries.

Relations

Anteriorly From above downwards, the abdominal aorta is related to:
1. The body of pancreas and splenic vein
2. The left renal vein
3. The horizontal part of duodenum
4. Root of mesentery
5. Coils of small intestine.

Posteriorly The abdominal aorta is related to the upper four lumbar vertebrae and the intervertebral discs between them.

To the right side are:
1. The inferior vena cava and the sympathetic chain
2. The right crus of the diaphragm.

To the left side are:
1. The left crus of diaphragm
2. The pancreas
3. The 4th or ascending part of duodenum
4. The left sympathetic chain.

Branches

Branches of abdominal aorta can be classified into the following groups:

1. *Ventral branches* (they supply the gut tube)
   a. Celiac trunk
   b. Superior mesenteric artery   *unpaired arteries*
   c. Inferior mesenteric artery
2. *Lateral branches*
   a. Right and left inferior phrenic arteries
   b. Right and left middle suprarenal arteries
   c. Right and left renal arteries
   d. Right and left gonadal (either testicular or ovarian) arteries.
3. *Dorsal branches:*
   a. 4 pairs of lumbar arteries
   b. An unpaired medial sacral artery
4. *Terminal branches*—They are the right and left common iliac arteries.

The Celiac Trunk (Fig. 11.20)

The celiac trunk arises from the front or ventral aspect of the abdominal aorta just below the aortic opening of the diaphragm. It is only about 1 cm long. It ends by dividing into its 3 terminal branches. They are:
1. The left gastric artery
2. The hepatic artery
3. The splenic artery.

Structures Supplied by Celiac Trunk

The celiac trunk supplies the following structures:
1. The lower end of esophagus, the stomach and upper part of duodenum up to the opening of the common bile duct (major duodenal papilla)
2. The liver
3. The spleen
4. Major part of pancreas
   (These structures are derived from the foregut; so, the celiac trunk is known as the artery of foregut).

Branches

The left gastric artery: It is the smallest branch of the celiac trunk. It supplies branches to the lower end of the oesophagus and continues along the lesser curvature of the stomach between the two layers of the lesser omentum. It ends by anastomosing with the right gastric artery. It sends numerous gastric branches along the lesser curvature of the stomach.

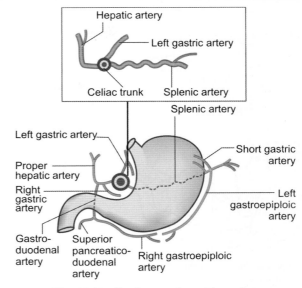

Fig. 11.20: Coeliac trunk and branches

The hepatic artery From its origin, it runs to the right and then ascends in the right free margin of the lesser omentum (along with the portal vein and common bile duct). Reaching the porta hepatis, it divides into right and left hepatic arteries.

Branches
1. *Gastroduodenal artery*—This large vessel runs downwards behind the 1st part of duodenum. It divides into right gastroepiploic artery and the superior pancreaticoduodenal artery (the right gastroepiploic artery runs along the right side of the greater curvature of stomach. It anastomoses with the left gastroepiploic artery and supplies the stomach along its greater curvature and also the greater omentum. The superior pancreaticoduodenal artery supplies the head of the pancreas and proximal part of duodenum – up to the level of the major duodenal papilla.
2. *The right gastric artery*—runs along the lesser curvature and anastomoses with the left gastric artery.

The splenic artery is the **largest** branch of the celiac trunk. It runs horizontally to the left along the upper border of pancreas behind the lesser sac. It enters the lienorenal ligament and reaches the hilum of the spleen where it divides into 5-7 splenic branches.

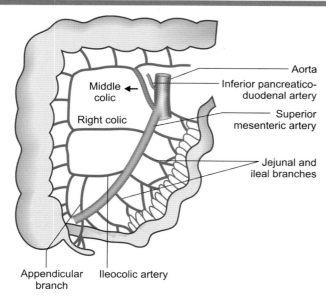

Fig. 11.21: Superior mesenteric artery

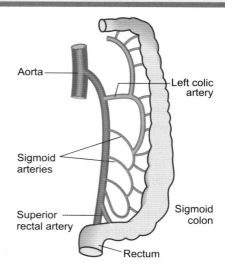

Fig. 11.22: Inferior mesenteric artery

Branches

   i. Pancreatic branches to the body and tail of the pancreas
  ii. 5-7 short gastric arteries to the fundus of the stomach
 iii. The left gastroepiploic artery to the greater curvature of stomach which anastomoses with the right gastroepiploic artery. It supplies the stomach and greater omentum.

The Superior Mesenteric Artery (Fig. 11.21)

It is the artery of the midgut. It supplies all derivatives of the midgut. They are (1) the lower part of duodenum, distal to the opening of common bile duct; (2) the jejunum; (3) the ileum (4) the cecum and appendix (5) the ascending colon and(6) the right 2/3 of transverse colon (7) the lower half of the head and uncinate process of pancreas.

The superior mesenteric artery is an unpaired branch of the abdominal aorta; it arises at the level of L1 vertebra, behind the body of the pancreas, 1 cm below the celiac trunk. It runs downwards and to the right, between the two layers of the root of the mesentery. It terminates in the right iliac fossa by anastomosing with a branch of the ileocolic artery.

Branches

1. *Inferior pancreaticoduodenal artery:* Anastomoses with the superior pancreaticoduodenal artery; it supplies the lower part of head of the pancreas and the duodenum.
2. 12-15 *jejunal and ileal branches* – arise from the left side of the artery.
3. *Ileocolic artery:* It supplies the cecum, appendix and the terminal part of ileum; it anastomoses with the terminal part of superior mesenteric artery and the middle colic artery.
4. *The right colic artery:* It passes to the right and divides into an ascending and descending branch to anastomose with the ileocolic and middle colic arteries; supplies the ascending colon.
5. *The middle colic artery:* It arises from the superior mesenteric artery, just below the pancreas. Its branches supply the transverse colon.

The Inferior Mesenteric Artery (Fig. 11.22)

This is the artery of the hindgut. It supplies the following structures (1) the left 1/3 of transverse colon (2) the descending colon (3) the sigmoid colon (4) the rectum, and (5) the anal canal, above the pectinate line.

This artery, an unpaired, ventral branch of the abdominal aorta, arises at the level of L3 vertebra, 3-4 cm above the termination of abdominal aorta.

Branches

Left colic artery is the first branch of inferior mesenteric artery. Its branches supply the left 1/3 of transverse colon and the descending colon.

Sigmoid arteries 2-3 in number; they supply the sigmoid colon.

Superior rectal artery It is the continuation of inferior mesenteric artery. It divides at the level of S3 vertebra into right and left branches, which descend on the right and left sides of the rectum. They divide into several branches, anastomose to form loops around the lower end of the rectum, in the submucosa. These branches communicate with the middle and inferior rectal arteries of the anal canal.

The Common Iliac Arteries (Fig. 11.19)

These are the terminal branches of the abdominal aorta, beginning at L4 vertebral level. The two common iliac arteries (right and left) pass downwards and laterally and end in front of the sacroiliac joint, by dividing into the external and internal iliac arteries.

The right common iliac artery is longer, it passes in front of the beginning of the inferior vena cava.

The External Iliac Artery

It is the larger of the two terminal branches of the common iliac artery. It starts in front of the sacroiliac joint; then passes deep to the midinguinal point where it becomes the femoral artery.

Branches
1. Inferior epigastric artery
2. Deep circumflex iliac artery
3. Branches to ureter, adjacent lymph nodes and muscles.

The Internal Iliac Artery

It supplies almost all the pelvic viscera. It starts in front of the sacroiliac joint and runs to the upper border of the greater sciatic notch, where it divides into two – an anterior division and a posterior division.

Branches from the Anterior Division

1. Superior vesical
2. Middle rectal
3. Vaginal artery (in female)
4. Uterine artery (female)
5. Umbilical artery.

Branches from the Posterior Division

1. Superior gluteal artery
2. Inferior gluteal artery
3. Obturator artery
4. Internal pudendal artery
5. Lateral sacral artery.

▌ ARTERIES OF LOWER LIMB (FIG. 11.23 AND FLOW CHART 11.2)

Femoral Artery

This is the **chief artery of the lower limb**. It is the continuation of the external iliac artery.

Flow chart 11.2: Arteries of lower limb

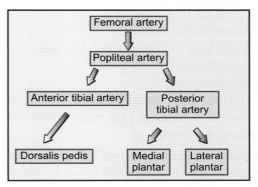

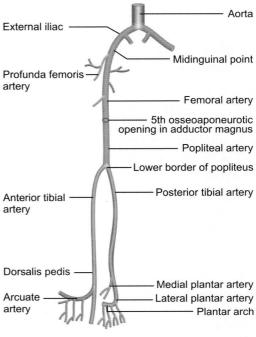

Fig. 11.23: Arteries of the lower limb

It begins behind the inguinal ligament at the midinguinal point (midway between anterior superior iliac spine and pubic symphysis). It passes downwards, first in the femoral triangle and then in the adductor canal. At the lower end of the adductor canal, it passes through an opening in the adductor magnus to continue as the popliteal artery.

Relations

In the femoral triangle (see Fig. 20.4) The femoral vein lies medial to it; the femoral nerve is lateral to the upper part of the artery. The artery is superficial; it lies deep to the skin, the fascia and the anterior wall of femoral sheath.

Branches in the femoral triangle
1. 3 superficial branches:
   a. Superficial epigastric
   b. Superficial external pudendal, and
   c. Superficial circumflex iliac
2. 3 deep branches:
   a. The profunda femoris artery
   b. The deep external pudendal artery
   c. Muscular branches.

The profunda femoris artery is the largest branch of the femoral artery. It is the chief artery of supply to the structures in the thigh. It arises from the lateral side of femoral artery, about 4 cm below the inguinal ligament.

Branches (1) medial and lateral circumflex femoral arteries (2) 4 perforating arteries (3) numerous muscular branches.

Relations in the adductor canal
Anteriorly: Vastus medialis
                 Subsartorial plexus of nerves
   The femoral artery is accompanied by femoral vein, saphenous nerve the branches of obturator nerve and the nerve to vastus medialis.

Branches in the adductor canal
1. Muscular branches
2. Descending genicular artery.

Applied Anatomy

1. The femoral artery can be compressed against the superior ramus of pubis, at the midinguinal point to control bleeding from the distal part of the limb.
2. Pulsations of femoral artery can be felt at the midinguinal point.

Contd...

Contd...

3. As the femoral artery is superficial in the femoral triangle, it can easily be exposed and cannulated
4. For left cardiac angiography and renal angiography, a long slender catheter is inserted into the femoral artery, in the femoral triangle.
5. The medial circumflex femoral artery is clinically important because it supplies most of the blood to the head and neck of femur.

The Popliteal Artery (see Fig. 20.9)

This artery is the continuation of the femoral artery. It begins at the opening in the adductor magnus (at the junction of middle 1/3 and lower 1/3 of thigh). At the lower border of popliteus muscle, it terminates by dividing into anterior and posterior tibial arteries.

   The popliteal artery is the deepest structure in the popliteal fossa. It lies on the popliteal surface of femur and the back of the knee joint. Superficial to the artery are the popliteal vein and tibial nerve.

Branches

(1) Muscular branches to the powerful muscles of the back of the thigh and leg (2) genicular branches to the knee joint.

Applied Anatomy

Blood pressure in the lower limb is recorded from the popliteal artery. The popliteal artery is more prone for aneurysm than other arteries in the body.

Arteries of the Leg

The anterior and posterior tibial arteries, the terminal branches of the popliteal artery, supply the various structures in the leg.

Anterior Tibial Artery

This is the main artery of the anterior compartment of the leg. It is the smaller terminal branch of the popliteal artery. It begins at the lower border of the popliteus muscle, on the back of the leg. It enters the anterior compartment of the leg through an opening in the upper part of the interosseous membrane.

   In the anterior compartment, it runs vertically downwards, to a point midway between the two malleoli, where, it ends by becoming the dorsalis pedis artery.

Branches (1) Muscular branches to adjacent muscles (2) Branches to the knee joint; (3) Branches to the ankle joint.

Posterior Tibial Artery

It is the larger terminal branch of the popliteal artery. It begins at the lower border of popliteus muscle. It enters the back of the leg (posterior compartment) deep to the tendinous arch of soleus muscle. The artery runs downwards to reach the posteromedial side of the ankle, midway between the medial malleolus and the medial tubercle of calcaneum. The posterior tibial artery terminates deep the flexor retinaculum by dividing into medial and lateral plantar arteries.

The artery is accompanied by the tibial nerve and 2 companion veins (venae commitantes).

Branches
1. *The peroneal artery*—it is the largest branch. It supplies the lateral compartment of leg.
2. Muscular branches.
3. A nutrient artery to tibia
4. Branches to the knee joint
5. Terminal branches – the medial and lateral plantar arteries.

Dorsalis Pedis Artery (Dorsal artery of the foot)
pes = foot

This artery, the continuation of the anterior tibial artery, is the chief artery of the dorsum of the foot.

The artery begins in front of the ankle between the two malleoli. It passes forwards on the dorsum of the foot to reach the proximal end of 1st intermetatarsal space. Here, it dips into the sole between the two heads of the first dorsal interosseous muscle and completes the plantar arch.

Branches
1. Lateral and medial tarsal arteries to the neighboring tarsal joints
2. *The arcuate artery* is a large branch, that runs laterally over the bases of the metatarsal bones. It gives off 2nd, 3rd and 4th metatarsal arteries; from which dorsal digital arteries arise.
3. First dorsal digital artery – gives branches to the adjacent sides of big toe and 2nd toe.

Applied Anatomy

Pulsations of the dorsalis pedis artery is easily felt midway between the two malleoli.

Medial and lateral plantar arteries These are the terminal branches of the posterior tibial artery. They begin deep to the flexor retinaculum.
1. *The lateral plantar artery* is the larger terminal branch of the posterior tibial artery. It continues as the plantar arch in the sole. It gives muscular branches, and anastomotic branches which anastomose with arteries of the dorsum of the foot.
2. *The medial plantar artery* is the smaller of the two terminal branches of posterior tibial artery. It gives muscular branches and 3 superficial digital branches that join 1st, 2nd and 3rd plantar metatarsal arteries.
3. *The plantar arch* is the direct continuation of the lateral plantar artery. It is completed medially by the dorsalis pedis artery.

The site of palpation of pulsations of various arterios is listed in Table 11.2.

■ THE EFFECTS OF ARTERIAL OCCLUSION

When the artery to any part of the body is occluded, necrosis or gangrene will ensue unless there is an alternative pathway to bypass the circulation around the obstructed vessel. This alternative pathway, present in most organs is known as "collateral circulation".

Ischemia

The word ischemia means a lack of blood flow. Commonly affected parts are the heart, the lower limbs, the brain, the kidneys and the intestine.

Atherosclerotic Obstruction

Atherosclerotic obstruction of vessels is most commonly seen in the arteries of the leg. The same patient will be having atherosclerotic obstruction in the coronary, carotid or cerebral vessels. The disease is more common in smokers. See Table 11.3 for the effects of arterial obstruction at various sites.

Gangrene

Gangrene means death and putrefaction of macroscopic portions of tissues. It commonly affects the

Table 11.2: Sites of palpation or pulsations of arteries

	Artery	Site of palpation
1.	Superficial temporal artery	Root of zygoma or preauricular point
2.	Facial Artery	A point on the base of mandible, at the antero-inferior angle of masseter muscle.
3.	Common/external carotid artery	A point on the anterior border of sternocleido-mastoid, at the level of upper border of the thyroid cartilage.
4.	Brachial artery	Front of the elbow, medial to the tendon of biceps.
5.	Radial artery	Against lower end of radius, just lateral to the tendon of flexor carpi radialis.
6.	Femoral artery	At midinguinal point, against the head of femur and tendon of psoas major.
7.	Popliteal artery	Middle of popliteal fossa; patient in prone position; knee slightly flexed.
8.	Anterior tibial artery	Midway between two malleoli
9.	Posterior tibial artery	Against calcaneum, 2 cm below and behind medial malleolus
10.	Dorsalis pedis artery	Proximal end of Ist Inter-metatarsal space, just lateral to the tendon of extensor hallucis longus OR on intermediate cunei-form.

Table 11.3: Relationship of clinical findings to site of arterial obstruction

1.	Aortoiliac	All pulses absent in both limbs (femoral, popliteal, posterior tibial and dorsalis pedis) *Claudication in buttocks, thighs and calves; Impotence
2.	Iliac obstruction	Unilateral absence of all pulses, unilateral claudication in thigh and calf
3.	Femoropopliteal obstruction	Unilateral absence of pulses below femoral artery Claudication in calf and foot
4.	Distal obstruction (popliteal, anterior or posterior tibial artery)	Ankle pulses absent claudication in calf and foot

distal part of a limb, the appendix, a loop of intestine or the testis.

(*Necrosis* means the death of groups of cells; *Slough* means, a piece of dead, soft tissue, like skin, tendon, etc).

Clinical Features of Gangrene

A gangrenous part lacks blood circulation, sensation, warmth and function. It has a characteristic greenish black appearance due to disintegration of hemoglobin and the formation of iron sulphide. *Dry gangrene* occurs by gradual slowing of circulation and *moist gangrene*, due to sudden occlusion; the epidermis is raised in blebs (moist gangrene can occur in acute appendicitis and strangulated bowel).

"Claudication" (claudicare = to limp. Claudius, a Roman Emperor, walked with a limp)

When an artery is occluded partially, the blood supply to the muscles of the leg, even though adequate at rest, becomes inadequate during exercise. On walking a certain distance (the claudication distance), the patient experiences severe pain, which forces him to stop. After a short rest, the pain disappears, allowing him to go on.

■ VEINS

Veins return blood to the heart from the tissues. Their walls are thinner than arteries. Contracting skeletal muscles compress the veins, directing the blood towards the heart.

The smallest veins are called venules. These tributaries unite to form larger veins which sometimes join to form venous plexuses or networks (e.g. dorsal venous arch of foot).

The veins that accompany muscular arteries, one on each side, are called venae comitantes (L. accompanying veins).

Many veins contain valves, which prevent the backflow of blood. The valves support the columns of blood superior to them (e.g. in the legs) and divide these columns into smaller amounts of less weight. This facilitates the return of blood to the heart and prevents pooling of blood.

Factors Affecting Venous Return

1. The pumping action of heart
2. The thoracic pump—A decrease in intrathoracic pressure during inspiration assists venous return
3. The skeletal muscle pump—the contraction of the powerful skeletal muscles of lower limb assist the venous valves in keeping the blood moving towards the heart.

In general, the veins carry deoxygenated blood (low in oxygen content); the pulmonary veins and portal vein are exceptions. The pulmonary veins carry oxygenated blood; the portal vein carries the products of digestion (of carbohydrates, fats and proteins from the intestine) and the products of red cell destruction from the spleen.

The veins of the body can be discussed under the following headings:
1. Veins of the head and neck
2. Veins of the upper limb
3. Veins of the thorax
4. Veins of the abdomen
5. Veins of the lower limb.

■ THE VEINS OF HEAD AND NECK

Veins of the Face (Fig. 11.24)

The external facial veins accompany the arteries of the face. They anastomose freely.

Supratrochlear Vein

The supratrochlear vein begins on the forehead, near the median plane, with the supratrochlear vein of the opposite side. Each supratrochlear vein unites with the supraorbital vein to form the facial vein near the medial angle of the eye.

Supraorbital Vein

The supraorbital vein begins near the zygomatic process of the temporal bone. It passes medially and joins the supratrochlear vein to form the facial vein near the medial angle of the eye.

Facial Vein

This vein provides the major venous return of the face. It begins near the medial angle of the eye, by the union of the supratrochlear and supraorbital veins. It runs downwards and posteriorly, posterior to the facial artery. It takes a straighter and superficial course than the artery.

Inferior to the margin of mandible, the facial vein is joined by the anterior division of retromandibular vein, to form the common facial vein, which ends by draining into the internal jugular vein.

The superior part of facial vein, near its origin, is often called the "angular vein" because of its relation to the medial "angle" (canthus) of the eye.

The Superficial Temporal Vein

This vein, draining the forehead and scalp, enters the parotid gland. Posterior to the neck of the mandible, it unites with the maxillary vein to form the *retromandibular vein*. This vein descends within the parotid gland, superficial to the external carotid artery. It divides into an *anterior division* and a *posterior division*.

The anterior division unites with the facial vein to form *the common facial vein,* which joins the internal jugular vein.

The posterior division of retromandibular vein joins the posterior auricular vein to form the external jugular vein which drains into the subclavian vein.

Applied Anatomy

The facial vein has clinically important connections with (1) the cavernous sinus, through the superior ophthalmic vein and (2) the pterygoid venous plexus.

The facial vein has no valves. It drains blood from the medial angle of eye, nose and lips (called the dangerous area or danger triangle of face) (Fig. 11.25). In patients with thrombophlebitis of facial vein, pieces of an infected clot may reach the cavernous sinus producing cavernous sinus thrombosis (Infection of facial vein may result from squeezing the pimples on the sides of the nose and upper lip, i.e. The dangerous area of the face).

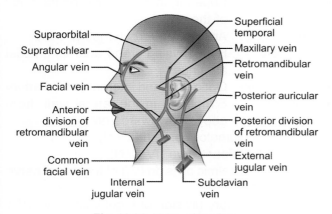

Supraorbital
Supratrochlear
Angular vein
Facial vein
Anterior division of retromandibular vein
Common facial vein
Superficial temporal
Maxillary vein
Retromandibular vein
Posterior auricular vein
Posterior division of retromandibular vein
External jugular vein
Internal jugular vein
Subclavian vein

Fig. 11.24: Veins of the face

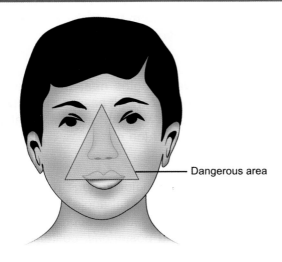

Fig. 11.25: Dangerous area of face

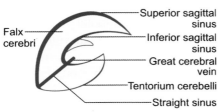

Fig. 11.26A: Dural venous sinuses

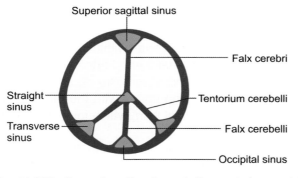

Fig. 11.26B: Coronal section through the posterior cranial fossa showing dural folds and dural venous sinuses

Dural Venous Sinuses (Figs 11.26 A and B)

The cranial dura mater has 2 layers—an outer endo-steal layer (which lines the inner aspect of skull bones and serves as the periosteum) and an inner meningeal layer, which surrounds the brain.

The two layers are fused with each other, except where the cranial venous sinuses are enclosed between them.

Peculiarities

The dural venous sinuses have an inner lining of endothelium. There is **no muscle** in their walls. They have **no valves**.

The dural venous sinuses receive venous blood from the brain, the meninges and bones of the skull. CSF is also drained into some of them (e.g. the superior sagittal sinus).

Classification

Classification of dural venous sinuses:
a. Paired dural sinuses
b. Unpaired dural venous sinuses.

Paired Dural Sinuses

1. Cavernous sinus
2. Superior petrosal sinus
3. Inferior petrosal sinus
4. Transverse sinus
5. Sigmoid sinus.

Unpaired Dural Venous Sinuses

These are median in position
1. The superior sagittal sinus
2. The inferior sagittal sinus
3. The straight sinus
4. The occipital sinus
5. The anterior and posterior intercavernous sinuses.

The Cavernous Sinuses (Fig. 11.27)

These are large venous spaces situated in the middle cranial fossa, one on either side of the body of the sphenoid. It is about 2 cm long and 1 cm wide.

Relations
- Anteriorly: Superior orbital fissure and the apex of the orbit
- Posteriorly: Apex of petrous temporal bone
- Medially: Hypophysis cerebri and sphenoidal air sinus
- Superiorly: Optic tract and internal carotid artery
- Laterally: Structures in the lateral wall, from above downwards are:
  1. The oculomotor nerve (III Cr.N)
  2. The trochlear nerve (IV Cr.N)
  3. The ophthalmic division (V_1) of trigeminal nerve (V Cr.N)
  4. The maxillary division of trigeminal nerve (V^2)

- *Structures passing through* the cavernous sinus are:
  i. The internal carotid artery
  ii. The abducent nerve (VI Cr.N)

Connections of Cavernous Sinus

1. The superior and inferior ophthalmic veins from the orbit
2. Facial vein, through the superior ophthalmic vein and pterygoid venous plexus
3. The right and left cavernous sinuses communicate with each other through the anterior and posterior intercavernous sinuses. All these communications are **valveless;** so, blood can flow through them in either direction.

Applied Anatomy

Cavernous sinus thrombosis It may be caused by spread of infection from the dangerous area of face, nasal cavities and paranasal air sinuses. It can give rise to severe pain in the area of distribution of the ophthalmic nerve (eye and forehead) and paralysis of muscles supplied by oculomotor, trochlear and abducent nerves.

Pulsating exophthalmos In fractures of the base of the skull, the internal carotid artery may tear within the cavernous sinus, producing an arteriovenous fistula. The arterial blood rushes into the sinus; from the sinus, blood reaches the superior and inferior ophthalmic veins. As a result, the eye protrudes (exophthalmos). The bulging eye pulsates, in synchrony with the radial pulse.

The Superior Sagittal Sinus

This unpaired dural venous sinus occupies the upper convex margin of the falx cerebri (The sickle shaped fold of dura mater, lying in the median plane, between the two cerebral hemispheres).

It begins anteriorly at the crista galli (a projection, resembling a cock's comb, from the ethmoid bone), runs upwards and backwards, to end near the internal occipital protuberance, by becoming continuous with one of the transverse sinuses, usually the right transverse sinus.

The superior sagittal sinus receives (1) venous blood from superior cerebral veins; (2) CSF from the subarachnoid space, which drains through the arachnoid villi.

Inferior Sagittal Sinus

This small sinus lies in the posterior 2/3 of the lower, concave margin of the falx cerebri. It ends by joining

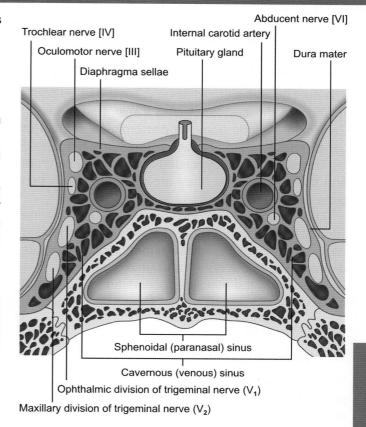

Fig. 11.27: Cavernous sinus

the great cerebral vein of Galen to form the straight sinus.

The Straight Sinus

This sinus lies in the median plane, at the junction of the falx cerebri and the tentorium cerebelli. It is formed by the union of the inferior sagittal sinus and the great cerebral vein. It ends at the internal occipital protuberance, by continuing with the left transverse sinus.

The Transverse Sinuses

These are large venous sinuses; the right sinus is usually larger than the left, because it receives blood from the superior sagittal sinus. They are situated in the posterior, attached margin of tentorium cerebelli; begin at the internal occipital protuberance and end at the posteroinferior angle of parietal bone, where it bends down to form the sigmoid sinus.

Confluence of Sinuses

The junction of the superior sagittal sinus, straight sinus, two transverse sinuses and the occipital sinus at the internal occipital protuberance is known as the confluence of sinuses.

Sigmoid Sinuses

The right and left sigmoid sinuses are the direct continuation of the corresponding transverse sinuses. Each one extends from the posteroinferior angle of parietal bone to the jugular foramen. It is S-shaped.

This sinus is closely related to the mastoid antrum and mastoid air cells; they are separated by a thin plate of bone.

The sigmoid sinus continues into the neck as the internal jugular vein. Usually the right internal jugular vein is larger, because the superior sagittal sinus drains blood to the right transverse sinus.

Infection of the middle ear can lead to the thrombosis of the sigmoid sinus.

Veins of the Neck

The Anterior Jugular Vein

This is usually the smallest of the jugular (jugular = pertaining to the neck) veins. It begins in the submental region (below the chin). It descends in the superficial fascia, about 1 cm from the median plane. About 2 cm above the sternum, it is connected to the anterior jugular vein of the opposite side, by a transverse channel, the jugular arch. The vein, then turns laterally and ends in the external jugular vein.

Variations The anterior jugular vein may be absent on one side. There may be a single median vein.

The External Jugular Vein (Fig. 11.24)

This vein begins near the angle of the mandible (just below the lobule of the pinna) by the union of the posterior division of retromandibular vein and the posterior auricular vein. It pierces the deep fascia about 5 cm above the clavicle. It ends by emptying into the subclavian vein, about 2 cm above the clavicle.

The size of the external jugular vein is inversely proportional to the size of the anterior jugular vein.

The Internal Jugular Vein

This vein, largest among the jugular veins, is a direct continuation of the sigmoid sinus. It begins at the jugular foramen and ends behind the sternal (medial)

end of the clavicle by joining the subclavian vein to form the brachiocephalic vein. The origin and termination of the vein are slightly dilated; they are called the superior and inferior bulbs respectively.

The internal jugular vein drains blood from the brain and superficial parts of the face and neck. This vein runs in the neck, in the carotid sheath, along with the internal carotid artery (above), the common carotid artery (below) and the vagus nerve.

The internal jugular vein is usually larger on the right side than on the left side because of the greater volume of blood entering it from the superior sagittal sinus, through the sigmoid sinus.

Deep cervical lymph nodes lie along the course of the internal jugular vein.

Tributaries Inferior petrosal sinus, the facial, lingual, pharyngeal, superior and middle thyroid veins drain into this large vein.

Applied Anatomy

When venous pressure is raised, (owing to heart failure), the external jugular vein becomes prominent throughout its course. Similarly, in obstruction of superior vena cava, enlarged supraclavicular lymph nodes, or increased intrathoracic pressure, the external jugular vein becomes prominent. So, when the external jugular vein becomes dilated and prominent, think about all these possibilities.

When the lower part of external jugular vein is cut, air may be sucked into the vein leading to venous air embolism. This is because, the vessel cannot retract owing to its attachment to the deep fascia.

In congestive heart failure or any other disease where venous pressure is raised, the internal jugular vein is markedly dilated and engorged. Pulsations of internal jugular vein resulting from contractions of right atrium become visible; they may also be palpable.

As the deep cervical lymph nodes lie very close to the internal jugular vein, removal of cancerous deep cervical nodes is difficult.

The Subclavian Vein

This large vein is the continuation of the axillary vein. It begins at the lateral border of the first rib and ends at the medial border of scalenus anterior muscle. Here, it joins the internal jugular vein, posterior to the medial end of clavicle to form the brachiocephalic vein (brachium = arm; kephale = Head).

It usually has one tributary, the external jugular vein.

The subclavian vein passes anterior and parallel to the subclavian artery, over the 1st rib. They are separated by the scalenus anterior muscle, and the scalene tubercle, to which the muscle is attached.

■ VEINS OF THE UPPER LIMB

The veins of the upper limb consists of 2 sets—superficial and deep. The deep sets accompany the arteries and are arranged in pairs as venae comitantes (except the axillary vein).

The superficial veins lie in the superficial fascia, communicate with the deep veins and finally drain into the axillary vein.

All these veins are provided with valves.

The Superficial Veins (Fig. 11.28)

These are the basilic vein, cephalic vein, median cubital and median vein of the forearm.

The venous blood from the hand drains mainly into the dorsal venous network on the back of the hand.

At the lateral end, the dorsal venous network joins with a vein from the index finger and veins from the thumb and continue upwards as the cephalic vein ("cephalic", means lying closer to the head).

At the medial end, the dorsal digital vein of little finger joins the dorsal venous network and continues upwards as the basilic vein.

The Cephalic Vein

It is formed in the anatomical snuff box (as explained earlier). It runs upwards along the radial border of forearm. In front of the elbow, it is connected to the basilic vein by the median cubital vein.

In the arm, the cephalic vein runs along the lateral side of biceps. In the groove between deltoid and pectoralis major, it pierces the deep fascia and drains into the axillary vein.

Basilic Vein

It begins from the ulnar side of the dorsal venous network, ascends along the ulnar (medial) side of the forearm. In front of the medial epicondyle, it is connected with the cephalic vein by the median cubital vein.

It ascends in the arm, medial to the biceps. In the middle of the arm, it goes deep, accompanies the

Fig. 11.28: Upper limb

venae comitantes of the brachial artery. Finally, the three veins join to form the axillary vein.

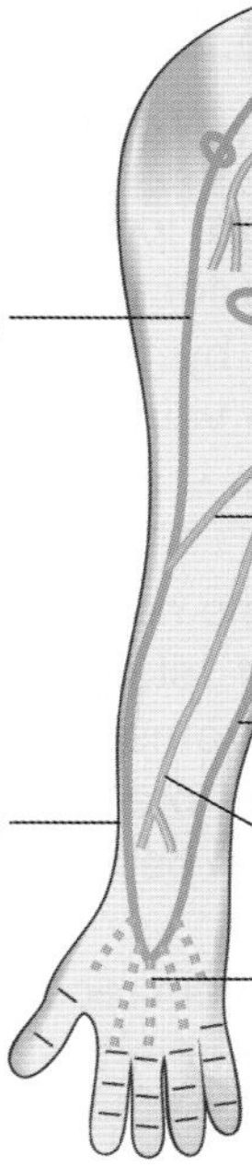

The Median Cubital Vein

This anastomosing vein lies in front of the elbow. It passes upwards and medially from the cephalic to the basilic vein. It lies in front of the bicipital aponeurosis, which separates the vein from the brachial artery and the median nerve.

The median cubital vein begins from the cephalic vein, 2.5 cm below the elbow; it runs obliquely upwards and medially, and ends in the basilic vein, about 2.5 cm above the medial epicondyle. It shunts blood from the cephalic to the basilic vein.

> **Applied Anatomy**
>
> The median cubital vein is connected to the deep veins by a perforator vein. The perforator vein fixes the median cubital vein and thus makes it ideal for intravenous injections, for withdrawing blood for biochemical and pathological studies, or for withdrawing blood from donors.

The Deep Veins

The deep veins accompany the arteries; they are arranged as venae comitantes.

Section C

The Organ Systems

■ VEINS OF THE THORAX (FIGS 11.29 AND 11.30)

The Brachiocephalic Veins

The brachiocephalic veins are located in the superior mediastinum. Each vein is formed posterior to the medial end of the clavicle by the union of internal jugular vein and subclavian veins (This is a union of the veins from the brachium = arm, and the kephale = head).

The brachiocephalic veins have no valves. At the level of the inferior border of 1st right costal cartilage, the two brachiocephalic veins unite to form the superior vena cava. The left brachiocephalic vein is twice as long as the right vein because it passes from the left to the right side.

The left brachiocephalic vein receives the thoracic duct, the largest lymph vessel in the body.

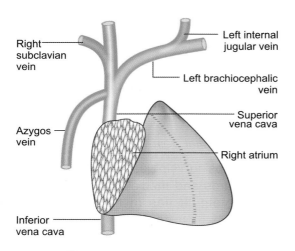

Fig. 11.29: Superior vena cava

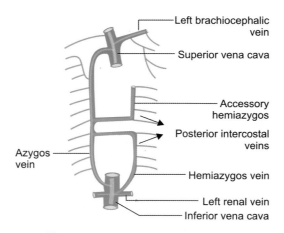

Fig. 11.30: Veins of the thoracic wall

Tributaries

Internal thoracic, vertebral, inferior thyroid and highest intercostal veins.

The Superior Vena Cava (Fig. 11.29)

This great vein lies in the superior mediastinum. It is about 7 cm long. The superior vena cava is formed posterior to the 1st right costal cartilage by the union of the right and left brachiocephalic veins. It passes inferiorly and at the level of the 3rd costal cartilage, it enters the right atrium.

The superior vena cava returns blood from all the structures superior to the diaphragm, except the heart and the lungs.

The Azygos System of Veins (Fig. 11.30) (Azygos = Unpaired)

The azygos system consists of veins on each side of the vertebral column that drains the back and the thoracic and abdominal walls.

The veins of this group are:
1. The azygos vein
2. The hemiazygos vein
3. The accessory hemiazygos vein.

The Azygos Vein

The azygos vein begins in the abdomen from the posterior surface of inferior vena cava. It enters the thorax through the aortic opening of the diaphragm or by piercing the right crus of diaphragm. It ascends in the posterior mediastinum, close to the right sides of the bodies of lower 8 thoracic vertebrae. It then arches over the right lung root and joins the superior vena cava.

The azygos vein connects the superior and inferior venae cavae directly or indirectly by the hemiazygos and accessory hemiazygos veins. It drains blood from the posterior walls of the thorax and abdomen.

Tributaries

1. Right superior intercostal vein (formed by the union of 2nd and 3rd posterior intercostal veins)
2. 4th–11th right posterior intercostal veins
3. The hemiazygos vein
4. The accessory hemiazygos vein
5. The right bronchial vein
6. Esophageal, mediastinal, pericardial veins
7. Ommunication with internal vertebral venous plexus.

The Hemiazygos Vein

This vein arises from the left renal vein. It ascends on the left side of the vertebral column, till it reaches T9 level. Here, it crosses to the right and enters the azygos vein.

Tributaries

1. Inferior 3 left posterior intercostal veins
2. Inferior oesophageal veins
3. Mediastinal veins.

The Accessory Hemiazygos Vein

This vein begins at the medial end of 4th or 5th inter-costal space, it descends on the left side of the vertebral column, from T5 to T8 level. Here it crosses to the right and drains into the azygos vein.

Tributaries Veins from 4th to 8th intercostal spaces and left bronchial veins.

The azygos, hemiazygos and the accessory hemiazygos veins show marked variation from one person to another. It forms an alternate channel of venous drainage from the thoracic, abdominal and back regions, when the inferior vena cava is obstructed.

Applied Anatomy

Infection and tumors from the abdomen and thorax can spread along the azygos vein, to different parts of the body.

Veins from the lower end of esophagus drain into (1) the left gastric vein, which drains into the portal vein and (2) the azygos vein, which drains into superior vena cava, a systemic vein. So, there is a porto- systemic anastomosis at the lower end of esophagus. In portal obstruction, these veins dilate, forming esophageal varices, which can give rise to massive hemorrhage.

Intercostal Veins

These veins accompany the intercostal arteries and nerves; they lie in the costal grooves (from above downwards vein, artery, nerve VAN). There are 11 pairs of posterior intercostal veins and one pair of subcostal veins. The posterior intercostal veins anas-tomose with anterior intercostal veins, which are tributaries of internal thoracic veins.

Most intercostal veins end in the azygos venous system.

■ VEINS OF THE ABDOMEN

1. The inferior vena cava
2. The portal vein.

The Inferior Vena Cava (IVC) (Fig. 11.31)

It is the largest vein in the body. The IVC returns blood from the lower limbs, most of the abdominal wall and the abdominopelvic viscera.

The inferior vena cava begins anterior to the L5 vertebra by the union of the two common iliac veins, to the right of the median plane, inferior to the bifurcation of aorta.

The IVC passes through the vena caval opening in the diaphragm at the level of T8 vertebra. It then pierces the fibrous pericardium and enters the inferior part of the right atrium of the heart.

Tributaries

These veins correspond to the branches of the aorta.

1. *The common iliac veins:* Each common iliac vein is formed by the union of the external and internal iliac vein.
2. The lumbar veins
3. The right testicular or ovarian vein (the left testicular or ovarian vein drains into the left renal vein)
4. The renal veins – (the left renal vein is longer as, it has to cross the midline to reach the IVC)

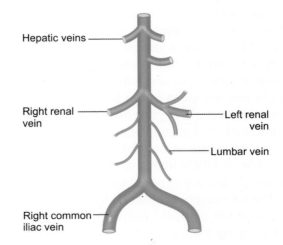

Hepatic veins

Right renal vein

Left renal vein

Lumbar vein

Right common iliac vein

Fig. 11.31: Veins of abdomen—inferior vena cava

5. The right suprarenal vein (the left suprarenal drains into the left renal vein)
6. The hepatic veins
7. The azygos vein – connects the SVC and IVC.

Applied Anatomy

1. Obstruction of the IVC due to thrombosis, below the level of the renal veins cause edema of the lower limbs.
2. Obstruction of IVC above the level of renal veins cause edema of lower limbs, albuminuria and ascites.
3. Tumors of the head of the pancreas can compress the IVC and can cause obstruction.
4. In inferior vena caval obstruction, the blood is bypassed along the azgous venous system and vertebral system of veins.

The Portal Vein (Fig. 11.32)

The portal vein is the main channel of the portal system of veins. It is peculiar in that it starts in capillaries and ends in venous sinusoids of the liver. It has no valves.

The portal vein collects blood from the abdominal part of the gastrointestinal tract, the gallbladder, the pancreas, the spleen and carries it to the liver. In the liver, it branches and re-branches to end in expanded capillaries or sinusoids. From the sinusoids, the blood is collected by hepatic veins, which drain into inferior vena cava.

Formation

It is formed by the union of the superior mesenteric vein and the splenic vein behind the neck of the pancreas, at the level of L1. Its average length is 8.5 cm.

The portal vein ascends to the liver in the free margin of the lesser omentum along with the bile duct and hepatic artery. At the porta hepatis, the portal vein divides into right and left branches, which enter the right and left lobes of the liver. They branch and end in sinusoids. The portal venous blood contains the products of digestion (carbohydrates, fats and proteins from the intestine) and the products of destruction of RBCs from the spleen.

The right branch of portal vein receives the cystic vein. The left branch receives (1) the ligamentum teres (obliterated left umbilical vein) (2) some small veins accompanying the ligamentum teres called paraumbilical veins; and (3) the ligamentum venosum (remnant of ductus venosus) extending from the left branch of portal vein to the IVC.

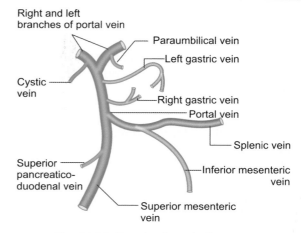

Fig. 11.32: Portal vein and tributaries

Tributaries

1. Left gastric vein
2. The right gastric vein
3. The cystic vein
4. Paraumbilical veins
5. *Superior mesenteric vein*—it receives tributaries from the areas supplied by superior mesenteric artery and also the right gastroepiploic vein.
6. *The splenic vein*—tributaries correspond to the branches of the splenic artery. The inferior mesenteric vein also drains into the splenic vein.

Communications Between the Portal and Systemic Veins—the Portosystemic Anastomoses

There are some areas, where the portal system of veins communicate with the systemic veins.

1. *At the lower end of the esophagus:* The venous plexus in the submucosa of lower end of esophagus drains inferiorly into the left gastric vein which ends in portal vein; and above, into the azygos venous system, which in turn, ends in the superior vena cava (systemic circulation).

   In portal obstruction, due to a backflow of blood through this plexus into systemic circulation, the submucous venous plexus of lower end of esophagus enlarge and become varicose. They are called esophageal varices – it can give rise to massive hemorrhage.

2. Superficial veins of anterior abdominal wall radiate from the umbilicus to drain into the axillary vein, saphenous vein or along the paraumbilical veins

into the portal vein. In portal obstruction, the radiating veins become dilated and tortuous, resembling "caput medusae" ("Medusa" is a character in Greek mythology, with serpents on the head).

3. Submucous venous plexus in the anal canal. These plexuses drain below into the inferior and middle rectal veins, which ultimately drain into the internal iliac and inferior vena cava (systemic veins). Superiorly, the plexus is drained by the superior rectal vein, which continues as the inferior mesenteric vein and drains into the splenic vein (portal circulation).

   In portal obstruction, the submucous venous plexus bulge to form hemorrhoids or piles.

Applied Anatomy

Portal obstruction may be due to:
1. *Cirrhosis of liver:* Damaged liver tissue is replaced by fibrous tissue, which constricts the branches of portal vein.
2. *Thrombosis of portal vein:* Portal obstruction leads to enlargement of spleen, ascites, collateral circulation through the portosystemic anastomoses and portal hypertension (normal pressure 5-15 mm of Hg; it exceeds 40 mm of Hg in portal obstruction)

Portocaval shunt is a surgical measure performed to relieve the portal vein obstruction by anastomosing the portal vein directly to the inferior vena cava.

■ VEINS OF LOWER LIMB (FIGS 11.33A and B)

Veins of lower limb can be classified into:
1. Superficial veins
2. Deep veins, and
3. Perforating veins.

The Superficial Veins

The superficial veins consist mainly of the great and small saphenous veins.

The Deep Veins

The deep veins accompany the arteries. They are surrounded and supported by the powerful muscles.

The deep veins include the tibial, popliteal and femoral veins. Below the knee, the deep veins are arranged as a pair of venae comitantes; above the knee, they form a single, large, vessel.

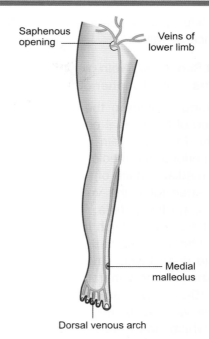

Fig. 11.33A: Great saphenous vein

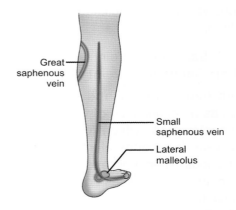

Fig. 11.33B: Veins of lower limb—small saphenous vein

The Perforating Veins

The perforating veins communicate the superficial with the deep veins.

All veins of the lower limb are provided with valves, to direct the blood flow to the heart, against gravity.

Factors Assisting Venous Return from Lower Limb

1. The main factor, is the contraction of calf muscles known as the "calf pump" or "peripheral heart"
2. Suction action of the diaphragm during inspiration

3. Transmitted pulsations of adjacent arteries
4. Presence of valves.

The Great Saphenous Vein or the Long (Fig. 11.33A) Saphenous Vein: (Saphenous = Easily Seen)

It is the longest vein in the body. It begins as a continuation of the medial end of the dorsal venous arch of the foot; the vein passes upwards about 2.5 cm anterior to the medial malleolus and ascends along the medial surface of tibia to reach the knee.

At the knee joint, the vein lies about a hand's breadth behind the patella. Then it ascends along the medial side of the thigh to reach the saphenous opening. The vein passes through this opening, pierces the deep fascia (the cribriform fascia) and finally drains into the femoral vein.

About 10-20 valves are present in this vein. Out of this, one valve is located at the saphenofemoral junction, which has great, functional importance. Incompetence of this valve will create great strain or pressure on the more distal valves and the great saphenous vein becomes dilated and tortuous within a short time.

The Small Saphenous Vein (The Short Saphenous Vein) (Fig. 11.33B)

This vein begins below and behind the lateral malleolus as a continuation of the lateral end of the dorsal venous arch, which joins with the lateral marginal vein. It ascends along the lateral side of tendocalcaneus. Then it ascends along the middle of the back of the leg; it passes between the two heads of gastrocnemius muscle, passes deep to end in the popliteal vein.

This vein, sometimes joins the great saphenous vein. It has 7-13 valves.

Applied Anatomy of Superficial Veins

1. The superficial veins are often dilated and become tortuous and become varicose. In the overlying skin, ulcers can develop due to lack of nutrition.
   Causes of varicosity
   a. Incompetency of valves
   b. Thrombosis of deep veins
   c. Persistent elevation of intra-abdominal pressure as in abdominal tumors or multiple pregnancy
2. A segment of great saphenous vein can be used as a graft in coronary bypass surgery.
3. Venesection in emergencies are carried out in the lower part of great saphenous vein.

Deep Veins of Lower Limb

These are the anterior and posterior tibial, peroneal, popliteal and femoral veins.

The Popliteal Vein

The popliteal vein begins at the lower border of the popliteus by the union of the veins accompanying the anterior and posterior tibial arteries. The vein continues as the femoral vein at the opening in the adductor magnus.

Tributaries (1) The small saphenous vein, and (2) the veins corresponding to the branches of the popliteal artery.

The Femoral Vein

This deep vein begins as an upward continuation of the popliteal vein at the lower end of adductor canal. It ends by becoming continuous with the external iliac vein behind the inguinal ligament, medial to the femoral artery.

Tributaries
1. The great saphenous vein
2. Profunda, deep external pudendal and muscular veins
3. Lateral and medial circumflex femoral.

The upper 4 cm of femoral vein is enclosed in a sleeve of fascia, called the femoral sheath (see Fig. 20.5). In the femoral sheath, the femoral vein occupies the middle compartment; the femoral artery, in the lateral compartment and the femoral canal is medial to the vein.

▊ VENOUS THROMBOSIS

Predisposing Factors—"Virchow's Triad"

1. Changes in the vessel wall, with damage to endothelium, e.g. inflammation
2. Diminished rate of blood flow, e.g. during or after operations
3. Increased coagulability of blood, e.g. infection, cancer.

Superficial Vein Thrombosis (Thrombophlebitis)

Commonly occurs in varicose veins or in veins which are cannulated for intravenous infusion. A painful, cord like inflamed area is diagnostic.

Deep Vein Thrombosis (Phlebothrombosis)

It may follow childbirth, operations, local trauma, immobility or any debilitating illness.

The thrombus may form in the tributary of a main vein; it gradually extends into the deep vein. The relatively faster blood stream may cause a portion of this thrombus to break off and cause a pulmonary embolus. Calf veins and pelvic veins are commonly affected.

Prevention of Deep Vein Thrombosis

a. *Before surgery* Overweight patients should reduce the weight.

b. *During surgery* It is essential that the venous return from the lower limb is not impeded. Pressure on the calf must be prevented; elevation and massage of legs are helpful.

c. *After surgery* Prevent conditions which predispose to a decreased circulation such as immobility of the lower limbs and dehydration.

To increase venous return from lower limbs, massage, leg movements, adequate hydration and early ambulation are advised.

■ DEVELOPMENT OF HEART

The heart develops from splanchnopleuric mesoderm of the cardiogenic area. Two endothelial heart tubes develop - the right and left heart tubes - which fuse to form one tube. It has an arterial end and a venous end (Figs 11.34A to C)

A series of dilatations appear on this tube (Figs 11.34A to C)

1. Bulbus cordis
   – Truncus arteriosus
   – Conus

2. Ventricle
3. Atrium
4. Sinus venosus
   – Right horn
   – Left horn

The bulbus cordis is further subdivided into a proximal conus and distal truncus arteriosus.

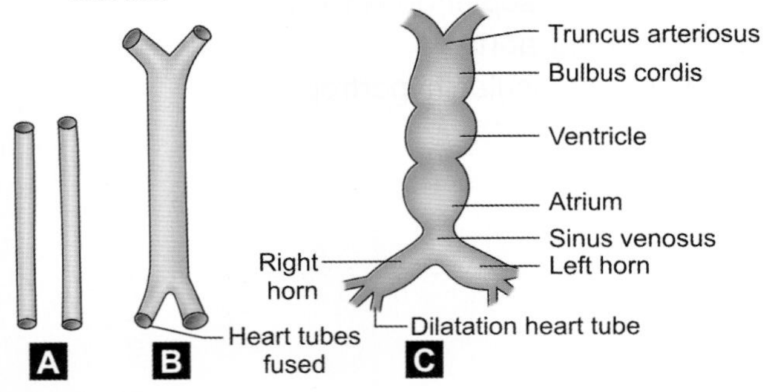

■ DEVELOPMENT OF ATRIA

The right atrium and the left atrium develop from the primitive atrium by two septa—the septum primum and septum secundum (Fig.11.35). A valvular passage, the foramen ovale, is present between these two septa. It allows the flow of blood from right atrium to the left atrium. After birth, there is a sudden increase in the volume of blood in the left atrium because it starts receiving oxygenated blood from the lungs. As a result, pressure in the left atrium increases, forcing the septum primum against septum secundum. The foramen ovale is obliterated. The fossa ovalis represents the septum primum and the annulus (limbus) ovalis represents the lower free edge of septum secundum (Fig. 11.35).

■ DEVELOPMENT OF VENTRICLE (FIG.11.36)

The conus and the primitive ventricle unite to form a common chamber. This is partitioned to form the right and left ventricles. The partition—the interventricular septum—is formed by;

1. Primitive interventricular septum that grows upwards from the floor of the primitive ventricle.
2. Bulbar septum
3. Proliferation of atrioventricular cushions—it fills the gap between (1) and (2)

Truncus arteriosus
Bulbus cordis
Ventricle
Atrium
Sinus venosus
Right horn
Left horn
Heart tubes fused
Dilatation heart tube
A **B** **C**
Right and left heart tubes

Figs 11.34A to C: Development of heart

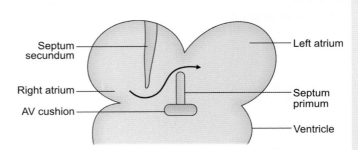

Septum secundum
Left atrium
Right atrium
AV cushion
Septum primum
Ventricle

Fig. 11.35: Development interatrial septum (arrow indicates direction of blood flow from right atrium to left atrium through foramen ovale)

Section C

The Organ Systems

PART 1

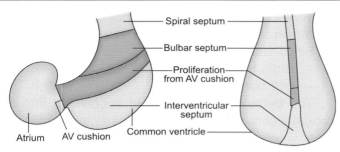

Fig. 11.36: Development interventricular IV septum

The ascending aorta and pulmonary trunk develop from the truncus arteriosus, which is divided by a spiral septum (Fig.11.36). After this partition, the aorta originates from the left ventricle and pulmonary trunk arises from the right ventricle.

Congenital Anomalies

Dextrocardia

It is the most frequent positional abnormality of the heart. The heart is displaced to the right and there is transposition in which the heart and its vessels are reversed left to right as in a mirror image.

Ectopia Cordis

The heart is partly or completely exposed on the surface of the thorax. It is usually associated with widely separated halves of the sternum and an open pericardial sac.

Atrial Septal Defects (ASD)

ASD is a common congenital anomaly. The most common form of ASD is patent foramen ovale.

Common Atrium

Common atrium is a rare cardiac defect in which the interatrial septum is absent, due to the failure of development of septum primum and septum secundum.

Ventricular Septal Defect

Ventricular septal defect (VSD) are the most common type of congenital heart defects (CHD). VSD occurs more frequently in males than in females. The defect may be small, which closes spontaneously during the first year. Patients with large VSD have a massive left-to-right shunt of blood.

Absence of interventricular septum - single ventricle or common ventricle - results from the failure of the IV septum to develop. The heart has 3 chambers - cor triloculare biatrium.

Transposition of Great Arteries

Transposition of great arteries (TGA) is the most common cause of cyanotic heart disease in newborn infants. TGA is often associated with ASD or VSD. The aorta arises from right ventricle and pulmonary trunk arises from left ventricle. Because of these anatomical abnormalities, deoxygenated blood from right atrium enters the right ventricle and then reaches the different parts of the body through the aorta. Oxygenated pulmonary venous blood reaches the left atrium, left ventricle and pulmonary trunk. Without surgical correction, these infants usually die within a few months.

Tetralogy of Fallot

Tetralogy of Fallot is characterized by four cardiac defects:
1. Pulmonary stenosis (narrowing or obstruction of right ventricular outflow)
2. Ventricular septal defect (VSD)
3. Overriding aorta
4. Right ventricular hypertrophy.

CHAPTER

12

Respiratory System

INTRODUCTION

The essential features of respiration are the transference of (1) oxygen from the atmosphere to the tissues, and (2) carbon dioxide from the tissues to the air.

EXCHANGE OF GASES

There are 2 phases in the exchange of gases:
1. External respiration, and
2. Internal respiration.

External Respiration

The *external respiration* takes place in the lungs. The oxygen is absorbed from air into the blood and CO_2 is excreted from the blood into the air.

Internal Respiration

The internal respiration or tissue respiration: O_2 is transferred from blood to the tissues, which give up CO_2.

The Air Passages

See Flow chart 12.1.

■ THE NOSE AND NASAL CAVITY

The nose has two functions to perform. They are (1) It is part of the respiratory passage; and (2) It has the olfactory function or the sense of smell.

External Features

The nose has a prominent ridge, separating the right and left halves, called the dorsum. The upper narrow end of the nose is called the root of the nose. The lower end of the dorsum is called the tip of the nose. At the lower end of the nose there are two openings—the right and left nostrils or anterior nares, which are

Flow chart 12.1: The air passages

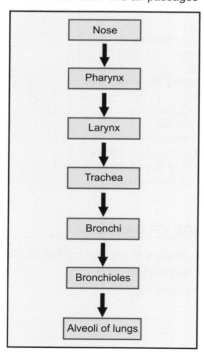

separated by a soft, median partition called the columella. Each nostril is bounded laterally by ala.

The nose has a skeletal framework, which is partly bony and partly cartilaginous. The bony part is contributed by the nasal bones and the frontal processes of maxillae. The cartilages are superior and inferior nasal cartilages and septal cartilage.

Nasal Cavity

It is pyramidal in shape. It extends from the nostrils (anterior nares) to the posterior nasal apertures (choanae). The nasal cavity is divided into right and left halves by a median septum.

Subdivision of Nasal Cavity

Vestibule: Vestibule is the dilated part, just inside the anterior nasal opening. There are numerous coarse hairs called vibrissae, which help to filter the air.

Olfactory region: The upper 1/3 of nasal cavity has the olfactory receptor cells. The mucosa is yellowish in color.

The respiratory region: The lower 2/3 of nasal cavity is lined by thick mucous membrane; it is highly vascular. It is lined by pseudostratified ciliated columnar epithelium.

Boundaries of Nasal Cavity

Each half of the nasal cavity has a roof, a floor, a medial wall and a lateral wall. Each half measures about 5 cm in height, 5-7 cm in length (antero-posterior), 1.5 cm in width (floor) and 1-2 mm at the roof.

The *roof* is mainly formed by the cribriform plate of the ethmoid bone. Anteriorly, roof is formed by the nasal bone and nasal part of the frontal bone; posteriorly, the roof is contributed by the body of the sphenoid bone.

The floor is formed by the palate, which separates the nasal cavity from the oral cavity.

The Medial Wall or the Nasal Septum (Fig. 12.1)

It is a median, osteocartilaginous partition between the two halves of the nasal cavity. It is covered on either side by mucous membrane (which is adherent to the underlying periosteum or perichondrium, forming mucoperiosteum or mucoperichondrium).

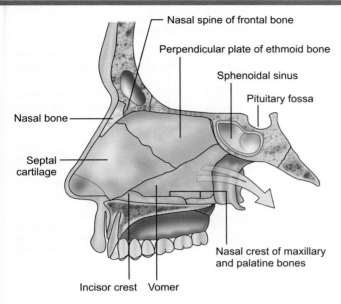

Fig. 12.1: Nasal septum

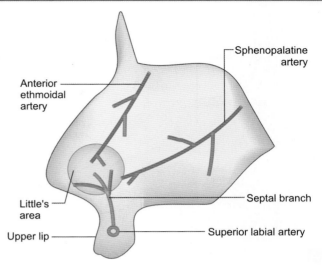

Fig. 12.2: Arteries supplying the nasal septum

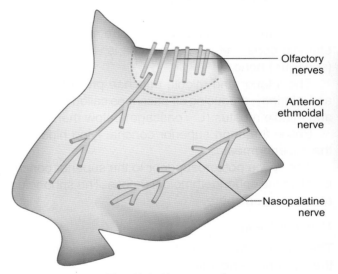

Fig. 12.3: Nerve supply

Components

The bony part: The bony part of the nasal septum is formed mainly by the vomer and the perpendicular plate of the ethmoid bone (The adjacent bones – frontal, maxilla, nasal and sphenoid – also contribute).

The cartilaginous part: It is formed by (1) the septal cartilage and (2) septal processes of inferior nasal cartilages.

The cuticular part: The cuticular part formed is by fibro-fatty connective tissue covered by skin. The lower margin of the septum is called the columella.

Blood Supply of Septum (Fig. 12.2)

The septum is supplied by branches of anterior ethmoidal artery (branch of ophthalmic artery), the superior labial branch of facial artery and the sphenopalatine branch of maxillary artery.

At the antero-inferior quadrant, there is a rich anastomosis between the septal branches of superior labial artery, anterior ethmoidal artery and the branches of sphenopalatine artery. This area, called the Little's area or Kiesselbach's area is a common site of epistaxis (bleeding from nose).

The corresponding veins drain into facial vein and the pterygoid venous plexus.

The nasal septum gets its sensory nerves from the branches of the trigeminal nerve; the upper 1/3 has olfactory receptor cells (Fig. 12.3).

Applied Anatomy

1. DNS (deviated nasal septum). The nasal septum is rarely median; it is often deflected or deviated to one side, more often, to the right side. It may require surgical correction.
2. Little's area is a common site of epistaxis. It can be due to trauma, infections or hypertension.

The Lateral Wall of Nose (Figs 12.4A and B)

The lateral wall of the nose is irregular, owing to the presence of 3 shelf-like bony projections called "conchae" (L.shells) or "turbinates". These elevations are called superior, middle and inferior nasal conchae according to their position on the lateral wall of the nasal cavity.

Section C

The Organ Systems

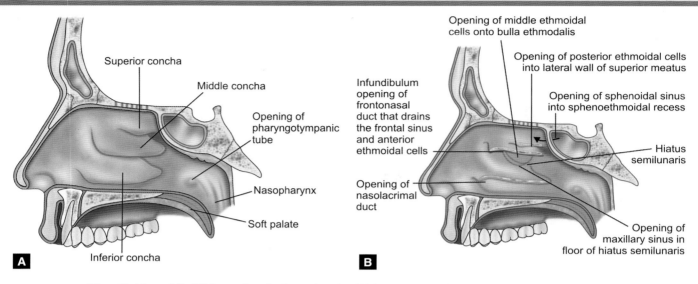

Figs 12.4A and B: (A) Lateral wall of nasal cavity (B) Lateral wall of nasal cavity (conchae are cut)

The superior and middle conchae are parts of the ethmoid bone; the inferior concha is an independent (separate) bone.

The inferior and middle conchae project medially and inferiorly; producing air passages called the inferior and middle meatus (L = passage) below them.

Below the short superior concha, lying hidden, is the superior meatus.

The space posterosuperior to the superior concha is called the sphenoethmoidal recess. The sphenoidal air sinus opens here.

Superior Meatus

The superior meatus is a narrow passage between the superior and middle conchae. The posterior ethmoidal sinuses open into the superior meatus.

Middle Meatus

The middle meatus is a longer and wider passage between the middle and inferior meatus. There is a funnel-shaped opening in its antero-superior part called the infundibulum. The frontal sinus opens into the infundibulum, through a narrow frontonasal duct.

There is a rounded elevation lying in the middle meatus, under cover of middle concha, called the bulla ethmoidalis ("bubble"). The middle ethmoid sinuses open on the surface of the bulla.

Inferior to the bulla, is a semicircular groove called the hiatus semilunaris. The maxillary sinus and the anterior ethmoidal sinuses open here.

The inferior meatus lies underneath the inferior concha. It is the largest of the 3 meatuses. The nasolacrimal duct opens into this meatus anteriorly. It is guarded by a mucosal fold called the valve of Hasner.

The lateral wall has a rich arterial supply from anterior ethmoidal artery and branches of maxillary artery. Most of the nerves are branches of the trigeminal nerve; the upper 1/3 is olfactory in function.

Paranasal Air Sinuses

These are airfilled extensions of the respiratory part of the nasal cavity into the following skull bones:

1. Frontal
2. Ethmoid
3. Maxilla
4. Sphenoid.

They are named according to the bones in which they are situated as follows:

1. The frontal air sinus (frontal sinus)
2. Ethmoidal air sinuses (ethmoidal sinuses)
3. Maxillary (air) sinus
4. Sphenoidal (air) sinus.

All of them open into the nasal cavity through its lateral wall. The sinuses are lined with mucous membrane, that is continuous with the mucous membrane of the nasal cavity. The epithelium is pseudostratified ciliated columnar.

The sinuses are absent or rudimentary at birth. They attain full size only after puberty.

Functions of Paranasal Sinuses

1. They reduce the weight of the skull
2. The inhaled air is warmed and humidified
3. They add resonance to the voice
4. The mucous secreted by the mucous glands help to trap dust particles in the inhaled air, the cilia remove the dust particles.

The Frontal Sinus

This large air sinus is situated in the frontal bone. It is divided into two halves by an asymmetrical septum. The frontal sinuses open into the middle meatus of the nose, at the infundibulum, through a frontonasal duct.

Relations: The frontal air sinuses are closely related to the anterior cranial fossa and the orbit. Infection of the frontal sinus—frontal sinusitis—can lead to abscess formation in the frontal lobe of the brain and osteomyelitis of the frontal bone.

The Ethmoidal Sinuses

The ethmoidal sinuses are a number of air filled cavities situated inside the ethmoid bone. They are divided into 3 groups—anterior, middle and posterior.

The anterior group of sinuses open into the hiatus semilunaris; the middle group, on the surface of the bulla ethmoidalis and the posterior group open into the superior meatus.

The Maxillary Sinus (Antrum of Highmore)

It is the largest of the air sinuses. It is situated inside the body of the maxilla. It is pyramidal in shape, with the apex directed laterally.

It opens into the hiatus semilunaris of the middle meatus of nose. This opening lies at a higher level than the floor of the sinus. Hence, fluids tend to collect in the sinus. It can be drained by tilting the head to one side (postural drainage) or by surgical puncture (antrostomy).

Sphenoidal Sinus

It is a cubical cavity, within the body of the sphenoid. It is divided into 2 by an asymmetrical septum.

Superiorly, it is related to the middle cranial fossa and the pituitary gland; laterally it is related to the cavernous sinuses, with the internal carotid artery and abducent nerve within the sinuses.

This sinus opens into the sphenoethmoidal recess of the nasal cavity.

Applied Anatomy

1. Infection from ethmoidal sinuses may spread to the orbit, because the medial wall of the orbit is thin and fragile (lamina papyracea). This can lead to orbital cellulitis, or optic neuritis.
2. The maxillary sinus is the one most commonly infected, probably because its opening is situated superior to its floor. Similarly, infected material from the frontal and ethmoidal sinuses flow into the maxillary sinus.
3. Since the floor of the maxillary sinus and the roots of the maxillary molar teeth are very closely related, removal of the molar teeth may lead to the formation of a communication between the oral cavity and the maxillary sinus.
4. As all the paranasal sinuses are lined by mucosa, which is an extension of the mucosa of nasal cavity, infection may spread easily, producing inflammation and swelling—this is called sinusitis.

■ PHARYNX

The pharynx is described under Chapter 13. "Digestive System"—*please refer.*

■ THE LARYNX

The larynx is the organ for the production of voice. It also serves as an air passage.

The larynx lies in the anterior midline of the neck, extending from the root of the tongue to the trachea. It lies opposite 3rd to 6th cervical vertebrae.

Size

Length	: 4.3 cm
Transverse diameter	: 4.2 cm
Anteroposteriorly	: 3.6 cm

The male larynx is larger than in females.

The larynx is made up of (1) a skeletal framework of cartilages; (2) joints, ligaments and membranes, which connect the cartilages and (3) muscles, which move the cartilages and the structures attached to them.

The cavity of the larynx is lined by respiratory epithelium, except the surface of vocal cord, which is lined by stratified squamous epithelium.

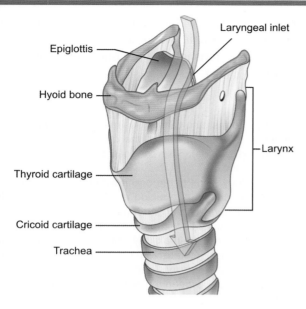

Fig. 12.5: Skeleton of larynx

The Skeleton (Cartilages) of Larynx (Fig. 12.5)

The larynx has 9 cartilages; 3 of them are unpaired; and 3 paired.

Unpaired Cartilages

1. Thyroid cartilage
2. Cricoid cartilage; and
3. The epiglottis.

The Paired Cartilages

1. The arytenoids
2. The corniculate; and
3. The cuneiform.

The Thyroid Cartilage

It is the largest of the laryngeal cartilages. It is a hyaline cartilage. It is made of two quadrangular plates. They have 2 surfaces (an inner and outer) and 4 borders (upper, lower, anterior and posterior).

The anterior borders of the two plates are united in their lower parts at an angle, which forms a prominence in the middle of the neck, called the laryngeal prominence or Adam's apple. It is more prominent in males.

In the upper part, the anterior borders are separated by a thyroid notch.

The posterior borders extend superiorly as the superior horns; and inferiorly, as the inferior horns.

To the upper border, the thyrohyoid ligament is attached.

The outer surface of the lamina presents an oblique line which gives attachment to strap muscles, inferior constrictor muscle of pharynx and pretracheal fascia.

The Cricoid Cartilage

It resembles a signet ring. It consists of a posterior quadrilateral lamina and an anterior arch. It is also a hyaline cartilage.

The Epiglottis

It is a leaf shaped, elastic cartilage. Its lower end is narrow and attached to the inner surface of the thyroid cartilage by a ligament.

The upper end is broad and free. Its anterior surface overhangs the posterior part of the dorsum of the tongue. Three folds of mucosa extend between this part and the tongue (one median and 2 lateral glossoepiglottic folds).

The posterior surface of epiglottis is covered by mucosa of the larynx.

Arytenoid Cartilages

These are placed on the upper border of the lamina of the cricoid cartilage. Each arytenoid is pyramidal in shape. It has an apex, a muscular process and a vocal process (which project from its base).

Corniculate and cuneiform cartilages are small cartilages, lying within the aryepiglottic folds.

Ligaments of Larynx

The extrinsic ligaments connect the cartilages. The intrinsic ligaments lie beneath the mucous membrane of the larynx. In the upper part, it is called the quadrangular membrane and in the lower part, it is called the conus elasticus.

The upper free border of the quadrangular membrane covered by mucous membrane, is called the aryepiglottic fold. Its lower border is thickened and called the vestibular ligament. With the mucosa covering it, it forms the false vocal cord or the vestibular fold.

The upper border of conus elasticus is attached anteriorly to the deep surface of thyroid cartilage, near the midline; posteriorly, it is attached to the tip of the vocal process of the arytenoid. Between these attachments, the upper border is free; it is thickened to form the vocal ligament. The vocal ligament, vocalis muscle and the stratified squamous epithelium covering them form the true vocal cord or vocal fold.

The Cavity of the Larynx (Fig. 12.6)

Extent

It extends from the laryngeal inlet above, to the lower border of cricoid. Superiorly, it opens into the laryngeal part of pharynx and inferiorly, it is continuous with the trachea.

The interior is lined by mucous membrane. Two pairs of folds project into the cavity from the lateral walls.

The upper pair of folds is the vestibular folds or the false vocal cords. The gap or fissure between them is called the rima vestibuli.

The lower pair of folds is the true vocal cords and the fissure between them is the rima glottis (glottidis).

The inlet of the larynx is an oval opening that is bounded by; (a) The free upper border of the epiglottis anteriorly (b) the aryepiglottic folds laterally; and (c) the interarytenoid mucous membrane posteriorly.

The cavity of the larynx is divided into 3 parts: (a) Vestibule (b) The sinus of the larynx, and (c) The infraglottic part.

The Vestibule

The vestibule extends from the inlet of the larynx up to the vestibular folds.

Sinus or the Ventricle of the Larynx

The sinus or the ventricle of the larynx is the smallest of the 3 parts of the larynx. It is situated between the vestibular folds and the vocal folds.

The infraglottic part extends from the level of true vocal folds to the beginning of the trachea.

Vocal folds or vocal cords or true vocal cords *(Fig. 12.7)*. They are two wedge-shaped, pearly white, folds of mucous membrane stretching from the angle of the thyroid cartilage to the vocal process of the arytenoid cartilages.

The fissure between the vocal cords is called the rima glottis or rima glottidis. They are concerned with voice production. Their average length varies in males and females; in the male, it is about 23 mm and in the female, 17 mm.

Each vocal cord consists of:
1. The vocal ligament
2. The vocalis muscle
3. The mucous membrane

The mucosa is firmly adherent to the underlying structures. There is no submucosa and blood vessels. Hence, the vocal cords are pearly white in color. The surface epithelium is stratified squamous, non-keratinizing.

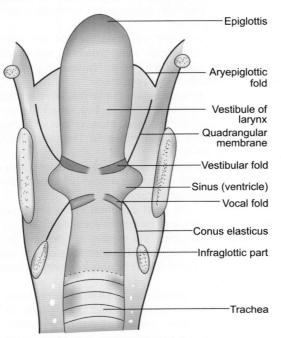

Fig. 12.6: Coronal section through larynx showing the cavity and subdivisions

Labels: Epiglottis; Aryepiglottic fold; Vestibule of larynx; Quadrangular membrane; Vestibular fold; Sinus (ventricle); Vocal fold; Conus elasticus; Infraglottic part; Trachea

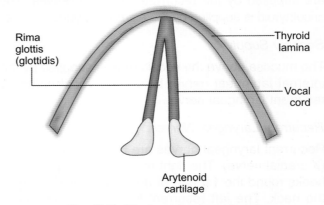

Fig. 12.7: Attachments of vocal cord

Labels: Rima glottis (glottidis); Thyroid lamina; Vocal cord; Arytenoid cartilage

Section C   The Organ Systems

	Table 12.1: Differences between vocal fold and vestibular fold		
Features	Vocal fold		Vestibular fold
1. Color	Pearly white		Pink
2. Epithelium	Stratified squamous non-keratinizing		Pseudostratified ciliated columnar epithelium
3. Submucosa	Absent		Present
4. Components	Vocal ligament, vocalis muscle and mucosa		Mucosa, submucosa and vestibular ligament
5. Function	Phonation (production of voice)		Prevents entry of food particles into larynx No role in phonation

The differences between vocal folds and vestibular folds are given in Table 12.1.

Muscles of Larynx

Classification

Extrinsic muscles: They are attached to the skeleton of the larynx and the bones above it. They move the larynx as a whole during respiration, deglutition, etc.

The intrinsic muscles: They are concerned with the movements of the vocal cords and hence, concerned with the production of voice. They are:
a. Cricothyroid
b. Thyroarytenoideus or thyroarytenoid
c. Vocalis
d. Posterior cricoarytenoid – the only abductor of vocal cords
e. Lateral cricoarytenoid
f. Transverse arytenoid
g. Oblique arytenoid
h. Aryepiglotticus
i. Thyroepiglotticus.

Innervation of the Muscles of Larynx

All the intrinsic muscles of larynx, except the cricothyroid, are supplied by the recurrent laryngeal nerves. The cricothyroid is supplied by the external laryngeal nerve.

Sensory Supply

The mucosa above the vocal cords is supplied by the internal laryngeal nerve; below the vocal cords by recurrent laryngeal nerve.

Recurrent Laryngeal Nerves

Recurrent laryngeal nerves are branches of the vagi (X cranial nerve). The right recurrent laryngeal nerve hooks round the 1st part of right subclavian artery in the neck. The left recurrent laryngeal nerve hooks around the ligamentum arteriosum, in the thorax.

Applied Anatomy

1. The term "glottis" refers to the vocal cords, the rima glottis (glottidis) and the narrow part of the larynx at the level of the vocal cords. The glottis is the part of the larynx most directly concerned with voice production and the rima glottis is the usual place, where aspirated food or other "foreign bodies" become lodged, causing laryngeal obstruction. No air can enter the trachea, bronchi and lungs. Emergency treatment should be given to the patients, to open the airway. If available, a wide bore needle is inserted into the larynx or trachea to permit fast entry of air; this procedure may be followed by tracheostomy and insertion of a metal tube into the trachea.

2. The recurrent laryngeal nerve is vulnerable to injury during thyroidectomy. Injury to the nerve results first in the paralysis of abductors of the vocal cord than adductors (Semon's law).
   When only one recurrent laryngeal nerve is paralyzed, phonation is possible, because the opposite vocal cord compensates for it.
   When both recurrent laryngeal nerves are affected, phonation is completely lost; breathing also becomes difficult. The vocal cords lie in "cadaveric position" (in between adduction and abduction).

3. When a foreign body enters the larynx, severe coughing is initiated to expel the object. But, when the internal laryngeal nerve is damaged, there is loss of sensation, and absence of the cough reflex. So, the foreign body can easily enter the larynx.

4. The interior of the larynx can be examined directly through a laryngoscope (direct laryngoscopy) or indirectly through a mirror (indirect laryngoscopy).

5. Hoarseness is the most common symptom of serious disorder of larynx, e.g. cancer of vocal cords.

6. As the age advances, the thyroid, cricoid and parts of arytenoid can undergo calcification. They can be broken, from blows received during boxing, or during automobile accidents.

7. Cancer of the larynx occurs between 40-60 years of age; males are 10 times more attacked than female. 70% of growths arise from the vocal cords. Huskiness or hoarseness is the first symptom; it is progressive, and finally gives place to aphonia. Diagnosis is made by laryngoscopy and biopsy.

Note: *To know the position of various structures in the thorax, a brief description of Mediastinum is given below.*

THE MEDIASTINUM
(L. MIDDLE SEPTUM) (FIG. 12.8)

The interval between the two pleural sacs is known as the mediastinum. The various organs of the mediastinum lie embedded in loose areolar connective tissue, nerves, blood vessels, lymph nodes and fat. In living individuals, the looseness of the connective tissue and fat, and the elasticity of lungs and pleura enable the mediastinum to accommodate movement and volume changes in the thoracic cavity.

Subdivisions of Mediastinum

For descriptive purposes the mediastinum is divided into superior and inferior parts.

The superior mediastinum extends from the superior thoracic aperture to a horizontal plane passing through the sternal angle and inferior border of T_4 vertebra.

The inferior mediastinum extends from the plane passing through sternal angle and lower border of T_4, to the T_{12} level. It is subdivided into 3 parts—anterior mediastinum, middle mediastinum and posterior mediastinum.

The Superior Mediastinum

The superior mediastinum lies above the horizontal plane passing through the sternal angle and the lower

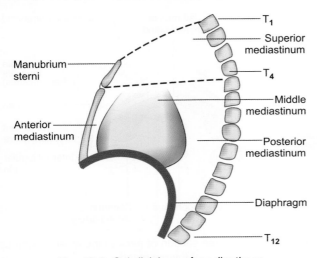

Fig. 12.8: Subdivisions of mediastinum

border of T_4. It is bounded anteriorly by manubrium sterni and posteriorly by T_1 to T_4 vertebrae.

Contents

1. Arch of aorta and its three branches
2. The esophagus
3. The trachea
4. The brachiocephalic veins
5. The upper part of superior vena cava
6. The thoracic duct
7. Vagi and recurrent laryngeal nerves.

Anterior Mediastinum

Lies between sternum and pericardium.

Contents

1. Mediastinal branches of internal thoracic artery;
2. 3-4 lymph nodes
3. Loose connective tissue.

Middle Mediastinum

Boundaries

- Superiorly – superior mediastinum
- Inferiorly – diaphragm
- Laterally – pleural sacs
- Posteriorly – posterior mediastinum.

Contents

1. Heart and pericardium
2. Ascending aorta
3. Lower half of superior vena cava
4. Bifurcation of trachea and the bronchi
5. Pulmonary trunk and its 2 branches
6. Four pulmonary veins
7. Tracheobronchial nodes
8. Nerves and nerve plexuses.

Posterior Mediastinum

Extends from T_5-T_{12}, posterior to middle mediastinum. Main contents are the (1) descending thoracic aorta; (2) the esophagus; (3) the azygos vein; (4) the thoracic duct; (5) nerves; and (6) lymph nodes.

Mediastinum—Surgical Importance

1. The posterior mediastinum is continuous with the neck, through the superior mediastinum. So,

Table 12.2: Effects of compression of mediastinal structures

Structures involved	Effect of compression
1. Superior vena cava	Engorgement of veins in the upper half of the body
2. Trachea	Dyspnea and cough
3. Esophagus	Dysphagia
4. Left recurrent laryngeal nerve	Hoarseness of voice
5. Phrenic nerve	Paralysis of diaphragm on that side
6. Vertebral column	Erosion of vertebrae

infections or fluid collections in the neck can spread to the superior and posterior mediastina.

2. *Mediastinal syndrome:* Compression of mediastinal structures by any growth (e.g. a tumor or an aneurysm) gives rise to a group of symptoms known as "mediastinal syndrome" (Table 12.2).

Common Causes

1. Bronchogenic carcinoma
2. Hodgkin's lymphoma
3. Aortic aneurysm.

THE TRACHEA (FIG. 12.9)

The trachea or windpipe is a wide, fibrocartilaginous tube, about 11-12 cm long. Its upper half is situated in the neck and the lower part, in the thorax.

Extent

It starts at the lower border of the cricoid cartilage (C6 vertebra) as the continuation of the larynx and ends at the level of the sternal angle (disc between T_4 and T_5, posteriorly) by dividing into right and left bronchi.

It is a tubular structure, with a flattened posterior wall.

The Cervical Part (Fig. 12.10)

Relations

- *Anteriorly:*
  1. Isthmus of the thyroid lies on the 2nd, 3rd and 4th tracheal rings
  2. Pre-tracheal layer of fascia
  3. Vessels of thyroid gland

4. The strap muscles – sternohyoid and sterno-thyroid.
- *Posteriorly:*
  1. The esophagus
  2. The recurrent laryngeal nerves lie in a groove between the trachea and the esophagus.
- *Laterally:*
  1. The lobes of the thyroid
  2. The common carotid arteries in the carotid sheath.

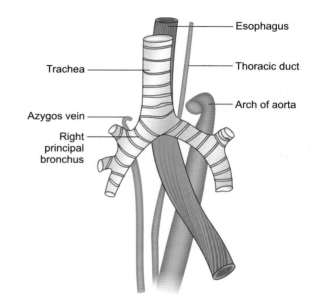

Fig. 12.9: Trachea and its relations

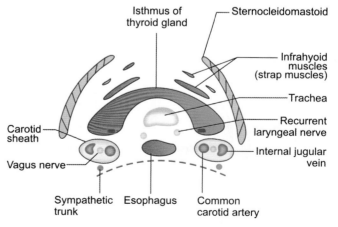

Fig. 12.10: Transverse section of neck at the level of isthmus of thyroid gland

Relations of Thoracic Part

- *Posteriorly esophagus.*
- *Anteriorly:*
  1. Manubrium sterni
  2. The arch of aorta
  3. The brachiocephalic and left common carotid arteries
  4. The thymus
  5. The left brachiocephalic vein crosses to the right side.
- *Left side:*
  1. Arch of aorta
  2. Left subclavian and left common carotid arteries
  3. Left recurrent laryngeal nerve
- *Right side:*
  1. Arch of azygos vein
  2. Right vagus nerve

Histology (Figs 12.11A and B)

1. *Mucosa:* Lined by pseudostratified ciliated columnar epithelium, with goblet cells in between.
2. *The submucosa:* Made of loose connective tissue, blood vessels, nerves and mucous glands.
3. *Cartilage rings* (16-20): They are C-shaped hyaline cartilage rings, deficient posteriorly, where the smooth muscle, trachealis, fills the gap.
4. The outer fibrous layer (Adventitia).

Carina

The last tracheal ring is thick and broad. From its lower border, a hook-like process curves downwards and backwards between the bronchi. This projection is called the carina (L. keel).

During bronchoscopy, the carina is observed between the orifices of the main bronchi. Normally the carina is in a sagittal plane. Enlarged tracheobronchial lymph nodes may alter the position of the carina; the carina may become distorted, widened and become immobile. Hence, changes in the carina is clinically important.

The mucous membrane at the carina is one of the most sensitive areas of the tracheobronchial tree. It is associated with cough reflex. For example, when a foreign body stimulates the mucous membrane at the carina (e.g. an aspirated peanut) the patient chokes and coughs.

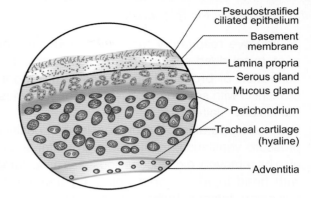

Fig. 12.11A: Trachea (TS)

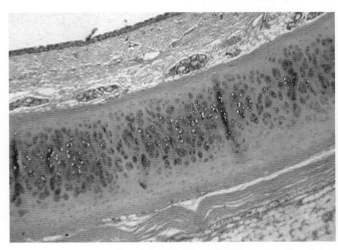

Fig. 12.11B: Trachea

Similarly, during postural drainage of the lungs, lung secretions pass to the carina; it causes coughing, which helps to expel the secretions.

Applied Anatomy

Foreign bodies entering the trachea have a tendency to go to the right bronchus, since it is wider and more in line with the trachea.

Tracheostomy

Tracheostomy is an emergency surgery, usually done to relieve complete obstruction of the respiratory tract at laryngeal level. An opening is made in the trachea, just below the isthmus of the thyroid and a tracheostomy tube is inserted, so that air enters into the distal part of respiratory tract through it.

Section C

The Organ Systems

Other Indications for Tracheostomy

1. To improve respiratory function by reducing the anatomical dead space, it also enables effective aspiration of bronchial secretions, e.g. chronic bronchitis with severe emphysema, chest injuries, particularly "flail-chest".
2. Respiratory paralysis: It allows assisted or positive pressure ventilation; inhaled foreign material can also be removed, e.g. unconsciousness associated with head injuries, coma, following barbiturate poisoning, tetanus, bulbar poliomyelitis.
3. As a preliminary to certain operations—e.g. laryngectomy.

■ THE BRONCHI

The Right Principal Bronchus

The right principal bronchus is approximately 2.5 cm long. It is wider and more vertical than the left. Hence foreign bodies which enter the trachea tend to fall into it. When this happens, the part of the lung distal to it collapses.

Relations

- Anteriorly
  1. The ascending aorta
  2. The superior vena cava
  3. Right pulmonary artery

  The right bronchus enters the hilum of the lung and divides into 3 lobar bronchi, superior, middle and inferior. The bronchus to the upper lobe lies above the pulmonary A (Eparterial); the other 2, below the pulmonary artery—Hyparterial.

The Left Principal Bronchus

The left principal bronchus is longer (5 cm), narrower and more oblique than the right. This bronchus, enters the hilum of the left lung and divides into 2 lobar bronchi—upper and lower.

It is related anteriorly to the left pulmonary artery and left pulmonary veins; it is related superiorly to the arch of the aorta. Posteriorly, it is related to the descending thoracic aorta and the esophagus.

Bronchoscopy

Bronchoscopy: By introducing a bronchoscope, the interior of the larynx, trachea and bronchi can be visualized. Foreign bodies can be removed and biopsies can be taken.

Further Division of Bronchi

Each principal bronchus (primary bronchus) divides into lobar bronchi (secondary bronchi) – 3 on the right side and 2 on the left side. Each lobar bronchus divides into tertiary or segmental bronchi, one for each bronchopulmonary segment. So, there are 10 segmental bronchi on each side. The segmental bronchi divide repeatedly to form very small branches called terminal bronchioles. Still smaller branches are called respiratory bronchioles.

■ THE LUNGS (FIGS 12.13A AND B)

The lungs or pulmones are the essential organs of respiration. Their main function is to oxygenate blood. They are situated in the thorax on either side of the middle mediastinum (See CP4 in Color Plate 1).

In healthy people, who live in a clean environment, the lungs are light pink in color; but, in people living in polluted area, lungs are dark and mottled due to the accumulation of dust or carbon particles which become trapped in the phagocytes.

Flow chart 12.2: Pattern of division of pulmonary bronchus

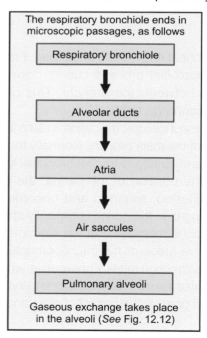

The respiratory bronchiole ends in microscopic passages, as follows

Respiratory bronchiole

↓

Alveolar ducts

↓

Atria

↓

Air saccules

↓

Pulmonary alveoli

Gaseous exchange takes place in the alveoli (*See* Fig. 12.12)

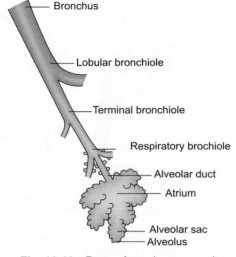

Fig. 12.12: Parts of a pulmonary unit

Each lung is conical in shape; it is contained in a pleural sac. The lungs are separated from each other by the heart and the great vessels in the middle mediastinum.

The lungs are attached to the heart and the trachea by the structures in the roots of the lungs, i.e. the pulmonary artery, the pulmonary veins and the main bronchi.

Each lung has
1. an apex, at the upper end
2. a base, resting on the diaphragm
3. three borders
   • anterior
   • posterior
   • inferior
4. 2 surfaces
   • costal
   • medial

Apex

The apex is the rounded superior end. It extends into the root of the neck, about 3 cm superior to the anterior end of the 1st rib, and medial end of the clavicle. It is covered by cervical pleura and suprapleural membrane. It is crossed by the subclavian artery.

Base

The base is the concave diaphragmatic surface, related to the dome of the diaphragm. The base of the right lung is deeper because the right dome rises to a more superior level, due to the presence of liver inferior to it.

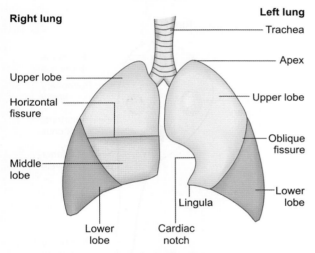

Fig. 12.13A: The lungs

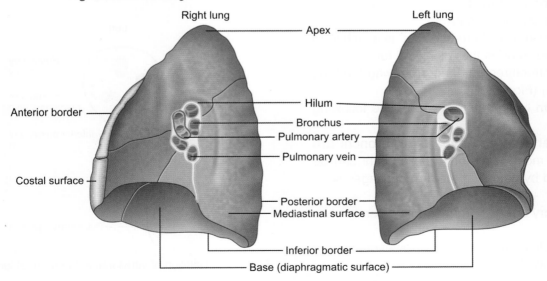

Fig. 12.13B: Lungs medial surface

Section C

The Organ Systems

Border

Anterior Border

The Anterior Border is very thin.

On the *right side,* it is vertical. The anterior border of the *left lung* shows a wide cardiac notch, which uncovers the heart and the pericardium. Below the notch is a tongue-shaped projection called *lingula.*

Posterior Border

The posterior border is thick. It extends from the C_7 to T_{10} level; related to the heads of ribs. The inferior border separates the base from the costal and medial surfaces.

Surfaces

Costal Surface

The costal surface is large and convex (the hardened lungs in an embalmed cadaver have the impression of the ribs on this surface). It is related to the pleura and the thoracic wall.

Medial Surface

The medial surface is divided into a posterior vertebral part and an anterior or mediastinal part.

The vertebral part is related to the vertebral column.

The mediastinal part is related to the mediastinal structures (heart and great vessels) and the hilum.

Mediastinal Surface

Mediastinal surface of right lung (Fig. 12.14): Structures and impressions seen are:

1. *The hilum* (Hilum is where the "root" is attached to the lung. The "root" is the "highway" for the transmission of structures entering or leaving the lung at the hilum) (Fig. 12.15).

    The hilum of right lung presents the following structures:

    a. The eparterial and hyparterial bronchi are situated most posteriorly (the bronchi can be identified by palpating the firm cartilages in its wall).

    b. Pulmonary artery in front and between the two bronchi.

    c. Upper pulmonary vein in front; lower pulmonary vein below.

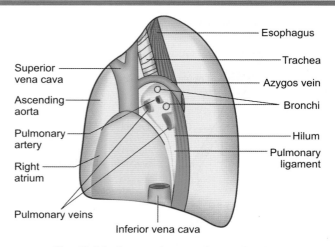

Fig. 12.14: Impressions on the mediastinal surface of the right lung

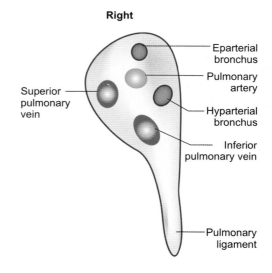

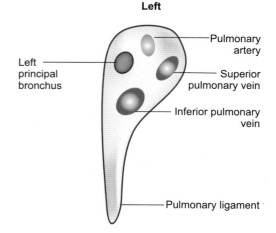

Fig. 12.15: Relation of structures at the hilum of lung

Other structures present at the hilum are the bronchial artery, hilar lymph nodes and pulmonary plexus of nerves.

2. *The cardiac impression* in front and below the hilum. This impression is produced by the right atrium (the heart is separated from the mediastinal surface of the lung by the pericardium).
3. *Superior vena cava:* It produces a vertical impression (groove) which runs up from the cardiac impression.
4. *Ascending aorta:* Area in front of the groove for superior vena cava.
5. Groove for inferior vena cava: It extends down, posteroinferior to the cardiac impression.
6. Azygos arch makes a curved groove just above the root of the lung.
7. Right surface of the trachea makes a groove just behind the groove of the superior vena cava.

The following markings can be seen on the mediastinal surface of the left lung (Fig. 12.16).
1. *The Hilum:*
   a. The bronchus, most posteriorly
   b. The pulmonary artery above the bronchus
   c. Upper pulmonary vein in front
   d. Lower pulmonary vein below.
      The bronchial artery, hilar nodes and pulmonary plexus of nerves are also seen.
2. *The cardiac impression* is seen in front and below the hilum. It is produced mainly by the left surface of left ventricle and left auricle (the pericardium separates the heart from the mediastinal surface of the lung).

3. The arch of the aorta makes an impression above the hilum and the descending aorta behind the hilum.
4. The left subclavian and left common carotid artery make an impression in the upper part.

Lobes and Fissures (Fig. 12.13)

The *right lung* is usually divided into 3 lobes.

Upper, middle and lower, by 2 fissures – the oblique fissure and horizontal fissure.

The *left lung* has only 2 lobes, upper and lower, divided by the oblique fissure.

The upper lobe of left lung has a cardiac notch and a tongue-shaped projection, the lingula.

The *oblique fissure* can be represented by a line connecting the 6th costochondral junction with a point between 3rd and 4th thoracic spines. This line passes along the 5th intercostal space.

The *horizontal fissure* is indicated by a transverse line drawn at the level of the 4th costal cartilage.

Differences between right and left lung are given in Table 12.3).

The Bronchopulmonary Segments
(Fig. 12.17 and Table 12.4)

Each primary bronchus (principal bronchus) on entering the hilum divides into branches, which, in turn, divide repeatedly.

The tertiary branches are referred to as segmental bronchi.

Similarly, the pulmonary artery also divides repeatedly and accompany the divisions of the bronchus.

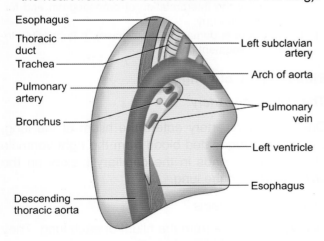

Fig. 12.16: Impressions on the mediastinal surface of left lung

Esophagus
Thoracic duct
Trachea
Pulmonary artery
Bronchus
Descending thoracic aorta
Left subclavian artery
Arch of aorta
Pulmonary vein
Left ventricle
Esophagus

Table 12.3: Differences between right and left lungs	
Right lung	*Left lung*
It has 2 fissures and 3 lobes normally	It has one fissure and 2 lobes normally
Anterior border straight	Anterior border shows a cardiac notch; there is a tongue shaped "lingula" below it
It is shorter and broader	It is longer and narrower
It is heavier—about 700 gm	Weighs about 600 gm
The arrangement at the hilum is different—refer Figure 12.18	

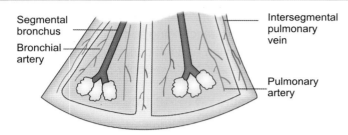

Fig. 12.17: Two adjacent bronchopulmonary segments

Table 12.4: Bronchopulmonary segments		
Lobe	Right lung	Left lung
Upper lobe	1. Apical	1. Apical
	2. Anterior	2. Anterior
	3. Posterior	3. Posterior
		4. Superior lingular
		5. Inferior lingular
Middle lobe	4. Medial	
	5. Lateral	
	6. Apical	6. Apical
	7. Medial basal	7. Medial basal
	8. Lateral basal	8. Lateral basal
	9. Anterior basal	9. Anterior basal
	10. Posterior basal	10. Posterior basal

Thus, each segmental bronchus and the accompanying division of the pulmonary artery supply a definite part of the lung and the entire unit is referred to as the bronchopulmonary segment.

The radicals of pulmonary veins are not segmental; they are intersegmental in position.

The bronchopulmonary segments are independent units; each unit is surrounded by connective tissue, which is continuous with the visceral pleura.

Bronchopulmonary Segments of Right Lung

There are 10 bronchopulmonary segments on the right side.

The upper lobe—has 3 bronchopulmonary segments
1. Apical
2. Anterior
3. Posterior.

The middle lobe has 2 bronchopulmonary segment
1. Medial
2. Lateral.

The lower lobe has 5 bronchopulmonary segments
1. Apical
2. Anterior basal
3. Medial basal
4. Posterior basal
5. Lateral basal.

Bronchopulmonary Segments of Left Lung

There are 10 bronchopulmonary segments in the left lung—5 each in upper and lower lobes.

Left upper lobe
1. Apical
2. Anterior
3. Posterior
4. Superior lingular
5. Inferior lingular.

Lower lobe
1. Apical
2. Medial basal
3. Lateral basal
4. Anterior basal
5. Posterior basal.

Applied Anatomy

1. Usually infections are confined to each bronchopulmonary segment; but tuberculosis and malignancy, breakthrough the septa and spread to the adjacent segments.
2. A knowledge of the bronchopulmonary segment is essential for the interpretation of bronchogram and for doing bronchoscopy
3. If a segment is damaged (by disease), it is possible to resect that segment only.

Blood Supply of Lungs

The Pulmonary Artery

One pulmonary artery enters the hilum of the lung, carrying deoxygenated blood from the right ventricle of the heart; it ends in the capillary plexus on the alveolar walls of the lung.

The Pulmonary Veins

Two veins emerge from the hilum of each lung. They carry oxygenated blood from the lung and empty into the left atrium.

The Bronchial Artery

This artery may arise from the thoracic aorta or one of the posterior intercostal arteries. It supplies the bronchial tree (as far as the respiratory bronchioles).

Histology (Figs 12.18A and B)

Alveoli

The alveoli are lined by a layer of simple squamous epithelium. There is a capillary network around the alveoli. The blood in the capillaries and the air in the alveoli are separated by the alveolar epithelium, the capillary endothelium and two layers of basement membranes in between them. At many places, the 2 basement membranes fuse together, thereby reducing the thickness of the barrier.

There are *various types of lining epithelial* cells in the alveoli.

1. The most numerous cells are the squamous cells – they are called Type I alveolar cells.
2. Type II alveolar cells – these cells secrete "pulmonary surfactant", which reduces surface tension and prevents collapse of the alveolus during expiration.
3. Type III alveolar cells : Exact function not known.

The total area of the alveolar surface of each lung is extensive – about 140 square meters.

Intrapulmonary Passages

1. Intrapulmonary bronchi: The structure is similar to that of the trachea.
2. Bronchioles: Cartilage is absent in the wall of bronchioles.
3. The amount of smooth muscle fibers in the wall of bronchus becomes more, as its size decreases. Spasm of these muscle fibers leads to bronchial asthma.
4. As the bronchi become smaller, the epithelium changes from simple ciliated columnar, then non-ciliated columnar and finally cuboidal.

The respiratory passages are lined by different types of cells—Goblet cells (mucous secreting), serous cells (secrete clear fluid), clara cells (surfactant- like secretion), lymphocytes and leukocytes.

Connective Tissue

The connective tissue of the lung has numerous elastic fibers, which plays an important role in providing the

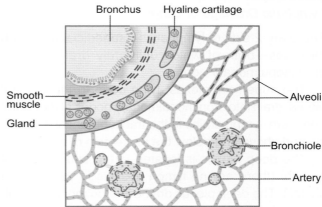

Fig. 12.18A: Histology of lung

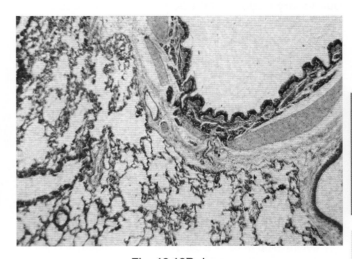

Fig. 12.18B: Lung

elastic recoil of lung tissue; it helps to expel the air from the lung during expiration.

Nerve Supply of Lungs

The lungs and the visceral pleura are innervated by the pulmonary plexuses, which are formed by the branches of vagus nerves (parasympathetic) and sympathetic trunks.

The innervation of parietal pleura is as follows: The costal pleura and the peripheral part of the diaphragmatic pleura are supplied by the intercostal nerves. The central part of the diaphragmatic pleura and the mediastinal pleura are innervated by the phrenic nerve. These nerves mediate the sensations of touch and pain.

Lymphatic Drainage of Lungs

There are 2 sets of lymphatics—superficial and deep plexuses—both of which ultimately drain into the bronchopulmonary nodes.

The superficial plexus, lying deep to the visceral pleura, drain into bronchopulmonary nodes at the hilum; from here, lymph drains to the tracheobronchial nodes, located above and below the bifurcation of trachea.

The deep plexus is located in the submucosa of bronchi (there are no lymph vessels in the walls of alveoli). The deep plexus drains into pulmonary lymph nodes of the lung, which are situated along the main bronchi; from here, lymph drains into the broncho-pulmonary nodes, then tracheobronchial nodes.

From the tracheobronchial nodes, lymph reaches the bronchomediastinal lymph trunks on the left side, this trunk may terminate in the thoracic duct.

The tracheobronchial node will be enlarged in tuberculosis, lymphomas or bronchogenic carcinoma. These nodes may be visible in the plain X-ray of the chest as opaque shadows.

Applied Anatomy

1. Fresh healthy lungs always contain some air. So, a part of the lung, when removed, will float in water. A fetal lung or lungs from a stillborn (baby who has never inspired air), which have never expanded, will sink when placed in water. This knowledge is of medicolegal importance in determining whether a dead infant was stillborn or whether it was born alive and had started to breath.
2. During the development of lung, in the fetus, the alveolar epithelium begins the secretion of surfactant at the 7th month. Hence the 7th month is considered as the viable age for the fetus.
   Lack of surfactant causes respiratory distress syndrome of the newborn. The alveoli collapse leads to a condition called Hyaline membrane disease.
3. The lung and the visceral pleura are supplied by the autonomic plexus of nerves. They are, therefore, not pain sensitive. The parietal pleura, is supplied by the intercostal or phrenic nerves. When the parietal pleura is involved, severe pain is experienced. Pain may be referred to the shoulder tip from the diaphragmatic pleura (along the phrenic and supraclavicular nerves, (C_3, C_4) or to the abdominal wall from the costal pleura (along the intercostal nerves).
4. *Postural drainage:* Secretions from a lobe has to be drained by altering the position of the patient, so that gravity will assist the process of drainage. For example, a patient with bronchiectasis (dilated bronchus) on the right side is positioned in bed on his left side, secretions from the right

Contd...

Contd...

lung and bronchi will flow toward the carina of the trachea. As this is a very sensitive area, the cough reflex is stimulated, bringing up the infected sputum. This clears the right bronchial tree.

Similarly, if the left lung and bronchi are to be drained, patient is made to lie on his right side.

5. In unconscious patients, infection from the nasal cavity or oropharynx may reach the lung, leading to pneumonia. So, the position of very ill and unconscious patients should be changed frequently to promote good aeration of lungs. Pillows are not advised in such cases.
6. Pulmonary thromboembolism is a serious condition; an embolus may reach the lung from a distant site (e.g. from the deep veins of the leg). It may partially or completely obstruct the arterial flow to the lung.
7. Bronchogenic carcinoma (cancer of a bronchus) is the most common cancer in men, the major cause being cigarette smoking and living in polluted industrial areas. This type of cancer can spread directly to the adjacent structures, along the lymphatics to the lymph nodes and along the blood-stream to distant sites like brain, bones, etc.
8. *Radiological anatomy:* In the plain X-ray chest, the lung fields are seen as translucent shadows because of the air they contain. The blood vessels, containing blood, stand out as radiating shadows from the hilum. Hilar lymph nodes may also be visible.

■ THE PLEURA (FIG. 12.19)

The pleura is a serous membrane, lined by a single layer of squamous cells (the mesothelium). There are 2 pleural sacs, one on either side of the mediastinum. Each pleural sac is invaginated from its medial side by the lung, so that it has an outer layer, called the parietal pleura and, the inner visceral layer or pulmonary pleura.

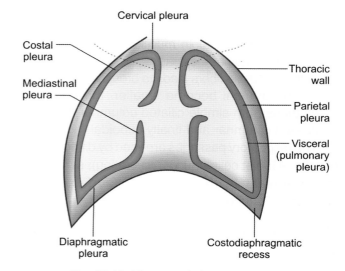

Fig. 12.19: Pleura and pleural recesses

These two layers are continuous with each other at the hilum of the lung. There is a potential space between the 2 layers, the pleural cavity, with a thin film of fluid, to prevent friction.

Visceral Layer

The visceral layer or pulmonary pleura covers the surfaces and fissures of the lung, except at the hilum. It is firmly attached to the surface of the lung; it cannot be separated from it.

Parietal Pleura

It has 4 subdivisions:
(a) The costal; (b) Diaphragmatic; (c) The mediastinal, and (d) Cervical.

The Costal Pleura

The costal pleura lines the inner side of the thoracic wall (the ribs and intercostal spaces).

The Mediastinal Pleura

The mediastinal pleura lines the mediastinal surface of the lung. It is reflected onto the root of the lung and becomes continuous with the pulmonary pleura around the hilum.

Cervical Pleura

The cervical pleura extends into the neck; it covers the apex of the lung. It is covered by suprapleural membrane. The subclavian artery is related to it.

Pulmonary Ligament

The parietal pleura surrounding the root of the lung extends downwards as a fold called the pulmonary ligament. This fold contains a thin layer of loose areolar connective tissue. It provides a space for the pulmonary vein to expand during increased venous return (during exercise).

Pleural Recesses (Figs 12.19 and 12.20)

There are 2 pleural recesses (pockets) of parietal pleura, which allow the lungs to expand during deep inspiration. These recesses are (1) the costomediastinal recess; and (2) the costodiaphragmatic recess.

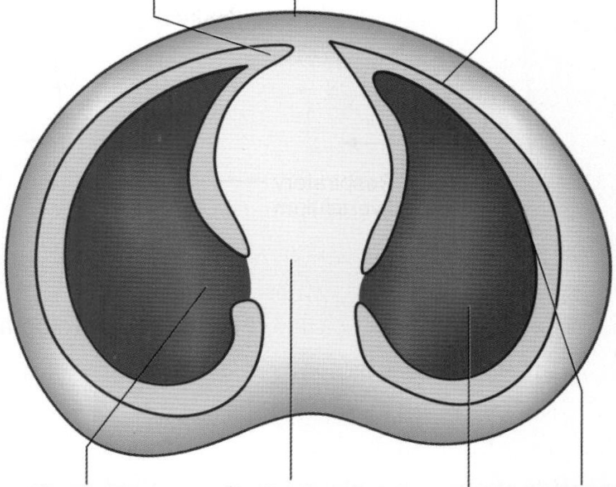

Fig. 12.20: Transverse section of thorax showing lungs and pleura

Costomediastinal Recess

The costomediastinal recess lies anteriorly, behind the sternum and costal cartilages, between the costal and mediastinal pleura.

Costodiaphragmatic Recess

The costodiaphragmatic recess lies inferiorly between the costal and diaphragmatic pleura. Vertically, it measures 5 cm; it extends from 8th to 10th ribs along the midaxillary line.

In pleural effusion, the costodiaphragmatic recess is the 1st part of pleural cavity to get filled; because it is the most dependent part of the chest.

The Nerve Supply of Pleura

The parietal layer is supplied by the intercostal nerves and the phrenic nerve. It is pain sensitive. The pulmonary pleura and the lungs are innervated by the autonomic nervous system.

Applied Anatomy

1. Inflammation of the pleura is called pleuritis or pleurisy. It may be associated with collection of fluid in the pleural cavity (pleural effusion).
2. Presence of air in the pleural cavity is called pneumothorax. Penetrating injuries of the thoracic wall can lead to this condition.
3. Presence of blood in the thoracic cavity is called hemothorax.

Contd...

Section C

The Organ Systems

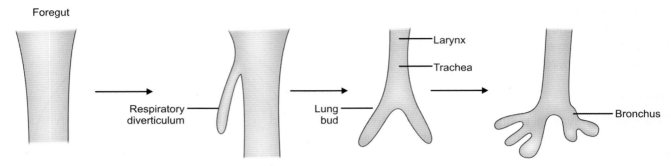

Fig. 12.21: Development of larynx, trachea and lungs

Contd...

4. Presence of pus in the pleural cavity is called empyema.
5. Aspiration of any fluid from the thoracic cavity is called paracentesis thoracis. It is usually done in the 6th intercostal space in the midaxillary line (in order to avoid injury to the liver).

Bronchogram

The bronchial tree can be visualized in X-ray after injecting radio-opaque compounds into the trachea and its branches. Only one side of the bronchial tree is usually injected at a time, because the contrast medium obstructs the air flow.

Nowadays, bronchography is replaced by CT or endoscopy (bronchoscopy).

■ DEVELOPMENT OF RESPIRATORY SYSTEM

The respiratory system develops from a median diverticulum of the foregut. Its lining epithelium is of endodermal origin. The connective tissue, cartilage and muscle are derived from splanchnopleuric mesoderm.

A groove appears in the floor of the developing pharynx - the tracheobronchial groove (Fig. 12.21). It elongates to form a diverticulum - the respiratory diverticulum. The free caudal end of the diverticulum becomes bifid; each subdivision is known as a lung bud.

The part of the diverticulum cranial to the bifurcation forms the larynx and trachea. The bronchi and lung parenchyma develop from the lung buds.

The communication between the pharynx and the respiratory diverticulum persists as the laryngeal inlet.

Alveoli are formed by expansion of the terminal parts of the bronchial tree.

The pleura and connective tissue of lung develop from mesoderm.

Within the respiratory passage, some cells become specialized for the production of surfactant. It forms a thin layer lining the alveoli. By reducing the surface tension, surfactant prevents the collapse of alveoli. By 7th month of intrauterine life, pulmonary circulation and surfactant are adequate. An infant born thereafter is viable. A deficiency of surfactant results in the failure of alveoli to remain open, resulting in respiratory distress syndrome (RDS) or hyaline membrane disease (HMD).

Glucocorticoid treatment during pregnancy accelerates fetal lung development and surfactant production. This finding has led to the administration of betamethasone to pregnant women for the prevention of RDS.

In premature babies, surfactant replacement therapy reduces the severity of RDS and neonatal mortality.

Congenital Anomalies

1. Stenosis (narrowing) or atresia (failure of canalization) of larynx.
2. Tracheo-esophageal fistula - an abnormal communication between trachea and esophagus. As a result, esophageal contents may enter the trachea or lungs.
3. Agenesis or failure of development of lung or lungs
4. Hypoplasia or under development of lungs
5. Absence of fissures and lobes of lungs

CHAPTER

13 Digestive System

■ INTRODUCTION

The purpose of digestion is to change the foodstuffs by mechanical and chemical action to simple forms, which can be easily absorbed into blood and utilized by various tissues in the body.

■ PROCESS OF DIGESTION (FLOW CHART 13.1)

The process of digestion takes place in the alimentary canal and is assisted by some accessory organs like salivary glands, liver and pancreas.

Food is processed within the body in 4 steps.
1. Ingestion
2. Digestion
3. Absorption
4. Excretion.

Ingestion

Ingestion or taking in of food and mastication (chewing) are functions performed by mouth and teeth, aided by tongue.

Pharynx and esophagus are concerned with swallowing.

Digestion

Digestion occurs in the stomach and upper part of small intestine.

Absorption

Absorption can occur from any part of alimentary canal (mainly by small intestine).

Excretion

Large intestine absorbs major quantity of water and the residue is excreted in the form of feces.

■ PARTS OF DIGESTIVE SYSTEM (FIG. 13.1)

Mouth

The mouth or oral cavity is the first part of the digestive tube. It is divided into mouth proper and a vestibule. The vestibule is the slit like space between the lips and cheek externally, and the gums and teeth internally. It communicates externally through the oral fissure. The parotid duct opens into the vestibule, on the inner surface of the cheek opposite the crown of the upper second molar tooth. Numerous mucous glands (buccal and labial) of lips and cheek open into the vestibule. The vestibule is lined by mucous

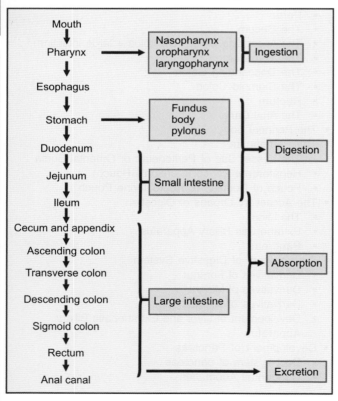

Flow chart 13.1: Process of digestion

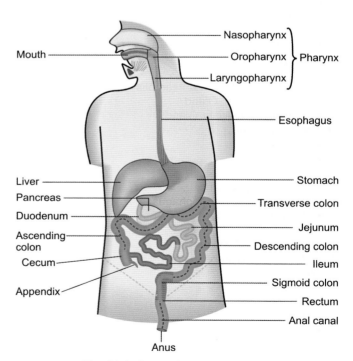

Fig. 13.1: Parts of digestive system

membrane, the lining epithelium of which is stratified squamous non-keratinizing.

Lips

Lips (labia) are fleshy folds lined externally by skin and internally by mucous membrane. Each lip is composed of skin, superficial fascia and orbicularis oris muscle, submucosa and mucous membrane.

Cheeks

Cheeks (buccae) are fleshy flaps, forming a large part of each side of the face. The cheeks are composed of skin, buccinator muscle, parotid duct, vessels and nerves, mucous glands and mucosa. In infants, a well-developed buccal pad of fat lies on the buccinator.

Oral Cavity Proper

Boundaries

- *Anterolaterally* by teeth, gums and alveolar arches of jaws.
- *Roof* is formed by hard and soft palate.
- *Floor* is formed by the tongue and sublingual region.
- *Posteriorly* the oral cavity communicates with the pharynx through the oropharyngeal isthmus, which is bounded superiorly by soft palate, inferiorly by tongue and on each side by palatoglossal arches.

Gums

Gums are the soft tissues which cover the alveolar processes of the upper and lower jaws and surrounded the necks of the teeth. The gums are made up of dense fibrous tissue, covered by stratified squamous epithelium.

Teeth

The teeth form a part of the chewing or masticatory apparatus and are fixed to the jaws. In man, teeth are replaced only once. The teeth of the first set are known as milk teeth or deciduous teeth and the second set, as permanent teeth.

The deciduous teeth are 20 in number. In each half of each jaw there are 2 incisors, 1 canine and 2 molars.

Permanent teeth are 32 in number, and consist of 2 incisors, 1 canine, 2 premolars and 3 molars in each half of each jaw.

The deciduous teeth are usually shed between 6 and 12 years of age and are replaced by permanent teeth. Eruption of permanent teeth is usually complete by 18 years of age, except for the third molars ("wisdom teeth").

Parts and Type of Teeth (Fig.13.2)

Each tooth consists of three parts, a crown, neck and root. The *crown* is the part that projects from the gum and meets (occludes) with one or more teeth in the opposite jaw. The *neck* is the part between the crown and the root . The *root* is fixed in the alveolus (socket) by a periodontal ligament. Each adult jaw consists of 4 chisel like incisors (cutters) two tearing teeth called canines (piercers) four premolars and six molars (grinders).

Structure of a Tooth (Fig.13.2)

Most of the tooth is composed of *dentin* that is covered by *enamel* over the crown and cementum over the root. The pulp cavity contains connective tissue, blood vessels and nerves. It is continuous with the perio-dontal tissue through the root canal and root foramen. The root canals transmit the nerves and vessels to and from the pulp cavity.

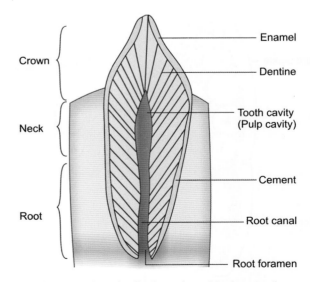

Fig. 13.2: Longitudinal section of incisor tooth

Section C

The Organ Systems

Vessels of Teeth

The branches of maxillary artery supply the upper and lower set of teeth. The veins join the pterygoid venous plexus. Lymph vessels drain mainly into submandibular nodes.

Nerves of Teeth

The teeth of upper jaw are supplied by maxillary nerve; the lower (mandibular)set of teeth are supplied by inferior alveolar nerve, a branch of mandibular nerve.

Palate (Fig. 13.3)

Palate forms the arched roof of the oral cavity and the floor of the nasal cavities. It separates the nasal cavities and nasopharynx from the oral cavity.

Parts

The palate consists of two regions— the anterior 2/3 or bony part, called the hard palate, and the mobile, posterior 1/3 or fibro-muscular part, known as the soft palate.

The Hard Palate (Fig.13.3)

The hard palate is formed by the palatine processes of maxillae and the horizontal processes of palatine

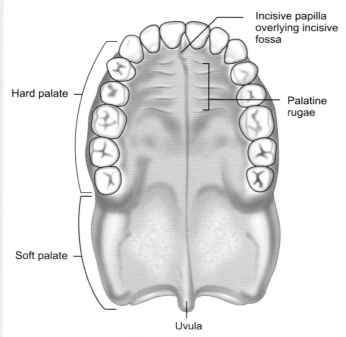

Incisive papilla overlying incisive fossa

Hard palate

Palatine rugae

Soft palate

Uvula

Fig. 13.3: Palate

bones. Anteriorly and laterally, the hard palate is bounded by alveolar processes and gums. Posterior to the central incisor teeth, there is an incisive foramen, which leads into incisive canal. This canal transmits nasopalatine nerve and sphenopalatine artery.

Medial to 3rd molar tooth is the greater palatine foramen through which greater palatine nerve and vessels emerge from the greater palatine canal. The hard palate is covered by mucous membrane that is firmly attached to the bone. There are numerous mucous-secreting palatine glands deep to the mucosa.

The Soft Palate

The soft palate is a movable, fibro-muscular fold, that is attached to the posterior edge of the hard palate. The soft palate shows a conical process called uvula (L. uva=grape) which hangs from its posterior, curved, free margin.

Laterally, the soft palate is continuous with the wall of the pharynx and is joined to the tongue and pharynx by the palatoglossal and palatopharyngeal arches, respectively. The palatine tonsil is located in the triangular interval between the palatoglossal and palatopharyngeal arches.

Muscles of soft palate The five muscles of soft palate pass from the base of the skull to the palate . They are levator palati, tensor palati, palatoglossus, palato-pharyngeus and musculus uvulae. The expanded tendon of tensor palati muscle forms the palatine aponeurosis which strengthens the soft palate.

Nerves of the Palate

- *Sensory*—Greater and lesser palatine nerves
- *Motor*—All the muscles of palate are innervated by pharyngeal plexus formed by cranial part of accessory and vagus nerves; the tensor palati is an exception, which is supplied by mandibular nerve.

Functions of Soft Palate

The soft palate controls and regulates the size of the pharyngeal isthmus during speech, swallowing, coughing, sneezing, etc. It prevents the entry of food from oropharynx into the nasopharynx.

Cleft palate: Palate develops from 3 palatal processes which unite along a y-shaped line. Nonunion of these processes result in cleft palate.

Paralysis of muscles of palate leads to nasal regurgitation and gives a "nasal sound" to the voice.

Tongue

The tongue (lingua, glossa) is a highly mobile muscular organ that can vary greatly in shape. It is situated partly in the mouth and partly in the oropharynx.

Parts

It has a root, body, tip, a dorsal surface, a ventral surface and two lateral borders.

Functions

Tongue is concerned with
1. Mastication (chewing)
2. Taste
3. Deglutition (swallowing)
4. Articulation/speech
5. Oral cleansing

Tongue is mainly composed of skeletal muscles, covered by stratified squamous epithelium.

The Dorsal Surface

The dorsal surface (Fig. 13.4) is covered by mucous membrane. A V-shaped groove, the sulcus terminalis, divides this surface into anterior 2/3 or palatine part and posterior 1/3 or pharyngeal part. At the apex of the V is a small foramen cecum. A shallow median groove extends from the tip to the foramen cecum, dividing the anterior 2/3 into right and left halves.

The mucous membrane of anterior 2/3 of dorsum of tongue is rough due to the presence of microscopic projections called papillae.

There are 3 types of papillae (Fig. 13.5). They are filiform, fungiform and circumvallate.

Filiform papillae are conical, most numerous, occur throughout the anterior 2/3.

Fungiform are red, mushroom shaped, scattered, provided with taste buds.

Circumvallate or vallate are large, drum-shaped papillae, easily seen with naked eye. They are 8-12

in number, arranged anterior to sulcus terminalis; each papilla is surrounded by a trench. Taste buds are present in the circumvallate papillae.

Mucous membrane of posterior 1/3 of tongue shows collections of lymphoid tissue, called lingual tonsil.

The Ventral Surface

The ventral surface of tongue is covered by smooth mucous membrane, devoid of papillae. There is a median mucous fold, the frenulum linguae, stretching from this surface to the floor of the mouth. On either side of the frenulum, the deep lingual vein can be seen.

Muscles of Tongue (Fig. 13.6A and B)

Muscles of the tongue are grouped into two—intrinsic and extrinsic.

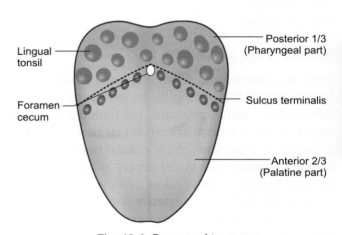

Fig. 13.4: Dorsum of tongue

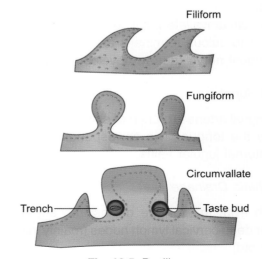

Fig. 13.5: Papillae

Section C   *The Organ Systems*

A median fibrous septum divides the tongue into right and left halves. Each half has 4 intrinsic muscles (Fig. 13.6A):
1. Superior longitudinal muscle
2. Inferior longitudinal muscle
3. Transverse muscle
4. Vertical muscle.

The extrinsic muscles of the tongue (Fig. 13.6B) are:
1. Genioglossus
2. Hyoglossus
3. Styloglossus
4. Palatoglossus.

Motor Nerve Supply (Fig. 13.7)

All the intrinsic and extrinsic muscles of tongue, except palatoglossus are supplied by the hypoglossal nerve. The palatoglossus is supplied by cranial accessory through vagus.

Sensory Nerve Supply (Fig.13.7)

Tongue is divided into anterior 2/3 and posterior 1/3. In the anterior 2/3 of tongue, general sensory (pain, temperature, touch, pressure, etc.) supply is by the lingual nerve, a branch of mandibular nerve (which is a branch of trigeminal, the V cranial nerve). Special sensory (gustatory or taste) supply is by the chorda tympani branch of facial nerve (VII cranial nerve).

Both the general and special sensations from the posterior 1/3 of tongue are carried by the glosso-pharyngeal nerve (IX cranial). The posterior most part is supplied by internal laryngeal nerve, a branch of vagus.

The circumvallate papillae, even though situated anterior to sulcus terminalis are supplied by gloss-pharyngeal nerve (due to developmental reasons).

Blood Supply

The lingual arteries (branch of external carotid artery) supply the tongue. The corresponding veins drain into internal jugular veins.

Lymphatic Drainage

Lymph from the tongue drains into superior and inferior deep cervical lymph nodes (mainly into jugulo-omohyoid).

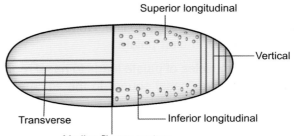

Coronal section of tongue (diagrammatic) to show the arrangement of intrinsic muscles

Fig. 13.6A: Intrinsic muscles of tongue

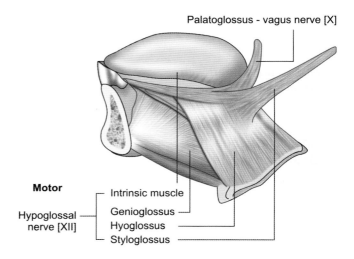

Fig. 13.6B: Extrinsic muscles of tongue

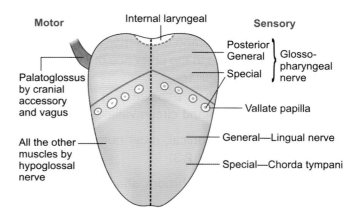

Fig. 13.7: Nerve supply

Histology (Fig. 13.8)

The tongue is covered by a mucous membrane. It has a stratified squamous non-keratinizing epithelium and lamina propria (the connective tissue layer.) There are 3 different types of papillae—filiform, fungiform and circumvallate. The fungiform and circumvallate papillae have taste buds.

Taste Buds

They are small, barrel-shaped structures, arranged along the sides of the grooves that surround the vallate or fungiform papillae.

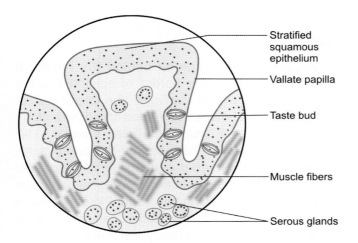

Fig. 13.8: Histology of tongue

Stratified squamous epithelium

Vallate papilla

Taste bud

Muscle fibers

Serous glands

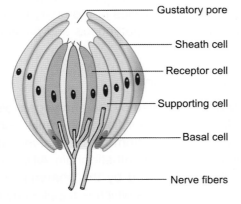

Gustatory pore

Sheath cell

Receptor cell

Supporting cell

Basal cell

Nerve fibers

Fig. 13.9: Taste bud

A taste bud (Fig. 13.9) has a layered appearance like an onion under the microscope. There are 2 types of cells—sustentacular(supporting) cells and neuro-epithelial taste cells. There is a pit, called inner taste pore, near the surface of the bud. The neuro-epithelial cells are intermingled with sustentacular cells. Their free ends, lying near the taste pore, have hairs.

Any substance to be tasted gets dissolved in saliva and passes through the taste pore into the pit of the taste bud, where, it stimulates the hairs of neuro-epithelial cells; these cells are richly innervated. So, a nerve impulse is set up.

Deep to the mucous membrane are interlacing striated (skeletal) muscle fibers. In between the muscle fibers are connective tissue fibers, adipose tissue and serous acini; there are large number of blood vessels and nerves also.

Applied Anatomy

1. *Congenital anomalies*
   a. Macroglossia (large tongue)
   b. Microglossia (small tongue)
   c. *Ankyloglossia or tongue tie*—the frenulum linguae extends up to the tip of the tongue, interfering with its protrusion. It may require surgical correction.
2. When quick absorption of a drug is required, e.g. nitroglycerin, a vasodilator used in angina pectoris, they are placed under the tongue where they dissolve and enter the deep lingual veins in less than a minute.
3. In unconscious patients, the tongue may fall back and obstruct the air passage. So, the patient can be made to lie on one side with head down.
4. Carcinoma of the tongue is common. Surgical removal of affected side is known as hemiglossectomy.
5. Injury to hypoglossal nerve leads to paralysis of muscles on the side of the lesion. When the patient is asked to protrude the tongue, the tongue deviates to the side of the lesion.
6. *Referred pain* – In cancer of the tongue, the lingual nerve may be involved. In such cases, pain is referred to the ear (pain radiates along mandibular nerve and its auriculotemporal branch).

■ PHARYNX (FIG. 13.1)

The pharynx is a wide muscular tube, situated behind the nose, mouth and larynx. It is about 13 cm long. Its upper part is widest (3.5 cm) and lower part is the narrowest part of gastrointestinal tract (except the appendix).

Parts of Pharynx

The cavity of pharynx is divided into:
1. The nasal part—Nasopharynx
2. The oral part—oropharynx, and
3. The laryngeal part—laryngopharynx

Nasopharynx

This part is situated behind the nose. It resembles the nose structurally and functionally. It is respiratory in function; it is lined by ciliated columnar epithelium. Anteriorly, it communicates with the nasal cavities, through the posterior nasal apertures.

Inferiorly, it becomes continuous with the oropharynx.

The lateral wall of nasopharynx shows the pharyngeal opening of the auditory tube or Eustachian tube. This opening is bounded by a tubal elevation. Behind this, there is a narrow vertical slit called the pharyngeal recess or fossa of Rosenmuller.

Near the junction of roof and posterior wall of nasopharynx, there is a collection of lymphoid tissue called the pharyngeal or nasopharyngeal tonsil. It is better developed in children. A pathologically enlarged pharyngeal tonsil is called adenoids.

Another collection of lymphoid tissue is present in the nasopharynx, behind the opening of auditory tube—it is called tubal tonsil.

Oropharynx

It is the middle part of the pharynx, situated behind the oral cavity. Superiorly it communicates with oral cavity through the oropharyngeal isthmus (isthmus of fauces). Inferiorly, it opens into laryngopharynx. The lateral wall of oropharynx presents the palatine tonsil, lying between the palatoglossal and palatopharyngeal arches.

Waldeyer's Ring (Fig. 13.10)

This is a "ring" of lymphoid tissue in relation to the oropharyngeal isthmus. The most important lymphoid tissue are the right and left palatine tonsils ("the tonsils"). Other aggregations forming the part of Waldeyer's ring are tubal tonsils, pharyngeal tonsil and lingual tonsil.

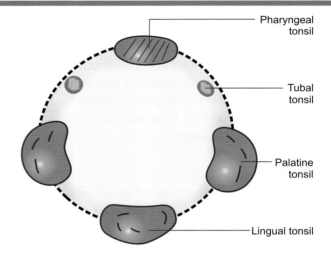

Fig. 13.10: Waldeyer's ring

Laryngopharynx

Laryngopharynx is situated behind the larynx. It extends from the upper border of epiglottis to the lower border of cricoid cartilage. The lateral wall of laryngopharynx shows a depression called the piriform fossa, one on each side of the inlet of larynx. This shallow depression is bounded by the aryepiglottic fold medially; by thyroid cartilage and thyrohyoid membrane laterally. The internal laryngeal nerve lies deep to the mucosa lining this fossa. Piriform fossa is a common site where foreign bodies can get arrested. While removing the foreign body, the internal laryngeal nerve may get damaged, leading to anesthesia in the supraglottic part of larynx.

Muscles of Pharynx

There are 3 pairs of constrictors—superior, middle and inferior. Other muscles of the pharynx are stylopharyngeus, salpingopharyngeus and palatopharyngeus.

Nerve Supply of Pharynx

The pharynx is supplied by the pharyngeal plexus of nerves, formed by motor fibers of cranial accessory nerve, sensory fibers from glossopharyngeal nerve and branches from sympathetic chain. All the muscles of pharynx are supplied by the pharyngeal plexus except stylopharyngeus which is supplied by glossopharyngeal nerve.

"Adenoids" or enlarged pharyngeal tonsil may cause difficulty in breathing in children. Surgical removal of adenoids is called adenoidectomy. Inflammation of mucosa of pharynx is "pharyngitis". Infection from nose can easily spread to the middle ear through the auditory tube.

Salivary Glands

The secretions of these glands help to keep the mouth moist, and provide a protective and lubricant coat of mucus.

The salivary glands can be classified into major salivary glands and minor salivary glands. The major salivary glands discharge their secretions into the oral cavity through a duct whereas, the minor glands open directly into the oral cavity.

Major Salivary Glands

Major salivary glands are (1) parotid, (2) submandibular and (3) sublingual.

Minor Salivary Glands

Minor salivary glands are labial glands (lips), buccal glands (cheeks), lingual glands (tongue) and palatine glands (palate).

Parotid Gland (Para = near; otis = ear)

Parotid glands are paired glands. They are the largest among the salivary glands. This gland is wedged between mandible and sternocleidomastoid muscle. It occupies the side of the face anterior and inferior to the auricle. It is an irregular, lobulated, yellowish mass, covered by a fibrous capsule. The parotid duct, or Stenson's duct passes horizontally from the anterior edge of the gland. This duct is 5 cm long and 5 mm in diameter. At the anterior border of masseter, this duct turns medially and pierces the buccinator muscle and buccal pad of fat to enter the vestibule opposite the crown of upper second molar tooth.

Relations (Fig. 13.11) within the Gland

1. The external carotid artery divides into its two terminal branches.
2. Retromandibular vein is formed; At the lower border of the gland, this vein divides into anterior and posterior divisions.

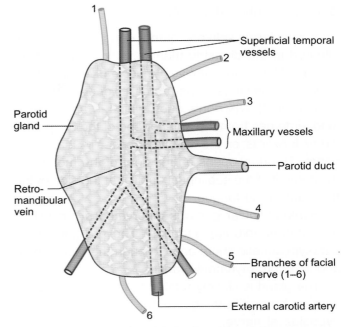

Superficial temporal vessels

Maxillary vessels

Parotid duct

Parotid gland

Retro-mandibular vein

Branches of facial nerve (1–6)

External carotid artery

Fig. 13.11: Parotid gland and its relations

3. Facial nerve enters deep to it and divides into 5 terminal branches.
4. Parotid lymph nodes.

The parotid gland is supplied by branches of external carotid artery. It is innervated by sympathetic and parasympathetic fibers. The postganglionic parasympathetic fibers reach it via auriculotemporal nerve from the otic ganglion. Stimulation of parasympathetic fibers results in copious, watery secretion; whereas sympathetic stimulation, as in fright, decreases the secretion and the mouth turns dry. Sensory nerve fibers pass to the gland through the great auricular and auriculotemporal nerves.

1. Mumps is an infectious disease caused by mumps virus, which affects the parotid gland.
2. Parotid swellings are very painful due to the unyielding nature of the parotid fascia. The pain radiates to the ear along the great auricular nerve.
3. Parotidectomy is the surgical removal of parotid gland. The facial nerve is preserved by removing the gland in two parts – superficial and deep – separately.
4. A tumor of parotid gland may infiltrate the facial nerve or its branches resulting in paralysis of muscles of facial expression, on that side.

Submandibular Gland (Fig. 13.12)

This is a large salivary gland, roughly J-shaped, situated in the digastric triangle, along the body of the mandible. The mylohyoid muscle divides it into a larger superficial part and a smaller deep part.

The duct(Wharton's duct) which is about 5 cm long, emerges from the anterior end of the deep part. It runs forwards; on the way, it is crossed by the lingual nerve. The duct opens on the floor of the mouth, on the summit of the sublingual papilla, at the side of the frenulum of the tongue.

Submandibular gland is innervated by branches from submandibular ganglion. The pre-ganglionic parasympathetic fibers pass through the facial nerve and its chorda tympani branch.

The gland is closely related to facial artery, facial vein, submandibular ganglion, lingual nerve and hypoglossal nerve.

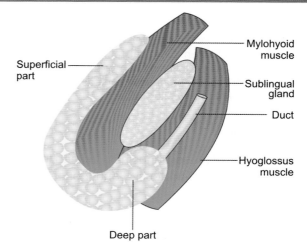

Fig. 13.12: Submandibular gland and its relations

Applied Anatomy

1. *Acute infections:* May be viral or bacterial. Mumps is the usual viral infection affecting all the salivary glands.
2. *Salivary calculi:* Submandibular calculi ("stones") are most common and can be demonstrated easily by plain radiography. The duct may be obstructed by the calculus. The characteristic symptom is recurrent painful swelling of the affected gland at meal times.

The calculus can be palpated by a "bidigital" examination.

Sublingual Salivary Gland (Fig. 13.13)

This is the smallest of the three major salivary glands. It weighs about 3 to 4 gm. This gland lies below the mucosa of the floor of the mouth. About 15 ducts emerge from the gland. Most of them open directly into the floor of the mouth. A few of them join the submandibular duct.

The gland receives secretomotor fibers from sub-mandibular ganglion. The pre-ganglionic parasympathetic fibers are carried by facial nerve and its chorda tympani branch.

Note: The parotid and submandibular salivary glands may be examined radiographically following injection of a contrast medium into their ducts. This special type of radiograph is called a sialogram.

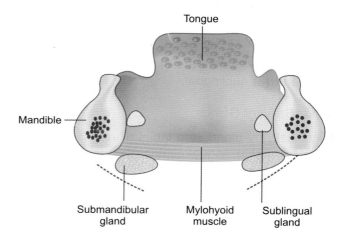

Fig. 13.13: Sublingual gland

Due to the small size of sublingual ducts, injection of contrast medium cannot be made into them.

■ THE ESOPHAGUS

The esophagus or gullet is a narrow part of the alimentary canal extending from the lower end of pharynx (at the level of 6th cervical vertebra or lower border of cricoid cartilage) to the cardiac orifice of stomach (at the level of 11th thoracic vertebra).

Length

Approximately 25 cm.

Diameter

About 1.5 cm.

Parts

(1) Cervical   (2) Thoracic   (3) Abdominal.

External Appearance

It is a tubular structure which remains collapsed anteroposteriorly. It shows anteroposterior curvatures which are similar to those of the vertebral column, to which it is closely applied (i.e. in the neck, it is convex forwards; in the thorax, it is convex backwards).

Constrictions of Esophagus

The esophagus presents 4 constrictions.
1. At the pharyngo-esophageal junction—15 cm away from the lower inicisors.
2. At the point where it is crossed by the arch of the aorta—21.5 cm away from the lower incisors.
3. At the point where it is crossed by the left bronchus—27.5 cm away from the incisors.
4. At the point where it passes through the diaphragm—40 cm away from the incisors.

Important Relations (See Figs 12.9 and 12.10)

Cervical Part

- Posteriorly—Vertebral column
- Laterally—Carotid sheath and its contents
- Anteriorly—Trachea.
   The recurrent laryngeal nerves ascend in the groove between trachea and esophagus.

Thoracic Part

- Posteriorly—Vertebral column
- Anteriorly—from above downwards
   1. Trachea
   2. Left bronchus
   3. Right pulmonary artery
   4. Left atrium
   5. Diaphragm.
   The esophageal plexus of nerves surround the esophagus on all sides.
- *On either side:* Mediastinal pleura of the lungs
- *Abdominal part:* This short segment of esophagus lies in a groove on the posterior surface of left lobe of liver.
- *Blood supply:* Esophagus is supplied by esophageal branches of descending thoracic aorta and esophageal branches of left gastric artery.

- *Veins* from the upper part drain into azygos venous system.
- Veins from the lower end drain into left gastric vein which drains into portal vein.
- *Important point:* In the submucous venous plexus at the lower end of esophagus, the systemic and portal veins meet. In cases of portal obstruction, due to the back flow of blood, esophageal varices are formed (i.e. the veins become dilated and tortuous). These varices may burst into the lower end of esophagus. Patient vomits blood (hematemesis).
- *Nerve supply:* Nerve supply by autonomic nervous system. Parasympathetic fibers from vagus nerve and sympathetic fibers from splanchnic nerves.

Histology (Figs 13.14A and B)

1. *Innermost mucosa:* Epithelium is stratified squamous, non-keratinizing. Beneath the epithelium is the lamina propria (connective tissue). Deep to this is the muscularis mucosa.
2. *Submucosa:* This layer contains large blood vessels and esophageal glands (mucous secreting glands).
3. *Outer muscular coat:* Made up of inner circular and outer longitudinal layers. In the upper 1/3, the muscle fibers are striated (skeletal); in the lower 1/3, smooth muscle; in the middle 1/3, a mixture of both types.
4. *Outermost fibrous coat* (adventitia). The lower end is covered by peritoneum also (Serosa).

Applied Anatomy

1. Esophagoscope is an instrument introduced into the esophagus to visualize the interior of esophagus. The sites of normal constrictions should be borne in mind while passing the esophagoscope (Difficulty may be experienced at these sites). These constrictions are the sites:
   a. Where foreign bodies get arrested
   b. Where carcinoma is usually seen
   c. Where strictures develop following swallowing of corrosive fluids such as acids.
2. In portal obstruction, esophageal varices develop at the lower end of esophagus; they may burst, leading to hematemesis.
3. Radiological anatomy: "Barium swallow" is the method of choice. The left atrial enlargement can be assessed by this method. The enlarged atrium presses on the esophagus and causes a depression in the esophageal shadow. Similarly, a malignant growth can also be detected.

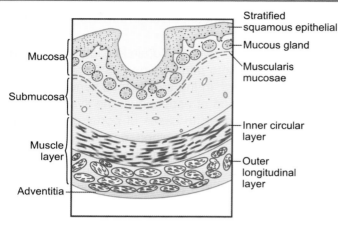

Mucosa
Submucosa
Muscle layer
Adventitia

Stratified squamous epithelial
Mucous gland
Muscularis mucosae
Inner circular layer
Outer longitudinal layer

Fig. 13.14A: Histology-esophagus

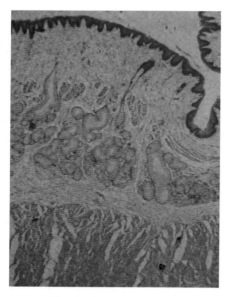

Fig. 13.14B: Esophagus

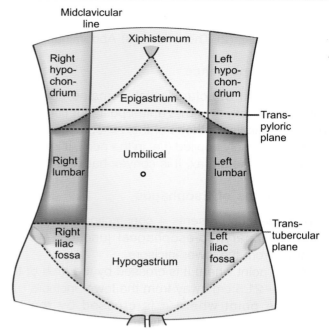

Midclavicular line
Xiphisternum
Right hypo-chon-drium
Left hypo-chon-drium
Epigastrium
Trans-pyloric plane
Umbilical
Right lumbar
Left lumbar
Right iliac fossa
Hypogastrium
Left iliac fossa
Trans-tubercular plane

Fig. 13.15: Regions of abdomen

Regions of Abdomen (Fig. 13.15)

For purposes of description, the abdominal cavity is divided into 9 regions by the two lateral vertical planes (mid inguinal or midclavicular) and 2 horizontal planes.
1. Transpyloric plane (TPP).
2. Transtubercular plane (TTP).

Transpyloric Plane of Addison (TPP)

Passes midway between the suprasternal notch and pubic symphysis, roughly a hand's breadth below xiphisternal joint. It passes through the tips of the 9th costal cartilages anteriorly and through the body of L_1 vertebra posteriorly. Structures lying at the level of transpyloric plane are:
1. Pyloric orifice of stomach
2. Gastroduodenal junction
3. Origin of superior mesenteric artery
4. Neck of pancreas
5. Upper part of hilum of right kidney
6. Lower part of hilum of left kidney
7. Fundus of gallbladder
8. Lower end of spinal cord.

Transtubercular Plane (TTP)

Transtubercular plane (TTP) passes through the iliac tubercles on the iliac crests. It passes through the body of L_5 vertebra.

The midclavicular lines or planes extend from the midpoint of clavicle to the midinguinal point (midpoint of a line joining anterior superior iliac spine and pubic symphysis).

The nine regions are:
1. Epigastric or epigastrium
2. Right hypochondrium
3. Left hypochondrium
4. Umbilical

5. Right lumbar
6. Left lumbar
7. Suprapubic or hypogastrium
8. Right iliac fossa or right inguinal region
9. Left iliac fossa or left inguinal region.

These regions are helpful clinically for describing the location of pain, site and size of a swelling, site of an incision or for locating the viscera (e.g. liver in the right hypochondrium, urinary bladder in the hypogastrium, etc).

Quadrants of the Abdomen

For simplicity, clinicians routinely divide the abdomen into 4 quadrants, right and left upper and lower quadrants. The median and transumbilical planes are used to divide the abdomen into these quadrants.

■ STOMACH

Stomach is the most dilated part of the alimentary tract. It extends from the cardiac end to the pyloric end. At its upper end, it is continuous with esophagus and at the lower end, it is continuous with the duodenum.

Position

In the supine position, the stomach occupies parts of the epigastric, umbilical and left hypochondriac regions.

Shape (Fig. 13.16)

It is j-shaped. Upper part is broader than the lower part. It has 2 ends – upper cardiac and lower pyloric; 2 surfaces – anterior and posterior; 2 borders or curvatures – the right border or lesser curvature and left border or greater curvature. The greater curvature is four to five times longer than the lesser curvature.

Capacity

A newborn infant's stomach can hold upto 30 ml of milk (It is about the size of a lemon). In adults, stomach is capable of considerable expansion. When empty, it is slightly larger than the large intestine; but it can hold upto 2 to 3 liters of food.

Functions of Stomach

1. It acts as a reservoir of food.
2. By its peristaltic activity, it makes the food particles smaller and softer and mixes the food thoroughly with gastric juice.

3. The gastric enzymes, produced by gastric glands have important role in digestion.
4. The hydrochloric acid secreted by gastric glands destroy many organisms present in food and drink.
5. Stomach secretes the intrinsic factor necessary for the absorption of vitamin B_{12}.
6. Some substances are absorbed in the stomach, e.g. some drugs, alcohol, glucose, etc.

Parts of Stomach (Fig. 13.16)

For descriptive purposes, stomach can be divided into 5 parts:
1. Cardiac part
2. Fundus
3. Body
4. Pyloric part
5. Pylorus.

Cardiac Part

The cardiac part or cardia lies around the cardiac orifice, which receives the opening of the abdominal part of esophagus (It was given this name because it lies near the diaphragm where the pericardial sac containing the heart rests – GK Kardia = heart).

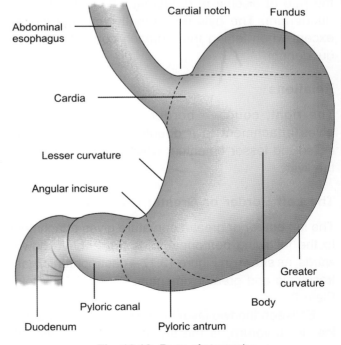

Fig. 13.16: Parts of stomach

Fundus

The fundus of the stomach, is the dilated portion, to the left and superior to the cardiac orifice. This is the most superior part of the stomach, which is related to the left dome of diaphragm. It usually contains a bubble of gas which can be seen in X-rays.

Body of the Stomach

The body of the stomach is the major portion of the stomach. It lies between the fundus and pyloric antrum.

Pyloric Part of the Stomach

There is a sharp angulation of lesser curvature of stomach called incisura angularis (angular notch) which indicates the junction of the body and pyloric part of the stomach. It has 2 parts—a wide portion, the pyloric antrum and a narrow portion, the pyloric canal. The pyloric canal is continuous distally with the pylorus.

Pylorus

The pylorus (GK. Gate keeper) is a sphincter, which guards the pyloric orifice. It has a thick wall because it contains extra smooth muscle fibers, which controls the rate of discharge of stomach contents into the duodenum. The pylorus normally remains closed except when letting the stomach contents into the duodenum.

Relations

The right, concave border or lesser curvature gives attachment to a double fold of peritoneum called the lesser omentum which stretches towards the liver.

The Left Border or Greater Curvature

The left border or greater curvature gives attachment to the following peritoneal folds from above downwards, as shown in Figure 13.17—gastrophrenic, gastrosplenic and greater omentum (See CP5 in Color Plate 2).

Between the two layers of lesser omentum, at the lesser curvature, is the anastomosis between left gastric and right gastric vessels.

Between the two layers of greater omentum at the greater curvature are the right and left gastroepiploic vessels.

Relations of Anterior Surface

1. Left lobe of liver
2. Diaphragm
3. Anterior abdominal wall

Relations of Posterior Surface

"*The stomach bed*" —— structures related to the posterior surface form the "stomach bed" (Fig. 13.18) They are:
1. Anterior surface of pancreas
2. Transverse mesocolon and transverse colon
3. Anterior surface of upper part of left kidney
4. Left suprarenal

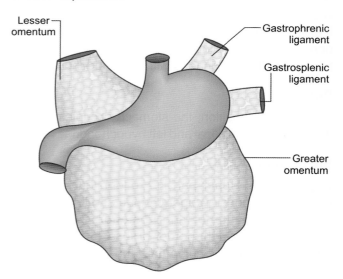

Fig. 13.17: Relations—peritoneal folds

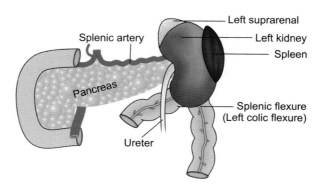

Fig. 13.18: Stomach bed—posterior relations of stomach

5. Splenic artery
6. Diaphragm
7. Gastric surface of spleen.

Blood Supply of Stomach (Fig. 13.19)

Arterial Supply

The stomach has a rich blood supply from the branches of celiac trunk.

The left gastric artery: It is a branch of the celiac trunk, which descends along the lesser curvature between the layers of lesser omentum.

Right gastric artery, a branch of hepatic artery, runs along the lesser curvature; it anastomoses with the left gastric artery and supplies the right half of the stomach.

Left gastroepiploic artery, a branch of splenic artery; it descends along the greater curvature between the layers of greater omentum.

Right gastroepiploic artery, a branch of gastroduodenal artery, which goes along the greater curvature between the two layers of greater omentum and anastomoses with the left gastroepiploic artery.

Short gastric arteries are 5-6 in number; they arise from the splenic artery and supply the fundus of stomach.

Venous Drainage

Right and left gastric veins end in the portal vein (Fig. 13.20), short gastric and left gastroepiploic veins drains into the splenic vein; right gastroepiploic vein drains into superior mesenteric vein. The prepyloric vein of Mayo connects the right gastric and right gastroepiploic veins and crosses the anterior surface of pylorus. So, it forms an important landmark for surgeons in the identification of location of pylorus.

Lymphatic Drainage (Fig.13.21)

The stomach is divided into 4 zones as shown in Figure 13.21. All the nodes finally drain into celiac nodes.

Nerve Supply

Nerve supply is by autonomic nervous system. Parasympathetic fibers come from the vagus. Sympathetic fibers arise from the celiac plexus. Parasympathetic stimulation increases the motility of stomach and increases the secretion of gastric juice rich in pepsin and hydrochloric acid. Sympathetic nerves are vasomotor and motor to pyloric sphincter and main pathway for pain impulses from the stomach.

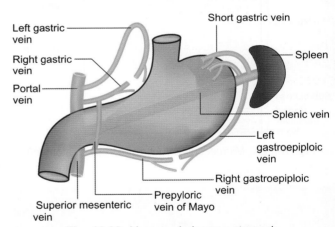

Fig. 13.20: Venous drainage—stomach

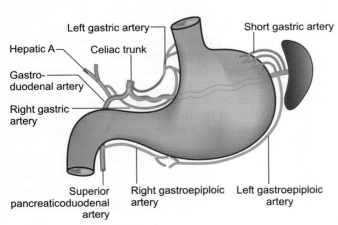

Fig. 13.19: Stomach-arterial supply

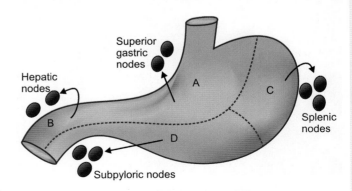

Fig. 13.21: Lymphatic drainage of stomach

Section C

The Organ Systems

Interior of Stomach

Mucosa of stomach is thrown into folds called gastric rugae, which disappear when the stomach distends. There are two longitudinal folds along the lesser curvature which form a canal called gastric canal; the swallowed liquids go along this passage to the lower end of stomach. Thus, the lesser curvature has a high risk of development of ulcers.

Histology (Figs 13.22A and B)

Stomach has 4 layers:
1. Mucosa
2. Submucosa
3. Muscular layer
4. Serosa.

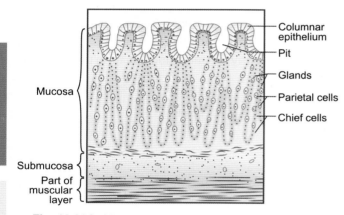

Fig. 13.22A: Microscopic structure of body of stomach

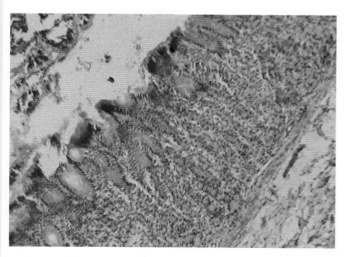

Fig. 13.22B: Stomach body

Mucosa

Mucosa is very thick; lined by simple columnar epithelium. Beneath the epithelium is the connective tissue layer—the lamina propria—which contains gastric glands. The glands open by ducts into pits, which open on the mucosa, releasing their secretions into the lumen of stomach.

The gastric glands are long and numerous in the fundus and body. They are lined by different types of cells (Figs 13.22A and B).
1. Chief cells or zymogenic cells, which secrete the enzyme pepsin.
2. Parietal or oxyntic cells are large, oval cells which secrete hydrochloric acid and intrinsic factor which is necessary for the absorption of vitamin B_{12}.
3. Mucous secreting cells.

Submucosa

Submucosa has loose areolar connective tissue and large plexuses of blood vessels.

Muscular Coat

Smooth muscle fibers are arranged in 3 layers, outer longitudinal, middle circular and inner oblique. The large food particles are broken into smaller ones and mixed thoroughly with gastric juice by the contraction of these muscles.

Serosa

Stomach is covered on all surfaces by peritoneum.

Applied Anatomy

1. Congenital hypertrophic pyloric stenosis (Fig. 13.23)— First born male infants are commonly affected. The musculature of pyloric antrum is hypertrophied. The symptoms are vomiting, visible peristalsis, constipation and loss of weight. Obstruction is relieved by Ramstedt's operation.
2. *Gastric ulcers: Acute gastric ulcers* are thought to be due to disruption of gastric mucosal barrier. They are frequently multiple. A common cause is ingestion of aspirin.
   Chronic ulcer: More common in men, around the age 45-55 years. Persons with blood group O are about 3 times more likely to develop an ulcer. Stress, anxiety, alcohol, smoking, endocrine disorders (like Zollinger-

Contd...

Contd...

Ellison syndrome) are some of the causes. Infection by *H.pylori* is said to be the main reason for the development of gastric and duodenal ulcers.

3. Gastric pain is felt in the epigastrium, because the stomach and skin of upper part of abdominal wall are supplied by T_6-T_{10} segments of spinal cord.

4. To study the interior of stomach, two methods are used
   (a) gastroscopy – direct visualization
   (b) radiological study – Barium meal

5. *Carcinoma of stomach:* The most common site is the prepyloric region. The cancer may spread directly to neighboring structures such as pancreas, liver, transverse colon or esophagus; cancer cells can spread to other parts of the body along the lymph vessels or blood vessels. The cells can sometimes pass from the stomach into the peritoneal cavity (transperitoneal implantation). In females, the cells get deposited on the ovaries giving rise to Krukenberg's tumors.

6. *Vagotomy* (Fig. 13.24): Stimulation of vagus nerve is known to cause increased secretion of HCl. So, as a method of treatment in ulcers, vagus nerve section is done.

7. *Achlorhydria* occurs when a stomach cannot produce juice with a pH of less than 7.1 even after maximal stimulation. Achlorhydria is associated with pernicious anemia because the parietal cell is responsible for the production of both HCl and intrinsic factor.

8. *Gastrectomy:* Removal of stomach, as in carcinoma, is called gastrectomy.

■ THE SMALL INTESTINE

Extent

The small intestine extends from the pylorus to the ileo-cecal junction.

It is about 6 meters long. Its structure is adapted for digestion and absorption.

Parts

It is divided into (a) an upper fixed part called the duodenum, approximately 25 cm long (10 inches) and (b) a lower (distal) mobile part. The proximal 2/5 of this long convoluted tube is continuous with the duodenum and is known as the jejunum. The distal 3/5 is called the ileum. The coils of jejunum and ileum are suspended by a fold of peritoneum called the mesentery from the posterior abdominal wall.

Important Features

1. The structure of small intestine is adapted for digestion and absorption. For absorption of digested

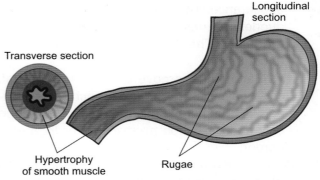

Fig. 13.23: Congenital hypertrophic pyloric stenosis

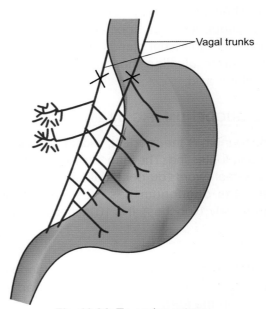

Fig. 13.24: Truncal vagotomy

food, a very large surface area is required. This is achieved by the following modifications:

a. Great length of small intestine (5½-6 m)

b. Presence of circular mucous folds or plicae circularis or valves of Kerckring. They are permanent mucous folds; (absent in the 1st part of duodenum), begin from the 2nd part of duodenum and become closely set in jejunum. Their number gradually decreases in the ileum. These folds help to increase the surface area for absorption and digestion and also help in slowing down the passage of food.

c. *Presence of villi and microvilli:* The intestinal villi are finger-like projections of mucous membrane, which give the intestinal surface a velvety appearance. They are large and numerous in duodenum; their number and size decrease gradually. They increase the surface area about 8 times.

Each villus is covered by a layer of tall columnar cells. The surface of each cell is provided with numerous microvilli (which are seen only under electron microscope) or the brush border.

2. *Intestinal glands* or crypts of Lieberkühn are simple tubular glands which open into the lumen, on the surface of mucous membrane, between the villi. They secrete enzymes and mucous.

The duodenum has special glands in the submucosa—the duodenal glands or Brunner's glands, which secrete alkaline mucous, to neutralize the hydrochloric acid.

■ THE DUODENUM (FIG. 13.25)

The duodenum is the shortest, widest and most fixed part of small intestine (The term "duodenum" is derived from a Greek word "du deka daktulos", meaning 12 fingers; the duodenum has 12-finger breadth—approximately 10 inches). It is 10 inches long (25 cm).

Extent

It extends from the pylorus to the duodenojejunal flexure. It is closely related to the head of the pancreas, in the form of letter "C".

Position

The duodenum is mostly retroperitoneal, lies above the level of umbilicus, opposite, the vertebrae L_1, L_2 and L_3.

Parts

Duodenum has 4 parts:
1. Superior or first part (5 cm)
2. Descending or second part (7.5 cm)
3. Horizontal or third part; (10 cm); and
4. Ascending or fourth part (2.5 cm).

First Part of Duodenum

It begins at the pylorus and passes upwards, backwards and to the right to meet the second part at the superior duodenal flexure.

Relations (Fig. 13.26)
1. Peritoneal relations: The first 2.5 cm, continuous with the pyloric end of stomach is attached to the lesser omentum above and greater omentum below. So, it is movable. The remaining 2.5 cm of first part is retroperitoneal and fixed.
   - *Anterior relations:* Liver and gallbladder
   - *Posterior relations:*
     1. Gastroduodenal artery
     2. Bile duct
     3. Portal vein.
   - *Inferiorly:* Head and neck of pancreas
   - Superiorly: Epiploic foramen.

Second Part of Duodenum

Second part of duodenum or the descending part of duodenum is about 7.5 cm long. It begins at the

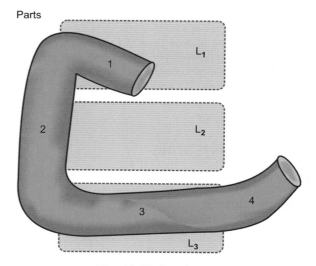

Fig. 13.25: Duodenum

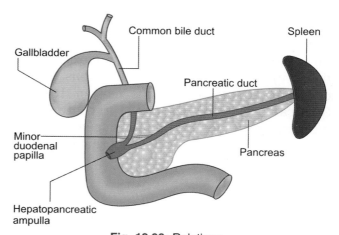

Fig. 13.26: Relations

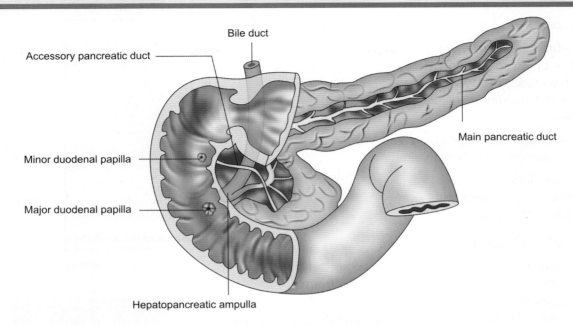

Fig. 13.27: Interior of 2nd part of duodenum

superior duodenal flexure, descends up to L_3 level and becomes continuous with the 3rd part at the inferior duodenal flexure. It is retroperitoneal.

Relations
- *Anteriorly:*
  1. Right lobe of liver
  2. Transverse colon
  3. Coils of small intestine.
- *Posteriorly:*
  1. Anterior surface of right kidney (near its hilum)
  2. Right renal vessels and pelvis of right ureter.
- *Medially:* The head of the pancreas and bile duct
- *Laterally:* Right colic flexure.

Interior of 2nd part
1. There are numerous plicae circularis.
2. About 8-10 cm distal to the pylorus, there is an elevation called major duodenal papilla. The hepatopancreatic ampulla of vater opens on the summit of the papilla. The papilla is guarded by a fold of mucous membrane referred to as the "hood of the monk"(Fig.13.27).
3. The minor duodenal papilla is present 6-8 cm distal to the pylorus; the accessory pancreatic duct, when present, opens on this papilla.

Third Part or Horizontal Part of Duodenum

This part is about 10 cm long. It begins at the inferior duodenal flexure, to the right of L_3 vertebra, runs horizontally and ends by joining the 4th part. It is also retroperitoneal and fixed.

Relations
- Anteriorly:
  1. Superior mesenteric vessels
  2. Root of the mesentery
- Posteriorly:
  1. Right ureter
  2. Right gonadal vessels
  3. Inferior vena cava
  4. Abdominal aorta
- *Superiorly:*
  1. Head of the pancreas and uncinate process
  2. Body of pancreas
- *Inferiorly:* Coils of jejunum

The Fourth or Ascending Part

It is only 2.5 cm long. It runs upwards to the left of the aorta up to L_2 level. It becomes continuous with the jejunum at the duodenojejunal flexure; here it is suspended by mesentery and is mobile.

It is related to the left kidney, its vessels and ureter. It is also related to the body of pancreas.

Arterial supply

1st part – branches from right gastric artery
2nd part – up to the major duodenal papilla, by superior pancreaticoduodenal artery.

Rest of the duodenum—inferior pancreaticoduodenal artery.

Venous drainage corresponding veins drain into superior mesenteric and portal veins.

Nerve supply: ANS
Sympathetic – T_9-T_{10} segments
Parasympathetic – Vagus.

Histology (Figs 13.28A and B) 4 Layers

Innermost *mucosa:* There are permanent mucous folds (plicae circularis). There are numerous villi which are long and spatula-shaped. Villi are lined by simple columnar epithelium. Beneath the epithelium is the lamina propria and muscularis mucosae – a thin layer of smooth muscle fibers which separates mucosa from submucosa.

There are numerous, simple tubular glands in the lamina propria – the intestinal glands or crypts of Lieberkühn. These glands are lined with Paneth cells which secrete enzymes; argentaffin cells secreting serotonin and undifferentiated cells which replace worn-out cells continuously.

Submucosa: Situated deep to mucosa; has numerous duodenal glands or Brunner's glands, secreting alkaline mucous. There are large plexus of vessels, nerves and lymphatics in this layer.

Muscular layer: Smooth muscle fibers are arranged in two layers—inner circular and outer longitudinal; control the peristaltic activity.

Outer serosa made of connective tissue and peritoneum.

Radiological Anatomy

(The alimentary canal can be demonstrated radiologically by giving barium meal (watery suspension of $BaSO_4$) and taking X-rays at regular intervals).

The barium meal passes from the stomach into the first part of duodenum, where it forms a homogenous triangular shadow called "duodenal cap" which resembles the "ace of spades". Due to the abseonce of mucous folds, it has a smooth outline. Persistent deformity of the duodenal cap is characteristic of duodenal ulcer.

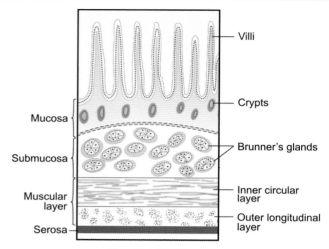

Fig. 13.28A: Histology

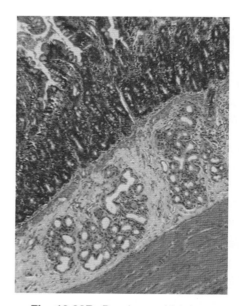

Fig. 13.28B: Duodenum histology

Applied Anatomy

1. *Duodenal ulcer:* Ulcers are more common in the first part of duodenum due to the following reasons:
   a. Stomach empties its acid chyme into the 1st part of duodenum.
   b. The arteries supplying the 1st part are said to be "end arteries" and a block of these vessels lead to necrosis of mucosa and ulcer formation.
2. A perforated duodenal ulcer may erode the gastroduodenal artery, producing severe bleeding.
3. *Superior mesenteric artery syndrome:* The superior mesenteric artery may compress the 3rd part of duodenum against the vertebral column producing duodenal stasis (See CP6 in Color Plate 2).

Table 13.1: Differences between jejunum and ileum

	Jejunum	Ileum
Cavity	Usually empty (L.jejunus = empty)	Loaded
Thickness of wall	Thicker	Thinner
Color	Red, due to increased vascularity	Paler
Position	Coils occupy the umbilical region	Hypogastric and right iliac fossa
Plicae circularis	Large and well developed, numerous	Small in the superior part and in the terminal part, absent
Presence of fat in the mesentery	Less	More
Presence of lymph nodes in the mesentery	Less	More
Arteries in the mesentery	Vasa rectae longer and less numerous	Shorter, closely placed and numerous
Villi	Long and abundant	Short, thin and less numerous
Peyer's patches	Absent	Present

Jejunum and Ileum

The coils of jejunum and ileum are suspended by the mesentery (a fan-shaped fold of peritoneum) from the posterior abdominal wall; so they are freely mobile. The jejunum begins at the duodenojejunal flexure. Together, the jejunum and ileum are about 5½ to 6 meters, long; (jejunum, forms 2/5 and ileum 3/5). Although there is no clear line of demarcation between the jejunum and ileum, there are some gross characteristics which are given in Table 13.1.

The ileum ends at the ileocecal junction in the right iliac fossa.

The Mesentery of Jejunum and Ileum

The jejunum and ileum are suspended from the posterior abdominal wall by a fan-shaped mesentery. The pleated, fan-shaped mesentery has 2-layers of peritoneum. Between these layers, there are jejunal and ileal vessels, lymphatics, nerves and extraperitoneal fatty tissue.

The root of the mesentery is about 15 cm long. It is directed obliquely; runs downwards and to the right. It extends from the left side of L_2 vertebra to the right sacroiliac joint. The root of the mesentery, crosses the following structures:
1. Horizontal part of duodenum
2. Aorta
3. Inferior vena cava
4. Psoas major muscle
5. Right ureter
6. Right gonadal vessels.

Arterial Supply of Jejunum and Ileum

They are branches of superior mesenteric artery. This artery, a branch of abdominal aorta, runs obliquely in the root of the mesentery to the right iliac fossa; it gives 15-18 branches from its left side to the jejunum and ileum.

These arteries unite to form loops or arches called arterial arcades from which vasa recta (L. straight vessels) arise. The vasa recta do not anastomose with each other. They pass to the mesenteric border of the intestine, where they pass alternatively to opposite sides. The vascularity is greater in the jejunum, but the arterial arcades are shorter and more complex in the ileum.

The corresponding veins drains into the superior mesenteric vein.

Nerve supply: By superior mesenteric plexus (an autonomic plexus) containing sympathetic fibers from T_9 and T_{10} segments and parasympathetic from vagus.

Lymphatic drainage The lymphatics in the intestinal villi, called lacteals (L. lactis = milk), empty their milk-like fluid

Section C

The Organ Systems

into a plexus of lymph vessels in the walls of jejunum and ileum – from here to mesenteric lymph nodes and ultimately into superior mesenteric lymph nodes.

Applied Anatomy

1. Volvulus: (L. volvo = to roll). Twisting of intestines and superior mesenteric vessels often occurs. This may lead to deficiency in blood supply (ischemia) to the intestines leading to necrosis, and gangrene (death of tissue).
2. An ileal diverticulum (Meckel's diverticulum) is the most common malformation of the digestive tract. It is a blind sac or finger-like pouch arising from the ante-mesenteric border of ileum about 2 feet proximal to ileocecal junction, it is approximately 2 inches long and present in 2% population. It is of clinical significance because it may get inflamed (ileal diverticulitis) and the symptoms mimic acute appendicitis (ileal diverticulum is the remnant of proximal part of embryonic yolk stalk).

Histology of Small Intestine (Figs 13.29A to C)

4 layers:

1. Innermost mucosa – epithelium – simple columnar
   - lamina propria—presence of crypts of Lieberkühn
   - Muscularis mucosae
   Mucosa has mucosal folds and villi, which increase the surface area. The villi are long finger like in jejunum; short and less numerous in proximal ileum and almost absent in the terminal ileum.
2. *Submucosa:* In the ileum, there are large collections of lymphoid tissue, projecting from the mucosa, called Peyer's patches.
3. *Muscular layer:* Inner circular and outer longitudinal layer of smooth muscles; they produce peristaltic movements.
4. Outer serosa: Made of peritoneum.

■ THE LARGE INTESTINE

The large intestine extends from the ileocecal junction to the anus (See CP7 in Color Plate 2). It is about 1.5 m long. Differences between large intestine and small intestine are given in Table 13.2.

Parts

Cecum, the ascending colon, transverse colon, descending colon, sigmoid colon, rectum and anal canal. In the angle between caecum and terminal ileum, there is a narrow diverticulum called the vermiform appendix.

The rectum and anal canal are situated in the pelvis; the remaining parts, in the abdomen.

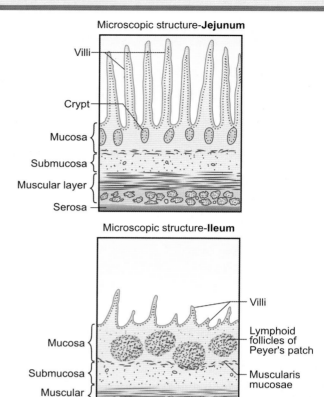

Fig. 13.29A: Histology—jejunum and ileum

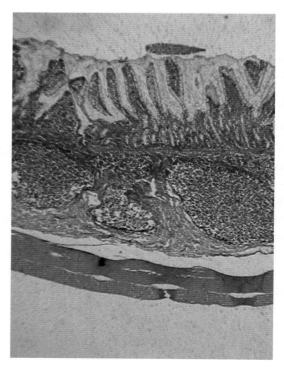

Fig. 13.29B: Ileum

Table 13.2: Differences between small and large intestines	
Large intestine	*Small intestine*
1. Large intestine has a greater diameter; the caliber decreases gradually towards the rectum	Smaller in diameter
2. Greater part of large intestine is fixed. Only the transverse colon, sigmoid colon and appendix are provided with mesocolon and are free to move	Greater part of small intestine is mobile (except duodenum)
3. The longitudinal muscle coat forms 3 ribbon – like bands called taenia coli	Taenia coli absent
4. Presence of appendices epiploicae – they are small pockets of peritoneum filled with fat and attached to the walls of large intestine	Appendices epiploicae absent
5. Haustrations are present (they are circular constrictions followed by saccular dilatations)	Haustrations absent
6. Villi are absent	Villi present

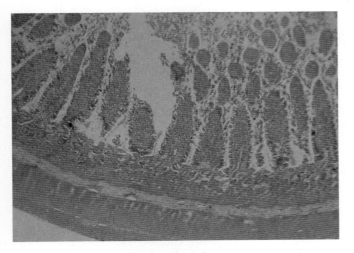

Fig. 13.29C: Jejunum

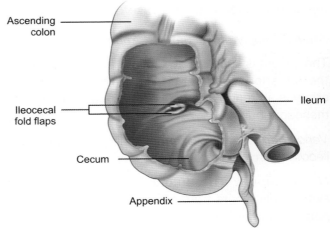

Fig. 13.30: Interior of cecum

The Cecum (Fig. 13.30)

The cecum is the first part of the large intestine. It is a broad, blind pouch (L. caecus = blind) about 5-7 cm in length. It is situated in the right iliac fossa. It is continuous with the ascending colon.

The ileum opens into the superior part of cecum. About 2.5 cm inferior to this ileocecal junction, the vermiform appendix opens into its medial aspect.

The cecum is almost completely covered by peritoneum and slightly movable. There are three peritoneal recesses (pockets) at the ileocecal junction—superior ileocecal, inferior ileocecal and retrocecal recesses. The retrocecal recess lies posterior to the cecum and in most people the appendix lies in it (See CP8 in Color Plate 2).

The ileum enters the cecum obliquely and partly invaginates into it. The ileocecal opening is seen as a transverse slit in the medial and upper wall of cecum. The upper and lower lips of the opening are prominent and are continuous with two ridges in the mucosa called the frenula of ileocecal valve. When the cecum is distended, the frenula are stretched and the lips of the opening close tightly. It is thought to prevent regurgitation to certain extent and to delay the passage of ileal contents into the large intestine.

The Vermiform Appendix
(L : vermis = worm; forma = form)

This narrow, worm – like, blind tube joins the cecum about 2.5 cm inferior to the ileocecal junction.

Appendix is longer and narrower children. The appendix has its own mesentery called mesoappendix, which suspends it from the mesentery of terminal ileum (Length = 3-5 cm).

Position (Fig. 13.31): The position of the body of appendix is variable. Usually it is retrocecal (posterior to the cecum) or pelvic (it hangs over the pelvic brim into the pelvis minor).

The base of the appendix is fairly constant in position. It lies deep to the McBurney's point—the junction of lateral 1/3 and medial 2/3 of a line connecting the anterior superior iliac spine and the umbilicus (Fig. 13.32).

The three taenia coli of the cecum converge at the base of the appendix and form a complete outer longitudinal coat for the appendix (The taenia coli help the surgeons to locate the base of the appendix).

Arterial Supply of Cecum and Appendix

The cecum is supplied by ileocolic artery, a branch of the superior mesenteric artery. The appendicular artery is a branch of ileocolic artery, which enters the mesoappendix.

Venous drainage: The corresponding veins drain into ileocolic vein, which drains into superior mesenteric vein.

Nerve supply: The autonomic nervous system; sympathetic fibers from superior mesenteric plexus, parasympathetic from vagus.

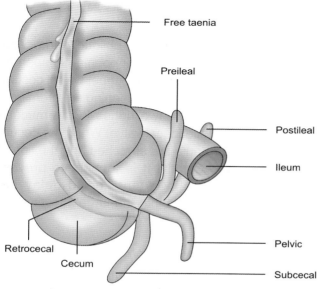

Fig. 13.31: Various positions of appendix

Histology of Large Intestine (Figs 13.33A and B)

4 layers – Mucosa, submucosa, muscular layer and serosa.

Mucosa: Innermost layer: epithelium has long columnar cells and plenty of *Goblet* cells which secrete mucous. Deep to the epithelium is the lamina propria

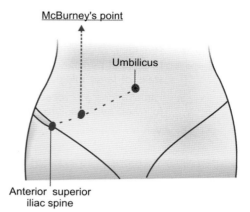

Fig. 13.32: McBurney's point

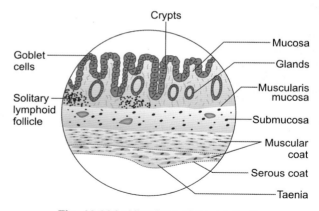

Fig. 13.33A: Histology of large intestine

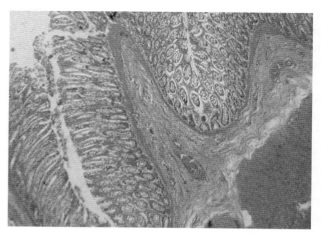

Fig. 13.33B: Histology of large intestine

with mucous glands and connective tissue. There is a thin layer of smooth muscle fibers—muscularis mucosa, which separates the mucosa from submucosa.

Submucosa: Consists of loose connective tissue, blood vessels, nerves and lymphatics.

The muscular layer: Smooth muscle fibers are arranged in two layers—inner circular and outer longitudinal.

Serosa: Consists of connective tissue fibers and peritoneum.

Appendix (Figs 13.34A and B) has similar layers; it has a stellate (star-shaped) lumen and dense collections of lymphoid tissue in the lamina propria of the mucosa and submucosa.

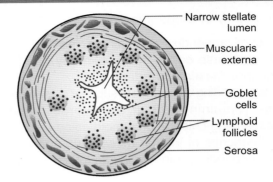

Fig. 13.34A: Appendix

Note: 1. No villi
        2. Numerous Goblet cells
        3. Solitary lymphoid follicle

Applied Anatomy—Vermiform Appendix

Inflammation of the appendix is called appendicitis. It is usually caused by obstruction of the appendix by fecal material. As a result, the appendix cannot drain the secretions; so it swells, obstructing its blood supply.

The pain of acute appendicitis is usually felt in the periumbilical region in the initial stage. This is because the appendix and skin of umbilical region are supplied by T_{10} segment. Later, the pain localizes in the right iliac fossa, due to the involvement of parietal peritoneum. Pressure over McBurney's point elicits maximum tenderness. Appendicectomy is the surgical removal of appendix.

Rupture of an inflamed appendix results in infection of part of or all of the peritoneum. This condition is called general peritonitis.

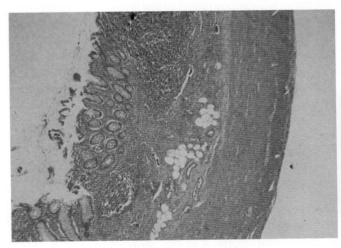

Fig. 13.34B: Histology of appendix

The Ascending Colon (GK kolos = large intestine)

It ascends on the right side of abdominal cavity from the cecum to the right lobe of liver where it turns to the left as the right colic flexure or hepatic flexure. Ascending colon is retroperitoneal. It is about 12-20 cm long.

The Transverse Colon

This part of the large intestine, actually hangs down as a loop. It is about 45 cm long. It is the largest and most mobile part of the large intestine. It crosses the abdomen from the right colic flexure to the left colic flexure. The transverse colon has a mesentery, the transverse mesocolon; it is double layered fold of peritoneum that suspends the transverse colon from the posterior abdominal wall.

The Descending Colon

Descends from the left colic flexure into the left iliac fossa, where it is continuous with the sigmoid colon. It is about 25-30 cm long. The diameter of descending colon is lesser than that of the ascending colon.

The ascending colon, transverse and descending colon form a frame for the small intestine.

The Sigmoid Colon

The sigmoid colon forms an S-shaped loop, usually 40 cm long; (it is called "sigmoid" because of its resemblance to the Greek letter sigma-S). Sigmoid colon is the part of large intestine situated between the descending colon and rectum. It has a long inverted V-shaped mesentery, called sigmoid mesocolon; so it is freely mobile. Feces are stored in the sigmoid colon until just before defecation.

Blood Supply of Large Intestine

1. Cecum, appendix, ascending colon and right 2/3 of transverse colon are supplied by branches of superior mesenteric artery.
2. Corresponding veins drain to superior mesenteric vein.
3. The remaining 1/3 of transverse colon, the descending colon and sigmoid colon are supplied by branches of inferior mesenteric artery.
4. Corresponding veins drain into inferior mesenteric vein.

Rectum

The rectum is the fixed terminal part of the large intestine. It begins anterior to the level of the third sacral vertebra; it follows the curve of the sacrum and coccyx and ends 3-4 cm anteroinferior to the tip of the coccyx,where it becomes continuous with the anal canal.

The U-shaped puborectalis muscle forms a sling at the junction of the rectum and anal canal producing an angle (90°) called anorectal angle.

The terminal part of the rectum has a dilatation known as the ampulla; the feces is held here just before defecation.

The Shape and Flexures of Rectum

The rectum (L.Rectus = Straight) was originally named in monkeys (by early anatomists) in which it is straight. In the human body, it is not straight. It has 3 sharp curves as it follows the sacrococcygeal curve. There is a sharp bend at the anorectal junction, the anorectal angle.

The rectum is S-shaped in the coronal plane. At each of the concavities, most of the wall of the rectum form transverse rectal folds, which partly close the lumen of rectum (Houston's valves).

Relations

Peritoneal relations (Fig. 14.13) Upper 1/3 of rectum is covered on the front and sides by peritoneum. The middle 1/3 is covered only on the front. The lower 1/3 is devoid of peritoneum.

In males, peritoneum is reflected from the rectum to the posterior of bladder; a rectovesical pouch is formed here.

In females, peritoneum is reflected from the rectum to the posterior part of the fornix of vagina and uterus forming the rectouterine pouch (pouch of Douglas).

Visceral relations

1. *Anterior relations—Males:* Upper 2/3 of rectum is related anteriorly to the rectovesical pouch. The lower 1/3 is related to the base of the urinary bladder, the terminal part of ureters, the seminal vesicles, vasa deferentia and the prostate gland.
2. *Anterior relations—in females:* Upper 2/3 of rectum is related to the rectouterine pouch. Lower 1/3 is related to the lower part of vagina.
3. *Posterior relations* are same in males and females. Rectum is related posteriorly to lower 3 pieces of sacrum, coccyx, piriformis, levator ani, coccygeus and vessels (such as median sacral and lateral sacral).

Arterial supply: The rectum is supplied by:

1. *Superior rectal artery:* This is the chief artery of rectum. It is the continuation of inferior mesenteric artery. It divides into right and left branches which pierce the muscular coats and run up to the anal valves; where they form loops.
2. *Middle rectal arteries: Branches of internal iliac artery.*
3. *Median sacral artery,* a branch of abdominal aorta.

Venous drainage: The veins form an internal rectal venous plexus and drain into superior rectal vein which continues upwards as the inferior mesenteric vein.

The middle rectal veins drain into internal iliac veins.

Nerve supply by autonomic nervous system.

Sympathetic L_1, L_2

Parasympathetic S2, S3, S4.

Applied Anatomy

Digital examination Per Rectum (PR) In a normal person, the following structures can be palpated by a finger (gloved and smeared with local anesthetic gel) passed per rectum:

a. In Males; the prostate, seminal vesicles and vasa deferentia.
b. In females; cervix, posterior fornix of vagina, occasionally the ovaries.
c. In both sexes – Sacrum, coccyx, ischial spines and on either side, the ischiorectal fossae.

Proctoscopy The interior of rectum and anal canal can be visualized directly using an instrument called proctoscope.

Prolapse of rectum Incomplete or mucosal prolapse and complete prolapse can occur through the anus. Predisposing factors are violent straining, laxity of pelvic floor and inadequate fixation of rectum.

The Anal Canal

It is the terminal part of large intestine. It is about 3.8–4 cm long. It extends from the anorectal junction to the anus. The anus is the surface opening of the anal canal, situated about 4 cm in front of the tip of the coccyx, in the cleft between the two buttocks. The surrounding skin is pigmented, thrown into folds and rich in apocrine glands.

Interior of Anal Canal (Fig. 13.35)

The interior can be divided into 3 parts—upper 15 mm; middle 15 mm and lower 8-10 mm.

Upper 15 mm: It is lined by columnar epithelium. The mucous membrane shows 6-10 vertical folds called anal columns. The lower ends of the anal columns are united with each other by short, transverse folds of mucous membrane called anal valves. Above each anal valve is a depression or a pocket called anal sinus.

The anal valves together form a continuous transverse line called pectinate line.

Middle 15 mm: Mucosa is lined with stratified squamous epithelium. The mucosa is bluish in color due to the presence of dense venous plexus lying deep to it. No anal columns are seen in this zone. The lower limit of this zone is whitish in color and is called the white line of Hilton.

Lower part (8-10 mm): It is lined with true skin containing sweat and sebaceous glands.

At the anus, the moist, hairless mucosa of the anal canal becomes dry, hairy skin.

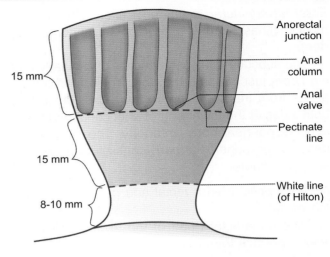

Fig. 13.35: Interior of anal canal

The anal canal, superior to the pectinate line differs from the part inferior to it in its arterial supply, innervation, venous and lymphatic drainage. This is due to their different embryological origins (Table 13.3).

Anal Sphincters (Fig. 13.36)

1. Internal sphincter
2. External sphincter.

Internal sphincter is formed by the thickened lower end of the circular layer of smooth muscle coat of anal canal. It is involuntary and supplied by autonomic nervous system. It surrounds the upper 30 mm of anal canal.

Table 13.3: Anal canal—differences above and below the pectinate line		
	Above	*Below*
Epithelium	Columnar	Stratified squamous
Nerve supply	Autonomic, hence, not sensitive to pain	Spinal nerve (inferior rectal), pain sensitive
Lymphatics	Drain into pelvic nodes	Drain into superficial inguinal nodes
Hemorrhoids	Internal hemorrhoids	External hemorrhoids
Blood supply	Superior rectal vessels (portal system)	Inferior rectal vessels (systemic system)
Development	From hindgut (Endodermal)	From proctoderm (ectodermal)
Sphincter	Controlled by internal sphincter	Controlled by external sphincter

External sphincter is made of skeletal (voluntary) muscle. It has 3 parts—subcutaneous, superficial and deep.

The subcutaneous part consists of circularly running fibers just beneath the skin. The superficial part arises from the tip of the coccyx, runs forwards on both sides of the lower end of anal canal and inserts into the perineal body anteriorly. The deep part is deep to the superficial part, consisting of circularly arranged fibers.

Nerve supply: Inferior rectal and perineal branch of 4th sacral.

(The puborectalis part of the levator ani muscle also surrounds the anal canal) (see Fig. 9.8).

Portacaval (Portosystemic) Anastomosis

There is a rich submucosal venous plexus in the rectum and anal canal, under the anal columns. Above the pectinate line, they end in the superior rectal vein and thus reach the portal system.

Below the pectinate line, the submucosal plexus drain into the inferior rectal veins and thus into the caval (inferior vena cava, a systemic vein) system.

Thus, there is a connection between the portal and systemic veins in the submucous plexus. In portal obstruction, these veins may become dilated and tortuous (varicose veins) and bulge – these are called piles or hemorrhoids.

Internal haemorrhoids are caused by the varicosities of tributaries of superior rectal vein covered with mucous membrane. The normal sites of these piles are 3,7 and 11 O'clock positions when seeing the patient in lithotomy position (Fig. 13.37). Internal hemorrhoids are associated with cirrhosis of liver, pregnancy and chronic constipation. They are painless and they bleed during straining.

The external hemorrhoids occur below the pectinate line. They are caused by varicosities of tributaries of inferior rectal vein. They are covered by the anal skin. The external hemorrhoids do not bleed; they are painful.

The Ischiorectal (Ischioanal) Fossae (Fig. 13.38)

On each side of the anal canal is a wedge-shaped space called ischiorectal (ischioanal) space. It is located between the skin of the anal region and the pelvic diaphragm.

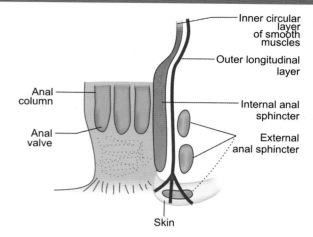

Fig. 13.36: Sphincters

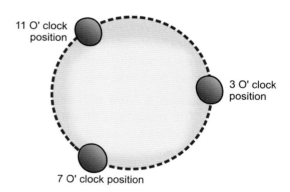

Fig. 13.37: Sites of internal hemorrhoids

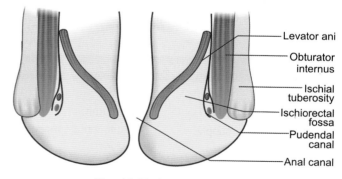

Fig. 13.38: Ischiorectal fossa

Boundaries of ischiorectal fossae
* *Laterally:* Ischium and obturator internus muscle.
* *Medially:* Anal canal, levator ani and sphincter ani externus muscles.
* *Posteriorly:* Sacrotuberous ligament and gluteus maximus muscle.

- *Anteriorly:* Base of urogenital diaphragm. The two ischiorectal fossae communicate with each other.

Contents

1. Ischiorectal pad of fat – It supports the anal canal.
2. Internal pudendal artery and vein.
3. Pudendal nerve.

    The vessels and nerve run in the pudendal canal on the lateral walls of the fossae. They give rise to inferior rectal vessels and nerves, which supply the sphincter ani externus and the perianal skin.

4. The perineal branch of S_4 also passes through the ischiorectal fossae.

Applied Anatomy

The ischiorectal fossae may get infected, leading to the formation of ischiorectal abscess (collection of pus), which is a painful condition. Infections may reach the ischiorectal fossae in the following ways:
1. Following inflammation of anal sinuses (cryptitis)
2. Following a tear in the anal mucous membrane
3. From a penetrating wound in the anal region.

    There is a fullness and tenderness between the anus and ischial tuberosity.

    An ischiorectal abscess may burst open, spontaneously, into the anal canal, rectum or perianal skin. An abscess in one fossa may spread to the other one because of their communication.

Anal fistula is an abnormal passage, that develops from the abscess, which, at one end, opens into the anal canal or rectum and at the other end on the perianal skin.

■ THE PERINEAL BODY (FIGS 13.39A AND B)

The perineum is the diamond-shaped area extending from the pubic symphysis, to the tip of the coccyx; it overlies the inferior pelvic aperture or pelvic outlet.

    The perineum is bounded by (1) the pubic symphysis; (2) the ischiopubic rami; (3) the ischial tuberosities; (4) the sacrotuberous ligaments and (6) the coccyx.

    For descriptive purposes, the perineum is divided into two unequal triangles by an imaginary transverse line joining the anterior ends of the ischial tuberosities. Its anterior part, called the urogenital triangle contains the external genitalia and the urethra. The posterior part or anal triangle contains the anal canal with ischioanal fossae on each side.

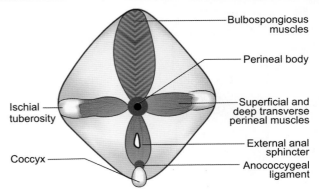

Fig. 13.39A: Perineal body

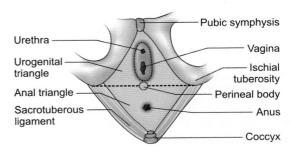

Fig. 13.39B: Boundaries of perineum

    The midpoint of the line joining the ischial tuberosities overlies the perineal body or the central perineal tendon. It is a mass of fibrous tissue, forming the landmark of perineum, where several muscles converge—the transverse perineal, levator ani, sphincter ani externus, etc.

    Perineal body is an important structure in women. Tearing or stretching of the perineal body, during childbirth, removes support for the vagina. As a result, prolapse of vaginal wall may occur.

    To prevent this tearing, an incision is made in the perineum, during delivery, to enlarge the vaginal orifice. This incision is called an *episiotomy* (see Fig. 9.9). Being a clean, surgical incision, it heals better. The incision does not extend to the external anal sphincter.

■ THE PERITONEUM (FIGS 13.40A AND B)

The peritoneum is a thin, transparent serous membrane. It consists of 2 layers—the peritoneum lining the abdominal wall is called parietal peritoneum, whereas, the peritoneum covering the viscera is called the visceral peritoneum. These layers consist

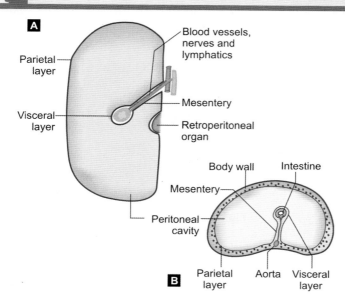

Figs 13.40A and B: Peritoneum—(A) sagittal section of abdomen, (B) TS of abdomen

of a single layer of squamous epithelium called mesothelium. These two layers are separated from each other by a thin capillary film of peritoneal fluid, which lubricates the peritoneal surfaces, reducing friction during movement of viscera.

In the early embryo, the peritoneum is a large sac that lines the walls of the abdominal cavity. As the developing organs enlarge, the peritoneal cavity is almost completely obliterated. Some organs become retroperitoneal (e.g. kidneys, pancreas); some organs protrude completely into the peritoneal sac (e.g. stomach, intestine).

As the fetal organs assume their final adult positions, the peritoneal cavity is divided into two peritoneal sacs, the greater and lesser sacs of peritoneum.

A surgical incision through the anterior abdominal wall enters the greater peritoneal sac.

The Lesser Sac of Peritoneum or Omental Bursa

It is a large compartment of the peritoneal cavity located posterior to the stomach and lesser omentum. It permits the stomach to slide freely.

The omental bursa communicates with the greater sac through the omental foramen or epiploic foramen of Winslow, located posterior to the free edge of lesser omentum.

Boundaries of Epiploic (omental) Foramen

- *Anteriorly:* Portal vein, hepatic artery, bile duct, in the free edge of lesser omentum.
- *Posteriorly:* Inferior vena cava
- *Superiorly:* Caudate lobe of liver
- *Inferiorly:* Superior part (1st part) of duodenum hepatic artery, bile duct and portal vein.

The peritoneal cavity is closed in males; but, in females, there is communication with the exterior through the uterine tubes, uterus and vagina.

Various terms are used to describe different parts of the peritoneum and peritoneal cavity. They are:
1. Mesentery
2. Omentum
3. Ligament
4. Recess.

Mesentery

This is a double layer of peritoneum that encloses an organ and connects it to the abdominal wall. It contains connective tissue, fat, vessels, nerves and lymphatics, passing to and from the organ or viscera. Mesentery of the stomach is called mesogastrium; mesentery of the transverse colon is called transverse mesocolon.

Omentum (Fig. 13.41)

It is a double-layered sheet of peritoneum attaching the stomach to the body wall or other abdominal organs.

The *lesser omentum* connects the lesser curvature of stomach and proximal part of duodenum to the liver.

The greater omentum—This apron-like, fat-laden fold of peritoneum hangs from the greater curvature of stomach and connects the stomach with the diaphragm, spleen and transverse colon.

The greater omentum is called "the policeman of the abdomen". It has considerable mobility and can migrate to any area of the abdomen and wraps itself around an inflamed organ such as appendix or a perforated duodenal ulcer. Thus it "walls off" and protects an infected or perforated organ from other viscera.

Peritoneal Ligament

Peritoneal ligament is a double layer of peritoneum that connects an organ with another organ or with the abdominal wall.

For example, gastrosplenic ligament—connects spleen to greater curvature of stomach.

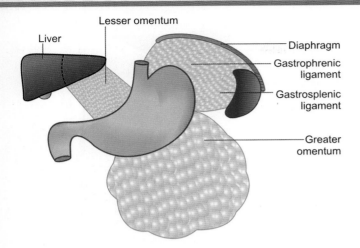

Fig. 13.41: Omenta

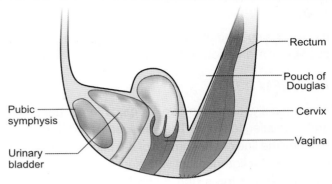

Fig. 13.42: Pouch of Douglas

Gastrophrenic ligament—connects stomach to diaphragm;

Lienorenal—connects lien (spleen) and rene (kidney).

Peritoneal Recesses

These are blind pouches or sacs (cul-de-sac) formed in some areas; they are closed at one end, the other end opens into the peritoneal cavity. For example, retrocecal recess.

The peritoneal recesses are clinically important because pus can collect in these regions. Similarly, loops of small intestine can slide into any of these recesses, leading to "internal hernias".

Hepatorenal Pouch (Morison's Pouch)

This space lies posterior to the right lobe of the liver; it lies anterior to the upper part of right kidney and right suprarenal. Inferiorly, it communicates with the general peritoneal cavity.

It is the lowest (most dependent) part of the abdominal cavity proper, in the supine (lying on the back) position. Fluids tend to collect here. This is the commonest site of subphrenic abscess; the site of infection may be an inflamed appendix.

Pouch of Douglas or Rectouterine Pouch (Fig. 13.42)

Pouch of Douglas or rectouterine pouch is the most dependent part of the peritoneal cavity when the body is in the upright (standing) position. It is bounded anteriorly by the uterus and posterior fornix of vagina; it is bounded posteriorly by the rectum.

This pouch is of great clinical importance. The floor of this pouch is only 5.5 cm away from the anus. By doing a per rectal (PR) examination or pervaginal (PV) examination, any abnormal collection in this pouch can be easily felt. Being the most dependent part, pus, blood or ascitic fluid tends to collect here; these fluids can be drained or a sample can be collected through the posterior fornix of vagina.

Applied Anatomy of Peritoneum

1. Inflammation of peritoneum is called peritonitis. It is a very serious condition.
2. Collection of free fluid in the peritoneal cavity is called ascitis; it can be due to cirrhosis of liver, congestive heart failure or tumors of abdominal viscera. Removal of this fluid is called paracentesis.
3. *Peritoneal dialysis:* In renal failure, a suitable glucose-electrolyte solution is introduced into the peritoneal cavity. Excess urea and electrolytes will enter into the dialysis fluids. This fluid is then removed.

■ THE ACCESSORY ORGANS OF DIGESTION

The liver, biliary system and pancreas form the accessory organs of digestion.

The Liver

The liver is the largest gland in the body (See CP9 in Color Plate 2). It is reddish brown in color, highly vascular and weighs about 1.5 kg in adults.

Position (Fig. 13.43)

It occupies the right hypochondrium and extends into the epigastrium and left hypochondrium.

Shape

It is wedge shaped. It has an apex, base or right lateral surface, an anterior, superior and inferior surfaces.

Lobes (Fig. 13.44)

Anatomically, liver has two lobes, a large right lobe and a smaller, left lobe. They are separated from each other by the line of attachment of the falciform ligament, the fissure for ligamentum venosum and by the groove for ligamentum teres (the round ligament of liver).

Quadrate lobe is seen on the inferior surface of the right lobe (Fig.13.45).

Caudate lobe It is chiefly situated on the posterior surface (It has a short tail-like extension – the caudate process) of the right lobe.

Functional or Physiological Lobe (Fig. 13.46)

Depending on the areas supplied by (or drained by) the branches of hepatic artery, portal vein and bile duct, the liver is divided into 2 functional lobes.

The caudate and quadrate lobes are parts of the right anatomical lobe.

But the functional right lobe is demarcated by an imaginary line passing through the inferior vena cava and fossa for the gallbladder. So, the quadrate and caudate lobes become parts of the left functional lobe of the liver.

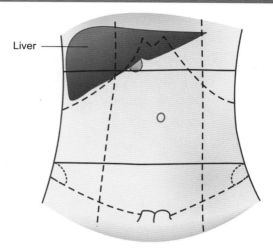

Fig. 13.43: Position of liver

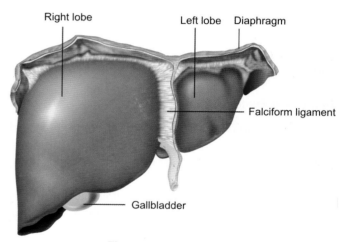

Fig. 13.44: Anterior view

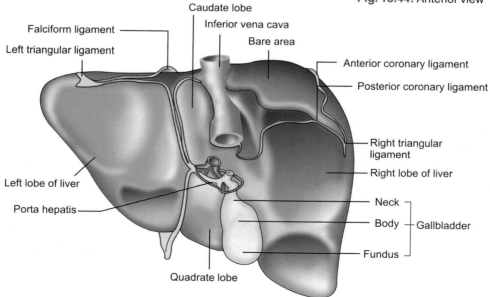

Fig. 13.45: Posterior view

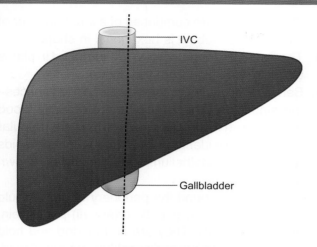

Fig. 13.46: Physiological lobes of liver

Peritoneal Folds of Liver

1. Falciform ligament (falx = sickle): It is a sickle-shaped fold of peritoneum stretching from the liver to the diaphragm and anterior abdominal wall. The lower free border of this ligament lodges the ligamentum teres or the round ligament of liver, which is the remnant of left umbilical vein.
2. Coronary ligament: This ligament, having an (upper) anterior and (lower) posterior layer, stretches between the right lobe of liver and diaphragm.
3. The anterior or upper layer and the posterior or lower layer fuse laterally, forming the right triangular ligament.
4. *Left triangular ligament:* It is a short triangular fold, stretching between the left lobe of liver and diaphragm.
5. *The lesser omentum*: This double fold of peritoneum connects the lesser curvature of stomach to the liver. In the liver, it is attached to the fissure for ligamentum venosum (obliterated ductus venosus) and to the lips of porta hepatis along an L-shaped line.

The right free margin of lesser omentum contains the portal vein, hepatic artery, the common bile duct, hepatic plexus of nerves and lymphatics. The right free border of lesser omentum forms the anterior boundary of epiploic foramen. To control bleeding from the cystic artery, during cholecystectomy (removal of gallbladder), the hepatic artery can be compressed between two fingers – the index finger inserted into the epiploic foramen, the thumb resting anteriorly on the right free border of lesser omentum – This is called Pringle's maneuver.

Relations of Liver

Superior surface: It is related to the diaphragm, which separates the liver from the bases right and left pleura and the fibrous pericardium.

Right lateral surface: It is related to diaphragm which separates this surface from right pleura and lung.

Anterior surface: It is related to the 5th to 10th ribs and costal cartilages on the right side, the anterior abdominal wall and 7th, 8th costal cartilages on left side.

Posterior surface (Fig. 13.45):
1. Groove for vertebral column.
2. Fissure for ligamentum venosum—It lodges the obliterated ductus venosus, which is a wide channel connecting left branch of portal vein with inferior vena cava in fetal life.
3. Groove for abdominal part of esophagus.
4. *Bare area of liver*—It is a rough, triangular area on the posterior surface of liver. It is bounded by the inferior vena cava at the base, by the right triangular ligament at the apex and the upper and lower layers of coronary ligament on the sides.

The bare area is not covered with peritoneum. It is directly related to the diaphragm. It is also related to the upper pole of right kidney and right suprarenal. Bare area forms one of the sites of portosystemic (portacaval) anastomosis.

Inferior Surface of Liver (Fig. 13.47) (Visceral Surface)

This oval surface is divided into right and left lobes by the fissure for ligamentum venosum and the fissure for ligamentum teres (obliterated left umbilical vein). The left lobe is related to the stomach:

On the right lobe is (a) *fossa for* the gallbladder. This fossa is not covered by peritoneum. It lodges the gallbladder. (b) *the hilum of liver or porta hepatis* – it is a transverse fissure, about 3.5 cm in length; it separates the quadrate lobe from caudate lobe. The lips of this fissure give attachment to the lesser omentum.

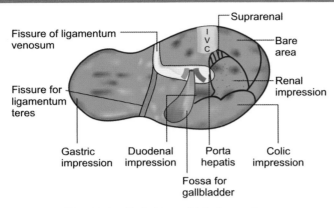

Fig. 13.47: Relations of inferior surface

Structures entering the hilum are the right and left branches of hepatic artery, the portal vein and hepatic plexus of nerves.

Structures emerging out of the hilum are the right and left hepatic ducts and some lymph vessels. The visceral surface is related to the upper end of second part of duodenum, right colic flexure, right kidney and right suprarenal and pyloric end of stomach (Fig. 13.47).

Blood supply Liver gets blood from 2 sources—
1. From the hepatic artery – a branch of celiac trunk. It carries oxygenated blood. It divides into right and left branches.
2. The portal vein—it carries products of digestion from the alimentary canal. It is formed by the union of superior mesenteric vein and splenic vein (RBCs are destroyed in the spleen. The splenic vein also receives blood from inferior mesenteric vein).

The right and left branches of hepatic artery and portal vein enter the porta hepatis and divide into 8 segmental branches. Each branch supplies a definite part of the liver called the hepatobiliary segment which is drained by a segmental branch of bile duct.

The branches of hepatic arteries and the portal vein open into the sinusoids of liver, which in turn are drained by the tributaries of hepatic veins (central veins). These intersegmental veins join to form the right and left hepatic veins and drain into the inferior vena cava.

Histology (Figs 13.48A and B)

1. The liver is covered by a connective tissue capsule called Glisson's capsule.

2. The two lobes are composed of a large number of lobules; each lobule is hexagonal in shape.
3. The liver cells (hepatocytes) are arranged in plates or cords, radiating from the central vein.
4. Between the plates of cells are blood filled spaces— the sinusoids (The sinusoids are actually blood vessels, with incomplete walls, which are irregular in shape and wider than capillaries). The sinusoids are lined by endothelium and phagocytes, known as Kupffer cells.

Arranged around the periphery of each lobule are branches of (i) hepatic artery, (ii) portal vein, and (iii) bile duct. They are surrounded and held by connective tissue and constitute the portal triad. The oxygenated blood from the hepatic artery, the products of digestion and destruction of RBCs from portal vein are drained into the sinusoids. Finally, the sinusoids drain into the central vein. The central veins join to form larger veins and ultimately form the hepatic veins, which drain into the inferior vena cava.

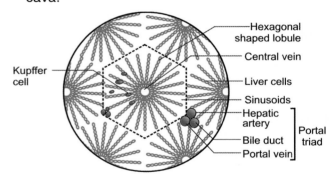

Fig. 13.48A: Histology of liver

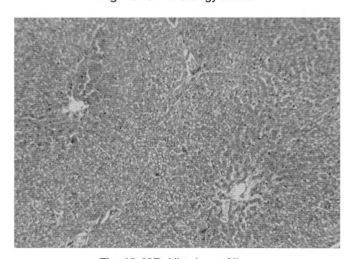

Fig. 13.48B: Histology of liver

5. The hepatocytes are polyhedral cells, these cells are in close contact with sinusoids. They absorb nutrients from sinusoids. The cells secrete bile. Lying between adjacent cells, there are minute channels called biliary canaliculi. Bile is secreted into these canaliculi. These canaliculi join together to form larger ducts, which finally emerge as right and left hepatic ducts.

Functions of Liver

1. Secretion of bile – this is the exocrine function of liver
2. Storage of glycogen
3. Gluconeogenesis
4. Metabolism of fat
5. Deamination of amino acids and production of ammonia
6. Production of plasma proteins
7. Storage of vitamins
8. Storage of iron, in the form of ferritin
9. Clotting factors—fibrinogen, prothrombin and factor VII are produced by liver.
10. Detoxification—liver modifies or destroys toxic substances.

The production of glucose, aminoacids, clotting factors, etc. are considered as endocrine functions of liver.

Applied Anatomy

1. Inflammation of the liver is called hepatitis.
2. Liver is a soft, friable tissue. Fracture of lower ribs or penetrating injury of upper part of abdomen can injure the liver. Bleeding is very profuse.
3. Because of its great vascularity, hematogenous spread of cancer from other parts of the body (liver secondaries) is very common.
4. Pain, in liver diseases, is referred to the right shoulder; this is because of the irritation of the diaphragm [the phrenic nerve (C_3, C_4 and C_5) and supraclavicular nerve supplying skin over the shoulder (C_3 and C_4) arise from the same spinal segments].
5. *Liver biopsy*—Hepatic tissue for diagnostic purposes may be obtained by inserting a liver biopsy needle through the right 10th intercostal space in the midaxillary line. Patient is instructed to hold his breath in full expiration to lessen the possibility of damage to pleura and lung.
6. *Hepatomegaly*: Enlargement of liver is associated with carcinoma, cirrhosis, congestive heart failure, etc.

Contd...

Contd...

7. Under certain conditions (e.g. chronic alcoholism) liver tissue undergoes fibrosis and it shrinks. This is called cirrhosis of liver.
8. Liver transplantation from living donors have good results.

Extrahepatic Biliary Apparatus (Fig. 13.49)

The biliary apparatus collects bile from the liver, stores and concentrates it in the gallbladder and transmits it to the second part of duodenum.

The apparatus consists of:
1. The right and left hepatic ducts
2. The common hepatic duct
3. The gallbladder
4. The cystic duct, and
5. The common bile duct.

The Right and Left Hepatic Ducts

The right and left hepatic ducts emerge at the porta hepatis from the right and left lobes of liver.

The Common Hepatic Duct

The common hepatic duct is formed by the union of right or left hepatic ducts near the porta hepatis. It runs downwards for about 3 cm and is joined on its

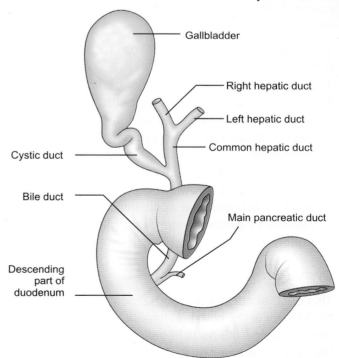

Fig. 13.49: Extrahepatic biliary apparatus

right side, at an acute angle, by the cystic duct to form the common bile duct (or "bile duct").

The Gallbladder

It is a pear-shaped reservoir of bile, situated in a fossa on the inferior surface of the right lobe of the liver. It is about 7-10 cm in length, 3 cm broad at its widest part; about 30 ml in capacity.

Parts of Gallbladder: It has a fundus, body and neck.

The fundus: It is the anterior, free, expanded portion of gallbladder. It projects beyond the inferior margin of liver. It comes in contact with the anterior abdominal wall at the junction of the 9th costal cartilage and linea semilunaris (on the right side).

Behind the fundus is the body; it is the main part of gallbladder. The body tapers posteriorly into a narrow neck. The neck is continued as the cystic duct.

The upper surface of the body is not covered with peritoneum. It is directly in contact with the fossa for the gallbladder in the liver. The lower surface of the body is covered with peritoneum.

The neck is funnel shaped. The right wall of the neck may present a dilatation called the Hartmann's pouch. This is the common site for gallstones to get lodged.

Histology of gallbladder (Figs 13.50A and B). Gallbladder has:

1. An outer serous coat made of peritoneum.
2. A fibromuscular coat deep to serosa – made up of fibrous connective tissue and interlacing smooth muscle fibers running in circular and longitudinal directions.
3. There is *no* submucous layer.
4. The mucosa is thrown into numerous folds. It is lined by tall columnar epithelium and adapted for absorbing excess water from bile and concentrating it.

Cystic Duct

This duct, about 3-4 cm long, emerging from the gall bladder, joins the common hepatic duct (at a variable distance below the porta hepatis) to form the common bile duct. The inner mucous lining of the cystic duct presents a spiral fold which acts like a valve. It is called the spiral valve of Heister.

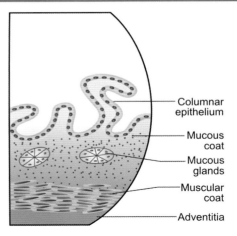

Fig. 13.50A: Histology of gallbladder

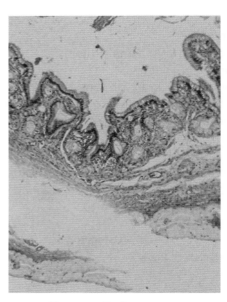

Fig. 13.50B: Gallbladder

Calot's Triangle or Triangle of Cystic Artery (Fig. 13.49)

This triangular space is bounded by the cystic duct, common hepatic duct and porta hepatis of liver. The cystic artery, a branch of right hepatic artery crosses this triangle to reach the neck of the gallbladder (The right hepatic artery may take a tortuous course in the triangle, which is described as a caterpillar turn or Moynihan's hump).

Common Bile Duct (CBD)

It is formed by the union of cystic duct and common hepatic duct. It runs downwards, within the free border of lesser omentum along with the portal vein and hepatic artery. Then it descends behind the 1st part of duodenum; then, behind the head of the pancreas and finally joins the main pancreatic duct to form the ampulla of Vater or hepatopancreatic ampulla. It pierces the posteromedial wall of the 2nd part (descending part) of duodenum about 8-10 cm from the pylorus.

The common bile duct, thus, has a supraduodenal part, retroduodenal (or duodenal) part, a pancreatic part and intramural part (The intramural part pierces the muscular wall of duodenum and opens on top of a mucous elevation called major duodenal papilla). The opening of ampulla of vater is partly covered by a mucous fold called the hood of Monk. It is guarded by a valve—the sphincter of Oddi. The terminal part of the bile duct, before it enters the ampulla is surrounded by the sphincter of Boyden.

Applied Anatomy

1. *Gallstones* are common in "fatty, fertile females of forty"; they may cause severe biliary colic when they get impacted in the cystic duct or the end of common bile duct. Impaction of gallstone in the ampulla may cause regurgitation of infected bile into the pancreatic duct causing acute pancreatitis.
2. Tumors of pancreas can cause obstruction to common bile duct. This leads to obstructive jaundice.
3. Inflammation of gallbladder, cholecystitis, may cause irritation of diaphragmatic peritoneum and the pain is referred to the right shoulder.
4. *Cholecystography:* Iodine containing compounds given orally or intravenously, get excerted in bile. In the gallbladder, it gets concentrated in bile. This casts a shadow of the gallbladder in X-rays. Gallstones can also be demonstrated.

Pancreas

Pancreas is a gland, that is partly exocrine and partly endocrine (Pan = all; kreas = fleshy; it is fleshy throughout). It lies transversely across the posterior abdominal wall at the level of L_1 and L_2 vertebrae. It is situated behind the stomach and retroperitoneally.

Dimensions

In the adult, it is about 15-20 cm long; weighing 90 gm; 3 cm broad.

Parts (Figs 13.51A and B)

Pancreas is divided, from right to left, into the head, neck, body and tail.

The head The head is enlarged; it lies within the concavity of the duodenum. From the lower and left part of the head, projects the uncinate process.

The bile duct is often embedded in the substance of the pancreas. It forms the most important, posterior relation of the head.

The uncinate process (hook-like process) is related to the superior mesenteric vessels.

Neck is the slightly constricted part between the head and body. The portal vein is formed behind the neck of the pancreas by the union of the superior mesenteric and splenic veins. Anteriorly, the neck is related to the gastroduodenal artery.

Body of Pancreas

Body of pancreas extends from the neck to the tail. It is elongated, passes upwards towards the left.

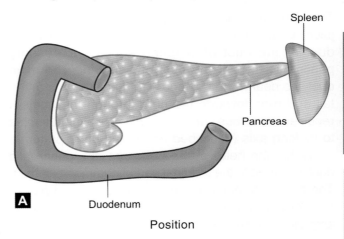

A Duodenum — Pancreas — Spleen

Position

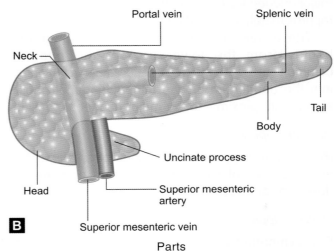

B Neck — Portal vein — Splenic vein — Tail — Body — Head — Uncinate process — Superior mesenteric artery — Superior mesenteric vein

Parts

Figs 13.51A and B: Pancreas

The superior border shows a projection (to the left of the neck) called tuber omentale. It is related to the celiac trunk. The superior border, to the left of tuber omentale, is related to the splenic artery.

The anterior surface of the body of pancreas is related to the lesser sac of peritoneum, which separates the pancreas from the stomach.

The posterior surface is related to the aorta, the origin of superior mesenteric artery, the left suprarenal, left kidney, left renal vessels and splenic vein.

Tail of the Pancreas

This is the narrow, left end of the pancreas. It reaches the hilum of the spleen. It lies along with the splenic vessels in the lienorenal ligament.

Ducts of Pancreas (Fig. 13.52)

The exocrine secretion of the pancreas (pancreatic juice) is drained by two ducts—the main pancreatic duct or the duct of Wirsung and the accessory pancreatic duct or the duct of Santorini. The main duct lies nearer the posterior surface; it begins at the tail and runs towards the right through the body. It receives numerous small ducts to join it at right angles to its long axis (described as herring-bone pattern).

Within the head of the pancreas, the pancreatic duct is related to the terminal part of common bile duct. The two ducts enter the muscular wall of 2nd part of duodenum and join to form the hepatopancreatic ampulla of Vater which opens on the summit of the major duodenal papilla, about 8-10 cm from the pylorus (approximately the middle of 2nd part of duodenum).

The accessory pancreatic duct (of Santorini) begins from the lower part of the head; it communicates with the main pancreatic duct. It opens into the duodenum on the minor duodenal papilla about 2 cm proximal to the major duodenal papilla.

Arterial Supply

Pancreas is supplied by branches of splenic artery and branches of superior and inferior pancreaticoduodenal arteries.

Nerve Supply

Parasympathetic (vagus) and sympathetic (lower thoracic) fibers innervate the pancreas; they reach it via blood vessels.

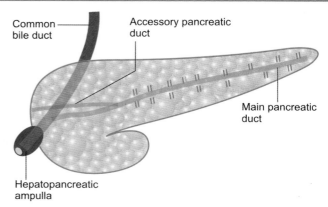

Fig. 13.52: Ducts

Histology (Figs 13.53A to C)

Pancreas is divided into numerous lobules. Each lobule is made up of a number of serous acini. Sections of interlobular and intralobular ducts can be seen. The ducts are lined by cuboidal epithelium.

Islets of Langerhans are seen as scattered groups of epithelial cells in between the serous acini. A number of blood capillaries are seen around the islets ("islands").

The islets of Langerhans are more numerous in the tail of pancreas. Each islet has (i) B cells or β-cells, which secrete insulin; (ii) A or alpha cells which secrete glucagon, and (iii) D cells or delta cells which secrete somatostatin. The islets form the endocrine part of the pancreas.

The serous acini form the exocrine part of pancreas. The cells lining the acini or alveoli are columnar or polyhedral; it has an inner granular zone and an outer clear zone. The outer zone stains deeply with basic dyes (blue). Inner zone stains with acidic dyes (pink). Thus the cells of the acini exhibit a "bipolar" staining. Filling the lumen of the acini are centroacinar cells.

Applied Anatomy

1. Deficiency of insulin causes diabetes mellitus.
2. Carcinoma of the head of the pancreas is common. Pressure over the bile duct causes obstructive jaundice; pressure over the portal vein leads to ascites; it can press upon the pylorus causing obstruction.
3. Anular pancreas (ring-like pancreatic tissue) is a congenital anomaly of pancreas which encircles the duodenum; it may obstruct the duodenum.
4. Accessory pancreatic tissue may be seen in Meckel's diverticulum.

Contd...

Contd...

5. Acute pancreatitis is a serious disorder in which there is autodigestion of pancreas followinig activation of trypsinogen. It is thought to be due to reflux of bile into the pancreatic duct system due to anatomical or mechanical causes or due to obstruction of pancreatic duct system. Patient has agonizing pain, vomiting, profound shock and faint jaundice. After 2 weeks, a collection of fluid or pus in the lesser sac will produce a mass in the epigastrium—this is known as pseudo-cyst of pancreas.

6. Cysts can be true cysts, may be solitary or multiple.

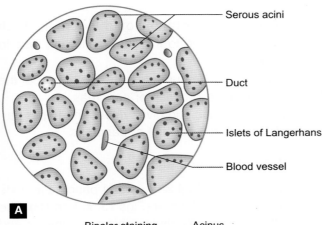

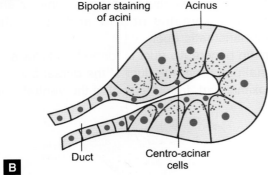

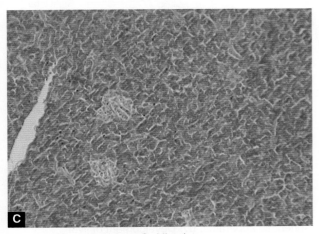

Figs 13.53A to C: Histology—pancreas

DEVELOPMENT OF DIGESTIVE SYSTEM

Three weeks after fertilization, the embryo is in the form of a disc, having 3 germ layers—outer ectoderm, middle mesoderm and inner endoderm. This is known as the trilaminar embryonic disc. However, there are two areas in this disc, where there is only ectoderm and endoderm. These are (1) the prechordal plate near the cranial end, and (2) the cloacal membrane, near the caudal end (Fig. 13.54).

Ectoderm forms the floor of the amniotic cavity. Endodermal cells multiply and form a lining for the yolk sac cavity.

The embryonic disc which is at first flat, undergoes folding at the cranial and caudal ends. These are the head and tail folds. Lateral folds also appear. With the formation of the head fold, tail fold and lateral folds, the flat endodermal sheet is converted into a tube. This is the gut tube.

The gut tube is subdivided into foregut, midgut and hindgut. The gut is closed cranially by the prechordal plate. It is now known as the buccopharyngeal membrane. Caudally the gut is closed by the cloacal membrane (Fig. 13.54).

Midgut communicates with the yolk sac through a channel called vitello-intestinal duct.

Derivatives of Foregut

- Part of mouth
- Salivary glands
- Pharynx
- Esophagus
- Stomach
- Duodenum—up to major duodenal papilla
- Liver
- Pancreas
- Extra hepatic biliary apparatus
- Respiratory system.

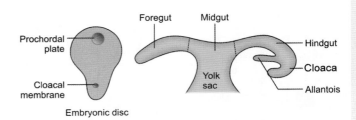

Fig. 13.54: Development of digestive system

Derivatives of Midgut

- Duodenum—distal to major duodenal papilla
- Jejunum
- Ileum
- Cecum
- Appendix
- Ascending colon
- Right two thirds of transverse colon.

Derivatives of Hindgut

- Left one third of transverse colon
- Descending colon
- Sigmoid colon
- Rectum
- Anal canal (except the terminal part, which is derived from proctodeum)
- Part of urogenital system.

Development of Liver and Extrahepatic Biliary Apparatus

The liver develops from an endodermal hepatic bud, that arises at the point of junction between foregut and midgut. It enlarges and divides into a large cranial part called pars hepatica (Figs 13.55A to D) and a smaller caudal part called pars cystica. Pars hepatica divides into right and left parts which develop into right and left lobes of the liver respectively.

Pars cystica gives origin to the gallbladder and cystic duct.

The part of the hepatic bud proximal to the pars cystica forms the bile duct.

■ DEVELOPMENT OF PANCREAS

Pancreas develops from two endodermal buds - the dorsal and ventral pancreatic buds. They arise from the gut tube which later forms the duodenum (Figs 13.56A and B).

As a result of rotation of duodenum, both pancreatic buds come to the same side and fuse to form a single mass.

Ventral pancreatic bud forms the lower part of the head and uncinate process of pancreas.

Dorsal pancreatic bud gives rise to the upper part of head, neck, body and tail of pancreas.

Duct System of Pancreas

The ducts of the dorsal and ventral pancreatic buds anastomose with each other (Figs 13.56A and B). The duct of the dorsal bud between this anastomosis and the duodenum remains narrow and forms the accessory pancreatic duct.

The main pancreatic duct is formed as follows:
1. Its distal part, from the duct of the dorsal bud,
2. Its proximal part, from the duct of the ventral bud.

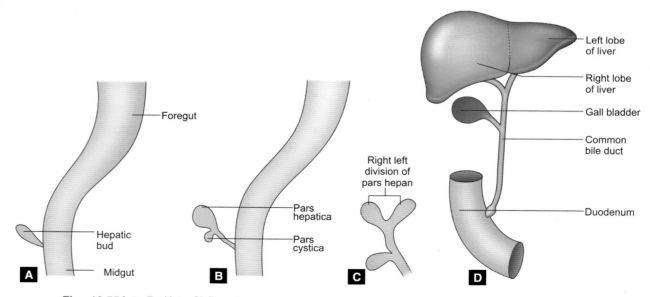

Figs 13.55A to D: (A to C) Development of biliary extraphepatic apparatus, (D) Development of liver

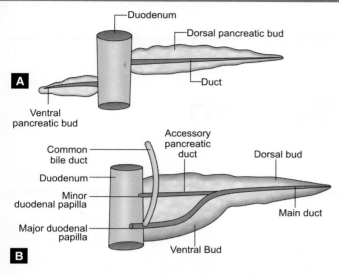

Figs 13.56A and B: Development of pancreas

The main pancreatic duct opens into the duodenum at the major duodenal papilla along with the bile duct.

The pancreatic acini and the islets of Langerhans are derived from the primitive duct system.

The connective tissue of liver, the gallbladder and pancreas are derived from mesoderm.

Congenital Anomalies

1. *Atresia:* Failure of development of lumen is common in esophagus and duodenum. Duodenal atresia is often associated with Down syndrome.
2. *Anular pancreas:* Dorsal and ventral pancreatic buds fuse to form a ring-like pancreatic tissue around the duodenum. May cause duodenal obstruction.
3. *Meckel's diverticulum or ileal diverticulum:* Persistence of vitellointestinal duct. It is seen in about 2 percent population, 2 inches long, 2 feet away from ileocecal junction on the antimesenteric border of ileum. May contain ectopic pancreatic or gastric tissue. Inflammation of this diverticulum may mimic features of acute appendicitis.
4. *Imperforate anus (absence of anal opening):* Caused by stenosis or atresia of lower part of anal canal.

14 Urinary System

■ INTRODUCTION

The urinary system is one of the four excretory systems in our body. The other three are the bowel, lungs and the skin.

■ COMPONENTS OF URINARY SYSTEM (FIG. 14.1)

1. 2 kidneys
2. 2 ureters
3. Urinary bladder
4. Urethra.

The kidneys remove waste products of metabolism, excess water and salts from blood and maintain the pH. The ureters convey urine from the kidneys to the urinary bladder. The urinary bladder is the muscular reservoir of urine and the urethra is the channel to the exterior.

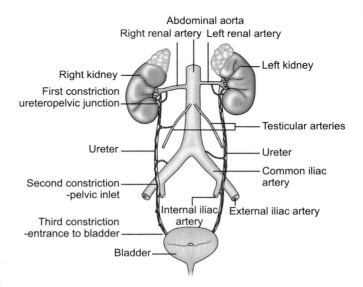

Fig. 14.1: Parts of urinary system

■ THE KIDNEYS (RENES; NEPHROS)

The kidneys are a pair of excretory organs, situated on the posterior abdominal wall, retroperitoneally, one on each side of the lumbar part of vertebral column.

The left kidney is slightly at a higher level than the right kidney since the massive liver occupies the right hypochondrium.

Shape and Size

Each kidney is bean-shaped; has 2 poles – upper and lower; 2 borders – medial and lateral; and 2 surfaces – anterior and posterior.

Measurements

- Length : 7.5-10 cm
- Width : 5 cm
- Thickness : 2.5 cm
- Weight : 150 gm

The upper pole is broad and is closely related to the suprarenal (adrenal) gland. The lower pole is pointed. The upper poles are nearer to the median plane than the lower poles.

The lateral border is convex. The medial border is concave. The middle part of medial border is depressed and is known as the hilum (hilus). The following structures are seen at the hilum (from anterior to posterior).

1. The renal vein
2. The renal artery
3. The renal pelvis—-the upper extended portion of ureter.

(The student can easily memorize this arrangement—VAU—vein, artery, ureter).

In addition to these 3 structures, the hilum also transmits nerves and lymphatics.

Coverings

The kidney has 3 coverings—
1. Innermost fibrous capsule or true capsule.
2. Middle fatty capsule or perinephric fat—it is a collection of fatty tissue.

It acts as a shock absorber and also helps to maintain the kidney in its position.
3. The false capsule—made of renal fascia. It has 2 layers—anterior and posterior. Superiorly the

2 layers enclose the suprarenal gland and then merge with diaphragmatic fascia.
(That is why the kidneys move with respiration)
4. Pararenal fat – forms a cushion for the kidney.

Surface Anatomy

The outline of kidney can be marked on the posterior aspect of the body by drawing the Morris' parallelogram as shown in (Fig. 14.2).

Relations (See CP10 in Color Plate 3)

1. *Upper pole* of each kidney is related to the corresponding suprarenal gland.
2. *The hilum* is related to renal vein, renal artery and ureter, with nerves and lymph vessels.

Posterior Relations (Fig. 14.3)

Posteriorly, the kidneys are related to (a) the diaphragm (b) psoas major (c) quadratus lumborum and trans-

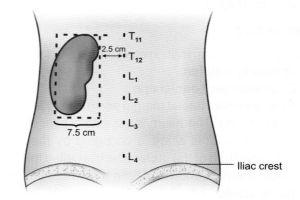

Fig. 14.2: Surface marking—Morris' parallelogram

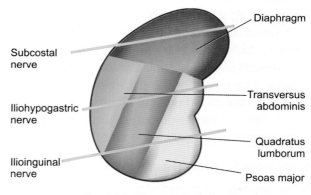

Fig. 14.3: Posterior relations

verses abdominis muscles. Posterior surface is also related to 3 nerves. From above downwards, they are: (i) subcostal; (ii) iliohypogastric and (iii) ilioinguinal. (d) In addition, the right kidney is related to 12th rib; left kidney is related to 11th and 12th ribs and the intercostal space between them.

Anterior Relations (Fig. 14.4)

Right kidney is related to (a) right suprarenal gland; (b) second part of duodenum (c) right lobe of liver; (d) hepatic flexure of colon, and (e) coils of small intestine.

Left kidney is related to:
a. Left suprarenal gland
b. Spleen and splenic vessels
c. Stomach
d. Pancreas
e. Left colic flexure or splenic flexure
f. Coils of small intestine.

Structure (Fig. 14.5)

Macroscopic

The naked-eye examination of a coronal section shows the following features:
a. Characteristic bean-shape
b. It has a convex lateral margin and a concave medial hilum (hilus). The hilum leads into a space called renal sinus. This sinus is occupied by the upper expanded part of the ureter which is called the renal pelvis (pelvis = basin). Within the sinus, the renal pelvis divides into 2 or 3 major calyces (calyx = cup). Each major calyx divides into minor calyces.
c. The kidney tissue, covered by a fibrous capsule, consists of an outer cortex and inner medulla. The cortex is light in color and the medulla, dark.
d. The medulla is made up of triangular areas of renal tissue called renal pyramids. The apex of the pyramid is called the papilla which fits into the minor calyx. The bases of the pyramids are capped by renal cortex called cortical arches.
e. The parts of the renal cortex projecting between the pyramids are called the renal columns.

Microscopic Structure (Histology) (Figs 14.6A and B)

The kidney may be regarded as a collection of millions of uriniferous tubules. Each uriniferous tubule consists

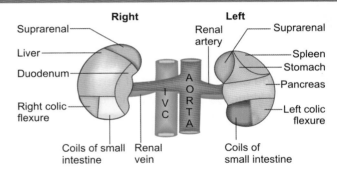

Fig. 14.4: Relations of kidney—anterior

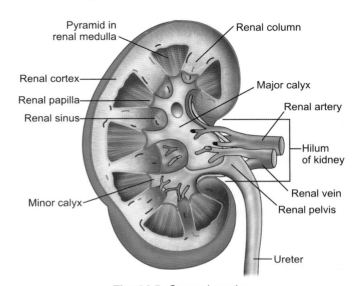

Fig. 14.5: Coronal section

of an excretory part called the nephron and of a collecting tubule. There are 1-2 million nephrons in a kidney.

Parts of a Nephron (Fig. 14.7)

Nephron, the functional unit of kidney, has the following parts:

A renal corpuscle or Malpighian corpuscle It is a rounded structure consisting of a tuft of capillaries called glomerulus and a cup-like double-layered covering called the Bowman's capsule or glomerular capsule.

The renal tubule is a long complicated tubule. It has different parts. They are:
  i. The proximal convoluted tubule (PCT)
 ii. The loop of Henle – having a descending limb, a loop and ascending limb
iii. The distal convoluted tubule (DCT)

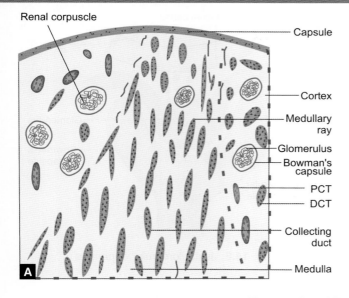

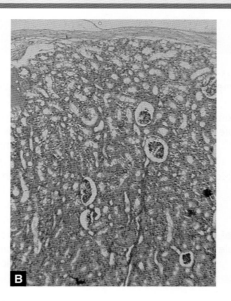

Figs 14.6A and B: Histology of kidney

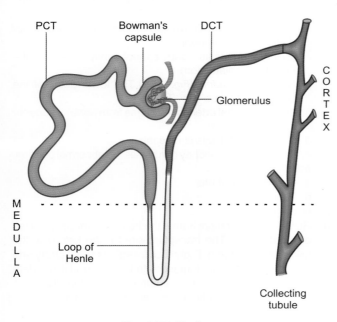

Fig. 14.7: Nephron

The DCT ends by joining a collecting tubule. The collecting tubules draining different nephrons join to form larger tubules called papillary ducts of Bellini, which open into a minor calyx at the apex of a renal papilla.

The renal corpuscles and major parts of PCT and DCT are located in the cortex of the kidney.

The loops of Henle and the collecting ducts lie in the medulla.

The inner surface of the Bowman's capsule is lined with highly specialized cells called podocytes. The outer layer is lined with simple squamous epithelium.

The PCT is lined with cuboidal or low columnar epithelium. The cells are provided with microvilli and numerous mitochondria. The loop of Henle is lined with low cuboidal or squamous epithelium.

The DCT is lined with cuboidal cells. These cells do not have microvilli.

The collecting tubules are lined with cuboidal or low columnar cells.

The renal artery after entering the kidney divides repeatedly. Functionally there are 2 sets of arterioles and capillaries. The first capillary system present in the glomeruli, is concerned exclusively with the removal of waste products from blood. It does not supply oxygen to the renal tissues. The second set of capillaries, present around the tubules are concerned with the exchange of gases.

Correlation of Structure and Function of a Nephron (Table 14.1)

The permeability of collecting ducts is controlled by antidiuretic hormone (ADH). This hormone can regulate the dilution of urine depending upon the functional requirements. Absence or deficiency of ADH is characterized by excretion of large volumes of dilute urine (Diabetes insipidus).

Section C

The Organ Systems

Table 14.1: Correlation of structure and function of a nephron	
Part	**Function**
1. Glomerular capsule	Filtration of molecules and water from blood
2. Proximal convoluted tubule	Selective reabsorption of many substances from glomerular filtrate. e.g. Glucose, amino acids, ions like Na^+, Cl^- etc.
3. Loop of Henle	Creates a hypertonic environment in the medulla to facilitate water reabsorption by collecting ducts
4. Distal convoluted tubule	Selective reabsorption of ions (mainly sodium bicarbonate) and secretion of hydrogen ions
5. Collecting ducts	Final concentration of urine

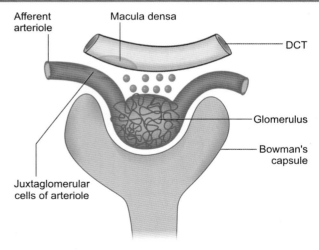

Fig. 14.8: Juxtaglomerular apparatus

Juxtaglomerular Apparatus (Fig. 14.8)

A part of the DCT and a part of the afferent arteriole of the glomerulus lying close together are modified to form a juxtaglomerular apparatus. The wall of the DCT, at this site shows densely packed columnar cells, forming the macula densa. The smooth muscle cells in the wall of the afferent arteriole are modified to form juxtaglomerular cells. The macula densa and juxtaglomerular cells together form the juxtaglomerular apparatus. It has an important role in Renin-Angiotensin mechanism, in the control of blood pressure.

Blood Supply of Kidney

The kidneys are supplied by renal arteries which are direct branches of abdominal aorta. They are large in size because the entire blood has to pass through the kidneys for filtration.

The renal vein emerges from the kidney at the hilum and ends in the inferior vena cava. The left renal vein is longer than the right.

Nerve Supply

Kidneys are innervated by autonomic nervous system. Sympathetic fibers are derived from T_{10}-L_1 segments and parasympathetic fibers are derived from the vagus nerves.

Lymphatics from the kidneys drain into lateral aortic nodes.

Applied Anatomy

1. *Congenital Anomalies:* These are developmental defects, present at birth.
   a. *Unilateral or bilateral agenesis—failure of development of kidney.*
      (Bilateral agenesis—baby cannot survive, Treatment – Renal transplantation)
   b. *Horse-Shoe kidney(Fig. 14.9)* The lower poles of the kidneys may fuse
   c. Abnormal renal arteries may arise from aorta or superior mesenteric artery
   d. Congenital polycystic disease
   e. Ectopic kidney—kidney can be seen in abnormal position.
2. *Infections:*
   a. Glomerulonephritis
   b. Tuberculosis.
3. Renal Failure
4. Renal transplantation: Forms the treatment for some cases of renal failure. The transplant site for the kidney is the lower abdomen (e.g. Right iliac fossa). Its renal artery and vein are joined (anastomosed) to the external iliac artery and vein, and the ureter is sutured into the bladder.
5. Surgical removal of kidney is called nephrectomy.

■ THE URETERS

The ureters are tubular structures which serve to conduct urine from the kidneys to the urinary bladder.

Length

25 cm; diameter 0.6 cm.

Extent

Each ureter starts from within the renal sinus as a funnel-shaped expanded part called the pelvis of

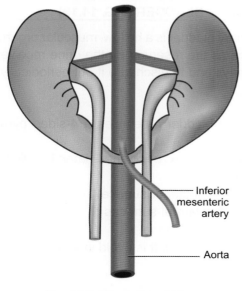

Fig. 14.9: Horse-shoe kidney

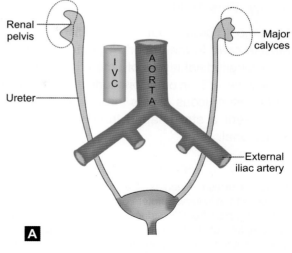

Fig. 14.10A: Ureter

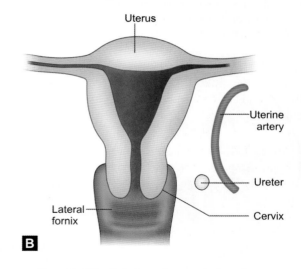

Fig. 14.10B: Relation of ureter in females

ureter. The ureter proper runs downwards and medially on the psoas major, crosses the pelvic brim to enter the pelvic cavity where it ends by opening into the lateral angles of bladder.

Parts

For purposes of description, ureter is divided into 2 parts.
1. From the site of origin to pelvic brim—abdominal part
2. From pelvic brim to entry into urinary bladder—pelvic part.

Constrictions

There are 3 natural constrictions in the ureter.
a. At the pelviureteric junction
b. At the pelvic brim
c. Just before it enters the bladder
The renal stones tend to get arrested at these sites.

Important Relations (Figs 14.10A and B)

In females, the pelvic part of ureter forms the posterior boundary of the ovarian fossa. As the ureter descends, it runs forwards on the lateral fornices of vagina, 1.5 cm lateral to the cervix of uterus. It is crossed by the uterine artery from lateral to medial direction. Near

its termination, the ureter runs down for a short distance on the anterior wall of vagina.

Blood Supply

Ureter is supplied by branches of renal artery, abdominal aorta, gonadal artery, common iliac artery, internal iliac artery and inferior vesical artery. These branches form a continuous arterial chain (There are corresponding veins).

Nerve Supply

Autonomic nervous system.

Section C

The Organ Systems

Histology (Figs 14.11A and B)

1. Outermost fibrous coat
2. *Muscle coat:* In the upper 2/3, outer circular and inner longitudinal layers of smooth muscle fibers. In the lower 1/3, an outer longitudinal layer is added.
3. Innermost mucous coat is thrown into folds so as to present a star-shaped lumen. Epithelium is transitional.

Applied Anatomy

1. Ureteric Calculi (L.Pebbles): Calculi may be located in the calyces of the kidneys, ureters or urinary bladder. A kidney stone may pass from the kidney to the ureter, causing partial or complete obstruction. It causes severe rhythmic pain called ureteric colic. The ureteric colic is a sharp, stabbing pain, which passes inferolaterally from loin to groin.
2. Ureter can be injured or accidentally tied during surgeries of ovary or during hysterectomy (removal of uterus).
3. Congenital anomalies: Bifid ureters or double ureters may be seen.

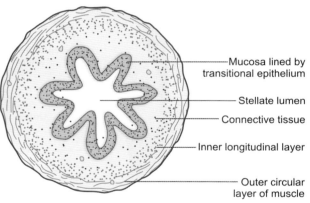

Mucosa lined by transitional epithelium

Stellate lumen

Connective tissue

Inner longitudinal layer

Outer circular layer of muscle

Fig. 14.11A: Histology—ureter

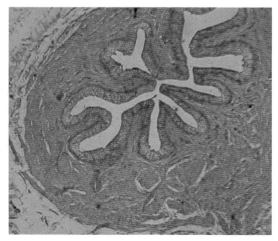

Fig. 14.11B: Histology of ureter

■ URINARY BLADDER (FIG. 14.12)

The urinary bladder is a hollow, muscular organ, which functions as the reservoir for the urine received from the kidneys and to discharge it out periodically.

Shape

The empty bladder resembles a 4-sided pyramid. It has:
1. 4 angles—apex, neck and 2 lateral angles
2. 4 surfaces
   a. Base (posterior surface)
   b. 2 inferolateral surfaces
   c. Superior surface
   When distended, it is ovoid in shape.

Position

Empty bladder, in the adult, is situated within the true pelvis. When distended, it rises up into the abdominal cavity and becomes an abdominopelvic organ.

In the newborn, it is abdominal in position.

Capacity

The normal capacity is 200-300 cc.

Relations (Fig. 14.13)

Apex or anterior angle: Connected to the umbilicus by a fibrous cord called urachus.

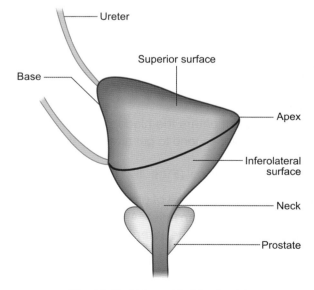

Ureter

Superior surface

Base

Apex

Inferolateral surface

Neck

Prostate

Fig. 14.12: Urinary bladder (male)

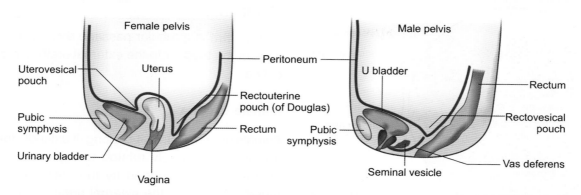

Fig. 14.13: Relations of urinary bladder (Sagittal section of female pelvis and male pelvis)

Neck or inferior angle is continuous with urethra. It rests on the upper surface of prostate in males and upper surface of urogenital diaphragm in the female.

1. *Superior surface*—is covered with peritoneum; it is related to the coils of small intestine.
2. *Inferolateral surface*—related to muscles of pelvic floor.
3. Posterior surface (Fig. 14.14)
   a. *Male: Upper part* covered by peritoneum and related to the rectovesical pouch of peritoneum with coils of ileum in it.
      The lower part of posterior surface is related to seminal vesicle and vasa deferentia.
   b. *Female:* The whole posterior surface is related to anterior wall of vagina (Fig. 14.13).

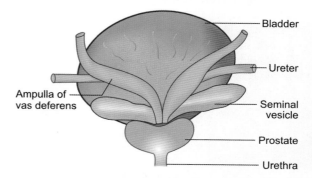

Fig. 14.14: Posterior relations—male

Interior of Bladder (Fig. 14.15)

The mucous membrane is straw-colored and is thrown into folds. When bladder is distended, these folds disappear.

The posterior wall shows a smooth triangular area called *trigone.* There are no mucous folds in this region. The mucosa is pink in color. It is richly innervated and highly sensitive.

At the upper lateral angles of the trigone are the ureteric openings. At its inferior angle is the internal urethral orifice.

Histology (Figs 14.16A and B)

1. Outer fibroelastic coat
2. The middle muscle coat made of smooth muscle fibers called detrusor muscle.
3. Inner mucous coat, lined by transitional epithelium. It rests on lamina propria, made chiefly of collagen fibers.

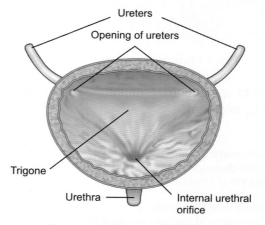

Fig. 14.15: Trigone of bladder

Blood Supply

Branches of internal iliac artery. The corresponding veins form a plexus and drain into internal iliac vein.

Nerve Supply

The vesical plexus, composed of sympathetic and parasympathetic fibers, innervate the bladder. Parasympathetic fibers are derived from S_2, S_3, S_4 seg-

Section C

The Organ Systems

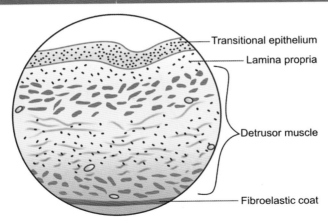

Fig. 14.16A: Histology—urinary bladder

- Transitional epithelium
- Lamina propria
- Detrusor muscle
- Fibroelastic coat

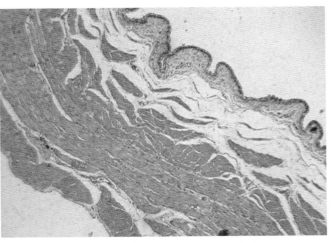

Fig. 14.16B: Histology of urinary bladder

ments of spinal cord. Sympathetic fibers are derived from L_1 segment.

Applied Anatomy

1. *Congenital anomalies:* Ectopia vesicae—there is a defect in the development of anterior abdominal wall, so that the trigone of bladder is exposed and the ureteric openings can be seen on it.
2. Infection of the bladder is called cystitis
3. Cystoscopy is done to visualize the interior of bladder
4. Neurological lesions of the bladder can occur due to injury of spinal cord at or above the level of $S_{2,3,4}$ segments of spinal cord.
5. Rupture of bladder—the distended bladder is in danger of rupture(as in fall from a height).
6. Cancer of bladder.

URETHRA

The urethra is a tubular passage extending from the neck of the bladder to the external urethral meatus or orifice.

Female Urethra

Female urethra is 3.75-4 cm long. It extends downward and forward, closely related to the anterior wall of vagina. It is surrounded by the sphincter urethrae muscle. It ends at the external urethral orifice in the vestibule. The mucosa is folded. There are a number of paraurethral glands in the submucosa which open by small ducts on the mucous membrane (These glands are said to be homologous to the prostate gland of male).

Male Urethra (Fig. 14.17A)

The male urethra is 18-20 cm long. In the flaccid state of penis, the urethra is S-shaped. When penis is erect, it becomes J-shaped. The male urethra forms a part of urinary system as well as reproductive system.

The male urethra is divided into 3 parts.

1. Prostatic part
2. Membranous part
3. Spongy part or penile part.

Prostatic Part (Fig. 14.17B)

Prostatic part is 3 cm long, passes through the prostate gland. It is the widest and most dilatable part of male urethra. It receives the openings of ejaculatory ducts on a raised area called verumontanum. It also receives the openings of glands of the prostate.

Membranous Part

About 1 cm in length, it is the narrowest and least distensible part. It passes through the urogenital diaphragm and is surrounded by the sphincter urethrae muscle.

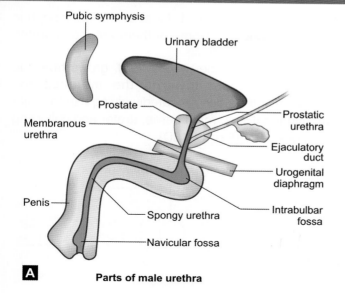

Parts of male urethra

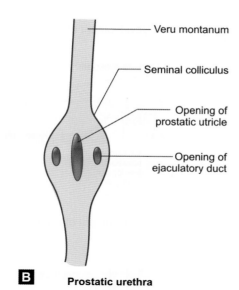

B   **Prostatic urethra**

Figs 14.17A and B: Male urethra

Spongy or Penile Part

About 15-16 cm in length, it passes through the bulb and corpus spongiosum of penis. It is narrow, with a diameter of 6 mm. It shows 2 dilatations:

1. At its beginning, to form the intrabulbar fossa
2. Within the glans penis, to form the navicular fossa.

There are a number of openings of urethral and bulbourethral glands into the urethra.

Sphincters

There are 2 sphincters in relation to the urethra. (1) Internal, and (2) External.

The internal sphincter is made of smooth muscle fibers and is situated at the neck of the bladder. It is innervated by ANS and involuntary.

The external sphincter is made up of striated (skeletal) muscle surrounding the membranous part of urethra. It is supplied by pudendal nerve (S_2, S_3, S_4) and voluntary.

The urethra is lined, internally by stratified columnar epithelium. Close to the external urethral orifice, it is lined by stratified squamous epithelium.

Applied Anatomy

1. *Congenital Anomalies:*
   a. *Hypospadias* – the urethra opens on the ventral surface of penis (normally opens at the tip)
   b. *Epispadias* – urethra opens on the dorsal side of penis
2. Infection – urethritis
3. Rupture of urethra can occur following a fracture of the pelvis. Usually, the membranous part is involved.
4. *Catheterization of bladder:* In some cases, the patient is unable to pass urine (retention of urine). In such cases, a rubber or metal tube (catheter) is introduced into the bladder through the urethra. While passing a metallic catheter in the males, the normal curvatures have to be considered. Forceful insertion may rupture the urethra. Catheterization of female urethra is easy because it is short and wide.

RADIOLOGICAL ANATOMY OF URINARY SYSTEM

1. Intravenous pyelography (IVP) or intravenous urography is done to visualize the calyces, pelvis and ureter. In IVP, an iodine containing compound is injected intravenously. It is excreted and concentrated by the kidneys. The shadows of calyces, pelvis, ureter and bladder can be seen.
2. Cystogram is a radiogram obtained with contrast medium instilled into the bladder.

DEVELOPMENT OF KIDNEYS, URETERS AND URINARY BLADDER

The urogenital system is developed from the intermediate mesoderm.

The intra embryonic mesoderm is subdivided into three (Fig. 14.18A)

a. Paraxial mesoderm
b. Intermediate mesoderm
c. Lateral plate mesoderm

Later, the intermediate mesoderm forms a bulge on the posterior abdominal wall, lateral to the dorsal mesentery. This is called the nephrogenic cord. A number of important structures develop from the intermediate mesoderm (Fig. 14.18 B) They are:

a. Excretory tubules associated with the development of kidney.
b. The mesonephric duct.
c. The paramesonephric duct, lateral to the mesonephric duct.
d. The gonad (Testis or ovary).

Development of Kidneys

Human kidney develops from 2 sources:

1. The excretory tubules or nephrons are derived from metanephros—The Bowman's capsule, proximal and distal convoluted tubules and Loop of Henle are derived from metanephros.
2. The collecting part of the kidney is derived from ureteric bud, a diverticulum arising from the lower part of mesonephric duct—The collecting tubules, minor calyces, major calyces, renal pelvis and ureter are the derivatives of ureteric bud.

During evolution, the vertebrate kidney has passed through three stages. The most primitive of these is called the pronephros. It is the functioning kidney in cyclostomes and fishes. The next stage, mesonephros, is seen in higher vertebrates (most anamniotes). The kidney of amniotes, including man is called metanephros.

During the development of human embryo, the human kidney passes through all these three stages. So, this is a classical example of the saying that "ontogeny repeats phylogeny".

The pronephros is formed in the cervical region of the embryo, in the nephrogenic cord. Later, mesonephros develops in the thoracolumbar region and metanephros, in the sacral region. Human pronephros is non-functional. A nephric duct formed in relation to the pronephros persists. Mesonephros consists of a series of excretory tubules which drain into the nephric duct. The nephric duct is now known

as the mesonephric duct. Most of the mesonephric tubules degenerate. Some of them are incorporated into the developing testis.

A diverticulum, the ureteric bud, grows from the mesonephric duct towards the metanephros (Fig. 4.18C). It divides repeatedly giving rise to major and minor calyces, renal pelvis, ureter and collecting ducts of kidney (Fig. 14.18D).

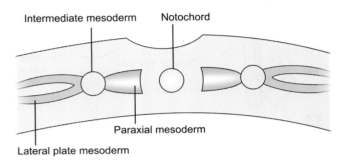

Fig. 14.18A: Development of kidney

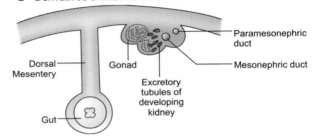

Fig. 14.18B: Development of kidney

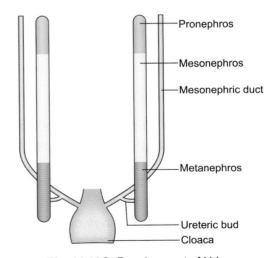

Fig. 14.18C: Development of kidney

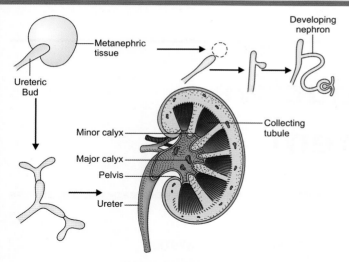

Fig. 14.18D: Development of kidney

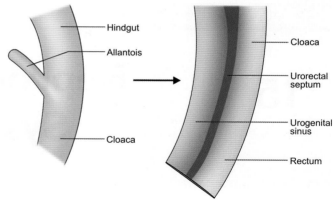

Fig. 14.19: Development of urinary bladder

The metanephros forms a cap around the ureteric bud—the metanephric blastoma. The Bowman's capsule, proximal convoluted tubule, loop of Henle and distal convoluted tubule are derived from the metanephric blastema. A tuft of capillaries invaginating the Bowman's capsule form the glomerulus.

The kidneys develop in the sacral region. Later, they ascend to the lumbar region.

Congenital Anomalies

1. Agenesis of kidneys—Unilateral or bilateral. Unilateral absence of a kidney often causes no symptoms because the other kidney undergoes compensatory hypertrophy and performs the function of both kidneys.

   Bilateral absence of kidneys occurs about 1 in 3000 births and is incompatible with postnatal life. During fetal life, electrolyte stability is not impaired because it is controlled by exchange through the placenta. Most infants with bilateral renal agenesis die shortly after birth.

   Horseshoe kidney (Fig. 14.9): The poles of the kidneys are fused - usually the lower poles. About 7% of persons with Turner syndrome have horseshoe kidneys.

Polycystic disease of kidneys is an autosomal recessive disorder. Both kidneys contain hundreds of small cysts, which result in renal insufficiency. Postnatal dialysis and kidney transplantation are the methods of treatment.

Development of Urinary Bladder (Fig. 14.19)

As a result of folding of the embryo, the flat endodermal sheet of the trilaminar embryonic disc is converted into a tube, the gut tube. It has 3 parts—foregut, midgut and hindgut. A small diverticulum called the allantoic diverticulum arises from the yolk sac and is absorbed into the hindgut.

The part of the hindgut distal to the attachment of allantoic diverticulum is called the cloaca. The urorectal septum divides the cloaca into a dorsal rectum and a ventral urogenital sinus.

The urogenital sinus gives rise to the urinary bladder and urethra.

The epithelium of the trigone of the urinary bladder develops from the absorbed part of mesonephric duct. Muscular wall develops from splanchnopleuric mesoderm.

The allantois gets constricted and becomes a thick fibrous cord, the urachus. It extends from the apex of the bladder to the umbilicus. In the adult, the urachus is represented by the median umbilical ligament.

Section C

The Organ Systems

15 Reproductive System

■ INTRODUCTION

One of the essential features of life is the power of reproduction. Reproduction, in man and higher animals is a complicated process, involving the existence of two sexes.

The reproductive organs of males and females differ structurally and functionally. The function of the male sex organs is to produce spermatozoa and the function of the female reproductive system is to produce ova.

If the ovum is fertilized by the spermatozoon, an embryo is formed, which remains in the cavity of the uterus, until the new individual is capable of an independent existence.

FEMALE ORGANS OF REPRODUCTION

The female sex organs, for the purposes of description, are classified as follows:
1. The external organs (Fig. 15.1) (External genitalia vulva or pudendum)
    a. The mons pubis
    b. The labia majora
    c. The labia minora
    d. The clitoris
    e. The vestibule of vagina
    f. Bulb of vestibule
    g. Greater vestibular glands.
2. The internal organs (Fig. 15.2)
    a. The uterus
    b. The ovaries
    c. The fallopian (uterine) tubes
    d. The vagina.

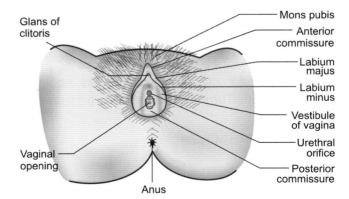

Fig. 15.1: Female reproductive system—external genitalia

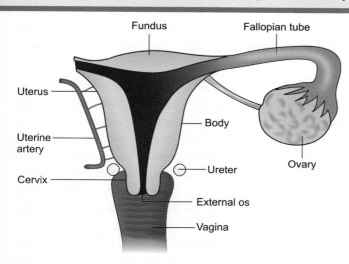

Fig. 15.2: Internal organs

3. The secondary organs—The breasts (mammary glands) (Figs 15.11A and B).

The female external genital organs are known collectively as the *vulva* (L. covering) or *pudendum*.

Mons Pubis (Mons = Mountain)

It is a rounded fatty elevation, situated anterior to the pubic symphysis. It consists mainly of a pad of fat deep to the skin. The mons pubis becomes covered with coarse pubic hairs during puberty. The typical female distribution of pubic hair has a horizontal upper limit.

Labia Majora (Labia Majora = Large Lips) (Singular Labium Majus)

They are 2 symmetrical folds of skin, which protect the urethral and vaginal openings, that open into the vestibule of the vagina. Each labium majus passes posteriorly from the mons pubis, to about 2.5 cm from the anus.

The slit between the labia majora is the pudendal cleft, into which the vestibule of the vagina opens.

The labia majora meet anteriorly at the anterior labial commissure. Posteriorly, they do not meet. A transverse skin fold, called the posterior commissure passes between them.

The thin outer surface of labia majora is covered with hair and their inner surface has large sebaceous glands.

Labia Minora (= Small Lips) (Singular Labium Minus)

They are thin, delicate folds of fat-free, hairless skin, which are located between the labia majora. They have many sensory nerve endings. They enclose the vestibule of the vagina, lying on each side of the orifices of urethra and vagina.

The labia minora meet just superior to the clitoris to form a fold of skin called the prepuce or clitoral hood. Posteriorly, they unite, forming the frenulum of labia minora; it is frequently torn during childbirth.

Vestibule of the Vagina (Vestibulum = Ante Chamber)

It is the space between the labia minora. The urethra, vagina and the ducts of the greater vestibular glands open into the vestibule.

The external urethral orifice is a median opening, located 2-3 cm posterior to the clitoris and immediately anterior to the vaginal orifice.

The vaginal orifice is a large opening located posterior and inferior to the urethral orifice.

The Greater Vestibular Glands (Bartholin's Glands)

They are located on each side of the vestibule of the vagina, posterolateral to the vaginal orifice, partly covered by the bulbs of the vestibule. Their ducts open into the vestibule of the vagina, on each side of the vaginal orifice. They secrete small amounts of lubricating mucus during sexual arousal.

Clitoris

It is 2-3 cm in length; homologous with the penis, it is also an erectile organ. But, it is not traversed by the urethra. It is located posterior to the anterior labial commissure. It is highly sensitive and very important in the sexual arousal of a female.

Bulbs of the Vestibule

These large, elongated masses of erectile tissue, about 3 cm long, lie along the sides of the vaginal orifice. They lie deep to the bulbospongiosus muscles.

Section C

The Organ Systems

Arterial Supply

Branches of external and internal pudendal arteries.

Venous Drainage

Internal pudendal veins drain the venous blood from the external genitalia.

Nerve Supply

1. Ilioinguinal nerve
2. Genitofemoral nerve
3. Perineal nerve.

Lymphatic Drainage

Mainly to the superficial inguinal nodes; from the clitoris, lymph drains into the deep inguinal lymph nodes.

Applied Anatomy

1. The female urethra is easily infected because it is open to the exterior via the vestibule of vagina. The resulting infection is called urethritis.
2. The short female urethra is highly distensible. So, the introduction of catheters and cystoscopes is much easier in females.

■ UTERUS (L. WOMB), (FIG. 15.2)

The uterus is a thick-walled, pear-shaped, hollow muscular organ, situated in the pelvic cavity, with the urinary bladder in front and the rectum, behind. It has thick muscular walls and a small central cavity.

Dimensions

The uterus is 7-8 cm long, 5-7 cm wide and 2-3 cm thick; and it weighs 30-40 gm.

Parts of the Uterus

The uterus has 2 major parts:
1. The expanded superior two-thirds is known as the body
2. The cylindrical, lower one-third is called the cervix (L. neck).

A slight constriction, called the isthmus, marks the junction between the body and the cervix.

Fundus

The fundus of the uterus, is the rounded upper part of the body. It is located superior to a line joining the points of entry of the fallopian tubes. The site of entry of the uterine tube is called the "cornu" (or horn).

Cervix

The cervix or neck of the uterus is the lowest portion, part of which projects like an inverted cone into the vault of the vagina. Thus, the cervix is divided into vaginal and supravaginal parts. The rounded vaginal part opens into the vagina through the external ostium or external os ("ostium" = door, entrance). The cavity of the cervix (cervical canal) communicates with the uterine cavity through the internal os.

The cavity of the uterus is a mere slit when viewed from one side; the cavity is flat and triangular when seen from the front.

Wall of the Uterus

The uterine wall has 3 layers:
1. Outer serous coat or *perimetrium,* made of peritoneum.
2. The middle, muscular coat or *myometrium.*

   It is 12-15 mm thick. Myometrium consists of smooth muscle fibers. The main branches of blood vessels and nerves are situated in this layer. During pregnancy, the thickness of myometrium increases greatly.
3. The inner mucous coat or *endometrium:* It is a highly specialized mucous membrane, which varies in thickness according to the phases of menstrual cycle. Except its basal layer (which lies close to the myometrium), the endometrium is shed off during menstruation. It regenerates after menstruation.

Position (Fig. 15.3)

In a nulliparous woman, uterus is completely within the pelvis. It is normally *anteverted* and *anteflexed* in position. The long axis of the cervix is bent forwards over the long axis of the vagina and this is called anteversion.

The body of the uterus is bent forwards over the cervix (at the isthmus). This is known as anteflexion. In some women, the uterus is angled backwards in a position of retroversion and retroflexion. When the urinary bladder fills, the uterus is physiologically retroverted.

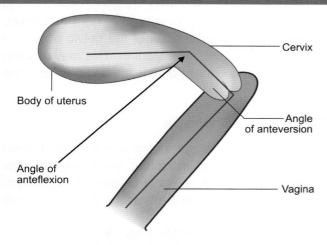

Fig. 15.3: Angulation of uterus and vagina

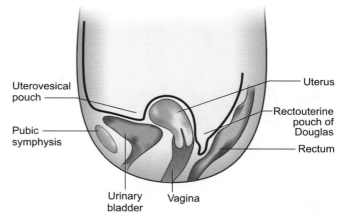

Fig. 15.4: Uterus—relations

Relations (Fig. 15.4)

Anterior Surface or Vesical Surface

The urinary bladder is situated anterior to the uterus. The uterovesical pouch of peritoneum separates the fundus and body of the uterus from the anterior 2/3 of the superior surface of the bladder. The supravaginal part of the cervix, which is not covered with peritoneum, is directly related to the posterior 1/3 of the superior surface of the bladder.

The Posterior Surface

The posterior surface of the fundus, body and supravaginal cervix are covered by peritoneum; it is related to the rectouterine pouch (pouch of Douglas) of peritoneum, which separates the uterus from the rectum. The pouch of Douglas is occupied by the coils of terminal ileum and sigmoid colon.

Lateral Borders

A double fold of peritoneum, the broad ligament, is attached to the lateral borders. It stretches from the lateral border of uterus to the side walls of the pelvis. Its upper free border encloses the fallopian tubes.

Other Structures in the Broad Ligament

Other structures between the 2 layers of the broad ligament are:
1. The ovarian ligament
2. The round ligament of uterus

3. The uterine vessels
4. Nerves
5. Lymphatics
6. Connective tissue.

The posterior layer of broad ligament is reflected backwards to form a fold called mesovarium, which encloses the ovary in its free border.

The part of the broad ligament between the mesovarium and the fallopian tube is called the mesosalpinx; the part below the mesovarium, is the mesometrium.

Supports of Uterus (Figs 15.5A and B)

The uterus is a mobile organ, which undergoes extensive changes in size and shape during pregnancy. It is supported and prevented from sagging down by a number of factors.
a. Muscular support
b. Fibromuscular ligaments
c. Peritoneal ligaments (of doubtful role)
A. *Muscular support:*
  1. The pelvic diaphragm
  2. The perineal body
  3. The urogenital diaphragm
B. *Fibromuscular ligaments:*
  1. Pubocervical ligament
  2. Transverse cervical ligament
  3. Uterosacral ligament
  4. Round ligament of uterus
C. *Peritoneal ligaments*: Broad ligaments

Section C

The Organ Systems

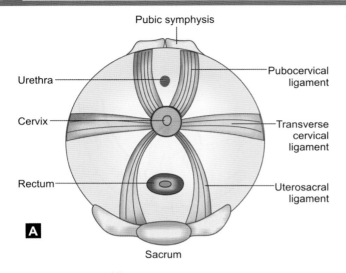

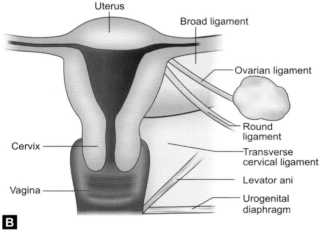

Figs 15.5A and B: Ligaments and supports of uterus

Muscular Support

Pelvic Diaphragm

The pelvic diaphragm is formed of the levator ani and coccygeus muscles (Fig. 9.8). They support the pelvic viscera. The pubococcygeus part and the pubovaginalis part of the levator ani form a strong support for the vagina, and indirectly, form a support for the uterus.

The Perineal Body (Fig. 13.39)

It is a fibromuscular body situated about 1.25 cm in front of the anal margin. It is also known as the central perineal tendon, to which several muscles of the perineum are attached.

The perineal body is very important in female, because, it provides support of the pelvic organs. It may be damaged during childbirth. This may result in prolapse of the urinary bladder, the uterus and the rectum.

The Urogenital Diaphragm

It is formed of the sphincter urethrae and deep transverse perineal muscles, lying between the superior and inferior layers of urogenital diaphragm. These fibers support the urethra as well as the vagina.

Fibromuscular Ligaments (Fig. 15.5A)

The Pubocervical Ligaments

These ligaments connect the cervix to the posterior surface of pubis.

The Transverse Cervical Ligaments
(or Mackenrodt's Ligaments or Cardinal Ligaments)

These ligaments fan out from the cervix and upper vagina to the lateral wall of pelvis.

The Uterosacral Ligaments

They connect the cervix to the sacrum.

The Round Ligaments of Uterus

They are two fibromuscular bands, about 10-12 cm long; Each ligament begins at the cornu of the uterus (anteroinferior to the uterine tube), run forwards and laterally, pass through the deep inguinal ring and traverse the inguinal canal; after leaving the inguinal canal through the superficial inguinal ring, the ligaments break up into thin filaments, which merge with the connective tissue of the labia majora.

The round ligaments pull the fundus anteriorly, thus keeping the uterus in anteverted position.

Peritoneal Ligaments

Broad Ligaments

The right and left broad ligaments, are double folds of peritoneum; rather than supporting the uterus, they convey blood vessels, nerves and lymphatics to the uterus (see page 253).

Section C

The Organ Systems

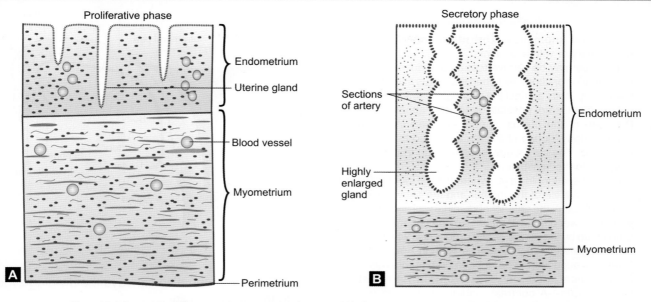

Figs 15.6A and B: Microscopic structure of uterus (A) Proliferative phase, (B) Secretory phase

Blood Supply of Uterus

Arterial Supply

The uterus is supplied chiefly by 2 uterine arteries and partly by ovarian arteries. The uterine artery is a branch of the internal iliac artery. It runs medially towards the cervix. 2 cm lateral to the cervix, it crosses the ureter and ascends along the side of the uterus, with a tortuous course. The uterine artery ends by anastomosing with the ovarian artery.

The uterine artery supplies (1) the uterus (2) the vagina (3) the medial 2/3 of fallopian tube; (4) the ovary; (5) the ureter and (6) the structures in the broad ligament.

Venous Drainage

The corresponding uterine veins drain into the internal iliac veins.

Lymphatic Drainage

To external iliac, internal iliac and aortic nodes; some lymphatics from the fundus, accompany the round ligament and end in the superficial inguinal nodes.

Nerve Supply

By autonomic nervous system—sympathetic (T12, L1) and parasympathetic (S2, S3, S4).

Histology of Uterus (Figs 15.6A and B)

1. Outermost perimetrium, made of peritoneum and connective tissue.
2. The middle myometrium is the thickest layer. The smooth muscle fibers are arranged in 3 layers—inner circular, middle oblique and outer longitudinal.
3. The inner endometrium: It has 2 zones—a deep basal zone and a superficial spongy layer. It is lined by a single layer of columnar epithelium. There are a number of uterine glands, which are tubular and lined by columnar cells.

The endometrium undergoes marked cyclical changes, under the influence of estrogen and progesterone. In the postmenstrual phase, the endometrium is thin; it is thickest at the end of secretory phase. The uterine glands, which were initially simple and tubular, become enlarged; their diameter increases and they become twisted; in sections, they acquire a saw-toothed appearance (Fig. 15.6B).

At the time of menstruation, the greater part of uterine glands are lost, along with the entire lining epithelium, leaving behind only their most basal parts. After menstruation, the lining epithelium regenerates from this basal part.

The branches of the uterine artery running through the myometrium and endometrium take a spiral course.

Applied Anatomy

1. Congenital anomalies: The uterus may be absent; rudimentary or with a septum.
2. The uterus may be retroverted; it can cause pain during menstruation or abortions (when the uterus enlarges during pregnancy, it hits against the sacral promontory; so further enlargement is not possible, leading to abortion).
3. Prolapse of the uterus: A perineal tear, involving the perineal body, during childbirth, may lead to prolapse of the uterus. The cervix descends through the vagina and bulges into the vestibule.
4. Cesarean section: In cases where normal labor is likely to endanger the life of the mother and the baby, an incision is made through the anterior abdominal wall and the uterine wall (the lower uterine segment, i.e. the upper 1/3 of cervix is taken up by the uterus during pregnancy, which is called the lower uterine segment). The Roman Emperor Julius Caesar is supposed to have been born in this way.
5. Radiological anatomy: Hysterogram and hystero-salpingogram—A contrast medium is injected into the uterine cavity and X-ray pictures are taken. The medium flows from the uterine cavity into the fallopian tubes also. The radiograph will show the cavity of the uterus and the lumen of the fallopian tubes.
6. Intrauterine contraceptive devices (IUCD) like Copper-T is introduced into the uterine cavity, which prevents implantation. This is an effective birth control measure.
7. Laparoscopy: By inserting a laparoscope through a small incision in the anterior abdominal wall, the external surfaces of uterus, uterine tubes and ovaries can be visualized (The laparoscope has illuminating and magnifying lens systems). Laparoscopic tubal ligation can be done with this instrument.
8. Fibroids are large masses of smooth muscle fibers arising from the myometrium. They may turn malignant and require surgical removal.
9. Surgical removal of uterus is called hysterectomy. The ureter may get damaged or accidentally ligated because of its close relation to the uterine artery.

UTERINE TUBES OR FALLOPIAN TUBES (FIG. 15.7)

The uterine tubes are long, tortuous tubes, extending laterally from the cornua of the uterus. Each tube is 10-12 cm long. These tubes lie in the free upper border of the broad ligament; the part of the broad ligament attached to the uterine tube is called the mesosalpinx (salpinx = tube).

Functions

1. They carry oocytes from the ovaries—to the site of fertilization.
2. They carry sperms from the uterus—to the site of fertilization.

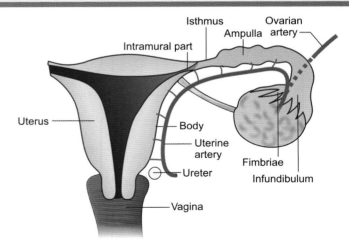

Fig. 15.7: Uterine tube

3. The uterine tube also conveys the dividing zygote to the uterine cavity.

Each tube opens proximally into the cornu of the uterus and distally, into the peritoneal cavity, near the ovary. Thus, the uterine tubes allow communication between the peritoneal cavity and the exterior of the body.

Parts

For purposes of description, the uterine tube is divided into 4 parts:

1. The infundibulum
2. Ampulla
3. The isthmus
4. The uterine or intramural part or interstitial part.

The infundibulum: (L: funnel) – It is the funnel-shaped lateral end (distal end) of the uterine tube, which is closely related to the ovary. Its opening into the peritoneal cavity is called the abdominal ostium (about 2 mm in diameter). Its margins have 20-30 finger-like processes, called the fimbriae. These processes are spread over the surface of the ovary. One of these fimbriae, is large and it is attached to the ovary. It is called the ovarian fimbria.

During ovulation, the fimbriae trap the oocyte and direct it through the abdominal ostium into the ampulla.

The Ampulla

The part of the uterine tube, medial to the infundibulum is called the ampulla. It is thin-walled, dilated and tortuous. It forms about two-third of the tube (about 6-7 cm). It arches over the upper pole of the ovary. This is the normal site of fertilization and most common site of ectopic pregnancy.

The Isthmus

It lies medial to the ampulla, it is the short (about 2.5 cm), narrow, thick-walled portion that enters the cornu of the uterus.

The Uterine or Intramural Portion

The uterine or intramural portion of the uterine tube is the short segment that passes through the thick myometrium of the uterus. It opens via the uterine ostium into the uterine cavity.

Arterial Supply

Derived from branches of uterine and ovarian arteries (medial 2/3 by uterine artery and lateral 1/3 by ovarian A, approximately). The corresponding veins drain into uterine and ovarian veins.

Lymphatic Drainage

Lymphatic drainage to aortic lymph nodes.

Nerve Supply

From autonomic system.

Sympathetic

Sympathetic, from T10-L2; mainly vasomotor

Parasympathetic

Parasympathetic from S2,S3, S4—they inhibit peristalsis and produce vasodilatation.

Histology

Histology (Fig. 15.8) fallopian tube has a mucous membrane, surrounded by a muscle coat and an external covering of peritoneum.

1. Mucous membrane: It shows numerous branching folds, which almost fill the cavity of the tube. These folds are very prominent in the ampulla. Mucosa is lined by columnar epithelium. Some of these cells are ciliated; their ciliary action helps to move the ova towards the uterus.

   Deep to the epithelium is a highly cellular connective tissue.

2. The muscle coat has an inner circular and outer longitudinal layer of smooth muscle. The circular muscle is thickest in the isthmus.

Applied Anatomy

1. One of the major causes of infertility in women is blockage of the uterine tubes from infection.

   Usually the oocyte is fertilized in the ampulla of uterine tube. The dividing zygote passes slowly along the tube and reaches the uterine cavity. Fertilization cannot occur when both tubes are blocked, because the sperms cannot reach the oocyte.

2. Patency of a uterine tube may be determined by:
   a. *Hysterosalpingography:* A radio-opaque material is injected into the uterus and X-ray is taken. The cavity of the uterus and the lumina of the tubes can be visualized.
   b. *Tubal insufflation test:* Normally, air pushed into the uterus passes through the tubes and leaks into the peritoneal cavity. This leakage produces a bubbling noise, which can be auscultated over the iliac fossae. Ideally, this test should be done soon after menstruation; otherwise, there is a remote possibility of dislodging a fertilized-ovum.

3. *Ligation of uterine tubes* is an effective method of birth control. The oocytes discharged from the ovaries reach the distal part of the tube and die; they disappear soon.

   Abdominal tubal ligation as well as Laparoscopic tubal ligation can be done.

4. *Salpingitis and peritonitis:* Because the female genital tract is in direct communication with the peritoneal cavity through the abdominal ostia of uterine tubes, infections of the vulva, vagina, uterus, and the tubes may result in peritonitis.

5. *Ectopic tubal pregnancy:* Infection of the uterine tube (salpingitis) may cause partial obstruction of its lumen due to adhesions. In these cases, zygote may not be able to pass to the uterus; so the blastocyst may implant in the mucosa of the uterine tube (commonest site is the ampulla). If not diagnosed early, ectopic tubal pregnancy results in rupture of the uterine tube and hemorrhage into the pelvic cavity. It may threaten the life of the mother and results in the death of the embryo.

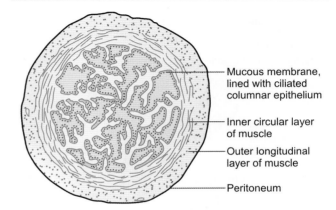

Fig. 15.8: Microscopic structure of uterine tube

- Mucous membrane, lined with ciliated columnar epithelium
- Inner circular layer of muscle
- Outer longitudinal layer of muscle
- Peritoneum

■ OVARIES (FIGS 15.9A AND B)

The ovaries are the female gonads. They are two in number, situated one on each side of the uterus in a fossa (the ovarian fossa) behind the broad ligament. The ovaries are situated on the lateral wall of the pelvis.

Boundaries of Ovarian Fossa

1. Behind—the ureter
2. In front—the obliterated umbilical artery
3. In the floor—obturator internus muscle, obturator nerve and vessels.

Size

Each ovary is 3 cm long, 1.5 cm wide and 1 cm in thickness.

Surface and Color

In young adults, the ovary is pink in color and its surface is smooth. Its long axis is nearly vertical.

In older women, due to repeated ovulations, the surface becomes irregular, puckered and the color becomes greyish. The long axis of the ovary becomes almost horizontal.

Each ovary is ovoid in shape. It has 2 poles – upper and lower; 2 borders – anterior and posterior; and 2 surfaces – Medial and lateral.

Relations

Poles

The upper pole is called the tubal pole. It gives attachment to the ovarian fimbria. A fold of peritoneum called the suspensory ligament of ovary or the infundibulopelvic ligament stretches from this end of the uterine tube to the lateral wall of the pelvis. It transmits the ovarian vessels and nerves.

The lower pole is tilted towards the uterus and hence called the uterine end. The ovarian ligament stretches from this pole to the lateral border of uterus.

Surfaces

The lateral surface of the ovary comes in contact with the ovarian fossa. It is related to parietal peritoneum, obturator internus muscle and its fascia, obturator nerve and obturator vessels.

The medial surface is overlapped by the uterine tube.

Border

The *posterior border* is called the free border.

The anterior border (mesovarian border) – a fold of peritoneum called mesovarium is attached to this border (it is continuous with the posterior layer of broad ligament). This border shows a cleft, called the hilum, through which the blood vessels and nerves enter the ovary; the lymphatics and veins come out.

Blood Supply

The ovarian artery, a branch of the abdominal aorta supplies the ovary. The veins, called pampiniform plexus, emerge from the ovary and join to form the

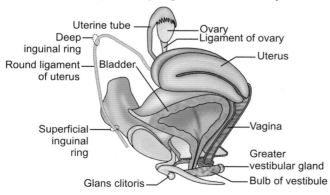

Fig. 15.9A: Ovary

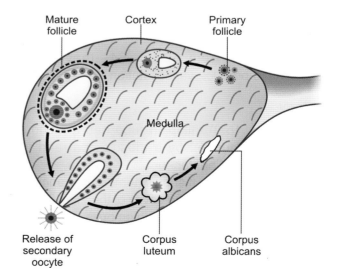

Fig. 15.9B: Ovary—structure

ovarian vein. The right ovarian vein drains into the inferior vena cava and left vein drains into the left renal vein.

Nerve Supply

Sympathetic fibers from T10, T11 spinal segments, parasympathetic from vagus.

Lymphatics

Lymphatics drain into lateral aortic nodes, lying near the origin of ovarian arteries.

Histology

The ovary has an outer cortex and an inner medulla. The free surface of the ovary is covered by a single layer of cuboidal epithelium called the germinal epithelium, which is continuous with the mesothelium of peritoneum.

Medulla

The medulla is made of loose connective tissue, connective tissue cells, blood vessels, nerves and lymphatics.

Cortex

The cortex has a dense collection of stromal cells. Different stages in the maturation of follicles may be seen in the cortex. They are:

Primordial follicle, with a central oogonium, surrounded by a single layer of cuboidal stromal cells.

Primary follicle, with a central oogonium, surrounded by a single layer of columnar cells (follicular cells).

*Secondary follicle,*with a central oogonium, surrounded by several layers of follicular cells (granulosa cells).

Mature ovarian follicle or Graafian follicle (Fig. 15.10) It is surrounded by a capsule, which has 2 layers – an inner vascular layer, the theca interna and an outer fibrous layer, the theca externa. The theca interna secretes estrogen.

The granulosa cells begin to secrete a fluid, liquor folliculi, which causes the oocyte to be displaced to one side of the follicle, where it becomes surrounded by a mass of granulosa cells called the cumulus oophorus. The oocyte is surrounded by a membrane, the zona pellucida.

The ovarian follicle is very small in the beginning. Later it becomes so big, that it reaches the surface of

Fig. 15.10: Mature ovarian follicle or Graafian follicle

the ovary and forms a bulge. The follicle ultimately ruptures and the oocyte is shed from the ovary. This process is called ovulation.

After ovulation, the remaining part of the follicle forms an important structure, called corpus luteum (yellow body). It has an endocrine function. It secretes the hormone progesterone, which prepares the uterus for implantation.

If the ovum is not fertilized, the corpus luteum degenerates and forms a mass of fibrous tissue called corpus albicans (white body).

The ovaries are under the control of 2 hormones – FSH and LH, secreted by the anterior pituitary. The formation of oocytes—*oogenesis,* is described below.

Oogenesis

The process of development of ova from the stem cells (oogonia) is known as oogenesis. It occurs in the cortex of the ovary as follows:

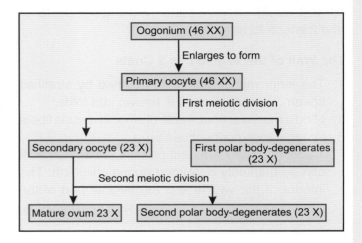

Section C

The Organ Systems

Applied Anatomy

1. At ovulation, some women experience pain, called "Mittelschmerz" (or ovulation pain). Pain radiates along T10 dermatome to paraumbilical region.
2. Ovulation occurs approximately 14 days before the next ovarian cycle (e.g. if the woman has a 28 day cycle, ovulation occurs on 14th day; if it is a 35-day cycle, ovulation occurs on 21st day).
3. Ovarian cysts: The developmental arrest of the ovarian follicles may result in the formation of ovarian cysts.
4. Carcinoma of the ovary is common.
5. Oocytes can be obtained from hyperstimulated ovaries for *in vitro* fertilization (IVF). Normally only one oocyte matures during the ovarian cycle. Several oocytes can be stimulated by the administration of clomifen citrate.

VAGINA (GK – KOLPOS)

The vagina is the female organ of copulation. It is a fibromuscular canal, forming the inferior portion of the female genital tract and the birth canal. It is 7-9 cm long. It extends from the cervix of the uterus to the vestibule of the vagina. In the anatomical position, it is directed anteroinferiorly.

The posterior wall of the vagina is 1 cm longer than the anterior wall. It is in contact with the external os.

The vagina communicates superiorly with the cervical canal and opens inferiorly at the vestibule, through the vaginal orifice.

Relations

The vagina is related anteriorly to the urinary bladder and posteriorly, to the rectum. It passes between the two levator ani muscles. It pierces the urogenital diaphragm. The sphincter urethrae muscle and the pubovaginalis muscle support the vagina.

The vaginal recess around the cervix is called the fornix (L. arch) – it has 4 parts – the anterior, posterior and 2 lateral fornices.

The Wall of the Vagina has 3 Coats

1. The inner mucous membrane, lined by stratified squamous epithelium; it is thrown into folds;
2. Middle muscular coat, made of smooth muscle fibers
3. Outer, dense connective tissue.

The vagina is the longest part of the birth canal. It can be markedly distended during childbirth. The interior of the vagina and the vaginal part of the cervix can be examined with a vaginal speculum.

Vaginal Examination

Vaginal examination (per vaginal examination or PV): Structures palpated by per vaginal examination are:
1. Anteriorly : Urethra
   Urinary bladder
   Pubic symphysis
2. Posterioly : Rectum
   Pouch of Douglas
3. Laterally : Ovary
   : Fallopian tube
   : Lateral pelvic wall
   : Ligaments
   : Ureter
4. Superiorly —cervix
5. It also helps in the measurement of diagonal conjugate, which extends from the lower border of symphysis pubis to the sacral promontory. It helps to assess the proportion between the fetal head and the pelvic inlet during the later part of pregnancy.
6. With one hand of the clinician on the lower part of anterior abdominal wall, and simultaneous per vaginal examination, the uterus can also be palpated. Pervaginal examination is done after emptying the urinary bladder. This is to differentiate between anatomical and physiological retroversion. The position of uterus changes from anteverted to retroverted as the bladder fills up.

This is called physiological retroversion. When the bladder is empty, the uterus regains its anteverted antelexed position. In some women, even after emptying the bladder, the uterus remains in retroverted position. This is anatomical retroversion. Patient is placed in lithotomy position before performing per vaginal examination (See CP11A and B in color plate 3).

Abnormal communication may exist between the vagina and bladder (vesicovaginal fistula) or between the vagina and rectum (rectovaginal fistula).

The lower 1/3 of vagina, supplied by the inferior rectal branches of pudendal nerve, is pain-sensitive, upper 2/3, supplied by autonomic plexus, is not sensitive to pain.

The Hymen

The lower end of the vagina is guarded by a crescentic fold of mucous membrane called the hymen. During infancy and childhood, the vaginal orifice is usually closed by the hymen but it normally ruptures before puberty, to allow the menstrual fluid to escape. In virgins, the vaginal orifice will be small, admitting the tip of one digit. After childbirth, very little of the hymen is left, except a few "tags" called hymenal caruncles.

Imperforate Hymen

The hymen, sometimes, is completely imperforate (without an opening). In such cases, the menstrual blood starts collecting in the vaginal canal after puberty (hematocolpos) and will require surgical intervention for letting the blood out.

▋ MAMMARY GLAND OR BREAST

The breast is present, bilaterally, in the pectoral region, in both sexes. After puberty, the female breasts are well-developed. The breast is a modified sweat gland.

The adult female breast has a base, nipple, areola and an axillary tail (Fig. 15.11A).

Base

The base is circular; vertically, it extends from the 2nd to 6th ribs, in the midclavicular line. Horizontally, it extends from the lateral border of sternum to the mid-axillary line. The base rests mainly on the pectoralis major muscle and the pectoral fascia (Fig. 15.11B). A retromammary space, containing loose areolar connective tissue intervenes between the base and the fascia covering the pectoralis major muscle.

Nipple

The nipple is a conical projection below the center of the breast. It is pierced by 15-20 lactiferous ducts. It has circularly and longitudinally arranged smooth muscle fibers. The circular fibers erect the nipple for sucking and the longitudinal muscle retracts the nipple. The nipple is richly innervated.

Areola

The areola is the pigmented circular area of skin around the base of the nipple. Outer margin of the areola has modified sebaceous glands. They are enlarged during pregnancy and lactation and are known as Montgomery's tubercles. Their oily secretions form a protective lubricant during lactation.

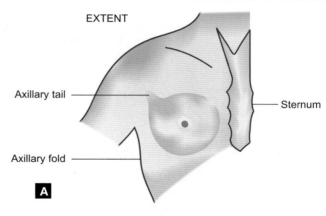

Fig. 15.11A: The breast

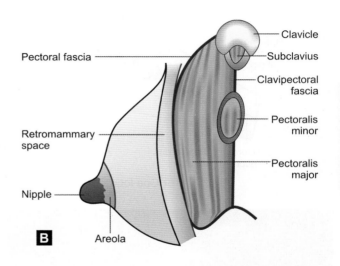

Fig. 15.11B: Deep relations

Axillary Tail of Spence

It is a tail-like projection from the upper outer quadrant of the breast into the axilla. When enlarged, it may be mistaken for an enlarged axillary node or a lipoma.

Structure (Figs 15.12A to C)

The breast is made up of 3 components:
1. Glandular tissue
2. Fibrous tissue
3. Interlobar fatty tissue.

Glandular Tissue

The glandular tissue consists of 15-20 pyramidal lobes. Each lobe is drained by a lactiferous duct. The lobes

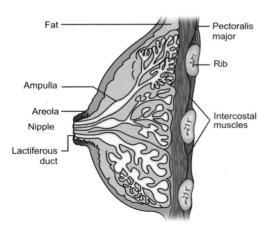

Fig. 15.12A: Lobes of mammary gland

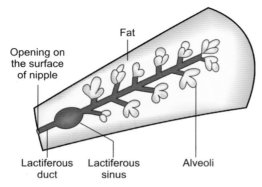

Fig. 15.12B: One lobe

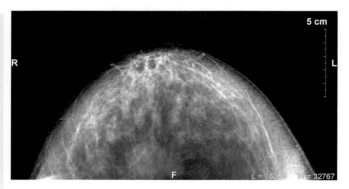

Fig. 15.12C: Mammogram showing tumor(F)

are arranged in a radiating manner from the nipple. Deep to the areola, each lactiferous duct dilates to form a sinus, the lactiferous sinus. The ducts open finally onto the nipple.

The glands are tubuloalveolar type.

The fibrous tissue supports the lobes; connects the skin to the pectoral fascia.

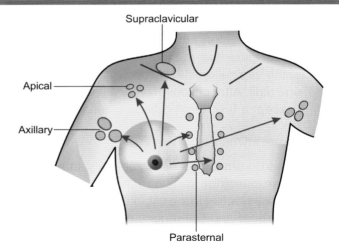

Fig. 15.13: Lymphatic drainage

The interlobar fatty tissue makes the breast rounded in contour. Fat is absent beneath the nipple and areola.

Blood Supply

Arterial Supply

Breast is supplied by lateral mammary branches (axillary artery), perforating branches from 2nd, 3rd, 4th intercostals arteries (internal thoracic artery), and superior thoracic artery.

The veins drain into the axillary, internal thoracic and intercostal veins.

Nerve Supply

The intercostal nerves and sympathetic fibers supply the skin and blood vessels respectively.

Lymphatic Drainage (Fig. 15.13)

About 75% of lymph from the mammary gland drains into axillary nodes.

About 20% into *parasternal* and the remaining 5% to posterior intercostal nodes. Some vessels drain from the upper part of the breast into the *supraclavicular* nodes also, and a few vessels may drain into *hepatic* nodes.

The cutaneous lymphatics of one breast communicate across the midline with those of the opposite side. In suspected cases of cancer of breast, lymph nodes of same side as well as opposite side should be examined.

BREAST CANCER

Breast cancer is one of the common malignant diseases in women. While any portion of the breast may be affected, the disease most commonly begins in the upper and outer quadrant. A blood-stained discharge from the nipple, presence of a "lump" with fixation to the underlying tissues, retraction of overlying skin, retraction of nipple, peau d'orange and enlargement of axillary lymph nodes are some of the important features.

Reason for Fixation of the Tumor

Infiltration of cancer cells along the fibrous tissue which connects the skin to the pectoral fascia (Ligaments of Cooper) and its fibrosis, results in the fixation of the tumor.

Retraction of Nipple

Retraction of nipple is due to extension of the growth along the lactiferous ducts and the subsequent fibrosis.

Peau D'Orange ("The Skin of an Orange")

The hair follicles overlying the tumor appear to be retracted. This is caused by obstruction of cutaneous lymphatics. This results in collection of lymph and edema of skin around the hair follicle. So, the skin resembles an orange peel.

Mammography (Fig. 15.12C)

The soft tissue radiography or mammography is useful in the early detection of breast cancer.

Treatment

Mastectomy, followed by radiotherapy or chemotherapy.

MALE REPRODUCTIVE SYSTEM

In males, the reproductive system is closely related to the urinary system. The urethra is shared by the urinary and genital systems.

The male reproductive organs include the external and internal genitalia.

1. The external genitalia or genital organs. The external genital organs include:
   a. The penis
   b. The scrotum
   c. The testis and epididymis
   d. The spermatic cord.
2. The internal genital organs: are the following, on each side
   a. The Vas Deferens or the ductus deferens
   b. The seminal vesicle
   c. The ejaculatory duct
   d. The prostate
   e. The urethra.

PENIS (L-TAIL) (FIG. 15.14)

The penis is the male organ of copulation. It has:
1. A root or attached portion
2. A body or free portion.

The root of the penis is situated in the superficial perineal pouch. It is composed of three masses of erectile tissue. They are (a) the bulb, and (b) The two crura (singular-crus).

Each *crus* is firmly attached to the margins of the pubic arch and is covered by the ischiocavernosus muscle. The crura, continue anteriorly into the body of the penis, as the corpora cavernosa.

The bulb is situated between the two crura. It is attached to the perineal membrane and is covered by the bulbospongiosus muscle. The bulb continues anteriorly, into the body of the penis, as the corpus spongiosum. The bulb is pierced by the urethra, which shows a dilatation called the intrabulbar fossa. The urethra then traverses the corpus spongiosum.

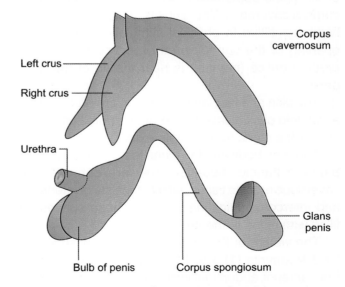

Fig. 15.14: Components of penis

Section C

The Organ Systems

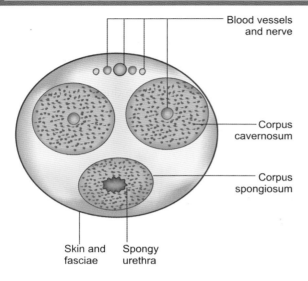

Fig. 15.15: Penis—transverse section

The Body of Penis (Fig. 15.15)

It is continuous with the root. It is composed of 3 cylindrical bodies (or corpora) of erectile tissue that are enclosed by dense fibrous tissue, deep fascia and skin. The skin is very thin and dark in color.

Two of the three erectile bodies, the corpora cavernosa, are arranged side by side, in the dorsal part of the penis. The corpus spongiosum lies ventrally, it contains the spongy urethra.

At the distal end of the body is the glans penis. It consists entirely of corpus spongiosum. The concavity of the glans penis covers the free blunt ends of the corpora cavernosa. The glans is richly innervated and highly sensitive to physical stimulation. The slit-like opening of the spongy urethra, called the external urethral orifice (meatus) is near the tip of the glans penis.

The skin and fasciae of the penis are prolonged as a free fold of skin called the prepuce (foreskin), which covers the glans to a variable extent.

The root, body and the glans feel spongy when the penis is flaccid, because they are composed of cavernous erectile tissue, which consists of interlacing and intercommunicating spaces. These spaces are filled with blood in the erect penis.

The weight of the body of the penis is supported by 2 ligaments (1) The fundiform ligament (extends from anterior abdominal wall) and (2) The suspensory ligament, (extends from the pubic symphysis).

Arteries of Penis

The internal pudendal artery gives off 3 branches to the penis.

Deep Artery

The deep artery runs in the corpus cavernosum. Its branches, having a spiral course are called helicine arteries (helix = spiral).

Dorsal Artery

The dorsal artery, runs on the dorsum, supplies the glans penis, the prepuce and distal part of corpus spongiosum.

Artery of the Bulb

The artery of the bulb supplies the bulb and proximal part of corpus spongiosum.

The Veins

Correspond to the arteries. Some of them drain into the prostatic venous plexus.

Nerve Supply

1. Sensory supply is derived from dorsal nerve of the penis
2. Motor supply to the muscles of penis are derived from perineal branch of pudendal nerve.
3. Autonomic nerves are derived from pelvic plexus. Sympathetic fibers are vasoconstrictors and parasympathetic (S2, S3, S4) are vasodilators.

Lymphatic Drainage

From glans penis – to deep inguinal nodes
From the rest of the penis – to superficial inguinal lymph nodes.

The Anatomical Basis of Erection

When a male is sexually stimulated, the smooth muscle in the helicine arteries relaxes owing to parasympathetic stimulation. As a result, the arteries straighten and their lumina enlarge. The cavernous spaces are filled with blood, they dilate and become rigid.

Meanwhile, the ischiocavernous and bulbospongiosus muscles compress the venous plexuses at the periphery of the corpora cavernosa, and prevent the return of venous blood. As a result, the three corpora become enlarged, rigid and the penis erects.

Following ejaculation and orgasm, the penis gradually returns to the flaccid state. This results from sympathetic stimulation that causes constriction of smooth muscle fibers in the helicine arteries. The bulbospongiosus and ischiocavernosus muscles relax, enhancing venous return, into the deep dorsal vein.

Applied Anatomy

1. *Congenital Anomalies:* A common congenital malformation of penis and urethra is *hypospadias.* The external urethral orifice opens on the ventral surface of the glans penis or the body.
2. *Phimosis:* The prepuce of the penis is usually sufficiently elastic to allow retraction over the glans penis; in some males, it cannot be retracted easily. The prepuce fits tightly over the glans. This condition is called phimosis.
3. *Circumcision:* Circumcision is the surgical removal of prepuce, in cases of phimosis.

SCROTUM

The scrotum is a loose cutaneous fibromuscular sac, that is situated posteroinferior to the penis and inferior to the pubic symphysis. It is composed of skin and dartos muscle. The dartos muscle, firmly attached to the skin, consists of smooth muscle fibers that contract under the influence of cold, exercise and sexual stimulation. Under these conditions, the wall of the scrotum becomes firm, and the skin becomes rugose (wrinkled).

Externally, the scrotum is divided into right and left halves by a ridge or raphe.

Layers of Scrotum

1. Skin
2. Dartos muscle
3. The external spermatic fascia
4. The cremasteric fascia
5. The internal spermatic fascia.

The scrotum is supplied by the superficial and deep external pudendal vessels.

Nerve Supply

The anterior surface (anterior 1/3) of the scrotum is supplied by L1 segment of spinal cord (through ilio-inguinal and genitofemoral nerves). The posterior surface (posterior 2/3) is supplied by S3 segment, through scrotal branches of pudendal nerve.

The dartos muscle is involuntary; it is supplied by sympathetic fibers.

Lymphatic Drainage

Lymphatic drainage: To superficial inguinal lymph nodes.

Contents

The scrotum contains the right and left testes, the epididymis and the lower part of the spermatic cords.

Function

Contraction of the dartos muscle and the cremasteric muscle causes the testes to be drawn against the body.

In hot weather, the scrotum relaxes and allows the testes to hang freely away from the body. This provides a larger surface area for the dissipation of heat.

These reflexes of the scrotum, in response to temperature, help to maintain a stable temperature; it is an important function because spermatogenesis will be impaired by extremes of heat or cold.

TESTES

Singular Testis

The testis is the male gonad. It is homologous with the ovary of the female.

It is suspended in the scrotum by the spermatic cord. It lies obliquely; the upper pole is directed forwards and laterally; the lower pole is tilted backwards and medially. Each testis is ovoid in shape. In adults, the left testis is lower in position than the right.

Size

Length—5 cm; thickness—2.5 cm; breadth—3 cm; weight—10 to 15 gram.

External Features (Fig. 15.16)

Testis has 2 poles—upper and lower; 2 borders—anterior and posterior; and 2 surfaces—medial and lateral.

The upper pole provides attachment to the spermatic cord. The anterior border is completely

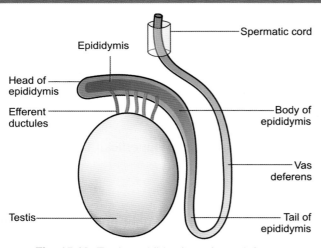

Fig. 15.16: Testis, epididymis and vas deferens

covered by the tunica vaginalis. The posterior border is related to the epididymis. The medial and lateral surfaces are convex and smooth.

Coverings of the Testis

The testis is covered by 3 coats:
1. The outermost tunica vaginalis
2. The middle, tunica albuginea, and
3. The inner tunica vasculosa.

Tunica Vaginalis

The tunica vaginalis is a closed, serous sac of peritoneum (It is a derivative of the processus vaginalis, which is a peritoneal sac surrounding the testis. It was cut off from the greater peritoneal sac before birth). The tunica vaginalis has 2 layers, the visceral layer is adherent to the testis and the parietal layer is adjacent to the internal spermatic fascia. A very small amount of fluid normally separates the visceral and parietal layers; this enables the testis to move freely within the scrotum.

Tunica Albuginea

The tunica albuginea is a dense fibrous coat covering the testis all around. It is covered by the visceral layer of tunica vaginalis, except posteriorly, where the testicular vessels and nerves enter the gland.

Posteriorly, the tunica albuginea is thickened to form the mediastinum testis. A number of septa extend from the mediastinum to the inner surface of the testis, dividing the testis into 200-300 lobules.

Tunica Vasculosa

The tunica vasculosa: Is the innermost vascular coat of the testis, lining its lobules.

The scrotum and its layers can also be considered as the coverings of testis. So, the testis is covered by skin, dartos muscle, the external spermatic fascia, the cremasteric fascia, internal spermatic fascia, tunica vaginalis, tunica albuginea and tunica vasculosa (These layers can be memorized with the help of a mnemonic "Some Dirty Englishmen Call It Testis. S = Skin; D – Dartos; E – External spermatic fascia; C = Cremasteric fascia; I – Internal spermatic fascia; T – the 3 tunicae).

Structure of Testis (Figs 15.17A and B)

1. The tunica albuginea consists of closely packed collagen fibers and elastic fibers. Posteriorly the tunica albuginea is thickened to form the mediastinum testis. A number of septa pass from the mediastinum testis to the tunica albuginea and divide the testis into 200-300 lobules.

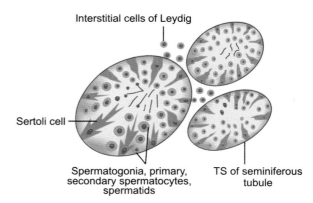

Fig. 15.17A: Structure of testis

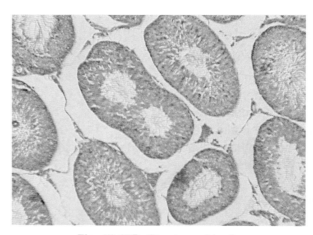

Fig. 15.17B: Structure of testis

2. Each lobule contains one or more highly coiled seminiferous tubules (each tubule is 70-80 cm long). These tubules are lined by cells, which are concerned with the production of spermatozoa (sperms).

3. Within the lobule, the spaces between the coiled seminiferous tubules are filled with loose connective tissue containing blood vessels, nerves and lymphatics.

4. Lying in the connective tissue, there are groups of cells having endocrine function – these cells are called the interstitial cells of Leydig. They secrete the male sex hormone (testosterone).

5. The ends of the seminiferous tubules join to form straight tubules, which ultimately form the efferent ductules. These ductules pass from the upper part of the testis into the epididymis. The epididymis has a head, body and a tail. At the lower end of the tail of the epididymis, the duct continues as the vas deferens.

6. Structure of the seminiferous tubule: The wall of each tubule is made up of fibrous tissue. Between this fibrous wall and the lumen (cavity) of the tubule, there are several layers of cells. Most of these cells represent stages in the formation of spermatozoa. These cells are:
   a. The spermatogonia (46 XY)
   b. Primary spermatocytes (46 XY)
   c. Secondary spermatocytes, and (23 X or 23 Y)
   d. Spermatids (23 X or 23 Y)

The process of formation of spermatozoa is called spermatogenesis. The steps are as follows:

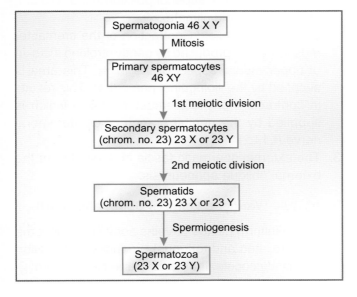

```
        Spermatogonia 46 X Y
                 │ Mitosis
                 ▼
        Primary spermatocytes
               46 XY
                 │ 1st meiotic division
                 ▼
        Secondary spermatocytes
        (chrom. no. 23) 23 X or 23 Y
                 │ 2nd meiotic division
                 ▼
             Spermatids
        (chrom. no. 23) 23 X or 23 Y
                 │ Spermiogenesis
                 ▼
             Spermatozoa
           (23 X or 23 Y)
```

The Spermiogenesis

The process by which a spermatid becomes a spermatozoon or sperm is called spermiogenesis. The spermatid is a rounded cell, with a nucleus, Golgi complex and mitochondria. The nucleus of the spermatid forms the head of the sperm. The Golgi complex forms the acrosomic cap. Mitochondria form the middle piece of the sperm.

At the end of spermatogenesis 2 types of sperms are formed—sperms with 23 X and sperms with 23 Y chromosomes.

In between the germ cells are supporting cells or Sertoli cells. They are tall, slender cells. They support the developing germ cells and provide them with nutrition. They may also act as macrophages.

Structure of a Spermatozoon (Sperm) (See Fig. 28.1)

The spermatozoon has a head, neck, a middle piece and a tail. The head is covered by a cap – the acrosomic cap. The neck is narrow, containing a spherical centriole. An axial filament begins just below this centriole.

The axial filament passes through the middle piece, into the tail. In the middle piece, the axial filament is surrounded by a spiral sheath of mitochondria.

The head of the sperm contains highly condensed chromatin (DNA). Each spermatozoon is an active, motile body. Its function is to fertilize the oocyte.

Arterial Supply of Testis

The testicular artery, a branch of the abdominal aorta, supplies the testis.

Venous Drainage

The veins emerging from the testis form a pampiniform plexus (resembling a vine). This plexus surrounds the testicular artery and the vas deferens. These veins condense to form a testicular vein. On the right side, the testicular vein drains into the inferior vena cava; on the left side, the testicular vein drains into the left renal vein.

Lymphatic Drainage

Lymphatics from the testes drain into preaortic and para-aortic nodes at the level of L2.

Section C

The Organ Systems

Nerve Supply

Testis is supplied by sympathetic fibers arising from T10 spinal segment.

Descent of the Testes

In the fetus, the testes develop in the lumbar region. Subsequently, they descend through the inguinal canals into the scrotum. They traverse the inguinal canal during the 7th month, and reach the bottom of the scrotum by 9th month.

Applied Anatomy

1. The testis may be absent on one side or both sides.
2. Undescended testis (cryptorchidism) – the testis fails to descend to the scrotum; it may lie in the lumbar, iliac, inguinal or upper scrotal region. The testes are undescended in 30% premature infants. Most undescended testes descend during the first few weeks after birth.
    Spermatogenesis may fail to occur in undescended testis and the chances of development of tumors is high.
3. Hydrocele is a condition in which fluid accumulates in the processus vaginalis.
4. An ectopic testis may be found in the perineum, at the root of the penis or in the femoral triangle.
5. Torsion of the testis: May result in obstruction of blood flow to the testis and lead to gangrene.

■ EPIDIDYMIS

This is a comma-shaped mass, made up of highly coiled tubes. It has 3 parts – a head, body and a tail. The epididymis is applied to the superior and posterolateral surfaces of testis.

The superior expanded part of the epididymis is called the head, which is composed of the coiled ends of the efferent ductules. They transmit the sperms from the testis to the epididymis.

The body of the epididymis consists of a highly coiled duct of the epididymis. The sperms are stored in the epididymis, where they undergo final stages of maturation. The tail of the epididymis is continuous with the ductus deferens (Vas deferens).

■ SPERMATIC CORD

The spermatic cord suspends the testis in the scrotum. It consists of structures running to and from the testis.

The spermatic cord begins at the deep inguinal ring, where its constituents assemble. The cord ends at the posterior border of the testis. The spermatic cord passes through the inguinal canal, emerges at the superficial inguinal ring and descends within the scrotum to the testis.

Constituents of Spermatic Cord

1. The ductus deferens or the vas deferens is a thick-walled tube, because of abundant smooth muscle in its wall.
2. Arteries:
    a. The testicular artery
    b. Cremasteric artery
    c. Artery to the ductus deferens
3. The pampiniform plexus of veins
4. Lymph vessels from the testis
5. The genital branch of genitofemoral nerve and sympathetic plexus, accompanying the arteries
6. Remains of processus vaginalis (a sleeve of peritoneum).

Coverings of Spermatic Cord

Three layers of fascia derived from the anterior abdominal wall, cover the spermatic cord.

1. Innermost, internal spermatic fascia (derived from fascia transversalis).
2. The cremaster muscle and fascia are derived from the internal oblique muscle of the abdomen and its fascia. This muscle, forming loops, reflexly draws the testis to a more superior position in the scrotum, especially in cold climate.
    Cremasteric Reflex: Contraction of the cremaster muscle can be produced by gently stroking the skin of upper medial aspect of the thigh. This area is supplied by the ilioinguinal nerve (L1). This results in contraction of the cremaster muscle which is supplied by the genital branch of genitofemoral nerve (L1,L2).
3. The external spermatic fascia is derived from the external oblique aponeurosis.

Varicocele (Varicose Veins of Spermatic Cord)

The pampiniform plexus of veins sometimes become varicose (dilated and tortuous) producing a condition known as varicocele, that feels like a "bag of worms".

Varicocele is more common on the left side which may be due to defective valves or due to compression produced by the loaded sigmoid colon.

THE VAS DEFERENS OR DUCTUS DEFERENS (FIGS 14.14 AND 14.17)

(Vas = vessel; ductus = duct; deferens = to carry down. The Latin word "ductus" is more apt than the Latin word "vas").

The ductus deferens is a thick-walled muscular tube. It is the continuation of the duct of the epididymis; it begins in the tail of the epididymis and ends by joining the duct of the seminal vesicle to form the ejaculatory duct. The ductus deferens is about 45 cm long; it transmits the spermatozoa from the epididymis to the ejaculatory duct.

In its course, the vas deferens lies in (1) the scrotum; (2) the inguinal canal; (3) the greater pelvis; and (4) the lesser pelvis.

The vas deferens leaves the spermatic cord at the deep inguinal ring, enters the greater pelvis, where it passes backwards and medially to enter the lesser pelvis. In the lesser pelvis, it lies deep to the peritoneum, crosses the ureter near the posterolateral angle of the bladder and reaches the base of the urinary bladder. At first, it lies superior to the seminal vesicle. Then it descends medial to the ureter and seminal vesicle. Here, it approaches the opposite duct and reaches the base of the prostate.

At the base of the prostate, the ductus deferens is joined by the duct of the seminal vesicle to form the ejaculatory duct.

The part of the ductus deferens lying behind the base of the bladder, is dilated and tortuous. It is known as the ampulla.

Structure (Figs 15.18A and B)

The wall of the ductus deferens has 3 layers (1) Innermost mucous membrane; (2) Middle muscular layer; and (3) The connective tissue.

The mucous membrane shows a number of longitudinal folds, so that the lumen appears stellate or star-shaped. Epithelium is simple columnar; in the extra-abdominal part, the cells are ciliated.

The muscle coat is very thick, consists of 2 layers, inner circular and outer longitudinal layers of smooth muscle. (May have an additional circular layer).

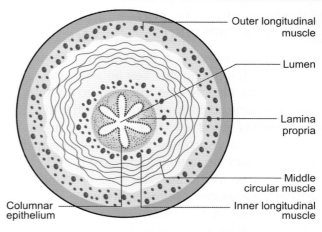

Fig. 15.18A: Structure of vas deferens

Outer longitudinal muscle
Lumen
Lamina propria
Middle circular muscle
Inner longitudinal muscle
Columnar epithelium

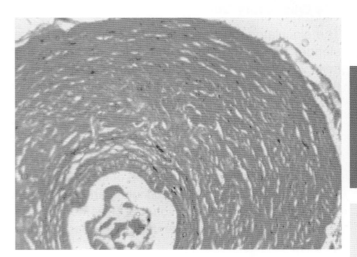

Fig. 15.18B: Structure of vas deferens

Vasectomy (Cutting the Vas Deferens)

Vasectomy is a common method of sterilizing the males. In this operation, small incisions are made in the superior part of the scrotum. The two deferent ducts are located. They are tied in two sites. The portion of the ducts between the sutures (ties) is then excised.

Although sperm production continues, sperms cannot pass into the ejaculatory ducts and urethra. The sperms degenerate in the epididymis and the proximal part of the vas deferens.

Recanalization can be done if required. The ends of the sectioned ducts are reattached under an operating microscope.

Section C

The Organ Systems

PART
1

SEMINAL VESICLES AND EJACULATORY DUCTS

The seminal vesicle is a thin-walled, pear-shaped structure, 3-5 cm long that lies between the fundus (base) of the bladder and the rectum. Actually, the seminal vesicle consists of a tube, which is 10-15 cm long, which is coiled to form a vesicle-like mass.

The seminal vesicles secrete a thick alkaline fluid, that mixes with the sperms. This fluid provides most of the volume of the seminal fluid or semen. The seminal vesicles do not store sperms.

The duct of each seminal vesicle joins the ductus deferens to form the ejaculatory duct, which opens into the posterior part of the prostatic urethra.

The Ejaculatory Ducts

The ejaculatory ducts are about 2.5 cm long. These slender tubes, formed by the union of duct of the seminal vesicle and the ductus deferens, pass through the prostate and open by slit-like openings into the posterior wall of the prostatic urethra, one on each side of the prostatic utricle.

PROSTATE (FIG. 15.19)

This is the largest accessory gland of the male reproductive system. It is a fibromusculo-glandular tissue. Its secretions add to the volume of semen (in the female, this gland is represented by the paraurethral glands).

The prostate is situated in the anteroinferior part of the pelvic cavity, below the neck of the bladder. It resembles an inverted cone. It has a base, apex, an anterior surface, posterior surface and 2 lateral surfaces. It weighs about 8 gm and surrounds the prostatic urethra.

Prostate has 2 capsules – the inner true capsule formed of connective tissue and the outer false capsule formed by the pelvic fascia. The prostatic venous plexus lies between these two capsules.

Lobes (Fig. 15.20)

The prostatic part of the urethra passes vertically through the prostate gland. The 2 ejaculatory ducts pierce the posterior surface of the prostate and open into the posterior wall of the prostatic urethra.

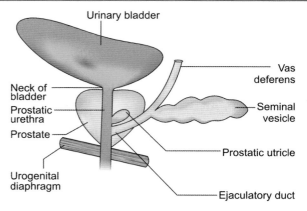

Fig. 15.19: Prostate gland

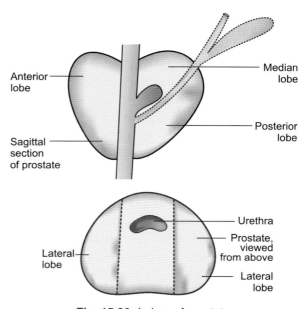

Fig. 15.20: Lobes of prostate

In relation to the urethra and the two ejaculatory ducts, the prostate can be divided into 5 lobes as seen in Figure 15.20.

The posterior lobe is commonly involved in cancer. The median lobe contains more glandular tissue than the other lobes. It is commonly involved in benign enlargement of prostate.

Relations of Prostate (see Figs 14.13 and 14.14– Urinary System, Fig. 15.19)

Superiorly, the base is related to the neck of the bladder. This surface is pierced anteriorly by the prostatic urethra as it enters the prostate.

Inferior, *apex*, is related to the urogenital diaphragm.

Anterior surface is related to the body of pubis. The prostatic urethra emerges out, at the lower part of this surface.

Posterior surface is related to the terminal portions of the vasa deferentia and the seminal vesicles. The ejaculatory ducts pierce this surface and enter the prostatic urethra.

The rectovesical septum or Denonvilliers' fascia lies between the posterior surface of prostate and the rectum.

The lateral surfaces are overlapped by levator ani. A few fibers of levator ani, called the levator prostati, are inserted into the capsule of the prostate.

The Prostatic Urethra

The prostatic urethra (see Fig. 14.17 Urinary system) is the widest and most distensible part of the male urethra. In its posterior wall is a vertical ridge called the urethral crest or the verumontanum. At the lower end of the urethral crest is a spherical swelling, the seminal colliculus. In the middle of the seminal colliculus is a blind recess called the prostatic utricle. On either side of the prostatic utricle are the slit-like openings of the ejaculatory ducts. The ducts of the prostate gland open directly by minute openings on each side of the urethral crest.

Applied Anatomy

1. Benign prostatic hyperplasia usually occurs after 45-50 years. It usually affects the median lobe, which enlarges upwards into the bladder, causing a reflex desire to pass urine frequently (frequency of micturition). The projection formed by the median lobe into the bladder is called "uvula vesicae". It obstructs the internal urethral orifice, when the patient strains during micturition.
2. Cancer usually affects the posterior lobe. The enlarged prostate can be palpated by doing a per rectal examination.
   The cancer of the prostate spreads via the blood stream and lymphatics. The prostatic venous plexus has communications with the valveless internal vertebral venous plexus. Straining to urinate may drive the venous blood from the prostatic venous plexus to the internal vertebral venous plexus. The cancer cells are carried along the blood stream and get deposited in the vertebral column or pelvis.
3. Partial or complete removal of prostate is called prostatectomy (TURP—Transurethral Resection of Prostate).

THE MALE URETHRA

For a description of the male urethra, please refer Chapter 14 "Urinary System".

DEVELOPMENT OF UTERUS AND FALLOPIAN TUBES

Uterus and fallopian tubes develop from paramesonephric ducts, which develop in the intermediate mesoderm.

The ducts of the two sides meet and fuse in the midline. The fused part of paramesonephric ducts give rise to the uterus. The unfused parts develop as the fallopian tubes (Figs 15.21A to C).

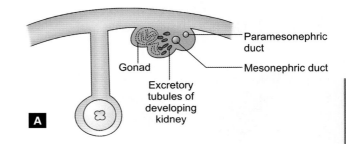

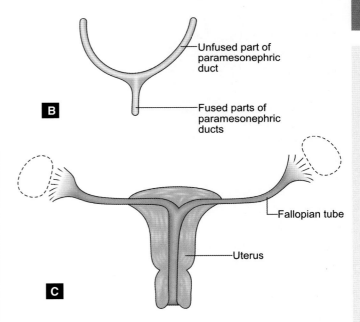

Figs 15.21A to C: Development of uterus and fallopian tubes

Uterus didelphys
(Duplication of uterus and vagina)

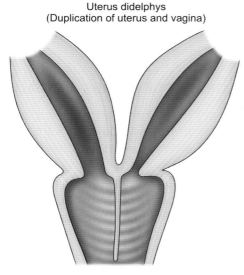

Fig. 15.22: Uterus didelphys

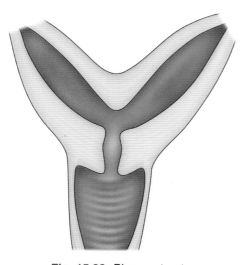

Fig. 15.23: Bicornuate uterus

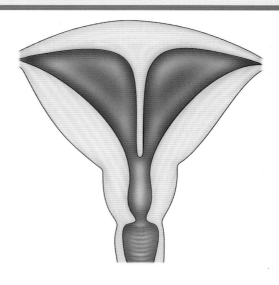

Fig. 15.24: Septum in uterus

Congenital Anomalies

1. *Agenesis:* Failure of development of uterus
2. *Complete duplication (Fig. 15.22) or uterus didelphys:* The inferior parts of paramesonephric ducts fail to fuse resulting in "double uterus". It may be associated with a double or single vagina.
3. *Bicornuate uterus:* If the duplication involves only the superior part of the body of uterus, the condition is called bicornuate uterus (Fig.15.23).
4. *Septum:* In some cases the uterus appears normal externally, but is divided internally by a thin septum (Fig.15.24).

CHAPTER

16 Nervous System

INTRODUCTION

The human body consists of numerous tissues and organs that are entirely different in structure and function. Yet, they function together, and in harmony, for the smooth functioning of the body as a whole. The nervous system is mainly responsible for this.

DIVISIONS OF NERVOUS SYSTEM

Nervous system may be divided into:
1. The central nervous system (CNS)—consisting of brain and spinal cord.
2. The peripheral nervous system (PNS)—consisting of peripheral nerves and ganglia associated with them.
3. The autonomous nervous system (ANS)—consisting of sympathetic and parasympathetic systems.

The Brain

The brain consists of:
1. The cerebrum—comprising two large cerebral hemispheres
2. The cerebellum
3. The midbrain
4. The pons
5. The medulla oblongata.

Brainstem

Midbrain, pons and medulla together form the brainstem. The medulla is continuous inferiorly with spinal cord.

Peripheral nerves attached to the brain are called cranial nerves; those attached to the spinal cord are called spinal nerves.

Tissues Constituting the Nervous System

1. Neurons
2. Neuroglia
3. Blood vessels.

The specialized cells that constitute the functional units of the nervous system are called neurons. Neurons are supported by a special kind of connective tissue called neuroglia (Refer Chapter 4 ,The Nervous Tissue).

Meninges (Fig. 16.1)

The brain and the spinal cord are enclosed by three membranous coverings – the meninges. These are, from the outside, the dura mater, arachnoid mater and pia mater. The dura mater is also called the pachymeninx; the arachnoid and pia are together known as leptomeninges.

The Dura Mater (Meaning—"Tough Mother")

The dura mater or pachymeninx is the outermost and toughest layer. It is a strong, thick, fibrous membrane consisting largely of white collagen fibers. The cerebral dura has 2 layers—an outer endosteal layer, which lines the interior of the cranium and an inner meningeal layer. Two large rigid folds of the inner layer of dura project into the cranial cavity and help to support the brain and to maintain it in position. These folds are (1) The falx cerebri, and (2) The tentorium cerebelli (Fig. 16.1).

Falx cerebri is a sickle-shaped fold lying vertically in the midline and separating the right and left cerebral hemispheres. Anteriorly, it is attached to the crista galli—a bony projection of the ethmoid bone. Posteriorly, it is attached to the tentorium cerebelli. The upper border is related to a dural venous sinus, the superior sagittal sinus. The lower border is related to the inferior sagittal sinus.

The tentorium cerebelli is a crescentic, arched sheet, which lies horizontally and forms a tent-like roof for

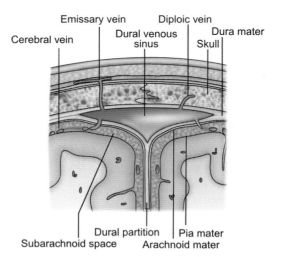

Fig. 16.1: Meninges and dural folds

the posterior cranial fossa, thereby separating the cerebrum above from the cerebellum below.

The arachnoid mater ("cobweb-like mother") is a delicate, transparent, avascular membrane situated between the dura mater and pia mater. It is closely applied to the dura mater but separated from pia by a subarachnoid space, which contains CSF. In some areas, the subarachnoid spaces are large, filled with large quantity of CSF—these spaces are called cisterns, e.g. lying between the undersurface of cerebellum and the medulla is the cisterna magna. This cistern is of importance, because, a needle can be introduced into this between the occipital bone and the atlas vertebra to get a sample of CSF. This procedure is called cisternal puncture.

Arachnoid is separated from the dura by a subdural space.

Arachnoid villi are small, finger-like processes projecting into the dural venous sinuses. They absorb CSF. With advancing age, these villi enlarge in size and form arachnoid granulations.

Pia Mater

Pia mater (Intimate or delicate mother) is the inner most of the meninges. It is closely applied to the brain and spinal cord. It is a very delicate, vascular membrane, which carries small capillaries to the surface of the brain and spinal cord. It closely follows the surface of the brain and dips into all fissures.

The pia mater, along with tufts of capillaries, form the choroid plexuses in the ventricles of brain, which secrete CSF.

Applied Anatomy

1. Inflammation of meninges is called meningitis – characterized by fever, headache, neck rigidity, and delirium. CSF pressure is raised; protein and cell content is increased in the CSF.
2. Intracranial hemorrhages can occur in head injuries. Different types of intracranial hemorrhages and the vessels commonly involved are given below:
   A. Extradural—tearing of middle meningeal vessels
   B. Subdural—cerebral veins
   C. Subarachnoid—rupture of aneurysms
   D. Intracerebral—rupture of arteries, following hypertension (raised BP).
3. *Tumors* arising from meninges are called meningeomas.

VENTRICLES OF THE BRAIN (FIG. 16.2)

The interior of the brain contains a series of cavities, filled with cerebrospinal fluid. These cavities are called the ventricles.

The cerebrum contains a median cavity, the *third* ventricle and *two lateral* ventricles, one in each cerebral hemisphere. Each lateral ventricle opens into the third ventricle through an interventricular foramen or foramen of Munro.

The third ventricle is continuous caudally with the *cerebral aqueduct,* which traverses the midbrain and opens into the *fourth* ventricle.

The 4th ventricle is situated dorsal to the pons and medulla and ventral to the cerebellum.

It communicates inferiorly with the central canal, which traverses the lower part of medulla and spinal cord. The entire ventricular system is lined by an epithelial layer called the ependyma.

The Lateral Ventricles

The lateral ventricles are two cavities, one situated within each cerebral hemisphere. Each ventricle is C-shaped; consists of a central part, which gives off 3 extensions called anterior, posterior and inferior horns.

Anterior horn lies in front of an imaginary line passing vertically through the interventricular foramen; it extends into the frontal lobe.

Posterior horn extends backwards into the occipital lobe.

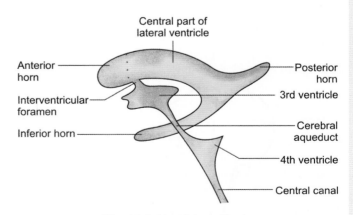

Fig. 16.2: Ventricles of brain

Inferior horn runs downwards and forwards into the temporal lobe.

Third Ventricle

Third ventricle is a slit-like median cavity situated between the right and left thalami. It communicates on either side with the lateral ventricle through the interventricular foramen.

Inferiorly (caudally), it continues into the cerebral aqueduct (of sylvius), which connects it to the 4th ventricle.

Fourth Ventricle

Fourth ventricle (Fig. 16.3) is a diamond-shaped space situated dorsal to the pons and to the upper part of medulla and ventral to the cerebellum.

The cavity of 4th ventricle is continuous superiorly with the cerebral aqueduct and inferiorly, with the central canal (of medulla and spinal cord).

It communicates with the subarachnoid space through 3 apertures (foramina)—one median aperture—the foramen of Magendie and two lateral apertures—foramina of Luschka.

The 4th ventricle has a floor called the rhomboid fossa (kite-shaped) and a roof. The floor is related to important structures like vestibular nuclei, vagal and hypoglossal nuclei, facial colliculus, etc.

Choroid Plexuses

Choroid plexuses are highly vascular structures that are responsible for the secretion of CSF.

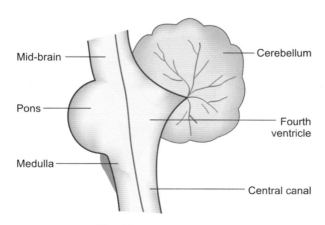

Mid-brain
Cerebellum
Pons
Fourth ventricle
Medulla
Central canal

Fig. 16.3: Fourth ventricle

Components

Components of choroid plexus are pia mater, ependyma and capillaries.

Cerebrospinal Fluid (CSF)

Cerebrospinal fluid (CSF) fills the subarachnoid space, ventricles of brain and the central canal of spinal cord. It is formed by the choroid plexuses of the ventricles.

Flow of CSF

CSF formed in the lateral ventricles reach the 3rd ventricle through the interventricular foramina of Munro. From here, through the aqueduct of Sylvius, CSF reaches the 4th ventricle. Through the median and lateral apertures, CSF is drained into the sub-arachnoid space. From here, through the arachnoid villi, CSF is finally drained into the dural venous sinuses, mainly the superior sagittal sinus.

The total volume of CSF is about 140 ml, of which 25 ml is in the ventricles. CSF is constantly replaced.

Constituents

Water, NaCl, potassium, glucose, proteins.

The arachnoid villi provide a valvular mechanism for the flow of CSF into blood, without permitting backflow of blood into CSF.

Functions of CSF

1. CSF provides a fluid—cushion which protects the brain from injury
2. Helps to carry nutrition to the brain
3. Removes waste products.

Applied Anatomy

1. *Hydrocephalus:* An abnormal increase in the quantity of CSF can lead to enlargement of head in children. This condition is called *hydrocephalus.*
   Causes
   a. Excessive production of CSF
   b. Obstruction to its flow
   c. Impaired absorption
   Meningitis or tumors may lead to the obstruction of flow of CSF.
2. *Lumbar puncture:* A needle is introduced into the lumbar subarachnoid space through the interval between 3rd and 4th lumbar vertebrae, for obtaining a sample of CSF for clinical diagnosis. (In meningitis, the CSF will be turbid and in subarachnoid hemorrhage, CSF will be blood stained).

Blood-brain Barrier

It has been observed that, while some substances can pass from the blood into the brain with ease, others are prevented from doing so. This has given rise to the concept of "blood-brain barrier."

Constituents (Fig.16.4)

1. Capillary endothelium
2. Basement membrane of endothelium
3. Processes of astrocytes.

Verticulography

Verticulography is the radiological study of the ventricles after injecting a radiopaque dye into the ventricular system.

Other Methods

Other methods of study include CT and MRI scans.

◼ THE BRAIN (FIG. 16.5)

The brain is that part of the CNS, which lies within the cavity of the skull. It consists of the following parts:

1. Forebrain — Cerebrum
              — Diencephalon
2. Midbrain
3. Hindbrain — Pons
             — Medulla oblongata
             — Cerebellum

Midbrain, pons and medulla together constitute the *brainstem.*

The Cerebrum

The cerebrum is the largest part of the brain. It consists of right and left cerebral hemispheres, which are partially separated by a sickle shaped fold of dura, the falx cerebri. The two cerebral hemispheres are connected by a bridge of white matter, the corpus callosum.

External Features

When viewed from the lateral aspect, each cerebral hemisphere has the appearance as shown in the Figures 16.6 and 16.7.

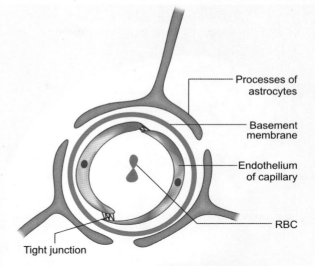

Fig. 16.4: Constituents of blood-brain barrier

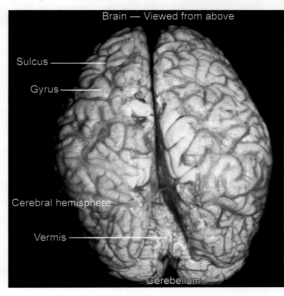

Fig. 16.5: Brain viewed from above

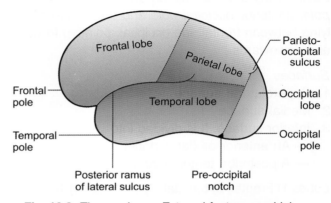

Fig. 16.6: The cerebrum: External features and lobes

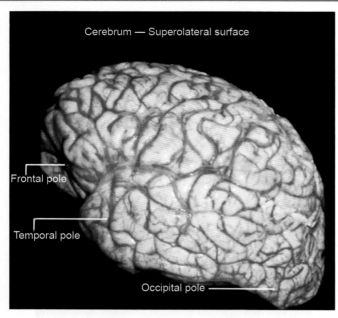

Fig. 16.7: Cerebrum—superolateral surface

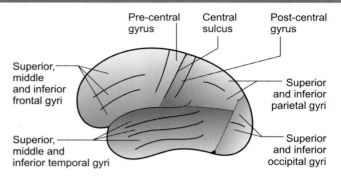

Fig. 16.8: Gyri and sulci

The surfaces of the cerebral hemispheres are not smooth. They show a series of grooves or sulci , which are separated by intervening areas that are called gyri. The sulci vary in depth from slight grooves to deep fissures. The gyri consist of a central core of white matter (nerve fibers running to and from the overlying cerebral cortex) covered by a layer of gray matter, the cerebral cortex. The cortex extends as a continuous sheet over the whole surface of the hemisphere and consists of nerve cells which are arranged in 6 layers.

Each cerebral hemisphere has 3 poles, 3 surfaces and 4 lobes.

Poles: They are the pointed ends – anterior *frontal pole,* posterior *occipital pole* and the *temporal pole* lying between frontal and occipital, pointing forwards and downwards.

Surfaces
1. Superolateral surface
2. Medial surface
3. Inferior surface: It is divided into
    — An anterior orbital surface, and
    — A posterior, tentorial surface

Lobes 1) Frontal, 2) parietal, 3) Occipital, 4) Temporal.

Important Sulci (Singular-Sulcus) (Fig. 16.8)

The Central Sulcus or the Fissure of Rolando

This sulcus runs downwards and forwards on the superolateral surface, a little behind the midpoint between the frontal and occipital poles. It ends a little above the lateral sulcus and separates the frontal and parietal lobes.

The Lateral Sulcus or the Fissure of Sylvius

It runs backwards and slightly upwards from the temporal pole, separating the frontal and parietal lobes from the temporal lobe.

The Parieto-occipital Sulcus

On the medial surface of the hemisphere, near the occipital pole, is the parieto-occipital sulcus. Upper end of this sulcus reaches the superomedial border.

Calcarine Sulcus

On the medial surface, situated below the parieto-occipital sulcus, is the calcarine sulcus.

Important Gyri (Fig. 16.8)

Lobes

Frontal lobe: The gyrus lying in front of the central sulcus is the precentral gyrus. There are 3 gyri that run in as anteroposterior direction. They are, from above downwards, superior, middle and inferior frontal gyri.

Temporal lobe shows 3 gyri; superior, middle and inferior temporal gyri.

Parietal lobe: Behind the central sulcus is the postcentral gyrus. The rest of the parietal lobe is divided into superior and inferior parietal lobules.

The occipital lobe shows superior and inferior occipital gyri.

Important Functional Areas (Fig. 16.9)

Brodmann has given different numbers (1 to 52) to different functional areas of brain.

Motor area or Area 4

It is situated in the whole of the precentral gyrus, slightly extending to the medial surface of cerebrum. Its function is to control the voluntary movements in the opposite half of the body. Man is represented in an upside down manner in this area. The trunk and upper limb are represented in the upper part of the gyrus, while the face and head are represented in the lower part. The area of cortex representing a part of the body is not proportional to the size of the part but proportional to the function of the part. Thus, relatively large areas of cortex are responsible for movements in the fingers and lips. This representation of the human body is referred to as the motor homunculus. (homunculus = miniature human).

Premotor Area (Area 6)

It is situated immediately in front of the motor area. This area is responsible for acts producing orderly series of movements, e.g. combing the hair.

Motor Speech Area of Broca (Area 44 and 45)

It is situated in the inferior frontal gyrus. It is present unilaterally in the dominant hemisphere, i.e. in the left cerebral hemisphere in right handed individuals. A destructive lesion in this area (e.g. total loss of blood supply) causes motor aphasia, in which, the patient loses his speech (even though the muscles of larynx, tongue and lips are not paralyzed).

Two other areas concerned with speech are speech area of Wernicke and supplementary motor area.

Sensory Area (Areas 3, 1 and 2)

These areas lie in the postcentral gyrus. A sensory homunculus represents the parts of the body in the post central gyrus. The digits, lips and the tongue have a disproportionately large representation. This area receives afferents from opposite (contralateral) side of the body (except larynx, pharynx, mouth and perineum).

Visual Area

It consists of visuosensory and visuopsychic areas.

Visuosensory Area or Area 17

It is situated in the occipital lobe, below the calcarine sulcus. The color, size and movements of an object are appreciated here.

Visuopsychic or Psychovisual Area (Areas 18 and 19)

It is situated, surrounding the area 17. These areas are concerned with identification of objects and assessing the distance or depth.

Auditory Area or Acoustic Area (Area 41-42)

It is situated in the temporal lobe; concerned with hearing.

Parietal Lobe or Parietal Area

It is concerned with learning process.

■ BLOOD SUPPLY OF CEREBRUM

Introduction

Two systems of arteries supply the brain. They are:
1. The vertebral system
2. The carotid system.

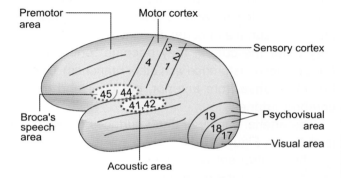

Fig. 16.9: Functional areas

Vertebral Arteries

The vertebral arteries, branches of the subclavian arteries, reach the cranial cavity through the foramen magnum. They unite at the lower border of pons to form the basilar artery. At the upper border of pons, this artery divides into two posterior cerebral arteries.

Carotid System

The cerebral parts of the internal carotid arteries enter the subarachnoid space at the medial end of lateral cerebral fissure of Sylvius. The internal carotid arteries give rise to the following branches:
1. Anterior cerebral artery
2. Middle cerebral artery
3. Posterior communicating artery, by which internal carotid artery is connected to the posterior cerebral artery
4. The anterior choroidal branches – to the choroid plexus of lateral and third ventricles.

The Circulus Arteriosus or the Circle of Willis (Fig. 16.10)

It is an arterial circle situated at the base of the brain. It is formed anteriorly by the anterior communicating artery, anterolaterally by anterior cerebral arteries, laterally by the posterior communicating artery and posteriorly by the posterior cerebral arteries.

The *cerebrum* is supplied by cortical and central sets of arteries, which are branches of the above-mentioned arteries.

The cortical branches supply the outer portions of cerebrum. These arteries anastomose freely and form a network in the pia mater on the surface of the cortex. From this network (plexus) branches arise and pierce the surface of cortex at right angles. But, once they enter the cortex, they become "end arteries" (they do not anastomose with other branches). Cortical branches arise from anterior, posterior and middle cerebral arteries.

The arterial supply of the 3 surfaces of cerebrum is shown in Figure 16.11.

The central set of arteries supply the centrally located parts of the cerebrum. They are end arteries.

Venous Drainage

Venous drainage is by cerebral veins. These veins are extremely thin due to the absence of muscle fibers in their walls. They cross the subarachnoid spaces and finally drain into dural venous sinuses.

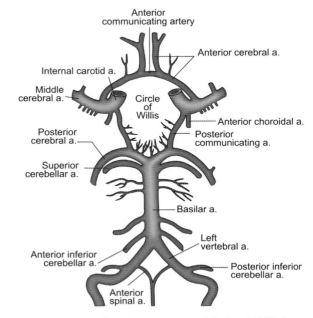

Fig. 16.10: Circulus arteriosus (circle of Willis)

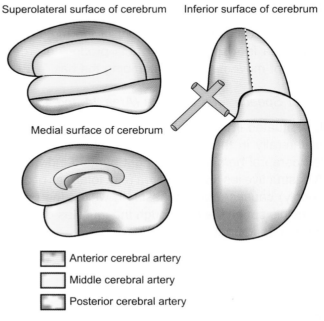

Superolateral surface of cerebrum   Inferior surface of cerebrum

Medial surface of cerebrum

☐ Anterior cerebral artery
☐ Middle cerebral artery
☐ Posterior cerebral artery

Fig. 16.11: Arterial supply of cerebrum

Functions of Cerebrum

1. Receives all sensory stimuli and conveys most of them to consciousness
2. Initiates all voluntary movements
3. Correlates and retains all impulses; this ability forms the basis of memory
4. Seat of intelligence
5. Controls other parts of the nervous system.

■ DIENCEPHALON

We have seen that the forebrain has two parts, the cerebrum and the diencephalon.

Diencephalon is a midline structure, which is largely embedded in the cerebrum and therefore hidden from the surface view. Its cavity is the third ventricle.

Subdivisions

The hypothalamic sulcus extending from interventricular foramen to the cerebral aqueduct divides each half of the diencephalon into dorsal and ventral parts.
1. *Dorsal part of diencephalon*
   a. Thalamus
   b. Metathalamus
      i. Medial geniculate body (Concerned with hearing)
      ii. Lateral geniculate body (Concerned with vision)
   c. Epithalamus.
2. *Ventral part of diencephalon*
   a. Hypothalamus
   b. Subthalamus.

Thalamus (Fig. 16.12)

Thalamus is a large mass of grey matter, situated one on either side of the lateral wall of the 3rd ventricle.

It is a sensory relay station with afferent fibers reaching it from the spinal cord and the brainstem.

Medial surfaces of the two thalami are connected across the midline by the interthalamic adhesion; but fibers from one thalamus never cross the midline.

Thalamus is divided into 3 parts – anterior, medial and lateral, by a y-shaped lamina of white matter. Each part has several important nuclei.

Anterior Group of Nuclei

Anterior group of nuclei are concerned with recent memory. Lesions in this region lead to Karsakow's syndrome, in which recent memory is lost.

Medial Group of Nuclei

Medial group of nuclei are concerned with integration of olfactory, visceral and somatic sensibilities.

Ventral Group of Nuclei

Ventral group of nuclei have 3 subdivisions— ventral anterior, intermediate and posterior.

Connections and Functions

Afferent fibers from different parts of CNS reach the thalamus. It is regarded as a great integrating center where information from different parts of brain and spinal cord are brought together. This information is projected to all parts of the cerebral cortex through thalamo-cortical projections or thalamic radiations.

The most important connections of the thalamus are those of the ventral posterior nucleus (VPN) , which receives terminations of the major sensory pathways coming from the brainstem and spinal cord. All these sensations are carried to areas 3,1 and 2 of the cerebral cortex.

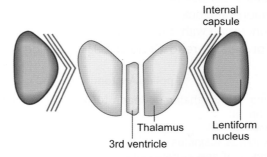

Fig. 16.12: Coronal section of cerebral hemispheres showing position and relations of thalamus

Applied Anatomy

1. Lesions of the thalamus cause impairment of all types of sensations.
2. "Thalamic syndrome" is a clinical condition characterized by disturbances of sensation, hemiplegia, hyper-aesthesia (even a gentle touch produces severe pain) and severe spontaneous pain in the opposite side of the body.

Hypothalamus

The hypothalamus is a part of the diencephalon. It lies in the floor and lateral wall of the third ventricle. It has been described as the chief ganglion of autonomic nervous system because of its role in the control of many metabolic activities of the body.

Important Nuclei

1. *The supraoptic nucleus* lies above the optic chiasma; secretes anti-diuretic hormone (ADH)
2. *Paraventricular nucleus lies* just above the supra-optic nucleus; secretes oxytocin.

Functions

Hypothalamus controls cardiovascular, respiratory and alimentary functions. It secretes releasing hormones that regulate thyroid, adrenal cortex, ovaries and anterior pituitary. Oxytocin and ADH (vasopressin), secreted by hypothalamus are transported and stored in the posterior lobe of pituitary.

The anterior part of hypothalamus is mainly concerned with regulation of parasympathetic system and posterior part, with the sympathetic system.

Other Functions

- Temperature regulation.
- Regulation of food and water intake; controlled by "feeding centre" and "thirst" centre respectively.
- Sexual behaviour and reproduction.
- Biological clock.
- Concerned with emotion, fear, rage, pleasure, etc. (For details, read Chapter 29, Physiology).

Applied Anatomy

Lesions of hypothalamus leads to one of the following syndromes:
1. Obesity
2. Diabetes insipidus
3. Abnormal sexual behaviour
4. Hyperglycaemia

Basal Nuclei (Ganglia) (Fig. 16.13)

The basal nuclei or basal ganglia are large masses of gray matter situated deep in the cerebral hemispheres (By definition, "nuclei" are collections of nerve cell bodies within the CNS and "ganglia" are collections of nerve cell bodies outside the CNS. So, the term "basal nuclei" is more apt).

The basal nuclei are:
1. Caudate nucleus ⎫ Corpus striatum
2. Lentiform nucleus ⎭
3. Amygdaloid nucleus; and
4. Claustrum

Caudate Nucleus

It is a C-shaped nucleus, which is surrounded by and closely related to the lateral ventricle. The concavity of the caudate nucleus encloses the thalamus and internal capsule. The caudate nucleus has a head, body and tail.

Lentiform Nucleus

Lentiform nucleus is a large, biconvex-lens shaped nucleus forming the lateral boundary of internal capsule.

Subdivision (Fig. 16.14) The lentiform nucleus is divided into 2 parts by a thin lamina of white matter.

The larger, lateral part is called the putamen and the medial part is called the globus pallidus.

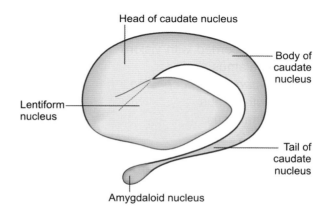

Fig. 16.13: Basal nuclei (ganglia)

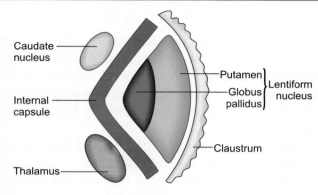

Fig. 16.14: Horizontal section of cerebral hemisphere showing relations of lentiform nucleus

Amygdaloid Body

Amygdaloid body or amygdala lies in the temporal lobe close to the tip of the tail of caudate nucleus.

Functions of Basal Nuclei
1. The corpus striatum (caudate nucleus + lentiform nucleus) regulates the muscle tone and makes the voluntary movements smooth.
2. Controls the coordinated movements of different parts of the body.

Applied Anatomy

Basal nuclei form a part of the extrapyramidal system, concerned with movement. Lesions of corpus striatum lead to hypertonicity, muscular rigidity, loss of facial expression and involuntary movements (chorea and athetosis).

Parkinsonism is a common syndrome characterized by marked rigidity, stooped posture, slow shuffling gait, difficulty in speech and a mask-like face. This condition is believed to be due to degenerative changes in the corpus striatum and a part of the midbrain, the substantia nigra. There is a reduction in the production of a neurotransmitter dopamine. It is treated with Levo-dopa.

■ WHITE MATTER OF CEREBRUM

The white matter of cerebrum consists chiefly of myelinated fibers, which connect various parts of the cortex to one another and also to the other parts of CNS. They are classified into 3 groups:
1. Association fibers
2. Projection fibers
3. Commissural fibers.

Association Fibers

Association fibers connect different cortical areas of the same hemisphere to one another. For example superior longitudinal fasciculus (Fig. 16.15). It connects the frontal, parietal, occipital and temporal lobes of the same hemisphere.

Projection Fibers

Projection fibers connect the cerebral cortex to other parts of CNS like brainstem and spinal cord. Many of the important tracts, e.g. corticospinal tract, are made up of projection fibers.

Commissural Fibers

Commissural fibers are fibers, which connect corresponding parts of two hemispheres. Largest commissural fibers of the cerebrum form the corpus callosum.

Corpus Callosum (Figs 16.16 and 16.17)

The corpus callosum is the largest collection of commissural fibers of the cerebrum. It connects the

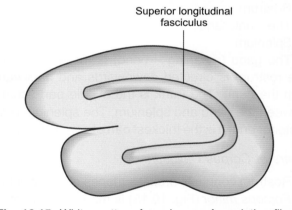

Fig. 16.15: White matter of cerebrum—Association fibers

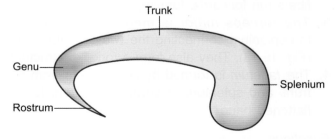

Fig. 16.16: Corpus callosum—parts

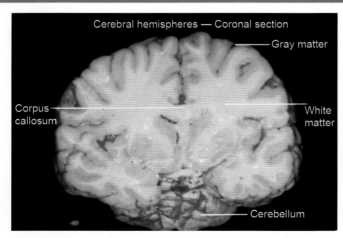

Fig. 16.17: Coronal section through cerebral hemispheres, showing corpus callosum

two cerebral hemispheres. It is 10 cm long. It connects all parts of cerebral cortex of the two sides except the lower and anterior part of temporal lobes. The corpus callosum is closely related to the lateral ventricle.

Parts

1. Genu
2. Rostrum
3. The trunk, and
4. Splenium

The genu (meaning – bend) is the anterior end. The rostrum is directed downwards and backwards from the genu. The trunk is the middle part situated between the genu and splenium. The splenium is the posterior end. It is the thickest part of corpus callosum.

Fibers of Corpus Callosum

1. Rostrum connects orbital surfaces of frontal lobes.
2. *Forceps minor* (Fig. 16.18) is made up of fibers of the genu that connect the two frontal lobes. These fibers run forwards, forming a fork-like structure.
3. *The forceps major* is made up of fibers of the splenium connecting the two occipital lobes (Fig. 16.18). They also form a fork-like structure.
4. The *tapetum* is formed by some fibers from the trunk and splenium of corpus callosum. It is a flattened band.

Functions

It co-ordinates the activities of the two cerebral hemispheres.

The Internal Capsule (Fig. 16.19)

The internal capsule is a large band of projection fibers, situated in the inferomedial part of each cerebral hemisphere. In horizontal sections of the brain, it appears V-shaped, with its concavity directed laterally. This concavity is occupied by the lentiform nucleus.

The internal capsule is composed of fibers going to and coming from the cerebral cortex.

When traced upwards, the fibers of the capsule diverge and form the corona radiata.

When traced downwards, these fibers form a compact bundle and continue as the crus of midbrain.

Parts of Internal Capsule (Fig. 16.20)

1. Anterior limb—short
2. Posterior limb—long

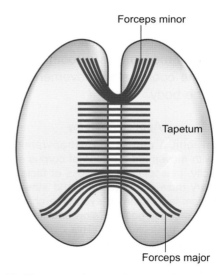

Fig. 16.18: Forceps minor and major of corpus callosum

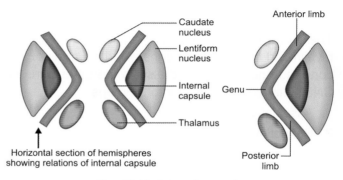

Fig. 16.19: Internal capsule

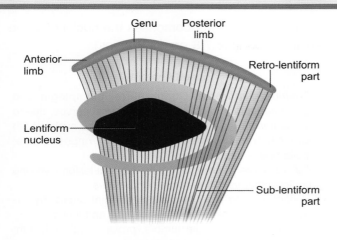

Fig. 16.20: Parts of internal capsule

3. Genu—junction of anterior and posterior limbs
4. Retrolentiform part (i.e. behind the lentiform nucleus)
5. Sublentiform part (i.e. below the lentiform nucleus).

Anterior limb lies between the head of caudate nucleus and lentiform nucleus (Fig. 16.19). It contains thalamocortical, frontopontine and corticobulbar (cortex to medulla oblongata) fibers.

Posterior limb lies between the thalamus and lentiform nucleus. It contains, mainly corticospinal fibers.

Genu ("bend") lies between anterior and posterior limbs; contains corticospinal and corticonuclear (nuclei of cranial nerves) fibers, mainly for the head and neck.

Retrolentiform part lies behind the lentiform nucleus; composed of optic radiations (fibers from lateral geniculate body to visual cortex).

Sublentiform part lies below the lentiform nucleus, contains auditory radiations (from medial geniculate body to auditory cortex).

Applied Anatomy

The internal capsule is made up of ascending (sensory) and descending (motor) fibers. It is a narrow area, where fibers are densely packed and hence, even a pinpoint lesion in the internal capsule gives rise to widespread clinical symptoms in the body.
Internal capsule is supplied by branches of the middle cerebral, posterior communicating and anterior cerebral arteries. A hemorrhage or thrombosis in one of these branches, leads to hemiplegia on the opposite half of the body, including the face.

Brainstem (Fig. 16.21)

The brainstem connects the forebrain and spinal cord.
It consists of 3 parts—
1. Midbrain
2. Pons
3. Medulla oblongata.

Midbrain

Midbrain or Mesencephalon is the shortest segment of the brainstem.

Parts (Fig. 16.22): The midbrain is traversed by the aqueduct of Sylvius, which connects the 3rd and 4th ventricles.

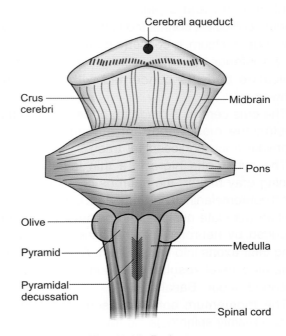

Fig. 16.21: Brainstem

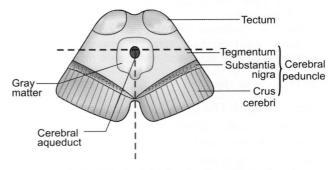

Fig. 16.22: TS of midbrain showing different parts

The part of the midbrain lying behind an imaginary transverse line drawn through the cerebral aqueduct is called the *tectum*.

The part lying in front of the transverse line is made up of right and left cerebral peduncles (Fig. 16.22). Each peduncle consists of 3 parts; from anterior to posterior, they are:

1. The crus cerebri
2. The substantia nigra
3. The tegmentum.

The aqueduct is surrounded by gray matter which contains the nuclei of trochlear nerve (IV cranial nerve) and oculomotor nerve (III cranial nerve).

The tectum is made up of a pair of superior colliculi (right and left) and a pair of inferior colliculi. The superior colliculus, connected to the lateral geniculate body by a brachium, is concerned with visual pathway.

The inferior colliculus, connected to the medial geniculate body, is concerned with the auditory pathway.

The crus cerebri is made up of descending fibers (tracts); the most important among these is the pyramidal tract (corticospinal tract).

The substantia nigra is a thin layer of deeply staining gray matter; the neurons contain a pigment called neuromelanin. It has important connections with striatum (caudate nucleus and putamen). Dopamine produced by neurons in the substantia nigra pass along their axons into the striatum. A reduction in the Dopamine level results in Parkinsonism (already described under "Basal nuclei").

The tegmentum contains ascending (sensory) fibers – mainly spinothalamic tract.

Applied Anatomy

1. A blockage of cerebral aqueduct may lead to hydrocephalus.
2. *Weber's syndrome*—thrombosis or hemorrhage in vessels supplying the midbrain may lead to oculomotor paralysis (same side) and hemiplegia (opposite side).

The Pons (Fig. 16.21)

(*Pons = bridge*). The pons lies between the midbrain and the medulla oblongata. On each side, it is connected to the cerebellum by the middle cerebellar peduncle. Pons is made up of gray and white matter. The white matter consists of ascending and descend-

ing tracts. Gray matter comprises the nuclei of some cranial nerves (5,6,7,8 cranial nerves).

Applied Anatomy

1. *Foville's syndrome*: Contralateral hemiplegia and ipsilateral paralysis of 6th and 7th cranial nerves, due to a lesion involving corticospinal tract and the nuclei of 6th and 7th cranial nerves (contralateral– opposite side; ipsilateral – same side).
2. *Millard Gubler syndrome* results due to a lesion involving corticospinal tract and facial nerve fibers.
3. *Pontine hemorrhage*: Results in pinpoint pupils, hyperpyrexia and bilateral paralysis of face and limbs.
4. Cerebellopontine angle tumors, press on 5th, 6th, 8th, 9th, 10th and 12th cranial nerves, and cerebellum. Depending on the structure compressed, symptoms vary.
[5th cranial nerve—facial anesthesia, loss of corneal reflex
6th cranial nerve—internal squint
8th cranial nerve—progressive deafness, vertigo
9th cranial nerve—dysphagia (difficulty in swallowing)
10th cranial nerve—syncope
12th cranial nerve—ipsilateral tongue paralysis
Cerebellum – Ataxia, vertigo, hypotonia].

The Medulla Oblongata (Medulla, "bulb") (Fig. 16.21)

The medulla, a part of the brainstem, is about 3 cm in length; it is continuous above with the pons and below, with the spinal cord.

Posteriorly, it is related to the 4th ventricle. Anteriorly, there are two swellings on either side of the anterior median fissure, caused by the pyramidal tracts. They are called the pyramids. They contain the corticospinal fibers. In the lower part of the medulla most of these fibers cross the median plane – the decussation of pyramids – and pass down to the opposite side of the spinal cord in the lateral corticospinal tract.

The vital centers (respiratory and vasomotor) are situated in the medulla. The last four cranial nerves (9,10,11 and 12) emerge from the medulla. Ascending tracts and descending tracts pass through it.

Damage to the vital centers lead to death. "Bulbar paralysis" is characterized by paralysis of muscles supplied by the last four cranial nerves.

Medulla is supplied by anterior spinal arteries and posterior inferior cerebellar arteries.

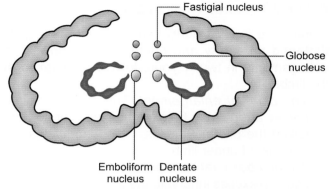

Fig. 16.23: Nuclei of cerebellum

Obstruction of Posterior Inferior Cerebellar Artery(PICA) leads to Wallenberg Syndrome characterized by dysphagia, dysphonia, dysarthria and Horner's syndrome due to the involvement of IX,X and XIth cranial nerves and sympathetic fibers.

The Cerebellum (Cerebellum = Little Brain)

The cerebellum is the largest part of the hindbrain. It is situated in the posterior cranial fossa. It lies below the occipital lobe of cerebrum, from which it is separated by the tentorium cerebelli (a fold of dura).

Cerebellum has 2 hemispheres joined together by a worm-like portion, the vermis. Each hemisphere is divided into 3 lobes—anterior, middle and flocculo-nodular lobes. These lobes are divided into numerous small parts by fissures. The cerebellum has an outer layer of gray matter and inner layer of white matter. Situated deep inside the white matter, there are 4 pairs of nuclei (Fig. 16.23). They are, the
1. Dentate
2. Emboliform
3. Fastigial, and
4. Globose (Mnemonic – "DEFG").

The cerebellum is connected on either side, to the brainstem, by bundles of fibers called superior, middle and inferior cerebellar peduncles (Flow chart 16.1).

Flow chart 16.1: Cerebellar peduncles

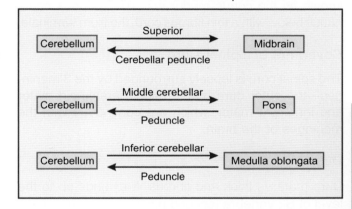

Functions

Main function is to coordinate the movements. To perform this function, the cerebellum receives information from cerebral cortex, vestibular, auditory, visual and proprioceptive receptors. Other functions include maintenance of balance and muscle tone.

Cerebellar dysfunction results in
1. Hypotonia or Atonia: A decrease or loss of muscle tone, due to involvement of cerebellar nuclei.
2. Ataxia: Uncoordinated movements.
3. *Dysarthria:* Loss of coordination between muscles of larynx, mouth and respiratory system. Lack of coordination of the muscles and inability to adjust the intensity and duration of sound, causes jumbled vocalization, with some words loud, some weak, some long or some short. So, the resultant speech is almost unintelligible.

4. *Cerebellar nystagmus* is the tremor or oscillations of the eyeball, when the patient tries to fix his gaze.
5. *Dysdiadochokinesia* is the loss of ability to perform alternating movements like pronation and supination rapidly and continuously.
6. *Intention tremor:* When the patient with cerebellar dysfunction tries to perform a voluntary act, the movements tend to oscillate, especially when they approach the intended mark (e.g. picking a pen from the table); this reaction is called intention tremor.
7. *Loss of balance*.

■ THE SPINAL CORD

The spinal cord is a part of the central nervous system, which lies within the vertebral canal.

The spinal cord begins as a continuation of the medulla oblongata. It extends from the level of the foramen magnum to the lower border of L1 vertebra. It is approximately 45 cm long. In the adult, it occupies only the upper 2/3 of the vertebral canal.

External Features (Fig. 16.24)

Spinal cord is a cylindrical structure that is slightly flattened anteriorly and posteriorly. It shows two enlargements in two regions, for the innervation of the limbs.

1. *The cervical enlargement:* The cervical spinal nerves arising from here form a plexus or network called the brachial plexus, which innervates the muscles of upper limb.
2. *The lumbosacral enlargement:* The lumbar and sacral plexuses innervate the powerful muscles of the lower limb.

   The lowest part of the spinal cord is conical and is known as the conus medullaris. The conus is continuous below with a thin fibrous cord, the filum terminale.

Coverings of Spinal Cord

The spinal cord is loosely surrounded by the 3 meninges—the outer dura mater, middle arachnoid mater and inner pia mater, which are continuous with the meninges of the brain.

Dura Mater

Dura mater is thick and fibrous. It extends up to the level of S2 vertebra (second piece of sacrum). At this level, it is pierced by the filum terminale.

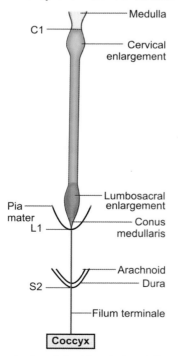

Fig. 16.24: Important vertebral levels in relation to spinal cord

Laterally, the dura is pierced by the roots spinal nerves around which, the dura form a tubular sheath.

Extradural or Epidural Space

Extradural or epidural space lies outside the dura, but inside the vertebral canal. It contains some areolar tissue, fat and internal vertebral venous plexus. This space is clinically important because anesthetic agents can be injected into this space – Epidural anesthesia.

Arachnoid Mater

Arachnoid mater is a delicate, avascular membrane situated between dura and pia mater. It ends at the level of S2 vertebra.

The subarachnoid space is continuous with the subarachnoid space surrounding the brain. It contains CSF.

Pia Mater

Pia mater is thin and transparent; it closely adheres to the surface of the spinal cord. Below the conus medullaris, the pia mater continues as a thin, silvery, thread-like filament called the filum terminale. It is 20 cm long.

The pia mater of the spinal cord presents 21 tooth-like projections on each side – the ligamenta denticulata – which are attached to the inner surface of dura and help to anchor the spinal cord.

31 pairs of spinal nerves arise from the spinal cord. Each nerve is attached to the spinal cord by dorsal and ventral roots. There are 8 cervical, 12 thoracic, 5 lumbar, 5 sacral and 1 coccygeal nerves on each side.

Internal Features (Fig. 16.25)

A transverse section of the spinal cord at the thoracic or lumbar region shows the following features:

1. There is an anterior median fissure and a posterior median sulcus.
2. The gray matter is situated internally. It forms an H-shaped mass. In each half of the spinal cord, the gray matter is divisible into a larger ventral mass—the anterior or ventral horn (or column) and a narrow, elongated posterior or dorsal horn (or column).

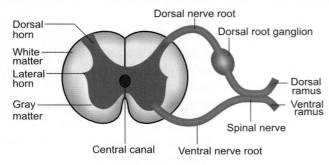

Fig. 16.25: The spinal cord—internal features

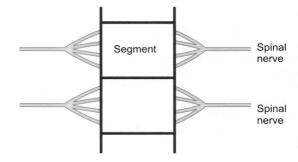

Fig. 16.26: Spinal segments

3. In the thoracolumbar region, the spinal gray matter shows a small lateral projection – the lateral horn.

   The gray matter of right and left halves are connected across the midline by a gray commissure, in the center of which is the central canal (lined by ependyma and contains CSF).

4. The white matter of spinal cord is external to the gray matter; it is divided into right and left halves. Each half contains many ascending and descending tracts.

   The spinal cord gives attachment, on each side to a series of spinal nerves. Each spinal nerve arises by two roots—anterior or ventral or motor root and posterior, dorsal or sensory root (Fig. 16.25). The ventral and dorsal roots join to form a spinal nerve (so, it has both motor and sensory fibers). Just proximal to this junction, the dorsal root is marked by a swelling—the dorsal root ganglion or spinal ganglion. Each spinal nerve passes through the intervertebral foramen, immediately dividing into two – a dorsal ramus and a ventral ramus.

   The part of the spinal cord giving origin to one pair of spinal nerves is called a spinal segment (Fig. 16.26). Spinal cord is made up of 31 spinal segments—8 cervical, 12 thoracic, 5 lumbar, 5 sacral and 1 coccygeal.

   In the fetus, the spinal cord and the vertebral canal are of the same length. But later, the length of the vertebral canal increases. So, the spinal nerves have to follow an oblique downward course to reach the appropriate intervertebral foramina. This obliquity is most marked in the lower nerves and many of these nerves occupy the vertebral canal below the level of spinal cord. These roots, resembling a "horse's tail" is called the cauda equina (Fig. 16.27).

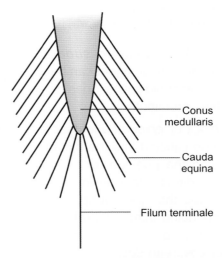

Fig. 16.27: Cauda equina

Blood Supply

The spinal cord is supplied by anterior and posterior spinal arteries (branches from the vertebral), intercostal and lumbar arteries. Veins drain into internal vertebral venous plexus.

Applied Anatomy

Lumbar puncture: The spinal cord ends at the level of the lower border of L1. The dura, arachnoid and subarachnoid space containing CSF extend up to the second sacral vertebra. Between L1 and S2, the subarachnoid space contains the spinal nerve roots of the cauda equina. In this region, a needle can be introduced into the subarachnoid space to withdraw a sample of CSF, without any harm to the spinal cord or spinal nerves. This procedure is called lumbar puncture. Infections of the meninges (meningitis) and diseases of CNS may alter the cells in the CSF or change the concentration of its chemical constituents. In subarachnoid hemorrhage, the CSF will show the presence of blood.

Contd...

Contd...

Lumbar puncture (LP) is performed with the patient leaning forward or lying on the side with the back flexed. Flexion of the vertebral column facilitates insertion of LP needle because the spinous processes of the lumbar vertebrae move apart (See CP12 in Color Plates 3).

Usually, the LP needle is inserted, under aseptic conditions, in the midline between the spinous processes of L3 and L4 or L4 and L5 vertebrae. (A plane passing through the highest point of iliac crests corresponds to L4 spinous process).

Lumbar puncture should not be done in patients with suspected high intracranial pressure. The release or decrease of pressure, following LP, may result in herniation of the brainstem and cerebellum, which could be fatal.

Spinal block or lumbar subarachnoid block (LSAB): An anesthetic agent can be injected directly into the subarachnoid space. Anesthesia usually takes effect within one minute.

Epidural block or epidural anesthesia An anesthetic agent can be injected into the extradural or epidural space (using the position for LP). The anesthetic asent has a direct effect on the spinal nerve roots, which traverse the epidural space, on their way to the corresponding intervertebral foramina.

An epidural block is commonly given during delivery or cesarean section to prevent pain.

To reduce postoperative pain, an in-dwelling catheter can be inserted into the epidural space, through which an anesthetic agent is injected periodically (8th hourly or 12th hourly).

Spinal cord injuries: Transection of the spinal cord results in loss of all sensations and voluntary movements below the lesion. If transection occurs above C5, patient is quadriplegic (4 limbs paralyzed). Patient may die of respiratory failure if spinal cord is transected superior to C4 (the phrenic nerve, supplying the diaphragm, arises from C3,4 and 5).

The control of rectum, anal canal and bladder is also lost. Injury to lower part of spinal cord (below T1 results in paralysis of lower limbs (paraplegia).

Compression of roots of spinal nerves may be caused by prolapse of intervertebral disc, osteophytes, a cervical rib or a tumor.

Study of spinal cord

1. *Myelography:* An injection of a radio-opaque contrast medium into the subarachnoid space, followed by X-rays can help in the visualization of the spinal cord and nerve roots.

2. *CT and MRI:* Computerized tomography differentiates between gray matter and white matter, detects a fracture of the vertebral column and the degree of compression of spinal cord.

   a. *Magnetic resonance imaging* (MRI) is a computer-assisted radiation-free procedure, which produces excellent images of vertebral column, spinal cord and subarachnoid space. Herniation or prolapse of intervertebral disc and its relation to the nerve roots are also well defined by MRI.

Poliomyelitis is an acute viral infection of the neurons of the anterior gray columns of the spinal cord and motor nuclei of cranial nerves. This causes death of the motor neurons resulting in paralysis and wasting of muscles. Muscles of the lower limb are more often affected than muscles of upper limb.

Paralysis of diaphragm and muscles of face, larynx and pharynx can also occur.

■ CRANIAL NERVES (CN)

Cranial nerves, attached to the brain, are bundles of processes of neurons that innervate muscles or glands or carry impulses from sensory areas. They are called cranial nerves because they emerge through foramina or fissures in the cranium (skull) and are covered by sleeves of meninges.

There are 12 pairs of cranial nerves. They are numbered from I to XII according to their attachment to the brain (Fig. 16.28A).

   I. The olfactory nerve (CN I)

  II. The optic nerve (CN II)

 III. The oculomotor nerve (CN III)

 IV. The trochlear nerve (CN IV)

  V. The trigeminal nerve (CN V)

 VI. The abducent nerve (CN VI)

 VII. The facial nerve (CN VII)

VIII. The vestibulocochlear nerve (CN VIII)

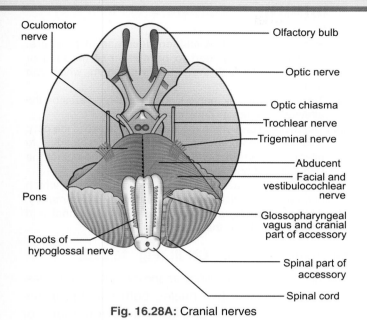

Fig. 16.28A: Cranial nerves

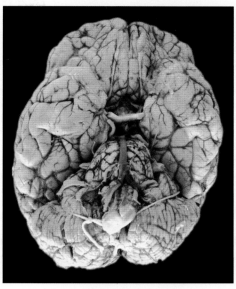

Fig. 16.28B: Base of brain

IX. The glossopharyngeal nerve (CN IX)
X. The vagus nerve (CN X)
XI. The accessory neve (CN XI); and
XII. The hypoglossal nerve (CN XII)

Meaning/Explanation

1. Olfactory—Concerned with the sense of smell or olfaction.
2. Optic—Concerned with the sense of vision.
3. Oculomotor—Motor to muscles which move the eyeball.
4. *Trochlear* (Trochlea = pulley)—The nerve which supplies the muscle, which passes through a pulley; i.e. superior oblique muscle of eyeball.
5. *Trigeminal*—Nerve having 3 roots (tri-three) ophthalmic, maxillary and mandibular.
6. *Abducent*—The nerve which supplies the muscle that abducts the eyeball, i.e. lateral rectus.
7. *Facial*—The nerve which supplies the muscles of facial expression.
8. *Vestibulocochlear*—The nerve concerned with vestibular (balance) and cochlear (hearing) functions.
9. *Glossopharyngeal* ("glossa"—pertaining to tongue)—Nerve which supplies tongue (posterior 1/3) and pharynx (stylopharyngeus).
10. *Vagus*—This word is derived from "vagabond", meaning "wanderer" Vagus nerve has a long course;

starting from the medulla, it comes out through the jugular foramen; reaches the neck, thorax and abdomen, supplying large number of structures.
11. *Accessory*: It helps the vagus nerve to form nerve plexuses to supply various muscles. So it acts as an "assistant" to vagus.
12. *Hypoglossal,* lies *beneath* (hypo) the tongue (glossa) and supplies all the muscles of tongue (except palatoglossus).

Olfactory Nerve—CN I (Ref. chapter 18, Fig. 18.1)

This is the first cranial nerve, associated with the sense of smell. "Olfaction" is the sensation of smells (odors) that results from detection of odorous substances mixed with air.

The olfactory cells or olfactory neurons are situated in the olfactory epithelium in the roof, upper part of nasal septum and lateral wall of nasal cavity. The central processes of the bipolar olfactory neurons join to form about 20 bundles of olfactory nerve fibers that pass through the cribriform plate of the ethmoid, pierce the dura and arachnoid, and end in the olfactory bulb in the anterior cranial fossa (Fig. 18.1, Fig. 16.28A).

Olfactory stimuli arouse emotions and enable us to appreciate the aromas of food. Smell induces salivation, increased gastric secretion and sometimes nausea.

Applied Anatomy

1. *Anosmia* is the loss of sense of smell. Commonest cause is common cold (rhinitis). It is a transitory stage. Loss of olfactory fibers occur with aging. Loss of smell can lead to loss of taste also. Injury to the nasal mucosa, olfactory nerve fibers or the olfactory bulb may also impair smell.
2. Fracture of the base of the skull—especially, the anterior cranial fossa—as a result of severe head injuries may tear the olfactory nerve fibers, leading to anosmia and CSF rhinorrhea (leak of CSF through nose).
3. A tumor or abscess (collection of pus) in the frontal lobe of brain may compress the olfactory bulb leading to anosmia.

Optic Nerve—CN II

The optic nerve or the "nerve of sight" is formed by the axons of the ganglion cells of retina (Chapter 18, Fig. 18.11). All these fibers converge at the optic disc, pierce the sclera and form the optic nerve. It is surrounded by the 3 meninges of brain and the spaces between them. These meninges are pierced by the central artery and vein of retina; these vessels course in the anterior part of the optic nerve.

The optic nerve, after piercing the sclera, passes posteromedially in the orbit and exits through the optic canal entering the middle cranial fossa. The two nerves, coming from the eyeballs, join here to form the optic chiasma (chiasm).

At the optic chiasma, fibers from the nasal or medial half of each retina cross over or decussate. They join uncrossed fibers from the temporal or lateral half of the retina to form the optic tract. This partial crossing of fibers is essential for binocular vision, allowing a three-dimensional vision.

Most fibers in the optic tract terminate in the lateral geniculate body. Fibres arising from here form the optic radiation, related to the retro-lentiform part of internal capsule, to end in the visual cortex of the occipital lobe of cerebrum. (Visual Pathway, Chapter 18, Fig. 18.11).

Applied Anatomy

1. *Papilledema* or the edema of the optic disc(or papilla) is the result of increased intracranial pressure. This causes a compression of the retinal vein, which traverses the subarachnoid space, containing CSF. The swollen optic disc can be visualized with an ophthalmoscope.

   Papilledema is a valuable indicator of raised intracranial pressure. Lumbar puncture in such patients

Contd...

Contd...

   is contraindicated. A sudden decrease in pressure, following LP may lead to herniation of medulla or cerebellum, through the foramen magnum, which can be fatal.
2. Transection of optic nerve leads to total blindness on the affected side (as in penetrating injuries)
3. The hypophysis cerebri (pituitary) can compress the optic chiasma, when it is enlarged (as in tumors) leading to bitemporal hemianopia (loss of vision of one-half of the visual field of both eyes).
4. *Optic neuritis* are lesions of the optic nerve which decreases the visual acuity, caused by inflammatory, degenerative or toxic disorders. Methyl alcohol and tobacco can injure the optic nerve.

Oculomotor Nerve—CN III (Fig. 16.28B)

It is the chief nerve to the ocular and extraocular muscles.

The nerve has two nuclei, both situated in the midbrain—(1) Motor and (2) Parasympathetic or Edinger-Westphal nucleus.

The oculomotor nerve, having motor and parasympathetic fibers, emerges from the midbrain, runs in the lateral wall of cavernous sinus (a dural venous sinus). It enters the orbit through the superior orbital fissure. Within the orbit, the nerve divides into a superior division and an inferior division.

Superior division—It supplies the levator palpebrae superioris and the superior rectus muscles. The *inferior division* supplies inferior and medial rectus and the inferior oblique muscles. The inferior division also carries parasympathetic fibers from the Edinger-Westphal nucleus to the ciliaris or ciliary muscle (accommodation of lens) and sphincter pupillae (constriction of pupil).

(Note: The extraocular muscles of the eyeball are:
1. Lateral rectus (LR)
2. Superior oblique (SO)
3. Medial rectus
4. Inferior rectus
5. Superior rectus
6. Inferior oblique, and
7. Levator palpebrae superioris.

Lateral rectus is supplied by 6th cranial nerve (abducent) and superior oblique by 4th cranial nerve. The rest of the muscles are supplied by the 3rd cranial(oculomotor) nerve. So, the innervation of the extraocular muscles can be easily memorized using the formula $(LR_6SO_4)_3$.

Applied Anatomy

Oculomotor nerve palsy: A lesion of the oculomotor nerve (e.g. hemorrhage or thrombosis of vessels supplying the midbrain, a tumor compressing the nerve, etc.) causes paralysis of all extraocular muscles (except lateral rectus and superior oblique), the sphincter pupillae and the ciliaris muscles.

Characteristic features of 3rd nerve palsy or oculomotor nerve palsy:

1. *Ptosis* or drooping of upper eyelid, caused by paralysis of levator palpebrae superioris (Note: A part of the levator palpebrae superioris—the Muller's muscle—is supplied by sympathetic fibers. These fibers are not paralyzed. So, there is only partial drooping of the upper eyelid).
2. Absence of pupillary reflex or light reflex, (i.e. constriction of pupil in response to bright light) due to paralysis of constrictor (sphincter) pupillae.
3. Dilation (dilatation) of pupil—due to the unopposed action of dilator pupillae (supplied by sympathetic fibers).
4. Absence of accommodation of lens because of paralysis of the ciliaris muscle.
5. Squint—The eyeball is abducted and directed inferiorly due to the unopposed actions of lateral rectus and superior oblique muscles.

Trochlear Nerve—CN IV (Fig. 16.28A)

This nerve supplies the superior oblique muscle of the eyeball. Its nucleus is situated in the gray matter surrounding the cerebral aqueduct of midbrain. The nerve emerges from the dorsal surface of the midbrain, winds round the brainstem, passes anteriorly in the lateral wall of the cavernous sinus.

The nerve then passes through the superior orbital fissure into the orbit, where it innervates the superior oblique.

Trochlear nerve is the only cranial nerve to emerge dorsally from the brainstem.

(*Note:* This nerve is called "trochlear nerve" because, the muscle which it innervates (superior oblique) passes through a fibrocartilaginous pulley (trochlea).

Lesions of trochlear nerve leads to paralysis of the superior oblique, leading to diplopia or double vision (because the direction of gaze is different for the normal eye and the affected eye).

Trigeminal Nerve—CN V (Figs 16.28A to C)

This nerve emerges from the pons by a small motor root and a large sensory root.

The nerve has 3 divisions—ophthalmic (V1) maxillary (V2) and mandibular (V3). The ophthalmic and maxillary are sensory nerves; the mandibular nerve has both motor and sensory fibers.

Fibres in the sensory root are mainly axons of neurons in the trigeminal ganglion. Peripheral processes of these neurons form the ophthalmic and maxillary nerves and the sensory component of the mandibular nerve.

Ophthalmic Nerve

The ophthalmic nerve enters the orbit through the superior orbital fissure. It supplies the skin overlying the forehead, anterior quadrant of scalp, skin of upper eyelid cornea and dorsum of nose.

Maxillary Nerve

The maxillary nerve leaves the cranial cavity through the foramen rotundum. It carries sensations from skin overlying the maxilla, maxillary teeth, maxillary sinus, upper lip and the palate.

Mandibular Nerve (Fig. 16.28C)

The mandibular nerve reaches the infratemporal fossa through the foramen ovale.

Branches of mandibular nerve are:

1. Muscular branches to muscles of mastication (Temporalis, masseter, medial and lateral pterygoid) through its anterior division.
2. Auriculotemporal—supplies the temporomandibular joint, sensory fibers to auricle (pinna) and temporal region, parasympathetic (secretomotor) fibers to parotid gland.
3. *Inferior alveolar nerve* enters the mandibular foramen and passes through the mandibular canal and supplies the mandibular teeth of the corresponding side; its branch, mental nerve—comes out of the mental foramen, supplies lower lip and chin.

Before entering the mandibular foramen, the inferior alveolar nerve gives a slender branch, the mylohyoid nerve, which supplies the mylohyoid and anterior belly of digastric muscles.

4. *The lingual nerve* lies anterior to the inferior alveolar nerve. It is closely related to the 3rd molar tooth. It is sensory to the anterior 2/3 of tongue, the floor of the mouth and gums.

The chorda tympani nerve, carrying taste sensation from anterior 2/3 of tongue joins the lingual nerve. The chorda tympani (a branch of facial nerve) also carries secretomotor fibers for submandibular and sublingual salivary glands.

The submandibular ganglion, a parasympathetic ganglion hangs by 2 roots from the lingual nerve.

Applied Anatomy

1. *Injury*-CN V may be injured in trauma, tumors or meningeal infections. Its nuclei, situated in the pons and medulla may be destroyed by vascular lesions or tumors. *Results of injury*
   • Paralysis of muscles of mastication, with deviation of mandible to the same side of lesion.
   • Loss of sensation from areas supplied by the nerve.
   • Loss of corneal reflex (blinking in response to the cornea being touched by a wisp of cotton).
2. *Trigeminal neuralgia (tic douloureux)*, the main disease affecting the trigeminal nerve, is characterized by attacks of severe pain in the areas of distribution of maxillary and mandibular divisions.
3. *Mandibular nerve block* a needle is introduced through the mandibular notch into the infratemporal fossa and the anesthetic agent is injected. This is to anesthetise the branches of mandibular nerve.
4. *Inferior alveolar nerve block* is commonly used by dentists while extracting or repairing the mandibular teeth. The injection is given around the mandibular foramen.
5. *Lingual nerve* can get injured while extracting the 3rd molar tooth, because the nerve lies just inferior to the 3rd molar tooth.

Abducent Nerve—CN VI or Abducens

This nerve supplies only one muscle—the lateral rectus, an extraocular muscle.

The abducent nucleus lies in the pons. The fibers of facial nerve wind round this nucleus and produces a swelling in the floor of the 4th ventricle—the facial colliculus.

This nerve, emerges from the brainstem between the pons and medulla; after running a long intracranial course (it has the longest intracranial course, of all cranial nerves), it enters the cavernous sinus, coursing through its venous blood; here it is closely related to the internal carotid artery.

Abducent nerve enters the orbit through the superior orbital fissure; it supplies the lateral rectus muscle (abductor of the eyeball).

Applied Anatomy

Because of its long intracranial course, it is often stretched in raised intracranial pressure. A tumor or hematoma can compress the nerve.

Paralysis of lateral rectus causes medial deviation of the affected eye—the eyeball is fully adducted because of the unopposed action of the medial rectus; this results in diplopia (double vision).

Facial Nerve—CN VII (Figs 16.28C III, IV)

The facial nerve emerges from the junction of pons and medulla. It has two divisions:
1. A large motor root; and
2. A small nervus intermedius

The larger motor root, or the facial nerve proper, arises from the motor nucleus, situated in the pons. These fibers wind round the nucleus of the abducent nerve, producing a swelling in the floor of the fourth ventricle, called the facial colliculus.

Muscles innervated by facial nerve are:
1. Muscles of facial expression
2. Posterior belly of digastric
3. Stylohyoid
4. Stapedius (of middle ear)
5. Occipitofrontalis (of scalp)
6. Platysma; and
7. Muscles of pinna.

The smaller root, the nervus intermedius, carries taste, parasympathetic and somatic sensory fibers.

Distribution of Nervus Intermedius

1. Sense of taste from anterior 2/3 of tongue, via chorda tympani branch
2. Sensation from the external ear (skin around external auditory canal)
3. Parasympathetic (secretomotor) fibers to the submandibular, sublingual salivary glands, the lacrimal gland and glands of the nasal cavity and palate.

Course

During its course, the facial nerve traverses the posterior cranial fossa, the internal acoustic meatus,

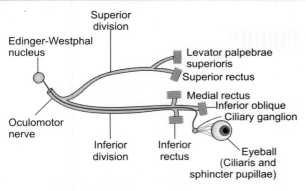

Fig. 16.28C-I: Distribution of oculomotor nerve

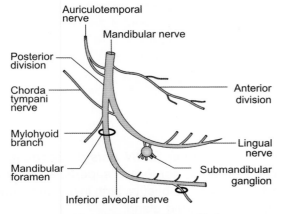

Fig. 16.28C-II: Branches of mandibular nerve

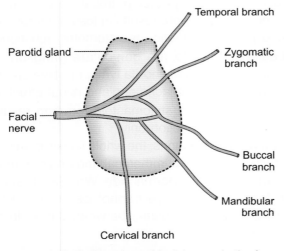

Fig. 16.28C-III: Branches of facial nerve in the face

the facial canal in the temporal bone, stylomastoid foramen and the parotid gland.

At the medial wall of tympanic cavity, the facial canal bends posteroinferiorly. The sensory ganglion

of facial nerve—the geniculate ganglion is located here.

Branches

Within the facial canal
1. Greater petrosal nerve
2. Nerve to stapedius
3. Chorda tympani nerve

Branches from the extracranial part (after emerging from the stylomastoid foramen – there are 6 branches.
1. Posterior auricular;
   Within the parotid gland
   1. Temporal
   2. Zygomatic
   3. Buccal
   4. Mandibular or marginal mandibular
   5. Cervical.

The names of the branches refer to the regions they supply.

Posterior Auricular

The posterior auricular, passes posterosuperior to the ear and supplies the occipital belly of occipitofrontalis muscle and auricularis posterior muscle of pinna.

Temporal Branch

The temporal branch emerges from the superior border of the parotid gland and supplies:
1. Superior part of orbicularis oculi
2. Frontal belly of occipitofrontalis
3. Muscles of auricle (anterior and superior).

Zygomatic Branch

The zygomatic branch divides into two or three branches, passing above and below the eye. Supply
1. Inferior part of orbicularis oculi
2. Facial muscles inferior to the orbit.

Buccal Branch

The buccal branch passes external to buccinator muscle. It supplies the—
1. Buccinator
2. Muscles of upper lip.

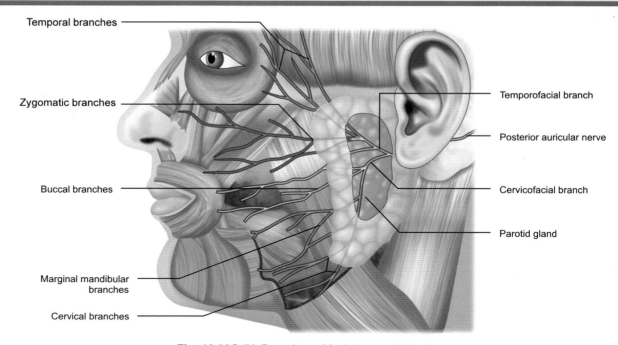

Fig. 16.28C-IV: Branches of facial nerve in the face

Labels: Temporal branches, Zygomatic branches, Buccal branches, Marginal mandibular branches, Cervical branches, Temporofacial branch, Posterior auricular nerve, Cervicofacial branch, Parotid gland

Marginal Mandibular

The marginal mandibular supplies the muscles of lower lip and chin. This nerve emerges from the inferior border of parotid gland and crosses the inferior border of mandible to reach the face.

Cervical Branch

The cervical branch passes inferiorly from the inferior border of the parotid gland, runs posterior to the mandible to supply the platysma, which is a superficial muscle of the neck.

Applied Anatomy

The lesions of facial nerve can be discussed under 2 headings:
- *Supranuclear lesions* or upper motor neuron lesions (i.e. from cerebral cortex to the level of facial nerve nucleus).
- *Infranuclear lesions* or lower motor neuron lesions (i.e. Nucleus of facial nerve, the facial nerve and its branches).

The supranuclear lesions of facial nerve (UMN palsy) is usually a part of hemiplegia (e.g. vascular lesions of internal capsule). In this type, only the lower part of face is paralyzed.

The upper part of the face escapes from paralysis. The frontal belly (frontalis) and upper part of orbicularis oculi escape due to the bilateral representation in the cerebral cortex.

Infranuclear lesions (LMN palsy): Because the branches of facial nerve are superficial, they are subject to injury by stab, cuts, birth injury and surgical procedures on parotid gland.

1. Lesions near the origin of facial nerve or proximal to the origin of greater petrosal nerve from the geniculate ganglion, results in loss of motor, taste and parasympathetic (secretomotor) functions.
2. Lesions distal to the geniculate ganglion, but before the origin of chorda tympani nerve produce same effects, but lacrimal gland is not affected.
3. Lesions near the stylomastoid foramen result in loss of motor function only.

In infranuclear lesions, the whole face is paralyzed. The face becomes asymmetrical, the angle of mouth is drawn towards the normal side. Wrinkles disappear from forehead. The eye cannot be closed. While chewing, food accumulates between the teeth and cheek (vestibule).

Bell's palsy is a common disorder involving the facial nerve. There is a sudden loss of control of muscles of entire right or left half of the face. Usually there are no other neurological defects. This type of LMN facial nerve palsy, without any known cause is called Bell's palsy. It often follows exposure to cold (e.g. travelling on a bus, sitting on the side seat).

The facial nerve can be compressed by a tumor of the parotid gland.

When the stapedius muscle is paralyzed, the ear becomes more sensitive to low tones (Normally stapedius dampens vibration of the stapes). So, even a small sound is appreciated by the patient as a loud noise (Hyperacusis).

Vestibulocochlear—CN VIII

This cranial nerve is a special sensory nerve having 2 functions—the functions of hearing and equilibrium.

It emerges from the junction of pons and medulla. It enters the internal acoustic meatus along with the facial nerve and labyrinthine vessels. Here, the CN VIII divides into vestibular and cochlear nerves.

The vestibular fibers, enter the maculae of utricle and saccule and ampullae of the semicircular ducts.

Cochlear fibers enter the spiral organ of Corti. They are concerned with hearing.

Applied Anatomy

Injury of the vestibulocochlear nerve may cause:
1. Tinnitus—"ringing" in ears
2. Vertigo—giddiness or loss of balance
3. Impairment or loss of hearing.

Sensorineural deafness is the result of disease in the cochlea or in the pathway from the cochlea to the brain.

Acoustic neuroma is a slow growing tumor of the neurolemma cells, covering the vestibular nerve. This results in compression of vestibular as well as cochlear nerves, resulting in loss of hearing, tinnitus and loss of equilibrium.

Glossopharyngeal Nerve—CN IX

This cranial nerve emerges from the medulla. It passes anterolaterally and leaves the cranium through the anterior part of jugular foramen.

Function

1. Sensory
   • General
   • Special (taste)
2. Motor
3. Parasympathetic.

The glossopharyngeal carries general sensation from the posterior 1/3 of tongue, mucosa of pharynx, palatine tonsil, auditory (Eustachian) tube, middle ear and from the carotid body and sinus.

The *taste* sensation from posterior 1/3 of tongue and the circumvallate papillae is carried by the glossopharyngeal nerve.

Motor fibers of CN IX innervates the stylopharyngeus muscle.

Parasympathetic or secretomotor fibers supply the parotid gland and glands of posterior 1/3 of tongue.

The nuclei of glossopharyngeal nerve are situated in the medulla, which are closely related to the nuclei of vagus and accessory nerves.

Applied Anatomy

Isolated lesions of glossopharyngeal nerve is rare. When affected, taste is absent on the posterior 1/3 of tongue, and the gag reflex is absent on the side of the lesion.

The glossopharyngeal nerve emerges from the cranium through the jugular foramen, along with vagus and accessory nerves. So, a tumor in this region can affect these three nerves—the jugular foramen syndrome.

Vagus Nerve—CN X

This nerve derives its name from the word "vagabond", meaning "wanderer", because of its long, complicated course.

The vagus nerve arises by a series of roots from the medulla and leaves the skull through the jugular foramen, along with glossopharyngeal and accessory nerves.

The nerve shows 2 swellings—superior and inferior ganglia—at and below the jugular foramen.

The vagus nerve has connections with the glossopharyngeal and the accessory nerves at the jugular foramen.

Vagus nerve continues inferiorly in the carotid sheath to the root of the neck (the carotid sheath is a tubular sheath of cervical fascia enclosing the internal carotid artery, vagus nerve and internal jugular vein. In the lower part, it encloses the common carotid artery).

The right vagus nerve enters the thorax anterior to the right subclavian artery, where it gives the right recurrent laryngeal nerve; this nerve hooks around the right subclavian artery and ascends in the groove between trachea and esophagus, to supply the larynx.

In the thorax, the right vagus gives several branches which form 3 plexuses, the pulmonary, cardiac and esophageal.

The left vagus descends in the neck posterior to left common carotid artery. At the inferior border of the arch of aorta, it gives off the left recurrent laryngeal nerve. This nerve hooks the ligamentum arteriosum (a fibrous ligament extending between arch of aorta and left pulmonary artery) and ascends to the larynx between the trachea and esophagus.

The left vagus nerve also contributes to the pulmonary, cardiac and esophageal plexuses.

The two vagi form the esophageal plexus along with branches from sympathetic trunks. This plexus follows the esophagus through the diaphragm into the abdomen, where the fibers of the vagi (now they are called anterior and posterior vagal trunks) innervate the esophagus, stomach, duodenum, jejunum, ileum, cecum, appendix, ascending colon and right 2/3 of transverse colon.

The nuclei of vagus are situated in the medulla, which are shared by glossopharyngeal and accessory nerves.

Functions

1. Sensory:
   a. General
   b. Special (taste)
2. Motor
3. Parasympathetic.

General sensory—from lower part of pharynx, larynx, thoracic and abdominal organs.

Sense of taste (special)— from posterior most part of the tongue and epiglottis.

Motor—to soft palate, pharynx, muscles (intrinsic) of larynx, palatoglossus.

Parasympathetic (secretomotor)—to the glands of trachea and bronchi, glands of digestive tract, parasympathetic to thoracic and abdominal viscera.

The vagi form the largest part of parasympathetic nervous system (cranio-sacral outflow) innervating many important structures including the heart.

Lesions of Vagus Nerve

Isolated lesions are rare.

Injury to pharyngeal branch of vagus results in dysphagia (difficulty in swallowing).

Injury of a recurrent laryngeal nerve can result from operations on thyroid gland or tumors of the neck or aneurysm of arch of aorta. Recurrent laryngeal nerve injury causes hoarseness and dysphonia due to paralysis of vocal cord.

In patients with tumors of posterior cranial fossa or raised intracranial tension, the cerebellum and medulla oblongata tend to herniate downward through the foramen magnum. This will produce symptoms of headache, neck stiffness and paralysis of glosso-pharyngeal, vagus, accessory and hypoglossal nerves due to traction.

The lateral medullary syndrome or Wallenberg syndrome results from thrombosis of posterior inferior cerebellar artery, in which the nuclei of 8th, 9th, 10th and 11th cranial nerves are affected.

Bulbar Palsy

In this degenerative motor neuron disease, the cranial nerve nuclei of the bulb (medulla oblongata) are invol-ved leading to dysarthria, dysphonia and dysphagia.

Accessory Nerve—CN XI

The 11th cranial nerve has two roots—cranial and spinal.

The cranial root arises by a series of rootlets from the medulla.

The spinal root emerges as a series of rootlets from the first five cervical segments of the spinal cord. It ascends through the foramen magnum and joins the cranial root at the jugular foramen. These two roots are united only for a short distance.

The cranial root of accessory joins the vagus and together, they innervate the skeletal muscles of the soft palate, pharynx, larynx and the esophagus.

The spinal part of accessory descends along the internal carotid artery, penetrates and innervates the sternocleidomastoid muscle. It then crosses the posterior triangle of neck and innervates the trapezius muscle.

Summary

1. Cranial root—motor to muscles of soft palate pharynx, larynx and esophagus.
2. Spinal root—motor to sternocleidomastoid and trapezius.

Applied Anatomy

The spinal part of accessory is likely to get injured in the posterior triangle of neck during surgical procedures such as lymph node biopsy. This leads to weakness of trapezius and sternocleidomastoid muscles.

Wry neck: Paralysis of sternocleidomastoid causes the neck to be flexed to the opposite side and the face turned to the paralyzed side by the unopposed action of the normal sternocleidomastoid muscle.

In paralysis of trapezius, patient cannot shrug the shoulder on the affected side.

Hypoglossal Nerve—CN XII

This nerve is purely motor. It arises as a series of rootlets from the medulla, and leaves the skull through the hypoglossal canal. After emerging from the canal, this nerve is joined by a branch from the first cervical nerve (C1).

The hypoglossal nerve passes medial to the angle of mandible. It then curves anteriorly to enter the tongue. It divides into several branches. *Hypoglossal nerve supplies all the intrinsic and extrinsic muscles of tongue except palatoglossus muscle.*

The hypoglossal nerve conveys fibers of C_1 to the ansa cervicalis (a loop of nerves in the neck), which supplies the infrahyoid muscles.

Applied Anatomy

Injury to the hypoglossal nerve paralyzes the muscles of tongue on the same side (ipsilateral); so, when the tongue is protruded, its tip deviates towards the paralyzed side because of the unopposed action of the muscles of normal side.

In medial medullary syndrome (due to thrombosis of a branch of vertebral artery supplying the medial part of medulla) the nucleus of hypoglossal nerve is affected.

▮ NERVI TERMINALIS—CN XIII

A minute bundle of nerves attached to the cerebrum posterior to the olfactory tract has been described as a 13th pair. They are distributed with the olfactory nerves to the nose; their function is unknown.

▮ SUMMARY OF CRANIAL NERVES

See Table 16.1.

▮ NERVE PLEXUSES

Nerves arising from some regions of the spinal cord join together to form networks called plexuses, which gives rise to important nerves. These plexuses are:
1. Cervical plexus
2. Brachial plexus
3. Lumbar plexus
4. Sacral plexus.

Cervical Plexus

This network of nerves is formed by the communications between the ventral rami of the superior four cervical nerves (C1-C4). This plexus lies deep to the internal jugular vein and the sternocleidomastoid muscle. Cutaneous nerves arising from this plexus supply the skin of the neck and scalp.

Nerves derived from the cervical plexus are lesser occipital, great auricular, transverse cervical, supraclavicular and phrenic nerves.

The supraclavicular nerve (C3 and C4) arises as a single nerve; then it divides into medial, intermediate and lateral branches and supplies the skin over the anterior aspect of chest and shoulder.

The *phrenic nerve,* the only motor supply to the diaphragm, arises from the ventral rami of C3,4 and 5. It is about 30 cm long. Injury to the phrenic nerve results in paralysis of the corresponding half of the diaphragm.

Most of the nerves arising from the cervical plexus are related to the posterior border of the sternocleidomastoid muscle, at the junction of its superior and middle thirds—this site is called the *"nerve point of the neck"*. For regional anesthesia before surgery, a "cervical plexus block" can be done, by injecting the anesthetic agent around the nerve point of the neck or along the posterior border of sternocleidomastoid muscle.

Brachial Plexus (Fig. 16.29)

This large network of nerves to the upper limb extends from the neck into the axilla. This important plexus is situated partly in the neck and partly in the axilla.

Section C

The Organ Systems

Section C

The Organ Systems

Table 16.1: Summary of cranial nerves

Cranial nerve	Location of nerve cell bodies (Nuclei)	Cranial exit/entry	Main action / Actions
1. Olfactory nerve	Olfactory epithelium—Nose	Foramina in cribriform plate of ethmoid	Carries the sense of smell from nose to olfactory bulb
2. Optic nerve	Ganglion cells of retina	Optic canal	Vision from retina
3. Oculomotor	Midbrain	Superior orbital fissure	1. Motor to superior, inferior and medial rectus, the inferior oblique and levator palpebrae superioris
			2. Parasympathetic fibers to sphincter pupillae and ciliaris muscle
4. Trochlear	Midbrain	Superior orbital fissure	Motor to superior oblique muscle of eyeball
5. Trigeminal			
Ophthalmic division CN V1	Trigeminal ganglion	Superior orbital fissure	Sensations from cornea, forehead, scalp, eyelids, nose, nasal cavity and paranasal sinuses
Maxillary division CN V2	Trigeminal ganglion	Foramen rotundum	Sensation from skin over maxilla, maxillary teeth, maxillary sinuses, palate
Mandibular division CN V3	Motor-Pons Sensory-Trigeminal ganglion	Foramen ovale	Muscles of mastication, anterior belly of digastric, tensor tympani levator palati muscles
			Sensation from skin over mandible, lower set of teeth, anterior 2/3 of tongue, etc.
6. Abducent	Pons	Superior orbital fissure	Motor to lateral rectus muscle of eyeball
7. Facial	a. Motor-Pons b. Special sensory (taste)—geniculate ganglion c. Parasympathetic fibers—geniculate ganglion	Stylomastoid foramen	Muscles of facial expression, stapedius, posterior belly of digastric muscles of scalp stylohyoid Taste from anterior 2/3 of tongue (chorda tympani branch) Parasympathetic fibers to submandibular, sublingual salivary glands, lacrimal gland and glands of palate and nose
8. Vestibulocochlear			
Vestibular	Vestibular ganglion	Internal acoustic meatus	Vestibular sensation from semicircular canals, utricle and saccule
Cochlear	Spiral ganglion	Internal acoustic meatus	Hearing – from spiral organ
9. Glossopharyngeal	Medulla	Jugular foramen	1. Motor to stylophargngeus
			2. Parasympathetic (secretory) to parotid gland
			3. General sensations from posterior 1/3 of tongue
			4. Taste from posterior 1/3 tongue
10. Vagus	Medulla	Jugular foramen	1. Motor to constrictor muscles of pharynx, muscles of larynx, palate (except tensor palati), muscles of upper 2/3 of esophagus

Contd...

Contd...

Cranial nerve	Location of nerve cell bodies	Cranial exit/entry	Main action / Actions
			2. Parasympathetic supply to smooth muscles of trachea, bronchi, digestive tract, cardiac muscle
			3. Sensation from posterior part of tongue, pharynx, larynx, trachea, bronchi, digestive tract
			4. Taste from palate, epiglottis
11. Accessory			
a. Cranial part	Medulla	Jugular foramen	Motor to skeletal muscle of soft palate and pharynx by forming a plexus with vagus
b. Spinal part	Upper 5 spinal segments of spinal cord	Jugular foramen	Motor to sternocleidomastoid and trapezius muscles
12. Hypoglossal	Medulla	Hypoglossal canal	Motor to muscles of tongue (except palatoglossus).

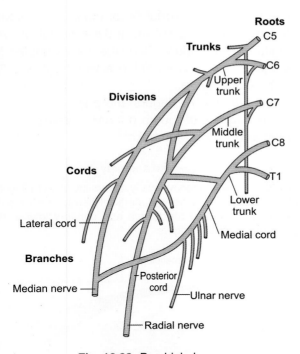

Fig. 16.29: Brachial plexus

The brachial plexus is formed by the union of the ventral rami of nerves C5 to C8 and T1. These ventral rami are called the "roots" of brachial plexus.

Typical Plan of Brachial Plexus

The ventral rami of C5 and C6 unite to form the upper trunk; the ventral ramus of C7 continues as the middle trunk; the ventral rami of C8 and T1 unite to form the lower trunk.

Each trunk divides into anterior and posterior divisions.

The anterior divisions of upper and middle trunk join to form the lateral cord.

Anterior division of the inferior trunk continues as the medial cord.

The three posterior divisions unite to form the posterior cord.

These divisions are of fundamental significance because the anterior divisions supply the anterior or flexor muscles and posterior divisions supply the posterior or extensor muscles.

Important Branches of Brachial Plexus

Only the larger, clinically important branches are mentioned here.

Lateral cord
1. Musculocutaneous nerve
2. Lateral root of median nerve

Medial cord
1. Ulnar nerve
2. Medial root of median nerve

Posterior cord
1. Radial nerve
2. Axillary nerve

■ BRACHIAL PLEXUS INJURIES

Injuries to different parts of brachial plexus may lead to characteristic clinical features.

Erb's Paralysis

Injury to the upper trunk of brachial plexus at the Erb's point (site of union of C5 and C6) leads to Erb's paralysis. Excess stretching of the head from the shoulder as in birth injury, or a fall on the shoulder, are the usual causes. The upper limb shows a characteristic position described as "waiter's tip hand" or "porter's tip hand". The arm hangs by the side, it is adducted and medially rotated; forearm is extended and pronated.

Klumpke's Paralysis

Results from injury to the lower trunk of brachial plexus. The cause for this type of paralysis is undue abduction of the arm. For example, forceful pull of an infants' shoulder during birth. It may also occur when a person grasps something during a fall. This results in an injury to the inferior trunk of brachial plexus (C8 and T1). The resulting paralysis and anesthesia usually affects the skin and muscles supplied by the ulnar nerve.

A cervical rib can also compress the lower trunk.

Deformity in Klumpke's Paralysis

Claw hand results from paralysis of intrinsic muscles of hand and flexors of wrist and fingers. There is loss of sensation over the ulnar side of forearm and hand. Injury to sympathetic fibers which leave the spinal cord through T1 results in Horner's syndrome (drooping of eyelid, loss of sweating of face on the affected side, and constriction of pupil).

Musculocutaneous Nerve (C5,6,7)

It supplies the muscles of the anterior aspect of arm. The 3 muscles supplied by this nerve are (BBC):
1. Biceps brachii
2. Brachialis
3. Coracobrachialis.

Just proximal to the elbow, it becomes superficial and forms the lateral cutaneous nerve of the forearm, supplying skin over the lateral aspect of forearm (This is why the nerve is called musculocutaneous—supplying "muscles and skin").

Median Nerve or "Labourer's Nerve" (Figs 16.29 and 16.30)

Formed by the union of medial root from medial cord of brachial plexus and lateral root from the lateral cord of brachial plexus. So it receives fibers from C5,6,7,8 and T1 (Root value of median nerve is C5,6,7,8 T1).

In the arm, it accompanies the brachial artery. Then it passes in front of the elbow and enters the forearm, where it lies deep to the flexor muscles.

Just proximal to the wrist, the nerve becomes superficial. Then, it passes through an osseofibrous tunnel called the carpal tunnel. This restricted space lies between the flexor retinaculum and the carpal bones.

Area of Distribution of Median Nerve

1. It supplies the powerful flexor muscles of forearm (except flexor carpi ulnaris and medial half of flexor digitorum profundus). Since the nerve supplies the muscles essential for lifting a weight, it is called Labourer's nerve.
2. Skin of lateral 3½ fingers and palm.
3. Short muscles of thumb (thenar muscles).
4. First and second lumbricals.

Applied Anatomy

Injury to the median nerve usually occurs just proximal to the wrist (because the nerve is superficial here), which results in inability to oppose the thumb and sensory loss in the palm and lateral 3½ fingers.

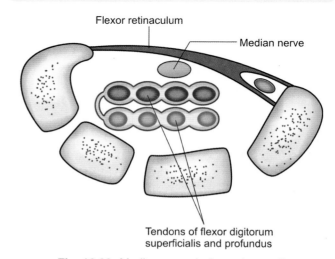

Fig. 16.30: Median nerve in "carpal tunnel"

Carpal tunnel syndrome: The median nerve lies in a restricted space between the flexor retinaculum and the carpal bones (the carpal tunnel). If this space gets reduced following bony injury or soft tissue swelling (e.g. fracture or dislocation of carpal bones, myxoedema, acromegaly) carpal tunnel syndrome results. There is weakness of short muscles of thumb, pain, tingling, impairment of sensation in the lateral 3½ fingers.

Ulnar Nerve

The ulnar nerve is formed in the axilla as a continuation of the medial cord of brachial plexus and conveys fibers from C8 and T1.

After its formation, the nerve lies on the medial side of the axillary artery and then on the medial side of brachial artery. It *passes behind the medial epicondyle of humerus.* It enters the forearm between the two heads of flexor carpi ulnaris muscle.

In the upper 1/3 of forearm, the nerve lies deep to the flexor muscles, but in the lower or distal 2/3 of forearm, the nerve becomes superficial.

At the wrist, the nerve divides into superficial and deep branches.

Branches of Ulnar Nerve (Motor)

It supplies:
1. The flexor carpi ulnaris
2. Medial half of flexor digitorum profundus,
3. All the intrinsic muscles of hand *except* the thenar muscles and the first and second lumbricals.

Since most of the intrinsic muscles of the hand are used by a musician (e.g. a violinist) the ulnar nerve is called the *musician's nerve.*

Sensory—It supplies the skin on the medial one and half of the fingers and skin over the medial side of palm.

Applied Anatomy

As the ulnar nerve is very superficial in the region of the elbow, the name "funny bone" is given to this part. A sharp tap behind the medial epicondyle of humerus produces stimulation of the ulnar nerve leading to pain and tingling sensation in the ring and little fingers.

Injury to ulnar nerve leads to paralysis of flexor carpi ulnaris, flexor digitorum profundus (medial ½) and most of the intrinsic muscles of hand leading to a condition called *"claw hand".*

Radial Nerve

The radial nerve is the largest branch of brachial plexus and begins in the axilla as a continuation of posterior cord. It conveys fibers from C5,6,7,8 and T1; it is the nerve of extensor compartment of the arm and forearm.

After leaving the axilla, the radial nerve winds round the posterior aspect of the shaft of the humerus in the *spiral groove* or *radial groove.*

The radial nerve supplies the triceps, the extensors of the back of the forearm, skin on the dorsal aspect of lateral 2/3 of dorsum of hand and fingers.

Applied Anatomy

The radial nerve is often injured in fracture of the humerus in the region of spiral groove or by the pressure of a crutch in the axilla. Compression of the nerve against the spiral groove by placing the outstretched arm on an arm chair (e.g. under drunken condition) may be associated with temporary radial nerve palsy. This is known as "Saturday night palsy".

When radial nerve is injured, the hand becomes flaccid, and it is flexed at the wrist—this condition is called *"wrist drop".*

Axillary Nerve (Fig. 16.31)

This is a branch from the posterior cord of brachial plexus. It conveys fibers from C5 and C6.

The nerve, accompanied by the posterior circumflex humeral vessels, winds round the posterior surface of the *surgical neck of humerus* (so, it is called the "circumflex nerve").

Axillary nerve supplies an articular branch to the shoulder joint, muscular branches to the deltoid and teres minor muscles and the skin overlying the deltoid muscle. So, the axillary nerve obeys the Hilton's Law, which enunciates that a nerve which supplies a joint

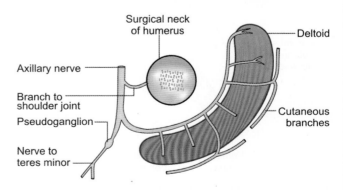

Fig. 16.31: Axillary nerve

also innervates the muscle and skin overlying that joint.

Applied Anatomy

Injury to the axillary nerve due to fracture of the surgical neck of humerus results in paralysis of deltoid muscle. So the patient cannot abduct the arm. There will be loss of sensation over the deltoid. The deltoid muscle atrophies and the rounded contour of the shoulder disappears.

Intramuscular injections are often given into the deltoid. The injection is given in the lower half of the muscle, in order to avoid injury to the axillary nerve.

Lumbar Plexus (Fig. 16.32)

The lumbar plexus lies within the substance of the psoas major muscle. It is formed by the ventral rami of upper 4 lumbar nerves (L1-L4). The first lumbar nerve gets a contribution from the subcostal nerve (T12) and the fourth lumbar nerve gives a contribution to the lumbosacral trunk (which takes part in the formation of the sacral plexus).

Branches

1. Iliohypogastric nerve.
2. Ilioinguinal nerve.
3. Genitofemoral nerve—it divides into 2 branches, genital and femoral. The genital nerve supplies the cremaster muscle in males. The femoral branch accompanies the proximal part of femoral artery and supplies the skin over the upper part of front of the thigh.

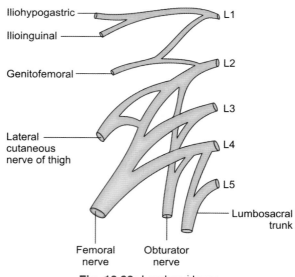

Fig. 16.32: Lumbar plexus

4. Lateral cutaneous nerve of thigh.
5. The femoral nerve.
6. The obturator nerve.
7. The lumbosacral trunk (L4, L5), takes part in the formation of the sacral plexus.

Femoral Nerve

It is the *largest branch* of the lumbar plexus. It is formed by the union of the dorsal divisions of L2, L3 and L4. It emerges at the lateral border of psoas major muscle, runs downwards and laterally. It passes behind the inguinal ligament to enter the femoral triangle in front of the thigh. Here, it lies lateral to the femoral artery. After a short course, of about 2 cm below the inguinal ligament, the nerve divides into branches.

Femoral nerve is the chief nerve of the anterior or extensor compartment of the thigh. It supplies the extensor muscles (mainly quadriceps femoris), hip and knee joints and skin. *Saphenous nerve*, the *longest cutaneous nerve* in the body is a branch of femoral nerve.

Injury to the femoral nerve causes paralysis of the quadriceps femoris and loss of sensation on the anterior and medial sides of the thigh.

Obturator Nerve

It is a branch of the lumbar plexus. It is formed by the ventral divisions of L2, L3 and L4 ventral rami. It is the chief nerve of the medial (adductor) compartment of thigh. The upper part of the nerve lies in the pelvis. It enters the medial aspect of thigh by passing through the obturator canal.

The obturator nerve supplies the following muscles—Pectineus, adductor longus and brevis, part of adductor magnus and obturator externus. The nerve supplies the hip joint and knee joint also. So, a disease of the hip joint may cause referred pain in the knee.

Sacral Plexus (Fig. 16.33)

The sacral plexus is formed by the lumbosacral trunk (L4 and L5) and the ventral rami of S1-S3. This plexus lies in front of the piriformis muscle; posteriorly it is related to the internal iliac vessels and the ureter.

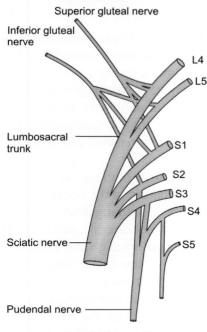

Fig. 16.33: Sacral plexus

The sacral plexus gives rise to two main branches, the sciatic and pudendal nerves.

Sciatic Nerve (Fig. 16.34)

This is the largest branch of the sacral plexus; it is the *thickest nerve in the body*. It is formed by the ventral rami of L4, 5, S1, S2 and S3. The sciatic nerve begins in the pelvis and terminates at the superior angle of the popliteal fossa by dividing into tibial and common peroneal nerves.

In the pelvis, the nerve lies anterior to (in front of) the piriformis muscle.

The sciatic nerve enters the gluteal region through the greater sciatic foramen, below the piriformis. It lies under cover of the gluteus maximus muscle.

The sciatic nerve enters the back of the thigh at the lower border of gluteus maximus. Then it runs vertically downwards up to the superior angle of popliteal fossa (the junction of upper 2/3 and lower 1/3 of the back of the thigh). Here, the nerve divides into its two terminal branches—the tibial and common peroneal nerves.

The sciatic nerve may divide into its terminal branches anywhere between the pelvis and thigh. When the division occurs in the pelvis, the tibial nerve passes through the greater sciatic foramen inferior

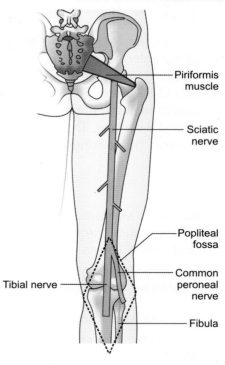

Fig. 16.34: Sciatic nerve

to the piriformis and the common peroneal nerve pierces the piriformis to enter the gluteal region.

Branches

1. Articular branches to hip joint
2. *Muscular branches*—the tibial part of sciatic nerve supplies the "hamstring" muscles and muscles of the flexor compartment of leg. The common peroneal supplies the short head of biceps, muscles of the anterior and lateral compartments of leg.

Applied Anatomy

1. Compression of the sciatic nerve (e.g. after sitting for a long time) against the femur may give rise to a "sleeping foot".
2. Shooting pain along the distribution of sciatic nerve and its terminal branches is known as sciatica. This may be due to an intervertebral disc prolapse or osteoarthritis.
3. The sciatic nerve may be injured by penetrating wounds or fracture of the pelvis. This results in loss of all movements below knee, including "foot drop" and loss of sensation over a large area of thigh, leg and foot.
4. While giving intramuscular injections in the gluteal region (buttock) the upper outer quadrant is chosen, so that injury to sciatic nerve can be prevented.

Pudendal Nerve

This branch of sacral plexus is the chief nerve of perineum and the external genitalia.

This nerve arises in the pelvis. It is derived from the ventral rami of S2,3 and S4. It leaves the pelvis through the greater sciatic foramen. It is accompanied by internal pudendal vessels. These structures enter the pudendal canal by passing through the lesser sciatic foramen.

Branches

Sensory branches to skin around the anus, penis, terminal part of anal canal, scrotum, labia majora and minora, terminal part of vagina.

Motor or muscular branches to urogenital muscles, external anal sphincter and levator ani.

Applied Anatomy

Pudendal nerve block has been used in obstetrics prior to forceps delivery to relax the pelvic floor muscles and to abolish sensation (pain) in the lower vagina and vulva (The anesthetic agent, e.g. lignocaine is injected near the ischial spine by a needle passed through the vaginal wall and then guided by a finger).

■ AUTONOMIC NERVOUS SYSTEM (ANS)

Definition

The portion of the nervous system that controls the visceral functions of the body is called the autonomic nervous system. This system controls visceral functions such as arterial pressure, gastrointestinal motility and secretion, urinary bladder emptying, sweating, body temperature and many other activities.

The ANS is activated mainly by centers located in the spinal cord, brainstem, hypothalamus and the limbic system of the cerebrum. Sensory signals from the visceral organs enter these centers and these, in turn, transmit appropriate reflex responses back to the visceral organs to control their activities (visceral reflex).

Subdivisions of ANS

1. Sympathetic system
2. Parasympathetic system

Sympathetic System (Thoracolumbar Outflow)

This system consists of two chains of ganglia on either side of the vertebral column, from which nerves extend to different internal organs.

The sympathetic nerves originate in the spinal cord between the spinal segments T1 and L2 and pass from here first into the sympathetic chain, thence to the tissues and organs that are stimulated by the sympathetic nerves.

Each sympathetic pathway is composed of two fibers—a preganglionic and a postganglionic neuron. The cell body of a preganglionic neuron lies in the lateral horn (intermediolateral horn) of the spinal cord and its fiber passes through the anterior or ventral root of the cord into the spinal nerve.

Immediatley after the spinal nerve leaves the spinal cord, the preganglionic sympathetic fiber leaves the nerve and passes through the white ramus into one of the ganglia of the sympathetic chain. Then the course of the fibers can be one of the following three (Fig. 16.35).
1. The fibers can synapse with the neurons in the ganglion that it enters.
2. The fibers can pass upward or downward in the sympathetic chain and synapse in one of the other ganglia of the chain.

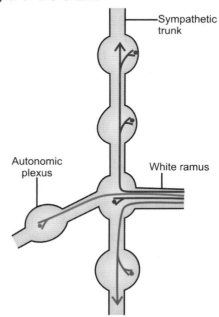

Fig. 16.35: Mode of termination of sympathetic preganglionic fiber

3. The fiber can pass for variable distances through the chain and then through one of the nerves radiating outward from the chain, finally ending in an outlying sympathetic ganglion.

Many of the postganglionic fibers pass back from the sympathetic chain into the spinal nerves through gray rami at all levels of the cord.

Segmental distribution of sympathetic nerves The sympathetic fibers from T1 generally pass up the sympathetic chain into the head; from T2 into the neck; T3 to T6 into the thorax; T7 to T11 into the abdomen; T12, L1 and L2 into the legs (This is an approximate distribution; there may be considerable overlap).

Parasympathetic Nervous System (Craniosacral Outflow)

The parasympathetic fibers leave the central nervous system through cranial nerves 3,7,9 and 10 and through S2, S3 and S4 sacral nerves. About 75% of all parasympathetic fibers are in the vagus nerves.

The vagus nerves supply the parasympathetic fibers to the heart, lungs, esophagus, stomach, the small intestine, large intestine (up to proximal 2/3 of transverse colon), the liver, gallbladder, pancreas and upper portions of ureters.

The 3rd cranial nerve (oculomotor) supplies the sphincter pupillae and ciliaris muscle of the eye. Fibres from the 7th (facial) nerve pass to the lacrimal, nasal, submandibular and sublingual glands; fibers from the 9th (glossopharyngeal) nerve pass to the parotid gland.

The sacral parasympathetic fibers join to form the nervi erigentes or pelvic nerves, which distribute their fibers to the descending colon, rectum, bladder, lower portions of ureters and external genitalia.

The sympathetic system stimulates activities that are performed during emergency and stress situations ("fight, fright and flight" situations).

The parasympathetic system stimulates activities that conserv and restore body resources.

For distribution of ANS and the effects of stimulation, see Table 16.2.

Tissue/Organ	Sympathetic stimulation	Parasympathetic stimulation
Heart	Increases the rate and strength of contraction. Coronary arteries dilated. Blood flow to heart muscle increases	Decreases the rate and strength of contraction. Heart beats slowly. Constricts coronary arteries; blood flow to myocardium decreases
Lungs	Dilates bronchi	Constricts bronchi
Arrector pili muscle of skin	Contraction of this smooth muscle, causes the hairs to "stand on end"	Not innervated by parasympathetic fibers
Iris	Pupils dilate due to contraction of dilator pupillae	Constricts pupil due to contraction of sphincter pupillae
Ciliary muscle	Not innervated by sympathetic fibers	Contracts ciliary muscle and accommodates lens for near vision
Glands		
Sweat, Lacrimal, Salivary	Expulsion of sweat	Not innervated by parasympathetic fibers
Gastric, Intestinal, Stomach	Decrease or inhibit secretion	Stimulate secretion
Intestine	Decreases motility	Increases motility
Pancreas	Inhibits secretion of enzymes	Promotes secretion of enzymes
Urinary bladder	Contracts internal sphincter, causing retention of urine	Relaxes internal sphincter, resulting in passage of urine

Table 16.2: Distribution of ANS and the effects of stimulation

Applied Anatomy

1. *Visceral Referred Pain:* Pain arising from an abdominal viscus (organ) radiates to the part of the body supplied by somatic sensory fibers associated with the same spinal segment of spinal cord that receives visceral sensory fibers from the viscus (organ) concerned. For example, pain from a gastric ulcer is referred to the epigastric region, because the stomach is supplied by T7 and T8 segments of spinal cord. Pain from an inflamed vermiform appendix is referred to the umbilicus along T10; (but later, due to the involvement of parietal peritoneum, the pain is referred to right lower quadrant of abdomen).

Contd...

2. *Lumbar Sympathectomy:* Surgical removal of one or two sympathetic ganglia and their rami communicantes, as a treatment for arterial disease of lower limb is called lumbar sympathectomy.
3. *Syringing:* Syringing the external auditory meatus to remove a "foreign body" may stimulate the auricular branch of vagus nerve and cause bradycardia.
4. *Horner's Syndrome:* Cutting of sympathetic trunk in the neck interrupts the sympathetic nerve supply to the head and neck on that side leading to Horner's syndrome characterized by (a) Pupillary constriction owing to paralysis of dilator pupillae (b) Ptosis or partial drooping of upper eyelid due to paralysis of smooth muscle component of levator palpebrae superioris muscle; (c) Vasodilation and absence of sweating on the affected side.

17 Endocrine System

INTRODUCTION

The functions of the body are regulated by two major control systems (1) the nervous system, and (2) the hormonal or endocrine system.

A hormone is a chemical substance that is secreted into the body fluids by one cell or a group of cells and that exerts a physiological control effect on other cells of the body.

In general, the hormonal system is concerned principally with control of the different metabolic functions of the body.

LOCAL HORMONES AND GENERAL HORMONES

Examples of local hormones are (1) acetylcholine released at the skeletal nerve endings; (2) secretin released by the duodenal wall and transported in the blood to the pancreas to cause a watery pancreatic secretion.

The general hormones are secreted by specific endocrine glands and are transported in the blood to cause physiologic actions at distant points in the body. A few of the general hormones affect almost all the cells in the body, e.g. growth hormone and thyroid hormone.

Chemically, the hormones are of two types:

1. Proteins, derivatives of proteins or amino acids
2. Steroids.

The hormones secreted by adrenal cortex, the ovary and the testis belong to the steroid group. All other hormones are basically proteins.

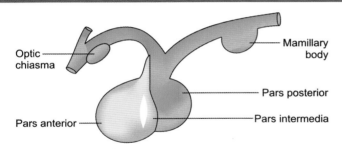

Fig. 17.1: Parts of pituitary gland

Major Endocrine Glands

The major endocrine glands (ductless glands) are:
1. The hypophysis cerebri or pituitary
2. The thyroid gland
3. The parathyroid glands
4. The adrenal or suprarenal glands.
   These glands are entirely endocrine in function.

Other Endocrine Glands

Groups of endocrine cells may be present in organs that have other functions. They include:
1. The islets of Langerhans of pancreas
2. Interstitial cells of testes
3. Follicle and corpus luteum of ovary
4. Hormones secreted by placenta.

THE HYPOPHYSIS CEREBRI OR PITUITARY (FIG.17.1)

The hypophysis cerebri is one of the most important endocrine glands. It is a small gland about 1 cm in diameter and 0.5-1 gram in weight that lies in the sella turcica, at the base of the brain. It is connected to the hypothalamus by the pituitary or hypophysial stalk.

Physiologically, the pituitary gland is divided into two distinct portions:
1. The anterior pituitary or adenohypophysis and
2. The posterior pituitary or neurohypophysis.

Between these, is a small, relatively avascular zone called pars intermedia, which is almost absent in human being (It is larger and more functional in lower animals).

Pars tuberalis is an upward prolongation from the anterior lobe , which covers the front and sides of the pituitary stalk.

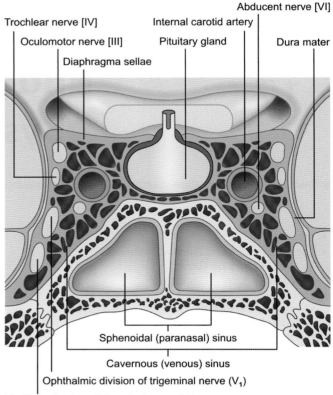

Fig. 17.2: Relations of pituitary gland

Important Relations (Fig.17.2)

- *Above:* A circular fold of dura called diaphragma sellae; and circulus arteriosus
- Above and in front: The optic chiasma
- Below: Sphenoidal air sinus
- Laterally: Cavernous sinuses with 3rd, 4th and 5th cranial nerves in the lateral wall of cavernous sinus.

Blood Supply

Pituitary is a highly vascular gland. It is supplied by superior hypophyseal arteries (branches arising from circulus arteriosus) and inferior hypophyseal arteries (branches of internal carotid arteries).

The anterior pituitary has extensive capillary sinuses among the glandular cells. Almost all the blood that enters these sinuses passes first through a capillary bed in the tissue of the lower most portion of hypothalamus, called the median eminence. Then, through the hypothalamic-hypophyseal portal vessels into the anterior pituitary sinuses.

Histology (Figs 17.3A to C)

Anterior Lobe (Pars Anterior)

Anterior lobe consists of thick irregular cords of cells, branching and freely anastomosing. These cells are separated by sinusoids.

The cells are of two types:
a. Chromophils (50%)
b. Chromophobes (50%)

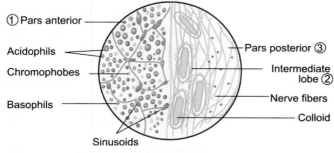

Fig. 17.3A: Histology of pituitary

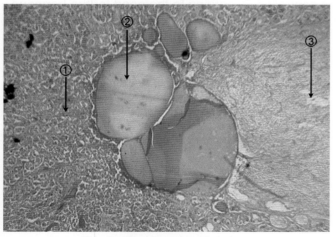

Fig. 17.3B: Histology of pituitary

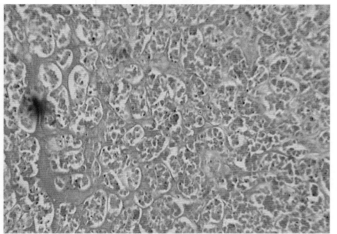

Fig. 17.3C: Pars anterior

Chromophobes are much smaller in size.

Chromophils are of two types:
a. Acidophils (75%) or alpha cells
b. Basophils (25%) or beta cells.

Intermediate Lobe (Pars Intermedia)

Intermediate lobe has an irregular row of follicles containing pale staining colloid material and clusters of cells.

Posterior Lobe (Pars Posterior)

Posterior lobe consists of hypothalamo-hypophyseal nerve fibers and pituicytes. The nerve terminals show vesicles called Hering bodies where the hormone is stored. Most of the nerve fibers arise in the supraoptic and paraventricular nuclei of hypothalamus.

The pituicytes are supporting cells, lying between the axons.

Types of acidophils, basophils and the hormones secreted by them are shown in Table 17.1.

Pars Posterior—Neurohypophysis

Hormones

The hormones released are as follows:

Vasopressin or antidiuretic hormone (ADH): This hormone is actually produced by the neurons of supraoptic nucleus of the hypothalamus. Then it passes down the axons of the neurons and stored in vesicles at nerve endings (Hering bodies) and released into the bloodstream, when required.

Oxytocin—It is synthesized by neurons located in the paraventricular nucleus of hypothalamus, pass along the axons to the posterior lobe of pituitary.

The ADH controls reabsorption of water by kidney tubules. The oxytocin controls the contraction of smooth muscle of the uterus and also of the mammary gland (the myo-epithelial cells surrounding the alveoli).

The cells of the pars intermedia secrete a hormone—Melanocyte stimulating hormone or MSH, which causes increased pigmentation of skin.

Control of Secretion of Hormones of the Adenohypophysis

The secretion of hormones by the adenohypophysis takes place under the control of neurons of the hypo-

Table 17.1: Parts of pituitary, cells and hormones	
I. Pars Anterior or adenohypophysis	
a. Acidophils	
1. Somatotrophs (somatotropes)	Somatotropin or growth hormone
2. Mammotrophs (mammotropes)	Prolactin
b. Basophils	
1. Corticotrophs (corticotropes)	ACTH or Adrenocorticotropin
2. Thyrotrophs (thyrotropes)	TSH or Thyrotropic Hormone
3. Gonadotropes	a. In females, FSH and LH
	b. In males, ICSH (Interstitial cell stimulating hormones)
II. Pars posterior or neurohypophysis	
1. Vasopressin or ADH	Actually produced by neurons of supraoptic nucleus of hypothalamus; stored and released by neurohypophysis
2. Oxytocin	Synthesized by neurons of paraventricular nucleus of hypothalamus; stored and released by neurohypophysis
III. Pars intermedia or intermediate lobe	
1. MSH—Melanocyte Stimulating Hormone	

*Chromophobes of anterior pituitary are thought to have depleted granules; they are also considered as "stem cells"

thalamus. These neurons produce specific "releasing factors" or "releasing hormones" for each hormone of adenohypophysis. These hormones or factors are released into the bloodstream and reach the adenohypophysis through the hypothalamic hypophyseal portal system. Some neurons of the hypothalamus secrete "inhibitory factors or inhibitory hormones" also. The hypothalamic releasing and inhibitory hormones that are of major importance are:

1. Thyroid stimulating hormone releasing hormone (TRH) which causes release of the TSH.
2. Corticotropin releasing hormone (CRH), which causes release of ACTH.
3. Growth hormone releasing hormone (GH-RH), which causes release of growth hormone, and growth hormone inhibitory hormone (GHIH) which inhibits the release of growth hormone.
4. Luteinising hormone releasing hormone (LRH), which causes release of both LH and FSH. This hormone is also called gonadotropin releasing hormone.
5. Prolactin inhibitory hormone.

"Negative Feedback" in the Control of Hormone Secretion

Each endocrine gland has a basic tendency to "over secrete" its particular hormone. But, once the physiological effect of the hormone has been achieved, information is sent-directly or indirectly-to the gland to inhibit further secretion.

On the other hand, if the gland under secretes, the physiological effects of the hormone decreases; this stimulates the gland to secrete adequate quantities of hormones once again.

In this way, the rate of secretion of each hormone is controlled, according to the need for the hormone.

Applied Anatomy

1. Tumors of pituitary may increase the depth of the hypophyseal fossa of the sphenoid—this can be seen in an X-ray of the lateral view of skull or in a CT scan.
2. A tumor of the pituitary may press on the middle fibers of optic chiasma, causing bitemporal hemianopia.
3. A pituitary tumor can press on the 3rd and 4th cranial nerves producing paralysis of the muscles innervated by them.
4. Abnormalities in hormone secretion—(Refer Part II, Physiology, Chapter 27).

■ THE THYROID GLAND (thyroid = Shield-like)

This is the largest endocrine gland in our body. It clasps the anterior and lateral surfaces of the pharynx, larynx, esophagus and trachea like a shield. It is yellowish brown in color and highly vascular.

Position

It is situated in front and sides of the lower part of the neck, opposite the levels of C5-C7 and T1 vertebrae.

Parts

It has 2 lateral lobes connected by a central isthmus (Fig.17.4).

Each lateral lobe is roughly conical in shape, measuring 5 cm in length, 3 cm in breadth and 2 cm in thickness. The isthmus is quadrilateral in shape.

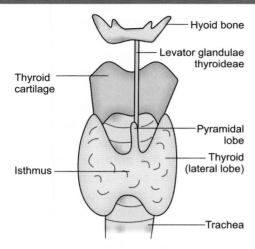

Fig. 17.4: The thyroid gland

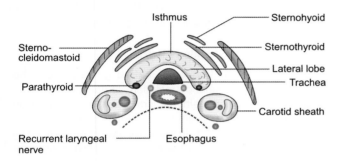

Fig. 17.5: Relation of thyroid gland (TS of neck at the level of isthmus)

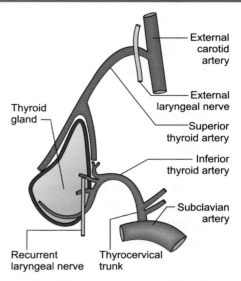

Fig. 17.6: Arterial supply of thyroid

The gland weighs approximately 30 gm. The gland is larger in females, which enlarges during menstruation and pregnancy.

Capsules

Thyroid has a true fibrous capsule, formed by the condensation of connective tissue fibers.

The false capsule is formed by the pretracheal layer of deep cervical fascia, which splits to enclose the gland. The pretracheal layer is attached to the hyoid bone and oblique line of the thyroid cartilage. Because of these attachments, the thyroid gland moves up with deglutition (swallowing).

Relations (Fig. 17.5)

*   *Anterolaterally*: The lobes are covered by the "strap muscles"—the sternohyoid and sternothyroid.

Surfaces

Posterior Surface

Posterior surface is related to the carotid sheath and its contents (internal jugular vein, common carotid artery and vagus nerve).

The superior and inferior parathyroid glands lie on the posterior surface of thyroid.

Medial Surface

Medial surface of the gland is related to *2 tubes* (trachea and esophagus), *2 nerves* (external laryngeal and recurrent laryngeal), *2 muscles* (cricothyroid and inferior constrictor of pharynx) and *2 cartilages* (thyroid and cricoid).

Isthmus

Isthmus is covered anteriorly by skin and fasciae. Posteriorly, it rests on 2nd, 3rd and 4th tracheal rings. From its upper border, a small projection may be seen—called the *pyramidal lobe*. A fibromuscular strand, called *levator glandulae thyroideae* may extend from the pyramidal lobe to the hyoid bone.

Blood Supply (Fig.17.6)

Arterial Supply

A pair of superior thyroid arteries—arise from external carotid arteries; they enter the upper poles of the lobes. These arteries are closely related to the external laryngeal nerves.

A pair of inferior thyroid arteries—they arise from the thyrocervical trunk of the 1st part of the subclavian arteries. The terminal part of each inferior thyroid artery is intimately related to the recurrent laryngeal nerve.

Arteria thyroidea ima—this unpaired branch may arise from the arch of aorta or brachiocephalic trunk.

Venous Drainage

The veins form a plexus beneath the true capsule. The plexus is drained by 3 pairs of veins, viz. superior, middle and inferior thyroid veins. Superior and middle veins drain into internal jugular vein; inferior veins drain into brachiocephalic vein.

Nerve Supply

Parasympathetic from vagi and sympathetic from cervical sympathetic ganglia.

Lymphatic Drainage

Pretracheal and upper and lower deep cervical nodes.

Histology (Fig.17.7)

1. *Capsule*: There is a thin fibrous capsule; thin septa arise from this capsule, which pass interiorly.
2. The parenchyma consists of thyroid follicles (of different size) filled with "colloid". Each follicle has a basement membrane and a single layer of cuboidal epithelium.
3. In between the follicles and deep to the basement membrane of follicles, there are clusters of cells called parafollicular cells or C cells.
4. There are connective tissue cells, fibers and plenty of blood vessels.

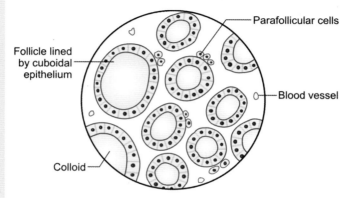

Fig. 17.7: Microscopic structure of thyroid

The follicular cells are concerned with the production of thyroid hormones. Thyroid gland stores large amounts of hormone in an inactive form called colloid in the follicles. The parafollicular cells secrete a hormone called calcitonin.

The functions of thyroid gland is controlled by anterior lobe of pituitary, which, in turn, is controlled by the hypothalamus. The thyroid stimulating hormone (TSH) secreted by the anterior lobe of pituitary stimulates the follicular cells to trap iodine from blood and in the synthesis of thyroglobulin. The cells then convert the thyroglobulin into thyroid hormones—Tri-iodothyronin (T3) and Tetra-iodothyronin (T4).

In turn, the release of TSH by anterior pituitary (adenohypophysis) is stimulated by thyrotropin releasing hormone (TRH) secreted by the hypothalamus.

The parafollicular cells, which secrete calcitonin is solely controlled by the level of calcium in blood.

Applied Anatomy

1. Enlargement of thyroid is called goitre. It can be physiological (pregnancy, lactation) or pathological.
2. Underactivity or hypothyroidism results in cretinism in children and myxoedema in adults. Overactivity or hyperthyroidism results in thyrotoxicosis, characterised by loss of weight, nervousness, increased sweating, tachycardia and protrusion of eyeballs.
3. Swellings of thyroid move with deglutition because of the attachment of pretracheal facsia to the hyoid bone and thyroid cartilage. Thus, asking the patient to swallow to see whether the swelling moves with deglutition or not, forms a simple test to differentiate a thyroid swelling from other swellings in the neck (midline swellings).
4. The external laryngeal nerve is intimately related to the superior thyroid artery close to the origin of the artery. Near the upper pole of the lobe of thyroid, the nerve moves away from the artery. So, during thyroidectomy, the artery should be tied very near the upper pole of the gland.

The terminal part of each inferior thyroid artery is intimately related to the recurrent laryngeal nerve. To avoid injury to the recurrent laryngeal nerve, the inferior thyroid artery is tied a little away from the gland during thyroidectomy. Injury to the nerve will result in paralysis of vocal cord and will cause hoarseness of voice and difficulty in breathing.

▮ THE PARATHYROIDS (FIG. 17.8)

There are 4 parathyroids—a pair of superior and a pair of inferior glands. Each gland is yellowish orange in color. *Size*: 6 × 4 × 2 mm; weight about 50 mg.

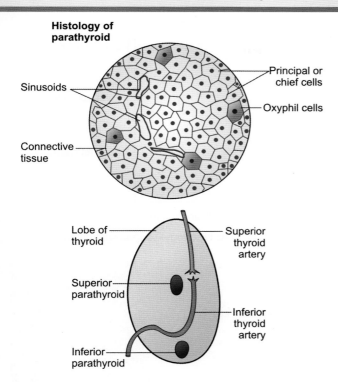

Histology of parathyroid

Sinusoids

Connective tissue

Principal or chief cells

Oxyphil cells

Lobe of thyroid

Superior parathyroid

Inferior parathyroid

Superior thyroid artery

Inferior thyroid artery

Fig. 17.8: Position of parathyroid glands on posterior surface of thyroid lobe

Position

The superior parathyroid lies on the posterior surface of lateral lobe of thyroid gland at the junction of the upper and middle thirds.

The inferior parathyroids are variable in position and lie on the posterior surface of lateral lobe of thyroid gland, close to the entrance of inferior thyroid artery.

Blood Supply

Mainly by inferior thyroid artery, they also receive blood supply from superior thyroid arteries.

Development

Superior parathyroid develops from endoderm of fourth pharyngeal pouch; the inferior parathyroid develops from endoderm of third pharyngeal pouch, along with the thymus. So, the inferior parathyroid may lie very close to the thymus.

Histology (Fig. 17.8)

1. There is a thin fibrous capsule from which septa extend into the substance of the gland.

2. The parenchyma consists of irregular cords of cells with blood capillaries in between. There are 2 types of cells (1) *chief cells* or principal cells; and (2) *oxyphil* cells. The chief cells are more in younger individuals; they are polygonal cells with centrally placed nucleus and pale-staininig acidophilic cytoplasm. The oxyphil cells are larger in size, with abundant granules and darkly staining cells. Their number increases with age.

These cells (mainly chief cells) secrete parathormone which raises serum calcium levels. The activity of parathyroid gland is controlled by alterations in the level of calcium in blood. The maintenance of delicate balance of blood calcium level is essential for the normal function of cells, mainly those of the heart, skeletal muscles and nerves.

Applied Anatomy

1. Accidental removal of parathyroids is possible during thyroidectomy. To avoid this, the posterior part of lateral lobes of the thyroid gland is left behind.
2. Hypoparathyroidism results in tetany—the blood calcium level decreases (hypocalcemia), leading to carpopedal spasms and convulsions.
3. *Hyperparathyroidism*—the parathormone stimulates osteoclastic activity, thereby increasing bone resorption by mobilizing calcium and phosphate. Hyperparathyroidism has been described as a disease of "bones, stones, abdominal groans and psychic moans".
   (Bones—there may be generalized decalcification
   Renal stones: Formation of renal stones commonly results from hyper-parathyroidism.
   Patient may have nausea, vomiting, anorexia, tiredness and abnormal behavior).

■ THE SUPRARENAL OR ADRENAL GLANDS

The suprarenals are paired endocrine glands. They are highly vascular; yellowish pink in color.

Position

They are situated on the superomedial aspect of kidneys (Fig.17.9).

Shape

The right suprarenal is pyramidal in shape (resembles a "cocked Roman hat"). The left suprarenal is semilunar in shape.

Dimensions: 50 mm × 3 mm × 10 mm; weight— 5 gm.

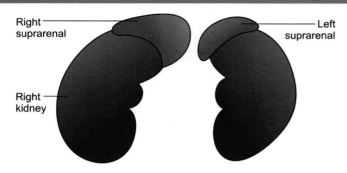

Fig. 17.9: Suprarenal gland

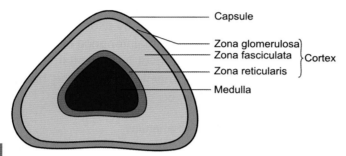

Fig. 17.10: Sub-divisions of suprarenal

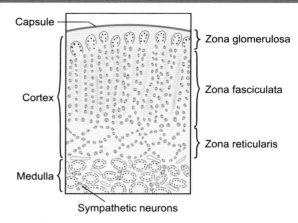

Fig. 17.11A: Histology—suprarenal gland

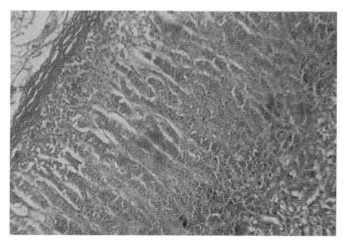

Fig. 17.11B: Histology of suprarenal

Sub-divisions (Fig.17.10)

Each gland has an outer cortex and inner medulla; they are entirely different in structure, function and development.

The cortex develops from mesoderm (coelomic epithelium) and medulla develops from neural crest.

Important Relations

Right Suprarenal

Medially it is related to the right edge of inferior vena cava; posteriorly it is related to the diaphragm and upper pole of right kidney; superiorly, related to bare area of liver.

The Left Suprarenal

Its anterior surface is overlapped by the body of pancreas. Its upper half is related to the posterior surface of stomach. Posterior surface rests on the diaphragm (part of stomach bed). Lateral surface is related to the left kidney.

Blood Supply

Suprarenal is a highly vascular gland. Each gland receives 3 sets of arteries.

1. Superior suprarenal arteries—they are branches of inferior phrenic arteries
2. Middle suprarenal arteries are direct branches of abdominal aorta
3. Inferior suprarenal arteries arise from the renal arteries.

These arteries pierce the capsule, supply the cortex and finally end in the medullary sinusoids. From the sinusoids, a *single* suprarenal vein comes out. On the right side, this vein ends in the inferior vena cava; the left suprarenal vein ends in the left renal vein.

Histology (Figs 17.11A and B)

1. *Capsule*: There is a thin, fibrous capsule for this gland
2. The gland has an outer *cortex* and inner medulla. Medulla forms 1/10th of the size of the cortex.

Cortex

The cortex: It has 3 zones:

The zona glomerulosa: This is the outermost zone; the polyhedral cells occur as small clusters.

The middle zona fasciculata: This is the thickest layer of the cortex. There are large polyhedral cells, arranged in long column, with sinusoids in between.

Inner zona reticularis: The cells are arranged in the form of a network; the cell cords branch and anastomose.

The Medulla

The medulla is made of irregular collection of large cells with large sinusoids in between. The medulla is dark in color; it receives sympathetic innervation from the coeliac plexus.

Functions of Cortex

The hormones of the suprarenal cortex are derived from cholesterol and belong to a class of hormones called "steroids".

1. The zona glomerulosa secretes "mineralocorticoids", mainly, aldosterone, which regulates sodium-potassium balance; it also regulates fluid-electrolyte balance.
2. Zona fasciculata secretes a group of hormones collectively called "glucocorticoids", which influence the carbohydrate metabolism. The principal glucocorticoid is cortisol (hydrocortisone).
3. The zona reticularis secretes small amounts of sex hormones (androgens and estrogens), which influence sexual growth and development.

   The suprarenal medulla secretes adrenaline and noradrenaline (catecholamines); their secretion is not under the control of adenohypophysis. Adrenal medulla is stimulated by the sympathetic nervous system.

   The adrenal cortex is controlled by the adenohypophysis, which secrete ACTH; the adenohypophysis, in turn, is controlled by the hypothalamus, which secretes the corticotropin releasing factor. Hypersecretion of cortical hormones is prevented by a negative feedback mechanism.

Control of Adrenal Cortex

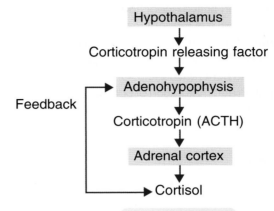

Applied Anatomy

1. Pheochromocytoma is a tumor of the medulla characterized by the following symptoms—hypertension, palpitation, excessive sweating and pale skin.
2. Tumors of the cortex (adenoma or carcinoma) may give rise to Cushing's syndrome—(hypogonadism, diabetes, obesity and hirsutism). In children, excessive sex hormones cause adrenogenital syndrome or precocious puberty.
3. *Addison's disease:* Insufficiency of adrenal cortex due to tuberculosis; characterized by hypotension, pigmentation of skin, anemia and muscular weakness.
4. Total loss of adrenocortical secretion usually causes death, unless the patient gets mineralocorticoid therapy. In the absence of mineralocorticoid, potassium ion concentration increases markedly; sodium and chloride concentration decrease; the total blood volume is reduced, cardiac output decreases, leading to a "shock-like" state, followed by death. Therefore, the mineralocorticoids are said to be the acute "life-saving" portion of adrenocortical hormones; the glucocorticoids are also equally necessary to meet "stress" (trauma, infection, intense heat or cold, surgical operations, etc).

■ PINEAL GLAND ("EPIPHYSIS CEREBRI")

The pineal gland was considered as a vestigial organ of no functional importance, but, now it is known to be an endocrine gland of great importance.

It is a small, reddish structure, about the size of a pea, situated in the midline of brain, immediately behind the 3rd ventricle. In adults, it may be calcified and identified on skull X-ray films, in which its displacement to one side would indicate the presence of a space occupying lesion within the cranium.

Function

It is thought to be an active endocrine gland. Its function is poorly understood. It is thought to modify the activities of other endocrine glands like adenohypophysis, neurohypophysis, the thyroid, parathyroids, the adrenal gland, gonads and pancreatic islets. One of its hormones inhibits the growth and maturation of gonads until puberty. Pineal gland has an extensive role in coordinating circadian and diurnal rhythms throughout the body.

■ PANCREAS

The pancreas, in addition to its digestive functions, secretes hormones.

The pancreas is composed of two types of tissues (1) the pancreatic *acini,* which secrete digestive juices into the duodenum (2) the islets of Langerhans, which release their secretions directly into the blood.

The islets of Langerhans, contain 3 main types of cells—alpha, beta and delta. The alpha cells secrete glucagons, the beta cells secrete insulin and the delta cells secrete somatostatin.

Insulin and glucagons, even though, concerned with carbohydrate metabolism, are diametrically opposite in their function. Insulin causes rapid uptake, storage and use of glucose by almost all tissues of the body; glucagons, on the other hand, has two major effects on glucose metabolism. They are (1) breakdown of liver glycogen (glycogenolysis), and (2) increased gluconeogenesis. *(Described in detail in Chapter 27).*

Somatostatin, secreted by the delta cells of pancreas, has the capability of inhibiting the secretion of both glucagons and insulin by the alpha and beta cells (Somatostatin is the same as growth hormone inhibitory hormone that is secreted by the hypothalamus, a hormone that might help to control growth hormone secretion).

■ INTERSTITIAL CELLS OF TESTES (FIGS 15.17A AND B)

The testes secrete several male sex hormones, which are collectively called androgens, which include testosterone, dihydrotestosterone and androstenedione. Testosterone is much more abundant and potent than the others.

Testosterone is formed by the interstitial cells of Leydig, which lie in the interstices (intervals) between the seminiferous tubules. They constitute about 20 percent of the mass of the adult testes. They are numerous in an adult male (any time after puberty). Even when the germinal epithelium of the testes is destroyed by X-ray treatment or by excessive heat, the interstitial cells continue to produce testosterone. Testosterone is responsible for the distinguishing characteristics of the masculine body.

Ovarian Hormones (Figs 15.9 and 15.10 Chapter 15 Reproductive System)

The developing ova, in the ovary, are surrounded by stromal cells. Ovarian follicles or Graafian follicles are derived from these stromal cells. The stromal cells surrounding the oocyte become flattened and are known as follicular cells. These cells, later, become columnar, then they proliferate to form several layers of cells, which constitute the membrana granulosa. The cells are now called granulosa cells.

Later, some of these cells partially separate, to enclose a cavity—the follicular cavity; which is filled with a fluid, the liquor folliculi. This cavity rapidly increases in size. As a result, the wall of the follicle becomes relatively thin. The stromal cells surrounding the membrana granulosa become condensed to form a covering called the theca interna. Outside this, some connective tissue fibers become condensed to form the theca externa.

The cells of theca interna secrete estrogen; now it is called a thecal gland.

Ultimately, the follicle ruptures, to release the ovum. After ovulation, the remaining part of the follicle undergoes changes that convert it into an important structure called the corpus luteum. It secretes a hormone, called progesterone.

Estrogen is mainly responsible for the development of secondary sexual characteristics of the female. Progesterone is concerned with the final preparation of uterus for pregnancy and the breasts for lactation.

Placenta

In pregnancy, placenta secretes large quantities of human chorionic gonadotropin, estrogens and progesterone. These are essential for the continuation of pregnancy. The large quantity of estrogen causes enlargement of uterus, enlargement of breasts and growth of duct system of breasts; estrogen also relaxes the pelvic ligaments, for easier passage of the fetus through the birth canal.

Progesterone decreases the contractility of uterine muscle, thus preventing uterine contractions from causing spontaneous abortion. Progesterone also helps to prepare the breasts for lactation.

18 Sense Organs

INTRODUCTION

The sensory nerves reaching the CNS have their beginnings in various peripheral structures such as:

1. Skin
2. Muscles
3. Joints
4. Special organs like eye and ear

Structures concerned in the production of sensations are:

1. An end organ or a sensory receptor
2. Afferent nerve fiber
3. Thalamus
4. Cerebral cortex.

CLASSIFICATION OF SENSATIONS

1. Special senses—sight, hearing, smell, taste and equilibrium.
2. Somatic senses—pain, touch, pressure, vibration.

SENSE OF SMELL OR OLFACTION

Mechanism

The mechanism of smell depends upon:

1. Olfactory receptor cells lying in the epithelial lining of the olfactory mucosa of nasal cavities.
2. Olfactory nerves, olfactory bulb and olfactory tract which convey the impulses to brain.
3. The limbic lobe, situated in the medial surface of cerebral hemisphere.

The olfactory receptor cells (Fig. 18.1) are situated in the epithelium lining the olfactory mucosa. Each receptor cell consists of a cell body and 2 processes (i.e. it is a bipolar cell). The dendrite reaches the surface of the olfactory epithelium and ends in a small swelling, to which a number of cilia are attached. The central process or axon forms one fiber of the olfactory nerve. These fibers form 13 to 20 bundles forming olfactory nerves; they pass through the cribriform plate of ethmoid to reach the anterior cranial fossa and

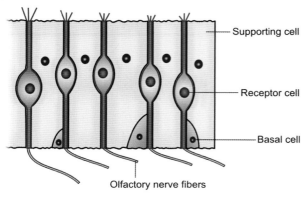

Fig. 18.1: Olfactory epithelium

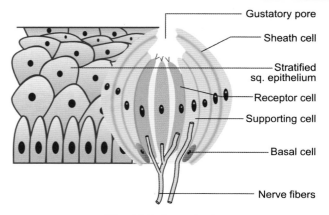

Fig. 18.2: Taste buds

terminate in the olfactory bulb. The olfactory bulb has important neurons called mitral cells. The axons of the mitral cells form the olfactory tract. These fibers pass to primary olfactory cortex. It has connections with the limbic system.

The stimulus for the sense of smell are usually minute particles which are soluble in the secretions of nasal mucous membrane.

The sense of smell is closely related to the sense of taste. So, in common cold, there will be loss of sense of smell and altered taste. Loss of sense of smell is called anosmia.

THE SENSE OF TASTE

The true sense of taste is localized in the tongue. There are 4 basic tastes—bitter, sweet, sour and salt (all other flavors are appreciated by the sense of smell).

The tongue is covered by a mucous membrane. There are numerous elevations called papillae, on the tongue (see, Digestive system—Tongue). The end organs for the sense of taste are called taste buds. They are situated most densely at the tip, sides and base of the tongue. Taste buds are also present in the walls of the circumvallate papillae, on the palate, epiglottis and oropharynx.

The taste buds (Fig. 18.2) consist of collections of receptor cells and supporting cells, together with the sensory nerve endings. The taste bud has a pore which opens on the epithelial surface of the tongue. A number of hairs project from the receptor cells into the pore.

Sensory Nerves of Tongue

The mucosa lining the dorsum of the tongue is divided by a V-shaped sulcus terminalis into anterior 2/3 and posterior 1/3. The anterior 2/3 is again divided into right and left halves by a median furrow.

Taste fibers from the anterior 2/3 of tongue travel in the chorda tympani branch of the facial nerve and those from the posterior 1/3 of tongue, travel along the glossopharyngeal nerve. The circumvallate papillae are innervated by the glossopharyngeal nerve.

The general sensations from anterior 2/3 are conveyed by the lingual branch of mandibular, a division of trigeminal nerve. The general sensations from posterior 1/3 are carried by glossopharyngeal nerve; from the posterior most part of the tongue, by internal laryngeal nerve.

The taste sensations ultimately reach the post-central gyrus of the cerebral cortex.

In order to appreciate the taste of substances, they must be soluble in the watery secretions of mouth and saliva.

SENSE OF HEARING

The hearing or auditory apparatus consists of:
1. The external ear
2. The middle ear
3. Cochlea of the internal ear
4. The cochlear nerve
5. Auditory area of temporal lobe.

Functions

The ear has two functions—balance and hearing. The external and middle ears are mainly concerned with conduction of sound to the internal ear; the inner (internal) ear contains the vestibulocochlear organ that is essential for equilibrium and hearing.

External Auditory Meatus

The external auditory meatus is 2.5 cm long. Its outer 1/3 is cartilaginous and inner 2/3 is bony. Lateral third is lined with skin; contains hair follicles and ceruminous glands (modified sweat glands). The ceruminous glands produce a waxy exudate called cerumen (wax). The medial 2/3 is lined with very thin skin, which is continuous with the external layer of tympanic membrane.

The external auditory meatus is not a straight passage. It shows an S-shaped bend. In order to make it straight (while examining the tympanic membrane), the pinna should be pulled upwards and backwards in adults (in children, pinna should be pulled downwards and backwards).

Nerve Supply

1. Auricular branch of auriculotemporal nerve
2. Facial nerve
3. Auricular branch of vagus.

Applied Anatomy

1. External auditory meatus is directed anteromedially. So, the tips of the stethoscope are angulated to conform to this shape.
2. While syringing the external auditory meatus to remove wax or foreign bodies, vagus nerve may get stimulated, leading to bradycardia.
3. Pain from the tongue (e.g. as in ulcer, cancer) may radiate to the ear along the auriculotemporal nerve.

Tympanic Membrane (Figs 18.3A and B)

This thin, semitransparent, pearly white, oval membrane is situated at the medial end of the external auditory meatus. It forms a partition between external ear and middle ear.

It has three layers:
1. Outer, thin skin (developed from ectoderm)
2. Middle fibrous layer (developed from mesoderm)
3. Inner mucous membrane (developed from endoderm).

The tympanic membrane shows a concavity towards the external auditory meatus with a central depression called umbo, on the inner side of which, the handle of malleus is attached. From the umbo, a bright area referred to as the "cone of light" radiates anteroinferiorly.

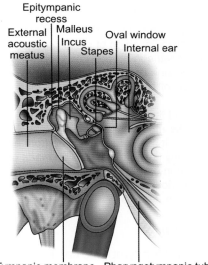

Fig. 18.3A: Parts of ear

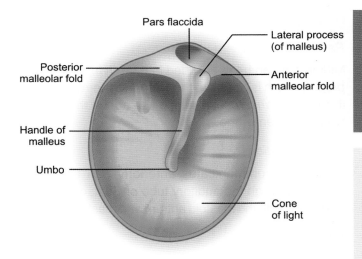

Fig. 18.3B: Tympanic membrane

Tympanic membrane moves in response to ear vibrations. From here, the vibrations are transmitted by the ear ossicles through the middle ear to the internal ear.

Nerve Supply

External surface is supplied by auriculotemporal nerve and auricular branch of vagus; internal surface is supplied by tympanic branch of glossopharyngeal nerve.

Section C   The Organ Systems

Applied Anatomy

1. Perforation or rupture of tympanic membrane may result from infection, insertion of sharp objects into the external auditory meatus in an attempt to remove wax or foreign bodies, or due to excessive pressure as in diving. A perforation of tympanic membrane leads to conduction deafness.
2. A fracture of the base of the skull may lead to perforation of tympanic membrane and CSF leak—known as CSF otorrhea; or it may cause bleeding from ear.
3. Myringotomy is a surgical incision made on the tympanic membrane in order to drain pus from the middle ear. The posteroinferior quadrant is chosen, because no important structures lie deep to it.

The Middle Ear

It is also called the tympanic cavity or tympanum. It is a narrow, air filled space, situated in the petrous part of temporal bone, between the external and internal ear. It is shaped like a cube; the cavity is biconcave as the medial and lateral walls lie close together at the center.

Parts

Middle ear can be subdivided into the *tympanic cavity proper,* which is opposite the tympanic membrane and the *epitympanic recess,* which lies above the level of tympanic membrane.

Communications

Middle ear communicates anteriorly with the nasopharynx through the auditory tube or Eustachian tube; posteriorly with the mastoid antrum and mastoid air cells.

Boundaries

It has a roof, a floor, lateral, medial, anterior and posterior walls.

Roof is called the tegmental wall; formed by tegmen tympani of petrous part of temporal bone. This thin plate of bone separates the middle ear from the middle cranial fossa and temporal lobe of cerebrum.

Floor or jugular wall: It is related to the superior bulb (swelling) of internal jugular vein.

Anterior or carotid wall: Upper part of anterior wall has the opening for tensor tympani; middle part has the opening of auditory tube; lower part is related to the carotid canal, through which internal carotid artery passes.

Posterior or mastoid wall: The tympanic cavity communicates with the epitympanic recess through an opening in the posterior wall called aditus. There is a conical projection called pyramid; it has an opening at its apex, through which the tendon of stapedius muscle emerges.

Lateral or membranous wall is formed by the tympanic membrane; it separates the middle ear from the external ear.

Medial or labyrinthine wall separates the middle ear from the internal ear. It presents the following features:
1. The promontory, a rounded bulging produced by the first turn of cochlea.
2. The fenestra vestibuli (oval window) is an oval opening, closed by the foot-plate of stapes; it leads to the vestibule of the internal ear.
3. The fenestra cochleae (round window) is a round opening, which leads to the scala tympani of cochlea and is closed by the secondary tympanic membrane.

Contents of Middle Ear

1. Three small bones or ossicles—the malleus, incus and stapes
2. Ligaments of ear ossicles
3. Two muscles—tensor tympani and the stapedius
4. Vessels of middle ear
5. Nerves—chorda tympani and tympanic plexus.

Ear ossicles (Fig. 18.4)

These 3 bones are situated in the middle ear: they are (1) Malleus; (2) Incus, and (3) Stapes.

Malleus: It resembles a hammer; it is the largest and most laterally placed ossicle.

Parts: (1) a head; (2) a neck (3) anterior and lateral processes and (4) a handle, which is attached to the tympanic membrane.

Incus ("Anvil") resembles a premolar tooth. It has a body and a long process. The body articulates with the head of the malleus. The tip of the long process articulates with the head of the stapes.

Stapes ("Stirrup") is the smallest and most medially placed ossicle of the ear.

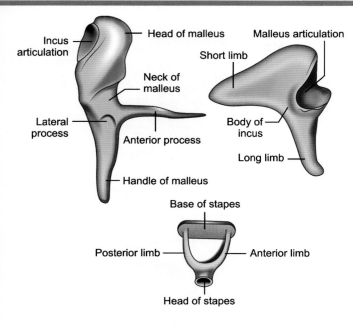

Incus articulation — Head of malleus

Malleus articulation

Short limb

Neck of malleus

Lateral process

Body of incus

Anterior process

Long limb

Handle of malleus

Base of stapes

Posterior limb — Anterior limb

Head of stapes

Fig. 18.4: Ear ossicles

Parts

1. Head: It has a concave facet, with which the tip of the long process of incus articulates
2. A narrow neck, which provides insertion for the stapedius muscle
3. Two limbs or crura
4. *Foot plate:* It is oval in shape; fits into the fenestra vestibuli (oval window) of the medial wall of middle ear.

Muscles of Middle Ear

There are two muscles:
1. The tensor tympani, and 2. The stapedius

The tensor tympani arise from the cartilaginous part of auditory tube and from the base of the skull. The muscle lies in a bony canal that opens on the anterior wall of middle ear; its tendon bends sharply around a bony process called processus cochleariformis and inserted into the handle of the malleus. It tenses the tympanic membrane and reduces the amplitude of its oscillations.

Nerve supply: By mandibular nerve (a division of trigeminal nerve—V cranial nerve).

The stapedius lies in a bony canal, on the posterior wall of middle ear. Anteriorly, the canal opens on the summit of the pyramid. Its tendon emerges through this opening and passes forwards, to get inserted into the neck of the stapes.

Nerve supply: A branch of the facial nerve (VII cranial nerve).

Function

Function of these muscles: They act simultaneously to damp down the intensity of high-pitched sound waves. Thus they help to protect the internal ear. Paralysis of stapedius as a result of lesion of facial nerve is associated with HYPERACUSIS (excessive intensity of hearing).

The Auditory Tube

The auditory tube or pharyngotympanic tube or Eustachian tube.

This tube connects the middle ear cavity with the nasopharynx. It is about 4 cm long. It is directed downwards, forwards and medially. The tube has two parts—bony and cartilaginous.

Bony part: It forms the lateral one-third of the tube. It is about 1.2 cm long and lies in the petrous part of temporal bone. Its lateral end opens on the anterior wall of the middle ear cavity. It is related superiorly to the bony canal for tensor tympani and medially to the carotid canal.

Cartilaginous part forms the medial two-thirds of the tube; approximately 2.8 cm long. It is made up of a triangular plate of cartilage, which is curled up to form a tube. The apex of this plate is attached to the medial end of the bony part. The base is free and forms the tubal elevation in the nasopharynx. The "tubal tonsil" is located near this opening. The cartilaginous part is related to the mandibular nerve, the otic ganglion, chorda tympani nerve and middle meningeal artery.

Lining

It is lined by mucosa, the epithelium is ciliated columnar (Respiratory epithelium). The mucosa is continuous with the mucosa of nasopharynx; laterally, it is continuous with the mucosa of middle ear cavity.

Section C

The Organ Systems

Function: It forms a channel of communication of the middle ear cavity with the exterior (The nasopharynx opens anteriorly into the nasal cavities), thus ensuring equal air pressure on both sides of the tympanic membrane, so that the tympanic membrane can vibrate freely.

The tube is usually closed. It opens during swallowing, yawning and sneezing by the action of tensor and levator palati.

Applied Anatomy: Middle Ear

Infections from the nose and throat can spread to the middle ear through the auditory tube resulting in otitis media. The pus, which collects in the middle ear may drain into the external ear by the rupture of tympanic membrane.

Other Complication of Otitis Media

1. Infection can spread to the meninges (meningitis) by the erosion of the tegmental wall.
2. It may erode the floor (jugular wall), causing thrombosis of internal jugular vein or sigmoid sinus.
3. Pus may spread to the mastoid air cells and cause mastoiditis and mastoid abscess.

Otitis media is more common in children because the auditory tube is shorter, wider and straight.

The Internal Ear (Figs 18.5A and B)

The internal ear contains the vestibulocochlear organ, which is concerned with the reception of sound and maintenance of balance. It is situated in the petrous part of temporal bone.

The internal ear consists of the sacs and ducts of the membranous labyrinth. This system is filled with a fluid, the endolymph. The membranous labyrinth is suspended within the bony labyrinth. Between the bony labyrinth and the membranous labyrinth is a clear fluid called the perilymph.

The Bony Labyrinth

The bony labyrinth is composed of three parts:
1. The cochlea
2. The vestibule
3. The semicircular canals.

The Cochlea (L = Snail Shell)

This shell-like part of the bony labyrinth contains the cochlear duct, which is concerned with hearing.

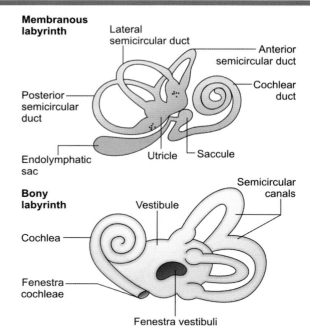

Fig. 18.5A: Lateral view of membranous labyrinth and bony labyrinth

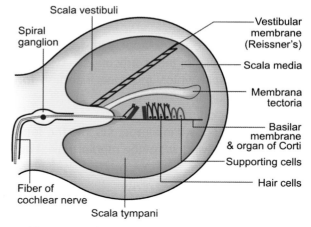

Fig. 18.5B: Schematic section through cochlea

The cochlea makes two and one-half turns around a bony core called the modiolus (L = The nave of a wheel). The modiolus is traversed by canals for nerves and blood vessels. The basal turn of the cochlea produces an elevation, the promontory, on the medial wall of tympanic cavity.

A small shelf of bone, the spiral lamina, protrudes from the modiolus similar to the thread of a screw. The basilar membrane is attached to the spiral lamina.

The spiral lamina and the basilar membrane divide the cochlear canal into the scala vestibuli above and the scala tympani below. The scala vestibuli communicates with the scala tympani at the apex of the cochlea by a small opening called helicotrema.

The scala vestibuli opens into the vestibule of the bony labyrinth. Through this opening, perilymph can be exchanged freely between the vestibule and scala vestibuli.

The scala tympani, is related to the tympanic cavity at the fenestra cochleae (round window) which is closed by the secondary tympanic membrane.

Both the scala vestibuli and scala tympani are filled with a clear fluid, the perilymph. It is similar in composition to CSF, because, a narrow perilymphatic duct connects the scala vestibuli to the subarachnoid space.

The Vestibule

The vestibule is an oval bony chamber which is continuous anteriorly with the bony cochlea and posteriorly, with the semicircular canals. The vestibule contains the utricle and saccule, which are parts of the balancing apparatus.

The Semicircular Canals

The anterior, posterior and lateral semicircular canals communicate with the vestibule of the bony labyrinth. The canals are set at right angles to each other. Each semicircular canal forms about two-third of a circle; there is a swelling at one end of each canal—the ampulla. The canals have only 5 openings into the vestibule because, the anterior and posterior canals have one stem common to both.

Membranous Labyrinth

The membranous labyrinth consists of a series of sacs and ducts; situated in the cavities of the bony labyrinth. It contains a fluid called endolymph, which differs in composition from perilymph.

Parts

1. Two small communicating sacs, the utricle and saccule are situated inside the vestibule
2. Three semicircular ducts, in the semicircular canals
3. The cochlear duct in the cochlea.

The parts of membranous labyrinth are suspended in the bony labyrinth by connective tissue. The cochlear duct is secured to the cochlear canal by a spiral ligament (thickened periosteum).

The various parts of the membranous labyrinth form a closed system; they communicate with one another.

The utricle and saccule: These dilatations have a specialized area of sensory epithelium called a "macula". The macula of the utricle is known as macula utriculi, which lies in its floor, parallel with the base of the skull.

The macula of the saccule, the macula sacculi, is vertically placed on the medial wall of saccule

The hair cells in the macula are innervated by fibers of the vestibular division of vestibulocochlear nerve. "Motion sickness" results from prolonged stimulation of maculae.

The semicircular ducts: Each duct has an ampulla at one end, containing a sensory area called "crista ampullaris". The hair cells of the cristae are supplied by vestibular branches of vestibulocochlear nerve.

The cochlear duct: This is a spiral, blind tube, triangular in transverse section, firmly fixed to the cochlear canal by a spiral ligament.

The cochlear duct lies between the spiral lamina and the external wall of the cochlear canal. Its roof is formed by the vestibular membrane and floor, by the basilar membrane and spiral lamina.

The spiral organ of Corti: The receptor of auditory stimuli is situated on the basilar membrane. There is an overlying gelatinous membrane, the membrana tectoria (tectum = roof). The spiral organ has hair cells, that respond to vibrations induced in the endolymph by sound waves. The hair cells are innervated by the cochlear part of vestibulocochlear nerve.

Mechanism of Hearing (Summary)

Sound waves in the air are collected by the pinna (auricle) and conducted into the external auditory meatus. They cause vibrations (oscillations) of tympanic membrane and the ear ossicles. Stronger waves are transmitted to the perilymph at the fenestra vestibuli by the foot plate (base) of stapes.

Vibrations of the perilymph are transmitted to the basilar membrane; its displacement causes bending of the hair-like projections of the sensory hair cells of organ of Corti. These cells are innervated by cochlear part of vestibulocochlear nerve. Different parts of the organ of Corti are stimulated by different frequencies of sound. The loudness of sound depends on the amplitude of vibrations.

The *auditory pathway:* Cochlear nerve → cochlear nuclei → cross over to the opposite side, in the trapezoid body of pons → inferior colliculus of midbrain → medial geniculate body → auditory radiation, through sublentiform part of internal capsule → auditory area (41,42) in the temporal lobe.

Applied Anatomy

1. Persistent exposure to loud noises (e.g. persons working around jet engines) causes degenerative changes in the spiral organs of corti, resulting in high tone deafness.
2. Divers and fliers are likely to develop otic barotraumas due to imbalance in pressure between the surrounding air and air in the middle ear cavity.
3. Ménière's syndrome is characterized by dizziness (vertigo), tinnitus ("ringing or jingling" or noises in the ear) and deafness. This is thought to be due to an increase in the amount of endolymph, which leads to degeneration of hair cells in the maculae of the vestibule and spiral organ of Corti.

THE SENSE OF VISION (SIGHT)

The eye or the organ of sight is situated in the orbital cavity of the skull and it is well protected by its bony walls. The orbit also contains the muscles of the eyeball, their nerves, vessels, the lacrimal gland and fat.

The Living Eye

When examined from the anterior aspect, most of the eyeball appears to be in the orbit, but from the side, it is seen to be protruding between the eyelids (palpebral fissure). The "white of the eye" is the anterior aspect of the sclera. The anterior, transparent part of the eye is the cornea. At the margins, the cornea is continuous with the sclera. The dark, circular aperture that is seen through the cornea is the pupil. This opening is surrounded by a circular, pigmented diaphragm, the iris. The sclera is covered by a thin, moist, transparent mucous membrane, the bulbar conjunctiva. It does not cover the cornea.

From the sclera, the bulbar conjunctiva is reflected onto the deep surface of the eyelids or palpebrae, lining them, up to their margins. This is the palpebral conjunctiva. When the eyelids are closed, the bulbar and palpebral conjunctivae form a closed sac.

Essential Organs

Essential organs of visual apparatus are;
1. The eyeballs
2. The optic nerves
3. The visual centers in the brain (occipital lobe).

Accessory Organs

Accessory organs for the protection and smooth functioning of the eyes are;
1. The eyebrows
2. The eyelids or palpebrae
3. The conjunctiva
4. The lacrimal apparatus
5. Muscles of the eye.

The Eyebrows

The eyebrows are formed by skin covering the orbital processes of the frontal bone. It has thick, short hairs which prevent sweat from the forehead from pouring into the eyes.

The Eyelids or Palpebrae

They protect the eyes from injury and excessive light. They help to keep the cornea moist.

The upper eyelid is larger and more mobile than the lower one. The upper eyelid partly covers the cornea.

The eyelids are covered externally by thin skin and internally by highly vascular palpebral conjunctiva. The palpebral conjunctiva is reflected on to the eyeball, where it becomes continuous with the bulbar conjunctiva. Between these two are deep recesses, known as superior and inferior conjunctival fornices.

Each eyelid is strengthened by a dense, connective tissue plate, the tarsal plate or tarsus. In between the skin and tarsal plate are the fibers of the orbicularis oculi muscle.

Embedded in the tarsal plates are numerous tarsal glands (Meibomian glands). Their oily secretion lubricates the edges of the eyelids and prevent them from sticking together.

The eyelashes (cilia) project from the margins of the eyelids. Ciliary glands are large sebaceous glands associated with the eyelashes (glands of Zeis). Between the hair follicles, there are modified sweat glands of Moll.

The two eyelids meet medially and laterally forming medial canthus (corner of eye) and lateral canthus respectively.

The space between the two eyelids is the palpebral fissure.

Functions of Eyelids

1. They prevent the entry of excess light
2. Protection from foreign bodies
3. By frequent blinking, tears are smeared over the cornea and conjunctiva, which prevent them from drying.

Applied Anatomy

1. Infection of the glands of the eyelids can occur following obstruction to their ducts. When the ducts of the ciliary glands become obstructed and inflamed, a painful red swelling known as "stye" develops.
   Cysts of the sebaceous glands of eyelids are called chalazia.
2. Palpebral conjunctiva is commonly examined in suspected cases of anemia; normally it is red and vascular; in anemia, it is pale.
3. Infection of conjunctiva is called conjunctivitis.
4. Slanted palpebral fissures are present normally in some Asian races. They are also present in persons with Down's syndrome (Trisomy 21) and the cri-du-chat syndrome (terminal deletion of chromosome 5).

The Lacrimal Apparatus (Fig. 18.6)

The lacrimal apparatus includes the parts concerned with the secretion and drainage of lacrimal fluid (tears). It is made up of the following parts:

1. The lacrimal gland and its ducts
2. The conjunctival sac
3. Lacrimal canaliculi
4. Lacrimal sac, and
5. Nasolacrimal duct.

The Lacrimal Gland

The lacrimal gland is a serous gland; it lies in the lacrimal fossa (on the anterolateral wall of the roof of the orbit) and partly on the upper eyelid. It has a larger orbital part and a smaller palpebral part.

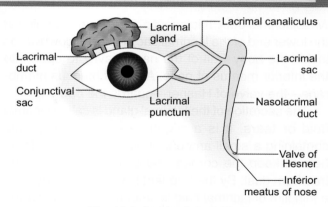

Fig. 18.6: Lacrimal apparatus

About 10-12 ducts pierce the conjunctiva of the upper eyelid and open into the conjunctival sac near the superior fornix.

The gland is supplied by lacrimal branch of ophthalmic artery. The secretomotor nerve fibers come along the facial nerve.

The Conjunctival Sac

The potential space between the palpebral conjunctiva and bulbar conjunctiva is the conjunctival sac. The lines along which the palpebral conjunctiva of upper and lower eyelids is reflected onto the eyeball are called the superior and inferior fornices (singular—fornix).

Lacrimal Puncta and Canaliculi

When the edge of the lower eyelid is everted, a small black pit can be seen at its medial end; this is the lacrimal punctum; it is placed on the summit of a small elevation called the lacrimal papilla. There is a similar punctum and papilla on the upper eyelid.

Each lacrimal canaliculus begins at the lacrimal punctum. Both canaliculi open in the lateral wall of the lacrimal sac.

The Lacrimal Sac

It is a membranous sac, about 12 mm long and 5 mm wide, situated in the lacrimal groove, behind the medial palpebral ligament.

The lacrimal sac may be regarded as the upper expanded portion of nasolacrimal duct. Its upper end is blind.

Inflammation of lacrimal sac is known as dacryocystitis.

Nasolacrimal duct It is about 1.8 cm long. It begins at the lower end of the lacrimal sac; it passes downwards inside the bony wall of the nasal cavity to open into the inferior meatus of nose. A fold of mucous membrane—the valve of Hasner—guards the opening.

The secretion of the lacrimal gland is called lacrimal fluid or tears. It is a slightly alkaline watery fluid containing a small amount of sodium chloride. Tears help to keep the cornea and conjunctiva moist and free from dust. By the frequent blinking of the eyelids, a thin film of lacrimal fluid is smeared over the cornea and conjunctiva.

The secretion of tears is increased by the presence of foreign bodies, inflammation of conjunctiva, irritating fumes, emotion, pain and very bright light.

Extraocular Muscles (Fig. 18.7)

These are muscles which move the eyeball.
1. *Four Recti* (Rectus = straight)
   a. Superior rectus
   b. Inferior rectus
   c. Medial rectus
   d. Lateral rectus
2. *Two oblique muscles*
   a. Superior oblique
   b. Inferior oblique
3. The Levator Palpebrae Superioris.

Rectus Muscles (Fig. 18.7A)

The rectus muscles are voluntary, ribbon-shaped muscles, which arise from a tough tendinous ring called the common tendinous ring, which surrounds the optic canal and the junction of superior and inferior orbital fissures.

From their origin, these muscles run anteriorly, close to the walls of the orbit; their position in relation to the eyeball is indicated by their names.

They are inserted on the eyeball (on the sclera) just posterior to the sclerocorneal junction.

The Oblique Muscles (Fig. 18.7B)

The superior oblique muscle This muscle arises from the body of the sphenoid bone; it passes anteriorly, superior and medial to the medial rectus muscle. It ends in a round tendon, which passes through a fibrocartilaginous pulley (trochlea), which is attached to the superomedial wall of the orbit.

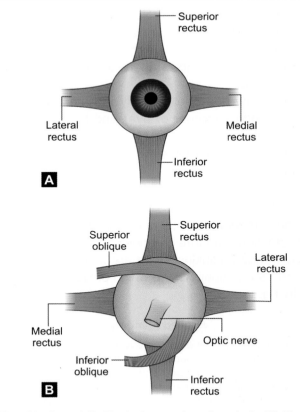

Figs 18.7A and B: Extrinsic muscles of eyeball. (A) Eyeball and insertions of 4 recti-front view (B) Eyeball and insertions of two oblique muscles—posterior view

After passing through the pulley, the tendon turns posterolaterally and inserts in to the sclera at the posterosuperior aspect of lateral side of eyeball.

The inferior oblique muscle is a thin, narrow muscle that arises from the floor of the orbit. It passes laterally (inferior to inferior rectus) and inserts into the sclera at the posteroinferior aspect of the lateral side of the eyeball.

Levator Palpebrae Superioris (Muscle which Elevates the Upper Eyelid)

It takes origin from the roof of the orbit, anterior to the optic canal.

Insertion

It is inserted partly into the skin of the upper eyelid and partly into the tarsal plate. The part which is inserted into the tarsal plate is involuntary; it is called the superior tarsal muscle or Müller's muscle.

Action

It elevates the upper eyelid. This muscle is continuously active, except during sleeping.

Nerve Supply of the Muscles of the Orbit

1. Lateral rectus—Abducent Nerve (VI cranial nerve)
2. Superior oblique—Trochlear Nerve (IV cranial nerve)
3. All the other muscles –Oculomotor Nerve (III cranial nerve)
4. Involuntary part of levator palpebrae superioris—Fibers from cervical sympathetic trunk.

[The following "formula" is an easy memory key (**LR$_6$ SO$_4$)$_3$**, i.e. Lateral rectus by VI, superior oblique by IV and all the others by III cranial nerves].

The Actions of the Extraocular Muscles

- Medial rectus — Adducts the eyeball
- Lateral rectus — Abducts
- Superior rectus — Elevates, adducts and rotates medially
- Inferior rectus — Depresses, adducts and rotates laterally
- Inferior oblique — Elevates medially rotated eye; abducts and rotates laterally.
- Levator palpebrae superioris — Elevates upper eyelid.
- Superior oblique — Depression, abduction, medial rotation

Applied Anatomy

Paralysis of one or more extraocular muscles due to injury of III, IV or VI cranial nerves results in diplopia or double vision.
   Paralysis of lateral rectus results in medial squint.
If the cervical sympathetic trunk is injured or compressed (by a tumor), the Müller's muscle is paralyzed, which causes drooping of the upper eyelid (ptosis); it is a part of Horner's syndrome.

The Eyeball (Fig. 18.8)

The eyeball is situated in the anterior part of the orbital cavity and is almost spherical in shape. Each eyeball is like a camera. It has a lens, which produces images of objects that we look at. The images fall on a light—sensitive membrane called the retina. Cells of the retina convert light images into nervous impulses which pass through the optic nerve and other parts of the visual

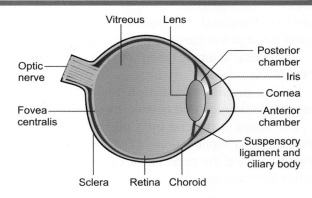

Fig. 18.8: Section of eyeball

pathway to reach the visual areas of the cerebral cortex.

Structure (Table 18.1, Fig. 18.8)

a. Coverings of eyeball
b. Light transmitting media or refractive media

Table 18.1: Coverings and refractive media of eyeball		
a. Coverings of eyeball (Fig. 18.8)		*b. Light transmitting media*
I. Fibrous coat	a. Sclera	Cornea
	b. Cornea	Aqueous humor
II. Vascular coat	a. Choroid	Lens
	b. Cilliary body	Vitreous body
	c. Iris	
III. Nervous coat	The retina	

Fibrous Coat

The Sclera (SKLEROS = HARD)

The posterior 5/6th of the outer coat of the eyeball consists of strong, opaque fibers and is called the sclera. It is protective in function and helps to maintain the shape of the eyeball. When viewed from the front, it is that portion, which is referred to as the "white of the eye"; in this position it is covered by a transparent mucous membrane, the bulbar conjunctiva.

   Posteriorly, the optic nerve passes through it from the retina. At this site, the sclera shows numerous perforations for the passage of optic nerve fibers. Because of its sieve-like appearance, this region is called the lamina cribrosa.

   The sclera is continuous anteriorly with the cornea at the sclerocorneal junction or limbus. Deep to the limbus is a circular canal called the sinus venosus sclerae or the canal of Schlemm. The aqueous humor drains through this canal to ciliary veins.

Posteriorly, sclera is fused with the dural sheath of optic nerve.

Sclera provides insertion to the extraocular muscles—(the recti and oblique muscles).

The sclera is pierced by a number of structures. They are: (1) Optic nerve; (2) Ciliary nerves and vessels, around the optic nerve (3) 4-5 choroid veins (venae vorticosae).

The Cornea

It forms the anterior 1/6th of the external coat of the eyeball. The cornea is transparent. Its junction with the sclera is called the sclerocorneal junction or limbus.

The cornea is more convex than sclera. It is separated from the iris by a space called the anterior chamber of the eye; it is filled with a clear fluid, the aqueous humor.

Cornea is innervated by branches of ophthalmic division of trigeminal nerve; it is sensitive to pain.

Histology (Figs 18.9A and B): Cornea consists of the following layers, from anterior to posterior:

1. *Corneal epithelium*—Stratified squamous non-keratinized type; continuous with the epithelium of conjunctiva.
2. *Bowman's membrane;* a homogeneous layer made of fine fibrils.
3. *The substantia propria,* containing parallel collagen fibers; the transparency of the cornea is because of the regular arrangement of collagen fibers, and because of the fact that the refractive index of the fibers and ground substance are the same. The fibroblasts present in the substantia propria are called keratocytes or corneal corpuscles.
4. *Descemet's membrane*—It is a thin homogeneous layer.
5. The posterior surface of the cornea is lined by a single layer of flattened cells—the *endothelium* of anterior chamber; it is in contact with the aqueous humor of the anterior chamber. The cornea has no blood vessels or lymphatics; it derives its nutrition from (1) vessels in its periphery, (2) by direct diffusion from aqueous humor.

The cornea is sometimes described as the "window of the eye."

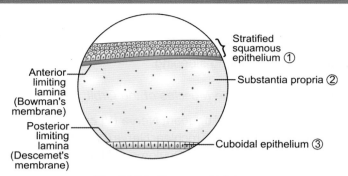

Fig. 18.9A: Cornea—histology

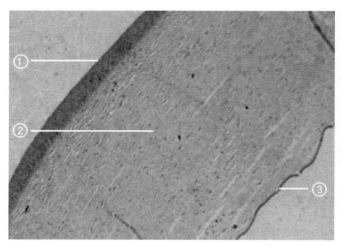

Fig. 18.9B: Histology of cornea

The Vascular Coat

This is the middle layer of the eye. It contains many blood vessels which are derived from the ophthalmic branch of internal carotid artery. The choroid, ciliary body and the iris together form the *uveal tract.* Its inflammation is called uveitis.

The Choroid

The choroid is a thin pigmented membrane, dark brown in color, which is situated between the sclera (externally) and retina (internally).

The Ciliary Body

The ciliary body is a circular structure continuous with the anterior part of the choroid. It contains smooth muscle fibers—the *ciliary muscle;* to the ciliary body is attached the suspensory ligament which helps to suspend the lens in position. The ciliary muscle helps in accommodation by adjusting the thickness of lens.

The Iris

This is the pigmented membrane which surrounds the pupil of the eye. It arises from the margin of the ciliary body and forms a diaphragm with a dark central opening (the pupil), immediately in front of the lens. The color of the eye is dependant on the pigment in the iris.

The space between the cornea (in front) and the lens (behind) is the anterior segment. It is divided into 2 by the iris; the space between the cornea and iris is the anterior chamber; the space between the iris and lens is the posterior chamber. These two chambers communicate with each other through the pupil. They are filled with a clear fluid, the aqueous humor.

Muscles of iris The iris contains a well-developed ring of muscles called sphincter pupillae, which lies near the margin of the pupil. The dilator pupillae is an ill-defined sheet of radially arranged fibers. Both are involuntary (smooth) muscles.

The sphincter pupillae and the ciliary muscle are supplied by parasympathetic fibers which come from Edinger-Westphal nucleus of midbrain, oculomotor nerve and the ciliary ganglion.

The dilator pupillae is supplied by sympathetic nerve fibers. Contraction of dilator pupillae causes dilatation (mydriasis) of pupil.

Contraction of sphincter pupillae causes constriction of pupil (miosis).

Nervous Coat

The Retina

It is the innermost coat of the eyeball. It is a thin, delicate layer, continuous posteriorly with optic nerve. The outer surface of the retina, formed by pigment cells is attached to the choroid. Its inner surface is in contact with the hyaloid membrane of the vitreous.

Opposite the entrance of the optic nerve is a circular area called optic disc.

The retina has three parts—Optic, ciliary and iridial parts. The optic part contains nervous tissue and it is sensitive to light. It extends from the optic disc to the posterior end of ciliary body. The thin, non-nervous, insensitive layer that covers the ciliary body and iris called the ciliary and iridial parts of the retina.

At the optic disc, there are no rods and cones. So, it is called the blind spot.

About 3 mm lateral to the optic disc, is a depression called macula lutea. (Due to its yellow color). The center of the macula is further depressed to form the fovea centralis. It contains cones only and is the site of maximum acuity of vision.

Histology (Figs 18.10A and B)

Retina has 10 layers, which includes the pigmented layer, layer of rods and cones, bipolar neurons, ganglion cell layer and nerve fiber layer. The fibers of the optic nerve begin from the ganglion layer of the retina, which converges at the optic disc: These fibers pierce the choroid and sclera and pass backwards as the optic nerve, covered by the 3 meninges; then through the orbit to the optic chiasma and brain.

The rods and cones are the actual receptors of light and sight. These cells contain photosensitive pigments (Rods–rhodopsin; cones–iodopsin)

The retina is supplied by a branch of the ophthalmic artery—the **central artery of retina**. It enters the eye with the optic nerve. Occlusion of this artery by thrombosis leads to total blindness.

Retina may become partially detached from the choroid—this is called retinal detachment. It can occur in hypertension or blunt injury to the eyeball.

The Light Transmitting Media or Refractive Media

The Aqueous Humor

This is a clear fluid which fills the space between the cornea and lens (anterior segment of the eye). Anterior segment is subdivided by the iris into an anterior chamber (between cornea and iris) and a posterior chamber (between iris and lens). Both these chambers, filled with aqueous humor, communicate with each other through the pupil.

The aqueous humor is secreted into the posterior chamber by the capillaries of ciliary processes. From here, the fluid reaches the anterior chamber, which ultimately reaches the canal of Schlemm. Interference with the drainage of aqueous humor results in an increase of intraocular pressure (glaucoma). This leads to atrophy of the retina, leading to blindness.

The aqueous humor has 2 functions: (1) It helps to maintain the intraocular pressure and thus maintain the shape of the eyeball, (2) It is rich in ascorbic acid, glucose and amino acids and nourishes the cornea and lens.

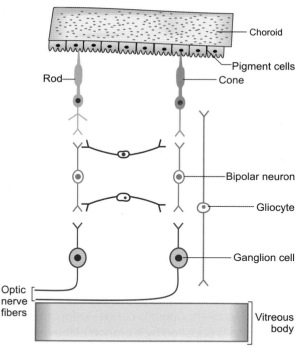

Fig. 18.10A: Structure of retina

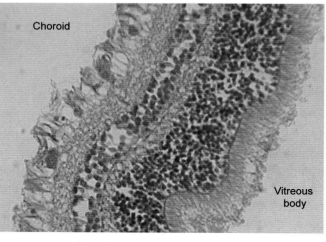

Fig. 18.10B: Histology of retina

The Lens

The lens is firm, transparent and biconvex. It is enclosed in a transparent, elastic capsule. Posterior surface of the lens is more convex than the anterior. The suspensory ligament of the lens retains the lens in position and its tension keeps the anterior surface of the lens flattened. The lens is placed immediately behind the iris and the pupil of the eye. Its function is to focus the light rays entering through the pupil on to the retina.

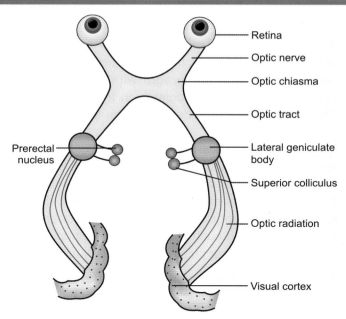

Fig. 18.11: Visual pathway

The Vitreous Body

The vitreous body is a colorless, transparent, jelly like substance, which fills the posterior 4/5 (posterior segment) of the eye. It helps to preserve the spherical shape of the eyeball and to support the retina. It is enclosed in a delicate, homogeneous hyaloid membrane.

Summary of the Sense of Light (Visual Pathway Fig. 18.11)

Light rays → cornea → aqueous humor → lens → vitreous body → retina → optic nerve → optic chiasma → optic tract → lateral geniculate body → optic radiation → areas 17, 18, 19 of occipital lobe of cerebrum.

Applied Anatomy

1. *Funduscopy:* Inspection of the optic fundus with an ophthalmoscope is an essential part of neurological examination. The central artery and vein of retina run within the anterior part of the optic nerve, with the subarachnoid space around them. An increase in CSF pressure slows venous return from the retina, causing edema. Edema of retina is seen as a swelling of the optic disc or optic papilla. It is called papilledema.
2. Detachment of the retina may follow a blow to the eye; it is a separation of pigment layer from the neural layer. (Because they develop from the outer and inner layers of optic cup, in the embryo).
3. Homologous corneal transplants can be performed for patients with opaque corneas.
4. An area of opacity of the lens is called cataract. Cataract extraction is the most common eye operation.
5. The retinal artery, a branch of the ophthalmic artery, is an end artery. Obstruction of this artery by an embolus or a thrombus leads to instant, total blindness.

19   Medical Genetics

It is the branch of Medicine dealing with inheritance, diagnosis and treatment of diseases due to–
a. Single gene mutation
b. Chromosome abnormalities, and
c. Multifactorial reasons
It also deals with genetic counseling and screening.

Chromosomes (Chroma = color; soma = body)

Chromosomes form the physical basis of inheritance. They are located in the nucleus. E Strasburger coined this term in 1875. Chromosomes are made up of genes. They become visible during cell division. Number of chromosomes is species specific.

Structure of Chromosome

Chromosome is composed of a double helix of DNA (deoxyribonucleic acid). Watson and Crick proposed the double helix model of DNA in 1953 and were awarded the Nobel Prize in 1962.

DNA is composed of two chains of nucleotides arranged in a double helix, which can be compared to a twisted ladder. Each chain has a sugar-phosphate backbone, which can be compared to the "side of a ladder". The two chains are held together by hydrogen bonds between the nitrogen bases.

Nucleotides

The nucleotides are composed of–
1. A nitrogenous base
2. A sugar molecule – In DNA, it is a deoxyribose sugar
3. A phosphate molecule

Nitrogenous Bases

There are 2 types of nitrogenous bases
A. Purines – Adenine and guanine
B. Pyrimidines – Cytosine and thymine (In RNA, thymine is replaced by uracil).

Salient Features of Watson and Crick Model of DNA (Fig. 19.1)

DNA consists of two polydeoxyribose nucleotide chains. These chains are twisted in a right handed double helix, similar to a spiral stair case. The sugar and phosphate molecules form the "handrail" of this ladder and the nitrogenous bases form the "steps" of the ladder or staircase. The two strands are always complementary to each other. Adenine always pairs with thymine and guanine in one chain always pairs with cytosine. A double hydrogen bond holds A with T; and a triple hydrogen bond holds G with C.

The two strands of DNA are antiparallel. One strand runs in the 5' to 3' direction whereas the other strand runs in the 3' to 5' direction. This arrangement can be compared to a road divided into two, each half carrying traffic in the opposite direction.

James Dewey Watson, Francis Harry Crick

Fig. 19.1: Structure of DNA

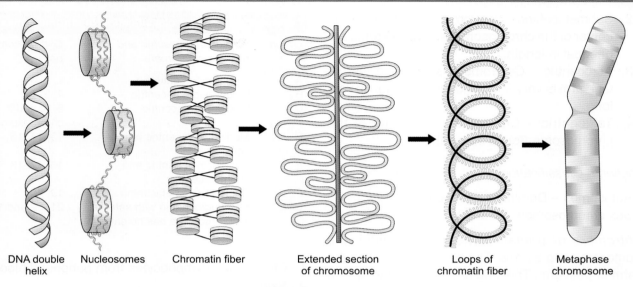

DNA double helix   Nucleosomes   Chromatin fiber   Extended section of chromosome   Loops of chromatin fiber   Metaphase chromosome

Fig. 19.2: Higher organization of DNA

Higher Organization of DNA (Fig. 19.2)

- DNA is wound over histones to form nucleosomes.
- A group of nucleosomes form DNA fibrils.
- 6 such fibrils are supercoiled to form chromatin fibers.
- Chromatin fibers are supercoiled to form loops.
- Loops are condensed to form chromosomes.
- DNA is condensed or compressed 10,000 folds to form a chromosome.
  Human diploid genome consists of 7×10^9 base pairs

Cytogenetics is the study of cell division and chromosomes. Human somatic cells have 46 chromosomes and gametes have 23 chromosomes. 2 types of cell divisions occur – Mitosis and meiosis (reduction division). Meiosis occurs during gametogenesis. Mitosis takes place in somatic cells. Mitosis results in each daughter cell having a diploid chromosome complement (46). During meiosis the diploid count is halved so that each mature gamete receives a haploid complement of 23 chromosomes.

Consequences of Meiosis

1. It facilitates halving of the diploid number of chromosomes so that each cell receives half of its chromosome complement from each parent.

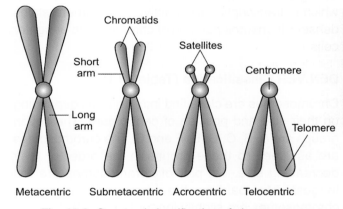

Fig. 19.3: Structural classification of chromosomes

Metacentric   Submetacentric   Acrocentric   Telocentric

2. It provides an extraordinary potential for generating genetic diversity due to independent assortment of chromosomes and "crossing over" of genetic material.

Classification of Chromosomes

Chromosomes can be classified into different groups based on structure and function.

Structural Classification (Fig. 19.3)

A. Based on the position of centromere, chromosomes are classified as follows—
   1. Metacentric – Centromere in the middle, so that the arms are of equal length.

2. Submetacentric – Centromere slightly below the center of the chromosome, so that the arms are unequal in length.
3. Acrocentric – Centromere is closer to the tip. One arm is very short and the other arm is very long.
4. Telocentric – Centromere at the tip of the chromosome (found in mice).

Functional Classification

A. Autosomes – Dominant and recessive.
B. Sex chromosomes – X and Y.

Centromere or primary constriction divides the chromosome into 2 arms. The short arm is called the 'p' arm or petit arm. The long arm is called the 'q' arm ('g' arm) or the grand arm.

The centromere is the site for spindle attachment during cell division. It has a special DNA sequence, which is transcriptionally inactive. Centromere stains densely. It ensures partition of chromatids to daughter cells.

DEN VER Classification (Table 19.1)

Chromosomes are classified into 7 groups depending on their size and position of centromere. The seven groups are A, B, C, D, E, F and G. The chromosomes are assigned to these groups in the order of their decreasing size. 1st pair of chromosomes are the longest. 21st pair are the smallest (22nd pair of chromosomes are slightly longer than 21st).

Satellite: It is a stalk – like appendage of acrocentric chromosomes, which form the nucleolus of the resting interphase cell and contain multiple repeat copies of the genes for ribosomal RNA.

Karyotyping is a method employed in genetic analysis for the study of chromosomes. It was first employed by Tjio and Levan in 1956. "Karyotype" is the photomicrograph of an individual's chromosomes arranged in a standard manner. Structural and numerical anomalies of chromosomes can be detected by this method (Figs 19.4A and B).

Any nucleated cell which can undergo cell division can be used for this study. Most commonly used cells are:

Table 19.1: Den Ver classification of chromosomes		
Group	*Size of chromosome and position centromere*	*Chromosome pair*
A	Large, metacentric	1, 2, 3
B	Large, submetacentric	4, 5
C	Medium metacentric	6-12 and X chromosome
D	Medium acrocentric with satellites	13, 14, 15
E	Shorter metacentric and submetacentric	16, 17, 18
F	Shortest submetacentric	19, 20
G	Short acrocentric with satellite Y chromosome has no satellite	21, 22 and Y

1. Circulating T-lymphocytes from peripheral blood
2. Skin
3. Bone marrow
4. Chorionic villi
5. Amniocytes.

Procedure (Fig. 19.5)

- 5 ml of peripheral blood drawn
- A sample of this blood added to a small volume of culture medium
- Add phytohemagglutinin (PHA) which induces T-lymphocytes to divide
- Cells are cultured under sterile condition at 37°C for about 72 hours
- Colchicine is added to the cell culture to arrest cell division (colchicine prevents spindle formation)
- Cell division gets arrested at metaphase, when the chromosomes are maximally condensed and most easily visible
- Hypotonic potassium chloride added. RBCs are lysed. Lymphocytes are swollen, spreading the chromosomes
- The cells are spread onto a slide by dropping, stained and banded (Fig. 19.6A).
- Photomicrographs of metaphase spread taken. Individual chromosomes are cut out from the prints and arranged in a systematic manner.
- Nowadays, a camera attached to the microscope transfers the picture to the computer, which analyzes the image and automatic karyotyping is done.

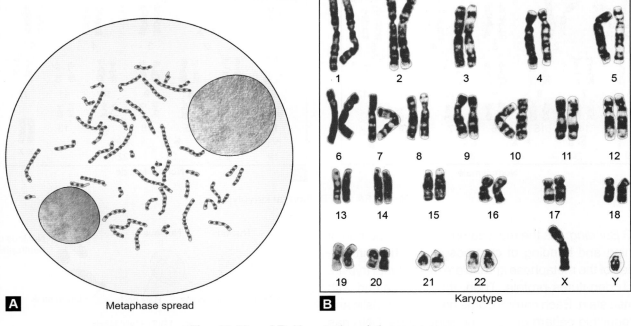

A Metaphase spread

B Karyotype

Figs 19.4A and B: Karyotyping of chromosomes

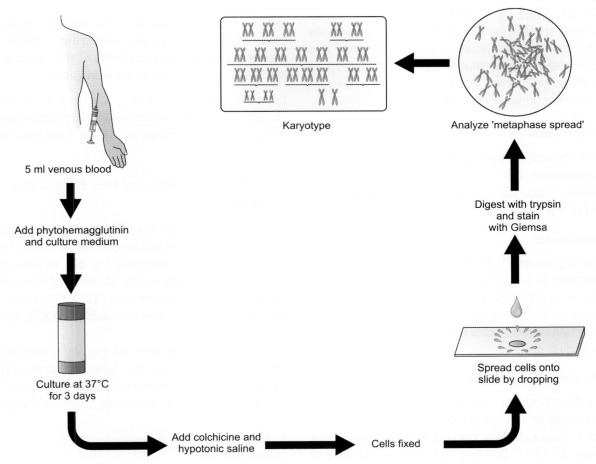

5 ml venous blood

Add phytohemagglutinin
and culture medium

Culture at 37°C
for 3 days

Add colchicine and
hypotonic saline

Cells fixed

Spread cells onto
slide by dropping

Digest with trypsin
and stain
with Giemsa

Analyze 'metaphase spread'

Karyotype

Fig. 19.5: Karyotyping – different steps

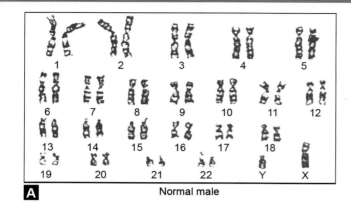

A   Normal male

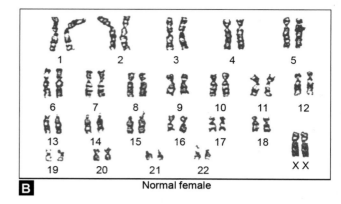

B   Normal female

Figs 19.6A and B: Normal karyotype

GTG Banding: It is the most commonly used method of staining and banding of chromosomes. The chromosomes of the metaphase spread are treated with trypsin, which denatures protein. Then, they are stained with giemsa stain. Each chromosome has characteristic and reproducible pattern of dark and light bands. DNA has large collections of AT rich areas, which stain deeply than GC rich areas, which take lighter stain.

A normal male somatic cell has 46 chromosomes which includes 44 autosomes, one X chromosome and a Y chromosome. A normal female somatic cell has 46 chromosomes which includes 44 autosomes and two X chromosomes (Fig. 19.6B).

Sex Chromosomes

X and Y chromosomes are the sex chromosomes. They determine the sex of the individual. These chromosomes also carry genes affecting structure and function of the body.

X chromosome

Belongs to group C. It carries genes for ovarian development. In females, one of the X chromosomes remains in a partially inactivated state called sex chromatin or Barr body. X chromosome may carry genes for hemophilia, color blindness and Duschenne muscular dystrophy.

Sex Chromatin or Barr Body (Fig. 19.7)

In 1949, Barr and Bertram discovered a small darkly staining mass in the interphase nucleus of somatic cells of females. It was named Barry body or sex chromatin. In 1961, Dr Mary Lyon noted that one of

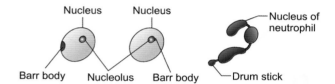

Fig. 19.7: Barr Body

the X chromosomes in mice differed from other chromosomes. She proposed that this X chromosome was inactivated. In recognition, the process of X inactivation is referred to as lyonization.

Lyon hypothesis: In a cell, all the X chromosomes except one undergo differentiation or partial inactivation early in embryogenesis and form Barr body or Barr bodies.

Calculation of Number of Barr Bodies in a Somatic Cell

"n-1 rule" states that the number of sex chromatin or Barr bodies in a cell is always one less than the total number of X chromosomes (n). So, in females, the number of Barr body is one. In males, there is no Barr body. In Klinefelter syndrome (47, XXY), there will be one Barr body. In super females (47,XXX), there will be two Barr bodies.

Sex chromatin studies are usually made on buccal smears. This test can be done for screening large populations for X chromosomal abnormalities.

Sex chromatin can be demonstrated in the following cells:

1. Buccal mucosa
2. Neutrophils

3. Bladder mucosa cells–A sample of urine is collected, centrifuged. The mucosa cells will sediment, in which Barr body is demonstrated.
4. Vaginal smears
5. Cells from the root of a single hair

The Barr body can be seen as a dark staining spot attached to the nuclear membrane or the nucleolus. In neutrophils of females, the Barr body appears as a "drum stick", an appendage from the nucleus.

The process of X inactivation in the somatic cell is initiated by a gene called XIST (X Inactivation Specific Transcript).

Genes

The chromosomes have genes along their length. A gene is nothing but a segment of the DNA that contains information in code for the synthesis of a polypeptide chain or a protein (gene product). Each gene codes for one specific polypeptide or a protein. There are about 25,000 genes within the human genome. That means, they can code for about 25,000 types of proteins or gene products. Genes normally exist in pairs, because each gene is present at corresponding sites on both homologous chromosomes (obviously, the gene pair for X and Y chromosomes will be absent in males).

Structure of gene (Fig. 19.8): Gene is made up of a long strand of DNA. There are coding sequences called EXONS with intervening non-coding sequences called INTRONS. The number and size of introns and exons in various genes is highly variable. Larger the gene, greater the number of exons.

Genetic Code

Genetic information is stored within the DNA molecule in the form of "triplet code", that is, a sequence of 3 bases determines one amino acid. Only 20 different amino acids are found in proteins. If three bases can code for one amino acid, the possible number of combinations of the four bases would be 4^3 or 64. This is more than enough to account for all the 20 amino acids and is known as the genetic code.

Regulation of Gene Expression (Flow chart 19.1)

The genes which are concerned with protein synthesis are called *structural genes.* The function of the

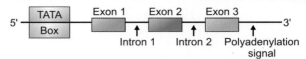

Fig. 19.8: Structure of a gene

Flow chart 19.1: Regulation of gene expression

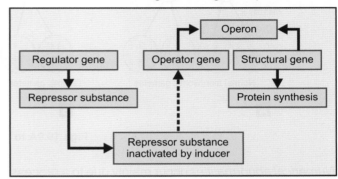

structural gene is controlled by an *operator gene.* Jacob and Monod coined the term *OPERON* for the unit of gene action which consists of an operator gene and the adjacent structural gene. The operator gene in turn, is controlled by a *regulator gene.* The regulator gene synthesizes a substance *repressor,* which inhibits the operator gene. Thus, when the regulator gene is functioning, proteins are not synthesized by the structural gene. The structural gene functions only when the regulator gene is "switched off" by the inactivation of the repressor by a specific metabolite referred to as *inducer.*

Chromosomal Aberrations or Abnormalities

Abnormalities of chromosome number and structure could seriously disrupt normal growth and development.

Classification

A. Numerical abnormalities
B. Structural abnormalities
C. Different cell lines (Mixoploidy)

Causes of abnormalities

1. Viruses
2. Irradiation
3. Autoimmune diseases
4. Late maternal or paternal age
5. Non-disjunction

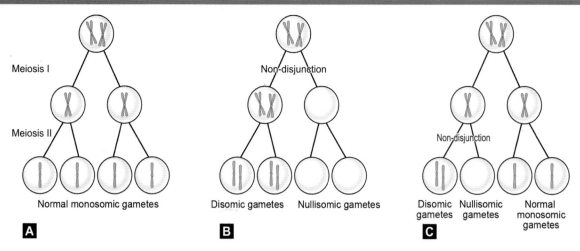

Figs 19.9A to C: Non-disjunction

Numerical abnormalities occur mainly due to a process called non-disjunction. Non-disjunction means failure of separation of homologous chromosome pairs during Meiosis I or Meiosis II. It can rarely occur in early mitotic divisions of the embryo. This results in an abnormality of number of chromosomes in the daughter cells as shown in Figures 19.9A to C.

Numerical Anomalies are of Two Types

A. Aneuploidy
B. Euploidy.

Aneuploidy: When there is an increase or decrease in the number of one chromosome, the abnormality is referred to as aneuploidy. The normal number of chromosomes in a human somatic cell is 46. When the chromosome number is less than the normal (diploid) number it is called hypoploidy. When the number is more than the normal, it is called hyperploidy.

Monosomy = 2n – 1, where 'n' is the haploid number (23). So, the chromosome number will be 46–1 = 45 e.g. Turner syndrome (45, X). Monosomy of an autosome is almost always lethal resulting in spontaneous abortion.

Trisomy = 2n+1, where 'n' is the haploid number (23). So, the chromosome number will be 46+1 = 47. Different types of chromosomal trisomies are shown in Figures 19.10A to C. One of the chromosomes will be present in triplicate e.g. Trisomy 21 or Down syndrome (Table 19.2).

Euploidy: When one set (23) of chromosomes are involved. The number of chromosomes will be in multiples of 23, as shown in Table 19.3.

Structural abnormalities of chromosomes occur due to chromosome breakage followed by loss of part of chromosome or reunion in a different configuration.

Effects of structural abnormalities are:

A. Balanced rearrangements
B. Unbalanced rearrangements

Table 19.2: Some common chromosomal anomalies		
No.	*Syndrome*	*Chromosomal abnormality*
1.	Down syndrome	Trisomy 21
2.	Edward syndrome	Trisomy 18
3.	Patau syndrome	Trisomy 13
4.	Turner syndrome	Monosomy of X
5.	Klinefelter syndrome	Trisomy of sex chromosome XXY

Table 19.3: Types of euploidy	
Chromosome number	*Type of euploidy*
n (23)	Haploidy
n (46)	Diploidy (normal)
n (69)	Triploidy
n (92)	Tetraploidy

Note: All the pregnancies, (except diploidy) end up in spontaneous abortion or the baby will die soon after birth.

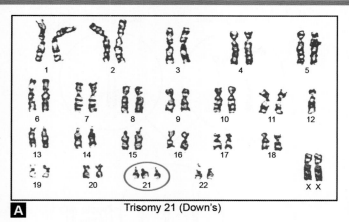

A Trisomy 21 (Down's)

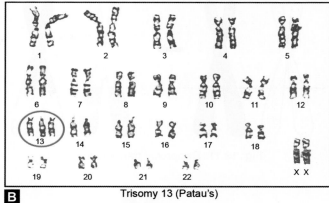

B Trisomy 13 (Patau's)

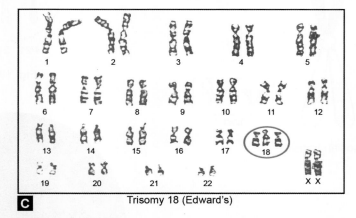

C Trisomy 18 (Edward's)

Figs 19.10A to C: Different types of chromosomal trisomies

Balanced rearrangements: Even though there is a rearrangement of parts of chromosomes, there is no loss or gain of genetic material. The chromosome complement is complete. The carriers of balanced rearrangement are at risk of producing children with unbalanced chromosome complement.

Unbalanced rearrangements: The chromosome complement contains an incorrect amount of chromosome material. The clinical effects produced by these abnormalities will be usually serious.

Types of Structural Abnormalities

1. Translocations
   (a) Reciprocal,   (b) Robertsonian
2. Deletions
3. Inversions
4. Insertions
5. Rings
6. Isochromosomes

Translocations

Refer to the transfer of genetic material from one chromosome to another non-homologous chromosome.

Reciprocal Translocation (Fig. 19.11)

A break occurs in each of the two chromosomes, segments are exchanged, two new chromosomes are formed. Most often, the condition is genetically harmless. But, a translocation between chromosome 9 and 22 may result in chronic myeloid leukemia.

Robertsonian Translocation (Fig. 19.12)

This phenomenon was first described by the American insect geneticist WRB Robertson. This is also known as centric fusion. It results from the breakage of two acrocentric chromosomes (13, 14, 15, 21 and 22) at or close to their centromeres, with subsequent fusion of their long arms. The arms are lost.

Section C

The Organ Systems

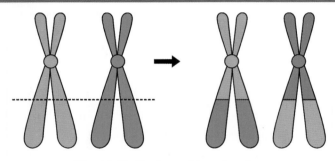

Fig. 19.11: Reciprocal translocation

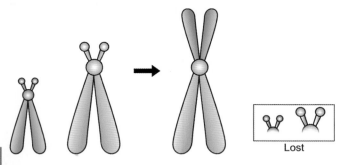

Fig. 19.12: Robertsonian translocation

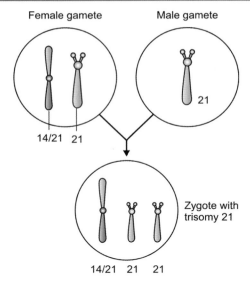

Fig. 19.13

Practical Importance

Robertsonian translocation can predispose to Down syndrome, if translocation occurs between chromosome 21 and one of the other acrocentric chromosomes, e.g. 14q 21q translocation. If the embryo inherits two normal 21 chromosomes (one from each parent), plus a translocation chromosome involving a 21 chromosome, Trisomy 21 results (see Fig. 19.13).

Deletions

Deletion is the loss of part of a chromosome. Results in monosomy for that segment e.g. **cri-du-chat syndrome** – Due to the loss of a segment from short arm of chromosome 5 (5p). The child has severe mental retardation, cat like cry (due to under development of larynx), congenital heart disease like atrial or ventricular septal defects (ASD or VSD).

Insertions (Fig. 19.14)

A segment of one chromosome is inserted into another nonhomologous chromosome.

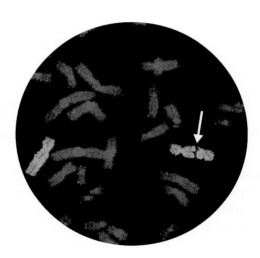

Fig. 19.14: Insertion

Inversions (Fig. 19.15)

An inversion is a two-point break rearrangement involving a single chromosome in which a segment is reversed or inverted in position.

There are two types of inversions:
a. Pericentric
b. Paracentric

In pericentric inversion, the centromere is involved. Paracentric inversion occurs in one of the arms, without involving the centromere. Inversions rarely cause clinical problems.

Inversions

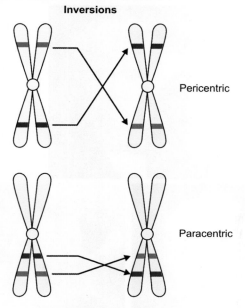

Pericentric

Paracentric

Fig. 19.15: Inversions

Ring chromosome

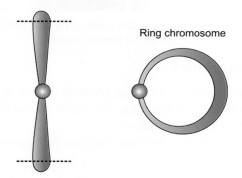

Fig. 19.16: Formation of ring chromosome

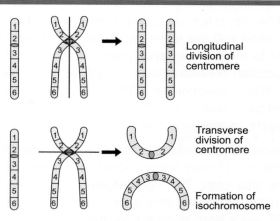

Longitudinal division of centromere

Transverse division of centromere

Formation of isochromosome

Fig. 19.17: Formation of isochromosome

Ring Chromosome (Fig. 19.16)

A break occurs on each arm of chromosome. The ends of the arms become sticky. They reunite to form a ring. If the deleted ends carry important genes, the effects will be serious. When one of the X chromosomes forms a ring, it can result in Turner syndrome.

Isochromosome (Fig. 19.17)

This is due to the loss of one arm and duplication of the other arm. So the chromosome carries identical information on both arms. This results from transverse division of centromere. Normally, during cell division, a longitudinal division occurs at the centromere and two sister chromatids are formed. Each sister chromatid has a p arm as well as q arm. If a transverse

division of centromere occurs, the p arm gets separated from the q arm (see figure). Isochromosome X can be a cause of Turner syndrome.

Different cell lines: In the same individual, cells with different genetic constitutions are found:
Two types:
a. Mosaicism
b. Chimerism
Mosaicism is defined as the presence in an individual of two or more genetically distinct cell lines derived from one zygote. It usually results from non-disjunction during mitotic divisions of zygote.

Chimerism: Chimera is a mythological Greek monster with head of a lion, body of a goat and tail of a dragon.
Two types of chimeras can occur—
1. Dispermic chimeras – 2 sperms fertilize two ova. The resulting zygotes fuse to form one embryo. If the zygotes are of two sex, the embryo will be a true hermaphrodite with XX/XY karyotype.
2. Blood chimeras result from exchange of cells via the placenta between non-identical twins in utero, e.g. 90% of one twin has XY and blood group B. 90% of cells of the other twin has XX and blood group A.

Clinical Features of Three Common Numerical Anomalies

Down syndrome (Trisomy 21)

The clinical features of this common disorder was first described by Dr Langdon Down in 1866 (Fig. 19.18A). The chromosomal basis was discovered by Lejuene et al.

The incidence of Down syndrome is 1 in 700 (i.e. when 700 live births occur, one of the babies is likely to have Down syndrome). There is a strong association between Down syndrome and maternal age. When maternal age is 20 years, the incidence of Down syndrome is 1 in 1500; when maternal age is 45 years, incidence is 1 in 30.

Reason: Oogenesis is a lengthy phenomenon. It begins in the female fetus, and gets completed several years after birth. This very lengthy interval between the onset of meiosis and its eventual completion, up to 50 years later, accounts for the well-documented increase in chromosome abnormalities in the offspring of older mothers. Non-disjunction of chromosomes can occur due to failure or abnormality of chiasmata formation or due to failure or abnormality of spindle formation. Exposure of oocytes to radiation and other environmental hazards is another reason.

Clinical Features (Phenotypic Anomalies)

Newborn period: A hypotonic, lethargic, sleepy baby with excess nuchal skin raises the first suspicion of Down syndrome.

Cranio-facial Features

- Brachycephaly (breadth of the skull is more than the antero-posterior measurement)
- Small nose, with shallow nasal bridge
- Protruding tongue, constantly open mouth (Fig. 19.18B)
- High arched palate
- Small ears
- Upward slanting palpebral fissures

Limb

- A single, prominent palmar crease—**Simian crease**—can be seen in 50% cases (Fig. 19.18C)
- Clinodactyly – short middle phalanx of 5th finger
- Wide gap between 1st and 2nd toes

Cardiac anomalies can occur in 40-50% patients with Down syndrome
- Atrial septal defect (ASD)
- Ventricular septal defect (VSD)
- Patent ductus arteriosus (PDA)
- Common atrioventricular canal

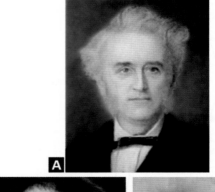
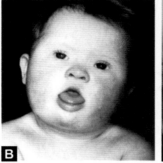
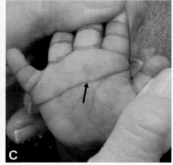

Figs 19.18A to C: (A) Langdon Down (B) A child with Down's syndrome (C) Simian crease

Other Anomalies

- Duodenal or anal atresia (Atresia means failure of canalization of the tube. So, there will be features of intestinal obstruction).
- Hirchsprung's disease (congenital aganglionic megacolon)
- Short stature
- Squint
- Mental retardation
- Adults develop Alzheimer's disease

In general, the children with Down syndrome are happy and quite affectionate.

Chromosomal Abnormality

95% - Trisomy 21
4%   - Translocation (Robertsonian)
1%   - Mosaicism

Prenatal Diagnosis

When the maternal age is more than 35 years, or when the previous child has a chromosome abnormality, rule out the possibility of Down syndrome or other chromosomal anomalies by the following methods:
1. Chorionic villus sampling (CVS)
2. Amniocentesis and culture

3. Triple or quadruple tests
4. Ultrasonography

Triple or quadruple tests are done in the maternal serum

1. Alpha fetoprotein (AFP) – Decreased in Down syndrome
2. Unconjugated estriol – Decreased in Down syndrome
3. HCG (Human chorionic gonadotropin) – Increased in Down syndrome
4. Inhibin A – Increased in Down syndrome

Ultrasonography around 12 weeks of pregnancy may reveal a fetal nuchal(neck) translucency due to accumulation of fluid behind baby's neck.

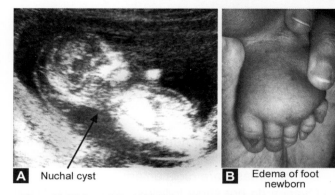

Figs 19.19A and B: (A) USG of fetus showing nuchal cyst (B) Edema of foot of newborn

Turner Syndrome (Ullrich-Turner Syndrome)

This clinical disorder was first described by Henry Turner in 1939. It is characterized by the absence of an X or a Y chromosome. The karyotype of the patient will be 45, X. Phenotypically, the patient will be a female.

Henry Turner

Incidence: 1 in 10,000.

Clinical Features

Turner syndrome can be suspected or based on certain clinical features before and after birth.

Clinical Features Before Birth

Ultrasonography may reveal generalized edema (hydrops) of the fetus and a nuchal cyst (collection of fluid behind the neck) (Figs 19.19A and B).

At birth: Puffy extremities.

Somatic Abnormalities

Craniofacial Features

• Epicanthal folds (fold of skin over the medial angle of eye)

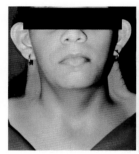 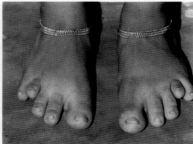

Figs 19.20A and B: (A) Webbed neck (B) Fourth short metatarsals and toes

• High arched palate
• Abnormal teeth
• Visual/auditory defects.

Neck (Fig. 19.20A)

Webbed neck or pterygium colli is one of the characteristic features of Turner syndrome. The skin and sometimes the trapezius or sternocleidomastoid muscle also extends on either side of the neck as a transverse fold.

Chest is described as a "shield chest". Nipples are widely placed and "pin point". Breasts are poorly developed.

Cardiovascular anomalies are also common. Ventricular septal defect and coarctation of aorta are the commonly encountered cardiovascular problems in Turner syndrome.

Skeletal Anomalies

• Short stature
• Cubitus valgus (prominent carrying angle)

Section C

The Organ Systems

- Spina bifida
- Short metacarpals/metatarsals – usually the 4th. So the fourth finger or toe appears to be shorter than the others (Fig. 19.20B)
- Osteoporosis.

Reproductive System

- Failure of development of secondary sexual characters – Axillary and pubic hairs are absent or scanty. Breasts are rudimentary
- Ovaries do not show follicles. They are replaced by fibrous tissue. The ovaries are described as "streak ovaries"
- Menarche occurs rarely (primary amenorrhea)
- Patients are usually sterile.

Karyotype: Usually 45, X. Rarely 46 XX/45X mosaics. Ring chromosome X or isochromosome X are also seen rarely.

Management of patients with Turner syndrome. Early detection is important in the management. A baby born with puffy extremities and a webbed neck will raise the first suspicion. Karyotyping will give the final diagnosis. Administration of growth hormone will enable the individual to attain almost normal height. Cyclical estrogen therapy can help to develop growth of breasts and other secondary sexual characters to a certain extent. If the patient insists on having a baby and if she can afford it, an in vitro fertilization (with donor's oocyte) and implantation into a hormonally prepared uterus is also possible.

Klinefelter Syndrome

The characteristic features of this clinical condition was first described by Dr Harry Klinefelter in 1942.

Dr Harry Klinefelter

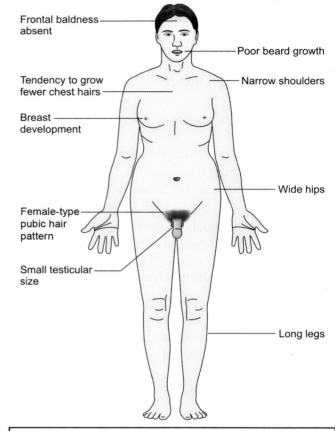

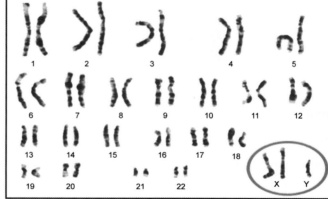

Fig. 19.21: Clinical features and karyotype Klinefelter syndrome

Incidence: 1 in 1000 male live births. High incidence in advance maternal age.

Karyotype: 47, XXY. Rarely 46 XY/47 XXY mosaic (Fig. 19.21).

Clinical Features (Fig. 19.21)

- Phenotypically male
- Barr body positive

- Azoospermia (absence of sperms in semen)
- Underdeveloped secondary sexual characters – Poor growth of facial, axillary and pubic hairs.
- Thin, tall individuals with long lower limbs.
- Gynecomastia (enlarged breasts)
- Narrow shoulders
- Female type of pelvis, wide hips
- Small size of testes

Patients are usually shy, clumsy and have low verbal IQ. They may show criminal traits and increased susceptibility to schizophrenia. These patients also show high incidence of leg ulcers, cancer of the breast and osteoporosis.

Management: Testosterone can be supplemented from puberty. Osteoporosis can be prevented by adequate calcium, vitamin D supplements.

Modes of Inheritance

Genetic information is inherited from one generation to the other by different ways –
- Dominant inheritance
- Recessive inheritance
  – Autosomal
  – Sex linked
- Sex linked inheritance

Dominant/recessive: Definitions—
When a gene produces its effects whether it is present either upon one or both chromosomes of the pair concerned, it is called dominant.

If the gene produces its effect only when present on both the chromosomes, it is called recessive.

Dominant Inheritance

If a gene is dominant for a given character, the individual will be definitely affected. Dominant conditions show up even when the abnormal gene is present in single "dose", i.e. in heterozygous state. The affected person has an affected parent (unless a mutation has occurred).

Autosomal dominant disorders: Examples are:
1. Marfan syndrome
2. Von Recklinghausen's neurofibromatosis
3. Dentinogenesis imperfecta
4. Achondroplasia

5. Brachydactyly
6. Osteogenesis imperfecta or Brittle bone disease

Marfan's syndrome is characterized by ectopia lentis (abnormal position of lens), long extremities, arachnodactyly (spider-like fingers), aneurysm of aorta, pectus excavatum (an abnormal, concave sternum).

Teeth are malformed in dentinogenesis imperfecta. Osteogenesis imperfecta or brittle-bone disease is characterized by malformations of skeletal system. The affected individual is prone for fractures even by trivial forces.

Von Recklinghausen's neurofibromatosis can present in mild to severe forms. In mild variety, only a dark pigmented patch ("*café au lait*" spots) or patches are seen on the skin. Some individuals may show numerous swellings on the body (small, rounded swellings arising from nerve sheaths). The swellings can arise from any nerve. Some of them may compress the spinal cord at different levels. Some of them can turn malignant (neurofibro-sarcoma) later in life.

Achondroplasia is a common cause of dwarfism. It is due to the abnormality in cartilage formation.

Brachydactyly: The fingers and toe are small (short).

Summary of Autosomal Dominant Traits

- Appear in every generation
- No skipping of generations
- Trait is transmitted by an affected individual to half of his or her offsprings
- Unaffected person cannot transmit the trait to their children
- Both the sexes have equal chances of having the trait and transmitting it.

Autosomal recessive inheritance: Fortunately these disorders are rare.
Examples
- Albinism
- Wilson's disease
- Microcephaly
- Phenylketonuria
- Porphyria
- Galactosemia
- Gaucher's disease
- Hemoglobinopathies

Characteristic features of autosomal recessive inheritance

- The disorder will manifest only when the gene is present in the homozygous state
- The individual has received the recessive gene from both parents
- Usually both parents are phenotypically normal and genotypically heterozygous carriers
- Parents are related (cousins) – i.e. consanguineous
- Males and females have equal chance of being affected and transmitting the trait
- Autosomal condition is seen typically in siblings – brothers and sisters. It is not seen in parents or other relatives
- On an average, the ratio of the affected, carrier and non-affected is 1:2:1 in the siblings.

Sex-linked Inheritance

The sex chromosomes have 2 functions:
a. Determination of sex
b. Genes for normal structure and function of the body

X-linked and Y-linked Genes

Genes carried by the X chromosomes are called the X-linked genes. They are exclusive to that chromosome. There is no corresponding locus on Y chromosome.

Genes carried by Y chromosome are called the Y-linked genes. These genes are unpaired. If present, it will be expressed. There is no question of dominance or recessiveness. Genes will be passed to all sons.

X-linked inheritance can be of two types—
a. X-linked recessive
b. X-linked dominant

Examples of X-linked Recessive Disorders

1. Hemophilia
2. Color blindness
3. Duschene muscular dystrophy (Pseudohypertrophy of calf muscles)

Some Facts About X-linked Inheritance

- Since male has only one X chromosome, whether recessive or dominant, an X-linked gene is always expressed in the male.
- X-linked trait cannot be transmitted from father to son.

- The disease can be transmitted to daughter only if the gene is dominant.
- The daughter will be a carrier if the gene is recessive.

Hemophilia is a congenital bleeding disorder. History's most famous carrier of the gene was Queen Victotria of England.

There are two types of hemophilia. Hemophilia A or the classical hemophilia and Hemophilia B or Christmas disease. Hemophilia A is more common (about 84%). The cause is Factor VIII deficiency. Even a minor injury like a tooth extraction or an intramuscular injection is sufficient to start prolonged bleeding. Hemophilia B or Christmas disease occurs due to the deficiency of Factor IX (Christmas factor or PTC). It was discovered in a family with surname "Christmas". The only effective therapy, when bleeding is uncontrollable, is intravenous administration of the deficient factor.

Prenatal diagnosis of hemophilia is possible. So, the birth of an affected male child can be prevented.

X-linked dominant conditions are very rare. Example is vitamin D resistant rickets. Affected females are twice as common as affected males. The trait is seen in females even if homozygous or heterozygous. The disorder is seen in males only if the single X chromosome carries the gene. The affected male passes on the trait to all his daughters. But none of his male children would be affected.

PRENATAL DIAGNOSIS OF GENETIC DISORDERS

For most of the genetic disorders, no specific treatment is available. Only symptomatic treatment can be given to the affected individual. The quality of life of these patients is not satisfactory. Prevention of such disorders by preventing the birth of a baby with serious congenital defects or metabolic disorders or bleeding disorders is far better than treating them. That is why prenatal diagnosis of genetic disorders has such an importance.

Definition

Prenatal diagnosis is the ability to detect abnormalities in an unborn child. It is an option chosen by many couples at high risk of having a child with a genetic disorder.

Indications

1. Advanced maternal age (> 35 years) is the commonest indication
2. Previous child with a chromosome abnormality
3. Family history of a chromosome abnormality
4. Family history of a single gene disorder
5. Family history of a neural tube defect
6. Family history of other congenital abnormalities
7. Abnormalities identified in pregnancy
8. Women who are carriers of X-linked recessive disorders, e.g. hemophilia
9. Carriers of inborn errors of metabolism

Standard techniques used for prenatal diagnosis can be broadly classified into two—
1. Non-invasive techniques
2. Invasive techniques

Non-Invasive Techniques

1. Maternal serum screening for
   a. AFP or Alpha fetoprotein
   b. Triple test
   c. Quadruple test
2. Ultrasonography (USG), 3D or 4D imaging
3. MRI (Magnetic Resonance Imaging)
4. CT (Computerized Tomography)

Invasive Technique

1. Amniocentesis
2. Chorionic villus sampling (CVS)
3. Fetoscopy
4. Fetal blood (cord blood) sampling

■ NON-INVASIVE TECHNIQUES

Maternal Serum Screening

Alpha fetoprotein level: AFP is a glycoprotein synthesized in the fetal liver, yolk sac and gut tube. Large amounts of AFP escape from fetal circulation into amniotic fluid in neural tube defects (NTD) and ventral wall defects (VWD). The level of AFP in maternal serum reaches a peak or maximum during the 16th week of pregnancy.

By estimating the level of AFP in maternal serum, nearly 75% of all case of open neural tube defects can be detected. 60-70% of all cases of Down syndrome can also be detected (in Down syndrome, the level of AFP in maternal serum is low).

Triple test

3 markers in maternal serum are evaluated
1. AFP
2. Unconjugated estriol (uE_3)
3. HCG (Human chorionic gonadotropin)
   Optimum time to do this test is 16 weeks.

Quadruple test

4 markers are evaluated in the maternal serum
1. AFP
2. uE_3
3. HCG
4. Inhibin

Ultrasonogram or USG is a valuable means of prenatal diagnosis. It also helps in
1. Placental localization
2. Detection of multiple pregnancy
3. Detection of structural abnormalities

Advantages of USG

- No known risk to the fetus or mother. So, it can be done routinely in all pregnant women around 18 weeks of gestation.
- It can be used as a screening procedure for NTD and cardiovascular anomalies
- USG can give a clue regarding underlying genetic disorder. For example, detection of a nuchal cyst (fluid accumulation behind the neck) can suggest Down syndrome or Turner syndrome or Rh isoimmunization.
- 3D or 4D USG can detect cleft lip, cleft palate and other structural abnormalities easily.

CT and MRI

- CT is helpful for differentiating between monoamniotic and diamniotic twins
- CT and MRI provide more accurate information about an abnormality detected in USG.

■ INVASIVE TECHNIQUES

Amniocentesis is the aspiration of 10-20 ml of amniotic fluid(through the mother's abdominal wall

Section C

The Organ Systems

and then the uterine wall) under USG guidance. Optimum time – 16 weeks

Technique (Fig. 19.22): Under sterile conditions, introduce a sterile needle (lumbar puncture needle), under ultrasound guidance into the amniotic cavity and withdraw 10-20 ml of amniotic fluid. This fluid will contain a mixture of cells shed from amnion, fetal skin, urinary tract epithelium and buccal mucosa of the fetus. Most of these cells are non-viable. Some cells will grow when added to a culture medium.

Centrifuge the amniotic fluid sample. The cells will sediment at the bottom of the tube and form a pellet. The cells are cultured. After about 14 days, biochemical, chromosomal or DNA studies can be carried out.

Informations that can be obtained from the cells are:
1. Fetal sex can be determined to rule out the possibility of severe sex-linked disorders like hemophilia or muscular dystrophy.
2. Detection of numerical anomalies like Down syndrome
3. Detection of inborn errors of metabolism

The supernatant fluid is used for assessing the level of AFP. It can also be used for spectrophotometric studies for assessing the degrees of erythroblastosis fetalis (HDN).

Advantages of Amniocentesis

Early detection of neural tube defects, chromosome abnormalities, metabolic disorders and molecular defects is possible.

Disadvantages: There is a 0.5-1% risk of miscarriage (abortion). Since, amniocentesis and culture of cells is a time consuming procedure, by the time the fetus is diagnosed to have a disorder, a mid-trimester termination of pregnancy will have to be considered.

Chorionic Villus Sampling (CVS)

Placenta is made up of two components – a maternal component called decidua basalis and fetal component called chorion frondosum or chorionic villi. The cells of the chorionic villi will have the same chromosome complement as that of the fetal cells because all these cells are derived from the zygote.

Optimum time: 10-12 weeks

Technique (Fig. 19.23): Under sterile conditions and ultrasound guidance, a sterile needle is introduced into the placenta through transabdominal or transvaginal route. A sample of chorionic villi is obtained. The chorionic villi cells can be cultured; karyotyping can be done. The DNA from the chorionic villi can be extracted for molecular studies.

Advantages of CVS

Prenatal diagnosis is possible in the first trimester.

So, if the fetus is found to have any serious chromosome or gene disorder, MTP can be carried out easily.

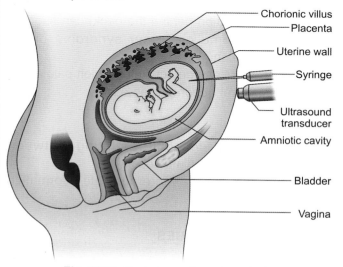

Fig. 19.22: Technique of amniocentesis

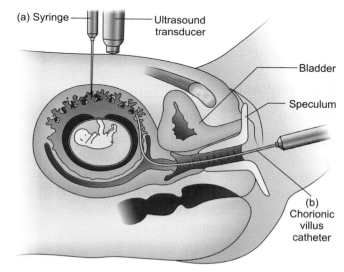

Fig. 19.23: Two techniques for getting samples of chorionic villi, (a) Transabdominal, (b) Transvaginal

Disadvantage

1. 2-3% risk of miscarriage (abortion)
2. Can cause limb abnormalities if carried out before 9 weeks of gestation.

Fetoscopy

Fetoscopy is the technique by which fetus can be visualized using endoscope. The fetoscope is introduced trans-abdominally. The procedure is done in the 2nd trimester of pregnancy. Detection of structural abnormalities, drawing of sample of fetal blood, getting a sample of fetal tissue for detailed study is possible by fetoscopy.

Disadvantage

3-5% risk of miscarriage.

Fetal Blood Sampling

Cordocentesis is a procedure by which a small sample of fetal blood is obtained from one of the umbilical cord vessels during fetoscopy. The blood is used for chromosomal analysis or DNA studies.

New Technique in Prenatal Diagnosis

Preimplantation genetic diagnosis (PGD) – *After in vitro* fertilization (IVF), the dividing zygote is cultured in the lab up to the eight cell stage. One of the blastomeres is removed and analyzed for chromosomal or gene abnormalities. If there are no defects, the embryo is implanted in the uterine cavity.

Human Genome Project (HGP)

Begun in 1990 by the US Department of Energy and the National Institute of Health. The project is sponsored by US, UK, Germany, France and Japan.

Goals

- Identify all the genes – approximately 26000 – in human DNA
- Determine the sequences of the 3 billion base pairs of DNA
- Improve tools for data analysis
- Address the Ethical, Legal, Social Issues (ELSI) that may arise from the project.

Aim

- Knowledge about DNA can help to diagnose, treat and prevent thousands of genetic disorders

Potential Applications of HGP are in the Following Field of Sciences

- Molecular medicine
- Microbial genomics
- Risk assessment
- Anthropology, evolution, human migration
- DNA forensics
- Agriculture, livestock, bioprocessing.

Applications of HGP in Molecular Medicine

- Improved diagnosis of diseases – Faster, cheaper, accurate methods are to be developed.
- Earlier detection of genetic predisposition to disease
- Gene therapy.

Applications of HGP in Microbial Genomics

- To develop new energy sources (biofuels)
- Protection from biological and chemical warfare
- Safe and efficient methods for toxic waste clean up.

Applications in DNA Forensics

- Identify potential suspects whose DNA may match evidence left at crime scenes
- Exonerate persons wrongly accused of crimes
- Establish paternity and other family relationships
- Match organ donors with recipients in transplant programs.

Applications in Agriculture/livestock/bioprocessing

- To develop disease resistant, insect resistant and drought resistant crops
- To develop healthier, more productive, disease resistant animals
- To develop more nutritive products
- To develop biopesticides
- To develop edible vaccines incorporated into food products.

Section C

The Organ Systems

Gene therapy is the ultimate cure in the correction of defective genes in genetic disorders

2 methods are being developed –

1. Somatic therapy
2. Germ line therapy

Somatic therapy: The functional gene is inserted into the somatic cell of the patient with the genetic defect which prevents or eliminates the disease. The drawback of this method is that, the defect is corrected for one generation only.

Germ line therapy: The gene is introduced into the gametes or zygote.

Practical Applications of Gene Therapy

1. In the treatment of cancer – A tumor necrosis factor is introduced into the lymphocytes, which will destroy the cancer cells
2. Treatment of diseases like cystic fibrosis, thalassemia
3. In the treatment of inborn errors of metabolism
4. Implantation of healthy myoblasts in patients with Duschene muscular dystrophy
5. Immunization
6. Protection of cells from HIV
7. Somatic gene therapy for *in vivo* production of insulin, growth hormone, factor VIII and erythopoietin.

20 Appendix

■ THE ADDUCTOR CANAL (FIG. 20.1)

It is a musculoaponeurotic tunnel, which occupies the middle one-third of the medial side of thigh. It is also known as the Hunter's canal or the subsartorial canal (Surgeon John Hunter utilized this region as the site of compression of femoral blood vessels by applying tourniquet against the linea aspera of femur, to arrest bleeding during the amputation of the lower limb below knee or in the surgery of aneurysm of the popliteal artery).

Extent

It extends from the apex of the femoral triangle to the fifth osseoaponeurotic opening of the adductor magnus.

Boundaries

It is bounded by:
1. Anterolaterally—vastus medialis
2. Posteriorly—Adductor longus and adductor magnus
3. Roof—Sartorius and fascia extending between the above muscles.

Contents

1. Femoral artery and femoral vein
2. Saphenous nerve—it pierces the roof
3. Nerve to vastus medialis
4, Terminal part of profunda femoris vessels.

■ THE AXILLA (FIG. 20.2)

The axilla or arm pit is a pyramidal-shaped area lying between the upper part of the arm and the lateral thoracic wall.

Boundaries

It has an apex, a base and 4 walls—anterior, posterior, medial and lateral.

Apex

The apex of the axilla is directed upward and medially to the root of the neck. It transmits structures to and from the neck to the upper limb.

Base

The base is directed below. It is concave; formed by skin, superficial fascia and axillary fascia.

Walls

Anterior wall is formed by the pectoralis major and pectoralis minor.

Posterior wall is formed by subscapularis, latissimus dorsi and teres major.

Medial wall is formed by (1) upper 4 or 5 ribs and the intercostal spaces between them; (2) the serratus anterior and the long thoracic nerve supplying it.

Lateral wall is formed by the humerus, biceps and coracobrachialis.

Contents

1. Axillary artery and its branches
2. Axillary vein and its tributaries
3. Cords of brachial plexus and its branches
4. Axillary lymph nodes
5. Axillary fat
6. Occasionally, the axillary tail of the breast.

■ THE CAROTID SHEATH (SEE FIG. 17.5)

The carotid sheath is formed by the contribution from the 3 layers of deep fascia of the neck—the pretracheal layer, the investing layer and deep layer of the cervical fascia. It extends from the base of the skull to the neck.

Contents

1. The common carotid artery (in the lower part)
2. The internal carotid artery (in the upper part)
3. The internal jugular vein
4. The vagus nerve.

Relations

1. Embedded in the anterior wall of the carotid sheath is a loop of nerves called the ansa cervicalis.
2. The cervical sympathetic chain lies behind the sheath.

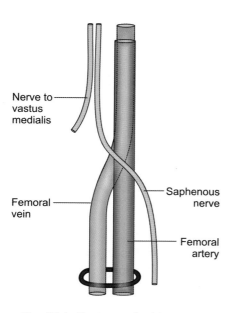

Fig. 20.1: Contents of adductor canal

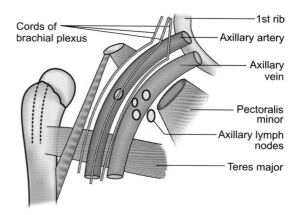

Fig. 20.2: Contents of axilla

THE CUBITAL FOSSA (FIGS 20.3A TO C)

It is a triangular space in front of the elbow.

Boundaries

- *Medially,* the lateral border of pronator teres
- *Laterally,* medial border of brachioradialis

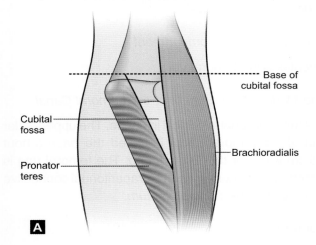

A

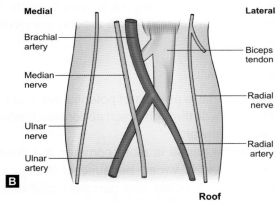

B

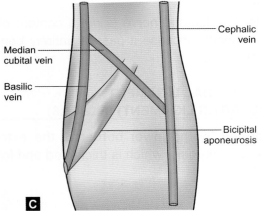

C

Figs 20.3A to C: Boundaries and contents of cubital fossa (left side)

- *Apex* is formed, where the above muscles meet
- *Base* is formed by an imaginary line joining both epicondyles of humerus
- *Floor* is formed by the brachialis and supinator muscles
- *Roof* is formed by deep fascia of the forearm and the bicipital aponeurosis.

Contents

From medial to lateral side, the contents are:
1. The median nerve
2. The brachial artery and its terminal branches
3. Tendon of biceps.
   Superficial to the bicipital aponeurosis lies the median cubital vein.

THE FEMORAL TRIANGLE (FIG. 20.4)

It is a triangular depression below the inguinal ligament, on the front of the upper one-third of the thigh.

Boundaries

- *Base* is formed by the inguinal ligament
- *Laterally* by the medial border of sartorius
- *Medially* by the medial border of adductor longus
- *Apex* is pointed donwards. It is formed, where the medial and lateral boundaries meet. It is continuous below with adductor canal.
- *Roof* is formed by (1) skin; (2) superficial fascia containing superficial inguinal lymph nodes, superficial branches of femoral artery and the upper part of great saphenous vein (3) the deep fascia.

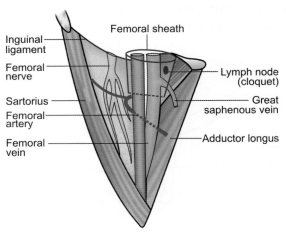

Fig. 20.4: Femoral triangle (right side)

- *The floor* of the triangle is formed, from medial to lateral by adductor longus, pectineus, psoas major and iliacus muscles.

Contents

1. Femoral artery and its branches
2. Femoral vein and its tributaries
3. The femoral sheath, which encloses the upper 4 cm of the femoral vessels
4. Femoral nerve and its branches
5. Deep inguinal lymph nodes.

■ THE FEMORAL SHEATH (FIG. 20.5)

It is a funnel-shaped sleeve of fascia, which encloses the upper 3-4 cm of the femoral vessels.

Formation

It is formed by the downward extension of two layers of fascia of the abdomen. They are:

The Fascia Transversalis

It forms the anterior wall of the femoral sheath (this fascia lines the deep surface of the transverses abdominis muscle of the anterior abdominal wall).

The Fascia Iliaca

It forms the posterior wall of the femoral sheath. This fascia covers the iliacus muscle.

Inferiorly, the femoral sheath merges with the connective tissue around the femoral vessels.

Function

The femoral sheath allows the femoral vessels to glide freely beneath the inguinal ligament during the movements of the hip joint.

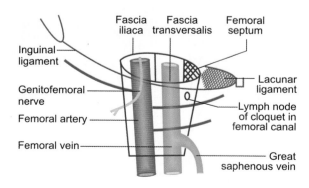

Fig. 20.5: Femoral sheath (right side)

Subdivisions

The sheath is divided into three compartments by two septa.

The Lateral Compartment

It contains the femoral artery and the femoral branch of genitofemoral nerve.

The Intermediate Compartment

It contains the femoral vein.

The Medial Compartment or the Femoral Canal

The femoral canal is conical in shape. The upper, wider end of the femoral canal is called the *base* (about 1.5 cm wide) or the *femoral ring.* The femoral ring is closed by a condensation of extraperitoneal connective tissue called the *femoral septum.*

Contents

The femoral canal contains a lymph node (of Cloquet), lymphatics and loose areolar connective tissue.

Femoral Hernia

The femoral canal is an area of potential weakness in the abdominal wall through which abdominal contents may bulge downward.

The femoral hernia is more common in females. The femoral ring is wider, because of the greater breadth of female pelvis and narrower diameter of femoral vessels.

Note: The femoral nerve is *not* a content of the femoral sheath, because it passes entirely beneath the fascia iliaca.

■ THE INGUINAL LIGAMENT (POUPART'S LIGAMENT) (FIG. 20.6)

It is formed by the lower border of the external oblique aponeurosis, which is thickened and folded on itself.

Extent

It extends from the anterior superior iliac spine to the pubic tubercle.

Section C

The Organ Systems

Attachments

1. The deep fascia of the thigh (fascia lata) is attached to its lower border. Traction of this fascia keeps the ligament convex downwards.
2. The internal oblique muscle of the abdomen takes origin from its lateral 2/3.
3. The transversus abdominis takes origin from its lateral 1/3.

Relations

1. The upper grooved surface of the medial half of the inguinal ligament forms the floor of the inguinal canal. It lodges the spermatic cord in males and the round ligament in females.
2. The femoral vessels, enclosed in the femoral sheath, lie beneath the inguinal ligament.

Extensions

Lacunar Ligament

Lacunar ligament is a triangular extension from its medial end, which is attached to the pecten pubis.

The Reflected Part

The reflected part of inguinal ligament or (ligamentum colles) consists of fibers, which are reflected from the pubic crest and pubic tubercle to the external oblique aponeurosis of the opposite side.

The Pectineal Ligament (Pectinate Ligament)

The pectineal ligament (of Astley Cooper) is a thick ridge which extends laterally along the pectineal line of pubis (It is used as an anchoring point for sutures while repairing the posterior wall of inguinal canal in hernias).

Middle Point of Inguinal Ligament

Middle point of inguinal ligament is midway between anterior superior iliac spine and pubic tubercle.

Mid Inguinal Point

Mid inguinal point is situated midway between the anterior superior iliac spine and symphysis pubis.

■ INGUINAL CANAL

The inguinal canal (Fig. 20.7) is an oblique intermuscular passage in the lower part of the anterior abdominal wall, situated just above the medial half of the inguinal ligament.
Length: About 4 cm.

Extent

It extends from the deep inguinal ring to the superficial inguinal ring. It is directed downward, forward and medially.

The Deep Inguinal Ring

The deep inguinal ring is an oval opening in the fascia transversalis, situated just above the midinguinal point and immediately lateral to the inferior epigastric artery.

The Superficial Inguinal Ring

The superficial inguinal ring is a triangular gap in the external oblique aponeurosis (The base of the triangle is formed by the pubic crest).

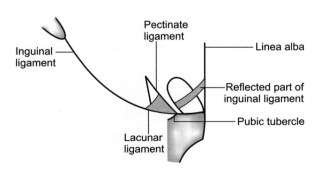

Fig. 20.6: Inguinal ligament

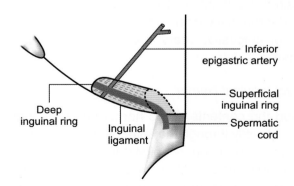

Fig. 20.7: Inguinal canal

Section C

The Organ Systems

Boundaries of Inguinal Canal

1. *Anterior wall:*   Skin
                      Superficial fascia
                      External oblique aponeurosis
                      Fibers of internal oblique
                      (lateral 1/3)
2. *Posterior wall:* Fascia transversalis
                      Conjoint tendon (medial 2/3)
                      Reflected part of inguinal ligament
3. *Roof:*   Formed by arched fibers of internal oblique and transversus abdominis
4. *Floor:*   Formed by the grooved upper surface of inguinal ligament

Inguinal canal is wider in males; it is a natural passage for the descent of testis.

Structures Passing Through Inguinal Canal

1. The spermatic cord, in males
2. The round ligament, in females
3. The ilioinguinal nerve.

■ INGUINAL HERNIA

Abnormal protrusion of abdominal contents, such as greater omentum or a loop of intestine, into the inguinal canal is known as inguinal hernia.

Types of Hernia

1. Indirect inguinal hernia
2. Direct inguinal hernia

Indirect Inguinal Hernia

When the contents of the hernia enters the inguinal canal through the deep inguinal ring and lateral to the inferior epigastric artery, it is said to be an indirect hernia.

Direct Inguinal Hernia

When the contents of the hernia enter the inguinal canal directly through the anterior abdominal wall and medial to the inferior- epigastric artery, it is said to be a direct hernia. Direct inguinal hernia occurs in old age due to the weakness of the abdominal muscles.

Hesselbach's triangle A direct inguinal hernia passes through the Hesselbach's triangle (Fig. 20.8).
   Boundaries of Hesselbach's triangle are:
   1. Lower 5 cm of lateral border of rectus abdominis muscle (medially)
   2. The inferior epigastric artery (laterally)
   3. Medial half of the inguinal ligament (inferiorly).

■ THE POPLITEAL FOSSA (FIG. 20.9)

It is a diamond-shaped space behind the knee.

Boundaries

- Above and medially—semitendinosus and semimembranosus
- Above and laterally—tendon of biceps femoris
- Below and medially—medial head of gastrocnemius
- Below and laterally—lateral head of gastrocnemius and plantaris
- *Floor* is formed by popliteal surface of femur, posterior part of upper end of tibia and popliteus muscle (with its fascia)
- *Roof* is formed by popliteal fascia (deep fascia).

Contents

1. Popliteal artery and vein
2. Tibial nerve
3. Common peroneal nerve
4. Termination of small saphenous vein
5. Popliteal lymph nodes
6. Pad of fat.

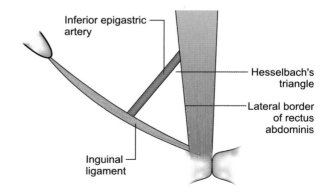

Fig. 20.8: Hesselbach's triangle

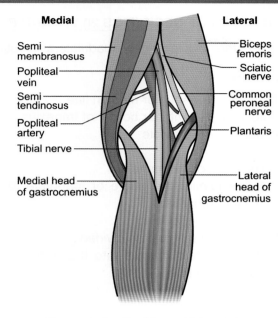

Fig. 20.9: Popliteal fossa (right side). Boundaries and contents

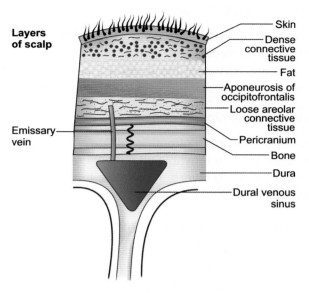

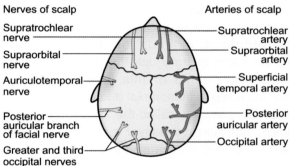

Fig. 20.10: Scalp

Popliteal Artery

Popliteal artery is the continuation of the femoral artery. It reaches the popliteal fossa through the 5th osseo-aponeurotic opening in the adductor magnus. It divides at the lower border of popliteus muscle into anterior and posterior tibial arteries. It is about 20 cm long.

Branches In the popliteal fossa, the popliteal artery gives muscular branches to the adjacent muscles and 5 articular branches to the knee joint (superior medial and lateral genicular, inferior medial and lateral genicular and middle genicular).

Aneurysm of the popliteal artery is common.

Popliteal vein continues above as the femoral vein.

The tibital and common peroneal nerves are the terminal branches of the sciatic nerve (Refer "Nervous system").

▮ THE SCALP (FIG. 20.10)

The soft tissues covering the cranial vault form the "Scalp".

Extent

• Anteriorly—supraorbital margins

• Posteriorly–External occipital protuberance and superior nuchal lines
• Laterally–Superior temporal lines.

Structure

Scalp has 5 layers:

S—Skin
C—Connective tissue (dense)
A—Aponeurosis of occipitofrontalis
L—Loose areolar connective tissue
P—Pericranium

Skin

Skin is thick and hairy, with plenty of sweat and sebaceous glands.

Section C

The Organ Systems

Dense Connective Tissue

Dense connective tissue layer is made up of thick bundles of collagen and elastic fibers. The blood vessels and nerves of the scalp are situated in this layer.

Aponeurosis of Occipitofrontalis

Aponeurosis of occipitofrontalis (Galea aponeurotica). This layer is freely movable on the pericranium along with the overlying layers. It receives the insertion of occipitofrontalis muscle (Refer Muscular System).

Loose Areolar Connective Tissue

This layer is made of fat and loose connective tissue. It extends anteriorly into the eyelids, because the frontalis muscle has no bony attachments. Hence, a collection of fluid (blood) is likely to spread into the upper eyelids (black eye).

Posteriorly it extends up to the superior nuchal lines and laterally upto the superior temporal line. Emissary veins of this layer connect the veins of the scalp to intracranial dural venous sinuses. Therefore, this layer is called the "dangerous layer of the scalp". Infection in this layer is likely to spread into the venous sinuses and cause thrombosis.

Pericranium

Pericranium or periosteum of the cranial bones is loosely attached to the surface of the bones, but it is firmly adherent to sutures where the sutural ligaments bind the pericranium to the endocranium.

Blood Supply of Scalp

See Figure 20.10.

Nerve Supply

Motor branches (to occipitofrontalis) from facial nerve; sensory fibers are branches of trigeminal nerve and cervical plexus.

Applied Anatomy

1. Sebaceous cysts are common in the scalp
2. Wounds of the scalp bleed profusely because the vessels cannot retract due to the abundance of fibrous tissue. Bleeding can be controlled by compression of the scalp against the skull bones.

■ TRIANGLES OF NECK (FIGS 20.11A AND B)

The side of the neck is roughly quadrilateral in outline.

Boundaries

- Anteriorly — Anterior median line.
- Posteriorly — Anterior border of trapezius
- Inferiorly — Clavicle
- Superiorly
  1. The base of the mandible
  2. A line joining the angle of mandible to the mastoid process
  3. Superior nuchal line.

This quadrilateral space is divided obliquely by the sternocleidomastoid muscle into the *anterior and posterior triangles* (Fig. 20.11A).

Posterior Triangle

Boundaries and Contents

- Anteriorly : Posterior border of sterno-cleidomastoid

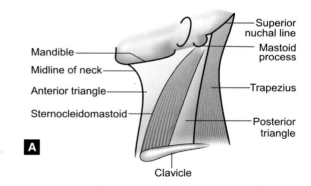

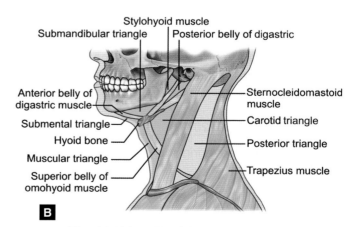

Figs 20.11A and B: (A) Triangles of neck (B) Subdivisions of anterior triangle

- Posteriorly : Anterior border of trapezius.
- Base : Middle one-third of clavicle

Apex

Lies on the superior nuchal line, where the sterno-cleidomastoid and the trapezius meet.

Contents

Nerves
a. Upper part of brachial plexus (Roots and trunks)
b. Spinal accessory nerve
c. Cutaneous branches of the cervical plexus (e.g. supraclavicular nerves).

Vessels
a. Subclavian artery and its branches (Transverse cervical and suprascapular)
b. Subclavian vein
c. External jugular vein
d. Occipital artery (near the apex).

Lymph nodes—Supraclavicular nodes.

The Anterior Triangle

Boundaries

- Anteriorly—by the median plane
- Posteriorly—sternocleidomastoid muscle
- Superiorly—base of the mandible and a line joining the angle of mandible to the mastoid process.

Subdivisions of Anterior Triangle (Fig. 20.11B)

The anterior triangle is subdivided by the digastric muscle and superior belly of omohyoid into 4 triangles.
a. Submental
b. Digastric
c. Carotid
d. Muscular triangles.

Submental Triangle

Boundaries On each side, the anterior belly of corresponding digastric muscle.
Base—Hyoid bone
Apex—Lies at the chin

Contents
1. Submental lymph nodes

2. Submental veins join to form the anterior jugular veins.

Digastric Triangle or Submandibular Triangle

Boundaries
- Anteroinferiorly—Anterior belly of digastric
- Posteroinferiorly—Posterior belly of digastric and stylohyoid.

Base: Base of the mandible and a line joining the angle of mandible to the mastoid process.

Contents
- 2 arteries—facial artery and its submental branch
- 2 veins– the corresponding veins of the arteries
- 2 nerves—hypoglossal nerve and nerve to mylo-hyoid muscle
- 2 glands—submandibular salivary gland and sub-mandibular lymph node.

The superficial part of the submandibular gland almost fill this triangle—that is why it is called the submandibular triangle.

Carotid Triangle

Boundaries
- Superiorly: Posterior belly of digastric and stylo-hyoid
- Anteroinferiorly: Superior belly of omohyoid
- Posteriorly: Anterior border of sternocleidomastoid

Contents
a. *Arteries:* Common carotid artery
Internal carotid artery
External carotid artery with its branches (superior thyroid, lingual, facial, ascending pharyngeal and occipital)
b. *Veins:* Internal jugular vein and its tributaries
c. *Nerves:* Vagus nerve and its superior laryngeal branch
Accessory nerve
Hypoglossal nerve
The sympathetic chain

Note: The internal jugular vein, the carotid arteries and the vagus nerve are enclosed in the carotid sheath (external carotid A is not a content).

Lymph nodes: The deep cervical lymph nodes are situated along the internal jugular vein.

Section C

The Organ Systems

Muscular Triangle

Boundaries

- *Anteriorly*—anterior median line of neck (from hyoid to sternum)
- *Posterosuperiorly:* superior belly of omohyoid
- *Posteroinferiorly:* anterior border of sternocleido-mastoid muscle.

Contents

1. *Infrahyoid muscles* (Ribbon muscles or strap muscles) Sternohyoid, Sternothyroid, Thyrohyoid and Omohyoid.

2. *Deeper structure* Thyroid gland.

MODEL QUESTION PAPERS
1st Year BSc (Nursing)
Degree Examination
Part I: Biological Science
Paper I: Anatomy Including Histology

MODEL QUESTION PAPER—I

Time : 2 Hours   *Maximum : 50 Marks*

Read the questions carefully. Your answers shall be specific to the questions asked.
Draw diagrams wherever necessary.
Answer section A and section B separately.

SECTION A

I. Name the parts of the digestive tract. Describe the gross anatomy of stomach. (8 marks)

II. Multiple choice questions (on separate sheet). (5 marks)

III. Write short notes on *any four*.
   a. Pituitary gland
   b. Mandible
   c. Thoracic duct
   d. Femoral sheath
   e. Mammary gland
   f. Median nerve   (4 × 3 = 12 marks)

SECTION B

IV. Classify epithelium. Describe the compound epithelium. (8 marks)

V. Write briefly on the extrinsic muscles of the eyeball. (5 marks)

VI. Write short notes on *any four*.
   a. Adductor canal
   b. Right coronary artery
   c. Female urethra
   d. Prostate gland
   e. Radius
   f. Talus
   g. Small saphenous vein
   h. Optic nerve   (4 × 2 = 8 marks)

VII. Match the following:

1. Kupffer cells	a.	Facial nerve
2. Lateral pterygoid	b.	Parathyroid
3. Pterion	c.	Stomach
4. Trapezius	d.	Vertebral artery
5. Oxyphil cells	e.	Spinal accessory nerve
	f.	Liver
6. Buccinator	g.	Elevation of mandible
7. Serratus anterior	h.	Mandibular nerve
8. Foramen transversarium	i.	Pancreas
	j.	Long thoracic nerve
	k.	Middle meningeal artery

(8 × 1/2 = 4 marks)

VII. Ans: 1-f, 2-h, 3-k, 4-e, 5-b, 6-a, 7-j, 8-d

SECTION A QUESTION NO. II
MULTIPLE CHOICE QUESTIONS

Encircle the alphabet corresponding to the most appropriate response: (10 × 1/2 = 5 marks)

1. "Wrist drop" results from injury to the:
   a. Axillary nerve
   b. Median nerve
   c. Radial nerve
   d. Ulnar nerve
2. Epithelium lining the vocal cord is:
   a. Ciliated columnar
   b. Columnar
   c. Stratified squamous non-keratinizing
   d. Stratified squamous keratinizing
3. The 'Pouch of Douglas' is situated between the:
   a. Liver and right kidney
   b. Rectum and uterus
   c. Rectum and urinary bladder
   d. Uterus and urinary bladder
4. The structure passing through a large foramen in the diaphragm at the level of T8 is:
   a. Aorta
   b. Inferior vena cava
   c. Esophagus
   d. Thoracic duct
5. Peyer's patches are seen in the:
   a. Duodenum
   b. Ileum
   c. Jejunum
   d. Vermiform appendix
6. The following nerves are branches of the brachial plexus EXCEPT:
   a. Axillary nerve
   b. Long thoracic nerve
   c. Phrenic nerve
   d. Ulnar nerve
7. The structure which passes deep to the flexor retinaculum of hand is:
   a. Median nerve
   b. Palmaris longus tendon
   c. Ulnar artery
   d. Ulnar nerve
8. The following structures are retroperitoneal EXCEPT:
   a. First 2.5 cm of duodenum
   b. Second part of duodenum
   c. Kidneys
   d. Pancreas
9. Normal position of uterus in the pelvic cavity is:
   a. Anteverted and retroflexed
   b. Retroverted and retroflexed
   c. Anteverted and anteflexed
   d. Retroverted and anteflexed.
10. Following are the contents of the posterior mediastinum EXCEPT:
    a. Azygos vein
    b. Thoracic aorta
    c. Trachea
    d. Esophagus

II. Ans : 1-c, 2-c, 3-b, 4-b, 5-b, 6-c, 7-a, 8-a, 9-c, 10-c

MODEL QUESTION PAPER—II

Time : 2 Hours   *Maximum: 50 Marks*

SECTION A

I. Name the parts of the female reproductive system. Describe the gross anatomy of uterus.   (8 marks)

II. Multiple choice questions (on separate sheet).   (5 marks)

III. Write short notes on *any four:*
   a. Sternum
   b. External carotid artery
   c. Cavernous sinus
   d. Anal canal
   e. Interior of larynx
   f. Left ventricle of heart   (4 × 3 = 12 marks)

SECTION B

IV. Describe the knee joint under the following headings:
   a. Type and articular surfaces
   b. Ligaments
   c. Movements and muscles producing them
   d. Applied anatomy   (8 marks)
V. Write briefly on liver.   (5 marks)
VI. Write short notes on *any four:*
   a. Thymus
   b. Cubital fossa
   c. Orbicularis oculi muscle
   d. Carpal bones
   e. Cornea
   f. Cerebellum
   g. Microscopic structure of suprarenal gland
   h. Ulnar nerve   (4 × 2 = 8 marks)
VII. Match the following:

1.	Superior rectus muslce	a.	Nasopharynx
2.	Interstitial lamellae	b.	Facial nerve
3.	Ciliated columnar epithelium	c.	Projection fibers
4.	Corpus callosum	d.	Pons
5.	Stapedius	e.	Oculomotor nerve
6.	Substantia nigra	f.	Midbrain
7.	Stratified squamous epithelium	g.	Gallbladder
8.	Hartmann's pouch	h.	Compact bone
		i.	Abducent nerve
		j.	Commissural fibers
		k.	Tongue

          (8 × 1/2 = 4 marks)

VII. Ans: 1-e, 2-h, 3-a, 4-j, 5-b, 6-f, 7-k, 8-g

SECTION A QUESTION NO. II
MULTIPLE CHOICE QUESTIONS

(10 × 1/2 = 5 marks)

1. The first carpometacarpal joint is a:
   a. Condyloid joint
   b. Hinge joint
   c. Pivot joint
   d. Saddle joint
2. The following structures lie in the free border of lesser omentum EXCEPT:
   a. Bile duct
   b. Hepatic artery
   c. Ligamentum teres
   d. Portal vein
3. The following are branches of the arch of aorta EXCEPT:
   a. Brachiocephalic trunk
   b. Left common carotid artery
   c. Left subclavian artery
   d. Thyrocervical trunk
4. Following viscera are related to the anterior surface of right kidney EXCEPT:
   a. Duodenum
   b. Jejunum
   c. Liver
   d. Stomach
5. Length of male urethra is about:
   a. 10 cm
   b. 15 cm
   c. 20 cm
   d. 25 cm
6. The muscle which unlocks the knee joint is:
   a. Biceps femoris
   b. Popliteus
   c. Quadriceps femoris
   d. Semimembranosus
7. The following structures form the "stomach bed" EXCEPT:
   a. Pancreas
   b. Right kidney
   c. Splenic artery
   d. Transverse colon
8. The carotid triangle is bounded by the following muscles EXCEPT:
   a. Anterior belly of digastric
   b. Posterior belly of digastric
   c. Sternocleidomastoid
   d. Superior belly of omohyoid
9. The portal vein:
   a. Is formed by the union of superior mesenteric vein and splenic vein
   b. Is formed anterior to the neck of the pancreas
   c. Drains the stomach and small intestine only
   d. Ends in the inferior vena cava
10. The following arteries are branches of the internal carotid artery EXCEPT:
   a. Anterior cerebral
   b. Middle cerebral
   c. Ophthalmic
   d. Posterior cerebral

II. Ans: 1-d, 2-c, 3-d, 4-d, 5-c, 6-b, 7-b, 8-a, 9-a, 10-d

Papers

Anatomy Including Histology

MODEL QUESTION PAPER—III

Time: 2 Hours   *Maximum: 50 Marks*

SECTION A

I. Name the major parts of brain. Describe the lobes, sulci, gyri, blood supply and functional areas of cerebrum. (8 marks)

II. Multiple choice questions. (5 marks)

III. Write short notes on *any four*:
   a. Duodenum
   b. Testis
   c. Portal vein
   d. Right atrium
   e. Male urethra
   f. Humerus (4 × 3 = 12 marks)

SECTION B

IV. Classify the joints. Name the synovial joints, giving one example each. Describe the ankle joint. (8 marks)

V. Write briefly on left lung. (5 marks)

VI. Write short notes on *any four*:
   a. Brachial artery
   b. Fallopian tube
   c. Retina
   d. Suprarenal gland
   e. Microscopic structure of tonsil
   f. Taste bud
   g. Tympanic membrane
   h. Biceps brachii (4 × 2 = 8 marks)

VII. Match the following:

1. Transitional epithelium	a. Celiac trunk
2. Haversian canal	b. Thyroid
3. Splenic artery	c. Thymus
4. Vermiform appendix	d. Esophagus
5. Hassal's corpuscles	e. Carpal tunnel syndrome
6. Median nerve	f. Musician's nerve
7. Epiglottis	g. Tonsil
8. Parafollicular cells	h. Compact bone
	i. Urinary bladder
	j. Elastic cartilage
	k. McBurney's point

(8 × 1/2 = 4 marks)

VII. Ans: 1-i, 2-h, 3-a, 4-k, 5-c, 6-e, 7-j, 8-b

SECTION A QUESTION NO. II
MULTIPLE CHOICE QUESTIONS

(10 × 1/2 = 5 marks)

1. Musculocutaneous nerve supplies the following muscles EXCEPT:
   a. Biceps brachii
   b. Brachialis
   c. Brachioradialis
   d. Coracobrachialis

2. The cerebellum has the following nuclei EXCEPT:
   a. Amygdaloid
   b. Dentate
   c. Emboliform
   d. Globose

3. The following muscles of the eyeball are supplied by the oculomotor nerve EXCEPT:
   a. Inferior oblique
   b. Medial rectus
   c. Superior oblique
   d. Superior rectus

4. The nerve likely to get injured while ligating the inferior thyroid artery during thyroidectomy is:
   a. External laryngeal
   b. Internal laryngeal
   c. Recurrent laryngeal
   d. Superior laryngeal

5. The following paranasal sinuses open into the middle meatus of nose EXCEPT:
   a. Anterior ethmoid
   b. Frontal
   c. Maxillary
   d. Posterior ethmoid

6. Lymphatics from palatine tonsils drain mainly into:
   a. Jugulodigastric nodes
   b. Retropharyngeal nodes
   c. Submandibular nodes
   d. Submental nodes

7. In the adult, lower limit of subarachnoid space coincides with:
   a. First lumbar vertebra
   b. Second piece of sacrum
   c. Fifth lumbar vertebra
   d. Coccyx

PART
1

8. Corpus callosum contains the following type of fibers:
   a. Association
   b. Commissural
   c. Projection
   d. Proprioceptive
9. In the fallopian tube, fertilization takes place at the:
   a. Ampulla
   b. Infundibulum
   c. Intramural part
   d. Isthmus

10. The following statements are true about esophagus EXCEPT:
   a. It begins at the level of 6th cervical vertebra
   b. It leaves the thorax at the level of 12th thoracic vertebra
   c. The venous drainage is into the azygos and portal systems
   d. It is lined by stratified squamous non-keratinized epithelium

II. Ans: 1-c, 2-a, 3-c, 4-c, 5-d, 6-a, 7-b, 8-b, 9-a, 10-b

MODEL QUESTION PAPER—IV

Time : 2 Hours   *Maximum: 50 Marks*

SECTION A

I. Describe the structure of the eyeball, with the help of a neat, labelled diagram.   (8 marks)
II. Multiple choice questions.   (5 marks)
III. Write short notes on *any four*.
   a. Vermiform appendix
   b. Midbrain
   c. Thyroid gland
   d. Scapula
   e. Hamstring muscles
   f. Ovary   (4 × 3 = 12 marks)

SECTION B

IV. Name the parts of the respiratory system. Describe the gross anatomy of lungs. (8 marks)
V. Write briefly on the muscles of mastication.   (5 marks)
VI. Write short notes on *any four*.
   a. Nephron
   b. Arch of aorta
   c. Submandibular salivary gland
   d. Median cubital vein
   e. Femoral triangle
   f. Flexor retinaculum of hand

g. Microscopic structure of compact bone
h. Atlas   (4 × 2 = 8 marks)
VII. Match the following:
   1. Pyramid   a. Cerebellum
   2. Hyaline cartilage   b. Abducent nerve
   3. Superior oblique   c. Deep peroneal nerve
   4. Splenium   d. Jejunum
   5. Foot drop   e. Epiglottis
   6. Vermis   f. Tibial nerve
   7. Villi   g. Thymus
   8. Lymphoid tissue   h. Trachea
                       i. Medulla oblongata
                       j. Trochlear nerve
                       k. Corpus callosum
                       (8 × 1/2 = 4 marks)

VII. Ans: 1-i, 2-h, 3-j, 4-k, 5-c, 6-a, 7-d, 8-g

SECTION A QUESTION NO. II
MULTIPLE CHOICE QUESTIONS

1. The nerve related to the surgical neck of humerus is:
   a. Axillary
   b. Median
   c. Radial
   d. Ulnar

Papers

Anatomy Including Histology

2. The following muscles are supplied by the median nerve EXCEPT:
   a. Flexor carpi ulnaris
   b. Flexor digitorum superficialis
   c. First and second lumbricals
   d. Pronator teres

3. The following structures drain into superficial inguinal lymph nodes EXCEPT:
   a. Anal canal, below pectinate line
   b. Labia majora
   c. Glans penis
   d. Scrotum

4. Brainstem consists of:
   a. Pons, medulla, cerebellum
   b. Midbrain, pons, cerebellum
   c. Midbrain, pons, medulla
   d. Midbrain, medulla, cerebellum

5. Which wall of the middle ear has the fenestra cochleae and fenestra vestibuli?
   a. Anterior
   b. Lateral
   c. Medial
   d. Posterior

6. Stylopharyngeus is supplied by:
   a. Facial nerve
   b. Glossopharyngeal nerve
   c. Hypoglossal nerve
   d. Vagus nerve

7. Pseudounipolar neurons are seen in:
   a. Dorsal root ganglion
   b. Olfactory epithelium
   c. Retina
   d. Sympathetic ganglion

8. The cells which produce myelin sheath in the CNS are:
   a. Astrocytes
   b. Microglia
   c. Oligodendrocytes
   d. Schwann cells

9. Which one of the following is a white fibro-cartilage ?
   a. Epiglottis   b. Intervertebral disc
   c. Septal cartilage   d. Thyroid cartilage

10. In the cerebral cortex visual area corresponds to area:
    a. 4   b. 17
    c. 41   d. 45

II. Ans: 1-a, 2-a, 3-c, 4-c, 5-c, 6-b, 7-a, 8-c, 9-b,10-b

Papers

Anatomy Including Histology

MODEL QUESTION PAPER—V

Time : 2 Hours   *Maximum: 50 Marks*

SECTION A

I. Name the primary tissues of the body. Classify connective tissue. Describe the cells of connective tissue.   (8 marks)

II. Multiple choice questions.   (5 marks)

III. Write short notes on any *four.*
   a. Lateral ventricle of brain
   b. Paranasal air sinuses
   c. Palatine tonsil
   d. Pancreas
   e. Temporal bone
   f. Great saphenous vein   (4 × 3 = 12 marks)

SECTION B

IV. Name the parts of the urinary system. Describe the gross anatomy of kidneys.   (8 marks)

V. Write briefly on right atrium.   (5 marks)

VI. Write short notes on *any four.*
   a. Neuroglia
   b. Rectus abdominis
   c. Radial artery
   d. Cornea
   e. Vas deferens
   f. Bell's palsy
   g. Inguinal ligament
   h. Popliteal fossa   (4 × 2 = 8 marks)

VII. Match the following: (8 × 1/2 = 4 marks)

1. Bipolar neurons   a. Trachea
2. Lentiform nucleus   b. Chorda tympani
3. Tongue   c. Urinary bladder
4. Lateral rectus   d. Elastic cartilage
5. Pseudostratified   e. Biceps brachii
   ciliated columnar
   epithelium
6. Intervertebral disc   f. Spinal ganglion
7. Radial tuberosity   g. Retina
8. Pinna   h. Triceps
      i. Abducent nerve
      j. Fibrocartilage
      k. Basal ganglia

VII. Ans: 1-g, 2-k, 3-b, 4-i, 5-a, 6-j, 7-e, 8-d

SECTION A QUESTION NO. II
MULTIPLE CHOICE QUESTIONS

(10 × 1/2 = 5 marks)

1. The nerve related to the neck of fibula is:
   a. Common peroneal
   b. Deep peroneal
   c. Peroneal communicating
   d. Superficial peroneal
2. Ligamentum arteriosum is the remnant of:
   a. Ductus arteriosus
   b. Ductus venosus
   c. Obliterated left umbilical vein
   d. Obliterated umbilical artery
3. The following muscles are attached to the coracoid process EXCEPT:
   a. Coracobrachialis
   b. Long head of biceps
   c. Pectoralis minor
   d. Short head of biceps
4. In the nasal cavity, the nasolacrimal duct opens into:
   a. Inferior meatus
   b. Middle meatus
   c. Sphenoethmoidal recess
   d. Superior meatus
5. All the following are branches of the first part of subclavian artery EXCEPT:
   a. Costocervical trunk
   b. Internal thoracic
   c. Thyrocervical trunk
   d. Vertebral
6. Following are the contents of carotid sheath EXCEPT:
   a. Common carotid artery
   b. External carotid artery
   c. Internal carotid artery
   d. Vagus nerve
7. Uveal tract consists of all the following EXCEPT:
   a. Choroid
   b. Ciliary body
   c. Iris
   d. Sclera
8. The nerve injured in "wrist drop" is:
   a. Axillary nerve
   b. Median nerve
   c. Radial nerve
   d. Ulnar nerve
9. The muscle which depresses the mandible is:
   a. Lateral pterygoid
   b. Masseter
   c. Medial pterygoid
   d. Temporalis
10. The only abductor of vocal cord is:
    a. Oblique arytenoid
    b. Posterior cricoarytenoid
    c. Thyroarytenoid
    d. Transverse arytenoid

II. Ans: 1-a, 2-a, 3-b, 4-a, 5-a, 6-b, 7-d, 8-c, 9-a, 10-b

Papers

Anatomy Including Histology

KERALA UNIVERSITY

MODEL QUESTION PAPER—I

Time : 2 Hours   *Maximum : 50 Marks*

SECTION A

1. Describe the structure of eyeball.   (10 marks)
2. Write short notes on:
   a. Knee joint
   b. Pancreas   (2 × 5 = 10 marks)
3. *Fill in the blanks:*
   a. Portal vein is formed by the union of ———— and ———— veins.
   b. Superior oblique muscle of the eyeball is supplied by ———— nerve.
   c. The biceps brachii muscle is inserted at the ————————— .
   d. Kupffer cells are seen in the ———— .
   e. The trachea is lined by ———— epithelium.   (5 × 1 = 5 marks)

SECTION B

4. Briefly describe the following aspects of kidneys:
   a. Position
   b. Relations
   c. Blood supply
   d. Microscopic structure
   e. Applied anatomy   (5 × 2 = 10 marks)
5. Write short notes on:
   a. Connective tissue cells
   b. Vas deferens
   c. Midbrain
   d. Clavicle
   e. Brachial artery   (5 × 2 = 10 marks)
6. Match the following:
   a. Coronary sinus   i. Bell's palsy
   b. Ampulla of Vater   ii. Intervertebral disc
   c. Corpus callosum   iii. Duodenum
   d. Facial nerve   iv. Commissural fibers
   e. Fibrocartilage   v. Right atrium
                          (5 × 1 = 5 marks)

   6. Ans: a-v, b-iii, c-iv, d-i, e-ii

MODEL QUESTION PAPER—II

Time : 2 Hours   *Maximum : 50 Marks*

SECTION A

1. Describe the gross anatomy of uterus:
                          (10 marks)
2. Write short notes on:
   a. External carotid artery
   b. Male urethra   (2 × 5 = 10 marks)
3. Fill in the blanks:
   a. The parietal cells of stomach secrete — and—.
   b. The nasolacrimal duct opens into the —— of nose.
   c. The lateral rectus muscle of the eyeball is supplied by the ———— nerve.
   d. ———— passes through a major foramen of diaphragm at the level of T_{10}
   e. ———— is the largest branch of celiac trunk
                          (5 × 1 = 5 marks)

SECTION B

4. Briefly describe the following features of the liver:
   a. Position and dimensions
   b. Lobes
   c. Blood supply
   d. Microscopic structure
   e. Applied anatomy   (5 × 2 = 10 marks)

5. Write short notes on:
   a. Thoracic duct
   b. Suprarenal gland
   c. Biceps brachii muscle
   d. Great saphenous vein
   e. Fetal skull   (5 × 2 = 10 marks)

6. Match the following:
   a. Trigone   i. Wrist drop
   b. Radial nerve   ii. Calcaneus
   c. Internal carotid iii. Patella
      artery
   d. Achille's tendon iv. Cavernous sinus
   e. Sesamoid bone   v. Urinary bladder
                          (5 × 1 = 5 marks)

   6. Ans.   a-v, b-i, c-iv, d-ii, e-iii

MODEL QUESTION PAPER—III

Time : 2 Hours   *Maximum : 50 Marks*

SECTION A

1. Describe structure of middle ear.   (10 marks)
2. Write short notes on:
   a. Cartilage
   b. Duodenum   (2 × 5 = 10 marks)
3. Fill in the blanks:
   a. The urinary bladder is lined by _____ epithelium.
   b. The brachiocephalic trunk divides into _____ and _____.
   c. The gluteus maximus is supplied by _____ nerve.
   d. The azygos vein drains into _____.
   e. The sciatic nerve divides into _____ and _____.   (5 × 1 = 5 marks)

SECTION B

4. Briefly describe the following aspects of the shoulder joint:

   a. Articular surfaces
   b. Ligaments
   c. Rotator cuff
   d. Movements and muscles producing them
   e. Applied anatomy   (5 × 2 = 10 marks)
5. Write short notes on:
   a. Lacrimal apparatus
   b. Talus
   c. Fallopian tube
   d. Small saphenous vein
   e. Palatine tonsil
                          (5 × 2 = 10 marks)
6. Match the following:
   a. Haversian canal   i. Femoral artery
   b. Adductor canal   ii. Mammary gland
   c. Cubital fossa   iii. Lymph channel
   d. Axillary lymph   iv. Brachial artery
      nodes
   e. Cisterna chyli   v. Compact bone
                          (5 × 1 = 5 marks)

   6. Ans. a-v, b-i, c-iv, d-ii, e-iii

MODEL QUESTION PAPER—IV

Time : 2 Hours   *Maximum : 50 Marks*

SECTION A

1. Describe the structure of skin.   (10 marks)
2. Write short notes on:
   a. Thyroid gland
   b. Muscles of mastication   (2 × 5 = 10 marks)

3. Fill in the blanks:
   a. Brunner's glands are seen in the _____.
   b. _____ is the largest lymphoid organ in the body.
   c. The two vertebral arteries join to form the _____.

Papers

Anatomy Including Histology

d. Fibers of the ganglion layer of the retina join to form the _____.
e. The terminal branches of brachial artery are _____ and _____. (5 × 1 = 5 marks)

SECTION B

4. Briefly describe the following aspects of stomach:
   a. Parts
   b. Peritoneal relations
   c. Arterial supply
   d. Histology
   e. Applied anatomy (5 × 2 = 10 marks)

5. Write short notes on:
   a. Humerus
   b. Simple epithelia
   c. Nephron
   d. Median cubital vein
   e. Glossopharyngeal nerve

6. Match the following:
   a. Medial rectus   i. Short head of biceps
   b. Coracoid process   ii. Ulnar nerve
   c. Volkmann's canal   iii. Deep peroneal nerve
   d. Musician's nerve   iv. Compact bone
   e. Foot drop   v. Oculomotor nerve
   (5 × 1 = 5 marks)

6. **Ans.** a-v, b-i, c-iv, d-ii, e-iii

MODEL QUESTION PAPER –V

SECTION A

1. Describe the gross anatomy of the right lung. (10 marks)

2. Write short notes on:
   a. Synovial joints
   b. Coronary arteries (2 × 5 = 10 marks)

3. Fill in the blanks:
   a. Common bile duct is formed by the union of _____ and _____.
   b. The tympanic membrane separates the _____ and _____.
   c. The terminal branches of external carotid artery are_____ and _____.
   d. The intrinsic muscles of larynx are supplied by the _____ and _____ nerves.
   e. The parotid duct opens in the vestibule of oral cavity at the level of _____.
   (5 × 1 = 5 marks)

4. Briefly describe the following aspects of the spinal cord:
   a. Extent and coverings
   b. External features
   c. Internal features
   d. Blood supply
   e. Applied anatomy (5 × 2 = 10 marks)

5. Write short notes on:
   a. Fibula
   b. Oculomotor nerve
   c. Femoral triangle
   d. Testis
   e. Hyaline cartilage (5 × 2 = 10 marks)

6. Match the following:
   a. Oxyphil cells   i. Liver
   b. Kupffer cells   ii. Ileum
   c. Hassall's corpuscles iii. Thyroid
   d. Parafollicular cells iv. Thymus
   e. Peyer's patches   v. Parathyroid
   (5 × 1 = 5 marks)

6. **Ans.** a-v, b-i, c-iv, d-iii, e-ii

CALICUT UNIVERSITY

MODEL QUESTION PAPER—2010

(Draw diagrams wherever necessary)

Time : 2½ Hours   *Maximum : 75 Marks*

I. MULTIPLE CHOICE QUESTIONS

Choose the most correct response:

1. The following is a ball and socket type of joint:
   a. Elbow joint
   b. Shoulder joint
   c. Ankle joint
   d. Knee joint

2. Mammary gland is a modified:
   a. Sweat gland
   b. Sebaceous gland
   c. Endocrine gland
   d. Holocrine gland

3. The following is the largest paranasal air sinus:
   a. Ethmoid
   b. Frontal
   c. Sphenoid
   d. Maxillary

4. Elastic cartilage occurs in ………….
   a. Articular cartilage
   b. Epiglottis
   c. Costal cartilage
   d. Intervertebral disc

5. Coronary artery supplies to the following organ:
   a. Liver
   b. Brain
   c. Heart
   d. Stomach

6. Average length of small intestine is:
   a. 2 meters
   b. 6 meters
   c. 4 meters
   d. 1 meter

7. Growth hormone is secreted by:
   a. Pancreas
   b. Pituitary
   c. Thyroid
   d. Suprarenal

8. Lateral ventricle is the cavity of:
   a. Cerebrum
   b. Cerebellum
   c. Diencephalon
   d. Mid brain

9. Surgical neck of humerus is related to the following nerve:
   a. Median
   b. Radial
   c. Ulnar
   d. Axillary

10. Vemiform appendix opens to:
    a. Cecum
    b. Ascending colon
    c. Sigmoid colon
    d. Ileum   (10 × 1 = 10 marks)

II. MATCH THE FOLLOWING

1. Odontoid process	a. Skin	
2. Epiglottis	b. Brain	
3. Meninges	c. Tongue	
4. Serous salivary gland	d. Heart	
5. Stratified squamous Keratinized epithelium	e. Elastic cartilage	
6. Taste buds	f. Parotid	
7. Coronary sinus	g. Bone	
8. Osteon	h. Scapula	
9. Acromion	i. Lymphoid organ	
10. Spleen	j. Atlas	
	k. Axis	

(10 × 1 = 10 marks)

III. DEFINE THE FOLLOWING

1. Cauda equina
2. Porta hepatis
3. Thyroid gland
4. Pericardium
5. Cavernous sinus   (5 × 2 = 10 marks)

IV. NAME THE FOLLOWING

1. Arteries supplying the uterus
2. Structures passing through foramen ovale
3. Bones forming the wrist joint
4. Structures in the stomach bed
5. Two types of ossification   (5 × 1 = 5 marks)

V. DIFFERENTIATE BETWEEN

1. Elastic cartilage and fibrocartilage
2. Adenohypophysis and neurohypophysis
3. Simple epithelium and pseudo stratified epithelium
4. Right atrium and left atrium
5. Duodenum and ileum   (5 × 2 = 10 marks)

VI. WRITE SHORT ANSWERS ON *ANY THREE*

a. Kidney
b. Tongue
c. Tonsil
d. Liver   (3 × 5 = 15 marks)

VII. ENUMERATE THE PARTS OF RESPIRATORY SYSTEM AND DESCRIBE THE LUNGS UNDER THE FOLLOWING HEADINGS

a. Lobes
b. Coverings
c. Bronchopulmonary segments
d. Applied anatomy   (15 marks)

MODEL QUESTION PAPER—I

[BSc (Hons) Nursing Phase I Professional Examination]

SECTION A

Answer all questions.

1. Draw a labeled diagram of the anterior abdominal wall showing various anatomical planes, lines and quadrants. Name the various organs present in the right iliac fossa. (3 + 2 = 5 marks)

2. Name the exact sites in the human body where the pulsations of the following arteries can be palpated distinctly: (2 marks)
   a. External carotid artery
   b. Brachial artery
   c. Arteria dorsalis pedis
   d. Femoral artery

3. Explain the anatomical basis of the following: (4 marks)
   a. Lumbar puncture is done below the level of L1 vertebra
   b. Upper and outer quadrant of gluteal region is chosen for deep intramuscular injection
   c. For bone marrow biopsy iliac crest is chosen
   d. Per vaginal examination is done after emptying the urinary bladder.

4. Write short notes on (4 marks)
   a. Synovial joint
   b. Down syndrome

5. Fill in the blanks with appropriate words (10 marks)
   a. Cavity between parietal and visceral peritoneum is known as _____.
   b. Hormones secreted by pancreas are _____ and _____.
   c. The superior vena cava is formed by the union of _____ veins.
   d. Taste sensation from the tongue is carried by _____ and _____ nerves.
   e. The cerebrospinal fluid is secreted by _____ of the ventricles of brain.
   f. The apex beat of the heart can be palpated in the _____ intercostal space.
   g. The bronchial muscles are examples of _____ muscles.

MODEL QUESTION PAPER—II

1. Define and classify synovial joints. Draw a labeled diagram of the coronal section of the hip joint showing its various parts.

2. Enumerate the following:
   a. Branches of posterior cord of brachial plexus
   b. Parts of extrahepatic biliary apparatus
   c. Branches of trigeminal nerve
   d. Dural venous sinuses

3. Write the vertebral level of the following structures of the body:
   a. Bifurcation of common carotid artery
   b. Bifurcation of trachea
   c. Termination of spinal cord
   d. Transpyloric plane

4. Draw a labeled diagram of the sagittal section of the female pelvis showing various parts of internal genital organs. Enumerate the supports of uterus.

5. Name the sites where the pulsations of the following blood vessels are felt
   a. Radial artery
   b. Superficial temporal artery

6. Fill in the blanks with appropriate words.
   a. The growth hormone is secreted by _____ cells of the pituitary.
   b) The parasympathetic nerve for the heart is _____ nerve.
   c. The space between the arachnoid and pia mater is called _____.
   d. The brain is supplied by ____ and _____ arteries
   e. The anterior fontanelle in a child is closed by _____ months after birth.
   f. _____ and _____ cells are photoreceptors in the retina.
   g. The portal vein is formed by _____ and _____ veins.

PART 2

PHYSIOLOGY

Introduction to Physiology

INTRODUCTION

Physiology is the branch of science that deals with various functions of living organism and the processes, which regulate them.

In multicellular organisms like man, the cells of the body are bathed in the extracellular fluid, which constitute the internal environment of the body. The volume, composition and temperature of this internal environment must be maintained within narrow limits for the smooth functioning of the cells. The concept of internal environment and its constant nature was first proposed by the French Physiologist Claude Bernard in 19th Century, who named it as 'milieu interieur'. According to him, the constancy of the extracellular fluid is the condition of 'free, independent life'.

Individual cell types in various organs act in concert to support the constancy of the internal 'milieu', and in turn, the internal 'milieu' provides these cells with a culture medium in which they can survive.

In humans and other vertebrates, there are specialized groups of cells for performing different functions, such as, (1) a gastrointestinal system to digest and absorb food; (2) a respiratory system to take up oxygen and eliminate carbon dioxide, (3) a urinary system for removal of waste products, (4) a

cardiovascular system to distribute oxygen, nutrition and products of metabolism, (5) a reproductive system for perpetuation of species, (6) a nervous system and an endocrine system for coordination and integration of other systems.

The mechanisms for maintaining the internal 'milieu' were studied by Walter B Cannon, who gave the term 'Homeostasis' to indicate the constancy of the internal environment. In short, all the physiological mechanisms are aimed at maintaining the homeostasis.

Several feedback control mechanisms maintain the constancy of the internal 'milieu'. Most common regulatory mechanism is the negative feedback mechanism, where a regulated variable is sensed, information is fedback to the controller and the effector acts to oppose the change, e.g. Hyperglycemia

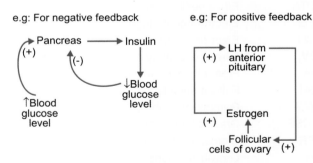

Fig. 21.1: Feedback control mechanisms

Table 21.1: Abbreviations and symbols commonly used in physiology

ABP	Androgen-binding protein	HbO$_2$	Oxyhemoglobin
ACh	Acetylcholine	hCG	Human chorionic gonadotropin
ACTH	Adrenocorticotropic hormone	HCS	Human chorionic somatomammotropin
ADH	Antidiuretic hormone		
ADP	Adenosine diphosphate	HIV	Human immunodeficiency virus
ANP	Atrial natriuretic peptide	5-HT	5-hydroxytryptamine (serotonin)
ANS	Autonomic nervous system	IDDM	Insulin-dependent diabetes mellitus
APUD cells	Amine precursor uptake and decarboxylation cells	IGF	Insulin-like growth factor
		IRDS	Infant respiratory distress syndrome
A-V difference	Arteriovenous concentration difference of any given substance	IU	International unit
		IUD or IUCD	Intrauterine device (contraceptive device)
AV node	Atrioventricular node	JG cells	Juxtaglomerular cells
aVR, aVL, aVF	Augmented unipolar ECG leads	k cal	Kilocalorie
AV valves	Atrioventricular valves	Kf	Glomerular ultrafiltration coefficient
BMI	Body mass index	LH	Luteinizing hormone
BMR	Basal metabolic rate	LH-RH	Luteinizing hormone-releasing hormone (same as GnRH)
cAMP	Cyclic adenosine 3' 5' – monophosphate		
CCK	Cholecystokinin	Mol	Mole; gram-molecular weight
CCK-PZ	Cholecystokinin-pancreozymin	mRNA	Messenger ribonucleic acid
CNS	Central nervous system	NIDDM	Non-insulin-dependent diabetes mellitus
CoA	Coenzyme A	NTS	Nucleus tractus solitarius
C peptide	Connecting peptide	Osm	Osmole
CRH	Corticotropin-releasing hormone	OVLT	Organum vasculosum of the lamina terminalis
CSF	Cerebrospinal fluid		
CVS	Cardiovascular system	PAH	Para-aminohippuric acid
DAG	Diacylglycerol	PBI	Protein-bound iodine
DNA	Deoxyribonucleic acid	PTA	Plasma thromboplastin antecedent
Dopa	Dihydroxyphenylalanine	PTH	Parathyroid hormone
ECG	Electrocardiogram	Rh factor	Rhesus group of red cell agglutinogens
EDTA	Ethylenediaminetetraacetic acid	RNA	Ribonucleic acid
EEG	Electroencephalogram	SA node	Sinoatrial node
EMG	Electromyogram	SCF	Stem cell factor
FEV$_1$	Forced expiratory volume in first second	SGLT	Sodium dependent glucose transporter
FFA	Free fatty acid, which is unesterified	SI units	Units of the system international
FSH	Follicle-stimulating hormone	SIADH	Syndrome of Inappropriate Antidiuretic Hormone
GABA	Gamma-aminobutyric acid		
G CSF	Granulocyte colony-stimulating factor	T3	3,5,3'- triiodothyronine
GFR	Glomerular filtration rate	T4	Thyroxine (Tetraiodothyronine)
GH	Growth hormone	TBG	Thyroxine-binding globulin
GIP	Gastric inhibitory peptide	TBW	Total body water
GLUT	Glucose transporter	TNF	Tumor necrosis factor
Gm	Gram	TSH	Thyroid-stimulating hormone
GM-CSF	Granulocyte- macrophage colony-stimulating factor	VIP	Vasoactive intestinal peptide
		VMA	Vanillylmandelic acid
GnRH	Gonadotropin-releasing hormone	WBC	White blood cell (corpuscle)
Hb	Hemoglobin (Deoxygenated)		

(increased blood glucose level) stimulates release of insulin from pancreas, which in turn causes a decrease in the blood glucose level (Fig. 21.1). Positive feedback mechanisms promote a progressive change in one direction, e.g. During the follicular phase of menstrual cycle, female sex hormone estrogen stimulates the release of luteinizing hormone, which in turn stimulates further estrogen synthesis by the ovaries (Fig. 21.1). This is called LH surge, which finally results in ovulation.

22

The Cell

INTRODUCTION

In order to understand how the body performs different functions, it is essential to know about the organization of the cell and functions of its components.

THE CELL

The smallest functional unit of our body is the cell. It was Robert Hook who first coined the term "cell" in 1665. The study of cells is called "Cytology" and the study of tissues is called "Histology". The size of a cell can vary from a minimum of 6 μm (that of a resting lymphocyte) to a maximum of 80 μm (that of a mature ovum). The chemical composition inside the cell is different from that outside.

STRUCTURE OF A CELL

A eukaryotic cell consists of the following (Fig. 22.1):
1. A cell membrane or plasma membrane, which is impermeable to large molecules like proteins and selectively permeable to small molecules like ions and metabolites.
2. A nucleus, which contains the genetic machinery.
3. Cytoplasm and the organelles.

CELL MEMBRANE OR PLASMA MEMBRANE OR UNIT MEMBRANE

The name "Unit membrane" was coined by Robertson in 1959. He conducted many biochemical studies on RBC membrane and came to the conclusion that the membrane is composed of two principal constituents – lipids and proteins, in the ratio 3:2.

Singer and Nicolson in 1972 proposed the "Fluid-Mosaic Model" of cell membrane, which has been confirmed by many independent researchers and is now generally accepted. According to this concept, cell membranes are fluid structures with a phospholipid bilayer interrupted by proteins. In addition, carbohydrate moieties are also present.

Lipid Bilayer

The phospholipids in the membrane possess hydrophilic heads, which are polar and water soluble, as well as hydrophobic tails, which dissolve readily in organic solvents (Fig. 22.2). The hydrophobic tail portions line up towards the center of the membrane, away from the water. Conversely, the hydrophilic heads orient themselves towards the polar water. Even though the phospholipid bilayer gives the idea of a

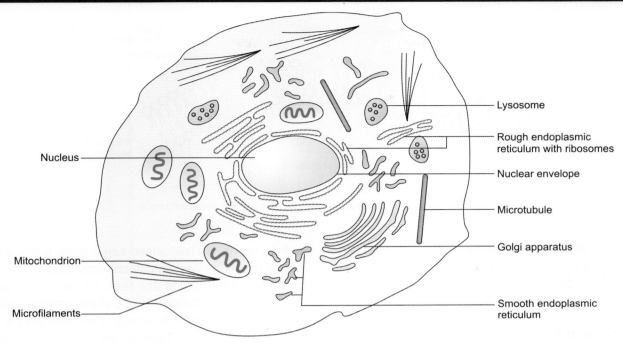

Fig. 22.1: Structure of a cell

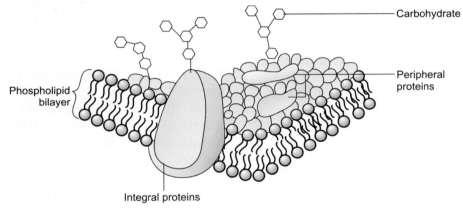

Fig. 22.2: Fluid—mosaic model of cell membrane

Part 2

Physiology

rigid appearance, it is a fluid structure, i.e., these lipid molecules can diffuse freely within the membrane, depending upon the surrounding temperature. At high temperatures, the phospholipid is said to be in the "sol" state and at lower temperatures, it attains the "gel" state.

In addition, there are also cholesterol molecules in the lipid bilayer, which stabilize and regulate the fluidity of the layer.

Membrane Proteins

They belong to two major classes of proteins – Peripheral proteins and Integral Proteins (Fig. 22.3). Peripheral or extrinsic membrane proteins are associated with the surface of the membrane. In contrast, integral proteins or intrinsic membrane proteins are seen embedded in the phospholipid bilayer. Some integral proteins span the entire

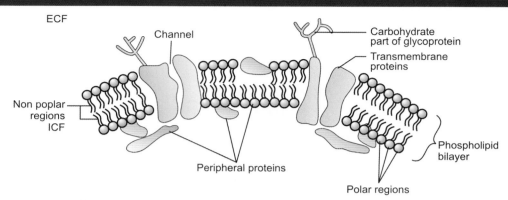

Fig. 22.3: Electron microscopic picture of a human cell membrane

thickness of the membrane once or several times and hence, are known as transmembrane proteins (Fig. 22.3). So those transmembrane proteins can diffuse freely within the surface of the membrane.

Functions of Membrane Proteins

Integral proteins serve as receptors which bind with various hormones. The interaction of a hormone with the extracellular protein receptor produce certain conformational changes within the receptor protein. As a result, the intracellular part of the integral protein becomes enzymatically active and initiate a number of reactions inside the cell.

Integral proteins help in the transport of substances across the cell membrane. They serve as "pores" and "channels" through which water and water-soluble ions can diffuse freely. Some of the integral proteins serve as "carrier proteins" which can facilitate transport of certain molecules across the membrane. Some of them act as "pumps" which can actively transport substances against their concentration gradient, e.g. $Na^+K^+ATPase$ pump, $Ca^{++}ATPase$ pump. "Pumps" are actually enzymes which can hydrolyze ATP for the release of energy. This energy can be utilized by the cell for various purposes like active transport. These membrane bound enzymes are richly seen in intestine.

Integral proteins can act as "cell-adhesion molecules" which are responsible for the particular shape, growth and differentiation of cells, e.g. Integrins, Cadherins, etc. They help to attach the cell with the surrounding extracellular matrix ("cell-matrix adhesion") and also with the neighboring cells (cell-

cell adhesion). In metastatic tumor cells, there is loss of cell-matrix as well as cell-cell adhesion.

Peripheral proteins are seen on the inner part of cell membrane, attached to one of the integral proteins (Fig. 22.3). They attach loosely to the lipid bilayer, but not embedded in it. Some of them act as enzymes, which can control intracellular functions. Some, along with peripheral proteins, form a part of cytoskeleton which give strength and structural integrity to the cell, e.g. Spectrin, Ankyrin, Actin, which are seen on RBC membrane (Fig. 22.4). Of these, most important is spectrin, that gives the biconcave shape and strength to the RBC. In persons with hereditary spherocytosis, there is mutation of genes coding for spectrin. As a result, their RBCs become extremely fragile and spherical in shape. These RBCs are unable to withstand shear stress, which leads to hemolytic anemia.

Membrane Carbohydrates (Glycocalyx)

Membrane carbohydrates are seen either attached to proteins ("Glycoproteins") or to lipids ("Glycolipids") (Fig. 22.3). They have different functions. The glycolipids seen on the RBC membrane form the ABO group substances. They give an overall negative charge on the outer surface of the cell. These negative charges on RBCs help to keep them dispersed in the plasma. Otherwise, they will get adhered to each other arranged like a pile of coins called Rouleaux formation (Refer "Blood"). Some of the membrane carbohydrate molecules act as receptors that combine with certain hormones like insulin.

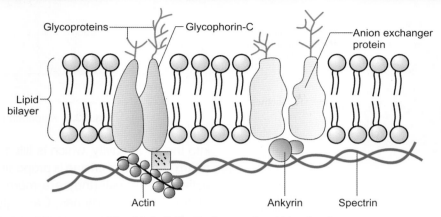

Fig. 22.4: Cell membrane of red blood cell

NUCLEUS (FIG. 22.1)

Nucleus is the structure that contains the machinery necessary to maintain, copy and transcribe DNA. This structure is visible under light microscope. The size of nucleus can vary from 2 to 20 µm in diameter. Usually each eukaryotic cell contains a single nucleus. Exceptions for this are skeletal muscle cells and certain cells of immune system.

Nucleus is surrounded by a double membrane called nuclear envelope, separated by a space. The rough outer membrane is continuous with that of rough endoplasmic reticulum and is studded with ribosomes. The inner membrane faces the nucleoplasm or intranuclear space. There are many circular openings or pores (nuclear pores) through which large molecules like RNA can pass in and out of the nucleus.

The nucleus is packed with chromatin fibres which, when the cells divide, coils and shortens into 4-6 rod shaped chromosomes. Chromosomes carry the "genetic message" responsible for inherited characteristics and are made up of segments of DNA. "Genes" which are the ultimate units of heredity, are segments of DNA molecule. The nuclei of all cells of an individual, except the germ cells of gonads, contain the same set complement of DNA, a quantity called the genome. This is present in the form of a fixed number of chromosomes – 46 in humans (the 'diploid' number, which consists of 44 autosomes and 2 sex chromosomes – either XX or XY. The germ cells of gonads contain only half the diploid number of chromosomes – 23 in humans (the "haploid" number

– 22 X or 22 Y). Many nuclei, especially those in the highly active cells, contain one or more dense structures called nucleoli, which are the sites of ribosomal RNA synthesis and ribosomal assembly.

Cytoplasm and its Organelles

Eukaryotic cells contain an intracellular matrix called cytoplasm which contains many organelles (Fig. 22.1). The main content of cytoplasm is water (70%). Also there are many organic and inorganic molecules and enzymes. The cytoplasmic organelles are described below:

ENDOPLASMIC RETICULUM (FIG. 22. 5)

Surrounding the nucleus, there is a web of membranous tubules, vesicles and flattened sacs (cisternae). This membrane divides the cytoplasm into two – the inner vaculoplasm and the outer cytosol or hyaloplasm. Two types of endoplasmic reticulum are there – granular or rough and agranular or smooth (Fig. 22.5). Granular endoplasmic reticulum is studded with ribosomes and can synthesize proteins. These proteins get accumulated in the cisternae, vesicles or lysosomes. Agranular endoplasmic reticulum lacks ribosomes and is concerned with the synthesis of lipids, steroids, cholesterol and carbohydrates. It acts as a surface for the attachment of many enzyme systems and helps in detoxifying drugs, alcohol, metabolic by-product, etc.

Highly specialized endoplasmic reticulum is present in some cells. In striated muscle cells, where it is called

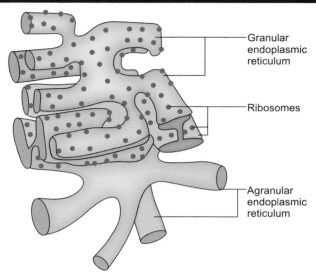

Fig. 22.5: Endoplasmic reticulum

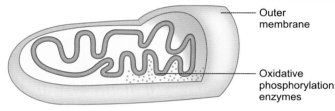

Fig. 22.6: Mitochondrion

sarcoplasmic reticulum (Fig. 25.3), it is involved in the storage and release of Ca^{2+} to initiate muscle contraction.

■ RIBOSOMES (FIG. 22.5)

Ribosomes are granules made up of proteins and RNA (rRNA). Each ribosome consists of a larger unit and a smaller unit (according to the rate of sedimentation during centrifugation) – 60 S and 40 S units. The larger subunit has 3 strands of highly coiled RNA and the smaller unit has one strand. When the cell is inactive, there is no protein synthesis, and the two units of ribosomes remain separate.

The principal function of ribosome is the synthesis of proteins from amino acids. The ribosomes can be seen either attached to the rough endoplasmic reticulum or free in the cytoplasm. The free ribosomes synthesize the structural proteins destined for the cytoplasm and also Hb in RBCs.

■ MITOCHONDRIA (FIG. 22.6)

Mitochondria are membrane bound organelles and are called the power-generating units of the cell. Each mitochondrion consists of two layers of membrane—the outer smooth membrane, which contains many pores, allowing free passage of small molecules, and the inner membrane, which is thrown into folds that increase the surface area. The folds or cristae project

into the inner cavity, which is filled with an amorphous matrix. Amount of cristae is proportional to the metabolic activity of the cell (they are more in cardiac muscle). Matrix contains enzymes, Ca^{2+}, glycogen, RNA, DNA etc. Mitochondria have several unique features among other cell organelles—they can move, change their size and shape and divide within the cells. Also they synthesize 37 of their own constituent proteins.

Functions

1. It is the chief site of TCA cycle, electron transport chain and fatty acid metabolism.
2. Release of energy from ATP and GTP.
3. It concentrates Ca^{2+}

Sperms do not contribute mitochondria to the zygote. Therefore it comes from the ovum and is completely maternal in origin.

Lysosomes

Lysosomes are membrane bound spheroidal bodies containing hydrolase enzymes capable of degrading a wide variety of substances. They are present in all cells except mature RBC. They are dominant in neutrophils. When one of the hydrolase enzymes is congenitally absent, the lysosomes become engorged with undigestable material normally degraded by that enzyme. This leads to lysosomal storage disorders, e.g. Fabry's disease, Gaucher's disease, Tay-Sachs disease, etc.

■ PEROXISOMES OR "MICROBODIES"

Peroxisomes are small, spherical, membrane bound organelle that closely resemble lysosomes. However, they contain entirely different set of enzymes – oxidases and catalases. Large peroxisomes are found in liver and kidney cells. They help in the detoxification and oxidation of a wide variety of compounds.

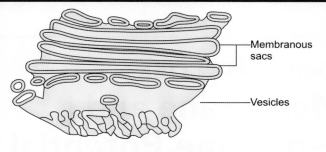
- Membranous sacs
- Vesicles

Fig. 22.7: Golgi apparatus

GOLGI COMPLEX OR GOLGI APPARATUS OR DICTYOSOMES (FIG. 22.7)

Golgi apparatus was discovered by an Italian optical microscopist – Camillo Golgi. It consists of a system of stacked, saucer shaped cisternae with the concave surface facing the nucleus.

23

Transport Across the Cell Membrane and the Resting Membrane Potential

- Passive Transport
  - Diffusion
  - Osmosis
- Active Transport
  - Primary Active Transport
  - Secondary Active Transport Process
- Vesicular Transport (Transcytosis)
  - Synaptic Transmitter Release
  - Resting Membrane Potential (RMP)
- Ionic Basis of RMP
  - Diffusion of Ions
- The Electrogenic Pump

Transport of various molecules through the cell membrane is essential for performing various functions of cell. The cell membrane is freely permeable to certain molecules while impermeable to some. Some molecules need certain carrier proteins for carrying them across the cell membrane. Lipid solid substances like N_2, O_2, alcohol, steroids, etc. can freely pass through the lipid bilayer of the cell membrane. But water soluble substances like glucose, ions, urea, proteins, etc. are not freely permeable and so require certain carrier proteins for their transport. Thus there are different types of transport systems for each molecule.

■ PASSIVE TRANSPORT

Passive transport includes diffusion and osmosis, which do not need energy.

Diffusion

Diffusion is the process of movement of molecules from an area of higher concentration to an area of lower concentration, down their concentration gradient, either chemical or electrical. The magnitude of diffusion is directly proportional to the cross-sectional area across which diffusion is taking place and also the amount of concentration gradient between them. This is according to the Fick's law of diffusion, which states that:

$$J = DA\frac{\Delta C}{t}$$

where J = *net rate of diffusion in gm/unit time*
        D = diffusion coefficient of the solute
        A = *Surface area of membrane*
   ΔC = Concentration difference (gradient) across the membrane
        t = thickness of membrane

Diffusion can be of two types—simple diffusion and facilitated diffusion.

Simple Diffusion (Fig. 23.1)

Movement of molecules across the cell membrane by simple diffusion occurs in two ways – through the lipid bilayer and through the transmembrane protein channels or ion channels. Simple diffusion through lipid bilayer occurs as in the case of lipid soluble substances like O_2, CO_2, N_2, fatty acids, steroids, alcohol, etc. Simple diffusion through transmembrane protein channels takes place in case of water and other lipid insoluble molecules. "Channels" means pathways from extracellular to intracellular region and are protein in nature. Na^+, K^+, Ca^{++}, Cl^-, water, glucose, etc. can diffuse directly through these channels. These protein channels are studied in detail by the Patch Clamp technique. According to such studies, there are leaky

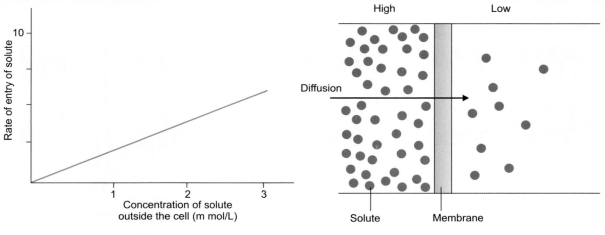

Fig. 23.1: Simple diffusion of solute across a plasma membrane

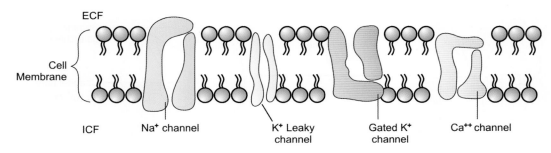

Fig. 23.2: Transmembrane protein channels with gates

or open channels which are continuously open and gated channels which have gates that can open or close (Fig. 23.2). "Gate" means the part of protein which closes these channels.

The opening and closing of the "gate" can occur in three different ways described below:

Voltage Gating: Variation in the electrical charges across the cell membrane can either open or close the channel, e.g. Na^+, K^+, Ca^{++}, and Cl^- channels (Fig. 23.3). Sodium channel is a tetramer protein with four subunits and has 0.3 to 0.5 nm in size. Inner surface of the channel is negatively charged which attracts Na^+ into it. In resting state, sodium channel is closed from outside (Fig. 23.3).

Potassium channel is 0.3 nm in size. It is also a tetramer like sodium channel, but inside is not negatively charged. So there is no attractive force which pulls K^+ into this channel. In some areas,

potassium channels are always open or leaky, whereas in other areas, they are closed from inside (Fig. 23.3). Permeability of the voltage gated channels depends on the charge of the ions, charges lining the interior of the channels, concentration difference between the two sides of membrane and the diameter of hydrated ion. When we compare the permeability of Na^+, and K^+ through the cell membrane, we can see that K^+ is more permeable than Na^+, even though the interior of sodium channel is negatively charged. This is because Na^+ are more hydrated than K^+. In other words, Na^+ are more surrounded by water molecules, which pull back the Na^+, from entering into the channel.

Ligand gating: "Ligand" is an ion or a specific substance like hormone, calcium, cyclic AMP, etc. Binding of the ligands to their corresponding receptors on the gate opens it (Fig. 23.4) and allows their passage.

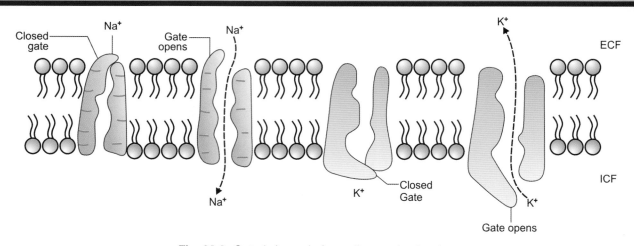

Fig. 23.3: Gated channels for sodium and potassium

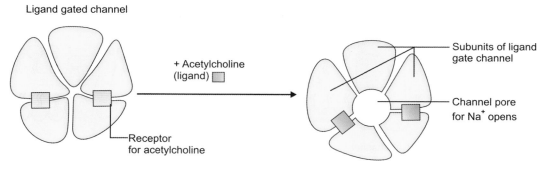

Fig. 23.4: Ligand-Gated channel (e.g. Acetyl choline receptor present at post-synaptic membrane)

Mechanical gating: The gate of these channels are opened by certain mechanical stimuli like stretch, sound waves, etc.

Facilitated Diffusion or Carrier Mediated Diffusion

Facilitated diffusion occurs via a carrier protein which helps in diffusion of the solute and does not involve energy. This type of diffusion is usually meant for those molecules which are too big or highly charged. Glucose and aminoacids are transported in this way.

The carrier protein has a specific receptor which is attached to the inner side of the channel. The molecule to be transported enters the channel and gets attached to the carrier protein. Now, this binding causes a conformational change in the carrier protein so that the channel opens (Fig. 23.5). This allows the passage of molecule according to its concentration gradient, across the membrane, e.g. GLUT (Glucose

Transporters) present on the intestinal epithelial cell help to transport glucose.

The rate of facilitated diffusion depends on the number of carrier proteins available, the temperature, the concentration gradient of substance and presence of any substance which can inhibit the carrier protein.

Osmosis

Osmosis is the net diffusion of water from a region of higher concentration of solutes to a region of lower concentration, across the cell membrane (Fig. 23.6). The shift of water between intracellular and extracellular compartments results from alterations in the effective osmolality of extracellular fluid (ECF). When the osmotic pressure of ECF is increased (as in hypernatremia), water leaves the cell by osmosis and the cell shrinks. Conversely, if the osmotic pressure of ECF is decreased (as in hyponatremia), water enters the cell and it swells.

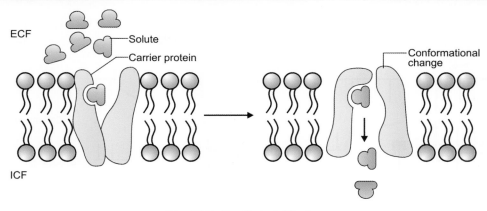

Fig. 23.5: Facilitated diffusion

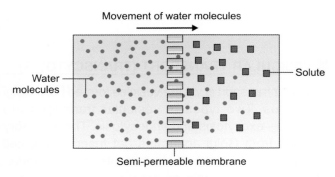

Fig. 23.6: Osmosis

ACTIVE TRANSPORT

The intracellular fluid has a different ionic composition from the extracellular fluid. It is essential to keep the concentration of certain ions like Na^+ less inside the cell than outside. Conversely, some other ions like K^+ are kept at a higher concentration inside the cell than outside. For this, we need a transport process which can transport these substances against their concentration gradient (either chemical or electrical), known as active transport. Active transport occurs at the expenditure of energy, derived by hydrolysis of ATP inside the cell. According to the source of energy involved, active transport can be classified into two – primary active transport and secondary active transport.

Primary Active Transport

In *primary active transport*, energy is derived directly from the hydrolysis of high energy phosphate compounds like ATP or creatine phosphate, e.g. Na^+ K^+ ATPase pump, H^+ K^+ ATPase pump, Ca^{++} pump, Na^+ H^+ ATPase pump, etc. "Pumps" are actually ATP hydrolyzing enzymes. Among these, Na^+ K^+ ATPase pump is present on almost all cells of body and is described below.

Na⁺ K⁺ ATPase Pump

This pump for primary active transport is present on almost all cells of our body. Fig. 23.7 shows the structural components of this pump in detail. It consists of two subunits—α and β.

α subunit has ATPase activity. It consists of intracellular binding sites for Na^+ and ATP along with a phosphorylation site. It has an intracellular binding site for K^+. β subunit is a glycoprotein which anchors the protein complex in the lipid bilayer. The pump functions as follows–Activation of the pump occurs when the concentration of Na^+ inside the cell increases. When activated, three Na^+ bind to the intracellular binding site of the pump while ATP binds to the ATP binding site (Fig. 23.7). Now, ATP is cleaved into ADP with release of energy. Using this energy, a conformational change occurs in the pump which results in extrusion of three Na^+ to the ECF. Simultaneously two K^+ bind extracellularly to the a subunit, which is released to the ICF. After this, the pump regains the previous state. In short, for every three Na^+ extruded, two K^+ are drawn inside (Fig. 23.7). Since this produces an electrical gradient across the membrane, this pump is also known as "electrogenic

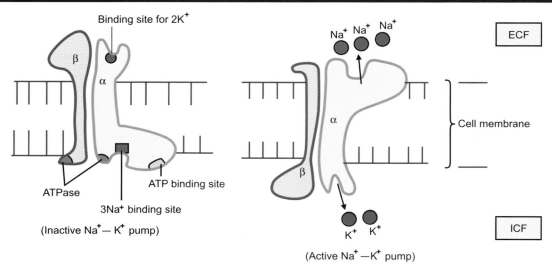

Fig. 23.7: Mechanism of operation of Na⁺-K⁺ ATPase pump

Binding site for 2K⁺

ECF

Cell membrane

ATPase

ATP binding site

3Na⁺ binding site

(Inactive Na⁺— K⁺ pump)

(Active Na⁺—K⁺ pump)

ICF

pump". Na⁺K⁺ ATPase pump helps to maintain the electrical gradient across the cell membrane. Also it maintains the volume of cells. In the absence of this pump, large amounts of Na⁺ along with proteins and other compounds tends to cause osmotic flow of water into the cell. The cell may swell up and burst. Thyroid hormone, insulin and aldosterone stimulate the activity of the pump. Digitalis, a drug used in heart failure, acts by inhibiting Na⁺K⁺ ATPase pump in the cardiac muscle.

Secondary Active Transport Process

In *secondary active transport process*, there is indirect involvement of energy. In many tissues, the transport of Na⁺ is coupled with that of other substances like glucose, amino acids, etc. Both the solutes bind to a common carrier protein on one side of the membrane and are transported together (symport or co-transport), e.g. SGLT in kidney transports both Na+ and glucose.

Sometimes, the carrier protein transports one substance in one direction and the other substance in opposite direction (Antiport or Counter transport), e.g. Na⁺ H⁺ antiport in kidney. In both symport and antiport there is indirect involvement of energy. The Na⁺ that is entering into the renal tubular cell has to be pumped into the circulation using Na⁺K⁺ATPase pump (*Refer* "Urinary system"). Otherwise, it will affect the proper functioning of the symport and antiport.

■ VESICULAR TRANSPORT (TRANSCYTOSIS)

Vesicular transport is via the vesicles formed from Golgi apparatus. Two types of vesicular transports are possible—exocytosis and endocytosis. The secretory vesicles from Golgi apparatus get fused to the cell membrane and empty their contents to the exterior. This process is known as exocytosis. Exocytosis is initiated by an increase in intracellular calcium concentration (Fig. 23.8), e.g. *Certain protein hormones like insulin, enzymes, secretions in GIT.*

Synaptic Transmitter Release

Endocytosis includes phagocytosis or "cell eating" by which extracellular microorganisms like bacteria, foreign particles, dead tissue, etc. are engulfed by the cells. At the region of contact of cell membrane with the above said substances, an invagination occurs on the cell membrane. Then gradually an endocytic vesicle is formed with the substance trapped inside it. Now, this endocytic vesicle fuses with a lysosome which releases various digestive enzymes that destroy that substance.

Resting Membrane Potential (RMP)

All the transport systems discussed above are co-ordinated in a living cell in such a manner that they maintain the particular ionic composition of intracellular and extracellular fluids compatible with life. Two types

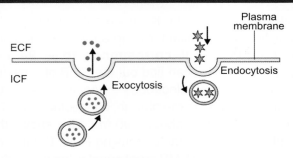

Fig. 23.8: Endocytosis and exocytosis

of ionic forces are existing across the cell membrane. The difference in concentration of a substance between ICF and ECF causes diffusion. Also there exists a difference in electrical potential between the two sides of the membrane because of difference in concentration of charged ions. This tends to move positive ions towards negative side and vice-versa. The sum total of these two driving forces is called electrochemical potential gradient for a specific solute. For uncharged solutes, only the chemical gradient exists. According to this gradient, net movement of solutes into or out of a cell occur till equilibrium is reached. At the equilibrium electrochemical gradient is zero and this is called equilibrium potential. The equilibrium potential is different for each solute.

The magnitude of the equilibrium potential at normal body temperature in the case of univalent ions like Na⁺, K⁺, Cl⁻, etc. can be calculated by Nerst equation, which is as follows:

$$E_{ion} = \frac{RT}{ZF} \ln \frac{C_{in}}{C_{out}}, \text{ where}$$

$$
\begin{aligned}
E_{ion} &= \text{Equilibrium potential of the ion} \\
R &= \text{Natural gas constant} \\
T &= \text{Absolute temperature} \\
Z &= \text{Valency of ion} \\
F &= \text{Faraday constant} \\
\ln &= \text{Symbol for natural logarithm} \\
C_{in} \text{ and } C_{out} &= \text{concentration of ion inside and} \\
&\quad \text{outside the membrane respectively}
\end{aligned}
$$

There is a difference in the electrical potential across the membranes of almost all cells of our body, with interior negative in relation to the exterior. This is called the resting membrane potential (RMP) or steady potential and is written with a negative sign which

means "inside is negative in relation to outside" of the cell. The value of RMP can widely vary from − 9 mV to −100 mV according to the function of the tissue. RMPs of excitable tissues are given in the (Table 23.1).

Excitable tissues include nerves and muscles (skeletal, cardiac and smooth muscles). They are called so because they can generate rapidly changing electrochemical impulses which are used to transmit signals along their membranes.

Cathode Ray Oscilloscope (CRO) is used to record the electric events taking place inside a cell which give rise to RMP (Fig. 23.9). Studies were first conducted in the axons of squid and frog's gastrocnemius muscle.

■ IONIC BASIS OF RMP

Diffusion of different ions across the cell membrane and the electrogenic pump are the two important factors responsible for the genesis of RMP in an unstimulated cell.

Diffusion of Ions

Because of large concentration gradient, K⁺ diffuse out of the membrane via K⁺ leaky channels (Fig. 23.2). This process is called K⁺ efflux. This upsets the electrical equilibrium between the two sides of cell membrane. To bring back the equilibrium, two things must occur—either the K⁺ efflux should be accompanied by an anion or a cation should come inside. Proteins and chloride ions are the main anions inside the cell. But they cannot accompany K⁺. Proteins are too big to come out through the pores of the cell membrane and the concentration of Cl⁻ is more outside the membrane. Therefore they come and line the interior of the cell membrane. At resting condition, Na⁺ channels are closed from outside. Also the size of hydrated Na⁺ prevents it from entering into the channel. This decreases the permeability of Na⁺ through the cell membrane about 100 times than that of K+.

Table 23.1: RMP of Different excitable tissues	
Tissue	*RMP (mV)*
Nerve	− 70
Skeletal muscle	− 90
Cardiac muscle	− 90
Smooth muscle	− 50 to − 60

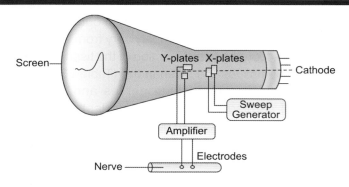

Fig. 23.9: Cathode ray oscilloscope

Therefore Na⁺ lines the exterior of the cell membrane.

As K⁺ moves out through the leaky channels, negativity is created inside the cell. Now the electrical gradient of K⁺ tends to prevent its efflux. In other words, the concentration gradient favors K⁺ efflux whereas the electrical gradient favors K⁺ influx. These two forces finally balance each other and net movement of K⁺ stops. The equilibrium thus reached is called the equilibrium potential for K⁺. If we calculate the value of equilibrium potential for K⁺ in a muscle cell (using Nerst equation) it corresponds to − 94 mV. Experimentally, we can find out the RMP of a muscle cell using two microelectrodes − one kept inside and one kept outside the membrane, connected to a CRO (Fig. 23.9), which is about − 90 mV. This shows that K⁺ efflux is the main factor causing RMP. But it is not the only factor since RMP is not equal to the equilibrium potential of K⁺ (− 94 mV).

THE ELECTROGENIC PUMP

Na⁺ K⁺ ATPase pump also contributes to the RMP. Whenever there is an increase in the Na⁺ concentration inside the cell, three Na⁺ are pumped out of the membrane in exchange for two K⁺ (Fig. 23.7), which again creates a negativity inside the cell.

Thus even though K⁺ efflux is the main factor in the genesis of RMP, small but significant contribution is done by the electrogenic pump.

24 The Nerve Physiology

The activities of all organ systems of our body are controlled and modulated by the nervous system. For this, the nerve cells or neurons rapidly transmit information from one cell to another by conducting electrical impulses and secreting neurotransmitters. The nerves sense all kinds of stimuli from both inside and outside our body, analyze and store them as well as respond to them. In this chapter, let us discuss the structure and properties of a neuron, initiation and conduction of action potentials or electrical potentials and synaptic transmission.

◼ NEURON (FIG. 24.1)

The nerve cell or neuron is the structural and functional unit of nervous system. There are about 10^{11} neurons in our central nervous system.

A typical neuron consists of a cell body or soma and several processes. Cell body or soma consists of a mass of cytoplasm surrounded by a membrane. The cytoplasm contains a large central nucleus, Nissl granules, numerous mitochondria, Golgi apparatus and other usual cell organelles. But the centrosome is absent, which means that neurons cannot divide. Nissl granules are granular material that are stained intensely with basic dyes like methylene blue. Electron microscopic studies reveal that they are composed of rough endoplasmic reticulum with ribosomes. Therefore, they are concerned with protein synthesis. The size and number of Nissl granules vary according to the physical condition of the cell. If the cell is active, fatigued or injured, these Nissl granules get disintegrated–a process called chromatolysis. A network of fibrils (neurofibrils), microfilaments and microtubules are also seen in the cytoplasm of the cell body. They provide support and shape to the cell.

The processes of neuron include several dendrites and a single axon (Fig. 24.1). Dendrites are irregularly thickened branches arising from the cell body. They carry impulses towards the cell body. Axon is another neurite arising from the cell body with uniform diameter and is unbranched. The portion of the cell body from which the axon arises is slightly thickened and is known as axon hillock (Fig. 24.1). Initial segment or trigger zone (Fig. 24.1) is the first portion of axon from which

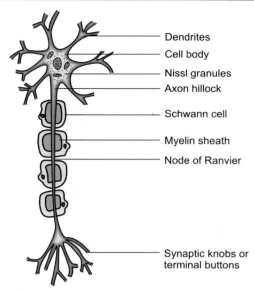

Fig. 24.1: Structure of a myelinated neuron

Labels:
Dendrites
Cell body
Nissl granules
Axon hillock
Schwann cell
Myelin sheath
Node of Ranvier
Synaptic knobs or terminal buttons

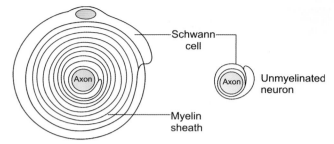

Fig. 24.2: Schwann cell covering of myelinated and unmyelinated neurons

Labels:
Schwann cell
Axon
Myelin sheath
Axon
Unmyelinated neuron

impulses arise. The cell membrane of axon is known as axolemma and the cytoplasmic fluid occupying the centre forms the axoplasm. The terminal part of axon ends as small knobby branches called terminal buttons or synaptic knobs, which synapse with dendrites of other neurons. These "knobs" contain vesicles in which synaptic neurotransmitter, secreted by the neuron, is stored. Axons carry impulses away from the cell body.

Axons may be covered by a myelin sheath or bare according to which neurons are classified as myelinated and unmyelinated respectively. Axons of myelinated nerves have a sheath of "myelin", made up of a protein-lipid complex, rich in sphingomyelin. The process of myelination (myelinogenesis) begins during fourth to fifth month of intrauterine life. In the central nervous system, specialized cells called "oligodendroglia" produce myelin, whereas in peripheral nerves, "Schwann" cells produce myelin. The membrane of Schwann cells wrap several times (about 100) around the axon of the myelinated peripheral nerves. In unmyelinated nerves also, Schwann cells are present, but they simply surround the axon once (Fig. 24.2). Myelinated nerves possess certain constricted areas at regular intervals, where myelin is absent. These are called the "Nodes of Ranvier" (Fig. 24.1). Myelination increases the velocity of conduction and decreases the energy expenditure for the same. Also it acts as an insulator, preventing cross stimulation of adjacent axons. Myelin gives a whitish appearance for the nerves.

APPLIED PHYSIOLOGY

Multiple Sclerosis

This is an autoimmune disorder, where patchy destruction of myelin sheath occurs in the central nervous system. This can cause delayed or blocked nerve conduction in those axons.

Properties of Neurons

Excitability and conductivity are the two prime properties of a neuron, which are discussed in detail below.

EXCITABILITY

Excitability is defined as the ability by which the nerve responds to a stimulus—either external or internal. If the stimulus is of adequate or threshold strength, the nerve responds by generating an electrical impulse or "action potential". The stimulus can be mechanical, chemical, thermal or electrical, which can cause a sudden change in the environment of a living organism. In the unstimulated state, the nerve cell exhibits an RMP of − 70 mV. Whenever there is a disturbance of RMP, ions start moving in or out to re-establish it. A decrease in RMP from its resting value towards zero is called depolarization (e.g. from − 70 mV to − 40 mV). Conversely, an increase in the negativity of RMP is called hyperpolarization (e.g. from − 70 mV to −110 mV).

When a stimulus is first applied to a nerve, two types of responses can be produced according to its strength:

1. If a sub-threshold stimulus (strength less than threshold stimulus) is applied, the cell depolarizes and a non-propagating local potential or graded potential is produced (Fig. 24.3).
2. If a stimulus of threshold strength is applied, the depolarized cell produces propagated impulses called action potential (Fig. 24.3).

■ CONDUCTIVITY

Conductivity is the self-propagating active process by which an action potential is conducted along the nerve. For this, the stimulus must be of threshold strength. Action potential can be recorded by monophasic recording, where one electrode is kept inside and the other kept outside the cell membrane of axon (Fig. 24.4). Both are connected to a CRO which accurately records the potential changes occurring when the axon is stimulated. If progressively larger depolarizing current or stimulus is applied to the membrane, a threshold membrane potential is reached at first followed by a brief and sudden explosive change to a positive potential. The phases of action potentials produced in different excitable tissues are different. The phases of action potential recorded from a squid axon is shown in Figure 24.5.

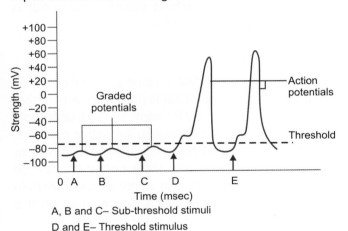

Fig. 24.3: Effect of different stimuli on nerve showing graded and action potentials

A, B and C– Sub-threshold stimuli
D and E– Threshold stimulus

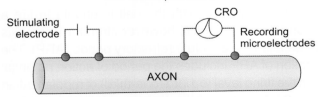

Fig. 24.4: Monophasic recording of action potential

Resting Phase (Latent Period)

This period denotes the time taken by the impulse to travel from the point of stimulus to the recording electrode during which the membrane potential is same as its RMP (– 70 mV). Duration of this period is directly proportional to the distance between the electrodes and inversely proportional to the conducting velocity.

Slow Depolarization Phase

During this phase, the axon is depolarized slightly, due to opening up of voltage gated Na^+ channels, through which Na^+ enters into the cell. In most neurons, the initial segment of the axon has a high density of voltage gated Na^+ channels and hence acts as the "trigger zone" for the generation of action potential. This phase lasts till the membrane potential reaches around –55 mV, called the firing level (Fig. 24.5) or the threshold level of excitation.

Rapid Depolarization Phase

When the membrane potential exceeds the threshold level of excitation (– 55 mV), the rate of depolarization sharply increases, overshoots the zero potential to about +35 mV. This is because, at the firing level, the voltage gated Na^+ channels start opening at an increased rate and there is an explosive (about 18-20 times increased) entry of Na^+ into the cell. This increase in Na^+ permeability reaches a peak value at a potential of + 35 mV, after which Na^+ influx stops.

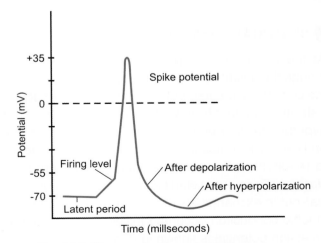

Fig. 24.5: Phases of action potential in a neuron

Part 2

Physiology

Rapid Repolarization Phase

Once the peak value of + 35 mV is reached, the membrane potential reverses and falls rapidly till 70% repolarization is completed (Fig. 24.5). Two reasons are there for this phase. Sudden closure of Na^+ channels at + 35 mV potential stops the Na^+ influx. Also there is opening up of voltage gated K^+ channels through which K^+ starts moving out. The net movement of positive charge (K^+) out of the cell along with stoppage of Na^+ influx speeds up the repolarization initially.

After Depolarization Phase (Slow Repolarization Phase)

After completing 70% repolarization, further decrease in membrane potential occurs slowly (Fig. 24.5). This is due to slowing down of K^+ efflux through the voltage gated K^+ channels.

After Hyperpolarization Phase

The slow repolarization phase does not stop once the RMP is reached (Fig. 24.5). The repolarization phase overshoots towards the negative side than the RMP for sometime. This small but prolonged increase in negativity of membrane potential is called after hyperpolarization. This is due to delayed closure of K^+ channels. Even after reaching – 70 mV, K^+ efflux is still occurring for a few more milliseconds.

The nerve action potential is also known as spike potential.

■ RESTORATION OF RMP

At the end of after hyperpolarization phase, even though the polarity of membrane reaches that of RMP, the concentration gradient of different ions are not re-established. In other words, at the end of after hyperpolarization phase, there is excess Na^+ inside and excess K^+ outside the membrane of axon. Hence to re-establish the previous concentration gradient, Na^+K^+ ATPase pump starts acting, which pumps three Na^+ out in exchange for two K^+. The overall changes in ionic conduction occurring inside a neuron during an action potential is shown in (Fig. 24.6).

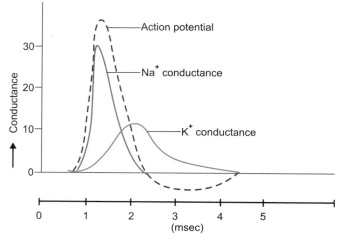

Fig. 24.6: Changes in ionic conductance during the action potential in a neuron

■ PROPERTIES OF ACTION POTENTIAL

Threshold Stimulus

A stimulus of threshold strength and duration is necessary to elicit an action potential. If you give a stimulus of adequate intensity for extremely short duration, action potential will not be produced. So both strength as well as duration of application of stimulus are significant. Sub-threshold stimulus can result in the formation of local potential.

All or None Law

Action potential obeys "All or None Law" which states that once an action potential is produced, any further increase in the strength of the stimulus (suprathreshold stimulus) produces no change in the action potential. In other words, action potentials cannot summate or fuse together. Conversely, local potentials can be summated. Application of a second sub-threshold stimulus within a fraction of a second can elicit a response (Fig. 24.7).

Refractory Period

The period during which the cell is unable to fire a second action potential, however strong the stimulus be, is called the absolute refractory period (ARP). The duration of ARP includes the period of action potential from the firing level to 1/3rd (one-third) of repolarization (Fig. 24.8). The cell is unresponsive or refractory because there is voltage inactivation of Na⁺ channels during this period. Therefore, no more Na⁺ influx can occur to evoke a second response. During the later part of repolarization, the cell is able to fire a second action potential, if a stronger stimulus is given. This period, which extends from 1/3rd (one-third) of repolarization phase to the end of action potential, is known as relative refractory period (RRP). With a stronger stimulus, some of the Na⁺ channels can be opened during this period.

Self-propagation (Conduction) of Action Potential

Action potentials are conducted along an excitable tissue by local current flow. The speed of conduction

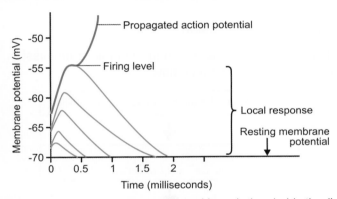

Fig. 24.7: Local responses produced by sub-threshold stimuli

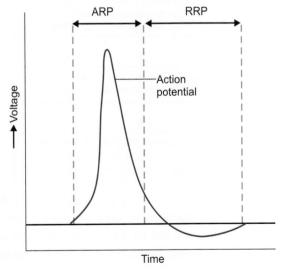

ARP– Absolute refractory period
RRP– Relative refractory period

Fig. 24.8: Refractory periods

through a nerve depends on the diameter and myelination of the nerve. The rate of conduction is directly proportional to the diameter of the nerve. In an unmyelinated axon, 'point-to-point' conduction of impulses takes place. Action potential elicited at one particular point excites the adjacent portions of the membrane, resulting in spread of action potential over the membrane (Fig. 24.9) shows the membrane of an axon where excitation is produced first at point 'A'. In this region, the external surface of the membrane is negative relative to the neighbouring areas. This potential difference causes a local circuit of current flow between the depolarized area and adjacent resting areas. Now, the neighbouring areas are depolarized (Fig. 24.9). These newly depolarized areas then cause current flows that depolarize other segments of the membrane (Fig. 24.9). Depolarization is followed by repolarization, which starts at the point of stimulus and spreads in both directions. Once initiated, the moving impulse does not depolarize the area behind it to the firing level, because this area is refractory. In other words, propagation of action potential occurs only in one direction.

In myelinated nerves, the type of conduction is known as "saltatory conduction". Here, action potentials are generated only at the Nodes of Ranvier (Fig. 24.10), as the myelin sheath is an insulator. Successive nodes are excited and the action potential appears to "jump" from one node to another and hence called "saltatory conduction".

■ SYNAPSE

Synapse is the region of contact between two neurons, where information from one neuron is handed over to

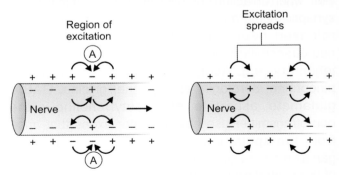

Fig. 24.9: Spread of action potential in an unmyelinated nerve

Part 2

Physiology

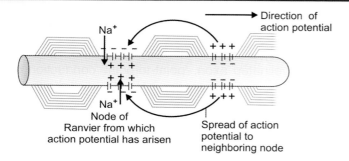

Fig. 24.10: Saltatory conduction in a myelinated nerve

another neuron without any structural contact with each other. Two types of synapses have been identified – Electrical and chemical. Electrical synapses are present in the central nervous system of animals, where there are "gap junctions" connecting the cytoplasm of adjacent neurons. These gap junctions permit the flow of ions from one neuron to another. Those present in the human beings are "chemical synapses", where the mode of transmission of impulses is via certain chemical neurotransmitters. Synapses can occur between an axon and dendrites (axo-dendritic) or between an axon and cell body (axo-somatic) or between an axon and axon of another neuron (axo-axonic). Among these, axo-dendritic synapse is more common. Structure of a typical axo-dendritic synapse is given in Figure 24.11.

The terminal button or synaptic knob of the axon forms the presynaptic membrane and the membrane of the dendrite forms the post-synaptic membrane. A narrow space called the synaptic cleft lies in between the pre-synaptic and post-synaptic membrane. The pre-synaptic terminal is packed with vesicles containing neurotransmitters which are released into the synaptic cleft, when an action potential enters the terminal. Post-synaptic membrane contains receptors, specific for the released neurotransmitter. There are several neurotransmitters, which can be either excitatory or inhibitory, e.g. Acetyl choline, dopamine, nore-pinephrine, serotonin, epinephrine, glycine, GABA, glutamate, aspartate, etc. Acetyl choline is the neurotransmitter used by all motor axons that arise from the spinal cord, pre-ganglionic and post-ganglionic parasympathetic fibres, and also in many of the central neural pathways.

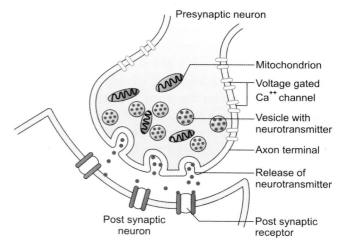

Fig. 24.11: Synapse

Synaptic Transmission

The events that occur during synaptic transmission can be summarized as below:

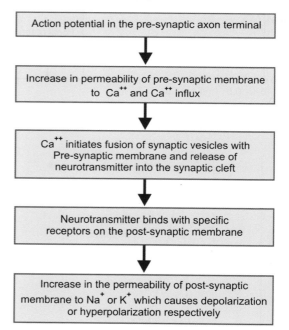

An increase in the membrane conductance to Na+ depolarizes the membrane and the result will be generation of an excitatory post-synaptic potential (EPSP) in the post-synaptic neuron. An increase in the membrane conductance that permits the efflux of K+ hyperpolarizes the cell and the result will be generation of an inhibitory post-synaptic potential (IPSP) in the post-synaptic neuron. Both EPSP and IPSP are local potentials. The depolarizing EPSP is not strong enough to trigger an action potential in the post-synaptic neuron. Thus the post-synaptic neuron will reach the threshold only when multiple EPSPs occur within a narrow time interval. Summation or fusion of these EPSPs occur and an action potential is generated.

Synaptic Delay

A small amount of time is required for each event of synaptic transmission explained above to occur, which comes to about 0.5 milliseconds (for a single synapse). This is called synaptic delay. By measuring the delay time between the input stimulus and the output response, we can estimate the number of synapses present in the neuronal circuit.

Part 2

Physiology

25 The Muscle Physiology

Muscle is an excitable tissue responsible for most of our functions like locomotion, posture, speech, respiration, pumping of blood, movement of food, expulsion of wastes, storage of energy, etc. Thus muscle is a highly diverse tissue which can perform a great variety of functions. In spite of the functional diversity, some basic similar features are there for all muscles. In this chapter, let us discuss the basic structure and mechanisms underlying muscle contraction.

■ TYPE OF MUSCLES

According to the Anatomical location, there are three types of muscles – Skeletal, cardiac and smooth muscles. Skeletal muscle controls the voluntary movement of skeleton. Smooth muscle controls the functions of our visceral organs and blood vessels. Cardiac muscle helps the heart to pump blood into the circulation. A common feature of all these muscles is that, when they are stimulated, they respond by contracting.

Skeletal Muscle

A single skeletal muscle is composed of numerous cells called muscle fibers. The plasma membrane of the muscle fiber is known as sarcolemma. The muscle fibers are multinucleated and contains numerous mitochondriae which supply energy in the form of ATP for contractions. Each muscle fiber contains several parallel myofibrils (Fig. 25.1). As revealed by electron microscopic studies, each myofibril is composed of about 1500 myosin filaments (thick filaments) and 3000 actin filaments (Light filaments), giving a "striated" appearance for the muscle fiber. These filaments are arranged alternatively and under polarized light, they appear as dark ("A") and light ("I") bands respectively (Fig. 25.1). "I"-bands are called so as they are isotropic to polarized light, and contain actin filaments. "A"-bands are anisotropic to polarized light and contain myosin filaments as well as ends of actin filaments, where they overlap the myosin filament (Fig. 25.1). "A"-band is divided at the center by a relatively lighter area called "H"-band. A prominent dark line is present at the center of the "H"-band called "M" line. Between the "A"-bands lie the "I"-bands. Crossing the center of "I"-bands, there is a dark line called "Z" line. Thin filaments of the "I" bands are attached to the "Z" line. This pattern of alternating bands is repeated over the entire length of a single muscle fiber. The portion of myofibril that lies between two successive "Z"-lines is known as the sarcomere (Fig. 25.1). Sarcomere is the functional unit of a muscle fiber.

When the muscle contracts, "Z"-lines move close together and the length of each sarcomere decreases. The width of "I"-bands also decreases. But the length of "A"-band remains the same (Fig. 25.2). When the muscle is relaxed, the thin filaments do not overlap, the thick filaments at the region of "H"-bands.

■ SARCOPLASMIC RETICULUM

Sarcoplasmic reticulum is the specialized endoplasmic reticulum found in other cells of our body. It forms an extensive network or "reticulum" around the myofibrils of each muscle fiber (Fig. 25.3). This reticulum plays an important role in the regulation intracellular Ca^{++} concentration. Inward extensions of sarcolemma (membrane covering the muscle fiber), called the T-tubules, pass into the muscle fiber at the junction between "A" and "I" bands. In other words, there are two T-tubules for each sarcomere. These T-tubules are in close contact with the extracellular fluid that surrounds the muscle fiber. The portion of sarcoplasmic reticulum on either side of T-tubule forms sac like dilatations called terminal cisternae (Fig. 25.3). They release as well as store Ca^{++} during muscle contraction and relaxation respectively. The two terminal cisternae on either side of the T-tubule in the center at the "A-I" junction form "the Triad".

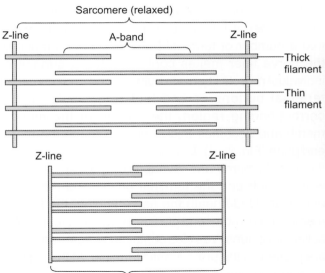

Fig. 25.2: Relaxed and contracted states of a sarcomere

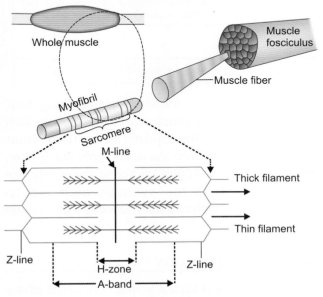

Fig. 25.1: Organization of skeletal muscle

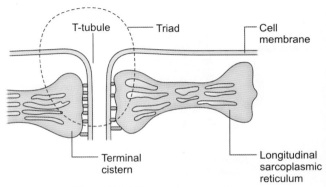

Fig. 25.3: Sarco-tubular system

NEUROMUSCULAR JUNCTION

The junction between a motor nerve and a skeletal muscle fiber is called neuromuscular junction or myoneural junction. Skeletal muscle fibers are innervated by large myelinated nerve fibers originating from the anterior horn cells of spinal cord. After entering the muscle, each nerve fiber divides into branches and thus can stimulate many muscle fibers. Each of the nerve ending or nerve branch makes one neuromuscular junction. All the muscle fibers innervated by a single nerve fiber is called a motor unit. The muscles which react rapidly and that require fine control have more motor units, e.g., laryngeal muscles. The muscles which do not require fine control have less number of motor units, e.g., Soleus (Fig. 25.4) shows the structure of a typical myoneural junction.

As the axon of the motor nerve approaches the muscle fiber, it loses the myelin sheath and divides into a number of terminal branches. Each branch of the nerve fiber makes contact with the surface of individual muscle fibers via bulb shaped endings called terminal buttons or end feet (Fig. 25.4). At the corresponding region, the portion of the muscle membrane becomes thickened and forms the motor end plate (Fig. 25.4). But the nerve ending lies outside the muscle membrane. Synaptic cleft – a small space, is there between the terminal button and the motor end plate. Underneath the nerve ending, the muscle membrane of motor end plate is thrown into many folds called pallisades. Vesicles containing neurotransmitter are present in the pre-synaptic terminal. At the skeletal muscle neuromuscular junction, acetyl choline is the excitatory neurotransmitter. There are about 2-3 lakhs of acetyl choline vesicles at the terminal button.

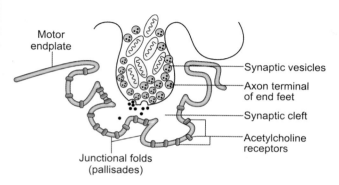

Fig. 25.4: Neuromuscular junction

Motor endplate
Synaptic vesicles
Axon terminal of end feet
Synaptic cleft
Acetylcholine receptors
Junctional folds (pallisades)

Specific receptors called nicotinic acetyl choline receptors are present in the folds of palisades of motor end plate.

Neuromuscular Transmission

Neuromuscular transmission begins when an action potential arrives at the pre-synaptic axon terminal. Summary of events that occur during the neuromuscular transmission is as follows:

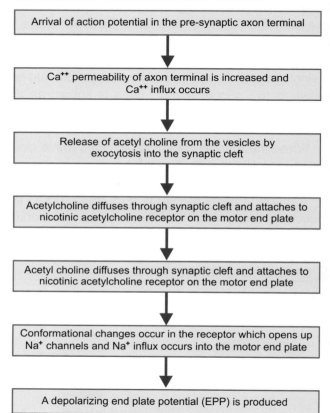

Arrival of action potential in the pre-synaptic axon terminal

↓

Ca^{++} permeability of axon terminal is increased and Ca^{++} influx occurs

↓

Release of acetyl choline from the vesicles by exocytosis into the synaptic cleft

↓

Acetylcholine diffuses through synaptic cleft and attaches to nicotinic acetylcholine receptor on the motor end plate

↓

Acetyl choline diffuses through synaptic cleft and attaches to nicotinic acetylcholine receptor on the motor end plate

↓

Conformational changes occur in the receptor which opens up Na^+ channels and Na^+ influx occurs into the motor end plate

↓

A depolarizing end plate potential (EPP) is produced

When EPP reaches the threshold level or firing level, an action potential is produced in the muscle fiber which is conducted to the adjacent areas. This initiates muscle contraction. Once the action potential is over, acetyl choline is destroyed by acetyl cholinesterase enzyme present in the synaptic cleft. Acetyl cholinesterase hydrolyzes acetylcholine into acetate and choline.

APPLIED PHYSIOLOGY

Myasthenia Gravis is an autoimmune disorder where antibodies are developed in the body against the nicotinic acetyl choline receptors. The receptors are thus destroyed, which affects the neuromuscular transmission. The end plate potentials produced in the muscle fibers are mostly too weak to initiate contraction, which results in generalized weakness and paralysis of muscles.

Skeletal Muscle Action Potential

The action potential of skeletal muscle fiber resembles that of a nerve fiber, except for a few differences. The RMP of skeletal muscle fiber is about -90 mV and the firing level or threshold is about -50 mV.

MOLECULAR MECHANISM OF SKELETAL MUSCLE CONTRACTION

During muscle contraction, the thin filaments slide into the spaces between the thick filaments so that the Z-lines of the sarcomere are pulled towards each other (Fig. 25.2). During muscle relaxation, the thin filaments do not overlap the thick filaments. Thus contraction of muscle occurs by "sliding filament mechanism". Certain "cross bridges" are formed between the overlapping thick and thin filaments (Fig. 25.5). In the following section, let us discuss about the contractile proteins of muscle and how the cross-bridging converts chemical energy from ATP into mechanical energy of contraction.

Contractile Proteins of Skeletal Muscle

Actin, troponin and tropomyosin are the three contractile proteins present in the thin filaments. Thick filaments contain only myosin. Actin molecules are arranged in the form of a double stranded α (alpha) helical polymer chain (Fig. 25.6) which forms the backbone of the thin filament. Tropomyosin molecules form long filamentous structure located in the groove between the two chains of actin (Fig. 25.6).

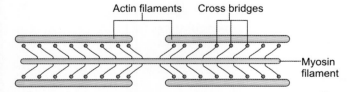

Fig. 25.5: Cross-bridge-interaction between actin and myosin filaments

Tropomyosin covers the active site of actin or the myosin binding site, at rest. Troponin is a heterotrimer globular protein located at intervals along the tropomyosin filaments (Fig. 25.6). It is made up of three subunits – Troponin-C (which binds to Ca^{++}), troponin-I (which binds other troponin subunits to tropomyosin). Polymers of myosin form the myosin filament. Each myosin molecule is composed of six polypeptide chains, of which two are heavy chains and four are light chains. Structurally, myosin molecule has two globular heads, a tail and a neck region (Fig. 25.6). Globular head contains both light and heavy chains. The head portion has an actin binding site and ATPase binding site. The heavy chains form a double helix and constitute the tail portion.

EXCITATION-CONTRACTION COUPLING

The process by which depolarization of muscle fiber initiates muscle contraction is known as excitation-contraction coupling. The process of generation of end plate potentials (EPPs) and formation of an action potential in the muscle fiber are already discussed. The T-tubular system helps in rapid transmission of action potential, causing release of Ca^{++} from the terminal cisternae into the sarcoplasm. Increase in the intracellular Ca^{++} initiates actin-myosin interaction and muscle contraction as follows:

The Ca^{++} released from sarcoplasmic reticulum binds to Troponin C. This causes a conformational change which facilitates the movement of the associated tropomyosin molecule resulting in the exposure of the active site of actin (Fig. 25.7). A single molecule of Ca^{++} binding to a troponin can expose

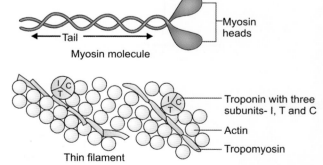

Fig. 25.6: Electron microscopic structure of thick and thin filaments

about seven myosin binding sites on actin filament. Now, myosin head attaches to the active site on actin. Once this binding happens, certain ATP-dependent conformational changes occur in the myosin molecule.

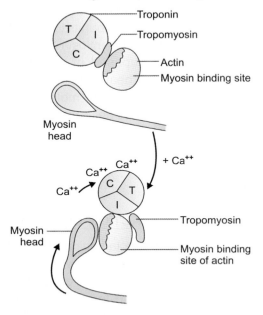

Fig. 25.7: The calcium 'switch' for controlling skeletal muscle contraction

Myosin head binds at the junction of head and neck, and pulls actin towards the center of the sarcomere (Fig. 25.8). Cross bridges from myosin extend towards the actin filament. Bending of myosin head is called "power stroke" and this requires energy. Bending of one myosin head exposes the other part of the head having ATPase activity. ATP combines with this site where it is hydrolyzed into ADP and high energy phosphate. This high energy phosphate remains attached to the myosin head till the next power stroke occurs, when this energy is utilized. Combination of myosin head with ATP also causes detachment of the head from actin. The head of myosin comes back to its original state with release of the cross bridges. This is called "cocking of head" (Fig. 25.8). If the intracellular Ca^{++} is still elevated, myosin will undergo another cross-bridge cycle, producing further contraction of the muscle. This process of attachment and detachment continues as long as ATP and Ca^{++} are available.

■ RELAXATION OF MUSCLE

The termination of contraction and beginning of relaxation of skeletal muscle requires the reuptake of

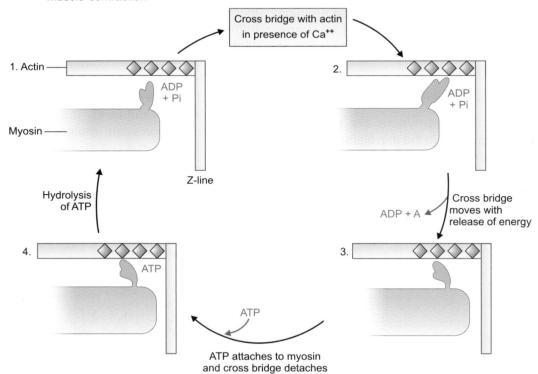

Fig. 25.8: Cross-bridge cycle of muscle contraction

intracellular Ca^{++} into the sarcoplasmic reticulum. Ca^{++}-Mg^{++} ATPase pump is present in the sarcoplasmic reticulum which pumps Ca^{++} back into the sarcoplasmic reticulum to be stored in the terminal cisternae. Decrease in intracellular Ca^{++} causes attachment of troponin and tropomyosin back to their original position, thus covering the myosin binding sites on actin. Hence the troponin-tropomyosin complex is also known as the "relaxing protein". Thus like contraction, relaxation of the muscle also requires energy.

APPLIED PHYSIOLOGY

Rigor Mortis

Several hours after death, all the muscles of our body go into a state of sustained contraction called rigor mortis. This is caused by depletion of ATP and creatinine phosphate after death. Here, all the myosin heads attach to actin molecules, but in an abnormal and resistant manner. Energy is not available for the separation of cross-bridges and relaxation of muscles.

Fibrillation

Destruction of a motor nerve supplying a skeletal muscle leads to increase in its sensitivity to the circulating acetylcholine. This property is termed as denervation hypersensitivity. This leads to abnormal excitation of the affected muscle and irregular contractions appear in individual muscle fibers. These contractions are not visible with naked eye and are known as fibrillations. They can be detected using an electromyogram (EMG).

Fasciculation

Lesions of spinal motor neurons can cause visible, jerky contraction of a group of muscle fibers. This is called fasciculation.

TYPES OF MUSCLE CONTRACTION

According to changes in length and tension, muscle contractions can be grouped into two – Isotonic and isometric contractions. An isotonic contraction is one in which the tone or force remains constant throughout the contraction and there is shortening of the muscle length, e.g. lifting up of a weight with hand, causing shortening of biceps muscle. Heat production is less during this process. An isometric contraction is one in which the muscle length is held constant, whereas tone or force generated during contraction is changed, e.g. contraction of postural muscles against gravity. Heat production is more in this case.

Types of Skeletal Muscle Fibers

According to the speed of activity, there are two types of skeletal muscle fibers – Slow twitch (Type I or Red muscle) fibers and Fast twitch (Type II or White muscle) fibers. Posture maintaining muscles like gastrocnemius belong to slow twitch group and the extraocular muscles and muscles of hand belong to the fast twitch muscles. The differences between the two types are summarized in the Table 25.1.

Energy Sources during Muscle Contraction

All muscles convert chemical energy into mechanical energy. ATP (adenosine triphosphate) is the immediate source of energy. ATP is formed from the metabolism of carbohydrates and lipids. Creatine phosphate present in the muscles can release energy by hydrolysis. Fatty acids represent an important source of energy for muscles during light exercises. Fatty acids release energy by β-oxidation within the mitochondria. Muscles store glycogen which can be metabolized during contraction for energy. If adequate oxygen is available, aerobic oxidation of glycogen occurs. In

Table 25.1: Types of skeletal muscle fibers		
	Type I (Slow fibers)	*Type II (Fast fibers)*
Myoglobin content	Less	More
Speed of response	Slow	Fast
Duration of muscle contraction	Long	Short
Myosin ATPase activity	Slow	Fast
Ca^{++} pumping capacity of sarcoplasmic reticulum	Moderate	High

severe and prolonged exercise, where oxygen is inadequate, anaerobic glycolysis occurs with accumulation of lactic acid. Depletion of energy sources along with accumulation of metabolic waste products like lactic acid makes the muscle fatigued.

OXYGEN DEBT

After a period of severe exercise, excess oxygen is consumed by the muscles to remove the accumulated lactic acid, replenish the ATP and creatine phosphate stores, and to replace the small amount of oxygen that was contributed by myoglobin. The consumption of extra oxygen by the muscles after severe exercise is called oxygen debt. Trained athletes are able to utilize free fatty acids more effectively for energy and therefore have smaller oxygen debts.

CARDIAC MUSCLE

The muscle mass of the heart, the myocardium, is an involuntary and striated muscle. Striations and structure of sarcomere are similar to that of a skeletal muscle. But there are certain unique features for cardiac muscle.

Cardiac muscle tissue consists of a branching network of cells called cardiac myocytes, interconnected at the intercalated discs (Fig. 25.9). Each cardiac myocyte contains a single nucleus, which is centrally placed. Intercalated discs are seen at the Z-lines. They maintain cell-to-cell adhesion and transmit the pull of one contractile unit to the next and thus to the whole cardiac muscle tissue. Near the intercalated discs, the sarcolemma of adjacent muscle fibers fuse for considerable distances forming gap junctions. They act as "low resistance bridges for the spread of action potential from one cell to another. Thus even though the individual cardiac

myocyte is single nucleated, the whole cardiac tissue is linked in a functional syncytium. The sarcoplasmic reticulum and the T tubules are not as extensive in cardiac muscle as in the skeletal muscle. Ca^{++} storage in sarcoplasmic reticulum is less when compared to the skeletal muscle. Hence Ca^{++} entry from ECF through the sarcolemma has an important role in the excitation-contraction coupling process in cardiac muscle.

Functionally, cardiac myocytes can be grouped into two:

1. Working cells
2. Pacemaker and conducting cells.

Working myocardial cells include atrial and ventricular muscles or myocardium. They are contractile in function. Pacemaker and conducting cells do not contract well, but generate and rapidly conduct impulses to different parts of heart. The conducting system of heart is discussed in detail in "Cardiovascular system". Innervation of cardiac muscle comes from sympathetic and parasympathetic nerves of autonomic nervous system which regulate the heart rate, conduction of impulses and strength of contraction.

SPECIAL PROPERTIES OF CARDIAC MUSCLE

Excitability or Bathmotropism

As in the other types of muscles, cardiac muscle also possesses both resting membrane and action potentials. The electrical potentials for working myocardial fibers are different from that of the pacemaker tissues.

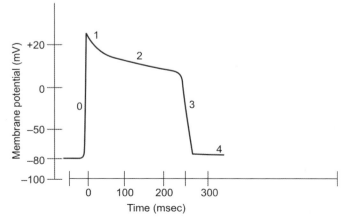

Fig. 25.10: Phases of action potential in ventricular muscle fiber

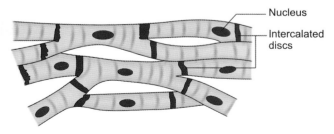

Fig. 25.9: Light microscopic appearance of cardiac muscle

Electrical Potential in Ventricular Musculature

Ventricular musculature represents the working myocardial cells, with a resting membrane potential of – 90 mV. (Fig. 25.10) illustrates the typical action potential in a ventricular muscle fiber. Note that the duration of action potential is quite long. There is a rapid depolarization for 2 milliseconds which overshoots to the positive side, followed by a plateau and then an abrupt repolarization. The plateau phase and repolarization last for about 200 milliseconds. There is no hyperpolarization phase. The different phases and ionic basis of ventricular action potential are discussed in detail below.

Phase 0 (Depolarization Phase)

There is a rapid depolarization from the RMP of – 90 mV which even overshoots to about + 20 mV. This phase is due to opening up of voltage gated Na^+ channels causing Na^+ influx into the myocytes.

Phase 1 (Initial Rapid Repolarization)

There is a rapid fall in the membrane potential from the peak value to about 0mV. This is due to the closure of the Na+ channels and opening up of K^+ channels causing increased K^+ efflux.

Phase 2 (Plateau Phase)

This unique phase of cardiac action potential takes about 200 milliseconds. Slower, but prolonged opening of the voltage gated Ca^{++} channels with Ca^{++} influx causes the membrane potential to remain at an almost constant level. The K^+ efflux is also continuing in this phase.

Phase 3 (Late Rapid Repolarization Phase)

After the plateau phase, the muscle rapidly repolarizes back to the RMP level due to the closure of slow Ca^{++} channels. K^+ efflux also helps in this.

Phase 4 (Restoration of RMP)

The Na^+K^+ ATPase pump works to re-establish the previous concentration gradient of ions and thus RMP is reached.

At a heart rate of 75/min, the duration of action potential is about 250 milliseconds. During the depolarization phase (Phase 0), initial rapid repolarization phase (Phase 1), Plateau Phase (Phase 2) and half of late rapid repolarization phase (Phase 3), the cardiac muscle cannot be excited by a second stimulus, however strong it may be. This is called the absolute refractory period (ARP) of cardiac muscle (Fig. 25.11). Later half of phase 3 and phase 4 are in relative refractory period, i.e., a second stimulus of stronger strength can evoke a response in the cardiac muscle (Fig. 25.11).

Autorhythmicity or Chronotropism

The contraction of cardiac muscle is not the result of stimulation by motor nerves. There are particular cells called pacemaker cells in the heart which can generate automatic and rhythmic action potentials that are conducted throughout the other parts. Pacemaker cells are spontaneously depolarized after every impulse and their membrane potential reaches the firing level of threshold which triggers the next impulse. This property is called autorhythmicity. SA node is the normal pacemaker of human heart. SA node can generate impulses spontaneously at a rate of 72 per minute. But if SA node is damaged, AV node or other parts of the conducting system of heart can act as pacemaker tissue, but the rate of impulse production is much less.

Pacemaker tissue is characterized by an unstable RMP of about – 50 to – 55 mV. As shown in Figure 25.12, the membrane potential, after each action potential, slowly raises to the firing level or the threshold (– 40 mV) to produce next action potential.

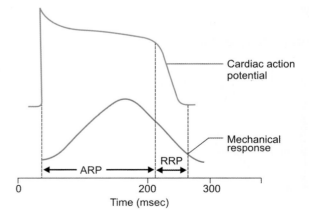

Fig. 25.11: Contractile response and action potential in cardiac muscle

The slow depolarization occurring in the pacemaker tissue till it reaches the firing level is called pre-potential or pacemaker potential.

Ionic Basis

Unstable RMP of pacemaker tissue is due to the leakiness of the membrane to Na^+. When the membrane potential reaches around -50 mV, certain voltage gated Ca^{++} channels called transient Ca^{++} channels or T Ca^{++} channels open, causing Ca^{++} influx. This completes the pre-potential and brings the membrane potential up to the firing level or the threshold. Once the firing level of -40 mV is reached, another set of Ca^{++} channels called long lasting Ca^{++} channels (L Ca^{++} channels) open which cause further increase in Ca^{++} influx and complete the depolarization. At the peak of each impulse, K^+ channels open, causing K^+ efflux which causes repolarization back to -55 mV. Since Na^+ leakiness occurs at this potential, the membrane potential becomes unstable and again a pre-potential is initiated. This process thus continues. Note that unlike in skeletal muscle, Ca^{++} plays an important role in the generation of pre-potential. The pre-potential or pace maker potential can be affected by the stimulation of sympathetic and parasympathetic nerves, which are explained in the "Cardiovascular System".

Conductivity or Dromotropism

The conducting system of heart and the mechanism of conduction are discussed in detail in "Cardiovascular System".

Contractility or Inotropism

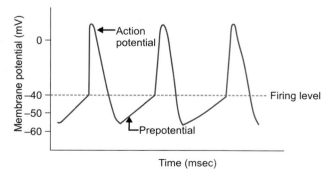

Fig. 25.12: Pacemaker potential

Along with the Ca^{++} released from sarcoplasmic reticulum, a significant amount of Ca^{++} enters the cardiac muscle from ECF during the plateau phase of action potential (Fig. 25.10). This helps in the excitation-contraction coupling of cardiac muscle. The force of contraction of cardiac muscle depends on the length of individual muscle fibers, which is explained by the Frank-Starling's law. This law states that the force of contraction of a muscle depends on the initial length of the muscle fiber, within physiological limits. Application of Frank Starling's law in heart is explained in the Chapter "Cardiovascular System".

■ SMOOTH MUSCLE

Smooth muscle differs from a skeletal muscle in its physical structure, electrical properties, excitation-contraction coupling, innervations and functions. Unlike skeletal muscles, smooth muscle consists of smaller muscle fibers, which are uninucleated, non-striated, spindle shaped and involuntary. Actin and myosin filaments are not regularly arranged as in the skeletal muscle. There are no Z-lines, instead of which "dense bodies" are present.

The dense bodies are attached to the cell membrane. There are no T-tubules and the sarcoplasmic reticulum is poorly developed. Also smooth muscles contain very few mitochondria. The contractile proteins present in the smooth muscle are actin, myosin, tropomyosin and calmodulin. Calmodulin is the Ca^{++} binding protein instead of troponin (seen in skeletal muscle).

Types of Smooth Muscles

Smooth muscles are broadly divided into two, according to their functional and structural differences – Multi-unit smooth muscles and single unit (visceral) smooth muscles.

Multi-Unit Smooth Muscles (Fig. 25.13)

Multi-unit smooth muscle is composed of discrete or separate spindle shaped muscle cells. Each one receives a separate synaptic input from a nerve terminal. Therefore, each of them can function independently. This allows fine motor control. Multi-unit smooth muscle is found in the ciliary body and iris of eye, piloerector muscles of skin and blood vessel walls. There is little intercellular communication among these smooth muscle fibers.

Single Unit (Visceral) Smooth Muscles (Fig. 25.13)

Single unit smooth muscle is the predominant smooth muscle type seen in the walls of visceral organs like uterus, gastrointestinal tract, urinary tract and many blood vessels. They consist of large sheets of closely packed muscle fibers which have extensive intercellular communications or gap junctions (as in cardiac tissue). These gap junctions permit easy spread of action potential and allow coordinated contraction of many cells. In other words, these cells contract as a "single unit" or as a "functional syncytium" (as cardiac muscles). No separate or independent innervations are present for each single unit muscle fiber.

Electrical Properties of Smooth Muscles

Smooth muscles possess an unstable resting membrane potential which varies from -50 to -60 mV. They can have different types of action potentials, which can either initiate or modulate contraction. Some smooth muscles exhibit prolonged action potentials with a prominent plateau, like cardiac muscle cells (Fig. 25.14). Here, the plateau phase can widely vary from 100 milliseconds to a maximum of 1000 milliseconds. This type of action potential is responsible for the prolonged contraction or tonus seen in uterus, ureters, etc. In some single unit muscles, action potentials can be of a simple spike type or a spike followed by a plateau or a series of spikes on the top of slow waves (Fig. 25.14). Some smooth muscle fibers are self-excitatory (like pacemaker cells of heart). Action potentials arise in them without any extrinsic stimulus. They are seen as slow wave potentials (Fig. 25.14). In any case, the upstroke or the depolarizing phase of action potential is due to opening of voltage gated Ca^{++} channels and the repolarization phase is due to opening of voltage gated K^+ channels and closure of Ca^{++} channels. The entry of Ca^{++} during the depolarization phase (especially in those with plateau type of potentials) plays an important role in the initiation of contraction of smooth muscles. (Note that there is no stored Ca^{++} in smooth muscles, as the sarcoplasmic reticulum is poorly developed). Unlike the skeletal muscle, smooth muscle can contract in response to a number of stimuli like electrical stimulation, hormonal stimulation and stretch.

Neuromuscular Junctions in Smooth Muscles

Neuromuscular junctions are not well-developed in smooth muscles as in the skeletal muscles. Smooth muscles are innervated by nerve fibers of autonomic nervous system. The axons of these nerves branch diffusely and innervate many smooth muscle cells (Fig. 25.13). Also the typical "end feets" seen in skeletal muscles are absent here. There are multiple varicosities along the length of these axons which contain vesicles with neurotransmitters (either acetylcholine or norepinephrine).

Excitation Contraction Coupling in Smooth Muscles

Arrival of an action potential at the end of pre-synaptic axon terminal causes release of neurotransmitter from the varicosities. The neurotransmitter binds with the specific receptors on smooth muscle causing opening up of many voltage gated Na^+ and Ca^{++} channels.

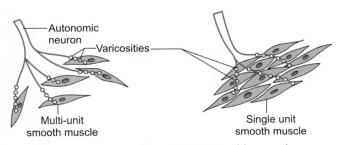

Fig. 25.13: Smooth muscle-types and innervation

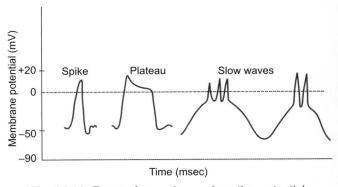

Fig. 25.14: Types of smooth muscle action potentials

Since the sarcoplasmic reticulum of smooth muscle is poorly developed, the entire Ca^{++} needed for muscle contraction comes from ECF during this time. Instead of troponin, Ca^{++} binds with calmodulin. The Ca^{++} - Calmodulin complex formed stimulates an enzyme – calmodulin dependent myosin light chain kinase (MLCK). Unlike in skeletal muscles, the myosin head develops the ATPase activity only when it is phosphorylated. This is catalyzed by the above enzyme-MLCK. Now the actin filament slides over the myosin filament, causing contraction.

There are a groups of enzymes called phosphatases in the smooth muscle, which cause dephosphorylation of myosin. But this may not stop the contraction. The dephosphorylated myosin head may still remain attached to actin. Cessation of contraction and initiation of relaxation of smooth muscle occurs only when Ca^{++} gets dissociated from Ca^{++}-Calmodulin complex. This is called "latch-bridge mechanism". By this mechanism, smooth muscles can remain in a contracted state for a longer time than skeletal muscles, with very little expenditure of energy.

26 Blood and Body Fluids

Blood is a vital transport system of the body which is continuously circulating in the vascular system. The unique fluid nature of this tissue is ideal for its function of transport of substances from one part of the body to another. Hematology is the branch of science that deals with the study of blood and blood forming tissues.

Blood is a red colored, opaque, viscid fluid. In the body of an adult male of 70 kg weight, there will be about 5600 ml of blood. Blood circulates in the cardiovascular system, which is composed of heart and blood vessels. Fresh blood, when taken from an artery, is scarlet red in color and when taken from a vein, is darkish red. Blood is fluid in circulation, but when taken out of the body, it has a tendency to solidify. This process is called coagulation or clotting.

COMPOSITION OF BLOOD

Blood consists of 55 percent plasma and 45 percent formed elements.

Plasma

Plasma is a clear, slightly yellow liquid, containing a large number of organic and inorganic substances dissolved in water. Plasma contains approximately 91 percent of water and 9 percent of solids. Of the solids, about 7 percent are the plasma proteins which consist of serum albumin, serum globulin and fibrinogen. Other organic substances present are glucose, amino acids, fatty acids, phospholipids, cholesterol, hormones, enzymes, antibodies, etc. Inorganic substances are sodium, potassium, calcium, magnesium, iron, copper, zinc, chloride, bicarbonate, etc. Also gases like oxygen, carbondioxide, Nitrogen etc. are present in the plasma.

Normal plasma volume is about 5 percent of total body weight (3500 ml). Plasma clots on standing. It remains in fluid state if an anticoagulant is added.

Formed Elements

They are called so, as they have definite structure and shape when viewed under microscope. There are three types of formed elements or cells. They are:
1. The red blood corpuscles (RBC) or erythrocytes,
2. The white blood corpuscles (WBC) or leukocytes
3. The platelets (thrombocytes).

If blood is collected without adding anticoagulant and if allowed to clot, after sometime, the clot will shrink and a fluid separates out. This supernatant fluid is called serum. It is plasma minus fibrin.

FUNCTIONS OF BLOOD

Transport of Respiratory Gases

Hemoglobin in the red blood corpuscles (RBC) carries oxygen from the lungs to the tissues for the oxidation of food and production of energy. From the tissues, carbondioxide is carried to the lungs, where it is exhaled.

Excretory Function

Various waste products of the tissue metabolism are carried by blood to the excretory channels – kidneys, skin and lungs.

Nutritional Function

The end products of digestion (glucose, aminoacids, lipids, etc.) are absorbed from the digestive tract and transported by blood to various tissues for growth and supplying energy.

Acid-base Balance

Normal pH of blood is 7.4. The enzymes of our body can act only within a narrow range of this pH. Large amounts of acids are produced daily as a result of metabolism. Blood contains various buffers , which can check the rise in H+ concentration.

Transport of Hormones

Hormones are secretions of endocrine or ductless glands, which are directly poured into the blood. Blood carries them to their target organs.

Protection or Defense

The white blood cells (WBC) especially the neutrophils and monocytes can attack the disease causing organisms like bacteria, virus fungus, etc. Blood also contains antibodies or immunoglobulins, which can act against the foreign antigens.

Temperature Regulation

Normal body temperature is 98.4°F or 37°C. Blood helps in easy dissipation of heat from warmer to cooler parts of body, thus helping to keep the temperature of the body at a constant. This helps in proper functioning of different enzymes.

Water Balance

Blood maintains and regulates the fluid contents in various body compartments.

Osmotic Pressure

Blood contains plasma proteins, which exert the osmotic pressure. This is responsible for the balance of fluid in the vascular system.

■ PROPERTIES OF BLOOD

Color

Red colored due to the presence of hemoglobin ("Red pigment").

Clotting

Blood, once taken out of the body, has a tendency to clot or solidify. This process is called clotting or coagulation. This property of blood helps to prevent excessive loss of blood from the body. Clotting can be prevented by adding substances called anticoagulants. For example, heparin, oxalates, citrate, etc. Clotting can also be prevented by defibrination.

Specific Gravity

The specific gravity of whole blood is 1.055 to 1.065. The specific gravity of plasma is 1.025 to 1.035 and that of RBC is 1.095 to 1.10. The specific gravity of blood decreases when there is fluid overload and increases when there is fluid loss, as in vomiting, diarrhea etc.

Viscosity

It is the resistance offered by a fluid to its flow. Normal viscosity of blood is 3.5 to 5.4, which is about 3-4 times that of water. Viscosity of blood is due to blood cells and plasma proteins. Viscosity of blood is slightly more in males than in females.

Osmotic Pressure

Normal osmotic pressure of blood is same as that of plasma and is expressed in terms of osmolarity. It is about 290 milliosmoles/liter. Plasma proteins exert the colloid osmotic pressure. Osmotic pressure exerted by the inorganic constituents is called crytalloid osmotic pressure. Isotonic solution is a solution having the same tonicity of blood. Hypotonic solution has tonicity less than plasma and hypertonic solution has tonicity more than plasma or blood.

■ PLASMA PROTEINS

Plasma proteins are the main constituents of plasma. Their normal level is 6-8 gm/100 ml. There are three major classes of plasma proteins – **Albumin** (3.5 to 5 gm/100 ml), **globulin** (2-3.5 gm/100 ml) and **fibrinogen** (0.2 to 0.3 gm/100 ml). The normal albumin-globulin ratio is 1.7:1. Of these, albumin and fibrinogen are homogeneous, but globulin is heterogeneous containing many different species, e.g. glycoproteins, ceruloplasmin, immunoglobulin, etc. They contain α, β or γ subfractions of globulin.

Synthesis: Plasma proteins are synthesized mainly in liver. However, the immunoglobulins are synthesized by plasma cells from the B-lymphocytes.

Plasma proteins can be separated from blood by:

- Electrophoresis
- Precipitation by using salt
- Fractional precipitation
- Sedimentation with ultracentrifugation.

Functions

Oncotic Pressure

The plasma colloid osmotic pressure (oncotic pressure) of about 25 mm of Hg, is due to plasma proteins. They help in keeping the fluid inside the vascular system and thus maintaining the normal blood volume and water balance. Albumin contributes maximally to the oncotic pressure because of its higher concentration and least molecular weight.

Immunity

The γ globulin fraction of the plasma proteins constitutes the immunoglobulins or the antibodies which protect the body from various infections.

Coagulation

Fibrinogen and other clotting factors (proteins) help in coagulation (clotting) of blood.

Acid-base Balance

Plasma proteins act as buffers since they contain a carboxyl and an amino group.

Viscosity

Plasma proteins provide viscosity to blood , which contributes to the peripheral resistance, that maintains the blood pressure.

Storage Proteins

Plasma proteins serve as reservoirs on which body can depend when there is depletion of tissue proteins, as in fasting.

Transport of Substances

They carry many hormones (e.g. Thyroid hormone), bilirubin, many drugs, etc.

Rouleaux Formation

Mainly fibrinogen, and also globulin can cause the red cells to form Rouleaux (Fig. 26.1).

Applied Physiology

Hypoproteinemia or decreased plasma protein concentration occurs in malnutrition, kidney damage (as there is increased excretion of proteins), and in liver damage. In severe liver damage, there is lower concentration of plasma albumin, but globulin fraction may not show significant fall. The albumin-globulin ratio is therefore altered or reversed in liver diseases.

■ RED BLOOD CORPUSCLES (RBCs)

Human RBC or erythrocytes are non-nucleated, biconcave, circular discs with a mean diameter of 7.5 μm, in adults. It is 2.2 μm thick at its periphery and 1 μm thick at its center (Fig. 26.2). Hence, the outer edge appears as a rim around a central depression, and when viewed from the side, it appears like a "dumb bell". The volume of a single RBC is about 90 μm^3 and surface area 120 μm^2. Each RBC contains 30 picograms of hemoglobin. RBC is red colored due to the presence of the pigment, hemoglobin.

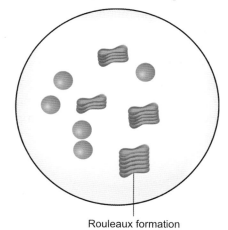

Rouleaux formation

Fig. 26.1: Frank blood viewed under microscope

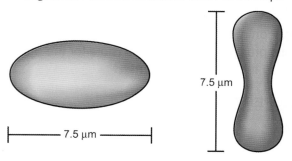

7.5 μm

7.5 μm

Fig. 26.2: Erythrocyte

Advantage of the Biconcave Shape of RBC

It increases the surface area of RBC for the diffusion of respiratory gases. Because of the biconcavity, RBC can squeeze itself through the capillaries, which have smaller diameter than that of RBC.

Composition of RBC

60-63 percent is water and 37-40 percent are solids.

Solids

Solids of which 34 percent is hemoglobin and the rest 3-6 percent is constituted by other proteins, K^+, phospholipids, cholesterol, ATP, many oxidative enzymes and co-enzymes, etc. RBC is devoid of a nucleus, mitochondria, ribosomes, endoplasmic reticulum and nucleic acids. So it cannot divide or synthesize proteins.

Hemoglobin is interwoven in the stroma of RBC, so that even if an RBC is cut, there is no extrusion of hemoglobin.

▮ STRUCTURE OF RBC

RBC is bound by a membrane, made up of a double layer of lipids and a protein layer in between (Fig. 26.3). It is selectively permeable with specific protein channels, pumps and gates. Lipid layer consists of phospholipids, cholesterol and small amounts of glycolipids. The glycolipids contain oligosaccharide sequences on the outer surface which form ABO blood group substances. The peripheral and integral membrane proteins of RBC provide structural integrity (e.g. Spectrin, actin, tropomyosin, glycophorin, ankyrin, etc). Spectrin is the important protein molecule which determines the structural integrity of RBC. These proteins form the skeleton of RBC, which provide its ability to change the shape as it passes through narrow capillaries and to resume the biconcave shape afterwards. The protein "glycophorin" is rich in sialic acid which contributes to the negative charge of the outer surface of RBC, at physiological pH.

Apart from these, a number of enzymes like $Na^+ - K^+$ ATPase, Ca^{++} ATPase, protein kinases, acetyl cholinesterases, etc. are also present in the RBC cell membrane. Hereditary spherocytosis is a genetic disease where the amount of spectrin is less. This renders the RBC fragile as they become spherical in

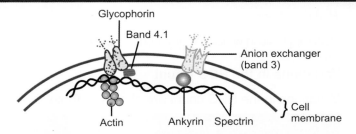

Fig. 26.3: Structure of RBC cell membrane

shape. Increased destruction of RBCs can cause hemolytic anemia also.

Since there is no protein or enzyme synthesis, RBCs metabolize glucose by anaerobic glycolysis for their energy.

▮ NORMAL RBC COUNT

4.5 to 5.5 millions /mm^3 in males
4 to 5 millions /mm^3 in females

Average **life-span** of an adult RBC is **120 days**. RBCs are counted on a special chamber called Neubauer's counting chamber.

Variations in Size and Shape of RBC

Under physiological conditions, the shape of the RBC changes considerably when it passes through the capillaries, but without any rupture. In certain diseases, the size, shape and structure of RBC changes resulting in its rupture. Some of the terms used to denote the abnormalities are:

Microcytosis

Microcytosis in which the size and volume of RBCs are reduced.

Hypochromia

Hypochromia in which the cells are pale and less deeply staining. This occurs in iron deficiency anemia and disorders of Hb synthesis.

Macrocytosis

Macrocytosis in which RBCs are large in size and volume, as in Megaloblastic anemia;

Anisocytosis

Anisocytos in which the cells are of different sizes.

Poikilocytosis

Poikilocytosis in which irregular shape of cells are seen.

Variations in the RBC Count

Variations in the RBC count are shown in Table 26.1.

Table 26.1: Variations in the RBC count		
	Physiological	*Pathological*
Increased RBC count	High altitude Neonates Infants Emotions	Polycythemia vera After hemorrhage Vomiting Diarrhea
Decreased RBC count	Pregnancy Women	Different types of anemia

■ PROPERTIES OF RBCs

Rouleaux Formation

When a freshly drawn drop of blood is observed under microscope, the RBCs are seen adhered to each other and arranged like a pile of coins. This is called rouleaux formatioin (Fig. 26.1). The reason for the Rouleaux formation is that the negative charges which keep the RBCs dispersed in plasma are diminished or lost when blood is kept outside.

Rouleaux formation is influenced by the plasma proteins, bacterial toxins, number of RBCs, size and shape of RBCs, etc.

Albumin contributes to the negative charge on RBCs and hence decreases Rouleaux formation, globulin and fibrinogen decrease the negative charge on RBCs and so favor Rouleaux formation.

Suspension Stability or Sedimentation

Normally the blood cells are uniformly suspended in plasma due to the streamline flow of blood. When a column of blood with anticoagulant is allowed to stand vertically in a tube undisturbed, after some time, the cells settle down in the bottom part and a clear plasma can be seen at the top. The length of the column of the clear supernatant plasma expressed in mm, after the end of one hour is called the Erythrocyte Sedimentation Rate (ESR) (Fig. 26.4). ESR can be estimated by two methods and its value varies with the method used.

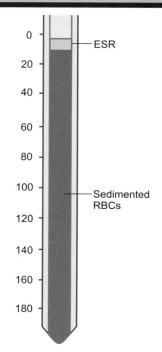

Fig. 26.4: Westergren's pipette

1. By Westergren's method : 3-5 mm at the end of 1st hour(males)
   5-12 mm at the end of 1st hour(females)
2. By Wintrobe's method : 0-9 mm at the end of 1st hour(males)
   0-20 mm at the end of 1st hour (females)

Variations of ESR are shown in Table 26.2. Though ESR is not diagnostic of any specific disease, it has a potent prognostic value.

Table 26.2 Variations of ESR		
ESR	*Physiological*	*Pathological*
Increased ESR	Pregnancy Exercise After a meal Females	Bacterial infections, Malignant diseases Burns, Hemorrhage Anemia
Decreased ESR	Infants Males	Polycythemia Sickle cell anemia Afibrinogenemia

Factors influencing ESR depend almost exclusively on the plasma factors rather than the red cells.

a. The most important factor is the concentration of fibrinogen to which ESR is directly proportional.

When fibrinogen concentration increases, the red cells conglomerate into larger Rouleaux and thus sink more rapidly

b. An increase in globulin can increase ESR. But an increase in albumin will decrease it

c. The biconcave disc shape of RBC favors sedimentation. Therefore anisocytosis and Poikilocytosis will decrease the ESR, since Rouleaux formation is not there

d. Increase in specific gravity or viscosity can decrease ESR

e. An increase in the number of red cells may decrease ESR

f. Technical errors like the slanting position of the tube used or inadequate anticoagulants may cause an increase in ESR.

Packed Cell Volume (PCV)

If a sample of anticoagulated blood is taken in a graduated tube and centrifuged at 300 rev/min for about 30 minutes, the blood gets separated into three layers – a tall bottom layer of red cells, a thin middle layer called "buffy coat" and a top layer of clear plasma. White blood cells and platelets constitute the "buffy coat" (Fig. 26.5). The above said method is called Wintrobe's method and the graduated tube is called the Wintrobe's tube. In this way, the percentage of total blood volume constituted by the red cells is determined, which is called the Packed Cell Volume (PCV) or the hematocrit. Normal PCV is about 45 percent in males and 42 percent in females. Variations in PCV are shown in Table 26.3.

Table 26.3 Variations in PCV		
PCV	*Physiological*	*Pathological*
Increased PCV	Neonates High altitude Exercise Males	Polycythemia Dehydration Burns
Decreased PCV	Pregnancy Females	Anemia

■ HAEMOGLOBIN (Hb)

Hemoglobin or "the red pigment" is the most important constituent of RBCs. It gives the blood its characteristic color. Besides serving as a carrier of oxygen and carbon dioxide, hemoglobin plays an important role as a buffer to maintain the acid-base balance of blood. Hemoglobin can combine with gases other than O_2 and CO_2 like carbon monoxide.

$$Hb + CO \rightleftharpoons COHb$$

COHb is about 250 times more stable than oxyhemoglobin. So even a small amount of carbon monoxide in the inspired air is dangerous, as the tissues will be deprived of oxygen.

Structure

Hb is synthesized by cells of erythroid series in the bone marrow.

Hb belongs to the class of conjugated proteins, with a molecular weight of 68,000. It consists of an iron containing "haem" portion and a colorless protein part called globiin. "Haem" is the pigment portion, belonging to the class of compounds called protoporphyrins. "Globin" belongs to the class of proteins called globulins.

Globin molecule has four polypeptide chains - 2α and 2β chains (in human adults). Therefore, the formula of globin of an adult Hb (HbA) is $\alpha_2\beta_2$. One molecule of haem is attached to each polypeptide chain of globin as in Figure 26.6. Therefore, one molecule of Hb contains 4 haem molecules.

Each haem molecule contains one atom of iron and the iron can combine with a molecule of oxygen (2 atoms of oxygen). Therefore, one Hb molecule can combine with 4 molecules of oxygen (Fig. 26.6). The

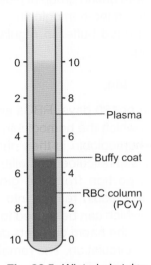

Fig. 26.5: Wintrobe's tube

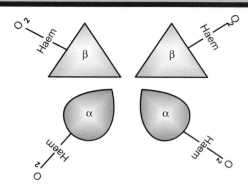

Fig. 26.6: Structure of hemoglobin

iron in haem is in the ferrous (Fe^{2+}) state. When Hb combines with 4 molecules of oxygen, oxyhemoglobin is formed. This occurs in lungs. If the oxyhemoglobin is then exposed to the tissues where oxygen pressure is low, the compound readily decomposes and oxygen is liberated, i.e. it becomes reduced Hb. One gram of pure Hb can combine with 1.39 ml of O_2 when fully saturated. This is called oxygen carrying capacity of hemoglobin.

VARIETIES OF Hb

1. Fetal Hb (HbF)
2. Grower 1 Hb ($Zeta_2$ $Epsilon_2$) and Grower 2Hb ($Alpha_2$ $Epsilon_2$) – These are normally found during the embryonic life.
3. Hb Barts (γ_4) – Homotetramer of γ chains, found in small amounts in the cord blood.

Fetal Hemoglobin

During fetal life, RBCs contain a different type of Hb called fetal Hb (HbF). HbF differs from HbA (adult Hb) in the fact that HbF has 2α and 2γ chains. This increases the oxygen affinity of HbF, which helps in transport of O_2 during fetal life. HbF concentration in blood decreases after birth.

DERIVATIVES OF Hb

1. Oxy and Deoxy hemoglobin are formed when Hb is oxygenated and deoxygenated respectively.
2. Carbamino haemoglobin: $Hb + CO_2 \rightleftharpoons$ Carbamino Hb
3. Carboxy haemoglobin : $Hb + CO \longrightarrow COHb$

4. Met haemoglobin: This is formed when Hb is treated with mild oxidizing agents like potassium ferricyanide.
5. Haematin: The iron present in Fe^{2+} state (Ferrous state) in haem is oxidized to haematin when treated with acid.
6. Glycosylated haemoglobin (HbA_1C)
   This is found in small amounts normally. Glucose combines with Hb non-enzymatically to form HbA_1C. This is the best index of long term control of blood glucose level.

NORMAL Hb CONTENT

Males	:	14-18 gm/100 ml
Females	:	12-16 gm/100 ml
Infants	:	18-23 gm/100 ml

FUNCTIONS OF Hb

O_2 transport

Hb combines loosely and reversibly with O_2 to form oxy Hb. 1 molecule of Hb can combine with 4 molecules of O_2. The iron stays in the Fe^{2+} state itself and so this reaction is called as oxygenation reaction. Thus O_2 is carried from lungs to the tissues, where the O_2 tension is much lower. Oxy Hb then becomes reduced Hb.

CO_2 Transport

CO_2 combines with amino group of globin and not with Fe^{2+} to form carbamino-hemoglobin.
Hb serves as a good buffer in regulating acid-base balance of blood.

Fate of Hemoglobin

After a life span of 120 days, RBCs are destroyed in the spleen from which the hemoglobin molecules are liberated. This hemoglobin is then phagocytosed by the macrophages of reticuloendothelial system, which split the haem portion from the globin part. The aminoacids in the globin returns into the aminoacid pool of plasma, which can be reused for the synthesis of Hb. The iron in the haem part gets detached and is released into the circulation. The remaining portion is converted into bilirubin. The released iron binds with a protein called transferrin and some of it is transported to bone marrow, which can be reused for the synthesis of new hemoglobin. Some of the iron is stored as Ferritin in liver, spleen and bone marrow. These stored forms can be used for the synthesis of hemoglobin, myoglobin or other enzymes which contain iron.

The bilirubin formed combines with albumin in the blood and is transported as a complex to liver. This form is called free bilirubin. In the liver, albumin is removed from bilirubin. Then it is conjugated with glucuronic acid to form conjugated products called bilirubin glucuronide (conjugated bilirubin). This is water soluble. A small amount of bilirubin conjugates with sulfate radicals forming bilirubin sulfate. Conjugated bilirubin is discharged into bile. It gives the particular yellowish green color to bile. Via bile, conjugated bilirubin reaches the intestine where it is reduced to urobilinogen. Some of it goes back into the circulation (enterohepatic circulation). A part of it is excreted through feces as stercobilinogen. Stercobilinogen on exposure to air gets converted into stercobilin, which gives a brown color to the feces. A part of urobilinogen passes through urine, which on exposure to air, gets converted into urobilin that gives the yellowish color to the urine. The whole process is summarized in Figure 26.7.

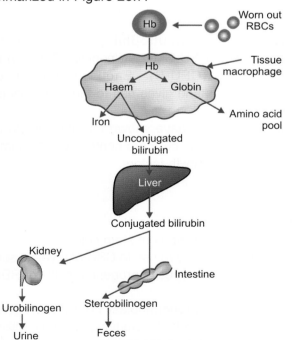

Fig. 26.7: Fate of hemoglobin

Applied Physiology

1. *Abnormal Hemoglobins:* Two major types of inherited disorders of Hb are: "Hemoglobinopathies and Thalassemia".
   - *Hemoglobinopathies:* Due to mutations in the structural genes controlling the production of globin chains, abnormal polypeptide chains are produced. For example, in HbS, the alpha chains are normal, but beta chains are abnormal. Here, valine residue replaces glutamic acid residue in the 6th position of beta chain. When deoxygenated HbS polymerizes inside the RBC and distorts its membrane resulting in "sickling" of the cells. These sickled RBCs are more fragile and give rise to a type of severe anemia called sickle cell anemia.
   - *Thalassemia:* Here the polypeptide chains are normal in structure, but produced in reduced amount or absent due to mutations occurring in globin genes. They can be named as a and b thalassemias when a and b chains are decreased (or absent) respectively.
2. *Jaundice:* Jaundice is the yellowish discoloration of skin and mucous membrane due to accumulation of excess bilirubin in plasma and tissue fluids.

   Normal serum bilirubin level is 0.2 to 1 mg percent. When it becomes >2 mg percent, it becomes clinically detectable. The ideal site for testing jaundice is the sclera of the eye because sclera contains the protein "elastin" which has high affinity for bilirubin. Also a yellow discoloration is more obvious on a white background. Tongue, palm, etc. are the other sites.
   - *Physiological Jaundice:* It is a mild form of jaundice that appears in a newborn after 2nd or 3rd day of birth. The causes of this can be:
     – Hepatic immaturity – Liver is not fully functioning and hence elimination of bilirubin is not adequately taking place.
     – Increased RBC destruction after birth.
       The jaundice is relieved spontaneously after 1 week.
3. *Pathological Jaundice – 3 Types*
   – *Hemolytic Jaundice:* Due to excessive destruction of RBCs, e.g. Malaria, hereditary spherocytosis.
   – *Obstructive Jaundice:* Due to obstructions in the pathway of bile, e.g. Tumor or stone in the common bile duct.
   – *Hepatic Jaundice:* Due to liver diseases, adequate conjugation and secretion of bilirubin into the bile does not take place, e.g. Viral hepatitis, toxins, etc.

■ DEVELOPMENT OF RED BLOOD CELLS

The process of formation of blood cells is called hemopoiesis and that of the red cells, as erythropoiesis.

Blood forming tissues of the body are the myeloid tissue and the lymphoid tissue. Myeloid tissue consists of the red bone marrow from which RBCs, granulocytes, monocytes and thrombocytes or platelets are formed. Lymphoid tissue consists of lymph nodes, thymus and spleen from which the lymphocytes are produced.

Site of Erythropoiesis

Primitive blood cells first appear in the third week of intrauterine life.
1. From third week to third month of intrauterine life – erythropoiesis occurs in the yolk sac. This is called intravascular erythropoiesis.
2. From third month to fifth month of intrauterine life – from liver and spleen (Hepatic erythropoiesis)
3. From fifth month of intrauterine life onwards and in the postnatal life – from the red bone marrow.

Formation of blood cells occurs in the red bone marrow only and not in the yellow bone marrow (as it contains fat). In infancy, all the bone marrow is red, but as age advances, more and more yellow bone marrow is formed due to fat deposition. After the age of 20, only the flat bones and the upper ends of long bones contain appreciable amount of red bone marrow.

■ ORIGIN OF BLOOD CELLS

All circulating blood cells originate from a single, most primitive cell called the pluripotent stem cell. The word "Pluripotent" means "capable of producing many". All the blood cells follow a common general pattern of stages of development (Fig. 26.8).
Steps are:
1. Formation of pluripotent stem cell (PSC)
2. Formation of committed stem cells (CSC)
   The pluripotent stem cells divide and give rise to the committed stem cells of myeloid series and the committed stem cells of lymphoid series. From the myeloid series – granulocytes, monocytes, platelets and RBC are formed. From the lymphoid series the T and B lymphocytes are formed (Fig 26.8).
3. Formation of colonies of progenitor cells or colony forming units (CFU) from the committed stem cells. There are separate CFUs which give rise to erythrocytes (BFU-E, CPU-E), megakaryocytes

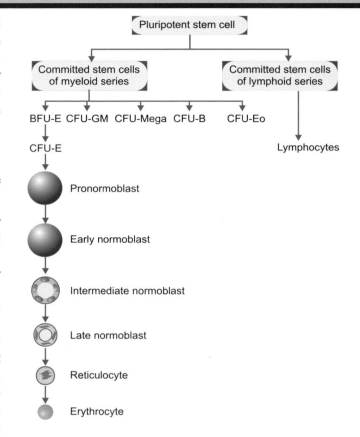

Fig. 26.8: Erythropoiesis

(CFU-mega), neutrophil granulocytes and monocytes (CFU-GM), eosinophils (CFU-Eo) and basophils (CFU-B). Each progenitor cell differentiates and gives rise to the corresponding cell.

In vitro studies have shown that the growth and differentiation of the cells are controlled by a number of hemopoietic growth factors.

Erythropoiesis

Steps
1. Formation of pluripotent stem cell
2. PSC divide and give rise to CSC of myeloid series
3. Formation of 2 types of progenitor cells for RBC – BFU-E and CFU-E.
4. Formation of pronormoblast from the progenitor cells, which is the first morphologically recognizable cell in the erythroid series (Fig. 26.8).

Features:

- The cell is large with a big nucleus, which is strongly basophilic
- Scanty, basophilic cytoplasm
- No Hb
- Mitosis present

5. Formation of early normoblast from pronormoblast
   Features:
   - Smaller than pronormoblast
   - Nucleus, smaller than that of pronormoblast
   - Scanty, basophilic cytoplasm
   - No Hb
   - Mitosis present

6. Formation of intermediate normoblast from early normoblast
   Features:
   - Smaller than early normoblast
   - Cytoplasm – moderate and polychromatophilic due to the presence of Hb
   - Mitosis present

7. Formation of late normoblast
   Features:
   - Nucleus very small and deeply stained
   - Cytoplasm – plenty and eosinophilic
   - Hb present
   - No mitosis (i.e. the cell cannot divide further)

8. Formation of reticulocyte
   Features:
   - Nucleus absent
   - Some DNA is still present which appears as "reticulum"
   - No mitosis. Hb present

9. Formation of mature erythrocyte
   - 7.5 μm diameter
   - Nucleus absent
   - Hb present
   - No mitosis

   All these processes are given in Figure 26.8.

■ FACTORS INFLUENCING ERYTHROPOIESIS

Erythropoietin

It is a glycoprotein hormone secreted mainly by the kidneys (85%) and also by liver (15%), in response to hypoxia. Tissue hypoxia can occur in anemia, chronic lung diseases, hemorrhage at high altitude, etc. Erythropoietin stimulates CFU-E to proliferate and

differentiate to form pronormoblast. It also increases the DNA and RNA contents, required for synthesis of Hb. Thus the precursors pass more rapidly through the different stages under the action of erythropoietin. In renal failure, there is erythropoietin deficiency and so anemia results.

Hemopoietic Growth Factors (Differentiation Factors)

They help in differentiation of blood cells and include the cytokines IL-1 (interleukine-1), IL-3, IL-6, GM-CSF (Granulocyte Monocyte Colony Stimulating Factor), etc. Cytokines are produced by macrophages, fibroblasts, activated T-lymphocytes and endothelial cells.

Maturation Factors

They include vitamins, proteins, minerals and trace metals.

Vitamin B_{12} and folic acid are essential for the synthesis of DNA and hence for the maturation of blast cells. In Vit B_{12} and folic acid deficiencies, the blast cells cannot divide or mature, but the cellular contents increase markedly giving rise to "megaloblast" (hence called "megaloblastic anemia"). Vitamin C is needed for absorption of iron from the gut. Other vitamins required are vitamin B6, thiamine and nicotinic acid. Proteins are essential for the synthesis of globin and stroma of RBCs.

Iron, copper, cobalt, nickel, manganese, zinc, etc. are also required for erythropoiesis.

Hormones

Androgens like testosterone, thyroid hormones, catecholamines (epinephrine and norepinephrine) and growth hormone stimulate erythropoietin, thus promoting the proliferation of RBCs in the bone marrow.

Fate of RBC

Total life span of an RBC is 120 days. As the age of RBC increases, the protective enzymes in it become inactive. So, oxidative damages appear and the RBC becomes rigid, spheroidal and fragile. When these fragile RBCs pass through small blood vessels of spleen, they are ruptured. Thus spleen is the main graveyard of RBC. But in certain diseases, liver can also destroy RBCs.

ANEMIA

Anemia is the condition in which either the RBC count or the Hb concentration or both are decreased.

Cut-off values

In adult female, Hb <11.8 gm percent and RBC count < 3.8 millions/mm^3 indicates anemia.

In adult male, Hb < 13.2 gm percent and RBC count < 4.5 millions/mm^3 indicates anemia.

Classification

Based on the causes of anaemia, it is classified into three:
I.   Due to decreased production of RBCs
II. Due to increased destruction of RBCs
III. Due to excessive loss of blood

I. Due to decreased production of RBCs: See Table 26.4

Table 26.4: Causes of decreased production of RBC	
Due to deficient supply of nutrients	*Due to inactive bone marrow*
• Iron deficiency anemia • Folic acid deficiency • Vit B$_{12}$ deficiency	• Irradiation • Anticancer drugs • Infiltrative diseases of bone marrow • Hypothyroidism

II. Due to increased destruction of RBCs:

• Sickle cell anemia
• Spherocytosis
• Hypersplenism
• Drugs

III. Due to excessive loss of blood:

• Acute hemorrhage as in accidents
• Chronic blood loss as in piles, hook worm infestation, heavy menstruation, etc.

Anemia may also be classified on the basis of changes in the appearance of RBCs as:
I.   Microcytic, hypochromic, e.g. Iron deficiency anemia

II. Normocytic, normochromic, e.g. anemia due to blood loss, aplastic anemia, etc.
III. Macrocytic anemia, e.g. Vit.B$_{12}$ and folic acid deficiency anemias.

Two common types of anemia that we come across most often are the Iron deficiency anemia and anemia due to Vit.B$_{12}$ and folic acid deficiency (Megalobalstic anemia).

IRON DEFICIENCY ANEMIA

Most common type, which occurs due to the following reasons – deficiency of iron in food, decreased intestinal absorption of iron, excessive loss of iron from the body and increased demand of the body (as in pregnancy and childhood).

Blood Picture

The Hb concentration is reduced (as iron is a part of Hb). Microcytic, hypochromic cells are seen in the peripheral smear.

Megaloblastic Anemia

It results due to the deficiency of Vit.B$_{12}$ and folic acid in our body. The deficiency occurs either due to decreased intake or dietary deficiency or decreased absorption from intestine. The latter cause is the commonest. Their deficiency results in maturation arrest during erythropoiesis. The precursors of RBCs cannot divide. In stead, they increase in size and form "megaloblasts", which can be seen in a peripheral smear of blood.

There is an intrinsic factor, produced by the stomach, which helps in the absorption of Vit.B$_{12}$ by binding with it. So in the deficiency of this intrinsic factor, there can be decreased absorption of Vit.B$_{12}$, resulting in "Pernicious anemia".

BLOOD INDICES

Blood indices are used for classifying anemias. They can be calculated from the values of PCV, RBC count and Hb concentration.

MCV (Mean Corpuscular Volume)

It is the mean volume of a single RBC.

$$MCV = \frac{PCV \times 10}{RBC \text{ count in million/mm}^3}$$

Normal value : 76-96 μ^3

A cell with normal MCV is named normocyte. MCV is increased in macrocytic anemias (e.g. Pernicious anemia) and decreased in microcytic anemias (e.g. Iron Deficiency Anemia).

MCH (Mean Corpuscular Hemoglobin)

It is the amount of Hb in a single RBC

$$MCH = \frac{Hb \text{ in gms/100 ml} \times 10}{RBC \text{ count in millions/mm}^3}$$

Normal value = 29 ± 2 picograms

MCH is decreased in hypochromic anemias (e.g. Iron deficiency anemia) and increased in macrocytic anemia (e.g. Pernicious anemia).

MCHC (Mean Corpuscular Hb Concentration)

It is the percentage concentration of Hb in a single RBC and is expressed in percent.

$$MCHC = \frac{Hb \text{ in gms/100 ml} \times 100}{PCV}$$

Normal value = 34 ± 2 percent

MCHC is decreased in iron deficiency anemia and is normal in pernicious anemia.

■ THE WHITE BLOOD CELLS (LEUKOCYTES)

The WBC or leukocytes are the body's "soldiers" that provide defense and immunity from invading organisms. They differ from the red cells in that they possess nuclei and do not contain hemoglobin. They are present in blood not because they have any respiratory or transport functions, but because they can reach the tissue easily through the blood and perform their functions. They are much less in number, as compared to the RBCs.

Types of Leukocytes

On the basis of morphology, the WBCs are broadly classified into 2 types: The granulocytes and the agranulocytes (See Flow chart 26.1).

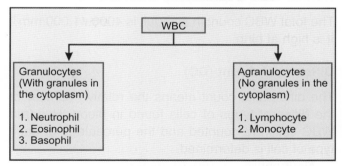

Flow chart 26.1: Classification of WBC

Development of WBCs (Leukopoiesis)

In the intrauterine life, the WBCs develop in the mesoderm and migrate into the blood vessels. In the postnatal life, the granulocytes and monocytes develop from the red bone marrow, while the lymphocytes develop from the lymphoid tissues mainly and to a lesser extent, from the red bone marrow.

Steps of leukopoiesis is concisely given in the Figure 26.9.

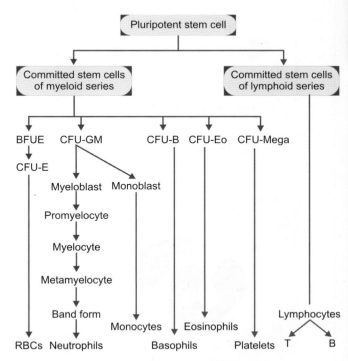

Fig. 26.9: Leukopoiesis

Part 2

Physiology

Normal WBC Count

The total WBC count in an adult is 4000-11,000/mm^3. It is high at birth.

Differential Count (DC)

The differential count means the relative number of the different types of cells found in blood. Hundred WBC cells are counted and the percentage of each type of cell is determined.

    Neutrophils   :   50-70 percent
    Eosinophils   :   1-4 percent
    Basophils   :   0-1 percent
    Lymphocytes   :   20-40 percent
    Monocytes   :   2-8 percent

Functions of WBCs

This includes all the functions of the different types of WBCs, which are separately dealt below.

■ NEUTROPHILS

They constitute about 50-70 percent of the total WBCs. They have a diameter of 10-12 μ and a multilobed nucleus (2-5 lobes), which stains purplish blue. Cytoplasm contains neutrally stained, fine granules (and hence the name "Neutrophil") (See Fig. 26.10B).

After its formation in the red bone marrow, the neutrophil spends 10 days there and after that it stays for 7 hours in the circulation. Then it migrates into the tissues where it spends 1-4 days, after which it dies. The dead neutrophil is removed by tissue macrophages.

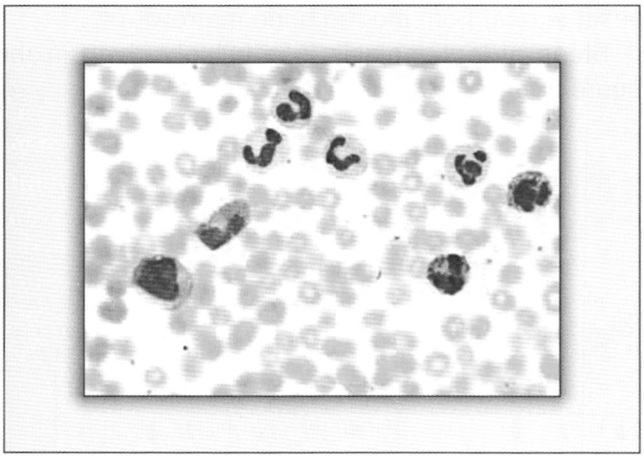

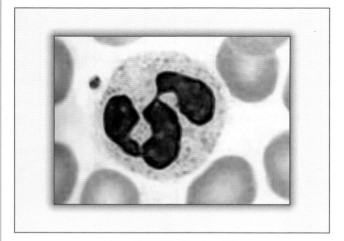

Fig. 26.10A: Peripheral smear

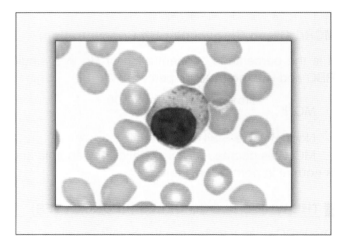

Fig. 26.10C: Large lymphocyte

Fig. 26.10B: Neutrophil

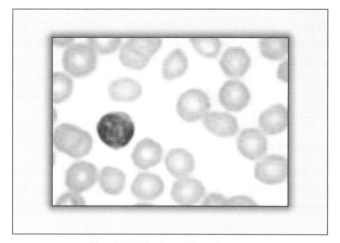

Fig. 26.10D: Small lymphocyte

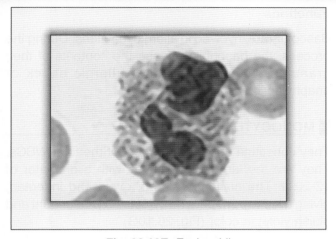

Fig. 26.10E: Eosinophil

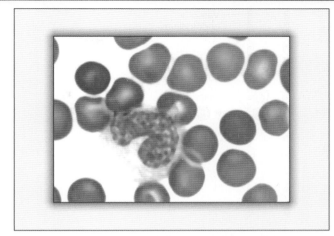

Fig. 26.10G: Monocyte

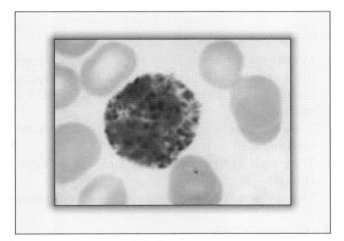

Fig. 26.10F: Basophil

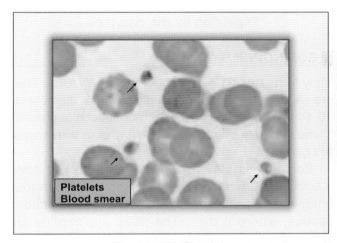

Platelets
Blood smear

Fig. 26.10H: Platelets

Part 2

Physiology

Functions of Neutrophils

The neutrophil constitutes the first line of defense against bacterial infections. Neutrophils are actively phagocytic.

The process of phagocytosis or engulfing includes the following events (Fig. 26.11).

When there is any bacterial invasion, neutrophils in the blood come and stick to the vascular endothelium by a process called margination. Then they move with the help of pseudopodia and reach the target tissue. This process is called diapedesis. The inflammatory processes produced by the bacteria will attract more and more neutrophils towards it and they get attached to the bacteria. This is called chemotaxis. Then they engulf or phagocytose the bacteria, by extending their pseudopodia. Neutrophil granules contain many enzymes, which are capable of digesting the engulfed bacteria. The whole process can be summarized as:

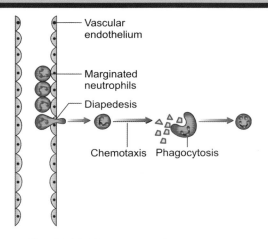

Fig. 26.11: Phagocytosis by neutrophils

Bacterial invasion → Margination → Diapedesis → Chemotaxis → Phagocytosis → Digestion by the enzymes.

EOSINOPHILS

They form about 1-4 percent of the total WBCs. They are slightly larger than neutrophils, with a diameter of 10-14 μ. The nucleus is bilobed or spectacle shaped. There is plenty of eosinophilic cytoplasm with coarse red granules (Fig. 26.10)

Eosinophils stay for 5-6 days in the red bone marrow, 8-12 hours in the circulation and then they enter the tissues, mainly skin and subcutaneous tissues, where it dies. The dead eosinophils are phagocytosed by the macrophages.

Functions

1. They prevent the spread of local inflammatory process
2. Larvicidal , in parasitic infections
3. Have a role in allergic inflammation.

BASOPHILS

Basophils are the rarest leukocytes which constitute only 0-1 percent of total WBCs. They have a diameter of 10-12 μ. The nucleus is often irregular shaped with a slight indentation. Cytoplasm contains coarse, bluish-black granules, which even mask the nucleus (Fig. 26.10F). These granules contain heparin, histamine and serotonin.

Functions

Basophils are involved in allergic reactions. During the process, the basophils release the contents of their granules, which bring about allergic shock or anaphylactic shock.

MONOCYTES

They constitute about 2-8 percent of the total WBCs. They are the largest leukocytes with a diameter of 12-20 μ. The nucleus is large, ovoid and indented, giving a kidney-shape. The cytoplasm is abundant and bluish violet in color (Fig. 26.10G)

After circulating in the blood for about 72 hours, the monocytes enter the tissues and become tissue macrophages (histiocytes), which form a component of the reticuloendothelial system .

Functions

1. Monocytes form the 2nd line of defense against the bacterial infections. They can phagocytose bacteria, cell debris and even dead RBCs or neutrophils.
2. They can lyse tumor cells and thus help to prevent malignancies.
3. They help in the development of cell mediated immunity.
4. They help in wound healing and tissue remodeling

LYMPHOCYTES

They constitute 20-40 percent of the total WBCs in adults. But in infants, their number is more than that of neutrophils. Structurally, they are of two types – large and small. About 80 percent of the circulating lymphocytes are small lymphocytes, which are the mature forms. Both have a nucleus – round or slightly indented, and occupying almost the entire cell. Entire nucleus stains deep blue. They contain only a very thin rim of cytoplasm, which have no granules (Figs 26.10C and D).

A group of the lymphocytes, formed from the bone marrow, go to the thymus for maturation into T-lymphocytes. Another set mature in the bone marrow itself to form B-lymphocytes. In the circulation, 80 percent are T-lymphocytes, 15 percent are B-lymphocytes and the rest 5 percent are null cells (neither T or B lymphocytes) or natural killer cells. From the blood, the lymphocytes enter the lymph nodes. There is a to and fro circulation of lymphocytes between the lymph nodes and blood.

The average life span of lymphocytes is 100-200 days.

There are different types of T lymphocytes . They are—

1. Cytotoxic T cells,
2. Helper T cells,
3. Suppressor T cells, and
4. Memory T cells.

The T-cells, by means of these different types of cells, destroy viruses and some bacteria, and provide cell mediated immunity (CMI).

B cells, on the other hand, differentiate, divide and form plasma cells, during infections by micro-organisms. These plasma cells secrete certain chemicals called immunoglobulins or antibodies against the antigen of invading organism. Thus B lymphocytes provide the type of immunity called humoral immunity.

Functions of Lymphocytes

T-lymphocytes provide cellular immunity and B lymphocytes provide humoral immunity. The T lymphocytes can recognize, attack and destroy foreign tissues and organs. Hence, they have a role in the rejection of transplanted organs.

■ IMMUNITY

The term immunity refers to the resistance exhibited by the host towards injury caused by microorganisms and their products. Some of the immune mechanisms are present from birth, by virtue of our genetic and constitutional make up. This is called *innate immunity*. On the other hand, the resistance that an individual acquires during life is called *acquired immunity*. 'Immunity' has been described in detail in chapter 27.

■ RETICULOENDOTHELIAL SYSTEM (RES)

RES or mononuclear phagocytic system means a system of cells with the following features:
1. They are powerful phagocytic cells
2. They can take supravital stains (i.e. stains used for living cells)

RES consists of the following cells:
• Monocytes
• Mobile macrophages
• Fixed tissue macrophages
• A few specialized cells in bone marrow, spleen and lymph nodes.

Since almost all the above cells are derived from the monocytic stem cells, they are called mononuclear phagocytic system (MPS). The cells of this system are distributed in various parts of our body. Monocytes, from the bone marrow, enter into circulation and remain for 72 hours, after which they enter the tissues. In the tissues certain changes occur in these cells. They swell and large number of lysosomes develop in them. These cells are now called macrophages. There are 2 types of macrophages – Mobile macrophages, which wander in circulating blood and Fixed macrophages in the tissues, e.g. for fixed tissue macrophages – Kupffer cells in liver, microglia in brain, pulmonary alveolar macrophages, reticulum cells in spleen, Langerhan's cells in skin, etc.

Functions of RES

1. They ingest and destroy old RBCs
2. They can phagocytose microorganisms
3. They assist to remove antigens in acute inflammations
4. They secrete various inflammatory mediators like interleukins, cytokines, etc.

■ PLATELETS OR THROMBOCYTES

Platelets are small, spherical, non nucleated mass of protoplasm with a diameter of 2-4 µm. The protoplasm contains mitochondria, granules, contractile elements, tubules, etc. (Fig. 26.12). The granules contain glycogen, platelet derived growth factors, ATP, histamine, serotonin, etc. The contractile elements contain actin and myosin, which help in amoeboid movements of platelets. Tubules are for the storage of calcium. Platelets do not have DNA and RNA. Therefore, there can be no protein synthesis inside the platelet.

Development of platelets: Platelets develop from the red bone marrow as follows:

Pluripotent stem cell → committed stem cells → CFU megakaryoblast → Promegakaryoblast →

Part 2

Physiology

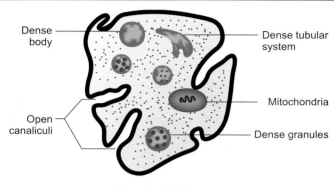

Fig. 26.12: Platelet

Dense body — Dense tubular system — Mitochondria — Dense granules — Open canaliculi

Megakaryoblast → Megakaryocyte → Platelets
(Refer Fig. 26.9)

Average life span of platelets is 7-10 days. After that, they are destroyed in spleen.

Normal platelet count is 1.5 to 2 lakhs/mm³ of blood.

Functions of Platelets

1. Help in vasoconstriction immediately after a vascular injury
2. Help in primary hemostasis, by forming a platelet plug and accelerating the process of coagulation.
3. Help in clot retraction, which is aided by the contractile elements in platelets.
4. Platelets can engulf some bacteria and carbon particles.
5. Platelets play an important role in inflammatory process, by the release of platelet growth factors from their granules. These platelet growth factors help in the process of chemotaxis.
6. Platelets store serotonin, histamine, etc.

■ HEMOSTASIS

The word hemostasis means "arrest of bleeding" by physiological process. Blood has got a natural, unique property of fluidity, when inside the blood vessels, and a tendency to set into a semisolid jelly, when shed. Both these properties are essential for our life. When there is a small injury to a blood vessel, a number of events are initiated that ultimately arrest the bleeding by the formation of a clot. Hemostatic events occurring are as follows:

I. *Immediate vasoconstriction* at the site of injury or vasospasm

II. *Formation of a platelet plug* or temporary hemostatic plug or primary hemostasis.

This process begins within a few seconds after the injury. The injury to the blood vessel causes the subendothelial tissue of the vessel to become exposed. The collagen fibers of the subendothelial tissue can attract platelets. Circulating platelets will come and adhere to these exposed collagen fibers. This binding is reinforced by the von Willebrand Factor (vWF). The whole process is called platelet adhesion. The adhered platelets can attract more and more circulating platelets to the injury site, resulting in *platelet aggregation*. After some time, a clump of platelets is formed at the site of injury called *the loose platelet plug*. Also, platelets secrete certain substances like serotonin, histamine, etc. which can cause local vasoconstriction. The platelet plug together with local vasoconstriction produce what is known as the *primary hemostasis*.

The time between the onset of bleeding and the primary hemostasis is called the bleeding time.

III. *Secondary Hemostasis*

The loosely aggregated platelets in the temporary plug are converted into a definitive clot by the deposition of fibrin. The formation of fibrin is by a process called clotting or coagulation. Coagulation mechanism includes a cascade of reactions in which some inactive enzymes are activated and the activated enzymes, in turn, stimulate other inactive enzymes.

Normally, blood contains two sets of substances called procoagulants and anticoagulants. Procoagulants are substances which help in clotting and anticoagulants are substances which prevent clotting. There is a balance between these two so that the anticoagulant action predominates over the other. But, when an injury occurs, procoagulants are activated.

Procoagulant Factors

Procoagulant factors or the clotting factors are 13 in number and are given below:

Factor I : Fibrinogen
II : Prothrombin
III : Tissue thromboplastin
IV : Calcium

V : Proaccelerin or labile factor

VI : The existence of this factor is not accepted

VII : Proconvertin or stable factor

VIII: Antihemophilic factor

IX : Christmas factor

X : Stuart-Prower factor

XI : Plasma thromboplastin antecedent (PTA)

XII : Hageman factor or Glass factor

XIII : Fibrin stabilizing factor

Mechanism of Coagulation

The fundamental reaction of coagulation is the conversion of the soluble plasma protein fibrinogen into insoluble fibrin threads. For this, the following reactions have to occur:

- Thrombin acts upon fibrinogen to form fibrin
- Thrombin is formed by activation of prothrombin
- Prothrombin to thrombin activation occurs in the presence of Factor X-a ('a' denotes activated)
- Factor X-a is produced by two major pathways – the intrinsic pathway and the extrinsic pathway

These steps are summarized as (explained afterwards) given in Flow chart 26.2

The Intrinsic Pathway (Occurring both *in vivo* and *in vitro*)

The intrinsic pathway is triggered by an injury to a blood vessel, leading to the contact of blood with the subendothelial collagen tissue and release of phospholipid from platelets.

1. The first reaction is the activation of Factor XII to XIIa, by the above said process, and in the presence of HMWK and Kallikrein.
2. XIIa, in the presence of HMWK and kallikrein, converts inactive Factor XI to XIa.
3. Factor IX is converted into IXa, in the presence of factor XIa, calcium.
4. Factor IXa, along with VIIIa, calcium and platelet phospholipids, converts X to Xa (Fig. 26.13).

The Extrinsic Pathway (Occurring only *in vivo*)

This pathway is triggered by a trauma to tissue, which causes the release of tissue factor and tissue thromboplastin from the injured tissues.

Flow chart 26.2: Steps of coagulation

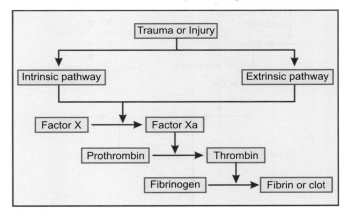

1. Tissue thromboplastin (Factor III) activates Factor VII to VIIa
2. Along with calcium, phospholipids and tissue thromboplastin, VIIa converts X to Xa (Fig. 26.13)

After the activation of Factor X to Xa by either intrinsic or extrinsic pathways, the common pathway for the formation of clot occurs as follows: (Fig. 26.12)

1. Factor Xa, along with factor Va, phospholipids and calcium, acts as the prothrombin activator which activates prothrombin to thrombin.
2. Formation of fibrin (clot). Thrombin converts fibrinogen to fibrin, in presence of calcium. Thrombin can also activate Factor XIII to XIIIa, which stabilizes the clot formed.
3. The fibrin threads form a network, entrapping RBCs, WBCs, platelets and plasma to form the clot. The clot becomes adherent to the injured vessel wall, plugging it permanently.
4. Clot retraction: Within a few minutes of formation of clot, it starts contracting to about 40 percent of its original size. Platelets help in this process. This is called clot retraction. This process can continue for about 1-4 hours. In this process, a clear fluid called serum is expressed from the clot. Clot retraction will draw the ends of the injured vessel closer and also helps in uninterrupted flow of blood through it.

Role of Vitamin K in Coagulation

Vitamin K is required for the synthesis of the procoagulant factors – II, VII, IX and X which are therefore called 'Vitamin K dependent procoagulants'.

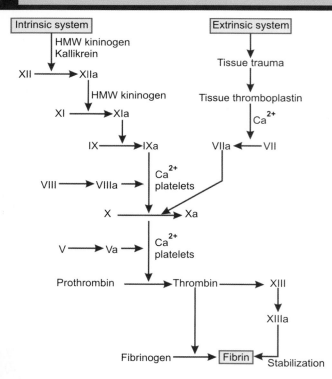

Fig. 26.13: The clotting mechanism

Vitamin K is present in many vegetables like cabbage and also synthesized to a lesser extent, by the bacterial flora of intestine.

Role of Liver

Liver synthesizes the procoagulants – Prothrombin, fibrinogen, factors V, VII, IX, X and XI. Liver failure can therefore lead to severe hemorrhagic disorders.

Role of vWF (von Willebrand's Factor)

VWF is a high molecular weight glycoprotein synthesized by the platelets and the endothelial cells. In the plasma, vWF serves as a carrier protein for the transport of clotting factor VIII. Hence it is important in clotting process. It is also required for the primary hemostasis, as it helps in platelet adhesion to the injured vessels.

Bleeding Time

It is the time between the onset of bleeding and the formation of loose platelet plug (primary hemostasis). Normal value lies within 8 minutes.

Clotting Time

It is the time between the onset of bleeding and the formation of a firm clot. Normally this occurs within 10 minutes.

ANTICOAGULANTS

Anticoagulants are substances, which prevent clotting. There are certain naturally occurring anticlotting mechanisms in our body.
1. Smoothness of normal endothelial lining of blood vessels
2. Heparin
3. Antithrombin III
4. The fast velocity of blood flow in larger blood vessels
5. *Fibrinolytic system:* This is the process which brings about dissolution of fibrin clot, thus preventing spread of intravascular clotting.

Clinically used Anticoagulants

Broadly classified into two : Used *in vivo*
                                  Used *in vitro*

Anticoagulants used in vivo

They include heparin and Vitamin K antagonists, which are used for therapeutic purposes, in patients with thrombotic disorders, myocardial infarction, pulmonary embolism, etc.

Heparin: It can inhibit the factors IIa, IXa, Xa, XIa and also potentiate the actions of antithrombin III. Heparin is given as intravenous or subcutaneous injections, as it is not absorbed orally. Commercially, it is prepared from the liver of animals. In case of heparin toxicity, protamine sulfate is used as the antidote.

Vitamin K antagonists: They are used as oral anticoagulant agents. Vitamin K is required for the synthesis of the factors II, VII, IX and X from liver. So Vitamin K antagonists can block their synthesis by inhibiting Vitamin K, e.g. Warfarin.

Anticoagulants used in vitro

They are not used therapeutically, but used in labs to prevent coagulation of stored blood (note that the intrinsic pathway can occur *in vitro* also), e.g. oxalates, citrate, EDTA, etc. All these can precipitate calcium from the blood and thus prevent clotting. Heparin can also be used to prevent *in vitro* clotting.

▎BLEEDING DISORDERS (FLOW CHART 26.3)

Hemophilia

Hemophilia is a congenital bleeding disorder. There are two types of hemophilia – Hemophilia-A or the classical hemophilia and Hemophilia-B or Christmas disease. The former is due to Factor VIII deficiency and the latter is due to Factor IX deficiency. A minor injury like falling on knees, tooth extraction, etc. is sufficient to start prolonged bleeding. At first the bleeding is less, due to the primary hemostasis, but then restarts severely within a few minutes (as there is no secondary hemostasis). Here, the clotting time is prolonged, but bleeding time is normal. The only effective therapy is intravenous injection of the deficient factor.

▎BLOOD GROUPS AND BLOOD TRANSFUSION

Several systems of classification of the types of blood are there, like ABO, Duffy, Rh, MNS, Kell, Lewis,

Lutheran, etc. Of these, the most important are the ABO and Rh systems, discovered by Karl Landsteiner in 1901. He was awarded the Nobel Prize for the same. The blood group of a person is genetically determined.

ABO System

According to this system, the blood grouping is done depending on the presence or absence of two agglutinogens (antigens), present on the surface of the red cells. The agglutinogens are named A and B and the blood group is determined according to the agglutinogen present. Agglutinins or antibodies against the agglutinogens are present in the plasma and are given the names α (anti-A) and β (anti-B) "A" agglutinogen will react with 'α' agglutinin and B agglutinogen with 'β' agglutinin. Human beings are divided into four main groups, whose features are given in Table 26.5.

Table 26.5: ABO System			
Blood group	Agglutinogen	Agglutinin	Percentage of population in India
A	A	β (anti-b)	25%
B	B	α (anti-A)	25%
AB	AB	-	5%
O	-	α and β	45%

Landsteiner's Law

It states that when a particular blood group antigen is present in the RBC, the corresponding antibody will be absent in the plasma.

Since people with *O group* have no agglutinogens (antigens), they are called as *Universal donors* and since there are no agglutinogens in people with *AB group*, they are called as *Universal recipients.*

Inheritance of Blood Groups

It depends on the inheritance of three genes – A, B and O (A and B being dominant). These genes give rise to the agglutinogens A, B and non-agglutinogenic O substance respectively. Genes A and B are co-dominant. The genotypes of different blood groups are shown in Table 26.6.

Flow chart 26.3: Types of bleeding disorders

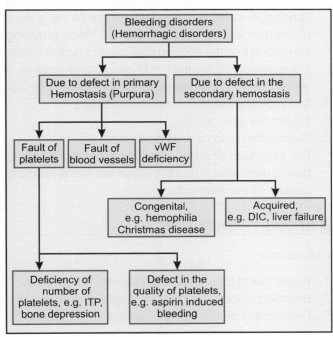

Part 2

Physiology

Table 26.6: Genotypes of blood groups

Blood group	Genotype
A	AA or AO
B	BB or BO
AB	AB
O	OO

THE Rh (RHESUS) SYSTEM

This system was first discovered in **Rhesus monkeys** and hence the name. In this system, there are several subgroups of Rh antigen – C,D,E,c,d,e, etc. but there are no naturally occurring antibodies. Out of these antigens, the D antigen is the most important. When Rh D is present in the RBC, the person is named Rh positive and when it is absent, he is named Rh negative. About 90 percent of Indian population are Rh positive and 10 percent are Rh negative. The Rh antibodies, or anti-D antibodies are absent in both Rh negative and Rh positive persons. But when Rh negative persons receive Rh positive blood or if Rh negative mother conceives Rh positive fetus, Rh antibodies or anti-D antibodies are formed. They belong to the IgG class of antibodies and therefore can cross the placenta.

Rh INCOMPATIBILITY

If an Rh negative individual receives Rh positive blood, there will not be any immediate reaction. However, the donor's red cells induce the formation of anti-D antibodies in the recipient. If he receives a subsequent Rh positive blood transfusion later, the anti-D antibodies immediately destroy the donor red cells and this is called Rh incompatibility.

Erythroblastosis Fetalis

Erythroblastosis fetalis or the hemolytic disease of the newborn occurs due to Rh incompatibility. This is a condition where the RBCs of the newborn are hemolyzed by the maternal antibodies which reach the fetus by crossing the placenta during pregnancy.

When an Rh negative mother conceives an Rh positive fetus, at the time of delivery , small amount of fetal blood that gets into the maternal circulation will induce the development of anti-D antibodies in the mother. Usually no reactions occur during the first delivery. But during the second pregnancy, these anti-D antibodies from mother cross the placenta into the Rh positive fetus, causing hemolysis. This will induce erythropoiesis in foetus, leading to the formation of more and more blast cells. Hence this condition is named erythroblastosis fetalis. If the hemolysis is severe, the fetus dies *in utero*. If it is mild, the newborn is born with the features of hemolytic anemia – low Hb concentration, increased reticulocyte count, nucleated RBCs (blast cells), jaundice, etc. Increased rate of hemolysis will increase the serum bilirubin level. When its concentration in the serum becomes high, bilirubin enters the brain and gets deposited. This condition is called kernicterus. This can be fatal and results in death of the newborn.

Prevention

Erythroblastosis fetalis can be prevented by giving a single dose of anti-D antibody injection to a Rh negative mother, immediately after her first delivery (provided the first child is Rh positive) and after each abortion.

USES OF BLOOD GROUPING

1. Blood transfusion
2. To diagnose or to predict Rh incompatibility
3. To investigate a case of disputed paternity. But this is not absolutely reliable. Here, if the blood groups of the mother and child are known, blood grouping can prove that the alleged man cannot be the father. For example: If mother is O group and baby is A group, then the alleged father cannot be of O group.
4. Medicolegal value – for criminal cases
5. Organ transplantation
6. Susceptibility to certain diseases
   For example: O group persons are more prone to peptic ulcer and A group people are more prone to cancer of the stomach.

BLOOD TRANSFUSION

Indications

1. Acute blood loss – as in accidents, surgery etc.
2. Bleeding disorders – as in hemophilia, purpura, etc.
3. Diseases of blood – as in severe anemia, leukemia, etc.

4. In poisoning – as in carbon monoxide poisoning
5. Shock

Before blood transfusion, the following should be done:
- Blood grouping
- Cross matching
- Screening for diseases like AIDS, hepatitis, malaria etc.

CROSS MATCHING

There are two types of cross matching – direct and indirect or major and minor respectively. In direct (major) cross matching, the serum of the recipient is matched with the RBCs of the donor and observed for any agglutination. In indirect (minor) cross matching, the recipient's RBCs are matched with the serum of the donor and observed for any agglutination. Direct method is more important than the indirect.

HAZARDS OF BLOOD TRANSFUSION

Can be grouped under two headings:

Due to Mismatched Blood Transfusion

This may be due to some clerical mistakes in labeling of blood group. If there is mismatching, the RBCs will be hemolyzed and clumped. The clumped RBCs may block small blood vessels in the heart and kidneys, causing ischemic changes. This can result in heart failure and kidney failure. Hemoglobin released from these hemolyzed RBCs can block the renal tubules causing anuria. Apart from this, there can be anaphylactic shock (due to increased release of histamine from massive lysis of basophils), fever and rigor. After a few days of mismatched transfusion, hemolytic jaundice may develop.

Due to Causes other than Mismatching

1. Circulatory overload – due to excessive blood transfusion, which can later give rise to pulmonary edema.
2. Transmission of diseases like AIDS, hepatitis B, malaria, etc.
3. Iron overload – due to repeated blood transfusions.
4. If not properly stored, the transfused blood may contain some pyrogens , which can produce fever.

BODY FLUID COMPARTMENTS AND LYMPH

Total body water constitutes a very high percentage – about 60 percent of the total body weight, which is distributed as the body fluid. Body fluids may be divided into two major compartments.

1. Intracellular fluid (ICF) – It is the fluid present inside the cells.
2. Extracellular fluid (ECF) – It is the fluid present outside the cells. ECF is again made up of 2 compartments – Plasma and interstitial fluid. Plasma means the fluid inside the vascular system, which surrounds the blood cells. Interstitial fluid means the fluid, which surrounds all cells except the blood cells.

About one-third of the total body water is constituted by the ECF and the remaining two-thirds, by the ICF. In addition, there is a transcellular fluid compartment made up of the fluids in synovial joint, pleural, pericardial and peritoneal spaces, cerebrospinal fluid, etc. But this forms only a small part of ECF (about 1 L).

Blood Volume

In healthy adults, 6-8 percent of total body weight is constituted by blood. The normal blood volume in an adult is 5600 ml. Other normal values related to formed elements of blood are shown in Table 26.7.

LYMPH

Lymph is a modified tissue fluid that enters the lymphatic vessels. It is an accessory route for the removal of excess interstitial fluid. The lymph channels begin in the interstitial spaces as thin walled capillaries, closed at one end. They join together to form large lymphatic vessels, which ultimately unite to form two large lymphatic ducts – the thoracic duct and the right lymphatic duct.

Lymphatic channels are present in all tissues except the superficial parts of skin, central nervous system, cornea and the endomysium of muscles and bones.

Formation of Lymph

It is a passive process. About 90 percent of the fluid filtered out of capillaries, at the arterial end to the tissues, is reabsorbed back into them at the venous end.

Table 26.7: Normal values (Blood, formed elements)	
Parameters	Values
pH of blood	7.4
Specific gravity	1.055-1.065
Osmotic pressure	290 milliosmoles/L
Viscosity	3.4-5.4
Plasma proteins	6.8 gm/100 ml
ESR	3-5 mm at the end of 1 hour (in males) 5-12 mm at the end of 1 hour (in females)
PCV	45 percent (males) 42 percent (females)
Diameter of RBC	7.5 µm
RBC count	4.4-5.5 millions/mm^3 of blood (in males) 4-5 millions/mm^3 of blood (in females)
Hb	14-18 gm percent (in males) 12-16 gm percent (in females)
Total WBC count	4000-11,000 /mm^3 of blood
Differential count	Neutrophils: 50-70 percent, Eosinophils : 1 to 4% Basophils : 0-1%; Lymphocytes : 20-40% Monocytes : 2-8%
Platelet count	1.5-4 lakhs / mm^3 of blood

Remaining 10 percent enters the lymphatic vessels to form lymph. Normal 24-hour lymph flow is 2-4 liters.

Functions of Lymph

1. *Transport of proteins:* Proteins and other large particles cannot be absorbed directly into the circulation. They enter the lymphatics and return to blood from the interstitial fluid.
2. Transport of fatty acids and cholesterol from intestine.
3. Lymphatics return the excess fluid from the tissue spaces into circulation. Thus it maintains interstitial fluid volume and pressure.
4. They can carry bacteria and other organisms to the nearest lymph node where they can be handled by lymphocytes and macrophages.
5. They can carry antibodies.
6. They transport Vitamin K from the intestine to the blood stream.

■ EDEMA

Edema is the accumulation of excess fluid in the interstitial spaces. It is due to some abnormality in the mechanism of transcapillary exchange.

Following can be the causes:

1. Increased capillary pressure causes the fluid to move from the capillaries to the tissue spaces.
2. Decrease in plasma proteins, due to malnutrition, liver diseases, burns, etc.
3. Increased capillary permeability, due to infections, inflammations, deficiency of Vitamin C, etc.
4. Blockage of lymphatics (lymphedema), due to malignancies, filariasis, surgery, etc.

27

Immunity

- Humoral Immunity
- Complement System
- Memory B Cells
- Antibody Titer
- Cell Mediated Immunity

- Natural Killer (NK) Cells
- Applied Physiology
  - Autoimmune Diseases
  - Acquired Immunodeficiency Syndrome (AIDS)
  - Graft Rejection or Rejection of Transplanted Tissue

The term "immunity" refers to the resistance exhibited by the host against foreign matter — both living and non-living. The immune defences provide:

1. Protection against infections caused by microorganisms like bacteria, fungi, viruses, parasites, etc.
2. Inactivation of non-microbial foreign substances like antigens, pollen grains, etc.
3. Destruction of cancer cells, formed inside the body.

Immune system of body recognizes the 'self' proteins and antigens from 'non-self' antigens. Against the non-self molecules, body responds by producing an 'effector response' and a 'memory response'. 'Effector response' tries to eliminate the non-self molecule. If our body is exposed to the same non-self molecule for the second time later, an immediate and stronger reaction is produced by the 'memory response'.

Immunity can be broadly classified into two categories—Innate or Natural or Non-specific Immunity; and Acquired or Specific immunity.

Innate or Natural immunity depends on the genetic make up of the individual and is present since birth. It is not specific to any particular foreign substance or cell, e.g. protection provided by skin and mucous membrane, gastric acid present in the stomach, tears, phagocytosis by neutrophils and macrophages, etc.

Acquired or specific immunity is developed by the body after an attack by a specific foreign substance or microorganism which acts as antigen. Antigens are high molecular weight proteins that can be either living or non-living thing. Pollen grains, egg white, transplanted tissue, incompatible blood cells, viruses, bacteria, fungi, parts of microbes like cell wall, capsule, flagella, toxin, etc. can act as antigens. The specific immunity can be passively acquired or actively acquired. The antibodies transported from mother to fetus through placenta form the passively acquired immunity. Actively acquired immunity involves the active participation of body's immune system by lymphocytes.

The B-lymphocytes provide 'humoral immunity' by producing antibodies and the T-lymphocytes provide 'cellular' or 'cell-mediated' immunity. The lymphocytes, after their maturation, get lodged in various lymphoid tissues. When the immediate defence by neutrophils and macrophages fail to control the invading foreign antigen, the immune systems by T and B lymphocytes are activated.

■ HUMORAL IMMUNITY

When the foreign antigen binds to the specific receptors on B-cell membrane, it gets activated. The activation of B-cells is also augmented by the helper-T-lymphocytes and the cytokines secreted by the macrophages. The activated B-cells proliferate and differentiate into plasma cells (Fig. 27.1), which can secrete antibodies or immunoglobulins. About 2000 antibody molecules per second can be produced from each plasma cell. These antibodies circulate through all body fluids, including plasma and saliva. Each immunoglobulin molecule consists of four polypeptide chains – two light and two heavy, interlinked by disulfide bonds (Fig. 27.2). According to the amino acid sequences in the heavy chains, they can be grouped into five classes – IgG, IgA, IgM, IgD and IgE with heavy chains γ, α, μ, δ and ε respectively. As given in (Fig. 27.2), each immunoglobulin molecule consists of a constant portion or Fc portion and a variable portion. The amino acid sequences of Fc portion are identical for all immunoglobulins of a single class.

Macrophages and the proteins of complement system have binding sites on Fc portion. The amino acid sequences of the variable portion differ from immunoglobulin to immunoglobulin in a given class. The variable portion can bind specifically to a particular type of antigen. The characteristics of different antibody classes are summarized in Table 27.1.

These antibodies combine with the foreign antigens and can render them nontoxic. The formation of antigen-antibody complex facilitates phagocytosis of the antigen by neutrophils and macrophages. Simultaneously the complex also activates the complement system which can directly lyse the foreign microorganism or the antigen.

COMPLEMENT SYSTEM

Complement system is a collective term given for a group of 30 proteins, many of which are enzyme precursors. Eleven of them form the main complement

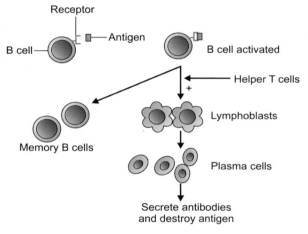

Fig. 27.1: Summary of humoral immunity by B cells

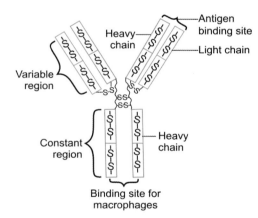

Fig. 27.2: Structure of immunoglobulin molecule

Table 27.1: Features of different antibody classes					
	IgG	*IgA*	*IgM*	*IgD*	*IgE*
Common location	Blood, Lymph, Intestine	Sweat, Saliva, Tears, Milk, Mucus and GI secretions	Blood Lymph	Blood	Mast cells and basophils lining respiratory and intestine tracts
Function	Protection from bacteria and viruses	Provide local immunity and protection to mucosa	First antibody to be secreted by the plasma cells	Activate B-lymphocytes	Involved in allergic reactions and provide protection from parasites
Transmission from mother to fetus via placenta	+++	–	–	–	–

proteins, viz; C to C9, B and D. Normally they remain inactive. There are two pathways by which they are activated–(1) the classical pathway, activated by the antigen-antibody complexes, and (2) the alternative (properdin) pathway, activated by cell wall polysaccharides and endotoxins of bacteria, yeast cell, etc. The activated complement proteins finally cause lysis of the foreign microorganisms.

MEMORY B CELLS

Some of the activated B-lymphocytes get converted into memory B cells which remain in our body. They respond more rapidly and markedly during a second attack by the same antigen. This is called as immunological memory by the B-lymphocytes.

ANTIBODY TITER

Antibody titer denotes the amount of a particular class of antibody in the serum. When a foreign antigen enters our body, there is a slow rise in the IgM antibody titer and then IgG titer, followed by a gradual decline. This is known as the primary response. When our body is exposed to the same antigen for the second time, the IgG antibody titer rises far greater than that during the primary response. This accelerated response is called secondary response, which is due to the immunological memory. This forms the basis of immunization.

CELL MEDIATED IMMUNITY

Cell mediated immunity or cellular immunity is mediated by T-lymphocytes. T-cells constitute a family that has two major subsets, according to the presence of certain protein receptors called CD4 and CD8 in their cell membranes. Functionally, they can be categorized into four – Killer T cells or Cytotoxic T cells, Helper T cells, Suppressor T cells and Memory T cells. Cytotoxic T cells and suppressor T cells have CD8 receptors and so are commonly known as CD8+ cells. Helper T cells have CD4 receptors and hence, also known as CD4+ cells. The T-cell receptor cannot combine with any antigen unless the antigen is first complexed with a particular self-antigen of the body called "Major histocompatibility proteins (MHC proteins) or "Human Leukocyte Antigen" (HLA). These proteins are unique for a person, except for identical twins. There are two classes of MHC proteins – Class I and Class II. Class I MHC proteins are found on the surface of almost all cells of our body except RBCs, while class II MHC proteins are found on the surface of lymphocytes and tissue macrophages. The different subsets of T cells bind to different classes of MHC protein. The CD8 receptors of cytotoxic T cells and suppressor T cells can bind with MHC class I proteins, whereas CD4 receptors on the helper-T cells can bind with MHC Class II proteins (Fig. 27.3). The binding will activate the T cells. The activated CD4 (helper) T cells secrete certain chemicals called cytokines which activate other T-cell subsets and also B cells. The activated CD_8 (cytotoxic) T cells secrete "perforins" ('pore' forming substances). Perforins form pores or holes in the cell membrane of the target cell, thus causing the cell to burst. Simultaneously, memory T-cells are also formed, like memory B-cells. The whole process is summarized in the Figure 27.3. Cell mediated immunity is responsible for delayed hypersensitivity reactions, immunity against viruses and fungi, transplantation immunity and immunity against tumors.

NATURAL KILLER (NK) CELLS

Natural killer (NK) cells constitute a distinct class of lymphocytes that are not T cells but are cytotoxic in nature. They attack, lyse and kill the target cells directly. Unlike cytotoxic T cells, NK cells are not antigen specific. In other words, NK cells can lyse the infected cells and tumor cells without recognition of a specific antigen on the part of the NK cells.

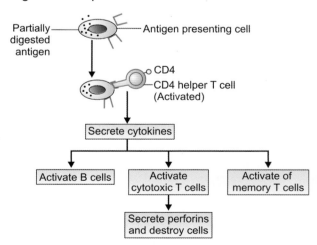

Fig. 27.3: Summary of cell mediated immunity by T cells

Table 27.2: Hormonal influences on immune response	
Hormones stimulating immune response	*Hormones inhibiting immune response*
Growth hormone	Glucocorticoids
Thyroid hormone	Estrogen
Insulin	Androgen
	Progesterone

The immune responses – cell mediated immunity and humoral immunity, are influenced by various hormones (see Table 27.2).

■ APPLIED PHYSIOLOGY

Autoimmune Diseases

The inappropriate immune attack triggered by the body's own proteins acting as antigens, leads to a group of disorders, together called autoimmune diseases. The activated T and B lymphocytes will attack some cells or organs of the host itself, e.g., In multiple sclerosis, auto-antibodies are formed against the myelin sheath of nerves. In rheumatoid arthritis, auto-antibodies are developed against the collagen tissue. Auto-antibodies attacking the acetylcholine receptors at the neuromuscular junction leads to myasthenia gravis.

Acquired Immunodeficiency Syndrome (AIDS)

AIDS is caused by HIV (human immunodeficiency virus) which affects the immune system. HIV infects the helper T cells and destroys them. This affects both cellular and humoral immunities, as helper T cells are needed for the stimulation of both. Thus the AIDS patient becomes prone to all types of infections and malignancies that otherwise would be readily handled by our immune system.

Graft Rejection or Rejection of Transplanted Tissue

Sometimes, the immune responses can become harmful or unwanted. "Graft" means the tissue or organ which is transplanted. Our immune system recognizes the graft as foreign antigen attacks it, resulting in its destruction. When a foreign graft is introduced into the body, the helper T cells get activated, which stimulate cytotoxic T cells and B-lymphocytes. The immune responses thus produced attack the graft which becomes necrotic and finally gets destroyed. This is because, the class I MHC proteins and class II MHC proteins on the cells of the graft differ from that of the recipient, except in the case of identical twins.

28 The Digestive System

The main functions of the digestive system or the gastrointestinal system are to digest food, to absorb the digested food and to eliminate the undigested food. It is also involved in maintaining the water and electrolyte balance of the body. The gastrointestinal tract consists of an alimentary canal of 8-10 meters length, extending from the mouth to the anus, the salivary glands, liver and exocrine part of pancreas. The process of digestion involves breaking up of large particles of foodstuff like polysaccharides, fats and proteins (which cannot be easily absorbed) into smaller particles like monosaccharides, fatty acids and amino acids, which can be easily absorbed. These absorbed products are metabolized for the production of energy for various body activities.

Parts of GIT

Refer Anatomy, Chapter 13 – Digestive System

General Histology

Histologically, the gastrointestinal tract (GIT) consists of the following layers from inside outwards (Fig. 28.1).

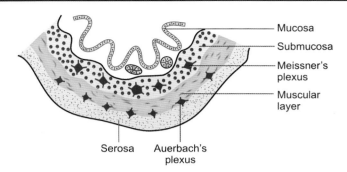

Fig. 28.1: Cross-section of GIT

Mucosa

It is lined with mucous membrane. It comprises (a) surface epithelium; (b) lamina propria with glands, blood vessels and nerves; (c) muscularis mucosa with smooth muscles.

Submucosa

It consists of loose connective tissue, blood vessels and lymphatics. It also contains a network of nerve fibers called the Meissner's plexus.

Muscular Layer

It consists of 2 thick layers of smooth muscles surrounding the submucosa – outer longitudinal and inner muscular layers. It contains a network of nerve fibers called the Myenteric plexus.

Serous Layer

It is the outermost layer with connective tissue.

Innervation of Gut

The special network of nerve fibers innervating the GIT consists of two groups of fibers.
   I. Enteric nervous system or intrinsic nervous system
   II. Extrinsic nervous system or autonomic nervous system

I. Enteric Nervous System (ENS) or Intrinsic Nervous System

It is formed by two different networks of nerve fibers –
1. Myenteric (Aurebach's) plexus, and 2. Meissner's (submucous) plexus.

1. Myenteric plexus lies between the longitudinal and circular muscle layers. They are responsible for the tone of the gut wall and the rhythmic contractions of gut. Thus they control the peristalsis of the gut.
2. Meissner's plexus lies between the submucous layer and the inner layer of muscular layer. It is mainly sensory in function and is responsible for the control of exocrine and endocrine secretions from GIT.

II. Extrinsic Nervous System (or ANS)

This is under the control of the parasympathetic and sympathetic nerves. Parasympathetic innervation to the gut arises from the vagus. The vagus nerve supplies the GIT up to the right 2/3 of transverse colon. The rest of the gut tube, receives parasympathetic supply from the pelvic splanchnic nerves. Stimulation of parasympathetic system results in an increase in the gastrointestinal secretions as well as the gastrointestinal motility.

Sympathetic fibers to GIT arise from T8 to L2 segments of the spinal cord. Most of these fibers terminate on the ENS, but a few go along with the blood vessels of the gut. Their stimulation causes inhibition of peristalsis and vasoconstriction of the blood vessels supplying GIT.

The Digestion in the Mouth

The food in the mouth, gets mixed with the saliva and chewed by the teeth, which converts the food into small pieces. Then it is swallowed. The chewing helps to increase the surface area of food on which the different digestive enzymes can act.

Salivary Glands

The secretions of these glands help to keep the mouth moist and provide a protective and lubricant coat of mucus.

Types of Salivary Glands and their Structure

(Refer Anatomy, Chapter 13)

■ SALIVA

Total amount of saliva secreted is 1500 ml/day.

Composition of Saliva

Saliva contains 99% water and 1% solids.

Solids include organic and inorganic constituents.

The organic constituents are:

1. The enzyme ptyalin or salivary amylase, secreted from the parotid gland.
2. Lingual lipase is another enzyme, which acts on triglycerides.
3. Mucin, which is a glycoprotein, secreted mainly from sublingual and a small amount from submandibular salivary glands. Mucin is responsible for the viscosity of saliva, which helps in the lubrication of food in mouth. The inorganic constituents in saliva include Na^+, K^+, Ca^{++}, HCO_3^- and Cl^-. Under resting conditions, the concentration of Na^+ and Cl^- in saliva is less than that in the plasma. But the concentrations of K^+ and HCO_3^- are more than that in the plasma. When there is an increased rate of salivary secretion, this ionic composition reverses.

Apart from the above substances, saliva also contains lysozyme and immunoglobulin A (IgA).

Functions of Saliva

Lubrication

Saliva helps in easy swallowing of food, as it lubricates food. Mucin is responsible for this action. Lubricating action of saliva also helps in speech.

Appreciation of Taste

Only those foods, which are dissolved in the saliva can stimulate the taste buds.

Digestive Function

Saliva contains salivary amylase or ptyalin, which mainly acts on cooked starch and converts it into maltose. After chewing, the food mixed with ptyalin is swallowed and its action continues in the stomach for some time. Lingual lipase has a weak action on fat digestion.

Defense

Saliva contains many substances like lysozyme, immunoglobulin A (IgA), etc. which can attack the microorganisms like bacteria which are present in food.

Other Functions

Saliva helps to cool the hot foodstuffs. Many metals like lead, mercury, etc. are excreted via saliva.

Innervation of Salivary Gland

Refer Anatomy, Chapter 13

Regulation of Salivary Secretion

There can be spontaneous as well as stimulated secretion of saliva.

Spontaneous secretion or basal secretion of saliva occurs without any stimulus and is present always. It is of small quantity (0.5 ml/minute) and is mucinous in nature. Stimulated secretion or reflex secretion of saliva is controlled by nervous stimulation, via secretomotor nerves. Two types of reflex secretion can occur.

Unconditioned Reflex (Inherent Reflex)

This is present since birth and needs no previous experience. Here, salivation occurs when food is placed inside the mouth. The receptors involved are the mechanoreceptors and taste receptors in the oral cavity. The afferent nerves involved are the V, VII, IX and X cranial nerves, which carry excitatory impulses to the "salivary center" constituted by superior and inferior salivary nuclei (*Refer Anatomy, Chapter 13*). The efferent nerves (VII and IX Cranial nerves) from the salivary center stimulate the salivary glands to secrete saliva.

Conditioned Reflex (Acquired Reflex)

This reflex is developed by previous experiences of the subject. Sight, smell or even thought of familiar food substances can initiate salivary secretion. This reflex is mediated by cerebral cortex. Stress, nausea, irritation of stomach and upper small intestine can also cause increased salivation by this reflex.

Applied Physiology

1. *Sialorrhoea:* Sialorrhea is increased salivation. It can occur in pregnancy or due to some irritation in the oral cavity, esophagus or stomach.
2. *Hyposalivation:* Hyposalivation is decreased salivation, which can occur due to fever, anxiety, drugs like atropine, etc.

Part 2

Physiology

DIGESTION IN THE STOMACH

The stomach is a hollow muscular organ. For "parts of stomach" and "functions of stomach", *(Refer Anatomy, Chapter 13).*

Gastric Glands

The glands are divided into three types depending on their location *(For figure – Refer Anatomy, Chapter 13, Figs 13.22A and B).*

Glands in the Fundus and Body of Stomach

Fundic or Oxyntic glands: They are elongated tubular glands which have 3 types of cells.
   i. Peptic, Chief or Zymogen cells – They are the predominant cells. They contain zymogen granules, which secrete the enzyme pepsinogen.
   ii. Parietal or oxyntic cells: They are large, oval cells, which secrete HCl and the intrinsic factor, necessary for the absorption of vitamin B_{12}.
   iii. The neck mucous cells, which secrete mucous.

Cardiac Glands

They are located near the cardio-esphageal junction. They secrete mucus.

Pyloric or Antral Glands

They are located in the pyloric antrum and the pyloric canal of stomach. They secrete soluble, mucous-rich, alkaline, viscid juice poor in enzymes. Pyloric glands also have G-cells, which secrete gastrin.

Gastric Juice

The cells of gastric glands secrete about 2500 ml of gastric juice daily. It has a pH of 1-2 at the luminal surface; i.e. it is strongly acidic. It contains 99% water and 1% solids. The important constituents of gastric juice are pepsinogen, hydrochloric acid, mucous, intrinsic factor, HCO_3^-, etc.

HYDROCHLORIC ACID (HCl)

HCl is secreted by the parietal (oxyntic) cells of the fundic glands. Functions of HCl:
- Activation of pepsinogen to pepsin
- Provides a medium necessary for the action of pepsin on proteins

- Has antiseptic and antibacterial actions
- Helps in absorption of iron and calcium by making them soluble
- Stimulates the secretion of pancreatic juice and bile.

Mechanism of Secretion of HCl

Purest specimen of HCl obtained has a pH as low as 0.87. But the pH of cytoplasm of parietal cell is 7 to 7.2. This means HCl is not formed as such inside the parietal cells, but in the canaliculi between the cells. Also the H^+ present in the parietal cells have to be pumped against the concentration gradient into the canaliculi. The whole process of HCl secretion was first described by Davenport in 1957 and is as follows (Fig. 28.2):
1. Generation of H^+ inside the parietal cells occurs from CO_2 and water formed as a result of various metabolic reactions.

$$CO_2 + H_2O \xrightarrow{\text{Carbonic anhydrase}} H_2CO_3$$
$$\rightleftharpoons H^+ + HCO_3^-$$

2. Pumping of H^+ thus formed into the canaliculi against concentration gradient, using energy (i.e. "active transport of H^+"). This is mediated by $H^+ - K^+$ ATPase present in the apical membrane of the cells which faces the lumen of gastric glands (Fig. 28.2). Using energy, $H^+ - K^+$ ATPase pumps H^+ outside the parietal cells in exchange for K^+.
3. To maintain intracellular neutrality, HCO_3^- are exchanged for Cl^- by '$HCO_3^- - Cl^-$ exchanger'. As a result Cl^- goes out of the cell and reaches the canaliculi.
4. H^+ and Cl^- react to form HCl at the apical membrane of canaliculus (Fig. 28. 2).

After a meal, when large amount of HCl is secreted into the gastric lumen, large amount of HCO_3^- also enters the systemic blood. As a result, the pH of systemic blood increases, i.e. it becomes alkaline. Some amount of this is excreted through urine. This is called "post-prandial alkaline tide".

Regulation of Acid Secretion

Vagal stimulation, Gastrin and Histamine can increase HCl secretion. Somatostatin, prostaglandin E, gastric inhibitory peptide (GIP) and vasoinhibitory peptide (VIP), decrease the acid secretion. Apart from this,

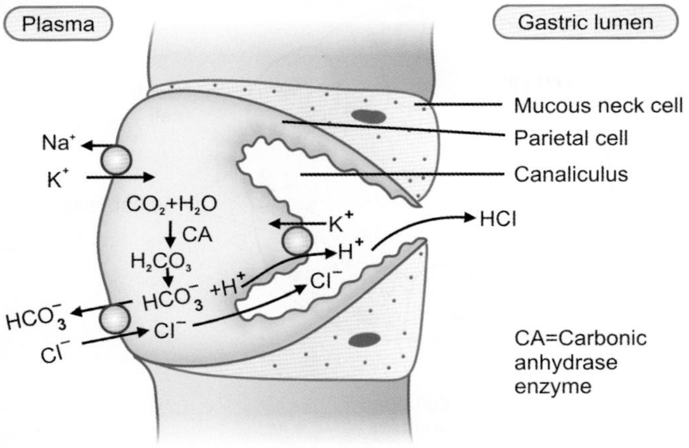

Fig. 28.2: Mechanism of HCl secretion by parietal cells in stomach

distension of stomach can stimulate HCl secretion and acids or fats in the duodenum can inhibit HCl secretion. There are also emotional influences on HCl secretion. Excitement, anger, etc. cause increased acid secretion while fear and depression decrease it.

Phases of Gastric Secretion

There are 3 phases: Cephalic, gastric and intestinal

Cephalic Phase

It is the psychic secretion that occurs even before food enters the stomach. Presence of food in the mouth, sight, smell or thought of food, etc. may stimulate gastric acid secretion, which is mediated via vagus nerve. This secretion is called appetite juice.

Gastric Phase

The presence of food in the stomach initiates this phase causing more than two-thirds of total gastric secretion. Gastric distention is a potent stimulus for the secretion of parietal and chief cells. Gastrin can also stimulate HCl production. When the PH of gastric contents becomes less than 2, there is a feedback inhibition of gastrin hormone and HCl. This protects the gastric mucosa from damage due to increased acidity.

Intestinal Phase

In contrast to the excitatory cephalic and gastric phases, the intestinal phase of gastric secretion is chiefly inhibitory in nature. The presence of acid or fat in

duodenum causes release of hormones like VIP, GIP somatostatin, etc. which inhibits gastric acid secretion.

▮ PEPSINOGEN

The chief cells secrete pepsinogen. This is an inactive precursor of pepsin, which is stored in the zymogen granules. Hydrochloric acid converts inactive pepsinogen into active pepsin by cleaving a peptide from the molecule. Actions of pepsin are:
- Splits proteins into proteoses, peptones and polypeptides.
- Unique ability to digest collagen.
- Splits the milk protein caesinogen into casein.

▮ SECRETION OF MUCUS

Mucus is secreted by the neck mucus cells of fundic gland, cells of pyloric gland and cardiac gland. Mucus forms a viscid, gel like coat over the mucosa. It is alkaline in nature and forms an extremely important protective layer, saving the stomach from the destructive action of HCl. Substances like alcohol, vinegar, aspirin, etc. can erode this barrier and produce gastritis.

Applied Physiology

1. *Peptic ulcer:* The term "Peptic ulcer disease" includes both gastric and duodenal ulcers. This occurs due to breakdown of gastric mucosal barrier or duodenal mucosal barrier respectively. Normal mucosal barrier of stomach and duodenum consist of the following:
   a. Mucus secreted by the surface mucous cells in the body and fundus of stomach.
   b. HCO_3^- secreted by the surface mucous cells
   Disruption of this "barrier" occurs in certain conditions like infection by *Helicobacter pylori* bacteria, alcohol, vinegar, long-term treatment with non-steroidal anti-inflammatory drugs, etc. which result in ulcer formation. In "Zollinger-Ellison syndrome" – a gastrin secreting tumor results in increased HCl secretion leading to peptic ulcer formation. These gastrin secreting tumors can occur in stomach and duodenum, but usually they are seen in the pancreas.
2. *Achlorhydria:* Achlorhydria is absence of acid secretion which occurs usually in pernicious anemia.
3. *Hypochlorhydria:* Hypochlorhydria is reduced acid secretion, seen in gastric carcinoma.

Exocrine Part of Pancreas

Pancreas consists of an exocrine and an endocrine part. The endocrine part consists of the Islets of Langerhans. The exocrine part consists of ducts and acini, which aid in digestion. So , we will discus about the exocrine part here.

The exocrine part is a highly lobulated structure with numerous acini. Acini consists of pyramidal shaped cells with their apices pointing towards the lumen of a ductile. Apex contains many zymogen granules. They secrete only enzymes. Acini drain into small ductules. The ductules unite to form bigger ducts which ultimately form the main pancreatic duct (of Wirsung).This duct joins with the common bile duct and this combined duct opens into the second part of duodenum. In some, there can be an accessory pancreatic duct also.

Ductal epithelium consists of cuboidal cells devoid of zymogen granules. At the beginning of a ductule (where it emerges from an acinus), the ductal epithelial cells invaginate into the acinus forming centriacinar cells (Fig. 28.3). Centriacinar cells and the epithelial cells lining of the ductules contain the enzyme carbonic anhydrase, which is required for the synthesis of bicarbonate.

Pancreatic Juice

Pancreatic juice is the name given to the secretions of exocrine pancreas. Normally about 1500 ml of pancreatic juice is produced everyday. It is alkaline (pH = 8) due to its high HCO_3 content. It consists of organic as well as inorganic constituents.

Organic constituents include various types of enzymes, which are given below:

  I. Proteases (proteolytic enzymes)
- Trypsinogen
- Chymotrypsinogen
- Proelastase
- Procarboxy peptidases.

  II. Amylolytic (Starch splitting) enzymes
- α-amylase

  III. Lipolytic (fat splitting) enzymes
- Pancreatic lipase and colipase for triglyceride digestion
- Phospholipase for phospholipid digestion,
- Cholesterol ester hydrolase for cholesterol digestion.

  IV. Nucleases (Nucleic acid splitting enzymes
- RNAase
- DNAase

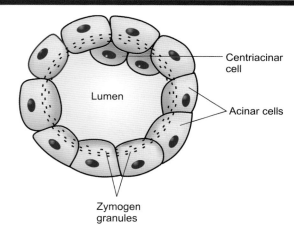

Fig. 28.3: Pancreatic acinus and centriacinar cells

I. Proteases

Pepsin of stomach can only partially digest the proteins and hence pancreatic proteases are essential for full protein digestion. Trypsin, chymotrypsin and elastases are endo-peptidases, i.e. they attack the peptide bonds in the interior of the proteins, forming peptide fragments. Carboxy peptidase is an exopeptidase, which splits off the terminal amino acid from proteins and peptides.

Trypsin

Trypsin is secreted as inactive trypsinogen. After reaching the second part of duodenum, it is activated to trypsin by an enzyme enterokinase (enzyme from proximal epithelial cells of small intestine). The active trypsin performs the following actions:

1. Trypsin converts other trypsinogen molecules to trypsin (autocatalytic action)
2. It activates chymotrypsinogen into chymotrypsin.
3. It activates proelastase to elastase.
4. It converts procarboxypeptidase to carboxypeptidase.
5. Trypsin can act directly on food particles.
6. It can digest other proteases and other enzymes present in the gastrointestinal lumen, thereby preventing their over action. There is a trypsin inhibitor in the pancreatic juice whose function is to check the actions of trypsin.

Activated carboxy peptidases act on polypeptides and split off their terminal amino acid residue so that free aminoacids are formed.

Proelastase becomes elastase, when activated. It breaks down elastin.

II. Amylolytic Enzyme – α-Amylase

There is no inactive form for this enzyme. Like the salivary amylase, alpha amylase acts on starch at an optimal pH of about 6.5. The starch is split into maltose, maltriose, dextrins, etc. Therefore, starch is not digested completely by exocrine pancreatic juice.

A part of this α-amylase enters into the systemic circulation and its serum concentration helps to diagnose acute pancreatitis – a condition where serum amylase is increased.

III. Lipolytic Enzymes

Triglycerides, phospholipids and cholesterol are the three main types of lipids present in our food. Pancreatic lipase splits triglycerides to monoglycerides. Cholesterol esters are split by cholesterol ester hydrolase enzyme. Phospholipases split phospholipids like lecithin. Of these, the most important one is pancreatic lipase. These enzymes are activated by the bile salts.

IV. Nucleases

DNAase splits DNA and RNAase splits RNA.

The inorganic constituents of pancreatic juice include Na^+, K^+, HCO_3^-, etc. The basal secretion (i.e. when there is no food for digestion) of pancreatic juice contains only Na^+ and Cl^-. When stimulated by certain gastrointestinal hormones like secretin, CCK-PZ, etc. or by vagal stimulation, there is increased rate of secretion of pancreatic juice rich in enzymes and HCO_3^-.

Regulation of Pancreatic Secretion

1. Neural Regulation
2. Hormonal Regulation.

1. Neural Regulation

Stimulation of the taste buds by food causes reflex increase in vagal discharge, which in turn increases the pancreatic juice secretion, rich in enzymes.

2. Hormonal

Secretin and cholecystokinin-pancreozymine or CCK-PZ are the two important gastrointestinal hormones influencing the secretion of pancreatic juice. These hormones are stimulated by the presence of acid rich chyme in the duodenum. Secretin, produced from the duodenal mucosa, stimulates pancreatic juice secretion rich in bicarbonate. CCK-PZ acts on pancreatic acini leading to secretion of enzyme-rich pancreatic juice.

Phases of Secretion

There are mainly 3 phases of secretion:

Cephalic Phase

It is brought about by chewing of food, sight of food, smell of food, etc. All these stimulate the taste buds, resulting in vagally mediated secretion of enzyme rich pancreatic juice (but poor in HCO_3^-).

Gastric Phase

It is brought about by distension of stomach by the food. This also evokes a vasovagal reflex, promoting secretion of enzyme-rich pancreatic juice.

Intestinal Phase

It is brought about by the presence of acid chyme in the duodenum. This stimulates the hormones – secretin and CCK-PZ, which promote the secretion of pancreatic juice rich in HCO_3^- and enzymes respectively. This phase of secretion is also initiated by the vasovagal reflex.

Autoprotection of Pancreas

Autodigestion of pancreas is prevented by the following:

1. Pancreatic proteases are not secreted in their active forms, but as inactive pro-enzymes, which may become active only in the presence of acid chyme.
2. Trypsin inhibitors are synthesized in the pancreas to check the overaction of trypsin.

Part 2

Physiology

Applied Physiology

1. *Acute pancreatitis:* Acute pancreatitis is a serious and fatal inflammatory disease of the pancreas. The enzymes leak out of the acini causing destruction of adjacent tissues. There is increased serum level of pancreatic amylase and trypsin.

2. *Chronic pancreatitis:* Usually seen in chronic alcoholics. Proteinaceous compounds and calcium compounds are deposited in the small ductules of pancreas, resulting in degeneration, atrophy and fibrosis of the concerned acinus draining into the ductule. So there is deficiency of pancreatic secretions. This mainly affects fat digestion and there is increased excretion of fat through feces— a condition called 'steatorrhea'.

The Liver and Gallbladder

For a detailed gross as well as microscopic structure of liver and the biliary system, (*refer Anatomy, Chapter 13*).

Bile

Bile is synthesized in the hepatocytes. Small amount of bile is continuously transferred to gallbladder through the cystic duct, because, during the interdigestive period, the tone of the sphincter of Oddi is so high that it cannot enter the duodenum. In the gallbladder, bile is stored and concentrated. After a meal, contraction of gallbladder transfers a large volume of bile into the duodenum.

About 700–1200 ml of bile is secreted by the liver daily.

Composition of Bile

Main component is water. A small amount of organic as well as inorganic constituents are also present in bile. The organic constituents are bile salts, bile pigments, cholesterol, lecithin, etc. The inorganic constituents are Na^+, K^+, HCO_3^-, etc.

In the gallbladder, water, sodium chloride and other inorganic constituents are absorbed. Therefore the organic components become 5–6 times concentrated.

Differences between hepatic bile and gallbladder bile are summarized in Table 28.1.

Table 28.1: Differences between hepatic bile and gallbladder bile		
	Hepatic bile	*Gallbladder bile*
Color	Golden yellow	Dark brown
Water	97%	89%
Bile salts	1.1 gm/dL	6 gm/dL
Cholesterol	0.1 gm/dL	0.6 gm/dL
pH	7.8 to 8.6	7 to 7.4
Na^+	145 mEq/L	130 mEq/L
K^+	5 mEq/L	12 mEq/L
Cl^-	100 mEq/L	25 mEq/L
HCO_3^-	28 mEq/L	10 mEq/L

Bile Salts

Bile salts are the most important constituent of the bile. They are sodium and potassium salts of bile acids conjugated to the aminoacids glycine or taurine.

Important bile acids formed in the liver are cholic acid and chenodeoxycolic acid. Both are derived from cholesterol. These bile acids are also called primary bile acids. Some of these primary bile acids get converted into secondary bile acids in the intestine, by the action of intestinal bacteria. The secondary bile acids are lithocholic acid and deoxycholic acid.

One of these bile acids combines with the aminoacids glycine or taurine to form a conjugated bile acid, which combine with Na^+, or K^+ to form Na^+ or K^+ salt of bile acid.

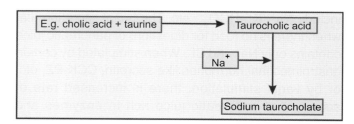

From ileum, 95% of bile salts are reabsorbed into the portal blood, which brings them back to the liver. These bile salts are almost entirely reabsorbed into the hepatic cells and then to the bile. In this way, bile salts circulate about 18 times before being excreted in the feces. This re-circulation of bile salts is called enterohepatic circulation.

Functions of Bile

All functions of bile are due to the bile salts in it.

1. Digestion and absorption of fat

    Pancreatic lipase is a water soluble enzyme and so it can only act on the surface of a lipid. Bile salts break up large lipid molecules into smaller ones by a process called emulsification, thereby facilitating the digestive action of water soluble lipase.

    Bile salts combine with products of hydrolysis of triglycerides to form small water soluble particles called micelles and transport them to the brush border of the epithelial cells of intestine.

2. Absorption of fat soluble vitamins – A, D, E and K.
3. Bile salts stimulate secretion of more and more bile from the liver – a process called choleretic action.
4. Bile salts keep cholesterol in solution in the bile.
5. Bile salts inhibit endogenous synthesis of cholesterol by liver.
6. Bile salts have mild antibacterial action.

Bile Pigments

Bile pigments are mono and di-glucorinides of bilirubin and biliverdin. They have no digestive functions. They are the excretory products of hemoglobin breakdown. They give the characteristic golden yellow color to the bile. Hepatocellular dysfunction or obstruction of the biliary passages cause accumulation of bile pigments in blood producing jaundice.

Regulation of Bile Secretion

Secretion of bile is influenced by the following.

1. *Bile salts:* Presence of bile salts in the liver or intestine stimulates more and more bile secretion – a processes called choleretic action. As the bile salts are recycled in the entero-hepatic circulation, they maintain a high level of bile secretion during digestion. As the digestive process completes, the tone of the sphincter of Oddi increases, recycling of bile salts decreases markedly and the choleretic action comes to an end.
2. *Secretin* stimulates bile secretion.
3. Vagal stimulation increases bile secretion.
4. *Cholegogues* are substances, which cause contraction of gallbladder and transfer of concentrated bile into the duodenum. The hormone CCK produces a powerful contraction of gallbladder. Fats and products of protein digestion are potent stimuli for the secretion of CCK and hence gallbladder contraction also.

Applied Physiology

Gallstones: Excessive formation of cholesterol and chronic inflammation of gallbladder result in precipitation of cholesterol in the form of sand like particles, which may later form gallstones. Sometimes, these stones may slip into the common bile duct and cause acute pain (biliary colic), and regurgitation of bile into the liver, thereby into general circulation causing obstructive jaundice.

Absence of bile causes impairment of digestion and absorption of fat and fat soluble vitamins.

Jaundice: *Refer Chapter 26 "Blood"*

■ THE SMALL INTESTINE

The small intestine extends from the pylorus to the ileo-cecal junction *(for a detailed gross and microscopic structure, refer Anatomy, Chapter 13).*

Small Intestinal Secretions

Succus Entericus

It is the characteristic alkaline fluid secreted by the crypts of Lieberkulin. About 2 liters of fluid is secreted per day, which is rapidly reabsorbed by the intestinal villi.

Secretion from the Brunner's Gland

It is alkaline as well as mucoid in nature and protects the duodenal mucosa from the acidic chyme.

Mucous Secretion by the Goblet Cells

Goblet cells secrete mucin, which combines with water to form the mucus, which protects the intestinal epithelium.

Succus Entericus or the Intestinal Juice

The digestive enzymes in the succus entericus are present in the brush border or micorvilli of the epithelial cells. The enzymes include:

Disaccharidases

They are chiefly sucrase, maltase and lactase enzymes. They act on the carbohydrates and split the disaccharides into monosaccharides.

Proteolytic Enzymes or Peptidases

They bring about the final breakdown of polypeptides into amino acids.

Intestinal Lipase

It splits neutral fats into diglycerides, monoglycerides, fatty acids and glycerol.

Enterokinase

This enzyme activates the conversion of trypsinogen into trypsin. Enterokinase is secreted in response to bile acids, secretin or CCK-PZ.

Regulation of Intestinal Secretion

Small intestinal secretions are regulated by neural mechanisms and hormonal mechanisms.

Neural reflexes are produced by the distension of intestine or irritation of the mucosa by the chyme. This stimulates the secretion from the crypts of Lieberkuhn as well as from the Brunner's glands. Vagal stimulation increases only the secretions of Brunner's glands and not that of the crypts of Lieberkuhn.

Gastrointestinal hormones like VIP, GIP and gastrin stimulate intestinal secretion. Acetylcholine increases and noradrenaline inhibits the secretion.

Applied Physiology

1. *Tropical sprue:* Tropical sprue is widely seen in Asia, especially in South India. This is a pathological condition in which there is atrophy of small intestinal villi. The cause is unknown. These patients suffer from malabsorption as the brush border enzymes are deficient or absent. Absence of enterokinase leads to deficiency of pancreatic proteolytic enzymes. Since there is no protein digestion, CCK is not stimulated causing decreased gallbladder contraction and deficiency of bile. This can decrease the digestion of fats. Ultimately, there is malabsorption of carbohydrates, proteins and fats.
2. *Lactose intolerance:* The brush border of small intestine contains the enzyme lactase, required for the digestion of lactose present in milk. Lactase is present since birth. But in some people, this enzyme may be absent or deficient. They develop diarrhea after consuming milk. This is called lactose intolerance.

   Thus, in the small intestine, there is complete digestion and absorption of food particles.

■ THE LARGE INTESTINE OR COLON

The large intestine extends from the ileocecal junction to the anus. *(Refer Anatomy, Chapter 13, to know about the parts, gross structure and histology of large intestine).*

Functions of Large Intestine

As digestion is completed in the small intestine, there is no role for the large intestine in the digestion of food.

1. *Secretory function:*

   Mucus is the chief secretion of the glands of large intestine. It helps to lubricate the fecal matter. It is alkaline in nature due to the presence of bicarbonate, which neutralizes the acids formed by the fecal matter. No digestive enzymes are secreted by the large intestine. Secretory activity of colon is stimulated by tactile stimuli of the mucosa. Stimulation of parasympathetic nerves cause marked increase in mucus secretion as well as motility of colon. That is why one gets diarrhea in intense emotional disturbances.

2. *Absorptive function:*

   The most important function of large intestine is the absorption of water and electrolytes (mainly in the proximal half of the colon). About 90% of fluids are absorbed. Certain vitamins like Vitamin K and B complex are synthesized by the normal colonic bacteria. They are also readily absorbed in the large intestine.

   After proper absorption, about 200-300 gm of solid feces are left.

3. *Storage function:*

   Feces are stored in the pelvic (sigmoid) colon until they can be expelled by the process of defecation.

Bacterial Flora in the Colon

Unlike small intestine, large intestine contains a very large number of bacteria; some bacteria have no particular effect on the host or vice-versa (called "commensals"). Some may be beneficial to the host and vice-versa ("Symbionts"), while some may cause diseases ("Pathogens"). All these bacteria are present in large amounts in the stool also. Some important beneficial aspects of these colonic bacteria are discussed below:

1. Some bacteria can synthesize Vit. B12, Vit.K, Thiamine, Riboflavin and Folic acid.
2. Brown color of feces is due to the action of bacteria on bile pigments.
3. Odor of the feces is due to amines formed by intestinal bacteria.
4. Certain short chain fatty acids are synthesized by these bacteria.
5. They have a role in cholesterol metabolism.

Blind Loop Syndrome

Bacterial overgrowth can occur when there is stasis of intestinal contents. This usually occurs when surgeons create a portion of small intestine that ends blindly. As a result, the person can have macrocytic anemia (due to malabsorption of Vit.B12) and Steatorrhea (malabsorption of fat).

Dietary Fibers

Dietary fiber means the undigestable part of food matter and includes cellulose, hemicellulose, pectin and lignin. They are abundant in green leafy vegetables, peas, etc. Dietary fibers are not digested by human digestive enzymes. Thus they increase the bulk of feces and prevent constipation. Dietary fibers bind with bile salts and the bound material is excreted via feces. To replenish the lost bile salts, more and more cholesterol is used up, thus decreasing the body cholesterol level. Also, dietary fibers decrease the chances for colonic cancer.

Feces consists of undigested plant fibers, inorganic materials (Ca^{2+}, PO_4^{2-}, etc.), fat, water and small amounts of nitrogenous materials. Also, there are large number of bacteria (most of which are dead), bile pigments and desquamated epithelial cells. pH of feces varies between 7 and 7.5. Its color is due to the oxidized bile pigment stercobilin. The characteristic odor of feces is due to the presence of indol, skatole and other related amines produced by colonic bacteria.

■ GASTROINTESTINAL HORMONES

GI secretions are regulated by a group of local hormones called GI hormones. They are synthesized by special cells in the GI mucosa called *APUD cells (*APUD = Amine Precursor Uptake and Decarboxylation)

Two characteristic features of GI hormones are:
1. Each hormone at physiological concentration may affect more than one target tissue.
2. Each target organ is usually responsive to more than one GI hormone.

The important GI hormones are discussed below:

Gastrin

Gastrin is secreted by the G-cells present in the antrum of stomach. It is also secreted from brain and peripheral nerves in a small amount, whose role is not fully known.

Actions

1. Stimulates secretion of HCl and pepsin
2. Increases gastric motility
3. Stimulates growth of mucosa of stomach and intestines.
4. Stimulates secretion of insulin, after a protein-rich meal.

Cholecystokinin-Pancreozymin (CCK-PZ)

CCK-PZ is secreted from the upper small intestinal mucosa.

Actions

1. Secretion of enzyme-rich pancreatic juice
2. Gallbladder contraction (cholegogue)
3. Potentiates the action of secretin
4. Increases the intestinal motility.

Secretin

It was the first hormone to be discovered. It is secreted from the S-cells of duodenal mucosa and upper jejunum.

Actions

1. Stimulates bicarbonate-rich pancreatic juice secretion.
2. Increases bile secretion (choleretic action)
3. Inhibits gastrin secretion in stomach
4. Potentiates the action of CCK-PZ.

GIP (Gastrointestinal Peptide)

It is secreted by the K-cells of mucosa of duodenum and jejunum.

Actions

1. Inhibits gastric secretions, mainly HCl.
2. Decreases gastric motility

Other GI hormones include vasoactive intestinal peptide (VIP), motilin, neurotensin, etc.

■ MOVEMENTS OF GIT

The gastrointestinal tract exhibits certain coordinated, purposeful series of movements, which help in the propulsion of food through different regions, where it is digested, absorbed and evacuated. The movements of GIT are broadly classified into 5.

I. Mastication
II. Deglutition
III. Movements of stomach
IV. Movements of small intestine
V. Movements of large intestine and defecation.

Mastication

Mastication or chewing is the first mechanical process to which the food is subjected to, on its entry into the mouth. It is a voluntary act. The food is broken into small pieces.

Purposes of chewing are:
1. To increase surface of food so that digestive enzymes can act on a greater area.
2. To help in mixing with saliva.
3. Helps in easy deglutition.
4. Helps to expose the inner digestible structure of certain vegetable foods containing cellulose, hemicellulose, etc.

Mechanism

The muscles of mastication are supplied by the motor part of trigeminal nerve. The presence of food in the mouth initiates reflex inhibition of the muscles of mastication → So the lower jaw drops → This stretches the jaw muscles → Rebound contraction of muscles occur → Jaw raises, compresses the food against tongue and teeth → This again inhibits jaw muscles (muscles of mastication) → Jaw drops and the process is repeated till a soft bolus of broken food is ready to swallow.

Deglutition

Deglutition or swallowing is the passage of masticated bolus of food from the oral cavity into the stomach. The entire process takes only a few seconds, but includes three different stages:
1. The oral or buccal stage.
2. The pharyngeal state.
3. The esophageal stage.

Of these, the oral stage is voluntary while the other two are involuntary.

Oral or Buccal Stage

This stage includes the passage of food bolus from the oral cavity into the oropharynx.

The bolus, after mastication and lubrication with saliva, is rolled posteriorly by the movement of tongue and palate. The soft palate rises upwards and closes the nasopharynx. This prevents nasal regurgitation of food. Ultimately, the bolus is pushed to the posterior part of tongue.

Pharyngeal Stage

This is an involuntary stage, concerned with the passage of bolus through the pharynx into the esophagus. As the bolus reaches the posterior part of oropharynx, it stimulates the "swallowing receptors" present in the pharynx. Impulses form these receptors pass to the brain stem to initiate the pharyngeal muscle contraction.

The food that has entered oropharynx has 4 possible outlets through which it can go (Fig. 28.4):
a. Back into the mouth
b. Into the nasopharynx
c. Pushed into the larynx and the respiratory tract
d. Travel down the esophagus

The re-entry of food into the mouth is prevented by the elevated position of the base of the tongue and by the contraction of the faucial pillars.

Regurgitation into the nasopharynx is prevented by elevation of the soft palate, which closes the posterior nares.

Entry of bolus into the larynx and the distal respiratory tract is prevented by the approximation of vocal cords and closure of the glottis (by swinging back

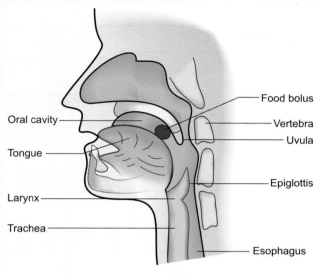

Fig. 28.4: Relation between mouth cavity, respiratory passages and esophagus

Labels in figure: Oral cavity, Tongue, Larynx, Trachea, Food bolus, Vertebra, Uvula, Epiglottis, Esophagus

of epiglottis). Also there is reflex inhibition of respiration (deglutition apnea), which lasts throughout this stage of deglutition.

Now, the only opening left is towards the esophagus. Involuntary contractions of the pharyngeal muscles propel the bolus into the esophagus. The upper esophageal sphincter relaxes, thus helping in the above act (This sphincter is usually contracted before swallowing in order to prevent the entry of air into the esophagus during respiration).

The pharyngeal stage of deglutition takes about 2 seconds.

Esophageal Stage

The bolus that had entered the esophagus is pushed to the stomach by peristaltic waves of esophagus. There are 2 types of peristaltic waves – primary and secondary. Primary peristalsis is the continuation of peristaltic wave that begins in the pharynx. If this wave fails to propel all the food downwards, secondary peristaltic waves start appearing. The secondary peristaltic waves are initiated by distension of esophagus by retained food and they continue till all the food had been emptied from esophagus.

As the bolus reaches the lower end of esophagus, the lower esophageal sphincter relaxes and the bolus enters the stomach. Once the bolus enters the stomach, this sphincter contracts again in order to prevent the regurgitation of the bolus.

Applied Physiology

1. *Achalasia or Cardiospasm:* This is a condition in which there is increased tone of lower esophageal sphincter. So the sphincter does not fully relax during swallowing and the food material gets accumulated just proximal to the contracted sphincter.
2. *Reflux esophagitis:* This is a condition where the lower esophageal sphincter does not contract fully, once the bolus enters the stomach. So there is reflux of gastric contents into the esophagus that damages the esophageal mucosa. This can cause diffuse ulceration, inflammation or stricture formation of esophagus.

Movements of Stomach

Physiologically or functionally, stomach can be divided into 2 parts – an orad part which consists of the upper region of stomach, and a caudad part, which consist of lower half of the stomach. The orad part acts as the reservoir which receives and stores food for some time, so that, it is mixed with the gastric juice. The caudad part helps in the evacuation of gastric contents from time to time.

Three major movements occur in stomach:
1. Receptive relaxation
2. Mixing – churning – breaking movements
3. Gastric emptying.

1. Receptive Relaxation

As the bolus passes through the lower esophageal sphincter, the orad part of stomach relaxes, helping in the easy entry of bolus. This process is called receptive relaxation. The stomach has the capacity to accommodate large volumes of food without appreciable increase in the intragastric pressure. Vasovagal reflex is responsible for this.

2. Mixing – Churning – Breaking (Gastric Peristalsis)

Gastric peristalsis helps to mix the food thoroughly with gastric juice and to breakup the food mass , which is softened by the digestive juices. It also helps to propel the food mass through the pylorus into the intestine.

The gastric peristalsis starts in the mid gastric zone. At first, this wave is weak and as it progresses from the body of stomach into the antrum, they become more intense. These strong waves force the antral contents towards the pylorus. The opening of the pylorus is so small that only a few milli liters or less of antral contents are expelled into the duodenum with each peristaltic wave. Also, as each peristaltic wave approaches the pylorus, the pyloric muscle itself contracts which further decreases the emptying through the pylorus. Therefore, most of the contents are squeezed backwards towards the body of the stomach and not through the pylorus. The moving peristaltic wave combined with this upstream squeezing action is called **retropulsion**. This helps in mixing, churning and breaking of food. The resultant food material leaving the stomach is called the **chyme**, which is acidic and semisolid.

Hunger contractions: This is a special type of contraction, which occurs when the stomach has been empty for several hours. These contractions are more intense in young and healthy persons. When this occurs, the person experiences mild pain called 'hunger pangs'.

3. Gastric Emptying

After a full meal, the stomach is emptied within 3-4 hours. The mechanism is by squeezing of a small quantity of chyme at each time through the pyloric canal to the duodenum. Following factors affect gastric emptying:

a. Increased food volume promotes gastric emptying
b. Emptying is delayed if there is increased fat in the food
c. Protein rich food also delays emptying
d. Fluids easily leave the stomach
e. Emotions can also affect gastric emptying, e.g. fear may prolong while excitement may hasten gastric emptying.

Applied Physiology

Vomiting: It is the powerful expulsion of gastric and upper GI contents to the exterior via the oral cavity.

Mechanism

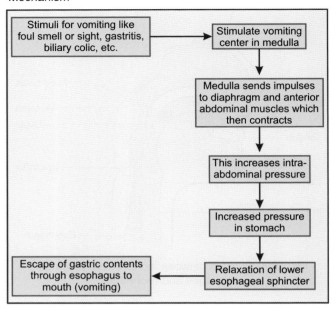

Movements of Small Intestine

Small intestinal movements helps in thorough mixing of the chyme with the digestive secretions and absorption. Also the unabsorbed residue is propelled downwards to be evacuated. There are 2 major movements and 2 minor movements. The two major movements are:

a. Mixing contractions (segmentation contractions)
b. Propulsive contractions (peristaltic movements)

a. Mixing Contractions

When a portion of small intestine becomes distended with chyme, the intestinal wall gets stretched which causes localized concentric contractions to take place (Fig. 28.5). They occur at regular intervals and last for a few seconds. Here only a small part of intestine is involved. In the next moment, the portions which are dilated become constricted and vice versa. This process continues. This helps in proper mixing of

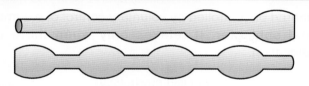

Fig. 28.5: Segmentation (mixing) contraction of small intestine

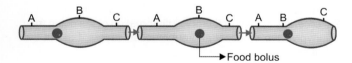

→Food bolus

Fig. 28.6: Peristaltic contractions of small intestine

chyme with small intestinal secretions. The segmentation contractions do not propel the chyme, but it is carried forward and backward.

b. Peristaltic (propulsive) Movements

Propulsive movements are for the propulsion of chyme. They can occur in any part of small intestine. Here, at first a ring of contraction occurs just behind the food bolus. The area in front of the bolus relaxes, thus resulting in forward movement of bolus (Fig. 28.6). This process repeats.

By peristalsis, the chyme progresses towards the ileocecal valve. On reaching the ileocecal valve, the chyme is blocked for some hours until the person takes another meal. At that time, a gastroileal reflex intensifies the ileal peristalsis and forces the remaining chyme through the ileocecal valve into the cecum.

The two minor movements are:
a. Ileal emptying
b. Movements of villi

a. *Ileal emptying*
Distension of ileum causes the ileocecal sphincter to relax and ileal emptying occurs. But when the cecum gets distended, this sphincter contracts, thus stopping further ileal emptying.

b. *Movements of villi*
On coming into contact with food material, the intestinal villi exhibit a to and fro lashing or swaying movements. They also shorten and elongate alternatively. The various movements of the villi help in the absorption of fat into the lacteals, which are located inside the villi.

Applied Physiology

1. *Malabsorption syndrome:* This occurs due to incompetence of ileocecal valve. This can cause reflux of the colonic material into ileum, which will lead to bacterial growth in the ileum.
2. *Small intestinal obstruction:*
   It can be due to:
   • Mechanical cause, e.g. tumor, band of adhesion, etc.
   • Non-mechanical cause, e.g. paralytic ileus occurring after surgery
   Here the part of intestine proximal to the obstruction, develops violent peristalsis, causing severe colicky abdominal pain called intestinal colic.

Movements of Large Intestine

Large intestine or colon serves as an organ of temporary storage for the unabsorbed ingested food and other waste products of digestion. During its stay in the colon, large bowel mucosa absorbs considerable quantity of water and converts the liquid chyme into semisolid feces. There are two types of movements in large intestine:
a. Mixing or segmentation or haustration
b. Mass peristalsis

a. Segmentation or Haustration

Segmentation or haustration is the major mixing movement of large intestine, which are much stronger than the segmented contractions of small intestine. So they divide the large intestine into very distinct segments called 'haustra'. This movement does not propel the colonic contents. But they facilitate water absorption.

b. Mass Peristalsis

One or two mass peristalsis may occur in a day. This is also a strong movement. At first, a ring of constriction starts appearing in the splenic flexure of the descending colon. This band of constriction then travels downwards and drives the colonic contents into the rectum. Normally, the rectum is empty. But when it is filled with the colonic contents by mass peristalsis, it becomes distended and urge for defecation develops.

Gastrocolic Reflex

After a meal, mass peristalsis becomes intensified and urge for defecation becomes stronger.

Part 2

Physiology

Defecation

The desire to defecate is felt only when the feces distends the rectum. Usually a pressure of 20-25 cm of water in the rectum is sufficient to initiate the desire for defecation.

There are two anal sphincters – external and internal.

External anal sphincter is under voluntary control and supplied by the pudendal nerve. Internal anal sphincter is involuntary. The urge to defecate can be ignored or controlled by voluntary control of external anal sphincter. It relaxes only when there is defecation.

Applied Physiology

1. *Crohn's disease* is characterized by chronic inflammation in the terminal part of ileum. It can affect all layers of the intestine.
2. *Ulcerative colitis* or Chronic inflammation of the colon may involve the mucosal and submucosal layers, but not the deeper structures.

The Respiratory System

The main function of our respiratory system is to provide oxygen to the body tissue from atmosphere and to remove carbon dioxide from tissues to the atmosphere. The lungs function as the organs of gas exchange between blood and atmosphere.

FUNCTIONAL ANATOMY OF RESPIRATORY SYSTEM

The components of respiratory system include the respiratory passages and the lungs. The respiratory passages are nose, pharynx, larynx, trachea, bronchi, bronchioles and alveoli.

TRACHEOBRONCHIAL TREE

The air passages between trachea and alveoli divided about 23 times and form the tracheo-bronchial tree (Fig. 29.1). This concept was first introduced by Weibel, a Swiss Anatomist. The main airway–'trachea'–branches into two bronchi. Each bronchus enters a lung and gives rise to many branches called bronchi, which further branch into smaller bronchioles. Trachea and the first 16 generations of tracheo-bronchial tree constitute the 'conducting zone', where no gas exchange occurs. The smallest airway in the conducting zone is the terminal bronchiole. The last 7

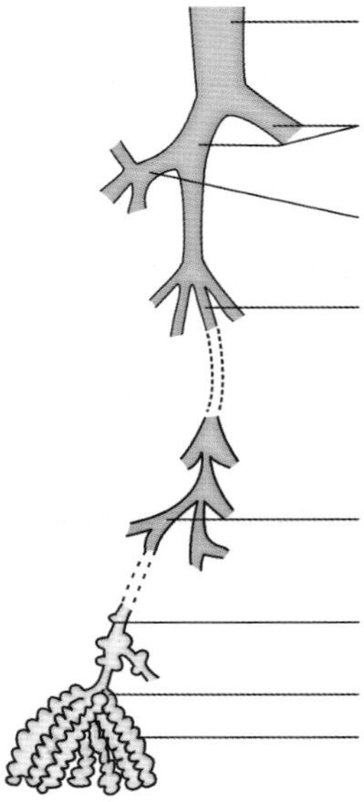
Trachea
Right and left principal bronchi
Lobar bronchus
Segmental bronchi
Terminal bronchiole
Respiratory bronchiole
Alveolar duct
Alveolar sacs

Fig. 29.1: Tracheobronchial tree

generations of tracheo-bronchial tree constitute 'the respiratory zone', where actual exchange of gases takes place. This zone consists of respiratory bronchioles, alveolar ducts and alveoli. The respiratory zone is surrounded by an extensive network of pulmonary capillaries. The alveolar ducts and the thin walled alveoli together form 'the respiratory unit'. The multiple divisions of tracheo-bronchial tree increase the cross-sectional area, i.e. cross-sectional area of trachea 2.5 cm^2 and that of alveoli is 11800 cm.^{2} There are about 300 million alveoli in human lungs. The pulmonary capillaries and alveoli form a 'blood-gas' interface.

LUNGS

Lungs occupy most of the thorax. Each lung is a spongy structure with the shape of a cone. It is covered by a thin, double layered membrane called pleura. The inner visceral pleura is adherent to the lung, while the outer parietal pleural is adherent to the thoracic cage. At the root of the lung (hilus), visceral pleura is continuous with parietal pleura. Pleura helps to transmit the force generated by respiratory muscles onto the lungs.

There is a potential space between the two layers of pleura called the pleural cavity, which is filled with a small amount of fluid (about 2 ml) called pleural fluid. Pleural fluid is formed by filtration of interstitial fluid and is continuously drained by the lymphatic channels of lungs. This continuous suction maintains a normal negative intra-pleural pressure. Pleural fluid acts as a lubricant which helps to reduce friction during movements of lungs. It provides a hydraulic traction making the two pleural layers inseparable.

Applied Physiology

In diseased conditions, the pleural cavity may be filled with blood ('Hemothorax'), fluid ('pleural effusion'), air ('pneumothorax') or pus ('pyothorax').

BLOOD SUPPLY

The conducting and the respiratory zones of tracheobronchial tree have their own distinct blood supply. The conducting zone is supplied by the bronchial artery, which originates from descending thoracic aorta and drains into the pulmonary vein. The bronchial circulation receives about 1-2% of cardiac output. The respiratory zone is supplied by the pulmonary artery that branches extensively to form a dense capillary network in the walls of alveoli. The pulmonary circulation receives whole of the cardiac output from right side of heart ('deoxygenated' blood) and therefore the blood flow is high (5-6 liters per min.). This blood is oxygenated in the alveoli and returned to the heart via the pulmonary vein. There is extensive anastomosis between bronchial and pulmonary circulation.

INNERVATION

Respiratory tract is innervated by sympathetic and parasympathetic divisions of autonomic nervous system, the stimulation of which causes bronchodilatation and bronchoconstriction respectively.

Non-respiratory Functions of Lungs

In addition to the respiration function, lungs subserve many other non-respiratory functions too.

Defence Functions of Lungs

- Bronchial secretions contain IgA which helps to resist infections

- Hair, cilia, pulmonary alveolar macrophages and the cough reflex prevent the entry of foreign particles into the alveoli. Foreign particles with size more than 10 μm are settled down by the hairs and mucous membrane of nose itself. Particles with size 2-10 μm can reach the bronchi. But they initiate reflex bronchoconstriction and cough which help to expel them. The ciliary movements also serve to expel them. Particles with size less than 2 μm can reach up to the alveoli, but are immediately phagocytosed by the pulmonary alveolar macrophages.

Functions Subserved by the Pulmonary Circulation

When the left ventricular output exceeds the systemic venous return, the excess blood can be stored in the pulmonary circulation, which helps in maintaining the cardiac output.

Small blood clots, cancer cells, fat cells, agglutinated RBCs, cell debris, masses of platelets, etc. are removed by certain lytic enzymes present in the pulmonary capillary endothelium.
Certain drugs can easily diffuse through the alveolar capillary barrier and enter into the systemic circulation, e.g. anesthetic drugs, aerosol.

Endocrine and Metabolic Functions of Lungs

Secretion of surfactant is the function of type II alveolar epithelial cells.

Certain substances like prostaglandins, histamine, kallikrein, etc. are stored in the lungs.

Certain substances are activated in the lungs. For example, angiotensin I is activated to angiotensin II (a potent vasoconstrictor) by the enzyme ACE (angiotensin converting enzyme) present in the lungs.

Certain substances like acetylcholine, serotonin, noradrenaline, etc. are inactivated in the lungs.

■ MECHANICS OF PULMONARY VENTILATION

Pulmonary ventilation is the process of movement of air from the atmosphere to the lungs and vice-versa, brought about by changes in the size of thoracic cage. Inspiration is an active process, brought about by contraction of inspiratory muscles while expiration is a passive process caused by elastic recoiling of lungs. In this section, let us discuss about the muscles involved in the process of breathing as well as the pressure changes that favor the movement of air into and out of the lungs.

■ MUSCLES CAUSING EXPANSION AND CONTRACTION OF LUNGS

The primary muscles of inspiration are diaphragm and external intercostals. In addition, scalene, sternocleidomastoids, neck and back muscles act as accessory muscles or secondary muscles of inspiration. Consequences of their contraction are given in (Fig. 29.2). When the diaphragm contracts, it moves downward towards the abdomen and the vertical diameter of thoracic cage is increased.

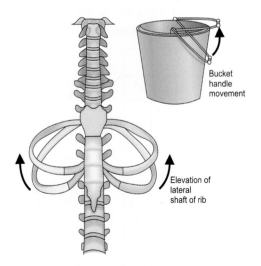

Bucket handle movement

Elevation of lateral shaft of rib

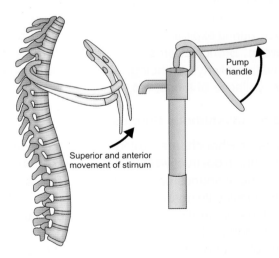

Pump handle

Superior and anterior movement of stirnum

Fig. 29.2: Actions of inspiratory muscles

Part 2

Physiology

Contraction of external intercostal muscles increase the transverse diameter ("bucket handle" effect) as well as the antero-posterior diameter ("water pump-handle" effect) of thoracic cage (Fig. 29.2). During quiet breathing, expiration occurs passively. Recoiling of lungs and chest wall (by relaxation of inspiratory muscles) cause compression of lungs and movement of air outwards. But during forced expiration (as in exercise or due to diseases like asthma and emphysema), expiration also becomes active by contraction of expiratory muscles. The abdominal recti and internal intercostals are the muscles concerned with expiration. Contraction of abdominal muscles increases the intra-abdominal pressure, which forces the diaphragm upwards. This decreases the vertical diameter of thoracic cage. Contraction of internal intercostal muscles reduce both antero-posterior and transverse diameters of thorax.

■ BREATH SOUNDS

Breath sounds can be heard clearly by auscultation. Breath sounds are produced due to passage of air through airways and alveoli. But the sounds are attenuated and filtered by the lung tissue. According to the quality of intensity, they are of two types – vesicular and bronchial. Vesicular breath sounds are the normal breath sounds heard all over the chest wall. They are heard mainly during inspiration and early expiration without a distinct gap. Bronchial breath sounds are high pitched sounds heard normally over trachea and bronchi, where there is less attenuation or filtration of sound. They are audible during inspiration as well as during expiration and there is a distinct pause in between. Bronchial breath sounds heard over the chest wall is always pathological (e.g. in pulmonary fibrosis, collapse of lungs, etc.).

Added or Adventitious Sounds

Wheezes or Rhonchi are prolonged musical sounds heard in certain obstructive lung diseases like bronchial asthma. These sounds are produced due to spasm of smooth muscles lining the airway.

Crackles or crepitations are short non-musical, bubbling sounds. These sounds are produced due to bubbling of air through secretions as in lung infections and interstitial lung diseases.

Pleural rub is the creaking or rubbing type of sound heard in inflammation of pleura or pleuritis.

Pressure Changes during Respiration

Intra-alveolar or Intrapulmonary Pressure Changes (Fig. 29.3)

Intra-alveolar pressure is the pressure exerted by the pleural fluid present between the two layers of pleura. Due to continuous suction of pleural fluid into lymphatic channels, a negative suction pressure is developed as the intra-pleural pressure. The normal intra-pleural pressure at the beginning of inspiration is – 2.5 to – 3 mm of Hg and at the end of inspiration, it becomes – 6 mm of Hg (Fig. 29.3). Intra-pleural pressure is exerted throughout the thoracic cavity, outside the lungs. However, this suction pressure is not uniform throughout the pleural space. It is more towards the apex of the lung (about –10 mm of Hg) and progressively falls to its lowest value near the base of the lung (– 2.5 mm of Hg), at the start of inspiration.

The pressure difference between intra alveolar pressure and intrapleural pressure is called the transpulmonary pressure. It gives an index of the elastic forces in the lungs that tend to collapse lungs at each stage of respiration.

■ SURFACE TENSION AND SURFACTANTS OF ALVEOLI

Surface tension is the force acting to pull a liquid's surface molecules together at an air-liquid interface

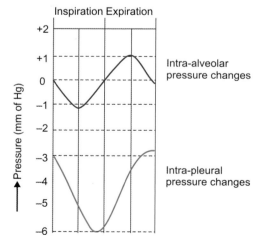

Fig. 29.3: Pressure changes in the alveolar and pleural spaces

(Fig. 29.4). Such an air-liquid interface exists in the alveoli also. This will tend to force air out of the alveoli causing them to collapse. But, normally the alveoli do not collapse due to the presence of surface tension lowering agent called surfactant. The presence of surfactant was first described by Pattle in 1955. Surfactant is a complex of phospholipids, proteins, carbohydrates and several ions. Dipalmitoyl phosphatidyl choline is the main surfactant. It is secreted by type II alveolar epithelial cells. It forms a layer at the fluid-air interface in the alveoli (Fig. 29.5). The air-surfactant interface has only one-twelfth to one-half the surface tension of air-fluid interface. Its advantages are given below:

1. It lowers the surface tension in the alveoli. This increases the compliance of lungs and reduces the work of breathing.

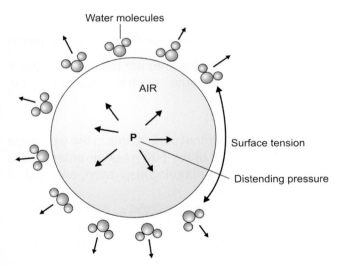

Fig. 29.4: Surface tension at the air-water interphase

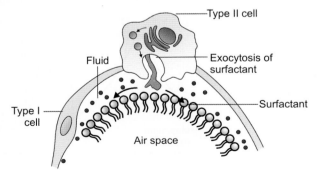

Fig. 29.5: Formation of pulmonary surfactant

2. The force of surface tension is inversely proportional to the concentration of surfactant present per unit area of alveolus. During expiration, the size of alveolus decreases and the concentration of surfactant per unit area increases. Thus surface tension is decreased which prevents collapse of lungs.

3. During inspiration, alveolar size increases and the concentration of surfactant per unit area decreases. This relatively increases the surface tension, thus preventing over distension of lungs.

4. By reducing the surface tension, surfactant prevents fluid accumulation in the alveolus and keeps it dry. In the absence of surfactant, fluid from interstitial space flows into the alveolar space due to increased surface tension.

Synthesis of surfactant starts in late fetal life. Its synthesis is facilitated by glucocorticoid and thyroid hormones. Insulin delays the synthesis of surfactant. There is an increase in fetal as well as maternal cortisol near term, which accelerates maturation of surfactant in lungs. In-utero, the fetal lungs remain collapsed. After birth, the infant makes several strong inspiratory efforts and the lungs expand. They are prevented from collapsing again by the surfactant.

▌ APPLIED PHYSIOLOGY

Hyaline Membrane Disease or Infant respiratory Distress Syndrome

This is a serious pulmonary disorder which develops in premature infants due to deficiency of surfactant. Surface tension of alveoli is high and many of them get collapsed. Also there can be fluid accumulation in the lungs ("pulmonary edema").

Measurement of Pulmonary Ventilation

For studying the volume of air moving into and out of the lungs, an instrument called spirometer is used. Most of the lung volumes and capacities can be easily determined by a water filled spirometer attached to a kymograph. The recording is called spirogram. Volume of air is plotted on Y-axis and time is plotted on X-axis of a spirogram (Fig. 29.6). A downward deflection in the recording obtained represents expiration and an upward deflection represents inspiration.

Part 2

Physiology

Lung Volumes (Fig. 29.6)

Tidal volume (TV): It is the volume of air that moves into the lungs with each inspiration or the amount that moves out with each expiration. It is about 0.5 liters in an adult.

Inspiratory reserve volume (IRV): It is the air inspired, with a maximal inspiratory effort, in excess of the tidal volume. It is about 1 liter in males and 0.7 liters in females.

Residual volume (RV): It is the air left in the lungs after a maximal expiratory effort. It is about 1.2 liters in males and 1.1 liters in females. This cannot be measured directly by simple spirometry.

Lung Capacities (Fig. 29.6)

Vital capacity (VC): It is the greatest amount of air that can be expired after a maximal inspiratory effort. It is about 4.8 liters in males and 3.1 liters in females. This is an index of pulmonary function.

Total lung capacity (TLC): It is the sum of vital capacity and residual volume. It is about 6 liters.

Inspiratory capacity (IC): It is the amount of air that a person can breathe, beginning at resting expiratory level and distending the lungs to the maximum amount.

    IC = IRV + TV

Functional Residual Capacity (FRC): It is the amount of air remaining in the lungs at the end of normal expiration .

    FRC = RV + ERV

Timed vital capacity (TVC): This is a dynamic lung function test where the subject is asked to inspire and expire forcefully and rapidly into a recording spirometer. The total volume of air thus expired is called timed vital capacity or forced expiratory volume (FEV). The forced expiratory volume in each second is calculated and plotted on a graph (Fig. 29.7) against time. In normal individuals, the forced expiratory volume in first second (FEV_1) is about 83%, in 2nd second (FEV_2) is about 95% and in 3rd second (FEV_3) is almost 100%. This measurement is used to detect obstructive and restrictive disorders of lungs. In an obstructive lung disorder (e.g. asthma), the forced vital capacity is significantly decreased and the FEV_1 : TVC ratio is low (Fig. 29.7). In a restrictive lung disorder (e.g. fibrosis of lungs), both TVC and FEV_1 are equally reduced and so the ratio of FEV_1 : TVC appear to be normal (Fig. 29.7).

Respiratory Minute Volume or Pulmonary Ventilation

It is the amount of air inspired per minute, i.e. 500ml/breath x 12 breaths/minute, which is about 6 L/minute.

Alveolar Ventilation

It is the volume of air that participates in the exchange of O_2 and CO_2. Out of the 500 ml of air entering into the lungs (tidal volume) during inspiration, only 350 ml reaches the alveoli.

    Therefore, alveolar ventilation = 350 ml x 12
                                        breaths/min
                                     = 4.2 liters/min.

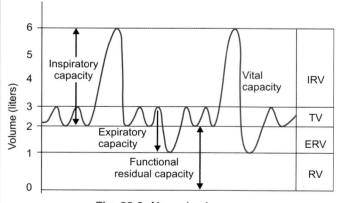

Fig. 29.6: Normal spirogram

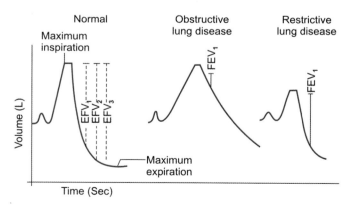

Fig. 29.7: Timed vital capacity in normal subject, obstructive and restrictive lung diseases

DEAD SPACE AIR

It is the volume of air present in the respiratory passage which does not take part in gaseous exchange. The volume of air in the "conducting zone" of respiratory passage, i.e. from nose to terminal bronchioles; constitute the 'Anatomical dead space'. Normal anatomical dead space volume is about 150 ml. Amount of air in this Anatomical dead space along with the air present in the alveoli which is not taking part in gaseous exchange constitute the Physiological dead space or total dead space of lungs. In healthy individuals, anatomical dead space is equal to physiological dead space, i.e.150 ml.

VENTILATION-PERFUSION RATIO

It is the ratio of the alveolar ventilation to the pulmonary blood flow. Normal alveolar ventilation at rest is 4.24 L/min. and pulmonary blood flow is 5 L/min. Therefore the ventilation-perfusion ratio is 4.2 L/5L or 0.84. The ventilation-perfusion ratio is high in the apex of the lungs and less in the base of the lungs (due to gravity).

PULMONARY DIFFUSION

Pulmonary diffusion is the gaseous exchange occurring at the alveolar level, i.e. the diffusion of O_2 from the alveolar air into the pulmonary capillary blood and diffusion of CO_2 from the capillary blood into the alveoli. Diffusion of gases takes place according to their partial pressure gradients (Table 29.1).

Respiratory Membrane (Fig. 29.8)

It is the blood-gas barrier through which diffusion of gases takes place. It is about 0.2 to 1 mm thick and consists of the following 6 layers, which separate the capillary blood from the alveolar air.

Table 29.1. Partial pressures of gases

	O_2 (mm of Hg)	CO_2 (mm of Hg)	N_2 (mm of Hg)
Inspired air	158	0.3	596
Expired air	116	32	565
Alveolar air	100	40	573
Arterial blood	95	40	573
Venous blood	40	46	573
Tissues	40	46	573

1. Lining of fluid, containing surfactant, around the alveolus
2. Alveolar epithelial cells
3. Basement membrane of alveolar epithelial cells
4. Space between epithelial and endothelial cells
5. Basement membrane of capillary endothelial cells
6. Endothelial cells.

Diffusion of a gas across the respiratory membrane depends on the Fick's law of diffusion, which states that:

$$V_{gas} = \frac{A}{d} \times D \times \Delta P, \text{ where}$$

A = Area of respiratory membrane
d = Thickness of respiratory membrane
D = Diffusion coefficient
ΔP = Different in partial pressures of the gas between alveolus and capillary blood
V_{gas} = Volume of diffusing gas

$$\text{Diffusion coefficient, D} = \frac{\text{Solubility of gas}}{\sqrt{\text{Molecular weight of gas}}}$$

Factors affecting the rate of diffusion of a gas across the respiratory membrane are:
1. The pressure difference (ΔP) across the membrane, i.e. diffusion is directly proportional to ΔP.
2. The thickness (d) of the membrane, i.e. rate of diffusion is inversely proportional to 'd'.
3. The surface area of the membrane (A), i.e. rate of diffusion is directly proportional to 'A'.
4. The molecular weight of the gas, i.e. rate of diffusion is inversely proportional to √molecular weight.
5. The diffusion coefficient (D) of the gas

Diffusion Capacity

The diffusion capacity of the lungs is the volume of the gas that diffuses through the respiratory membrane

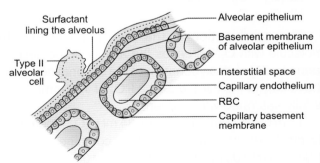

Fig. 29.8: Respiratory membrane

per minute for a pressure difference of 1 mm Hg. For O_2, it is normally 20-30 ml/min/mm Hg at rest. This value can increase markedly after exercise. The diffusion capacity of CO_2 is about 20 times that of O_2 under resting condition.

TRANSPORT OF GASES BETWEEN LUNGS AND TISSUES

Transport of gases is one of the most important functions of the blood.

Transport of O_2

It depends primarily on the amount of Hb in the RBCs. The partial pressure of dissolved O_2 (PO_2) in the pulmonary venous blood is 40 mmHg and that in the alveolar air is 104 mmHg. Because of this marked pressure gradient, O_2 diffuses from the alveolar air into the pulmonary capillary blood. However, the PO_2 becomes 100 mmHg when the blood reaches the aorta. This is due to some mixing of blood occurring in the basian vessels and physiological shunts.

Methods by which O_2 is carried by the blood:
1. About 97% of the O_2 transported from the lungs to the tissues is by chemical combination with Hb of RBCs.
2. The rest 3% is carried in the dissolved state in plasma.

Role of Hb in Oxygen Transport

Each of the four iron atoms of Hb can bind loosely and reversibly with one oxygen molecule, by oxygenation reaction (Not oxidation), i.e.

$$Hb_4 + O_2 \rightleftharpoons Hb_4O_2$$
$$Hb_4O_2 + O_2 \rightleftharpoons Hb_4O_4$$
$$Hb_4O_4 + O_2 \rightleftharpoons Hb_4O_6$$
$$Hb_4O_6 + O_2 \rightleftharpoons Hb_4O_8$$

The combination of the first heme in the Hb molecule with O_2 increases the affinity of the second heme for O_2 and oxygenation of the second increases the affinity of the third, etc. At a PO_2 of 100 mmHg, Hb is fully saturated with oxygen.

1 gm of Hb combines with 1.34 ml of O_2. Therefore, 15 gm (average normal Hb) of Hb combine with 15 x 1.34 $\approx$ 20 ml/100 ml of blood, i.e. each 100 ml of arterial blood can carry 20 ml of oxygen to the tissues. The

maximum amount of O_2 that can be carried in the blood by Hb is called the **oxygen carrying capacity of Hb (20 ml/100 ml of blood)**. About 5 ml of O_2 is given off to the tissue.

O_2-Hb Dissociation Curve

This curve represents the relationship between the PO_2 and the degree of oxygen saturation of Hb (Fig. 29.9). The curve is sigmoid shaped. This particular shape of the curve is due to the biological properties of Hb. When O_2 combines with Hb, it assumes the relaxed or R-state, which favors O_2 binding, and additional uptake of O_2 is facilitated. When all the four heme moieties have combined with O_2, it assumes the tense or T-state. The sigmoid shape is due to this T-R interconversion. The combination of the first heme in the Hb molecule with O_2 increases the affinity of the second heme for O_2 and oxygenation of the second increases the affinity of the third, etc.

At PO_2 values of 100 mmHg or above, Hb is 100% saturated (Fig. 29.9). Even when the PO_2 falls to 60 mmHg, the Hb saturation is 90%. This provides a margin of safety when one ascends up to high altitudes, where there is decreased alveolar PO_2. Even then he will suffer only a small decrease in the arterial O_2 content.

Factors Affecting O_2 Dissociation Curve

The important factors that affect the affinity of Hb for O_2 are:
1. pH
2. Temperature
3. 2,3-DPG concentration in the RBC

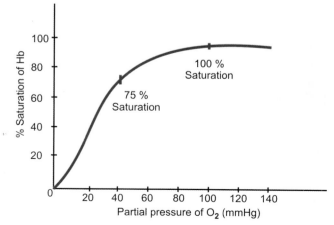

Fig. 29.9: The O_2-Hb dissociation curve

A rise in temperature, a fall in pH and an increase in 2,3-DPG concentration can shift the O_2-Hb dissociation curve to right, i.e. there is decreased affinity of Hb for O_2 and more of O_2 is liberated or off-loaded.

Bohr effect: The affinity of Hb for O_2 decreases when the pH of blood falls. This is called Bohr effect. At the tissue level, there is increased PCO_2 and decreased PO_2. This causes the Hb to liberate more O_2 into the tissues (Bohr effect).

The curve is shifted to left by a fall in temperature, a rise in pH and a fall in 2,3-DPG concentration, i.e. the affinity of Hb for O_2 is more and their binding is favored. HbF or fetal Hb contains $\alpha_2\gamma_2$ chains. γ chains have little affinity for 2,3-DPG. Hence HbF-O_2 dissociation curve is shifted to left when compared with the adult curve. The higher affinity of HbF for O_2 facilitates the extraction of O_2 by the fetus from the maternal circulation.

About 3% of O_2 is transported as dissolved form in blood. The dissolved O_2 exerts the partial pressure, which actually determines the amount of O_2 that combines with hemoglobin. Thus even though a small amount of O_2 is transported in dissolved form, it is very important in determining the total O_2 transport.

Transport of CO_2

Carbon dioxide is carried in the blood in 3 forms:
a. *As physically dissolved form:*
   Accounts for about 10% of the total CO_2 in blood. CO_2 is 20 times more soluble in water than O_2.
b. *As bicarbonate*
   About 80% of the CO_2 is transported as bicarbonate CO_2 diffuses into the red cells where, in the presence of carbonic anhydrase enzyme, it rapidly reacts with water to form carbonic acid (H_2CO_3). The carbonic acid dissociates into HCO_3^- and H^+

$$H_2O + CO_2 \underset{\text{Carbonic anhydrase}}{\rightleftharpoons} H_2CO_3$$
$$\uparrow\downarrow$$
$$H^+ + HCO_3^-$$

H^+ ion formed is buffered by Hb, while HCO_3^- enters the plasma. Deoxy Hb is a weaker acid than the oxyHb and therefore it binds to H^+ more than the oxyHb.

As the HCO_3^- which is formed in red cell diffuses out to the plasma, Cl^- diffuses from the plasma into the red cells to maintain electrical neutrality. This phenomenon is called the *chloride shift* or *Hamburger shift*. This is the cause of the increased content of Cl^- in venous blood.

c. *As carbamino hemoglobin*
   About 10% of CO_2 is transported as carbaminohemoglobin CO_2 combines with the amino ($-NH_2$) group of Hb to form carbaminoHb.

$$R\text{-}NH_2 + CO_2 \rightleftharpoons R\text{-}NH\text{-}COOH$$

No enzyme is required for this process. DeoxyHb can rapidly form carbamino compound than oxyHb. Therefore, in the tissues, where there is deoxyHb, the reaction shifts to right. In lungs, oxyHb shifts the reaction to left.

Haldane Effect

De-oxygenated Hb combines with more CO_2 compared to oxygenated Hb at any PO_2. This is called Haldane effect. Binding of O_2 to Hb thus reduces its affinity for CO_2.

CO_2 Dissociation Curve

This curve shows the relationship between PCO_2 and total CO_2 content of blood (Fig. 29.10). The relationship is linear over a wide range of PCO_2, than in case of O_2-Hb dissociation curve. When Hb is oxygenated, the CO_2-dissociation curve shifts to the right, i.e. CO_2 is removed (Haldane effect) (Fig. 29.10).

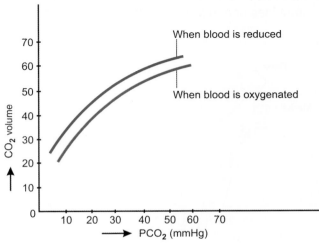

Fig. 29.10: CO_2 dissociation curve

REGULATION OF RESPIRATION

Respiration is an automatic process that occurs without any conscious effort while we are asleep or awake. Rhythmical excitation of respiratory muscles occurs as a result of multiple neuronal interactions involving all levels of the nervous system. Various neural and hormonal stimuli as well as chemical changes in blood (i.e. changes in blood PO_2, PCO_2 and pH) can influence the central control of respiration. Hence, regulation of respiration can be broadly classified as "Neural regulation" and "Chemical regulation".

Neural Regulation

The neural mechanism includes:
1. Automatic control by medulla and pons
2. Voluntary control by cortex.

Automatic Control by Medulla and Pons

Automatic centers in medulla and pons drive the respiratory muscles rhythmically and subconsciously. Certain pacemaker cells called "Pre-Botzinger complex" (Fig.29.11), situated on either sides of medulla, initiates rhythmical excitation of respiratory muscles. Medulla also contains special groups of neurons called dorsal and ventral groups, which are concerned with inspiration and expiration respectively (Fig. 29.12). Both dorsal and ventral groups are connected to the pre-Botzinger complex and can influence its activity.

a. The dorsal respiratory group is located in the nucleus of tractus solitarius. They receive afferent impulses coming from the lungs through IX and X

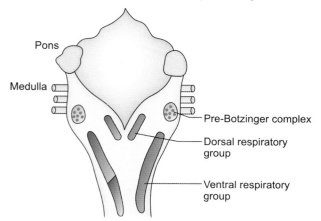

Fig. 29.11: Medullary respiratory neurons

cranial nerves. The inspiratory neurons (I-neurons) generate the basic rhythm of respiration. They also project to and drive the ventral respiratory group. These inspiratory neurons fire periodically and the frequency of impulse discharge gradually increases to maximum before coming to an abrupt end. This pattern of discharge is called ramp signals (Fig. 29.13)

Inspiration occurs as follows:

Dorsal group of neurons stimulated → I.neurons fire

- neurons supplying inspiratory muscles stimulated
- inspiratory muscles contract and inspiration starts.

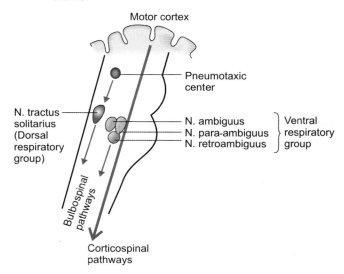

Fig. 29.12: The neural regulatory centers of respiration

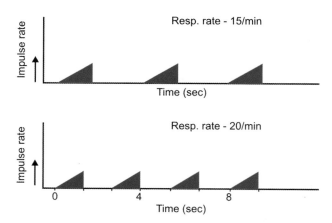

Fig. 29.13: The ramp signals in the medullary respiratory neurons

b. The ventral respiratory group is located in the ventral part of the medulla. They contain both expiratory (E-neurons) and inspiratory (I-neurons) neurons. Both the E and I neurons fire together only during deep breathing. The ventral group of neurons innervate the ipsilateral accessory muscles of respiration via the vagus nerves. The inspiratory neurons of the ventral group are controlled by the I-neurons of the dorsal group.

The pneumotaxic center or the pontine respiratory center:

The center is located on the upper pons bilaterally. These neurons have an inhibitory effect on the dorsal respiratory group of neurons. The pneumotaxic center, when stimulated "switches off" or inhibits inspiration, resulting in a shallow and more rapid respiration.

The sequence of events occurring during normal respiration can be summarized as follows:

At regular intervals, I-neurons fire and inspiration (given above) takes place.

Cessation of discharge from I-neurons → Beginning of expiration, which occurs passively.

In forceful expiration, the E.neurons fire during expiration.

Hering-Breuer Reflex

Stretch receptors are present in the smooth muscles of bronchial wall. They are stimulated when the lung inflates above the tidal volume. Afferent impulses from these stretch receptors reach the nucleus of tractus solitarius of medulla, via the vagus nerve. These impulses inhibit inspiratory neuron discharge and limit the tidal volume, thereby increasing the respiratory rate. This phenomenon is called Hering-Breuer reflex.

Other than the stretch receptors, there are also J receptors (Juxtacapillary receptors) and irritant receptors in the alveolar walls. J-receptors, when stimulated, decrease the depth of inspiration and increase the frequency of breathing.

Irritant receptors, when stimulated (by tobacco smoke, irritant gases, etc.), cause bronchoconstriction, cough, etc. to prevent that irritant substance from reaching the lungs.

Voluntary Control

Both inspiratory and expiratory muscles can be voluntarily controlled for a short period of time (breath-holding time) after which the chemical control overrides the voluntary inhibition and breathing takes place involuntarily. The voluntary system of control is located in the cerebral cortex and sends impulses to the respiratory neurons via the corticospinal tracts.

Other neural reflexes like visceral reflexes (like deglutition, vomiting, defecation, etc.) and proprioceptive afferent impulses from the muscles and joints can also regulate the respiration.

Chemical Control of Respiration

Chemical control of respiration is exerted through the central and peripheral chemoreceptors. The chemoreceptors respond to changes in the pH, PO_2 and PCO_2 of blood. A rise in PCO_2, a rise in H^+ concentration or a decrease in PO_2 can stimulate respiration.

The Central Chemoreceptors

They are located in the medulla, near the origin of IX and X cranial nerves.

Mechanism: The medullary chemoreceptors are sensitive to any change in pH of the CSF, including the interstitial fluid of the brain when there is increased PCO_2 in the blood, CO_2 penetrates the blood brain barrier and enters the CSF and interstitial fluid of brain. Now, the following reactions occur:

$$CO_2 + H_2O \rightarrow H_2CO_3 \rightleftharpoons H^+ + HCO_3^-$$

Thus an increase in H+ occurs, which stimulates the medullary chemoreceptors, which in turn stimulate ventilation. Also a decrease in H+ concentration can inhibit ventilation.

However, if an acid is injected into the blood, it may take a longer time to raise the ventilation, than the increased PCO_2. This is because CO_2 can cross the blood brain barrier more easily than the H^+.

The Peripheral Chemoreceptors

They are the carotid and aortic bodies. Carotid bodies are located at the bifurcation of the common carotid arteries. The aortic bodies are located in the arch of aorta (Fig. 29.14). The peripheral chemoreceptors are sensitive to both arterial PO_2 and PCO_2 levels. But a fall in pH of the blood can be detected by the carotid bodies only. The carotid bodies have a rich blood supply. The carotid bodies send afferent impulses to the respiratory centers through a branch of glosso-pharyngeal nerve or Hering's nerve, whereas, the aortic bodies send impulses through the vagus nerve.

RESPIRATORY ADJUSTMENTS IN HEALTH AND DISEASE

This topic deals with the following aspects:
- Respiratory adjustments to different stresses in day-to-day life (e.g. during exercise, at high altitude, etc).
- Abnormal respiratory patterns
- Respiratory disturbances like hypoxia
- Artificial respiration.

Respiratory Adjustments during Exercise

During exercise, many complex adjustments are occurring in our body regarding the blood flow, basal metabolism, cardiovascular system, respiratory system, temperature regulation, secretion of hormones, etc. Of these, the important adjustments that occur in respiratory system are discussed below.

Oxygen Uptake of the Body

In mild and moderate exercises (e.g. Steady and slow running or jogging), the increased O_2 uptake can be met by the body metabolism itself. But in severe exercise, the O_2 uptake or requirement shoots up to be >2 liters/min. 85% of energy required for severe exercise comes from anaerobic metabolism. So, there occurs an O_2 debt. O_2 debt includes the extra amount of O_2 consumed after exercise in order to remove excess lactate accumulation and to restore the ATP and creatine phosphate stores as well as to restore the small amount of oxygen that comes from myoglobin. To meet the O_2 debt, following changes in respiration occur:
1. Increased ventilation, due to stimuli like increased body temperature, accumulation of K^+, increased sensitivity to CO_2 during exercise, hypoxia, etc. There is increase in depth as well as rate of ventilation.

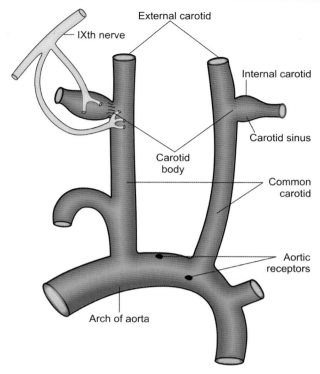

Fig. 29.14: Peripheral chemoreceptors (diagramatic)

2. Increased diffusion of O_2 into the pulmonary capillaries, i.e. increased pulmonary oxygen diffusing capacity.
3. There is no change in vital capacity.

Respiratory Adjustments in High Altitude

At high altitudes, the percentage of oxygen in the atmosphere is the same as at the sea level, but the barometric pressure is low and so the partial pressure of oxygen is proportionately decreased (see the Table below).

Altitude (feet)	Barometric pressure (mmHg)	Partial pressure of O_2 in air (mmHg)	Partial pressure of O_2 in alveoli (mmHg)
0 (sea level)	760	160	100
10,000	523	100	60
20,000	350	73	40
30,000	225	47	18

The highest permanent human habitation has been detected up to 18000 feet on Peruvian Andes.

Mountain Sickness

Mountain sickness or high altitude sickness occurs when a person, residing at sea level, ascends to a high altitude. Initially, he experiences the symptoms of hypoxia like breathlessness, headache, dizziness, palpitation, nausea, partial deafness, etc. These symptoms may disappear after a few days or weeks, due to a number of adjustments occurring in the body. This phenomenon by which they develop compensatory mechanisms is called *acclimatization*

The compensatory mechanisms are:
- Hyperventilation to raise the arterial PO_2
- Increased erythropoietin secretion
- Polycythemia
- Increase in myoglobin and cytochrome oxidase enzyme
- Increase in 2, 3-DPG, causing the O_2-Hb dissociation curve to shift to right.
- Renal compensatory changes like renal alkalosis, leading to excretion of excess HCO_3^- in urine.
- Increase in heart rate as well as cardiac output.

Decompression Sickness or Caisson's Disease

Typically, this condition can occur when a diver, who was under the sea, is brought to the surface rapidly; or when an aeroplane without pressurized cabin ascends suddenly,

When the diver is beneath the sea for a long time, large amount of nitrogen get dissolved in his body fluids. As he suddenly ascends to the surface of the sea, the dissolved N_2 diffuses out in the form of bubbles. These N_2 bubbles can block the blood vessels of heart, brain and kidney. The symptoms which result, constitute the decompression sickness, also called the "bends" or "Caisson's disease"

Symptoms: Pain in the muscles and joints; Dizziness and unconsciousness; Dyspnea or "chokes" (due to blocking of pulmonary capillaries by bubbles)

Prevention and Treatment

Caisson's disease can be prevented by gradually and slowly bringing the diver to the sea surface. Treatment is by immediate recompression in a pressurized tank followed by slow decompression to normal atmospheric pressure.

Abnormal Respiratory Pattern

Eupnea means the normal respiratory pattern

Apnea means temporary stoppage of breathing

Dyspnea means increased pulmonary ventilation or difficult labored breathing where the person is aware of shortness of breath.

A normal person is not aware of his respiration, until it is doubled. Dyspnea is one of the most important sign of lung diseases. Causes of dyspnea are as follows:
1. Respiratory Diseases – like bronchial asthma, pneumonia, pleural effusion, pneumothorax, emphysema, etc.
2. Cardiovascular Diseases – like congestive cardiac failure
3. Metabolic disorders – like acidosis, ketosis, hyperthyroidism, etc.
4. Anemia

Dyspnea occurring in lying down position is called orthopnea, usually seen in cardiac failure.

Periodic Breathing is an abnormal pattern of respiration which is characterized by alternate periods of respiratory activity and apnea. Two types of periodic breathing are usually seen:
1. Cheyne-Stokes breathing
2. Biot's breathing.

Cheyne-Stokes Breathing (Fig. 29.15)

It is a form of periodic breathing, which is characterized by a series of respirations that gradually increase in depth, reach a maximum and then progressively decline to stop in a state of apnea lasting for a few seconds. Again, breathing gradually begins and the whole cycle is repeated. This type of periodic breathing is normally seen in infants and in high altitudes. It is pathologically seen in uremia, heart failure, after head injury, etc.

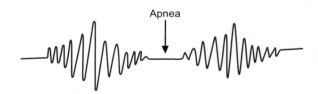

Fig. 29.15: Cheyne-Stokes breathing

Part 2

Physiology

Biot's breathing: It is characterized by irregular episodes of respiratory activity and apnea. There is no waxing and waning pattern of respiration as in Cheyne. Stokes breathing. This type of periodic breathing is seen in meningitis, pontine hemorrhage, cerebral ischemia, etc.

Hypoxia

Hypoxia means oxygen deficiency at the tissue level. Hypoxia can be of 4 types:
a. Hypoxic hypoxia or anoxic hypoxia
b. Anemic hypoxia
c. Stagnant or ischemic hypoxia
d. Histotoxic hypoxia.

Hypoxic Hypoxia

Most common type; it is due to a low PO_2 level in arterial blood. Conditions causing hypoxic hypoxia are high altitude, obstruction of airways, pulmonary oedema, asthma, congenital heart diseases, etc.

Anemic Hypoxia

Here, the arterial PO_2 is normal, but the O_2 carrying capacity of the blood is reduced, i.e. the amount of Hb available for O_2 transport is reduced. Apart from severe anemia, this can also occur in carbon monoxide poisoning, where there is increased formation of carbonmonoxy Hb compound and there is less available Hb for O_2 transport.

Stagnant Hypoxia

This occurs when the tissue blood flow is markedly reduced as in shock, heart failure or local obstruction. Here PCO_2 and Hb concentration are normal.

Histotoxic Hypoxia

This occurs in cyanide poisoning, where cyanide inhibits the enzyme cytochrome oxidase required for tissue oxidation. Therefore, the cells cannot utilize the O_2 supplied to them.

Cyanosis

Cyanosis means bluish discoloration of skin and mucous membranes due to excessive amount of reduced or deoxygenated Hb in blood ($\geq$ 5gm%). A person with severe anemia cannot become cyanotic as there is no adequate Hb to be reduced. Cyanosis is clinically diagnosed by looking under the tongue or nail beds of the patient. According to the region affected, cyanosis can be ***peripheral and central***. Central cyanosis is due to heart diseases, lung diseases, hypertension, warm periphery, etc. Peripheral cyanosis is due to circulatory shock, exposure to severe cold, etc. where there is severe peripheral vasoconstriction.

Artificial Respiration

Indications: Drowning, electric shocks, carbon monoxide poisoning, anesthetic accidents, respiratory failure, etc.

Artificial respiration should always be attempted in above said conditions because, respiration stops before the heart stops.

Mouth to mouth breathing: It is the most simple and effective means of resuscitation. The victim should be placed in supine position. Place a hand beneath the neck so that the airways are opened. The operator should blow into the victim's mouth about twice the tidal volume, 12 times a minute, and then permit the elastic recoil of the victim's lungs to produce passive expiration. Any entry of gas blown into the stomach is expelled by applying pressure on the abdomen. This method is more effective than other methods in producing effective ventilation. Also the CO_2 blown out from the operator's mouth is an effective respiratory stimulant.

Mechanical respirators or ventilators are useful in chronic ventilatory failure. They provide intermittent negative pressure to produce excursions of the chest wall resembling normal respiration.

The Cardiovascular System

A well functioning cardiovascular system is essential for the maintenance of the internal environment of the body. It carries nutrient substances absorbed from the intestine to different parts of our body and metabolic waste products to be excreted through kidneys and other organs. It is involved in the transport of gases and various hormones. It helps in the regulation of temperature, pH and water. The cardiovascular system has to perform all these functions in resting conditions as well as during a variety of challenging situations like exercise, stress, hot and cold climates, pregnancy and child birth, high altitudes, etc. It was William Harvey who first described the circulation of blood in 1628.

The cardiovascular system consists of heart, blood vessels and blood. Heart acts as the "pump" that pumps blood into the blood vessels so that it reaches all parts of our body. Blood vessels include: (a) an arterial system which distributes blood throughout our body; (b) the capillaries, where exchange of substances between blood and tissues occur; and (c) the venous system, which collects blood from different tissues back into the heart.

There are two components of circulatory system – pulmonary or lesser circulation; and systemic or greater circulation. The right ventricle pumps blood into the lungs via pulmonary artery for oxygenation. This forms the pulmonary circulation (Fig. 30.1). The left ventricle pumps oxygenated blood to all parts of our body via aorta. This forms the systemic circulation (Fig. 30.1).

HEART

Functional Anatomy

Heart is a hollow muscular organ, situated obliquely in the mediastinum – 2/3rd towards the left and 1/3rd towards the right of midline. It weighs about 300-350 gms. For a detailed structure of heart, different chambers and blood supply, *refer chapter 11(Anatomy)*.

Coverings of Heart

Heart is enveloped inside a fibro-serous covering called pericardium. Pericardium consists of an outer tough inelastic layer of dense connective tissue called fibrous pericardium and an inner double layered serous pericardium. Serous pericardium has a parietal layer and a visceral layer or epicardium, which is adherent to the surface of myocardium (Fig. 30.2). The pericardial layer anchors and protects heart. It prevents over distension of heart and at the same time, it allows free movement of heart during contraction. Between the two layers of pericardium, there lies the pericardial cavity, filled with 5-15 ml of pericardial fluid. Pericardial fluid reduces the friction during contractions of heart.

Applied Physiology

Pericarditis: Infection or inflammation of pericardium is known as pericarditis.

Pericardial effusion: Accumulation of excess fluids, pus or blood in the pericardial space is called pericardial effusion.

Musculature of Heart

Myocardium is the muscular coat of heart. It consists of three layers:

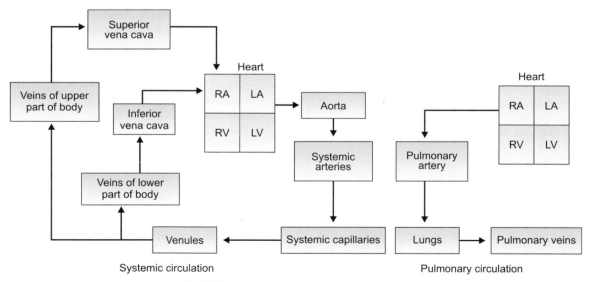

Fig. 30.1: Components of circulatory system

- Epicardium
- Myocardium proper
- Endocardium.

Epicardium is the outer most layer which is adherent to the visceral layer of pericardium. Myocardium includes two types of fibers – (1) the contractile fibers, constituted by the atrial and ventricular muscles, and (2) the specialized excitatory and conducting fibers which are less contractile. They include the pacemaker cells and conducting system of heart.

Atria and ventricles of both sides of heart are not anatomically or physiologically alike. Some important differences are given below.

Atrial walls are thinner than ventricular walls.

Left ventricular wall is thicker than that of right ventricle in the ratio 3:1.

Cavity of left ventricle is smaller than that of right ventricle.

Endocardium is the inner most layer of heart, which includes a thin layer of connective tissue and small amounts of muscular tissue, forming the flaps of valves, valve rings, etc.

Valves of Heart

There are two atrio-ventricular valves connecting atria and ventricles, and two semilunar valves guarding pulmonary and aortic orifices.

AV Valves (Fig. 30.3)

The atrio-ventricular valves (AV valves) are tricuspid valve on right side and mitral valve on left side. Tricuspid valve is larger than mitral valve and is guarded by three cusps or leaflets. Mitral or bicuspid

valve is guarded by two cusps or leaflets. AV valves are held in position by the papillary muscles, arising from the ventricular myocardium (Fig. 30.4). Papillary muscles are attached to the edge of valve leaflets by tough connective tissue strands called chordae tendinae (Fig. 30.4). Papillary muscles prevent undue bulging of AV valves into the atria during ventricular contraction.

Semilunar Valves (Fig. 30.3)

Semilunar valves guard the openings of pulmonary and aortic orifices. They consist of three cusps. Dense connective tissue covering these valves prevents their outstretching when blood passes through them. These valves are so arranged that only forward flow of blood opens them while backward flow of blood closes them. This prevents regurgitation of blood. The pressure changes occurring inside the cardiac

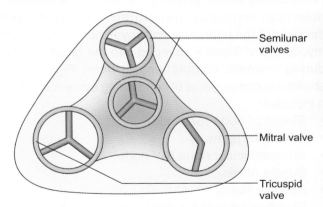

Fig. 30.3: Valves of heart

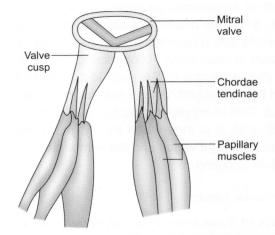

Fig. 30.4: Papillary muscles attached to atrioventricular valve

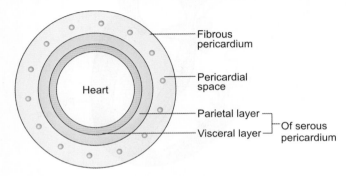

Fig. 30.2: Diagramatic representation of coverings of heart

chambers determine the opening and closing of valves.

CARDIAC MUSCLE

The gross as well as the microscopic structural peculiarities of cardiac muscle are already been discussed in "Muscle Physiology". Properties of cardiac muscle include the following:
- Excitability or bathmotropism
- Rhythmicity or chronotropism
- Conductivity or dromotropism
- Contractility or inotropism.

The first three properties are discussed in detail in "Muscle Physiology". In this portion, the conductivity of cardiac muscle is discussed in detail.

Conductivity or Dromotropism

Conductivity is the property by which the rhythmic electrical impulses generated are conducted throughout the heart, causing rhythmic contraction of myocardium. Blood is pumped into the great vessels during rhythmic contractions of heart Figure 30.5 shows the components of conducting system of heart. It includes:
- Sinoatrial node (SA node)
- Internodal pathway
- Atrioventricular node (AV node)
- Bundle of His
- Purkinje fibers.

Sinoatrial node (SA node) is situated at the junction of superior vena cava with right atrium. SA node contains specialized pacemaker type of myocardial cells. It has the property of self-excitation and so can generate action potentials. Such pacemaker cells are present in other parts of conducting system also. But the discharge rate of SA node is considerably more than their rates and hence it acts as the pacemaker of heart. Rate of impulse production by SA node is about 60-80/min. It is innervated by the parasympathetic fibers from right vagus and the sympathetic fibers via stellate ganglion.

Internodal Fibers

Bands of fibers connecting SA node to AV node are called internodal fibers. They include mainly three bands of fibers: Anterior internodal tract of Bachmann, middle internodal tract of Wenckebach, and posterior internodal tract of Thorel. The electrical impulses from SA node are conducted at about 1m/sec to the AV node.

AV Node (Fig. 30.5)

AV node is situated just beneath the endocardium on the right side of interatrial septum, near the tricuspid valve. It is the only conducting pathway between atria and ventricles. Pacemaker cells seen in AV node can generate electrical impulses just like SA nodal tissue, but at a lower rate of 40-50/min. The conduction velocity of AV node is about 0.05 m/sec.

Impulse from AV node takes about 0.08 to 0.1 sec for its transmission to 'Bundle of His". This delay is called AV nodal delay. This delay provides enough time for emptying of blood from atria into the ventricles so that effective pumping can be done. This enables atrial contraction to be completed before the starting of ventricular contraction. AV nodal delay is shortened by sympathetic stimulation and lengthened by stimulation of parasympathetic fibers of vagus.

Bundle of His

Bundle of His or atrio-ventricular bundle is the continuation of AV nodal fibers and is located beneath the endocardium on the right side of the interventricular

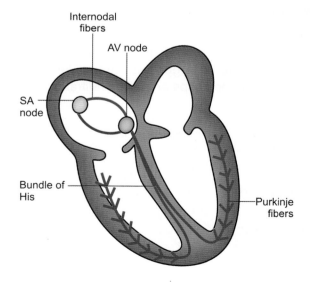

Fig. 30.5: Conducting system of heart

septum (Fig. 30.5). It gives off a left bundle branch and a right bundle branch. Left bundle branch again divides into anterior and posterior fascicles. All the branches run subendocardially on either sides of the septum. This bundle does not allow retrograde conduction of impulses back into the atria. The speed of conduction in the Bundle of His is about 1 m/sec.

Purkinje Fibers

Purkinje fibers arise from the branches of Bundle of His. They branch extensively and run subendocardially. They are fast conducting which conduct impulses at a rate of about 4m/sec. Fast conduction helps in instantaneous spread of impulses to both right and left ventricles so that they contract simultaneously.

In the interventricular septum, the depolarization wave of electrical impulse spreads rapidly through the Purkinje fibers to all parts of ventricles in the following orders:

- Septal activation from left side to right side
- Activation of outer septal region of ventricles
- Activation of major portion of ventricular myocardium from endocardium to epicardium
- Posterobasal portion of left ventricle, pulmonary conus and upper most part of interventricular septum, which together form the last portion of heart to be depolarized. The spread of the

electrical potential in the conducting system is summarized in (Fig. 30.6).

■ APPLIED PHYSIOLOGY

Ectopic Pacemaker

If the natural pacemaker or SA node becomes defective, the next fastest pacemaker tissue takes over (i.e. the AV node), which is then known as the "ectopic pacemaker".

Idioventricular Rhythm

If the atrio-ventricular bundle is destroyed, the atria and the ventricles get completely dissociated and they beat independently. Atria beat at a rate of 72/min ("SA nodal rhythm") while the ventricles beat at a rate of 15-40/min, which is called the "idioventricular rhythm".

Heart block or Conduction Block

Any disturbance which blocks the transmission of impulses from SA node downwards is known as heart block or conduction block. Depending on the extension of block, heart block can be of three types. If the conduction of impulses from SA node to AV node is just delayed by a partial obstruction, it is known as "first degree heart block". Diagnosis is possible only by taking an ECG where the PR interval will be prolonged (more than 0.2 sec). Second degree heart

Part 2

Physiology

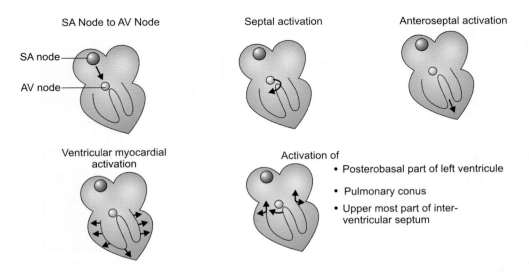

Fig. 30.6: Spread of electrical potential in the conducting system of heart

block is also an incomplete type where all the impulses from SA node are not conducted to the ventricles. Every second or third QRS complex is missed in the ECG. If the conduction from atria to ventricles is completely interrupted, it is called "third degree heart block" or complete heart block. Atria and ventricles beat at different rates, and idioventricular rhythm is present.

Innervation of Heart

Heart is innervated by sympathetic and parasympathetic fibers from ANS, which contain both afferent and efferent nerves. Parasympathetic supply is via vagus. Right vagus supplies SA node while AV node and the Bundle of His are supplied by left vagus. Atrial muscles also receive branches from vagal fibers, but very few branches go to the ventricles. Vagus is cardio-inhibitory in action. Sympathetic fibers arise from the lateral horn cells of spinal cord, relay in the stellate ganglion and supply SA node, AV node, Bundle of His, atrial and ventricular muscles. Sympathetic fibers are cardio-stimulatory in action. (Fig. 30.7) shows the effects of stimulation of sympathetic and parasympathetic nerves on the pacemaker potential.

Normally a minimal amount of continuous vagal discharge occurs in an adult at rest. This keeps the heart rate at about 70-80/min. This is called vagal tone. In the absence of vagal tone, the heart rate would be about 160-180/min.

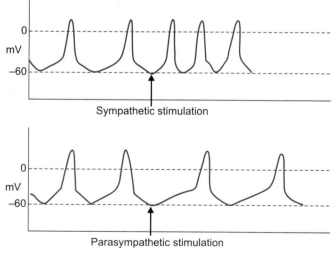

Fig. 30.7: Effects of sympathetic and parasympathetic stimulation on pacemaker potential

ELECTROCARDIOGRAM (ECG)

Electrocardiography is the method of recording electrocardiogram. ECG is defined as the graphical recording of the electrical activities of heart. The electrical changes occurring with each heart beat are conducted all over the body and can be recorded as ECG, by placing electrodes on the surface of the body. The potential difference between two electrodes is measured by the electrocardiograph.

Fluctuations in the potentials represent the algebraic sum of action potentials of individual myocardial fibers during cardiac cycle.

ECG was first recorded by Waller in 1887, using a capillary electrometer. William Einthoven, in 1903, used a string galvanometer for recording ECG and the recording was done on a photographic paper. Modern electrocardiograph machine consists of metallic leads and amplifier. The recording is done on a moving strip of specially waxed paper. Speed of movement of the paper is usually adjusted for about 25 mm/second (Fig. 30.8).

ECG Leads

'Lead' means a combination of wires and electrodes which are used to pick up electrical activity from the surface of body. Usually 12 leads are used for recording ECG – 3 bipolar (standard) limb leads, 3 unipolar augmented limb leads and 6 unipolar chest leads.

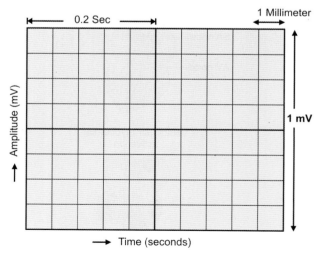

Fig. 30.8: ECG paper

Bipolar (Standard) Limb Leads

Bipolar limb leads were originally designed by Einthoven, which are still in use. Here, the potential difference between the two active or exploring electrodes placed on any two limbs is recorded.

Einthoven considered that three points in our body viz., right and left shoulders and mid point of left inguinal region, when joined constitute an equilateral triangle. Heart occupies the center of the triangle. This is called Einthoven's triangle (Fig. 30.9). Heart is considered as a dipole, situated at the center of a volume conductor. The sum of potentials recorded from the points of an equilateral triangle with a dipole in the center is zero at all times.

For convenience, limb leads are placed at right wrist, left wrist and left foot. The effect got is same as recorded from the points of Einthoven's triangle (Fig. 30.9). A right limb lead is kept for earthing. According to the law of circuits,

Lead$_I$ + Lead$_{II}$ + Lead$_{III}$ = 0

But Einthoven connected the leads in such a way that recording from all leads showed an upright deflection in the recording galvanometer. To obtain this, he reversed the polarity of lead II.

i.e. $L_I - L_{II} + L_{III} = 0$

or $L_{II} = L_I + L_{III}$ (Einthoven's Law).

The sum of potentials recorded from L_I and L_{III} is equal to the potential in L_{II}.

Unipolar Limb Leads (Fig. 30.10)

Unipolar leads were first introduced by Wilson in 1932. Here only one electrode is active or exploring and is connected to the positive terminal of the machine. The other electrode is called indifferent or inactive electrode, which is connected to the negative terminal. The potential of indifferent electrode is made zero by connecting the electrode on the limb leads through a resistance of about 5000 ohms. This is called "Wilson's Central Electrode" or "Terminal". The leads are designed as V_R, V_L and V_F. But they are not used now a days.

Unipolar Augmented Limb Leads (Fig. 30.11)

Unipolar augmented limb leads were first designed by Goldberger. Here, the active electrode is placed on one limb and only the other two limbs are connected to the negative terminal through high resistance. In other words, the electrode of central terminal to the limb, on which the active electrode is placed, is disconnected. When this is done, the amplitude of tracing got is about one and a half (1½) times that obtained with Wilson's central terminal. Since the amplitude is greater, it is called "augmented" unipolar limb leads.

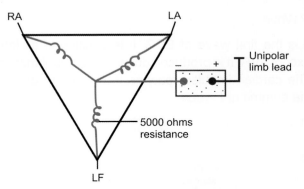

Fig. 30.10: Unipolar limb leads

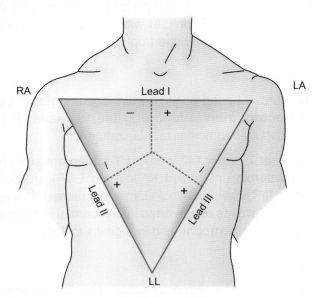

Fig. 30.9: Einthovan's triangle

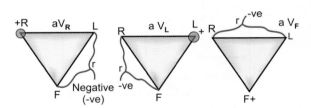

Fig. 30.11: Unipolar augmented limb leads

Part 2

Physiology

Unipolar Chest Leads

There are 6 chest leads – V_1 to V_6. The chest leads have an exploring electrode on the surface of the chest. The indifferent electrode is connected to the right arm (RA), left arm (LA) and left lower limb (LL) through a high resistance. The positions of active electrode in the six unipolar chest leads are given in (Table 30.1).

Normal ECG Pattern

The configuration of waves obtained during recording of ECG from different leads depends on the order in which each part of heart is getting depolarized. The direction of depolarizing wave in relation to the exploring electrode also affects the pattern got during the recording. If the depolarizing wave is passing towards the exploring electrode, an upward deflection is obtained. If the depolarizing wave is passing away from the exploring electrode, a downward deflection is got. No potential is recorded in the ECG when the ventricular muscle is completely depolarized (Fig. 30.12) shows the normal ECG with different waves, intervals and segments. The duration and amplitude of each wave are summarized in Table 30.2.

P Wave

It is the first wave of ECG. It is positive in all leads except aVR. It is produced due to atrial depolarization. The cardiac impulse reaches the AV node at about the summit of P-wave.

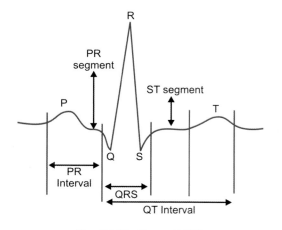

Fig. 30.12: Normal ECG

Table 30.1: Positions of active electrode in unipolar chest leads	
Lead	Position of Active Electrode
V1	4th Intercostal space at right border of sternum
V2	4th Intercostal space at left border of sternum
V3	Midpoint between V2 and V4
V4	Left 5th intercostal space on mid-clavicular line
V5	Left 5th intercostal space on anterior axillary line
V6	Left 5th intercostal space on mid-axillary line

Table 30.2: Duration and amplitude of ECG waves		
ECG wave	Duration	Amplitude
P wave	0.1 sec	0.5 mV
QRS complex	0.8 to 0.1 sec.	0.8 to 3 mV
T wave	0.27 sec.	0.5 mV

Abnormal P-wave: Tall prominent P-waves are seen in atrial hypertrophy. P-wave may be absent in ectopic beat as well as in atrial fibrillation.

QRS Complex

It consists of 3 waves – Q, R and S. It is produced by ventricular depolarization. The atrial repolarization wave appears at about the same time as that of QRS complex. Hence it is always totally masked by the large QRS complex.

Abnormal QRS complex: Wide QRS complexes of more than 1-2 seconds durations are seen in bundle branch block of heart, ectopic beats, hypertrophy of ventricles, etc. QRS complex with increased amplitude is seen in ventricular hypertrophy.

T-wave

It is a broad-based positive wave occurring after the iso-electric interval following the 'S'-wave. It is produced by ventricular repolarization.

Abnormal T wave: Tall, peaked T-waves are seen in hyperkalemia (increased K^+ level in ECF) and low amplitude T-waves in hypokalemia (decreased K^+ level in ECF). T-waves are inverted in myocardial ischemia, ventricular hypertrophy and in digitalis toxicity.

U-wave

U-wave is due to slow repolarization of papillary muscles of heart. This wave may not be present always.

Intervals

PR Interval

It is the interval between the beginning of P-wave to the beginning of QRS complex. It represents atrial depolarization and conduction through the AV node. PR interval is prolonged (more than 0.2 seconds) in heart block, hyperkalemia and hypercalcemia. PR interval is shortened (less than 0.12 seconds) if the impulse is arising from AV node or due to the presence of an abnormal conducting tissue between atria and ventricles.

QT Interval

It is the interval between the beginning of Q wave to the end of T wave. It represents ventricular depolarization and repolarization.

QT interval is prolonged in hypokalemia and hypocalcemia. It is shortened in hyperkalemia and hypercalcemia

R-R Interval

It is the interval between two successive R-waves. It is used for calculating the heart rate.

Heart rate = 60/RR interval

If RR interval is 0.8 seconds, then the heart rate is 60/0.8 or 75/min.

Segments

ST Segment

It is the iso-electric segment between the end of QRS complex and the beginning of T-wave.

ST segment is depressed in myocardial ischemia and hypokalemia. It is elevated in myocardial infarction and pericarditis.

PR Segment

It is the iso-electric segment between the end of P-wave and the beginning of T-wave.

Uses of ECG

ECG is used clinically for the diagnosis of suspected cardiac disorders especially myocardial ischemia and infarction, heart block, arrhythmias, etc. Analysis of ECG gives information about the orientation of heart, relative size of chambers, effects of altered electrolyte concentration and toxicity of certain drugs like digitalis, quinidine, etc.

▋ THE CARDIAC CYCLE

The cardiac cycle refers to a series of electrical and mechanical events that occur cyclically from the beginning of one heart beat to the beginning of the next. This also includes changes in pressure, blood flow and volume in the heart chambers. Duration of one cardiac cycle is 0.8 seconds (at a heart rate of 75/min). Of this, 0.1 sec. is for "atrial systole" (atrial contraction), 0.7 sec – for "atrial diastole" (atrial relaxation), 0.3 sec. – for "ventricular systole" (ventricular contraction) and 0.5 sec. for "ventricular diastole" (ventricular relaxation). From Figure 30.13, it is clear that atrial systole occurs in the last part of ventricular diastole and all the heart chambers are in "diastole" (relaxed state) for a period of 0.4 sec. of the cardiac cycle. This is called quiescent period.

Phases of Cardiac Cycle

The different phases of cardiac cycle and their duration are summarized in Table 30.3 below and the major events are given in (Fig. 30.14).

Isovolumetric Contraction Phase

In the beginning of ventricular contraction, the semilunar valves are closed while the AV valves are still open. As the ventricles start contracting, the intra-ventricular pressure gradually rises causing the AV

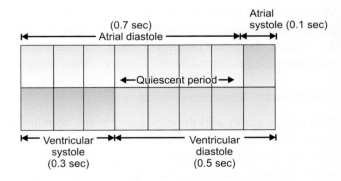

Fig. 30.13: Schematic representation of cardiac cycle

Part 2

Physiology

Table 30.3: Different phases of cardiac cycle	
Phase	Duration (sec)
Isovolumetric contraction phase	0.05
Phase of rapid ejection	0.10
Phase of reduced ejection	0.15
Protodiastole	0.04
Isovolumetric relaxation phase	0.06
First rapid filling phase	0.10
Reduced filling phase (diastasis)	0.2
Atrial systole	0.1

valves to close firmly. The ventricle now becomes a closed chamber. Further contraction of ventricles causes the intra-ventricular pressure to rise steeply, resulting in bulging of the AV valves into the atria. The left ventricular pressure rises from 0 to 80 mm of Hg. This phase lasts for 0.05 sec.

Phase of Rapid Ejection

When the intra-ventricular pressure exceeds the hydrostatic pressure of the aorta, the aortic valve opens and ejection of blood into the aorta occurs. Major part of ventricular ejection takes place during this phase, which lasts for 0.1 sec. Intra-ventricular pressure reaches a peak during this phase.

Reduced Ejection Phase

This phase follows the rapid ejection phase. During this phase, the ventricular pressure is slightly lower than that of aorta, but the blood flow continues due to momentum. This phase lasts for 0.15 sec.

Protodiastole

This phase lasts for only 0.04 sec. During this phase, ventricles start relaxing, allowing the intra-ventricular pressure to fall rapidly. But still the aortic valve is open. As a result, the blood column in the aorta tends to fall backwards, causing the aortic valve to close sharply.

Isovolumetric Relaxation Phase

This phase starts with the closure of aortic semilunar valve. Now the ventricle is a closed chamber, which is relaxing. The intra-ventricular pressure falls rapidly and when it is below the atrial pressure, the AV valves open. This phase lasts for 0.06 sec.

Rapid Filling Phase

As the AV valves open, blood rushes from the atrium to the ventricle. This causes the volume of ventricle to rise steeply. This phase lasts for 0.1 sec. Majority of ventricular filling occurs during this phase.

Reduced Filling Phase or Diastasis

After the initial rapid filling, the blood flow gradually decreases. This phase lasts for 0.2 sec.

Atrial Systole

This phase lasts for 0.1 sec. The ventricular inflow is increased again by the atrial systole. But this contributes to only 20% of the total ventricular inflow. Atrial systole functions as "primer pump" which increases the effectiveness of ventricular pumping.

The cardiac muscle rests and gets the coronary blood flow during the period of diastole. If the heart rate is increased (more than 150/min), duration of ventricular diastole, especially the phase of reduced filling is affected. This can lead to heart failure.

END DIASTOLIC VOLUME (EDV)

The volume of blood in the ventricles at the end of diastole is called end diastolic volume (EDV). It is about 120-130 ml. It acts as a pre-load which increases the initial length of cardiac muscle and hence the force of contraction. The relationship between the ventricular end diastolic fiber length and the force of contraction is called Starling's law of heart.

END SYSTOLIC VOLUME (ESV)

Some amount of blood remains in the ventricle even after complete ejection. This forms the end systolic volume (ESV) which is about 30-50 ml.

The difference between the end diastolic and end systolic volumes forms the "stroke volume" that is pumped out by the ventricles in each cardiac cycle.

Ejection Fraction

It is the fraction of EDV that is ejected by the ventricles in each cardiac cycle.

$$\text{Ejection fraction} = \frac{EDV - ESV}{EDV} \times 100$$
$$= 60 \text{ to } 70\% \text{ of EDV}$$

Wigger's diagram (Fig. 30.14) shows the correlation between ECG, pressure and volume changes in the right and left heart chambers and aorta as well as the heart sounds during different phases of cardiac cycle.

PRESSURE CHANGES DURING CARDIAC CYCLE

■ ATRIAL PRESSURE CHANGES (FIG. 30.14)

During atrial systole, atrial pressure starts increasing. It continues to increase during the iso-volumetric contraction phase also, because of the bulging of AV valves into the atria. Atrial pressure falls during ventricular systole as the AV valves are pulled downwards. Then it rises when atrial filling starts until AV valves open early in the ventricular diastole. After the opening of AV valves, atrial pressure again decreases, as blood flows from atria to ventricles.

Pressure changes in right atrium are reflected in jugular venous pulse (JVP). Right atrial pressure changes are transmitted to the right internal jugular vein in the neck, which is therefore selected to record JVP. JVP shows three positive waves (a, c and v waves) and two negative waves (X and Y descends) as given in (Fig. 30.14) 'a'-wave is due to increased atrial pressure during atrial systole. 'c'-wave is due to increase in atrial pressure during isovolumetric contraction phase, when AV valves bulge into the atrium. 'X'-descend is due to drop in the atrial pressure when the AV valves are drawn downwards during ejection of blood from ventricle. 'v'-wave is due to rise in atrial pressure during isovolumetric relaxation phase before the opening of AV valves. Atrial filling takes place at this time. 'Y'-descend is due to drop in the atrial pressure as the AV valves open at the end of isovolumetric relaxation phase.

Clinical Importance of JVP

In atrial fibrillation, there is no active atrial contraction and hence 'a'-wave will be absent. In tricuspid insufficiency, giant 'c'-wave is seen with each ventricular systole. Giant 'a'-waves or 'cannon' waves are seen in complete heart block.

■ VENTRICULAR PRESSURE CHANGES

Atrial systole causes a small increase in the left ventricular pressure, which becomes 2-3 mm of Hg at the end. In the isovolumetric contraction phase, the pressure increases, and when it exceeds that of aorta (80 mm of Hg), semilunar valves open. Rapid ejection of blood along with powerful ventricular contraction increases the left ventricular pressure to a maximum of 120-130 mm of Hg. In the isovolumetric relaxation phase, there is a sharp decrease in the left ventricular pressure to almost zero, during which the AV valves open. The whole process is shown in (Fig. 30.14).

When compared to left ventricle, pressure changes in the right ventricle is less during different phases of cardiac cycle (Fig. 30.14). The right ventricular pressure becomes 10 mm of Hg during isovolumetric contraction phase and rises to a maximum of 20-25 mm of Hg during the rapid ejection phase.

Table 30.4 shows the pressure changes in the atria and ventricles during systole and diastole.

■ PRESSURE CHANGES IN THE AORTA (FIG. 30.14)

The aortic pressure at the start of isovolumetric contraction phase is 80 mm of Hg. When the ventricular pressure exceeds this value, semilunar valves are opened. During the rapid ejection phase, aortic pressure starts increasing to a maximum of 120 mm of Hg. In the next phase (reduced ejection phase), the amount of blood pumped into the aorta gradually decreases and so does the aortic pressure. During ventricular diastole, ventricular pressure becomes less than aortic pressure. Backflow of blood from aorta into left ventricle causes 'eddy currents' at the root of aorta, causing the semilunar valves to close. This causes a momentary increase in the aortic pressure, which is recorded as an 'incisura' in the aortic pressure curve

Chamber	Systolic pressure	Diastolic pressure
Right atrium	5-6 mm of Hg	0
Left atrium	7-8 mm of Hg	0
Right ventricle	25 mm of Hg	0
Left ventricle	120 mm of Hg	0

Table 30.4: Pressure changes in the cardiac chambers during systole and diastole

(Fig. 30.14). Again the aortic pressure falls throughout the ventricular diastole to about 80 mm of Hg. Note that the aortic pressure fluctuates between systolic pressure of 120 mm of Hg and diastolic pressure of 80 mm of Hg. Aortic pressure never falls to very low levels during diastole because of the property of elastic recoiling of vessel wall. This maintains the forward flow of blood in the aorta during diastole also. This is called Windkessel effect.

Pressure changes occurring in the pulmonary artery are also similar to that in the aorta. But the maximum pressure is only 25 mm of Hg and the minimum pressure is about 10 mm of Hg (Fig. 30.14).

ARTERIAL PULSE

Pulse is the expansile pressure wave transmitted along the arterial wall, produced due to ejection of blood from left ventricle into the already filled aorta. During each heart beat, the left ventricle pumps about 70-80 ml blood into an already filled aorta. This sets up a pressure wave in the aorta, which moves forward along the arteries which can be palpated as the 'pulse'. The pulse can be conveniently felt at the radial artery, at the wrist, which is felt 0.1 sec after the peak of systolic ejection to aorta. Normal pulse rate in an adult is 72/min. The pulse rate increases in fever, exercise, anemia, etc. In addition to the pulse rate, the following aspects of the pulse should also be noted:

1. *Rhythm:* It is the interval between two successive pulse waves. When it is constant, the pulse is said to be regular. Irregular pulse is seen in arrhythmias, atrial fibrillation, heart block, etc.
2. *Volume:* It is the degree of expansion of the arterial walls during each pulse wave. The amplitude of pulse depends on the stroke volume. Low volume pulse is felt in shock and high volume pulse in atrial incompetence.
3. *Character:* The character of the pulse is best noted by palpating the carotid artery in the neck. Collapsing pulse is felt in atrial incompetence.
4. *Condition of vessel wall:* Normally, the arterial wall cannot be palpated. The wall is thickened by atherosclerosis and so it becomes palpable in old age.
5. *Radiofemoral delay:* Compare the radial pulse with the femoral pulse. Normally there is no radio-

femoral delay. Both pulses are obtained simultaneously. Delay is typically seen in co-arctation of aorta.

Movement of Valves

Opening and closing of AV valves and semilunar valves occur passively according to the pressure changes. When there is a forward pressure, the valves open and they are closed by a back-pressure. Opening of valves produces no sounds. But closure of valves occur suddenly or abruptly which produce audible sounds. The AV valves close at the beginning of isovolumetric contraction phase, which produce a soft sound. Mitral valve closes a few milliseconds before tricuspid valve. Semilunar valves are thicker than the AV valves. They are not attached to any fibrous strands. So their closure produces a sharp and loud sound. They close at the end of protodiastole. Aortic valve closes before pulmonary valve.

HEART SOUNDS

During each cardiac cycle, 4 sounds are produced, of which the first and second heart sounds can be normally heard using a stethoscope.

First Heart Sound

It is soft, low pitched and prolonged. It is expressed as "lubb" and lasts for 0.15 sec. It is produced due to the closure of the AV valves (mitral and tricuspid valves) at the beginning of ventricular systole. Although the first heart sound is heard all over the precordium, it is best heard in the mitral and tricuspid areas.

Second Heart Sound

It is sharp, loud, high pitched and lasts for 0.12 sec. It is expressed as "dub" and is due to the closure of the semilunar valves (aortic and pulmonary valves) at the end of protodiastole. This sound is best heard in the aortic and pulmonary areas. During deep inspiration, there is increased negative intrathoracic pressure, which increases the venous return to the heart. As a result, there is a delay in the closure of pulmonary valve than the aortic valve. This is called physiological splitting of the second heart sound.

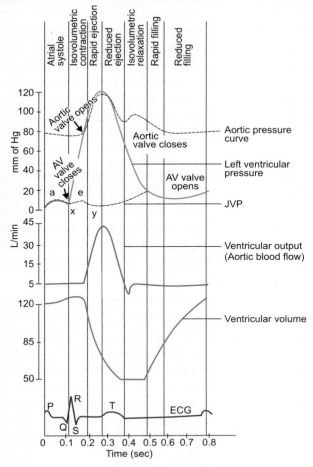

Fig. 30.14: Various events in cardiac cycle
(Wigger's diagram)

Third Heart Sound

It is a soft, low pitched sound heard at the end of first 1/3rd of diastole and is due to the rapid filling of ventricles. It can be sometimes normally heard in young individuals.

Fourth Heart Sound

This sound is never heard in normal persons, using a stethoscope. It occurs just before the first heart sound. This is due to rapid filling of ventricles during atrial systole. In certain conditions like ventricular hypertrophy, the fourth heart sound is heard using a stethoscope.

Apex Beat or Cardiac Impulse

Apex beat is defined as the outermost and lowermost point on the precordium, where a definite cardiac

pulsation can be seen or felt. Normally, apex beat lies in the *5th left intercostal space, 1 cm medial to the midclavicular line.* The position of the apex beat is shifted in fibrosis, pneumothorax, pleural effusion or collapse of lungs.

Applied Physiology

Murmurs: The normal blood flow is named as the laminar flow, which is silent. Murmurs can be heard when the blood flow becomes turbulent. This occurs when the blood flows through a narrow orifice or when the blood flows backwards through a valve which does not close fully. The former condition is called ***stenosis*** and the latter, ***incompetence***. Murmurs may occur in any phase of the cardiac cycle.

A systolic murmur is heard in aortic/pulmonary valve stenosis. A diastolic murmur is heard in aortic/pulmonary valve incompetence and mitral/tricuspid valve stenosis.

■ THE CARDIAC OUTPUT

Cardiac output is defined as the quantity of blood pumped by each ventricle into the aorta, per minute. In adults, the resting cardiac output is 5 L/min. It is about 10-20% less in females (4-4.5 L/min.)

Stroke volume: It is the volume of blood pumped out by each ventricle per heart beat. Normally, it is about 70 ml.

Therefore,

Cardiac output = Stroke volume x Heart rate
                     = 70 x 72 = 5.04 L

Cardiac output varies with body surface area of individuals. The cardiac output per square meter of body surface area is called *cardiac index,* which is about *3 L/m³ in an adult.*

End diastolic volume: It is the quantity of blood remaining in the ventricles at the end of diastole. The normal value is 120 ml.

Variations in Cardiac Output

The cardiac output is altered by the following factors:

I. Physiological Variations

1. Sex: Cardiac output is 10-20% less in females.
2. Age: At birth, cardiac output is 2.5 L/min/m². Body surface area, which gradually increases thereafter and becomes about 4 L/min/m² at 10 years of age. Cardiac output declines in old age also.

3. Exercise – Cardiac output is increased several folds (even up to 35 L/min).
4. After food intake, it is increased to about 30-40%.
5. Emotions like excitement, anxiety, etc. increase cardiac output.
6. A high environmental temperature can increase the cardiac output.
7. Posture: A change in posture from lying to standing produces a slight decrease in cardiac output.

II. Pathological Variations

Cardiac output is increased pathologically in fever, hyperthyroidism etc. and decreased in myocardial infarction, hemorrhage, shock, cardiac failure, arrhythmias, etc.

Distribution of cardiac output to various organs in our body is as follows:

Liver	1500 ml/min
Kidney	1250 ml/min
Brain	750 ml/min
Skeletal muscle	840 ml/min
Skin	460 ml/min
Heart	250 ml/min
Rest of the body	350 ml/min

Factors Affecting Cardiac Output

Cardiac output is affected by mainly stroke volume and heart rate. Stroke volume is affected by the following factors:
1. Preload acting on heart (End-diastolic volume)
2. Afterload acting on heart
3. Myocardial contractility.

Preload Acting on Heart (End–diastolic Volume)

It is the load acting on the cardiac muscle even before it starts contacting, i.e. the degree to which the myocardium is stretched before it contracts. According to Frank Starling's law, the force of contraction is directly proportional to the initial length of muscle fiber, within physiological limits. This effect is independent of innervation of heart. This type of regulation that involves change in length of muscle fiber is called heterometric regulation. In heart, preload is directly proportional to the end-diastolic volume, i.e. the

amount of blood remaining in the ventricle at the end of diastole (130 ml). The more the end-diastolic volume, more will be the stretching of myocardium and therefore the force of contraction will be increased (Fig. 30.15). Venous return, atrial contractility, intrapericardial pressure changes and ventricular compliance can influence the heterometric regulation.

Venous Return

It is the quantity of blood flowing from great veins into the right atrium per minute. At rest, it is same as cardiac output (5 L/min). After strenuous exercise, it can increase up to 35 L/min. Any factor that increases the venous return will increase the end-diastolic volume, which in turn, will increase the stroke volume. Factors affecting venous return are:

a. *Respiratory pump or intrathoracic pump:* Normal intrapleural pressure at the end of expiration is – 2.5 mmHg. During inspiration, it becomes more negative (–6 mm Hg). This negativity causes distension of thoracic veins and thus increased venous return to heart. The mechanism of increased blood flow during inspiration is called respiratory pump. Descent of diaphragm during inspiration increases the intra-abdominal pressure and squeezes blood from abdominal veins into thoracic veins. This also supports the "respiratory pump".

b. *Muscle pump or skeletal muscle contraction:* During exercise, skeletal muscles contract and the veins

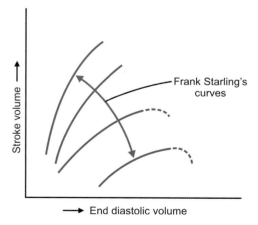

Fig. 30.15: Relationship between end diastolic volume and stroke volume, proving Frank Starling's law

in them are squeezed. Blood from these veins is driven towards the heart which increases the venous return. Backward flow of blood is prevented by valves present in these veins.

c. *Venous tone:* Sympathetic stimulation can increase the venous tone or venoconstriction which decreases the capacity of venous system and increases the venous return.

d. *Blood volume:* Changes in total blood volume can affect the venous return. Venous return is increased when blood volume is increased and vice-versa. For example, during pregnancy or intravenous fluid administration, the total blood volume increases and hence the venous return also increases.

e. *Posture:* Venous return is decreased during standing due to pooling of blood in the lower extremities.

Other Factors Affecting Preload

Atrial contractility: Atria act as "booster pumps" whose contraction propel some additional blood into the ventricles, after ventricular filling. This helps to increase end-diastolic volume and ventricular pumping effectiveness.

Intrapericardial pressure: It is inversely proportional to the ventricular filling and hence the end-diastolic volume decreases when intrapericardial pressure increases, e.g. Pericardial effusion can decrease the end diastolic volume and hence the cardiac output.

Ventricular compliance: This is also important in determining the force of contraction of ventricles. For example, in myocardial infarction, damaged myocardium becomes nonfunctional and ventricular compliance decreases.

After Load Acting on Heart

After load is the load acting on a muscle after it starts contracting. It has no effect on the muscle before it contracts and therefore there is no initial stretching of muscle fiber. In heart, afterload is the resistance against which ventricles have to pump blood. This depends mainly on the mean systolic pressure in aorta and pulmonary artery.

Stroke volume is inversely proportional to afterload. When peripheral resistance of blood vessels and viscosity of blood increase, aortic pressure increases and afterload increases. This can reduce the stroke volume. As a result, blood accumulates in the ventricle and according to Frank Starling's law, this accumulated blood will increase the force of contraction of ventricles. But this phenomenon occurs only during initial period. If the factors causing increase in after load persist, cardiac muscle undergoes hypertrophy leading to decreased ventricular compliance and heart failure.

Myocardial Contractility

The intrinsic level of myocardial contractility is independent of length of muscle fiber and hence independent of preload and afterload acting on heart. Therefore, this type of regulation is called homometric regulation. Myocardial contractility can be increased by increasing the availability of calcium in the cell. Factors affecting myocardial contractility was discussed before.

Heart Rate

Normal heart rate in man is 60-100/min, average 72/min (in adults). Heart rate can vary in many physiological conditions. Heart rate is more in infants, during exercise, emotions, pain, pregnancy and after meals. During sleep and in athletes, heart rate is decreased. Pathologically, fever, hyperthyroidism, hemorrhage, cardiac failure, cardiac arrhythmias, etc. cause increase in heart rate or tachycardia. Heart block, myxedema, viral infections, etc. produce a decrease in heart rate or bradycardia. Heart rate is regulated via neural as well as chemical and thermal stimuli. Neural regulation is through autonomic nervous system (ANS). Sympathetic stimulation increases heart rate while parasympathetic stimulation decreases heart rate. Higher centers, viz., vasomotor center in medulla, cerebral cortex, hypothalamus and limbic system also influence the heart rate. Peripheral neural reflexes like chemoreceptor and baroreceptor reflexes can also affect heart rate.

Hormones like adrenaline and thyroxine increase the heart rate while noradrenaline and vasopressin decrease the heart rate. Changes in blood gas composition can also affect heart rate. Hypoxia, hypercapnia and acidosis cause tachycardia. Increased body temperature causes tachycardia by a direct action on SA node.

Measurement of Cardiac Output

Direct Method

It is an invasive procedure done in experimental animals only, by placing electromagnetic flow meter in the ascending aorta.

Indirect Methods

Mainly 2 indirect methods are used in humans to determine the cardiac output. They are:
1. Fick's Principle method
2. Indicator dilution method

1. Fick's principle: According to this principle, the amount of substance taken up by an organ or whole body per unit time is equal to the arteriovenous concentration difference of the substance multiplied by the blood flow,

i.e. consumption of the substance in 1 min =
Arteriovenous difference of substance x Blood flow in 1 min.
(cardiac output)

Therefore cardiac output

$$= \frac{\text{Consumption of substance/min}}{\text{Arteriovenous difference of substance}}$$

$$= \frac{O_2 \text{ consumption of body in 1 min}}{\text{A-V difference of } O_2 \text{ in I L of blood}}$$

$$= \frac{200 \text{ ml/min}}{(200\text{-}160) \text{ ml/L}} = 5 \text{ L/min}$$

Arterial blood is obtained for this purpose from any peripheral artery. Venous blood is taken from the pulmonary artery by cardiac catheterization. The rate of O_2 consumption by the lungs is measured by using oxygen meter.

2. Indicator dilution method: In this method, a known amount of indicator or dye is injected into a large vein and the concentration of the dye in serial samples of arterial blood is determined. A graph is plotted with the concentration of dye in arterial samples on Y axis

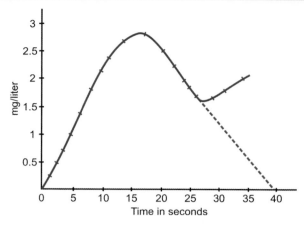

Fig. 30.16: Dye-dilution method of measuring cardiac output

and the time (in seconds) in the X-axis as in (Fig. 30.16). At zero time, none of the dyes has passed into the arterial tree. After 3 seconds, the concentration of dye in the serial arterial samples gradually increases and reaches a maximum in 6-7 sec. After that the concentration of dye in arterial samples falls rapidly. Before reaching the zero, it rises again due to recirculation of some of the dye. The down slope is extrapolated to the time scale to know the time taken for the first passage of dye through circulation.
From all the above data:
Cardiac output =

$$\frac{\text{Amount of dye injected (mg)} \times 60}{\begin{array}{c}\text{Average concentration of} \times \text{Duration of 1st} \\ \text{dye in serial arterial} \quad \text{circulation (sec)} \\ \text{sample (mg/L)}\end{array}}$$

The injected dye should have the following qualities – nontoxic, remain in circulation during the test, not metabolized, does not cause circulatory changes.

Cardiovascular Changes during Exercise

During rest, some of the capillaries in the muscle may remain collapsed with little blood flowing through them. During exercise, all the capillaries open up. Strenuous exercise can cause O_2 and nutrient deficiency in the muscle, which can be provided by these capillaries. They contribute to a 2 to 3 fold increase in O_2 and nutrient supply to the muscle tissue. Also, the O_2 deficiency can stimulate the release of certain vasodilator substances locally like adenosine, ATP, lactate, CO_2, etc. which also contribute to the rise in blood flow to the exercising muscle. By this time, certain circulatory readjustments occurs. They are:

1. Mass discharge of sympathetic nervous system throughout the body, which increases the heart rate, force of contraction and cardiac output. Muscle walls of the veins are strongly contracted, which increases the venous return.

   Arterioles of peripheral circulation are strongly contracted (except cerebral and coronary circulation).

2. An increase in blood pressure, which is due to vasoconstriction of arterioles, constriction of veins and increased cardiac output.

3. Increase in cardiac output, which is essential to provide large amounts of O_2 and nutrients for the exercising muscles.

■ ARTERIAL BLOOD PRESSURE

Arterial blood pressure is defined as the lateral pressure exerted by the column of blood on the walls of the arteries. It is expressed in mm of Hg. The arterial pressure fluctuates during systole and diastole.

Systolic Blood Pressure (Systolic BP)

It is the maximum pressure produced during the systolic phase of cardiac cycle. Normal value is 100-140 mmHg.

Diastolic BP

It is the minimum pressure produced during diastolic phase of cardiac cycle. Normal value is 60-90 mmHg.

Therefore the normal blood pressure is:

Systolic / Diastolic OR 120/80 mmHg.

Pulse Pressure

It is the difference between the systolic and the diastolic blood pressure (30-40 mmHg).

Mean Arterial Pressure

It is the average pressure produced during the cardiac cycle. It is calculated by adding diastolic pressure with one-third pulse pressure, i.e. Diastolic BP + 1/3 pulse pressure.

Methods of Measurement of BP

BP is measured by 2 methods—(1) direct and (2) indirect methods.

Direct Method

It is by introducing a cannula in the lumen of an artery and connecting the cannula to a manometer. Since this is an invasive procedure, it is used only in experimental animals.

Indirect Method (Sphygomanometry)

This is the safe, convenient and most widely used method, using the instrument called sphygmo-manometer.

Principle: The pressure in the brachial artery is balanced with the pressure inside the BP cuff, and is measured directly from the manometer.

Procedure: The subject should be relaxed mentally and physically before. Keep the arm and the apparatus in such a way that it is almost at the level of the heart. Tie the BP cuff around the upper arm, 1 inch above the cubital fossa. The blood pressure can be measured by two methods – palpatory and auscultatory methods.

Palpatory method: Palpate the radial artery at the wrist. Using the rubber bulb, the pressure inside the cuff is elevated until the pulse disappears. The cuff is now slowly deflated. Note the manometer reading at which the pulse reappears. That reading corresponds to systolic BP. But the diastolic BP cannot be determined by this method.

Auscultatory method: This method, introduced by Korotkoff, can be used to find out both systolic and diastolic BP. Here the brachial pulse is auscultated. After recording the systolic BP by palpatory method, the cuff is inflated 20-30 mmHg above the former value (e.g. If the palpatory BP is 130 mmHg you have to inflate the cuff up to 150-160 mmHg). Slowly deflate the cuff and note the reading when a tapping sound becomes audible. This corresponds to the systolic BP. Then slowly deflate the cuff, listening to the serial changes in the sound of the pulse. Note the manometer reading at which the sound disappears completely. This corresponds to the diastolic BP. The series of sounds heard in between are named Korotkoff sounds. They are divided into 5 phases:

Phase I : Sudden appearance of clear tapping sound
Phase II : Sound becomes soft and murmuring
Phase III : Murmuring character, but louder and clearer than phase II
Phase IV : Sound becomes soft and muffled
Phase V : Complete disappearance of sound

Cause for the Korotkoff sounds: Normally blood flow is streamlined and laminar and so no sound is heard on auscultaton. When blood flows through a partially occluded vessel, sound is produced, which is heard on auscultation.

Auscultatory gap: It is the phenomenon, usually seen in hypertensive patients. Sometimes, in the auscultatory method, the sound after the first appearance may get disappeared and reappear again. The reappearing sound may accidentally be taken as the systolic BP. This auscultatory gap can be avoided by doing the palpatory method to estimate the systolic BP, and the cuff should be inflated above this level during auscultatory method.

Variations in BP

Physiological

1. *Age:* BP is minimum at birth. It increases gradually with age, highest in old age.
2. *Sex:* BP is comparatively less in females of reproductive age group. The difference disappears after menopause.
3. *Diurnal variation:* BP is lowest in the morning and highest in the afternoon.
4. *Eating:* BP increases after a meal.
5. *Sleep:* Systolic BP decreases in sleep. But if the sleep is disturbed with dreams, systolic BP rises.
6. *Emotions:* Excitement, anxiety, etc. can increase the sympathetic discharge and hence the BP.
7. *Posture:* On immediate standing from the supine position, BP increases, but returns to normal soon. In lying down position, there is a decrease in BP.
8. *Exercise:* Due to sympathetic stimulation, during exercise, the systolic BP shows marked increase. There is only negligible change in diastolic BP.

Pathological

Hypertension or Increase in BP

Hypertension is defined as sustained elevation of the systemic arterial BP. Hypertension is classified as primary or essential hypertension and secondary hypertension. Primary hypertension or essential hypertension is the commonest, the cause of which is unknown.

Secondary hypertension is secondary to certain diseases like Cushing's syndrome, hyperaldosteronism, glomerulonephritis, pheochromocytoma, etc.

When the BP >140/90 mmHg, the condition is called hypertension.

Decreased BP or Hypotension

Systolic BP < 90 mmHg in adults is called hypotension. Hypotension can occur in hemorrhage, shock, cardiac disorders, Addison's disease, etc.

Factors Maintaining Blood Pressure

BP = Cardiac output x total peripheral resistance.

Therefore, factors that affect the cardiac output and peripheral resistance also affect BP. Including these two factors, there are altogether 4 important factors that maintain the BP. They are:

1. *Cardiac output:* Cardiac output influences the systolic pressure. Increase in cardiac output increases systolic pressure and vice-versa. Therefore, all those factors *(discussed previously in "cardiac output")* that affect cardiac output may alter the BP also.
2. *Peripheral resistance:* BP is directly proportional to the peripheral resistance. The peripheral resistance to the flow of blood depends on (a) lumen of blood vessels; (b) viscosity of blood; (c) velocity of blood flow. All these relations are in the Poiseuille's formula, which is:

$$Q = \frac{\pi \Delta P r^4}{8 L \mu}$$

Where Q is the rate of flow of fluid in a rigid tube, 'ΔP' is the pressure gradient between two ends of the tube; 'r' is radius of the tube; "L" is the length of the tube and "μ" is the viscosity of the fluid

$$Q = \frac{\Delta P}{R}$$

where R is the resistance to the flow of fluid

Combining the previous equation with the above one, we get

$$R = \frac{8L\mu}{\pi r^4}$$

i.e. Resistance to the flow (peripheral resistance) varies inversely with the fourth power of radius of the vessel, i.e. In aorta, the peripheral resistance is lower than that in the arterioles. The radius, in turn, depends on the tone of the vessel. The greater the tone, the smaller is the diameter (lumen) of the blood vessel and therefore, greater is the peripheral resistance.

Peripheral resistance mainly affects the diastolic BP. In conditions like anemia, hypoproteinemia, etc. peripheral resistance is decreased (viscosity of blood is decreased) and so the BP falls. In conditions like polycythemia (viscosity of blood is more), atherosclerosis (lumen of the vessel narrows), etc. peripheral resistance is more and hence BP tends to rise.

Elasticity of the Vessel Wall

The elasticity of the arterial wall is responsible for the maintenance of the diastolic BP. Because the arteries are elastic, they stretch when blood enters during systole and prevent a sudden rise in systolic pressure. During diastole, when the entry of blood stops, the arterial wall recoils, which is responsible for the maintenance of diastolic BP. Recoiling also helps to push the blood forwards. The elasticity of the vessels decreases in old age and in atherosclerosis, when the systolic BP rises causing hypertension.

Blood Volume

A reduction in blood volume, due to hemorrhage, burns, etc. can decrease the BP. In fluid overload (after IV infusion of large amounts of fluids), BP increases.

Regulation of Arterial BP

Our circulatory system is subjected to various forms of stress that varies from simple change of posture to severe hemorrhage. A number of mechanisms are there in our body to face such situations occurring in our daily life. Some mechanisms get activated fast and some slow. According to the time taken, they can be classified as follows:

I. Short-term Regulation Mechanisms

They get activated within seconds to minutes. They include:
1. Baroreceptor mechanism
2. Chemoreceptor mechanism
3. CNS ischemic response
The duration of action is less.

II. Intermediate-term Regulation Mechanisms

They become active in about 30 minutes to 1 hour. They include:
1. Renin-Angiotensin mechanism
2. Stress relaxation mechanism
3. Capillary fluid shift mechanism.
The duration of action is more.

III. Long-term Regulation Mechanisms

They take days, months or years to get activated. They include:
1. Mechanism for regulation of blood volume by kidneys
2. Hormonal mechanisms – by ADH, aldosterone, Renin-Angiotensin, and ANP (atrial natriuretic peptide)
   The duration of action lasts for a long time.

Short-term Regulation Mechanisms

1. Baroreceptor Mechanism

Baroreceptors are stretch receptors located in the heart and walls of large systemic arteries, especially in the carotid sinus and aortic arch. Carotid sinus is a small dilatation in the internal carotid artery seen immediately after the bifurcation of the common carotid artery (Fig. 30.17). Aortic body is present in the arch of aorta (Fig. 30.17).

Baroreceptors respond to changes in BP and are concerned with rapid control of BP. Baroreceptors are spray-type of nerve endings, which are numerous in number. The afferent fibers from the carotid sinus form the sinus nerve, a branch of 9th cranial nerve. Similarly, the afferent fibers from the aortic body form the aortic

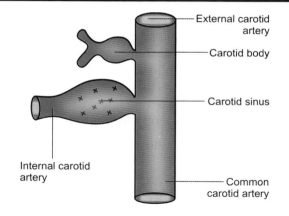

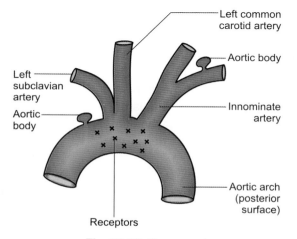

Fig. 30.17: Baroreceptors

The increased baroreceptor discharge, which has reached the vasomotor center through the buffer nerves, inhibits the vasoconstrictor area and excites the vagal innervation of the heart. Inhibition of vasoconstrictor area can decrease the vascular tone resulting in a drop in BP. Also, vagal stimulation of the heart causes bradycardia and a decrease in cardiac output, which can help in reducing the BP.

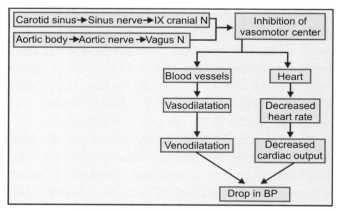

Similarly, when there is a fall in BP as in hemorrhage, there is less stimulation of baroreceptors, which will take away the inhibitory effects of baroreceptors on the vasomotor center. The vasomotor center or more accurately, the vasoconstrictor area, is activated which will cause a generalized increase in the tone of the blood vessels, which will raise the BP. Also the vagal tone is inhibited, that will decrease the cardiac output and increase the BP.

It is the baroreceptor action, that regulates the changes in BP occurring during different postures of the body. When a person stands up from a lying down position, the BP in the upper part of the body and head tends to fall when compared with the lower part of the body. If this persists or becomes marked, loss of consciousness can occur. This situation is prevented by the baroreceptors as follows:

nerve, a branch of 10th cranial nerve. Both these nerves are collectively called as Buffer nerves, because they act to prevent any rise or fall in arterial BP. At normal BP levels, buffer nerves discharge at a less rate.

Baroreceptor Reflex

When there is increased BP, there is distension of the arterial walls which stimulate the baroreceptors. They discharge at an increased rate. These impulses are carried via the afferent buffer nerves to the medulla, where the vasomotor center is located.

The vasomotor center consists of a vasoconstrictor area and a vasodilator area. The vasoconstrictor area sends excitatory impulses to the sympathetic nerves and the vasodilator area is concerned with vasodilatation of blood vessels by inhibiting the vasoconstrictor area.

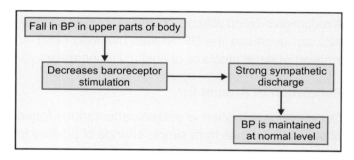

Baroreceptors respond very rapidly to a rapid change in BP than to a steady or stationary pressure. They respond maximally between 60-150 mmHg BP. They are of no use in long-term regulation of BP or in chronic hypertension.

Cardiopulmonary Baroreceptors

Baroreceptors are also located in the walls of right and left atria at the entrance of the venae cavae and the pulmonary veins respectively. They are called cardiopulmonary receptors.

In the atria, there are two types of stretch receptors, which respond during atrial systole (Type A baroreceptors) and atrial diastole (Type B baroreceptors). At the peak of atrial filling, Type B receptors are stimulated. Increased discharge from these receptors cause vasodilatation and decrease the BP.

Bainbridge Reflex

Rapid infusion of intravenous saline or blood in anaesthetized animals produces tachycardia (increase in heart rate), if the initial heart rate is low. This effect is called Bainbridge reflex. It is thought that this reflex is due to the stimulation of atrial baroreceptors. The significance of this is still under study.

Chemoreceptor Mechanism

Chemoreceptors are present in the aortic and carotid bodies. The carotid body is located on the ascending pharyngeal or occipital artery; and the aortic body, near the aortic arch. They respond to changes in the chemical composition of blood that occurs in hypoxia, hypercapnia (increase in CO_2 in blood) and acidosis. They exert their main effects on respiration. However, the afferents from these receptors also converge on the vasomotor center in the medulla. The cardiovascular response produced by the stimulators of chemoreceptors include peripheral vasoconstriction and bradycardia.

In hemorrhage, the BP falls and the blood flow decreases leading to hypoxia of carotid and aortic bodies. This hypoxia stimulates them to bring up the BP to normal. When the BP is normalized, there is no more hypoxia and so no stimulation of chemo-receptors. They respond maximally in the range of 40-70 mmHg BP.

CNS ischemic Response

In conditions like hemorrhage, there can be cerebral ischemia due to reduced blood flow to brain. There is decreased blood flow to the vasomotor center resulting in hypoxia and hypercapnia. This directly stimulates the center producing increase in BP and reflex bradycardia. This response is called CNS ischemic response and is a life-saving mechanism when BP falls below 40 mmHg.

Cushing's reflex: It is different from the CNS ischemic response. When the cerebral ischemia is the result of raised intracranial pressure, the response is called Cushing's reflex.

↑ Intracranial pressure → ↓ Blood flow in brain → Hypoxia and Hypercapnia →
→ Stimulation of vasomotor center → ↑ in arterial BP and ↓ in heart rate

Therefore, in patients with raised intracranial tension, bradycardia is seen.

Intermediate-term Regulation Mechanisms

Renin-angiotensin Mechanism

A fall in BP stimulates the release of renin from the Juxtaglomerular cells (JG cells) of the kidney. Renin converts the inactive angiotensinogen to angiotensin I; Angiotensin converting enzyme in the lungs convert antgiotensin I to angiotensin II, which is a potent vasoconstrictor. Therefore, angiotensin II increases the tone of the blood vessels, increases the venous return to heart, increases the cardiac output and thus increases the BP. Angiotensin II also stimulates the synthesis and secretion of aldosterone, which increases Na+ and water reabsorption from the kidney. This also assists in the restoration of pressure.

This mechanism takes about 20 minutes to become effective.

Stress Relaxation Mechanism

When there is an increase in BP, the walls of the blood vessels get stretched. If this remains for some time, the stretched vessel walls start relaxing, thereby decreasing the vascular tone. As a result, the pressure in the vessels falls to normal. This is due to the property of plasticity of smooth muscles. Even though there is an initial increase in tension on stretching, after some time, the smooth muscle tension gradually decreases. This is called stress relaxation.

Capillary Fluid Shift

When there is a decrease in BP, the capillary pressure becomes less → fluid is absorbed from the surrounding tissues into the capillaries by osmosis → increases blood volume in circulation → increases the BP.

When there is a rise in BP → capillary pressure increases → fluid leaves the capillaries → decrease blood volume → Brings down the BP.

Long-term Regulation Mechanisms

Mechanisms for Regulation of Blood Volume by Kidneys

When there is alteration of BP, the kidneys try to restore it by changing the excretion of sodium and water.

Pressure diuresis: When there is increased blood pressure, kidneys excrete more water thereby decreasing the ECF volume and the BP.

Pressure natriuresis: When there is increased ECF volume (due to increased BP), the salt excretion (Na^+ excretion) by kidney is raised. Conversely, when BP falls, kidneys reduce their rate of Na^+ excretion.

These mechanisms are very slow to act and hence not significant in the acute control of BP. However, it is by far the most potent among the long-term BP regulators.

Hormonal Mechanisms

a. The Renin-Angiotensin mechanism (already explained)
b. ADH (vasopressin) – decrease in ECF volume stimulates ADH release from posterior pituitary. ADH increases water reabsorption from kidney.
c. *ANP:* Increased blood volume causes distension of atria and stimulates the release of atrial natriuretic peptide (ANP). ANP induces natriuresis, bringing down the BP.

■ CORONARY CIRCULATION

(Refer "The blood supply of the heart", Anatomy, Chapter 11)

Normal coronary blood flow, at rest, is 250 ml/min, which is about 5% of total cardiac output. This may increase up to four folds during exercise. The oxygen consumption by cardiac muscle is very high even at rest. Hence, when the cardiac activity is increased, blood flow must increase to provide the much higher oxygen requirement.

Coronary blood flow changes according to the phases of cardiac cycle (Fig. 30.18). Most of coronary blood flow (about 70%) occurs during the diastole. During systole, the tension developed in the ventricles compresses the branches of coronary vessels passing through them. Hence the blood flow through the coronary vessels is less during systole. In severe tachycardia, the diastolic phase is shortened and therefore affects the coronary blood flow.

Regulation of Coronary Blood Flow

1. Pressure changes in aorta
2. Chemical factors
3. Neural factors

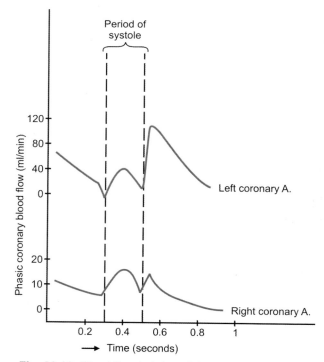

Fig. 30.18: Blood flow in left and right coronary arteries during different phases of cardiac cycle

Pressure Changes in Aorta

Coronary blood flow changes with alternations in systemic arterial pressure. However, over a small range of arterial BP, the coronary blood flow is maintained at a constant. This is called autoregulation, which fails when the BP is less than 70 mmHg.

Chemical Factors

Hypoxia, increased CO_2 (hypercapnia) or increased H^+ concentration can cause vasodilatation of coronary vessels. These metabolites are formed when the myocardial activity increases. The vasodilatation, thus produced, increases the coronary blood flow, thereby satisfying the O_2 demand.

Neural Factors

Autonomic nerves affect the coronary blood flow, either, directly or indirectly. *Sympathetic nerves* act via α or β receptors, both of which are present in all parts of coronary vessels. The action produced will depend on the type of receptor stimulated. The α receptor stimulation causes vasodilatation. However, usually the sympathetic stimulation is followed by coronary vasodilatation. This is due to an indirect effect. Sympathetic stimulation also increases the heart rate and the force of contraction. Therefore there is increased metabolic activity → accumulation of metabolites → vasodilatation

Parasympathetic supply is through the vagus nerve. Vagal stimulation dilates the coronary vessels.

Applied Physiology

Coronary artery disease: Long-standing atherosclerotic changes in the coronary blood vessels can decrease their lumen.

During conditions requiring increased O_2 demand (e.g. exercise), the coronary blood flow fails to meet the metabolic demands, resulting in accumulation of lactic acid and other metabolites. They can stimulate the pain nerve endings in the myocardium and produce severe pain called angina pectoris. The pain is usually felt beneath the sternum and often referred to the left shoulder, left arm or neck. The pain of angina pectoris is relieved on taking rest. When there is complete occlusion of the artery by the atherosclerotic lesion, that part of myocardium undergoes ischemic necrosis. This is called myocardial infarction. This condition can occur even at rest.

■ CIRCULATORY SHOCK

Shock is a clinical condition characterized by inadequate cardiac output and tissue perfusion. The tissue becomes deprived of oxygen and nutrients, resulting in its damage. According to the cause, shock can be broadly classified into four — hypovolemic, distributive, cardiogenic and obstructive shocks.

Hypovolemic Shock

As the name indicates, this type of shock is due to loss of body fluids as in hemorrhage, surgery, burns, vomiting and diarrhea. As a result, the cardiac output becomes inadequate. Massive hemorrhage is the commonest cause of hypovolemic shock. Clinical features include hypotension, rapid thready pulse, cold, pale and moist skin, decreased urine output, etc. Various regulatory mechanisms of blood pressure try to compensate for this and hence there will be intense vasoconstriction of the cutaneous blood vessels.

Distributive Shock

This type of shock is characterized by marked vasodilatation with warm extremities (alike hypovolemic shock). Sudden autonomic activity, producing vasodilatation and pooling of blood in the extremities, results in fainting episodes ('syncope'). Rapid severe allergic reactions can also cause marked vasodilatation ('anaphylactic shock'). Here, there is release of large quantities of histamine from basophils and mast cells. Certain endotoxins released from bacteria can cause severe vasodilatation and increased capillary permeability, resulting in shock ('septic shock').

Cardiogenic Shock

In this type of shock, cardiac output is decreased due to inadequate pumping of heart. Common causes include myocardial infarction, congestive cardiac failure and arrhythmias.

Obstructive Shock

Any obstruction to the outflow of blood from left ventricle or right ventricle can decrease the cardiac output, resulting in obstructive shock. Common causes include tension pneumothorax, massive pulmonary embolism and extreme cases of pericardial effusion.

Management of Shock

Care must be taken not to apply heat or to re-warm the patient, as this can worsen the shock, by causing further vasodilatation. Patient can be made to lie down in a "head-down" position; i.e. by keeping head at least 12 inches lower than the feet. This will increase the venous return and thus the cardiac output also. Oxygen inhalation can be started to prevent tissue damages. An early and rapid blood transfusion is given for hemorrhagic shock. In the case of shock due to vomiting, diarrhea and burns, plasma expanders should be given. Injections of norepinephrine and antihistamine can be given to patients with anaphylactic shock to counteract vasodilatation. Appropriate antibiotics are administered in septic shock.

31

The Endocrine System

INTRODUCTION

The endocrine glands are ductless glands whose secretions are called "hormones". The word "hormone" came from the Greek word "hormaein", which means "I arouse" or "I excite", and the word "endocrine" means "I separate". Secretin was the first hormone to be recognized by a name and was discovered by Bayliss and Starling in 1905. Hormones are also called chemical messengers and according to their mode of action, the endocrine glands can be grouped into different types as follows:

Classical Endocrine Glands

They secrete hormones into the blood that act at specific target tissues situated at different parts of the body. The different endocrine glands and their secretions are enlisted in Table 31.1.

For example, Thyroid gland, Parathyroid gland, Pancreas, Adrenal gland, Testes, Ovary, etc.

Paracrine Glands

They secrete hormones, which act on adjacent cells. The secreting cells are called paracrine cells and the hormones from paracrine cells pass through the extracellular fluid to reach the adjacent target cell. For example, Effect of somatostatin (secreted by D cells of islets of Langerhans) on A (alpha) and B (beta) cells of islets of Langerhans.

Autocrine Glands

They secrete chemical messengers that act on receptors present on the same cells, e.g. prostaglandins, certain chemicals produced by cancer cells.

Table 31.1: The different endocrine glands and their secretions		
Gland		Hormones
1. Pituitary gland	Anterior lobe	GH, TSH, ACTH, FSH, LH, Prolactin
	Intermediate lobe	MSH (in lower animals only)
	Posterior lobe	ADH (vasopressin), oxytocin
2. Thyroid gland		Thyroxine (T_4), Triiodothyronine (T_3), Calcitonin
3. Parathyroid gland		Parathormone
4. Adrenal gland	Cortex	Glucocorticoids, mineralocorticoids, sex hormones
	Medulla	Epinephrine, norepinephrine, dopamine
5. Endocrine pancreas		Glucagon, Insulin, Somatostatin, Pancreatic polypeptide
6. Gastrointestinal hormones		Gastrin, Secretin, CCK, VIP, GIP
7. Others		Vitamin D, ANP, Renin, Angiotensin, Erythropoietin, Melatonin

Neurosecretory Glands

The chemical messengers are secreted by neurons, e.g. epinephrine, norepinephrine, ADH, oxytocin, acetylcholine, etc.

Functions of Endocrine Glands

1. They integrate and coordinate various activities of the body along with the CNS.
2. They help in the growth and development of the body.
3. They help in proper digestion and absorption of food by controlling the secretions of digestive exocrine glands.
4. They help in reproductive functions.
5. They help a person to meet stressful situations and emergencies.

Hormones can be classified according to their chemical nature as below:

Steroid hormones: For example, Corticosteroids, Sex steroids, Mineralocorticoids, Vitamin D.

Proteins or polypeptide hormones: For example, Pituitary hormones, Parathyroid hormone, Insulin, Glucagon.

Amino Acid Derivative Hormones

For example, thyroid hormones, catecholamines (epinephrine, norepinephrine, dopamine).

Mechanism of Action of Hormones

All hormones act through specific binding sites on target tissues called receptors. Receptors are protein molecules which may be situated on the cell membrane, cytoplasm or nucleus of the target cells. Ligand is any substance that binds with the receptor and so hormone is a ligand.

All protein and polypeptide hormones combine with the cell membrane receptors, while the steroid hormones bind with the cytoplasmic receptors. Thyroid hormones combine with the nuclear receptors. The combination of the hormone with the specific receptor brings about many changes inside the target cell that is necessary for the action of that particular hormone (Fig. 31.1).

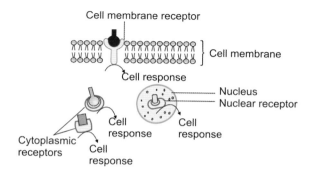

Fig. 31.1: Mechanism of action of different hormones, by binding with cell membrane, cytoplasmic and nuclear receptors

Part 2

Physiology

Regulation of Hormone Secretion

The amount of various hormones present in our body is thoroughly regulated according to their requirement by the following mechanisms:

Negative Feedback Mechanism (Fig. 31.2)

This is the main form of regulation, where the plasma level of hormone determines the secretion of hormone from the gland. When the plasma concentration of hormone exceeds its requirement, it inhibits the secreting gland and vice-versa. Majority of the hormones are regulated by this method.

Positive Feedback Mechanism

This type of control is less common, where the presence of hormone in the blood stimulates its secretion from the gland, thus amplifying its biological action, e.g. GnRH (Gonadotropin-releasing hormone) from hypothalamus stimulates pituitary gland to produce FSH and LH. FSH and LH stimulate ovary to produce estrogen. This estrogen will exert a positive feedback effect on anterior pituitary so that more LH is secreted. This is called LH surge. LH surge is essential for ovulation (Explained in detail in Reproductive System).

THE HYPOTHALAMUS AND THE PITUITARY GLAND

The hypothalamus and the pituitary gland form a single unit by which they influence the secretion of various other endocrine glands in our body.

The pituitary gland or hypophysis is situated in the sella turcica of the sphenoid bone and is attached to the hypothalamus by a thin stalk called the pituitary stalks. The word pituitary is derived from the Latin word "pituita" which means "mucus", as it was once thought to be secreting mucus into the nasal cavity. Later this was proved to be wrong and the word "hypophysis", which means "overgrowth" was introduced. But the term "pituitary" is still commonly used. It is a small gland, weighing about 0.5 to 1 gm.

Developmentally, it consists of two parts (Fig. 31.3):
1. Adenohypophysis or anterior lobe – developed from the Rathke's pouch and is made up of glandular cells.
2. Neurohypophysis or posterior lobe – developed from the neural tissue in the floor of the third ventricle.

Functionally, there is an intermediate lobe also, which is active in lower animals only.

Relations of Pituitary Gland and Hypothalamus

The posterior lobe of pituitary gland has neural connections between hypothalamus (hypothalamo-hypophysial tract) while the anterior lobe of pituitary has vascular connections with hypothalamus (Portal

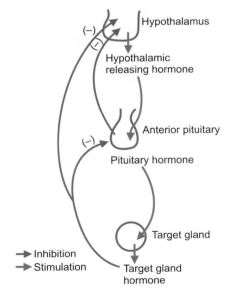

Fig. 31.2: Negative feedback mechanism

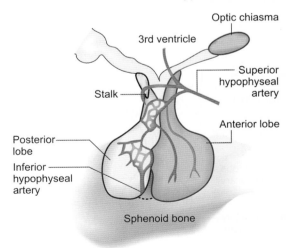

Fig. 31.3: Pituitary gland

Part 2

Physiology

hypophyseal vessels) (Fig. 31.4). In other words, the secretions from posterior pituitary is controlled by neural signals originating from hypothalamus while the secretions of anterior pituitary is controlled by hormones of hypothalamus, viz.hypothalamic releasing and hypothalamic inhibitory factors (or hormones). The hypothalamic releasing and inhibiting hormones are:

1. Growth hormone releasing hormone (GRH)
2. Growth hormone inhibiting hormone (GIH) or somatostatin
3. Corticotropin releasing hormone (CRH)
4. Thyrotropin releasing hormone (TRH)
5. Luteinizing hormone releasing hormone (LHRH) or Gonadotropin releasing hormone (GnRH)
6. Prolactin releasing hormone (PRH)
7. Prolactin inhibiting hormone (PIH)

The Adenohypophysis

The adenohypophysis or the anterior lobe of pituitary is a highly vascular structure, consisting of 5 types of cells:

1. *Somatotrophs:* They are acidophilic cells. They constitute about 50% of the total cells and secrete growth hormone (GH).
2. *Lactotrophs:* They form about 15-20% of total cells. They are also acidophilic and secrete prolactin.
3. *Corticotrophs:* They are basophilic cells, secreting ACTH and beta-lipoprotein. They constitute about 20% of total cells.

4. *Thyrotrophs:* They are basophilic cells, secreting thyrotropic hormone or thyroid stimulating hormone (TSH). They form 5% of the cells.
5. *Gonadotrophs:* They are also basophilic and secrete follicle stimulating hormone (FSH) and luteinizing hormone (LH). They form 5% of the cells.

The important adenohypophyseal hormones are explained below:

Growth Hormone (GH)

GH or somatotropic hormone is a polypeptide hormone secreted by the acidophilic somatotroph cells.

Actions of GH (Fig. 31.5)

The major actions of GH are on growth of the body and metabolism.

Actions on Growth

GH accelerates the skeletal as well as soft tissue growth. It promotes chondrogenesis and calcification of the long bones. Soft tissues like muscles, viscera, connective tissue, etc. grow under the stimulation of GH.

GH stimulates the production of somatomedins from liver, which promote the growth of skeletal and soft tissues. But GH has no role in the growth of brain tissue, hair and sex organs.

Actions on Metabolism

a. GH promotes protein synthesis and inhibits protein catabolism. It thus increases collagen synthesis.

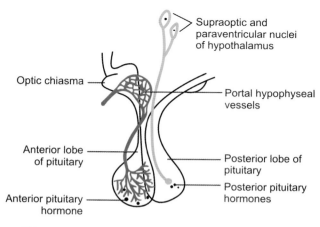

Fig. 31.4: Neural and vascular connections between hypothalamus and pituitary gland

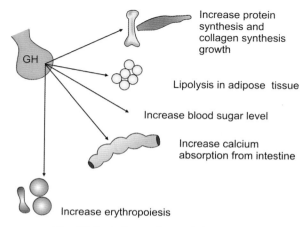

Fig. 31.5: Actions of growth hormone

b. GH increases the blood sugar level by reducing the cellular uptake of glucose and by increasing peripheral utilization of glucose. Therefore, GH is said to have "anti-insulin" effect or diabetogenic effect.

c. GH lipolysis. GH causes mobilization of free fatty acids from the adipose tissue, resulting in an increase in their level in plasma. Therefore, the free fatty acids become the main fuel for energy production.

d. GH increases plasma phosphorus level, increases Ca^{2+} absorption from the intestine. It favors the excretion of K^+ and Na^+.

e. GH promotes erythropoiesis.

Regulation of GH Secretion (Fig. 31.6)

Stimuli affecting GH secretion are enlisted in Table 31.2.

In children and adolescents, the GH level is more when compared to the adults. The secretion of GH is under the control of growth hormone releasing hormone (GHRH) and growth hormone inhibiting hormone (GHIH) or somatostatin, which are secreted from the hypothalamus.

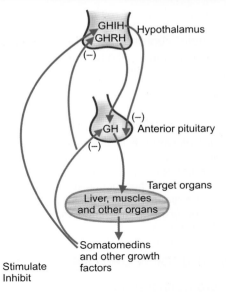

Fig. 31.6: Regulation of growth hormone secretion

Table 31.2: Stimuli affecting GH secretion

Increased GH secretion	Decreased GH secretion
Hypoglycemia	Hyperglycemia
Exercise	Free fatty acids
Fasting	Hormones like proges-
Stress	terone and cortisone
Increased levels of amino acids	High levels of GH itself
Sleep	
Hormones like estrogen, glucagon and vasopressin	

Applied Physiology

1. *Pituitary dwarfism:* This occurs due to deficiency of GH in the childhood. There is arrest of skeletal development. The growth is stunted, but proportionately. The puberty is delayed, as there is no sufficient gonadotropic hormone. In isolated GH deficiency, they may mature sexually and can have children. There is no mental retardation.

2. *Gigantism:* It is due to hypersecretion of GH occurring in the childhood, before the epiphyseal closure. Hypersecretion can be due to any tumors of anterior pituitary. All the body tissues, including the bones, grow rapidly. The stature is greatly increased. The limbs are disproportionately long. They tend to be mentally abnormal. The "giant" suffers from hyperglycemia.

3. *Acromegaly:* It occurs in adults, due to the hypersecretion of GH. Since the epiphyseal closure has already completed, the bones cannot grow in length and therefore the height of the person is not affected. But the soft tissues continue to grow under the influence of increased GH. Due to this, there is thickening of skin and subcutaneous tissues. The bones can grow in thickness. So there is broadening and thickening of hands and feet (acral parts and hence the name "acromegaly"). There is also prognathism, due to protrusion of lower jaw. Nasal bones, frontal and other skull bones and portions of vertebrae are affected. There can be hyperglycemia. There is enlargement of the viscera (splanchnomegaly). Gynecomastia and lactation can occur, as the GH excess is usually accompanied by hypersecretion of prolactin.

Prolactin

Prolactin or lactogenic hormone is a polypeptide hormone secreted from the lactotrophs of anterior pituitary.

Functions

Prolactin acts as the main hormone for the development of mammary gland and for secretion of milk during pregnancy and lactation respectively. During pregnancy, prolactin secretion increases which helps in development of tubuloalveolar system of mammary gland. But lactation does not occur. After delivery, prolactin causes secretion of milk from breast. The high prolactin levels will suppress LH secretion and accounts for amenorrhea (absence of menstruation) and sterility during the postpartum period.

Regulation

The prolactin secretion is influenced by the prolactin releasing hormone (PRH) and the prolactin release inhibiting hormone (PIH), which are secreted from the hypothalamus.

Applied Physiology

Hyperprolactinemia: It is the hypersecretion of prolactin occurring due to a tumor of anterior pituitary or due to certain drugs like chlorpromazine and alpha methyl dopa. This condition can cause hypogonadism. As a result, in females, there can be amenorrhea, hirstism, infertility and in males, there can be oligospermia, impotence, sterility and female type of fat distribution.

THE NEUROHYPOPHYSIS

The neurohypophysis or posterior pituitary is made up of modified glial cells called pituicytes, and nerve fibers. There are no glandular cells. The nerve fibers originate in the supraoptic and paraventricular nuclei of hypothalamus. They descend through the pituitary stalk and terminate in the posterior pituitary gland. Hence they are also called hypothalamohypophyseal fiber tract. The hormones formed in the above hypothalamic nuclei are stored in the expanded nerve endings of the hypothalamohypophyseal tract (Fig. 31.4).

The hormones of posterior pituitary are:
1. Vasopressin or antidiuretic hormone (ADH)
2. Oxytocin.

ADH or Vasopressin

ADH is formed primarily in the supraoptic nuclei of hypothalamus. It is a nonapeptide hormone.

Actions

The main function of ADH is to regulate the volume of ECF and plasma osmolality. The chief site of action is kidney. ADH increases water reabsorption by increasing water permeability in the renal collecting ducts. The mechanism of action is by combining with 'V_2' receptors in the epithelial membrane of collecting ducts. This results in opening up of water channels called aquaporins, which allows free passage of water.

In large doses, ADH causes contraction of smooth muscles of the vessels (Hence the name "vasopressin"). It can cause constriction of blood vessels,

thus decreasing cardiac output and cerebral blood flow. It also causes contraction of gastrointestinal smooth muscles.

Regulation of ADH Secretion

1. *Plasma osmolality:* An increased plasma osmolality can stimulate the osmoreceptors, present in the supraoptic nuclei of hypothalamus, to secrete more ADH. This causes retention of water and restores plasma osmolality.
2. *ECF volume:* A decreased ECF volume is a potent stimulus for ADH secretion. When the ECF volume increases, this inhibits ADH secretion.
3. *Low blood pressure:* Low blood pressure also stimulates ADH secretion.

Applied Physiology

1. *Diabetes Insipidus:* This is a condition due to deficiency of ADH or absence of ADH. It is characterized by polyuria and polydypsia (increased thirst). There is no concentration of urine and hence a large volume of dilute urine is excreted.
   Classification: According to the cause, diabetes insipidus can be of two types –
   a. Central diabetes insipidus
   b. Nephrogenic diabetes insipidus
   a. *Central diabetes insipidus:* When the deficiency of ADH is due to any lesions of hypothalamus or posterior pituitary, it is called central diabetes insipidus.
   b. *Nephrogenic diabetic insipidus:* Here, the secretion of ADH is normal, but the collecting ducts are unresponsive to the ADH.
2. *SIADH (Syndrome of inappropriate secretion of ADH):* In this condition, there is hypersecretion of ADH, leading to water retention, and hyponatremia. As the ADH secretion is not appropriate to the sodium concentration, it is called "SIADH".

Oxytocin

Oxytocin is a non-peptide hormone secreted from paraventricular nucleus of hypothalamus and stored in the posterior pituitary gland.

Actions

1. *Contraction of uterine smooth muscles during labor:* Estrogen increases and progesterone decreases the sensitivity of uterine smooth muscles to oxytocin. Hence during gestation oxytocin sensitivity is low due to a high progesterone level.

But at the end of pregnancy, the estrogen levels are much increased while there is no change in the progesterone level. Hence uterine muscle becomes highly sensitive to oxytocin. Oxytocin causes powerful contraction of uterus, pushing the fetus down. Dilatation of cervix and stretching of the birth canal as fetus passes down, sets up afferent impulses to the hypothalamus that stimulate the release of more and more oxytocin from posterior pituitary. Oxytocin also helps in expulsion of placenta and vasoconstriction of endometrial vessels which minimizes the bleeding. Dilatation of cervix and stretching of the birth canal as fetus passes down, sets up afferent impulses to the hypothalamus that stimulate the release of more and more oxytocin from posterior pituitary.

This is an example for positive feedback mechanism as well as for neuro- endocrine reflex. Oxytocin also helps in expulsion of placenta and vasoconstriction of endometrial vessels, which minimizes the bleeding.

2. *Milk ejection in lactating mothers:* In the lactating mammary gland, oxytocin causes contraction of myoepithelial cells in the alveoli and small ducts of mammary gland. As a result, there is expulsion of milk from the breast. This phenomenon is called milk ejection reflex, which is also an example for neuroendocrine reflex. The reflex occurs as follows (Fig. 31.7).

Stimulation of tactile receptors around the nipple, by suckling action of infant, sets up afferent impulses to the hypothalamus and stimulate release of oxytocin.

Applied Physiology

Oxytocin has wide clinical application. It can be used for induction of labor, to control uterine bleeding after delivery or abortion, and for the treatment of impaired milk ejection in lactating mothers.

■ THE THYROID GLAND

The thyroid gland is the largest endocrine gland, situated in the anterior aspect of neck, at the level of thyroid cartilage *(See CP13 in Color Plate 3)*. It consists of two lateral lobes connected together by a medial isthmus.

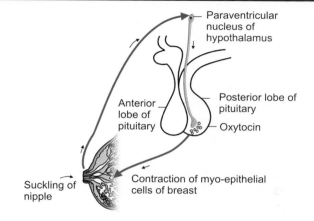

Fig. 31.7: Milk ejection reflex

It is a highly vascular gland. Microscopically, thyroid gland consists of a connective tissue capsule which sends septa dividing the gland into lobes and lobules. There are numerous colloid filled follicles. The epithelial lining of the follicle varies with the amount of colloid in it. The epithelial cells are tallest when the gland is active (for more details, refer Anatomy, part of this book).

The main constituent of the thyroid follicle is a glycoprotein – thyroglobulin.

Hormones of Thyroid Gland

The follicular cells secrete two principal iodine containing hormones—thyroxine (T_4) or tetraiodothyronine and triiodothyronine (T_3). In addition, the parafollicular cells, present in between the thyroid follicles, synthesize and secrete another hormone—calcitonin (dealt in detail later).

Biosynthesis and Secretion of Thyroid Hormone

T_3 and T_4 are synthesized from iodination of the aminoacid tyrosine. The steps involved in the synthesis of thyroid hormones are:

1. *Iodide trapping:* It is the transfer of iodide from the circulation into the follicular cells of thyroid gland. For this, an active transport mechanism is present in the follicular cells called the iodide pump. This transport mechanism is stimulated by TSH. In a normal thyroid gland, the iodide pump can concentrate about 30 times of iodide than its concentration in blood.

2. *Synthesis and secretion of thyroglobulin:* Thyroglobulin, a derivative of Tyrosine, is synthesized in the follicular cells and secreted into the colloid by exocytosis.

3. *Oxidation of iodide into iodine:* The iodides are oxidized into iodine by the peroxidase enzymes.

4. *Iodination of tyrosine:* The iodine is immediately bound to the tyrosine residues in thyroglobulin. In this process, the first product is monoiodotyrosine, which again combines with iodine to form diiodotyrosine. This step is also catalyzed by the peroxidase enzyme.

5. *Coupling:* Two molecules of diiodotyrosine couple to form a molecule of T_4. One molecule of monoiodotyrosine couple with a molecule of diiodotyrosine to form a molecule of T_3. But majority of T_3 is formed by deiodination of T_4. These reactions are also catalyzed by peroxidase enzyme.

6. *Storage:* The iodinated thyroglobulins are stored in the colloid present in the follicles. Thyroid is the only endocrine gland, which stores large amount of hormone extracellularly. The total amount of T_3 and T_4 thus stored is sufficient for 2-3 months.

7. *Secretion:* The release of T_3 and T_4 into the circulation requires proteolysis of the thyroglobulin. Small droplets of colloid are taken up by the follicular cells by endocytosis. In the cells, the globules of colloid merge with lysosomes. Lysosomes contain proteolytic enzymes that can cleave the thyroglobulin molecule, releasing T_3 and T_4 into the circulation. Diiodotyrosine and Monoiodotyrosine molecules are also formed during this cleavage. They are de-iodinated by the iodinase enzyme and the iodides and tyrosine residues are recycled for the synthesis of T_3 and T_4.

All these processes can be summarized as follows:

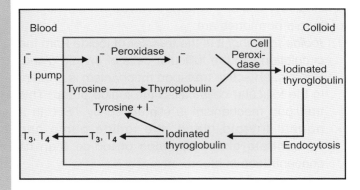

Normal level of total T_4 in plasma is about 8 μg/dl and that of T_3 is 0.5 μg/dL. T_3 is more potent and more rapidly acting than T_4, but has a short half-life. Only a small fraction of T_3 and T_4 are circulated in the free form. Majority are bound to the plasma proteins – thyroxine binding globulin (TBG) and thyroxine-binding pre albumin (TBA).

Actions of Thyroid Hormones

The actions of thyroid hormones are manifested in almost all tissues of the body.

1. *Growth:* Thyroid hormones are essential for normal growth and development. They promote skeletal growth, ossification of cartilage, eruption of teeth, brain development, etc. Thus thyroid hormones are essential for both mental and physical growth. They are also required for the normal secretion of growth hormone and they enhance the effect of growth hormone on tissues.

2. *Calorigenic action:* T_4 and T_3 increase the O_2 consumption of all tissues except adult brain, testes, uterus, spleen, lymph nodes and anterior pituitary. As a result, they increase the BMR and heat production in the tissues (calorigenic action).

3. *Brain and nervous system:* Thyroid hormones are essential for proper brain development. A hypothyroid infant may therefore develop irreversible mental retardation. Even though thyroid hormones do not increase O_2 consumption in nervous tissue, they are important for the normal activity of CNS. In hypothyroid patients, there is slowness of thought, somnolence, etc. and in hyperthyroidism there is irritability, anxiety and restlessness. In the peripheral nervous system, the deficiency of T_3 and T_4 leads to delay in the speed of conduction of impulses and so the stretch reflexes are prolonged.

4. *Cardiovascular system:* Thyroid hormones increase the number and affinity of beta-adrenergic receptors in the heart, thus increasing its sensitivity to the effects of catecholamines. They increase heart rate, force of contraction, cardiac output and systolic BP. The diastolic BP may be decreased due to cutaneous vasodilatation and reduced peripheral resistance, caused by T_3 and T_4. In hyperthyroidism, the skin is warm and moist.

5. *Metabolism of proteins, carbohydrates and lipids:* Physiological amounts of thyroid hormone promote protein synthesis, nitrogen retention and a positive nitrogen balance. But large amounts can cause protein catabolism, with increased excretion of N_2 in urine, causing negative nitrogen balance.

Thyroid hormones increase the rate of absorption of carbohydrates from the small intestine. They increase glycogenolysis and gluconeogenesis in liver. They also increase peripheral utilization of glucose. The overall effect results in a rise in the blood sugar level. Diabetes mellitus is aggravated by excess thyroid hormones. Thyroid hormones affect both synthesis and degradation of lipids, the degradation being more than the synthesis. T_3 and T_4 increase the cholesterol synthesis. At the same time, they enhance fecal excretion of cholesterol and promote its conversion into bile acids. Thus, there is a net reduction of plasma cholesterol concentration.

6. *Vitamin metabolism:* Thyroxine is essential for the conversion of carotene to vitamin A in the liver. Carotenemia occurs in hypothyroidism, giving a yellow color to the skin, but not sclera (unlike jaundice).

7. *Skeletal muscle:* Optimum amount of thyroxine is required for efficient muscle function. So in both hyper and hypothyroidism, muscle weakness occurs.

8. *Reproduction and fertility:* Normal gonadal functions require an optimum amount of thyroxine. Hypothyroidism in childhood delays the onset of puberty. In females, the rhythmicity of ovarian cycle and normal lactation can occur, only if an optimum amount of circulating thyroxine levels are present. Hypothyroidism causes increased menstrual bleeding and hyperthyroidism causes scanty or absent menstrual bleeding.

9. *Other effects:* Thyroxine is necessary for normal erythropoiesis. GI motility, secretions and appetite may be increased by the thyroid hormones.

Regulation of Secretion

The regulation of thyroid hormone secretion is controlled by the *"Hypothalamus - Anterior pituitary – Thyroid gland axis"* as well as the *negative feedback mechanism.*

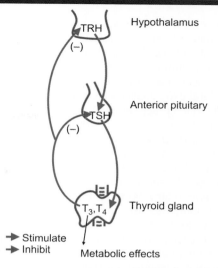

Fig. 31.8: Regulation of thyroid hormone secretion

a. *Role of TSH:* TSH from the anterior pituitary accelerates the secretions of thyroid gland by stimulating the steps of thyroid hormone synthesis. TSH also increases the size of the gland (hypertrophy) as well as the number of follicular cells (hyperplasia). The vascularity of the gland is also increased by TSH.

b. *Role of hypothalamus:* Thyrotropin releasing hormone (TRH) from hypothalamus stimulates TSH, via the hypothalamo-hypophyseal tract. In response to certain stimuli like stress, cold, etc. hypothalamus increases the secretion of TSH. Hence plasma levels of TSH and T_4 are slightly more during winter than in summer. The circulating levels of T_3 and T_4 exert a negative feedback by inhibiting TRH secretion from the hypothalamus (Fig. 31.8).

Applied Physiology

1. *Hyperthyroidism* (Thyrotoxicosis or Grave's Disease)
   This disorder is due to increased secretion of thyroid hormone as a result of hyperplasia and hypertrophy of the gland. There is increased circulating levels of free T_3 and T_4 and reduced TSH level (due to negative feedback). Grave's disease is an autoimmune disorder in which there are antibodies that bind to TSH receptors. They mimic the actions of TSH and stimulate T_3 and T_4 secretions from the thyroid gland. This leads to enlargement of the gland. The characteristic features seen are:
   a. Increase in BMR and O_2 consumption
   b. Weight loss in spite of hyperphagia
   c. Heat intolerance with a warm moist skin
   d. Nervousness, muscle weakness, etc.

e. Fine tremor of outstretched hands
f. Increased heart rate, palpitations, increased cardiac output and systolic hypertension
g. Sexual functions may be affected. In females, oligomenorrhea or amenorrhea can occur.
h. Bilateral exophthalmos causing protrusion of the eye ball ("staring" appearance). This sign is considered to be the hallmark of Grave's disease.

Blood picture: In most patients, TSH is reduced. But there is marked rise in T_3 and T_4 levels. Also long acting thyroid stimulator (LATS) is present in the plasma. LATS are the IgG antibodies against TSH receptors that mimic TSH.

Treatment: Antithyroid drug, that block the synthesis of thyroid hormones, are the first line of treatment, e.g. carbimazole. Destruction of the thyroid tissue by using radioactive iodine or surgical removal of the thyroid gland (subtotal thyroidectomy) are also used as the mode of treatment if antithyroid drug fails to control the situation.

2. *Hypothyroidism:* The hypofunction of thyroid gland in infants is called cretinism and in adults, it is called myxedema (myxoedema) .

Cretinism: It is due to lack of thyroid hormones at birth or congenital deficiency of thyroid hormones. The child is both physically and mentally retarded. The milestones in his development like crawling, sitting, standing, speech, etc. are delayed.

The characteristic features are dwarfism, coarse, thick and wrinkled skin, scanty hair, port belly, protruding tongue and thick lips. Puberty is arrested. BMR is reduced and therefore, there is subnormal temperature.

The commonest cause for cretinism is maternal iodine deficiency. Hence this condition is preventable, if mother is given sufficient iodine during pregnancy.

Myxoedema: It is due to severe hypothyroidism in adults, which can be caused by autoimmune diseases (e.g. Hashimoto's thyroiditis).

The characteristic features are: Puffy face, increased body weight, cold and dry scaly skin, decreased appetite, somnolence and muscular weakness. The BMR is low and there is intolerance to cold. The pulse rate is less and BP is low. Anemia may be present (as erythropoiesis requires thyroxine).

Blood picture: RBC count and Hb are reduced. Serum cholesterol is high.

T_3 and T_4 levels reduced, but TSH may be increased.

Treatment: Replacement therapy with thyroid hormones is effective.

3. Goiter:

Goiter is the visible enlargement of the thyroid gland. Goiter occurs due to iodine deficiency (simple goiter) or associated with hypofunction or hyperfunction of thyroid gland. Simple goiter occurs in regions where the iodine content in the soil is very less, and are called endemic goiters. In iodine deficiency, inadequate thyroid secretions occur, which lead to a feedback increase in TSH from anterior pituitary. This results in hypertrophy and enlargement of the thyroid gland (*See CP14 in Color Plate 3*). Later, there is hyperplasia and hypervascularity. Simple goiters may become huge, but remain symptomless.

Excessive consumption of certain substances can produce enlargement of thyroid gland or goiter. They are called goitrogens. Naturally occurring goitrogens are present in vegetables like cabbage, turnips, etc.

HORMONES REGULATING CALCIUM METABOLISM

Normal adult human body contains about 1100 gms of calcium. About 99% of it is present in the skeleton. The normal level of calcium in the blood is 9-11mg/dl. Calcium is present in the plasma as ionized form, as well as unionized form (protein bound form). Ionized calcium is freely permeable through the capillary membrane and is responsible for most of the functions.

Functions of Calcium

1. Mineralization of bone
2. Transmission of nerve impulses
3. Clotting of blood
4. Muscle contraction
5. Activation of certain enzymes, like lipase, succinate dehydrogenase, etc.
6. Required for secretion of certain hormones – insulin, calcitonin, vasopressin, etc.
7. Calcium decreases neuromuscular excitability and provides membrane stability.

About 30-80% of the ingested calcium is absorbed from the upper small intestine, for which Vitamin D is essential. Bones are the reservoir of calcium. When the serum calcium level rises, the excess calcium gets deposited on bones and when it falls, bones supply calcium. Calcium is excreted from the body through feces, sweat, breast milk, urine, etc. Majority of the ionized calcium filtered in the glomerulus is reabsorbed in the tubules, which is under the influence of parathyroid hormone. The hormone calcitriol also inhibits excretion of calcium. So, only a very small amount of calcium is excreted through urine. Serum calcium has got a reciprocal relationship with serum phosphate level.

Hormones associated with calcium metabolism are:
i. Parathyroid hormone or parathormone or PTH
ii. Calcitriol or 1,25-dihydroxy cholecalciferol
iii. Calcitonin.

Parathormone (PTH)

PTH is secreted from the chief cells of thyroid gland. It is a polypeptide hormone.

Regulation of PTH is by the serum calcium level. If serum calcium is low, the synthesis of PTH is stepped up and vice-versa. Also high calcitriol level inhibits PTH production.

PTH is degraded in the liver by the Kupffer cells and is excreted by kidney.

Actions

The main function of PTH is to raise the serum calcium level by the following actions (Fig. 31.9).
1. *Bone:* PTH increases the osteoclastic activity in the bone causing bone resorption and transfers the bone calcium into the ECF.
2. *Kidney:* PTH increases reabsorption of calcium from the distal tubules of kidney. At the same time, it decreases the reabsorption of phosphate in the proximal convoluted tubules of kidney, i.e. PTH causes phosphate excretion.

PTH also increases the synthesis of 1-α-hydroxylase enzyme, which is required for the conversion of Vitamin D to calcitriol.
3. *Intestine:* PTH increases the absorption of calcium from upper small intestine.

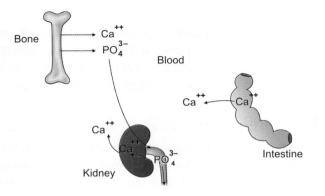

Fig. 31.9: Actions of parathormone

Abnormalities of Parathormone Synthesis

Hyperparathyroidism

It is characterized by an increase in PTH level in blood. *Physiologically*, increased PTH secretion occurs in pregnancy and lactation. *Pathological* hyperparathyroidism can be:
1. Primary (due to tumors of parathyroid glands) or
2. Secondary (due to renal diseases, rickets, etc).

Mechanism

$\uparrow$ PTH $\rightarrow$ $\uparrow$ osteoclastic activity in bone $\rightarrow$ $\uparrow$ bone resorption $\rightarrow$ $\uparrow$ serum Ca^{++} level and $\downarrow$ PO_4^{2-} level.

Because of increased bone resorption, large punched out cystic areas are seen in the X-rays of bones. There can be multiple fractures. There is depression of the central as well as peripheral nervous system, because the increased calcium level in blood decreases excitability of the neurons. There can be constipation, peptic ulcer and abdominal pain in such patients. Increased load of filtered calcium predisposes the patient to renal stone diseases. Metastatic calcification of soft tissues can also occur.

Hypoparathyroidism

It is characterized by a decrease in serum PTH level. PTH deficiency can occur due to inadvertent removal of the parathyroid glands during thyroidectomy or due to inadequate blood supply to the gland.

Mechanism

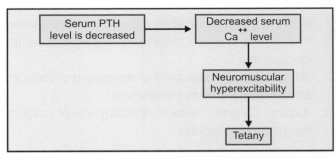

Tetany occurs when the serum calcium level falls below 50% of the normal. Tetanic muscle spasms first appear in the hands and feet (carpopedal spasm). Laryngeal spasm can give rise to a characteristic sound during inspiration called laryngeal stridor. In addition to these features, the following signs can be elicited in such patients:

Part 2

Physiology

a. *Chvostek's sign:* When the facial nerve of one side is tapped at the angle of jaw, a quick contraction of facial muscles occurs on the same side.
b. *Trousseau's sign:* Spasm of forearm, leading to flexion of wrist and thumb and extension of fingers.

1,25,dihydroxy Cholecalciferol or Calcitriol

Vitamin D belongs to the group of sterols.

Sources of Vitamin D

• Dietary – from fish liver oil, milk, eggs, etc.
• Cutaneous synthesis, from 7-dehydrocholesterol Vitamin D_3 is the active form of Vitamin D.

Synthesis of Calcitriol

7-dehydroxy cholesterol, present in the epidermal layers of skin, is converted into Vitamin D_3 by the ultraviolet sun rays. Vitamin D_3 is transported in the plasma bound to the Vitamin D binding protein. In the liver, Vitamin D_3 gets converted into 25-hydroxy cholecalciferol. In the kidney, the 1-α hydroxylase enzyme, converts 25-hydroxy cholecalciferol into 1,25 dihydroxy cholecalciferol or calcitriol. PTH stimulates the action of 1-α hydroxylase.

Since Vitamin D is synthesized in the skin, released into the circulation and has actions at a different site, it is considered as a steroid hormone.

Daily, requirement of vitamin D – 400 IU

Actions

1. *Bone:* Calcitriol mobilizes calcium and phosphate from bones to blood, thus to increasing the serum calcium level.
2. *GIT:* Calcitriol increases the absorption of calcium and phosphate from the jejunum.
3. *Kidney:* Calcitriol helps in reabsorption of calcium from the renal tubules.

Abnormalities of Calcitriol Levels

Rickets: Deficiency of Vitamin D in children, before the closure of epiphysis, causes rickets. Deficiency occurs due to lack of exposure to sunlight, decreased dietary intake of Vitamin D or decreased absorption of Vitamin D from the intestine (due to certain diseases like tropical sprue, short bowel syndrome, etc.)

Mechanism

↓ Vitamin D in blood → ↓ calcium in blood → Compensatory ↑ in PTH secretion → Osteoclastic bone resorption → Weakening of bones

Due to poor mineralization, bones are weakened and softened, and the leg bones "bow" (bend) on pressure ("Bow legs" or "knock knees"). Other characteristic bony abnormalities seen are rickety rosary, pigeon chest, bowing of femur and tibia.

Osteomalacia is the counterpart of rickets in adults.

Renal rickets or vitamin D resistant rickets – It occurs in prolonged renal diseases, where the conversion of 25-hydroxy cholecalciferol to 1,25 dihydroxy cholecalciferol is affected. Since this disease occurs not due to Vitamin D deficiency, it is called Vitamin D resistant rickets.

Calcitonin

Calcitonin is secreted by the parafollicular cells of the thyroid gland. These cells are also called clear cells or C-cells. It is a hypocalcemic hormone.

Actions

Calcitonin lowers the serum calcium level by favoring calcium deposition in the bones and by inhibiting osteoclastic activity of bones. Thus it prevents bone resorption. Calcitonin protects the bones of the mother from weakening during pregnancy and lactation. Also it helps in the skeletal development in young children.

Abnormalities of calcitonin secretion: Large quantities of calcitonin are secreted in medullary carcinoma of thyroid.

Other hormones regulating calcium level in blood: Apart from the 3 main hormones explained above, growth hormone, thyroxine, insulin and estrogens also play their roles in the regulation of calcium metabolism.

■ THE ENDOCRINE PANCREAS

The endocrine part of pancreas consists of a collection of cells called the Islets of Langerhans. There are about 1 to 2 million Islets of Langerhans, which have a rich blood supply. Histologically, there are 4 types of cells in the Islets. They are:

1. A cells (α cells) – Secrete glucagons; constitute about 25% of total Islets
2. B cells (β cells) – Secrete insulin, constitute about 60-65% of total islets
3. D cells (δ cells) – Secrete somatostatin, constitute about 10% of total islets
4. F cells (PP cells) – Secrete pancreatic polypeptide, constitute about <5%

There are connections or gap junctions between these 4 types of cells, through which, they can directly influence the secretion of each. This is called paracrine effect.

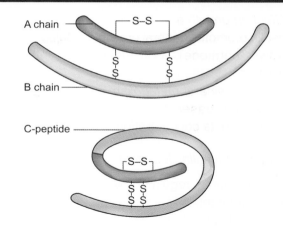

Fig. 31.10: Insulin

Insulin

Insulin was first isolated from pancreas by Banting and Best. It is the most abundant hormone among those secreted by the Islets. Insulin is a polypeptide hormone containing two amino acid chains (chains A and B), linked by disulfide bonds (Fig. 31.10). Chain A contains 21 amino acids and chain B contains 30 amino acids. Insulin circulates freely in blood. Therefore, its half-life is less (about 5-6 minutes). It is metabolized in liver and kidney.

Actions (Fig. 31.11)

Insulin is the only hypoglycemic hormone in our body. Insulin favors anabolism and causes storage of carbohydrates, proteins and fats. The main target organs are liver, muscle, adipocytes or fat cells in adipose tissue.

I. *Effect on carbohydrate metabolism*

Insulin causes rapid uptake, storage and utilization of glucose by almost all tissues of our body especially muscles, liver and adipocytes.

In liver, insulin stimulates glycogenesis and inhibits gluconeogenesis. Immediately after a meal, insulin secretion increases which causes absorption of glucose into liver and storage in the form of glycogen (glycogenesis). Insulin inhibits enzymes required for gluconeogenesis.

In muscles, insulin increases the transport of glucose into the cell. The normal resting membrane of muscle is impermeable to glucose. Only in presence of insulin or during exercise that the membrane becomes permeable to glucose. The glucose which has entered the muscle is stored as

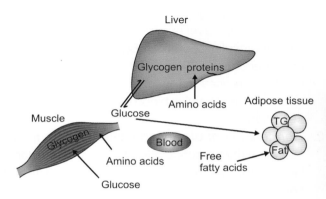

Fig. 31.11: Metabolic actions of insulin on carbohydrate, fat and protein

glycogen, by increasing glycogenesis. This stored glycogen can be used for energy later.

In adipocytes, insulin promotes conversion of excess glucose into fatty acids, which are deposited as fat.

II. *Effect on fat metabolism*

Insulin helps in fat storage.

Insulin *promotes fatty acid synthesis* in the liver from the excess glucose. These fatty acids are then transported from liver to the adipose tissue by lipoproteins and are stored there. Insulin *inhibits lipolysis.*

iii. *Effect on protein metabolism*

Insulin promotes protein synthesis and storage. Insulin stimulates transport of amino acids into the cells where they are used for synthesis of proteins. Insulin inhibits catabolism of proteins.

iv. *Effect on growth and development*
Insulin promotes general growth of body (along with growth hormone), as it favors the synthesis of proteins.

v. *Effect on mineral metabolism*
Insulin increases K^+ entry into the cell, thus decreasing its concentration in the ECF.

Regulation of Insulin Secretion

Insulin secretion is regulated by:
- Substrate control
- Hormonal control
- Neural control

1. *Substrate control:* Glucose is the most important substrate that regulates insulin secretion. When the blood glucose level is <70 mg/100 ml there is very minimal insulin secretion. But when blood sugar level becomes >70 mg/100 ml, the secretion of insulin increases many folds.
   Amino acid level in blood also stimulates insulin secretion.

2. *Hormonal control:* Somatostatin inhibits insulin secretion (by paracrine effect). Some GI hormones like gastrin, secretin, CCK, etc. promote insulin secretion.

3. *Neural control:* Islets of Langerhans receive both sympathetic and parasympathetic supply. Sympathetic stimulation suppresses insulin secretion while parasympathetic stimulation increases it.

Normal Blood Sugar Level

Fasting blood sugar level (FBS) is 70-110 mg%. Post-prandial blood sugar level (PPBS) is <140 mg%. So, if a random sample of blood is taken, the blood sugar level (RBS) comes up to 80-120 mg%.

Applied Physiology

Diabetes mellitus (DM): This disease is characterized by hypoinsulin secretion causing hyperglycemia. According to WHO, a person is said to be diabetic if his FBS is >140 mg% and PPBS is >200 mg%. There are generally 3 types of diabetes mellitus:
a. Type I diabetes (IDDM)
b. Type II diabetes (NIDDM)
c. Secondary diabetes

a. *Type I diabetes* or IDDM or Insulin dependent diabetes mellitus or juvenile diabetes is usually seen in young age. There is complete absence of insulin by destruction of beta cells. There may not be any family history of diabetes.

b. *Type II diabetes (NIDDM)* or Non-Insulin dependent diabetes mellitus or maturity onset diabetes. It becomes manifest after 40 years of age. It is caused by decreased sensitivity of target tissues to the effects of insulin. The persons are often obese and have a family history of diabetes mellitus.

c. *Secondary diabetes mellitus:* It is due to certain diseases like acromegaly, Cushing's syndrome, pheochromocytoma, etc.

Classical Features of Diabetes Mellitus

All the signs and symptoms are due to hyperglycemia and decreased utilization of glucose by cells. The features are – polyuria, polydipsia (or polydypsia = excess thirst), weight loss in spite of polyphagia (excess eating), hyperglycemia, glycosuria, dehydration, predisposition to infections (like boils, urinary infections, etc.), ketosis, acidosis, ketonuria and coma.

Metabolic Changes Occurring in Diabetes Mellitus

I. *Carbohydrate metabolism:* There is decreased entry of glucose into the peripheral tissues, even though there is excessive extracellular glucose ("starvation in the midst of plenty"). There is also impaired glucose tolerance leading to glycosuria, osmotic diuresis and polyuria.

II. *Protein metabolism:* There is increased rate of catabolism of proteins and amino acids (increased gluconeogenesis) leading to emaciation of muscles.

III. *Fat and cholesterol metabolism:* There is increased catabolism of lipids. This will increase the plasma levels of free fatty acids and ketone bodies. Some of these free fatty acids are converted into cholesterol, which is also increased in diabetes mellitus.

Complications

1. *Infections:* Patient may get repeated infections like boils, gangrene, etc.

2. *Dehydration, hypotension and coma:* Osmotic diuresis and polyuria can cause dehydration. Also there is loss of electrolytes in the urine → hypotension → coma.

3. *Diabetic ketoacidosis:* This is a medical emergency and is due to uncontrolled diabetes mellitus.

   Mechanism
   Increased lipolysis → Increased release of free fatty acids → Increased formation of acetyl Co-A → Acetyl Co-A gets converted into: a. Acetone, b. Acetoacetate, c. Beta-hydroxybutyric acid → Ketosis and acidosis
   Acetone, acetoacetate and beta-hydroxy butyric acid are the **ketone bodies**. They stimulate the respiratory center and cause deep sighing respiration called Kussmaul's breathing.
4. Atherosclerosis of blood vessels leading to myocardial infarction, stroke, etc.
5. Microangiopathy: This occurs in long-standing cases and is due to thickening of capillary basement membrane, e.g. retinopathy, nephropathy.
6. Diabetic neuropathy: There is degeneration of nerve fibers leading to weakness and numbness of skeletal muscles.

Hypoglycemia or Hyperinsulinism

This is a condition where the blood sugar level falls far below the normal limits. The causes of hypoglycemia may be one of the following:
   i. Patients in good diabetic control doing exercise
   ii. Over dosage of injection insulin or oral hypo-glycemic drugs
   iii. Insulinoma

Clinical Features

The signs and symptoms are due to the effect of hypoglycemia on CNS. They are giddiness, sweating, palpitation, mental confusion, convulsions and coma, finally leading to death.
   The condition can easily be reversed with immediate glucose administration.

Glucagon

Glucagon is a polypeptide hormone, secreted by the α(A) cells of Islets of Langerhans. It is also produced from the gastric and intestinal mucosa.

Actions (Fig. 31.12)

It is a hyperglycemic hormone. Liver is the principal target "organ".

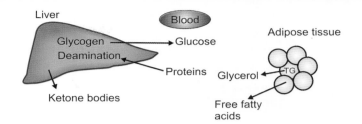

Fig. 31.12: Metabolic effects of glucagon on carbohydrate, fat and protein

Glucagon increases hepatic glycogenolysis and gluconeogenesis. It increases the ketone body formation as it promotes lipolysis. It stimulates the secretion of insulin, GH and somatostatin.

Regulation of Glucagons Secretion

1. *Substrate level control:* A reduced blood sugar level is the most potent stimulus for glucagon secretion (as in starvation, exercise, etc). An increase in amino acid level in blood also stimulates glucagons.
2. *Hormonal level:* CCK, gastrin, and GH stimulate glucagon secretion, while insulin and somatostatin inhibit it.
3. *Neural control:* Sympathetic stimulation promotes glucagon secretion.

Somatostatin

It is a polypeptide hormone secreted by the D (δ) cells of Islets of Langerhans. It is also secreted from stomach and the nerve endings in brain.

Actions: Multiple inhibitory actions. It inhibits both insulin and glucagon. In the GIT, it decreases GI motility, GI secretions, GI hormones and decreases the gallbladder contraction.

■ THE ADRENAL GLANDS

The adrenal or suprarenal glands are a pair of triangular glands, situated one on each side of the upper pole of the kidney. Each gland consists of two distinct parts—an outer cortex and an inner medulla. The cortex forms the bulk of the gland and secretes corticosteroids. The medulla is functionally related to the sympathetic nervous system and secretes epinephrine, norepinephrine and dopamine.

Part 2

Physiology

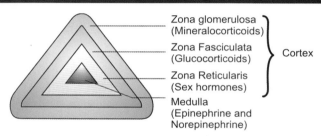

Fig. 31.13: Histology of adrenal gland with hormones
from each part

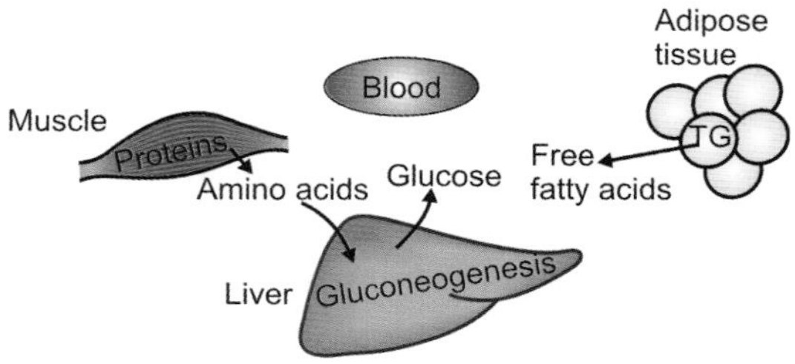

Fig. 31.14: Metabolic actions of cortisol

Adrenal Cortex

Histologically, there are 3 distinct layers in the cortex (Fig. 31.13):

1. Outermost zona glomerulosa, which secretes mineralocorticoids
2. Middle zona fasciculata, which secretes glucocorticoids
3. Innermost zona reticularis, which secretes sex hormones (androgens)

A small amount of sex hormones are secreted from zona fasciculata also and the zona reticularis secretes a small amount of glucocorticoids also.

Biosynthesis of Corticosteroids

All the corticosteroids is synthesized from cholesterol (Refer a standard Textbook of Biochemistry to know the biochemical pathways involved in the biosynthesis of corticosteroids)

Actions of Glucocorticoids

Glucocorticoids are cortisol (mainly) and corticosterone.

1. *Effects on the metabolism of carbohydrates (Fig. 31.14):* Glucocorticoids tend to increase the blood sugar level by stimulating gluconeogenesis and glycogenesis (synthesis of glycogen) in the liver, and decreasing the peripheral utilization of glucose (anti-insulin effect), but spares the brain and heart. They also decrease insulin sensitivity of the tissues. Therefore, they may aggravate diabetes.
2. *Effects on protein metabolism (Fig. 31.14):* Glucocorticoids favor protein catabolism (protein breakdown) and decrease protein synthesis in the extra hepatic tissues, especially skeletal muscles.

As a result, there is a rise in plasma aminoacid levels, which are diverted to form glucose (gluconeogenesis)

3. *Effects on fat metabolism (Fig. 31.14):* Glucocorticoids favor mobilization and utilization of fat from the adipose tissue, thereby increasing plasma free fatty acid levels. Excess cortisol causes re-distribution of fat in the body, especially in the face ("moon face") and shoulders ("Buffalo hump")
4. *Effects on electrolyte and water metabolism:* Cortisol causes retention of Na$^+$ and excretion of K$^+$. Cortisol maintains the ECF volume by providing an adequate glomerular filtration rate.
5. *Role in stress and inflammation:* Different types of stress like trauma, surgery, inflammations, infections, debilitating diseases, etc. tend to increase the secretion of cortisol (as stress stimulates ACTH secretion from anterior pituitary). Glucocorticoids mobilize fat and aminoacids, thus making them available for production of energy and other compounds, needed by different tissues of the body.

Cortisol inhibits the mediators of inflammation like prostaglandins, leukotrienes, bradykinins, etc. It increases the speed of healing process, thus helping in resolution of inflammation. Cortisol reduces fibroblastic activity and amount of fibrous tissue formed, due to which they are widely used for the treatment of connective tissue disorders like rheumatoid arthritis.

6. *Anti-allergic action:* High levels of glucocorticoids in plasma prevent histamine release. Hence they are very useful in allergic conditions, and delayed hyper-sensitivity reactions. They also relax bronchial smooth muscles, reduce the inflammatory changes and mucosal edema, and

therefore used in the treatment of bronchial asthma.

7. *Permissive action:* Small amounts of cortisol must be present in the plasma for the calorigenic effects, lipolytic effects and pressor responses of catecholamines.

8. *Effects on CNS:* Cortisol produces a sense of well-being. In deficiency, there are personality changes, irritability, apprehension and inability to concentrate.

9. *Effects on blood cells and lymphoid tissues:* Cortisol decreases the number of circulating basophils, eosinophils and lymphocytes. It reduces the size of lymph nodes and causes atrophy of thymus. But it increases the number of neutrophils and RBCs.

10. *Immunosuppressive effects:* Cortisol inhibits the formation of lymphocytes and plasma cells and therefore supresses immunity. This action occurs only after prolonged therapeutic use of glucocorticoids.

11. *Actions on Bone:* Glucocorticoids favor the osteoclastic activity and reduce the osteoblastic activity in bones. This leads to destruction of bones and osteoporosis.

12. *Effects on gastric secretions:* Glucocorticoids increase pepsin and HCl secretions in the stomach. Hence prolonged steroid therapy can cause peptic ulcer.

Regulation of Glucocorticoid Secretion

The secretion of glucocorticoids is regulated by the following hypothalamo-pituitary-adrenal axis or CRH – ACTH – Cortisol axis (Fig. 31.15)

1. *Anterior Pituitary:* ACTH, from the anterior pituitary, stimulates glucocorticoid secretion.

2. *Hypothalamus:* The circadian rhythm of glucocorticoid secretion as well as the response to stress are regulated by hypothalamus.

   The biological clock present in the suprachiasmatic nucleus of hypothalamus is responsible for the diurnal ACTH rhythm, and consequently that of glucocorticoids. The secretion is maximum during early morning hours and minimum during night.

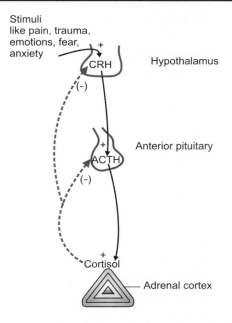

Fig. 31.15: Hypothalamo-pituitary-adrenal axis

Different types of stress stimuli increase the secretion of corticotropin releasing hormone (CRH) from hypothalamus, which activates ACTH secretion. Thus corticosteroids are increased during stress.

3. *Feedback control:* High plasma levels of glucocorticoids exert a negative feedback action on the secretions of ACTH and CRH.

Applied Physiology

Cushing's syndrome: This condition occurs due to excess cortisol production.

The **causes** can be – prolonged use of high doses of cortisol, glucocorticoid secreting tumors of adrenal cortex, ACTH secreting tumors of adrenal cortex or ACTH secreting tumors of anterior pituitary.

The **characteristic features** of Cushing's syndrome are:

a. Centripetal fat distribution with obesity of abdomen, buttocks, face and back of neck ("moon face" and "buffalo hump").

b. Excess protein catabolism leads to thin skin and subcutaneous tissue, and poorly developed muscles.

c. Osteoporosis, due to excessive osteoclastic activity

d. Hyperglycemia and hyperlipidemia

e. Purplish stria on the abdominal wall.

f. Hypertension, due to salt and water retention

g. Increased susceptibility to peptic ulcer and infections

h. Delayed wound healing.

Actions of Mineralocorticoids

Aldosterone is the chief mineralocorticoid secreted from the zona glomerulosa of adrenal cortex. Its actions are:

1. Increases the reabsorption of Na^+ in the distal convoluted tubules and collecting ducts of the kidney. As a result, water is also reabsorbed leading to an increase in the ECF volume.
2. Excretion of K^+ and H^+ are favored.
3. Increases Na^+ reabsorption from sweat, saliva and gastric juice.
4. Mild alkalosis is produced, as H^+ are excreted through urine.

Regulation of Aldosterone Secretion

1. Renin-Angiotensin mechanism (Fig. 31.16) (Explained in " *Chapter 6- Cardiovascular system*")
2. K^+ concentration in plasma : A high K^+ in plasma increases aldosterone secretion.

Applied Physiology

Conn's syndrome (Primary hyper aldosteronism): It is the hypersecretion of aldosterone from zona glomerulosa due to tumors. It is characterized by hypernatremia, increased ECF and plasma volume and hypertension. There is potassium depletion, alkalosis and severe muscle weakness, leading to tetany. When the ECF volume exceeds certain limit, the atrial natriuretic peptide is stimulated, which causes Na^+ excretion, in spite of continuous action of aldosterone. This is called aldosterone escape. Therefore, edema does not occur in primary hyper aldosteronism.

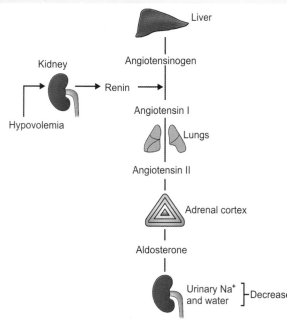

Fig. 31.16: Renin-Angiotensin-Aldosterone mechanism

Sex Hormones (Adrenalandrogens)

Dehydro-epiandrosterone is the principal sex hormone secreted from the adrenal cortex. It is secreted in both males and females and has a weak androgenic activity.

Large amounts of dehydroepiandrosterone is secreted during fetal life, which forms the precursor for the synthesis of estrogen by placenta.

In adult females, it contributes to increased muscle mass, growth of pubic and axillary hairs, acne, etc. Its action in males is overshadowed by testicular testosterone.

Applied Physiology

1. *Adrenogenital Syndrome:* It is due to excessive secretion of sex hormones (androgens). This can cause precocious pseudopuberty in boys and girls. In adult females, excessive secretion can cause virilization with appearance of beard, moustache, deepening of voice, breast atrophy, cessation of menstruation, etc. In adult males, no significant symptoms occur.
2. *Addison's disease* (Primary adrenocortical insufficiency): In this condition, there occurs deficiency of both glucocorticoids and mineralocorticoids. The cause for this disease can be tuberculosis of adrenal cortex or autoimmune diseases.

The characteristic features are:

1. Hyponatremia, hyperkalemia and dehydration occur due to reduced mineralocorticoids. As a result, there is low BP, reduced blood volume, acidosis, etc.
2. Weight loss, muscle weakness and muscle cramps.
3. Depression of sexual functions.
4. There is excessive secretion of ACTH from the anterior pituitary (due to feedback mechanism) and there is pigmentation of skin, due to melanocyte stimulating action of ACTH. Pigmentation is seen more over the exposed surfaces, palmar creases, scars and on mucous membranes of lips and mouth.
5. Decreased ability to withstand stressful situations.

ADRENAL MEDULLA

The hormones of adrenal medulla are the catecholamines viz. (1) Adrenaline (epinephrine); (2) Nor adrenaline (norepinephrine) and (3) Dopamine.

Dopamine is a precursor of noradrenaline and adrenaline. Adrenaline is the predominant hormone. Adrenal medullary hormones are synthesized from the amino acids phenylalanine and tyrosine (for the steps of catecholamine synthesis, please refer a standard Textbook of Biochemistry).

Actions of Adrenaline and Noradrenaline

Adrenaline to noradrenaline ratio is 4:1. Their actions resemble the effects of sympathetic stimulation.

1. *Effects on metabolism:* Calorigenic action occurs in presence of thyroid hormone (Permissive action) with increased O_2 consumption and BMR.
   Adrenaline tends to increase the blood sugar level by stimulating liver and muscle glycogenolysis.
   Both adrenaline and noradrenaline cause lipolysis in the adipose tissue.

2. *Effects on CVS:* Adrenaline increases the heart rate and the force of contraction (positive chronotropic and inotropic effects). Cardiac output and systolic BP are increased. Adrenaline causes vasoconstriction of cutaneous and splanchnic blood vessels except the blood vessels of skeletal muscles, coronary vessels and those of liver where it causes vasodilatation. The peripheral resistance is slightly reduced.

   Noradrenaline causes reflex bradycardia and positive inotropic effects. Noradrenaline may cause generalized vasoconstriction, but dilates the coronary vessels. The peripheral resistance is increased and hence, both the systolic and diastolic BP are increased. The hypertension thus produced, stimulates the baroreceptors producing a reflex bradycardia.

3. *Effects on muscles:* Adrenaline and noradrenaline cause relaxation of the smooth muscles of GIT, urinary tract, gallbladder, etc. but cause constriction of sphincters of GIT. They cause splenic contraction (not significantly effective in humans). They stimulate uterine contraction in nonpregnant females, but inhibit during late pregnancy.

   Adrenaline and noradrenaline increase the skeletal muscle excitability and contractility.

4. *Effects on respiration:* Adrenaline causes vasoconstriction of pulmonary vessels, reducing congestion. Adrenaline causes bronchodilatation, relieving bronchospasm.

5. *Effects on CNS:* They stimulate the ascending reticular activating system, giving rise to the feeling of general arousal and alertness - the "Fight or Flight" response.

6. *Effects during exercise:* Mild to moderate exercise activates only the sympathetic nerves. But severe exercise stimulates adrenal medullary secretion also. Both these provide appropriate cardiovascular responses for massive blood flow to the exercising muscles.

Applied Physiology

Pheochromocytoma: It is a tumor of adrenal medulla, which secretes both epinephrine and norepinephrine. The characteristic features are paroxysms (or bouts) of alarming hypertension, anxiety, palpitation, sweating, headache, blurred vision, decreased body temperature, cold and moist skin, muscular weakness and hyperglycemia.

The condition is diagnosed by detecting the presence of VMA (Vanillyl Mandelic Acid) in 24-hr urine sample. VMA is the metabolic product of catecholamines.

32 Reproductive System

INTRODUCTION

Reproduction is the process by which the species survive. In multicellular organisms like humans, reproduction includes the union of two gametes—one male and one female, to form the zygote. The gametes are formed by meiotic division and hence contain half the total number of chromosomes (haploid). The zygote is diploid, as it contains the original number of chromosomes.

THE MALE REPRODUCTIVE SYSTEM

The male reproductive system consists of the following parts:

1. A pair of testes (primary male sex organs)
2. A duct system, from testes to the penis
3. Accessory glands viz: prostate, seminal vesicles and bulbourethral glands.
   (For details, refer Anatomy, Chapter 15)

Spermatogenesis

Spermatogenesis is the process by which the sperms are formed. Spermatogenesis takes place in the seminiferous tubules. It begins at puberty and continues throughout adult life and declines in old age.

Two types of cells are seen in the seminiferous tubules:

a. Those cells which are in the process of becoming sperms.
b. Sertoli cells, which are large supporting cells, that extend from the basement membrane to the lumen of the tubules.

Steps of Spermatogenesis

1. The sperms develop from the primitive germ cells called spermatogonia, which are seen near the basement membrane of the tubules. The sper-

matogonia divide by mitosis and give rise to the primary spermatocytes, which are diploid (i.e. containing 44 XY chromosomes).

2. The primary spermatocytes divide by meiosis and give rise to two secondary spermatocytes, which are haploid (i.e. containing 22X and 22Y).

3. Secondary spermatocytes divide by mitosis and give rise to spermatids. Therefore, from one primary spermatocyte, 4 spermatids are formed, which are haploid.

4. The spermatids become embedded in the cytoplasm of the sertoli cells and mature to form spermatozoa or sperms, which are then released into the lumen of the tubules (spermiogenesis).

It takes about 74 days for a spermatogonium to form a spermatozoon. The steps can be summarized in Flow chart 32.1.

The mature sperm in the seminiferous tubule are non-motile and propelled into the epididymis.

In the epididymis, sperms gain motility. They then enter the vas deferens, ejaculatory duct and urethra).

Factors Controlling Spermatogenesis

Effect of Temperature

Spermatogenesis needs an optimum temperature of about 2 to 4°C lower than the body temperature. The testes are maintained at a temperature of about 32°C in the scrotum, which is kept outside the body. The thin skin and less subcutaneous fat in the scrotum enable the air circulating round scrotum to keep the

testes cool. The counter current heat exchange between the testicular arteries and pampiniform plexus of veins also helps to maintain the low temperature. In addition, when the environmental temperature increases, the dartos muscle relaxes.

Failure of testicular descent (cryptorchidism) causes defective spermatogenesis and infertility.

Effect of Hormones

Following hormones are essential for spermatogenesis:

Testosterone—It is required for the growth and division of testicular germinal cells. A high local concentration of testosterone is necessary for spermatogenesis.

FSH : It is essential for spermatid maturation and production of androgen binding protein by sertoli cells.

LH : It stimulates the Leydig cells to secrete testosterone.

GH : Growth hormone promotes the early divisions of spermatogonia.

Role of Sertoli Cells

(a) Sertoli cells are large supporting cells with rich glycogen content. They provide nutrition to the germ cells; (b) At the basement membrane, they form tight junctions with the adjacent cells that prevent passage of large proteins and other substances from the interstitium to the lumen. This is called blood-testis barrier; (c) They secrete ABP (androgen binding protein), which helps to maintain a high local concentration of testosterone; (d) In adults, they secrete a hormone called inhibin, which inhibits FSH secretion from the pituitary.

Spermatozoon/Sperm (Fig. 32.1)

A spermatozoon or sperm is a motile cell with a head, neck, body and tail. Head is elliptical in shape and has a large nucleus, rich in DNA. The head is "capped" by acrosome, made up of mucopolysaccharides and acid phosphatase enzyme. They are involved in the entry of sperm into the ovum.

The short neck of the sperm contains centrioles. The body or the middle piece contains large number

Flow chart 32.1: Spermatogenesis

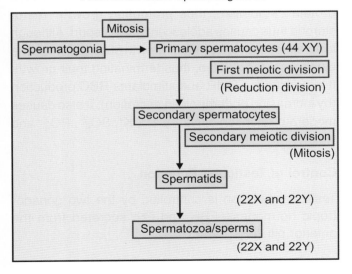

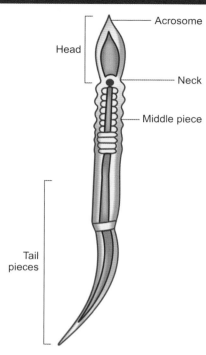

Fig. 32.1: Spermatozoon

of mitochondria, which are spirally arranged. They are the sites of energy production. The tail contains ATP and is concerned with movement.

Semen or Seminal Fluid

It is a milky opalescent mucoid fluid, which contains sperms and the secretions of seminal vesicles, prostate, Cowper's glands and bulbourethral glands. The average volume of semen is 2.5 to 3.5 ml per ejaculation. The normal pH is 7.5. Each ml of ejaculate contains 60 to 100 million sperms, of which 80 per cent or more are having normal morphology. After ejaculation, the sperms survive only for about 48 hours.

Applied Physiology

A reduction of sperm count to less than 25 million per ejaculate or 10 million/ml is called oligozoospermia. Azoospermia is the absence of sperms.

Vasectomy

It is a permanent method of contraception. It is a simple procedure in which the vas deferens is ligated and sectioned, so that the person is rendered sterile.

■ ENDOCRINE FUNCTIONS OF TESTES

Apart from its role in spermatogenesis, testis has got endocrine functions also as it secretes the following hormones:
- Androgens, mainly testosterone
- Estrogen
- Inhibin
- Activin.

Actions of Testosterone

Testosterone has androgenic as well as anabolic effects.

Androgenic Effects

1. Testosterone secreted during fetal life is responsible for the development of internal and external genitalia.
2. Testosterone is responsible for the development of secondary sexual characteristics and accessory sex organs (prostate, seminal vesicles, penis etc.) during puberty. The secondary sexual characters include—(a) Increased growth of hair on the face, chest and other parts of the body; (b) Baldness; (c) Masculine voice; (d) Broad shoulders; (e) Muscular body; (f) Thickening of sebaceous gland secretion, predisposing to acne formation; (g) Aggressive behavior.

Anabolic Effects

Testosterone favors protein synthesis and decreases protein catabolism, thus promoting growth. Testosterone thus causes adolescent growth spurt. Although it increases the rate of linear growth of bones, it causes epiphyseal fusion also, thus terminating their growth. It increases the BMR and stimulates RBC production (by increasing erythropoietin secretion). It also causes moderate retention of Na^+, Ca^{++}, K^+, SO_4^{2+}, PO_4^{2-} and water.

Control of Testicular Function

Testicular function is controlled by the two gonadotropic hormones—FSH and LH, secreted from the anterior pituitary.

FSH

Stimulates sertoli cells actively and increases the secretion of ABP (androgen binding protein). FSH acts on the germinal epithelial cells of seminiferous tubules and promotes spermatogenesis.

LH

LH or ICH (Interstitial cell stimulating hormone) acts on the Leydig cells and promotes testosterone secretion. Along with FSH, LH also promotes spermatogenesis.

The potent inhibitory feedback effect on anterior pituitary by inhibin (stimulated by FSH) provides the negative feedback mechanism for the control of testosterone.

■ THE FEMALE REPRODUCTIVE SYSTEM

The female reproductive system consists of uterus, fallopian tubes, a pair of ovaries and vagina.
(For details, refer Anatomy, Chapter 15)

Sexual Cycle in the Female

Cyclic activity is a characteristic feature of female sex functions. At puberty, the ovaries develop gametogenic and endocrine functions under the influence of pituitary gonadotropic hormones. The two major events in the female sexual cycle are menstruation and ovulation. "Menarche is the appearance of the first menstrual period, which occurs between 11 to 15 years of age."

Oogenesis

It is the process of gametogenesis in the female. The ova develop from the primitive germ cells called oogonia. About 6 million oogonia are present during 6th month of intrauterine life. The oogonium enters meiotic division, but halts at the stage of prophase, and becomes a primary oocyte. The first meiotic division (reduction division) is completed just before ovulation (after puberty) and gives rise to two daughter cells. One of them is larger and contains more cytoplasm, which is called the secondary oocyte. The other small cell is called the polar body, which fragments and disappears. The secondary oocyte is haploid (22+X). The secondary oocyte now begins the second meiotic division, but stops at the metaphase and will be completed only after the entry of the sperm into the oocyte. The secondary oocyte with 22+X chromosome matures to form the ovum and the second polar body. The steps can be summarised in Flow chart 32.2.

Thus, one primary oocyte gives rise to one ovum.

The Ovarian Cycle

It mainly consists of the follicular phase and the luteal phase.

The Follicular Phase

Some cells of the ovarian stroma become flattened and surround the primary oocyte, and form the primordial follicle (Fig. 32.2). The flattened cells are called granulosa cells. Every month, in each ovary, more than one primordial follicle start undergoing maturation process, but only one reaches maturity and the others regress. The maturation of the primordial follicle involves the following steps (Fig. 32.2):

1. The granulosa cells multiply into many layers and there is accumulation of fluid inside the follicle. The fluid is rich in estrogen.

2. The follicle enlarges with accumulation of large amounts of fluid so that it surrounds the ovum all around, except at one point, called cumulus oophorus. The cells covering the ovum called corona radiata remain in contact with the granulosa cells at the cumulus oophorus.

3. A transparent homogeneous membrane—zona pellucida starts appearing between the ovum and the corona radiata.

4. The mature primordial follicle is now called graafian follicle, which become surrounded by two layers—an inner rim of secretory cells called the theca

Flow chart 32.2: Oogenesis

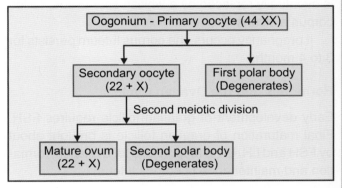

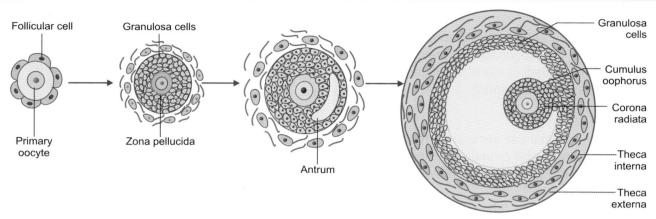

Fig. 32.2: Development of a graafian follicle

interna and an outer rim of fibrous tissue called the theca externa (Fig. 32.2).

5. The distended graafian follicle ruptures on the 14th day of the cycle, and the ovum is released into the peritoneal cavity, near the open end of fallopian tube. This process is called ovulation. Ovum is the largest cell in our body.

The Luteal Phase

Following ovulation, the wall of the graafian follicle collapses and it is filled with blood—corpus hemorrhagica. The granulosa and theca cells promptly begin to proliferate and get modified into large polyhedral cells. Their cytoplasm is rich in lipid and a yellow pigment called lutein. Hence this process is called luteinization. The whole structure, thus formed, is called the corpus luteum. The corpus luteum attains its maximum size by 22nd day of the cycle.

If pregnancy does not occur, the corpus luteum starts regressing rapidly after 24th day of the cycle. It is eventually replaced by a whitish scar tissue, called corpus albicans.

If pregnancy occurs, the corpus luteum persists for 3 to 4 months.

Hormones and the Ovarian Cycle

Early development of graafian follicle requires FSH. Final maturation of graafian follicle is brought about by FSH and LH. LH is responsible for ovulation, formation and maintenance of corpus luteum.

Two hormones are secreted from the ovary—Estrogens, by the graafian follicle and progesterone, by the corpus luteum. Corpus luteum also secretes estrogens.

The Menstrual Cycle or Uterine Cycle (Fig. 32.3)

The ovarian hormones (oestrogen and progesterone) produce cyclic changes in the endometrium of the uterus, terminating in sloughing of the endometrium and bleeding per vaginum, called menstruation. The duration of a menstrual cycle is 28 days, except during pregnancy and lactation, but it is highly variable. The phases of menstrual cycle are:

1. Proliferative phase.
2. Secretory phase.
3. Menstruation.

Proliferative Phase

Proliferative phase or pre-ovulatory phase or follicular phase begins on 6th day after menstruation and ends with ovulation (on 14th day).

At the beginning of this phase, only a thin basal layer of endometrium is present. Under the influence of estrogens from the developing follicle, the endometrium increases rapidly in thickness from the 6th day onwards. During this phase, the glands lengthen and become convoluted. The endometrium becomes more vascular, with coiled arteries.

At the end of the proliferative phase, the thickness of endometrium becomes 3 to 4 mm.

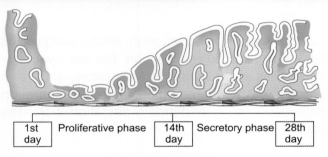

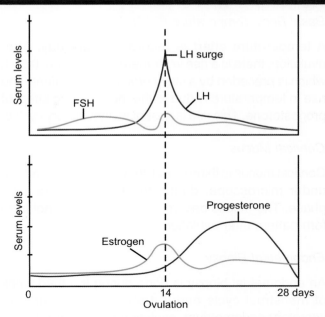

Fig. 32.3: Changes in the endometrium during the menstrual cycle

Fig. 32.4: Hormonal changes in different phases of menstrual cycle

Secretory Phase

Secretory phase or luteal phase begins from the 15th day onwards and lasts up to 28th day. This phase coincides with the luteal phase of the ovarian cycle (when corpus luteum is formed).

The endometrium shows marked thickness (hypertrophy), under the influence of estrogens. Progesterone produces marked enlargement of endometrial glands, which become coiled and distended with mucus. The endometrium becomes highly vascular. There is increase in the number of spiral arteries and the vessels are congested. Spiral arteries supply superficial 2/3rd of the endometrium, while the deeper layer is supplied by straight basilar arteries.

When the corpus luteum regresses, hormonal support for the endometrium is withdrawn and the next phase begins. But if pregnancy occurs, the thickened endometrium with the secretory glands provide a suitable site for the implantation of the fertilized ovum.

By the end of secretory phase, the thickness of endometrium becomes 5 to 6 mm. The length of secretory phase is remarkably constant (14 days).

Menstrual Phase

In the absence of pregnancy, the corpus luteum regresses, which decreases the level of the hormones. The superficial parts of the thickened endometrium are shed off, which is accompanied by menstrual bleeding. A few hours before the onset of menstrual bleeding, the spiral arteries undergo intense spasm and foci of necrosis appear in the endometrium and walls of spiral arteries gradually. The necrotic tissue gets separated from the remaining part and is shed off along with the

blood in the spinal arteries per vaginum. Duration of menstruation is about 3 to 5 days, during which an amount of 30 to 35 ml of blood is lost. Loss of more than 80 ml is abnormal.

Anovulatory Cycles

Sometimes ovulation does not occur and the corpus luteum is not formed; so, there is no effect of progesterone on the endometrium. However, under the influence of estrogen, the endometrium proliferates enough to breakdown and slough during menstruation.

Cyclical Changes in Cervix

Under the influence of estrogen, the cervical mucus becomes thinner and alkaline, which promotes the survival and transport of sperms. Progesterone makes it thick, tenacious and cellular.

Indicators of Ovulation

A knowledge about the time of ovulation is important in the treatment for infertility as well as the family planning. Although it is difficult to know the exact time of ovulation, the following methods indirectly help:

Part 2

Physiology

Basal Body Temperature (BBT)

A temperature chart is recorded every day. After ovulation, there is a rise in temperature by about 0.5°C, which is preceded by a slight drop in temperature. The rise in temperature is due to the thermogenic effect of progesterone. This rise is absent in anovulatory cycles.

Cervical Mucus

Cervical mucus is thinner and shows a fern-like pattern, under microscope, during the end of proliferative phase. The mucus becomes thick and does not form fern-pattern after ovulation.

Endometrial Biopsy

An endometrial biopsy is done during the third week of menstrual cycle and if it shows the features of secretory endometrium, ovulation has occurred.

Ultrasonography (USG) of Ovaries

Serial USGs can reveal the development and the day of rupture of graafian follicle.

Hormonal Control of Menstrual Cycle (Fig. 32.4)

a. The gonadotropin releasing hormone (GnRH) from hypothalamus stimulates the release of FSH from anterior pituitary.
b. FSH stimulates the development of the graafian follicle, which secretes estrogen. Estrogen is responsible for the proliferative changes in the endometrium. When estrogen level increases, it inhibits FSH secretion.
c. The rapid rise of estrogen occurs about 24 hours before ovulation, after which it drops. The high estrogen level induces a burst of LH secretion, which causes ovulation ["LH surge"]. LH helps in the formation and maintenance of corpus luteum.
d. Corpus luteum secretes progesterone as well as estrogen. The increased levels of progesterone and estrogen inhibit LH and FSH secretions. Progesterone is responsible for the secretory endometrial changes.
e. Regression of corpus luteum drops the levels of estrogen and progesterone, causing menstrual bleeding.

Menopause

It is the cessation of menstruation, which usually occurs between 45 and 50 years of age. The last few menstrual cycles are irregular and anovulatory, which stop gradually. After menopause, estrogen secretion decreases and gonadotropin secretion increases markedly.

Applied Physiology

1. *Amenorrhea:* It is the absence of menstruation. If menarche does not occur even after 18 years of age, it is called primary amenorrhea. Physiological amenorrhea occurs during pregnancy and lactation.
2. *Menorrhagia:* It is excessive bleeding during menstruation
3. *Metrorrhagia:* It is bleeding in between menstrual periods.
4. *Dysmenorrhea:* It is painful menstruation.

PREGNANCY

Fertilization and Implantation

In humans, fertilization of the ovum by the sperm usually occurs in the ampulla of the fallopian tube. Although many sperms surround the ovum, only one can penetrate it and fertilize it, which is aided by the proteolytic enzymes present in the acrosome of the sperm.

The developing embryo, called blastocyst, moves down the tube into the endometrium of uterus. This takes about 3 days. Now, the blastocyst becomes surrounded by two layers of cells—an outer syncytiotrophoblast layer and an inner cytotrophoblast layer. The syncytiotrophoblasts erode the endometrium, implanting the blastocyst in it. This process is called implantation. The usual site of implantation is on the dorsal wall of uterus. A placenta then develops, and the trophoblast remains associated with it.

Functions of Placenta

Endocrine Function

Placenta is the temporary endocrine organ of pregnancy and secretes a large number of hormones like human chorionic gonadotropin (hCG), estrogen, progesterone, human placental lactogen (hPL), relaxin, etc.

hCG It is a glycoprotein hormone synthesized by syncytiotrophoblast. It maintains the corpus luteum of pregnancy and stimulates fetal androgen secretion. Its presence in urine forms the basis of pregnancy tests.

Estrogens After the first 3 months of pregnancy, placenta takes up the function of estrogen secretion. Estrogens are synthesized in the syncytiotrophoblast and are responsible for the enormous growth of uterus and increased vascularity.

Progesterone It is synthesized from the syncytiotrophoblast. It is the hormone of pregnancy. It inhibits ovulation and menstruation during pregnancy, inhibits uterine motility and causes development of the alveolar system of breast.

Human placental lactogen (HPL) or human chorionic somatomammotropin: It has lactogenic and growth stimulating effects. It stimulates retention of calcium, nitrogen and phosphorus, lipolysis and inhibits gluconeogenesis. Therefore, it acts as a maternal growth hormone.

Respiratory Function

Placenta can be referred to as "fetal lung", as gas exchange occurs between maternal and fetal blood across the placenta.

Nutritive Function

Nutritive substances like glucose, iron, amino acids, vitamins, fatty acids, water, Na^+, Cl^-, etc. are transported from the maternal blood to the fetal blood via the placenta.

Excretory Function

Waste products like urea, uric acid, creatinine, etc. are removed from the fetal blood by diffusion into the maternal blood.

Other Functions

Maternal IgG antibodies can cross the placenta and reach the fetus, thus providing immunity. Placenta acts as a barrier protecting the fetus from the harmful effects of many drugs and infective agents.

Diagnostic Tests for Pregnancy

They are based on the detection of hCG in the urine of the pregnant female. hCG starts appearing in urine from 14th day after conception. There are biological as well as immunological tests for diagnosing pregnancy.

The Immunological Tests

The immunological tests are widely used nowadays, which can diagnose pregnancy as early as 2 weeks after conception. The basis of this method is antigen-antibody reaction. Here, the urine sample is mixed with an antiserum against hCG. To this, add latex particles coated with hCG, and watch for any agglutination. If there is no agglutination, the test is positive for pregnancy. This is because, the hCG present in urine neutralizes the antibodies.

In addition to pregnancy, the above tests are positive in certain uterine tumors secreting hCG, e.g. choriocarcinoma, hydatidiform mole.

Physiological Changes during Pregnancy

The average duration of normal pregnancy in human is 280 ± 20 days which is counted from the first day of last menstruation. The first 12 weeks are called first trimester, 13 to 28 weeks—second trimester and 29 to 40 weeks—third trimester. During these periods, the maternal system undergoes many changes to meet the increasing demands.

Changes in the Uterus and Ovary

Uterus enlarges considerably from 7.5 cm length in the non-pregnant female to about 35 cm at term. The weight also increases and can be about 1 kg at term. Enlargement of uterus during first trimester is due to hypertrophy of myometrial smooth muscles. Enlargement of uterus during second and third trimesters is due to stretching of the muscle fibers by the rapidly growing fetus.

In the ovary, the corpus luteum enlarges and secretes oestrogens and progesterone during first 3 months of pregnancy. Throughout the pregnancy, the menstrual cycle does not occur (amenorrhea).

Breast

There is breast enlargement, with proliferation of the alveolar ducts. The areola and nipple become pigmented.

Blood Volume

The blood volume starts rising after 12 weeks of pregnancy. Plasma volume is increased more than RBC volume (hemodilution). ESR is raised.

CVS

Cardiac output increases progressively as the pregnancy advances. The heart rate is slightly increased. BP remains normal in spite of increased blood volume. This is because of decreased peripheral resistance caused by progesterone.

The venous pressure in the femoral vein is increased by the effect of gravid uterus on the inferior vena cava. As a result, the pregnant woman is prone to varicose veins and venous thrombosis.

Weight Gain

An average weight gain of 10 kg occurs in a pregnant female, which may be due to the enlarged uterus with fetus, and water retention.

Skin

There is excessive pigmentation of face, areola, nipple and midline of abdomen.

Respiratory System

Tidal volume, O_2 consumption and CO_2 output are increased, but there is no change in the respiratory rate.

GIT

Nausea and vomiting occurs in the morning, during the first 2 to 3 months, which subsides thereafter. This is called "morning sickness". There is decreased intestinal motility, giving rise to constipation.

Kidney

Renal blood flow and GFR are increased (by 25%).

Endocrine Glands

The activities of anterior pituitary, adrenal cortex, thyroid, islets of Langerhans, etc. are increased in pregnancy.

Parturition or Labor

This is the process by which the products of conception (fetus, membranes and placenta) are expelled out of the uterus per vaginum. The onset of labor pains may usually start between 37th to 40th weeks of gestation.

The role of oxytocin in the initiation of labor pains has been already discussed in Chapter 26 "Endocrine System".

LACTATION

During pregnancy, prolactin levels increase steadily until term; under the influence of this hormone plus the high levels of estrogen and progesterone, full tubuloalveolar development of breast takes place.

Initiation of Lactation after Delivery

Prolactin and estrogen promote breast growth, but estrogen antagonizes the milk producing effect of prolactin on breast. After expulsion of placenta, there is a sharp decline in the levels of estrogen and progesterone. The drop in circulating estrogen level initiates lactation by prolactin.

After a few weeks, the basal level of prolactin falls. But when the infant sucks the nipple, neural stimulations are produced, which reach the hypothalamus and cause marked secretion of prolactin from the anterior pituitary. Therefore during breastfeeding, plasma prolactin level rises sharply.

Ejection ("Let down") of Milk

Even if the breast is filled with milk, it will not enter the ducts and come out unless oxytocin is present. Oxytocin causes contraction of myoepithelial cells around the alveoli, causing ejection of milk.

The milk ejection reflex is a neuroendocrine reflex, which occurs as follows (Flow chart 32.3).

Thus suckling not only causes formation of large quantities of milk by prolactin, but also helps in milk ejection by oxytocin.

During the period of lactation, menstruation and ovulation do not usually occur. This is due to the inhibitory effect of prolactin on GnRH secretion, that decreases FSH and LH.

Flow chart 32.3: Milk ejection reflex

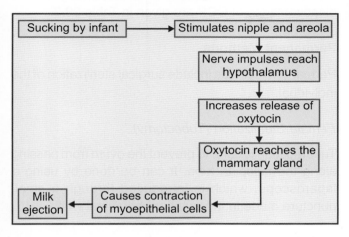

Colustrum

It is the thick, yellowish fluid that comes out of the breast during the first 2 to 3 days after delivery. Colustrum is rich in proteins, and sodium chloride. However, the lactose and fat content are less. It is an important source of antibodies for the newborn.

■ CONTRACEPTION

Contraception is the prevention of unwanted pregnancy. The methods used for contraception may be:
  I. Temporary methods
  II. Permanent methods

Temporary Methods

Barrier Methods

The "barriers" prevent the entry of sperms into the vagina. Most widely accepted barrier method is the use of condoms. They are cheap, easily available and have no side effects. In addition to contraception, they provide protection from sexually transmitted diseases like AIDS, hepatitis-B and other venereal diseases.

  "Diaphragm" is a similar mechanical barrier to be fitted in the vagina over the cervix, in females.

Natural Methods

Periodic abstinence of sexual intercourse during fertile period or the period of ovulation is one of the natural methods to prevent conception. This is called "safe period". The possibility of conception is minimum, if intercourse is avoided 5 days before and 3 days after

Table 32.1: Advantages and disadvantages of IUCD	
Advantages of IUCD	*Disadvantages of IUCD*
• No hospitalization required • Easily inserted • Cheap • Fertility is reversible on removal of IUCD	• Bleeding • Pain and uterine perforation • Pelvic inflammatory disease can occur • Ectopic pregnancy may occur

the predicted day of ovulation. But, the ovulation date can be predicted only if the menstrual cycle is of 28 days and regular. Hence this method is unreliable.

Intrauterine Contraceptive Devices (IUCDs)

There are mainly three types of IUCDs:
  First generation IUCD: Lippe's loop, which is a non-medicated device.
  Second generation IUCD: Copper-T, which is a medicated device.
  Third generation IUCD: Progestasert, which is a progesterone-filled device.
  The IUCD is introduced and left in the uterine cavity.

Mechanism of action The IUCD acts as a foreign body, causes "aseptic inflammation", of the uterine endometrium, and makes it unfit for the implantation of the fertilized ovum. The advantages and disadvantages of IUCD are given in Table 32.1.

Oral Contraceptive Pills (OCP)

They are hormonal pills and are of different types.

Combination pills or combined pills contain synthetic estrogens and progestins. These pills have to be taken regularly for 21 days, from the 5th day of menstrual cycle.

  Mechanism of action: High concentration of progesterone inhibits ovulation, by inhibiting the LH surge (negative feedback action). Progesterone also makes the cervical mucus thick and unfavorable for sperm penetration. They also prevent implantation of blastocyst, even if ovulation has occurred. Combined OCP are widely used and contain synthetic estrogens like ethinyl estradiol and progesterones like norethisterone or norethynodrel.

Part 2

Physiology

Progesterone only pill (Mini Pill) Contains small doses of progesterone alone and is given throughout the menstrual cycle.

The "sequential pill" Administration of high doses of estrogen alone for first 15 days, followed by 5 days of estrogen + progesterone.

Adverse Effects of OCP

Nausea, vomiting, breakthrough bleeding, weight gain are common side effects. Continuous use can cause hypertension, thromboembolic disorders, stroke and ischemic heart disease. The advantages and disadvantages of OCP are given in Table 28.2.

Permanent Methods

Permanent methods include surgical sterilization of the individual.

Female Sterilization (Tubectomy)

Tubal ligation is done to prevent the ovum from passing along the fallopian tube. It can be done by using a laparoscope, which is introduced through a small puncture made in the anterior abdominal wall.

Male Sterilization (Vasectomy)

Here, a portion of the vas deferens is excised and the open ends are ligated. Bilateral ligation of vas is a simple procedure, which can be done under local anaesthesia. The sperms are prevented from passing beyond the point of tying.

Table 32.2: Advantages and disadvantages of OCP	
Advantages of OCP	*Disadvantages of OCP*
• Easy to use • Reduced blood loss • Relief from dysmenor-rhea • Regular periods • Reversible fertility	• Failure can occur due to forgetfulness • Many adverse effects on continuous use

33 The Nervous System

The nervous system integrates and coordinates various activities of other organ systems. It controls muscle contraction, secretion of hormones from glands, rate and depth of respiration, cardiac activities and gastrointestinal activities. It is also involved in modulating and regulating a multitude of other physiological processes.

The nervous system can be broadly classified into two – the central and the peripheral nervous systems. The central nervous system consists of brain and spinal cord. The peripheral nervous system consists of the nerves lying outside the central nervous system, i.e., the somatic and the visceral groups of nerves. Somatic nerves supply structures other than viscera and include motor and sensory somatic nerves. Visceral nerves supply the structures of viscera and include sympathetic and parasympathetic nerves.

The structural and functional unit of the nervous system is the neuron or the nerve cell. The structure and basic mechanisms of sympathetic transmission are already discussed in "Nerve Physiology".

PROPERTIES OF SYNAPSE

1. One-way conduction : Synaptic conduction takes place always in one direction, i.e. from pre-synaptic to post-synaptic neuron, and never in the opposite direction.
2. Synaptic delay: (already discussed in "Nerve Physiology")
3. Convergence and Divergence: Convergence means that the axon terminals of multiple neurons end on a single neuron (Fig. 33.1A and B). This helps to summate the information from different

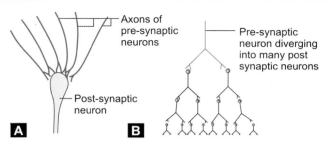

Figs 33.1A and B: Diagrammatic representation of convergence (A) and divergence (B)

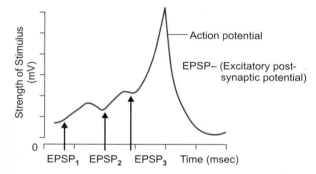

Fig. 33.2: Summation of EPSP's generating an action potential

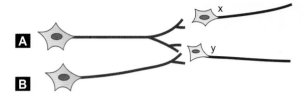

Fig. 33.3: Diagrammatic representation of subliminal fringe

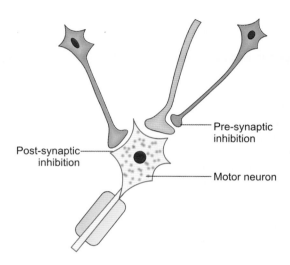

Fig. 33.4: Synaptic inhibition

sources. If a single neuron synapse with multiple post-synaptic neurons, it is called divergence (Figs 33.1A and B). Here an impulse arriving in the pre-synaptic neuron excites greater number of nerve fibers that leave the synapse.

4. Summation: Summation means fusion of electrical impulses which produce an enhanced response (Fig. 33.2). Only graded potentials can be summated to produce an action potential.

5. Subliminal Fringe: This is seen when a pre-synaptic neuron is shared by two or more post-synaptic neurons (Fig. 33.3). If the neuron "A" in the (Fig. 33.3) is stimulated, it causes excitation of both "X" and "Y" post-synaptic neurons. "X" and "Y" neurons are now said to be in the "subliminal fringe", i.e. the threshold for generation of action potentials in them is decreased. Now, even a subthreshold stimulation of the neuron "B" can produce an action potential in the post-synaptic neuron "Y", as "Y" is already in the "subliminal fringe".

6. Synaptic Plasticity: Certain changes can occur in the conduction of impulses through synapse on the basis of experience. This is called synaptic plasticity. If the pre-synaptic terminal is very frequently stimulated, it can result in an enhanced post-synaptic potential, which lasts for a longer time. It is due to increase in the intracellular Ca^{++} in the pre-synaptic and post-synaptic terminals. This is the basis of learning and memory. In certain synapses, repeated stimulation may result in decrease in the response. This is due to decreased release of neurotransmitter from the pre-synaptic terminal.

7. Synaptic inhibition: Inhibition of synaptic conduction can be of two types, post-synaptic inhibition and pre-synaptic inhibition.

Post-synaptic inhibition is due to the production of an inhibitory post-synaptic potential (IPSP) by an inhibitory neurotransmitter like glycine. This hyperpolarizes the post-synaptic neuron (Fig. 33.4). This type of inhibition, seen in motor neurons of spinal cord, cerebral cortex, limbic system and cerebellum, helps to check the over excitability of motor neurons.

Pre-synaptic inhibition is mediated by axons of certain neurons making synapses with pre-synaptic terminals of excitatory neurons (axo-axonal synapse) (Fig. 33.4). An inhibitory neurotransmitter is released at the axo-axonal synapse.

▉ THE CENTRAL NERVOUS SYSTEM

Spinal Cord

Spinal cord is the downward extension of medulla which extends from foramen magnum in the skull to the lower border of L1 vertebra. It extends to the upper border of L3 vertebrae in children. It consists of 31 segments and 31 pairs of spinal nerves, viz.8 pairs of cervical nerves, 12 pairs of thoracic nerves, 5 pairs of lumbar nerves, 5 pairs of sacral nerves and 1 pair of coccygeal nerve. For a detailed structure of spinal cord and the formation, drainage and functions of CSF, *refer "Anatomy"* part of this book.

Bell-Magendie law: The principle that in the spinal cord, ventral (anterior) nerve roots are entirely motor and dorsal (posterior) nerve roots are entirely sensory is called Bell-Magendie law.

Nerve Cells of Gray Matter of Spinal Cord

Anterior or ventral horn cells (motor neurons) Fig.33.5C

They include α (alpha) motor neurons (which innervate the contractile or extrafusal fibers of skeletal muscles), γ (gamma) motor neurons (which innervate the intrafusal fibers of muscles) and many interneurons (which make connections with α cells).

Posterior or Dorsal Horn Cells (Sensory neurons)

They receive sensory fibers from posterior or dorsal nerve root (Fig. 33.5). They give origin to the ventral spinothalamic tracts.

Lateral Horn Cells (Fig. 33.5)

They are present only in thoracic and upper lumbar segments. They give origin to sympathetic and parasympathetic fibers.

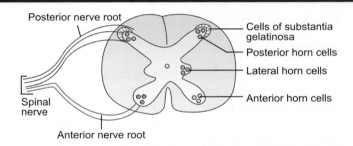

Fig. 33.5: Nerve cells in the spinal cord

Cells of Substantia Gelatinosa Rolandis (Fig. 33.5)

They consist of small group of cells which contain a gelatinous material, situated at the apex of dorsal horn. They receive fibers from dorsal nerve root and give origin to the lateral spinothalamic tract.

Nerve Fibers in the White Matter of Spinal Cord

The white matter has groups of nerve fibers called tracts. There are ascending (afferent) tracts, descending (efferent) tracts, association fibers (connecting adjacent segments of spinal cord) and commissural fibers (connecting opposite halves of same segments of spinal cord).

Ascending (Afferent) Tracts

They are constituted by a group of sensory fibers ascending from the receptors to the central nervous system. They include:
1. Dorsal column tracts of Goll and Burdach or dorsal column fibers.
2. Spinothalamic tracts (lateral and anterior)
3. Spinocerebellar tracts
4. Spinotectal tracts
5. Spino-olivary tracts

Descending (Efferent) Tracts

They convey efferent impulses from various parts of brain to the spinal cord and end in the anterior horn cells. They include:
1. Corticospinal (pyramidal) tracts
2. Extrapyramidal tracts—vestibulospinal, rubrospinal, reticulospinal, tectospinal and olivospinal tracts.

The tracts are described in detail later.

Part 2

Physiology

The Sensory System

Receptors

They are structures, which receive the sensations and form the first structure in the sensory pathway. They receive stimuli from both internal and external environments of our body. When these receptors are stimulated, the nerves emerging from them also get stimulated. An action potential is thus developed, which travels to the CNS.

Types of Receptors

1. *Exteroceptors:* They respond to immediate external environment, e.g. cutaneous receptors for touch, pain, temperature, etc.
2. *Interoceptors:* They respond to changes in internal environment, e.g. Baroreceptors, chemoreceptors, etc.
3. *Proprioceptors:* They detect the position of limb in space (posture), degree of contraction of muscles, etc, e.g. muscle spindle, Golgi tendon organ.
4. *Teleceptors:* They respond to stimuli coming from a distance, e.g. audition, vision.

Cutaneous receptors (Fig. 33.6): They respond to 4 cutaneous senses – touch, pain, pressure and temperature. They are—

- Merkel's disc (nerve endings with disc-like expanded tips)
- Meissner's corpuscles (encapsulated nerve endings)
- Pacinian corpuscles – for deep touch and pressure
- Ruffini's end organ
- Hair end organ
- Free nerve endings – for pain.

Proprioceptors

The somatic sensory receptors described above provide information about the external environment. Proprioception provides knowledge of positions of our limbs and joints as they move, and the degree of muscle contraction. Muscle spindle and Golgi tendon organ are the two mechano-sensitive receptors present in the skeletal muscles, which help in proprioception.

Muscle spindle (Fig. 33.7) is located in the modified skeletal muscle fibers called intrafusal fibers and are

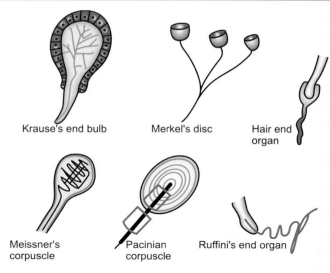

Krause's end bulb   Merkel's disc   Hair end organ

Meissner's corpuscle   Pacinian corpuscle   Ruffini's end organ

Fig. 33.6: Cutaneous receptors

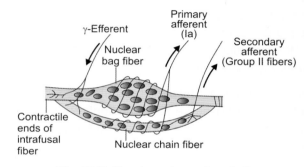

Fig. 33.7: Structure of muscle spindle

aligned in parallel with the contractile or extrafusal fibers. It has both afferent and efferent innervations. It contains two types of intrafusal fibers – nuclear bag and nuclear chain fibers, with two types of sensory endings surrounding them – primary and secondary endings (Fig. 33.7). The efferent innervation is via γ motor neurons, the stimulation of which causes the intrafusal fibers to contract in parallel with extrafusal fibers. The primary sensory endings of group Ia axons wind around and strongly innervate individual bag fibers (and also chain fibers). They are very sensitive to changes in muscle length or dynamic length. The secondary endings of group II axons mainly innervate nuclear chain fibers and are sensitive to static length of muscle. Both the afferent neurons discharge when the muscle spindle is stretched, resulting in contraction of extrafusal fibers. Now, the muscle spindle may be no more stretched which makes it insensitive to further changes in length. To avoid this situation and to maintain control over the sensitivity of muscle spindle, γ motor neurons discharge and contract the contractile elements present at the edges of intrafusal fibers (Fig. 33.7). Thus the stretch over the muscle spindle is maintained.

The Golgi tendon organ is situated near the tendon of the skeletal muscle. They are aligned in series with extrafusal fibers. When tension is developed in the muscle due to either passive stretch or active contraction, the mechano-sensitive nerve endings (group Ib axons) of Golgi tendon organs fire action potentials. The group Ib axons make connections with the anterior horn cell supplying the same muscle via an inhibitory neuron. Therefore, when group Ib axons are stimulated, they stimulate the inhibitory interneuron, which in turn inhibit the contractile action of the muscle (Fig. 33.8). Thus Golgi tendon organ controls the tension or force developed in a muscle.

Properties of Receptors

1. *Weber Fechner's law:* Receptors obey this law, which states that the magnitude of sensation felt is directly proportional to the log of intensity of stimulus.
2. *Specificity of response:* A receptor is maximally excited when it is subjected to a specific stimulus. For example, sound waves can stimulate auditory receptors and not the visual receptors.
3. *Adequate stimulus:* The particular form of energy to which a receptor is most sensitive is called its adequate stimulus. For example, rods and cones present in the eye can be stimulated by light as well as by pressure. But the threshold of stimulus required is less in the case of light and so, rods and cones are more sensitive to light than pressure.
4. *Adaptation:* When repeatedly stimulated, the receptor response decreases gradually, and finally they stop to respond. This is called adaptation (But we do not get adapted to pain), e.g. we get used to ear rings, bad smell, clothes, etc.
5. *Law of projection:* If a sensory pathway is stimulated anywhere along its course to sensory cortex, the conscious sensation is always felt at the site of receptors and not at the point of stimulation. This is called law of projection. For example, a person who has undergone amputation (removal of a limb), usually experiences pain and proprioceptive sensations in the limb which is removed ("phantom limb"). Irritation of damaged nerve endings at the stump of the removed limb generates impulses in the nerve fibers. According to the law of projection, these sensations are projected to where the receptors used to be located.

Reflex

Reflex is the basic unit of integrated neural activity. It is the mechanism by which a sensory impulse is automatically converted into a motor effect through the involvement of central nervous system.

Reflex arc: It is an arc consisting of the receptor, afferent nerve fibers, the center, efferent nerve fibers and the effector organ. The "Center" means that part of the CNS where the afferent nerve fibers synapse with the efferent nerve fibers, either directly or indirectly via an interneuron. A typical example for a monosynaptic (involves only one synapse) reflex arc is the stretch reflex (Fig. 33. 9). In stretch reflex, the

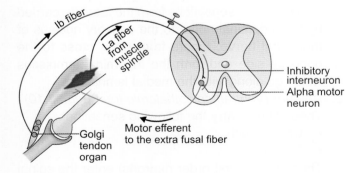

Fig. 33.8: Action of Golgi tendon organ ("inverse stretch reflex")

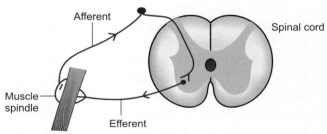

Fig. 33.9: Stretch reflex

receptor is the muscle spindle, which when stimulated, sends an action potential that travels through the afferent nerve to the center. The center in the stretch reflex is the spinal cord, where the impulse is processed and sent through the efferent nerve to the effector organ, which is the muscle. Now, the muscle contracts.

Classification of Reflexes

I. According to the number of synapses:
    a. Monosynaptic, e.g. stretch reflex
    b. Polysynaptic, e.g. abdominal reflex, withdrawal reflex, etc.
II. Clinical classification:
    a. Superficial e.g. corneal reflex, conjunctival reflex, etc.
    b. Deep, e.g. Tendon jerks
    c. Visceral, e.g. Baroreceptor reflex, micturition reflex
    d. Pathological, e.g. Babinski's sign
III. According to the development:
    a. Conditioned reflexes – which develop after birth and depend on previous experiences, e.g. Salivation at the sight of delicious food.
    b. Unconditioned reflexes – which are present since birth, e.g. Micturition reflex.

Sensory Tracts (Ascending Tracts)

The main ascending or sensory tracts are:
1. Dorsal column fibers or the posterior column fibers or the tract of Goll and Burdach
2. Anterior spinothalamic tract
3. Lateral spinothalamic tract.
1. *Dorsal (posterior) column tract or fibers (Fig. 33.10)*
    These fibers carry the following sensations:
    •   Fine touch or light touch
    •   Vibration sense
    •   Tactile localization
    •   Tactile discrimination or two-point discrimination
    •   Proprioception
    The fibers carrying the above sensations enter the spinal cord through the dorsal funiculus (dorsal white column) and ascend upwards. Hence they are called the dorsal column fibers. They form the first order neurons. They relay in the dorsal column nuclei in the medulla. The dorsal column nuclei are

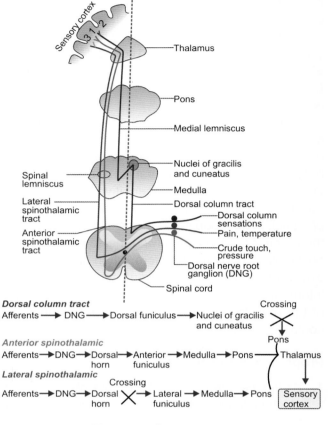

Fig. 33.10: Sensory tracts

the nucleus cuneatus and nucleus gracilis. Second order neurons from the medulla cross to the opposite side and form the medial lemniscus, which ascend upwards and relay in the posterolateral nucleus of thalamus (also called specific sensory relay nucleus). Third order neurons arise from the thalamus and are projected to the primary sensory cortex (Brodmann's area 3,1,2)

The above sensations from the face are carried by the trigeminal nerve to the sensory nucleus of trigeminal nerve. Fibers from here, cross to the opposite side and form the trigeminal lemniscus which go along with the medial lemniscus.

2. *Anterior (Ventral) spinothalamic tract (Fig.33.10)*
    These fibers carry the following sensations:
    •   Crude touch
    •   Pressure
    These fibers (first order neurons) enter the spinal cord via the dorsal nerve root ganglion and terminate in the dorsal horn (dorsal gray column).

Second order neurons arising from the dorsal horn cross to the opposite side and ascend upwards through the anterior (ventral) funiculus or white column. Hence, they are called the anterior spinothalamic tract. They ascend upwards to the medulla, where they join the lateral spinothalamic tract to form the spinal lemniscus. These fibers then relay in the specific sensory nucleus of thalamus. The third order neurons from the thalamus are projected to the primary sensory cortex (Brodmann's area 3,1,2).

The above sensations from the face are carried via the trigeminal nerve to the sensory nucleus of trigeminal nerve → cross to opposite side → trigeminal lemniscus → join the spinothalamic tracts → reach the sensory cortex.

3. *Lateral spinothalamic tract (Fig. 33.10)*
The fibers of this tract carry the pain and temperature sensations.

The first order neurons enter the spinal cord via the dorsal nerve root ganglion and relay in the dorsal horn. Second order neurons from here cross to the opposite side and ascend upwards through the lateral funiculus (white column) forming the lateral spinothalamic tract.

They ascend upwards to the medulla where they join the anterior spinothalamic tract to form the spinal lemniscus. The spinal lemniscus ascends upwards to relay in the specific sensory nucleus of thalamus. The third order neurons, arising from here, are projected to the primary sensory cortex (Brodmann's area, 3,1,2).

The above sensations from the face are carried via the trigeminal nerve to the sensory nucleus of trigeminal nerve → cross to the opposite side → join the lateral spinothalamic tract → ascend to the cortex.

Types of Pain

According to the quality of sensation felt, pain can be grouped into two – fast pain and slow pain. As the name indicates, fast pain is felt with in 0.1 sec after a painful stimulus and is of sharp quality. It is elicited by either mechanical or thermal stimulation of skin and is transmitted via Aδ type of nerve fibers. Slow pain is a diffuse type of pain felt after 1 sec or more. It is always

associated with tissue destruction and involves deeper structures along with skin. It is poorly localized and is associated with autonomic responses like vomiting, increase in BP, increase in heart rate, etc. It is transmitted via type C unmyelinated nerve fibers.

According to the site of production, pain can be classified into somatic and visceral types. Somatic pain involves pain arising from skin and subcutaneous tissue, muscles, joints and bones. Visceral pain is the deep pain felt in visceral organs and is of 'slow type' in nature. Any one of the following stimuli can cause visceral pain – ischemia, overdistension, obstruction or spasm of hollow viscera, inflammation, traction of mesentery, chemical stimuli, etc. Visceral pain is usually 'referred' to a distant somatic structure and hence also known as 'referred pain'. The visceral pain is referred to a structure that has developed from the same embryonic segment as the structure in which the pain originates. This principle is called "dermatomal rule". e.g. pain due to appendicitis is felt around the umbilical region; pain due to myocardial infarction is felt in the inner aspect left arm; pain due to stones in the gallbladder is felt at the tip of the shoulder.

Somatosensory Cortex

From the specific sensory nuclei of thalamus, the nerve fibers of all sensory pathways project to mainly two somatic sensory areas of the cerebral cortex viz.primary (S_1) and secondary (S_{II}) somatosensory areas. Primary sensory area (S_I) is located in the post-central gyrus in the Brodmann's area – 3,1,2. Secondary somatosensory area (S_{II}) is located in the parietal cortex. For a detailed diagram of the above described areas, refer (Chapter 16, Anatomy).

Thalamus

There are two thalami, which are ovoid masses of gray matter lying close to each other, separated only by the third ventricle. For the detailed structure and different types of nuclei present in the thalamus, refer Anatomy part of this text.

Functions

1. *Relay station:* Thalamus is the relay station for pathways coming from all peripheral sensory systems, except olfaction. The sensations are

relayed and projected to appropriate cortical areas. Pain sensations are perceived in the medial nucleus of thalamus, which is also associated with subjective emotional feelings ("affect").

2. Thalamus controls wakefulness, alertness and consciousness.

3. *Motor influences:* Thalamus has connections with cerebral cortex, cerebellum and basal ganglia. Along with basal ganglia, thalamus controls planning and programming of motor activities by cerebral cortex. Along with cerebellum, it helps in error detection and correction of motor activities by cerebral cortex.

4. Personality and emotional control: Thalamus is concerned with social behavior and emotions of a person via its connections with hypothalamus and pre-frontal cortex.

Applied Physiology

Thalamic Syndrome

This is a clinical condition caused by vascular lesion of thalamus. It is characterized by over reactions to some sensations, along with hemiplegia. There is hyperesthesia, i.e. sensations like touch, warmth, cold or even sound on the affected side are associated with pain and unpleasant emotions.

■ MOTOR SYSTEM

This involves the descending cortical pathways which control the actions of skeletal muscles of our body. Muscles of limb and trunk are innervated by motor nerves, originating from α motor neurons of spinal cord. Muscles of head and neck region are controlled by motor neurons of cranial nerve nuclei in the brain stem. Basal ganglia, thalamus and cerebellum provide feedback to the motor control areas of the cerebral cortex and brain stem (Fig. 33.11).

Motor Unit

A motor unit consists of an α-motor neuron and the group of extrafusal fibers it innervates.

The muscles, joints and ligaments possess sensory receptors that inform the central nervous system about body position and muscle activity. Skeletal muscles contain muscle spindles, Golgi tendon organs, free

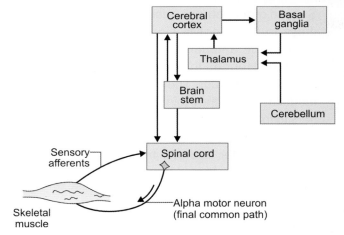

Fig. 33.11: Schematic representation of control of motor system

nerve endings and pacinian corpuscles. Muscle spindle and Golgi tendon organ are the proprioceptors. Their functioning were described earlier in this chapter.

Cerebral cortex has the highest control over motor activities which is exerted via the "Corticospinal tract" or the "pyramidal tract". The actions of pyramidal tract are influenced by the extrapyramidal system. Both pyramidal and extrapyramidal tracts form the descending tracts of spinal cord, which are described below.

1. Pyramidal Tract or Corticospinal Tract (Fig. 33.12)

The motor cortex controls the anterior horn cells of spinal cord through this tract. It is named so, because it forms a prominent elevation ("pyramid") on the ventral surface of the medulla.

The corticospinal tract originates from the following regions of cortex:

1. Primary motor cortex (Brodmann's area – 4)
2. Pre-motor cortex (Brodmann's area – 6)
3. Supplementary motor area
4. Somatosensory area

About 30% of the fibers constituting the pyramidal tract originate from the primary motor cortex, 30% from the pre-motor area and supplementary motor area together, and 40% from the somatosensory areas.

Course of Pyramidal Tract

The fibers originating from the above mentioned areas descend through the subcortical structures as radiations called "Corona radiata", which converge and

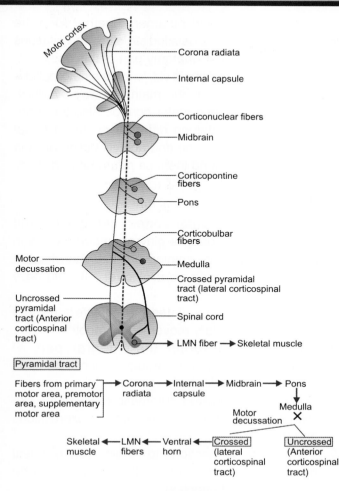

Fig. 33.12: Pyramidal tract

post-synaptic neurons from the anterior horn cells supply the effector organ through the spinal nerves.

Some pyramidal tract fibers, while traveling down the brain, terminate on the nuclei of cranial nerves. They are called the corticonuclear fibers. Some other fibers terminate on the nuclei other than the cranial nerve nuclei on pons and medulla and are called the corticopontine and the corticobulbar fibers respectively.

The activities of pyramidal system is under the control of cerebellum, basal ganglia and sensory cortex.

Functions of Pyramidal Tract

This tract is concerned with the voluntary motor movements, skilled motor movements and reflex activity of limbs and trunk. The corticonuclear, the corticobulbar and the corticopontine tracts are concerned with the movements of head and neck muscles.

Through the pyramidal tract, the motor cortex controls the opposite half of the body (as the tract crosses in the medulla).

2. The Extrapyramidal Tracts or System

This system includes all other descending fibers arising from the motor cortex, apart from the pyramidal tract fibers. These fibers arise from the same cortical areas as for the pyramidal system, but instead of going to the medulla, they relay in the basal ganglia, red nucleus and brainstem reticular formation. Ultimately, they form separate tracts and descend through the spinal cord. For example, for the extrapyramidal tracts – reticulospinal tract, rubrospinal tract, vestibulospinal tract, olivospinal tract, etc.

The extrapyramidal system is concerned with the regulation of involuntary motor activities, muscle tone and posture.

Upper motor neuron or UMN means the descending nerve fibers from the cerebral cortex to the anterior horn cells of the spinal cord (excluding the anterior horn cells).

Lower motor neuron or LMN includes the anterior horn cells of spinal cord and the efferent fibers arising from the anterior horn cells to the effector organ or muscle.

go through a compact "V" shaped mass of grey matter called the internal capsule. The corticospinal tract occupies the genu and the anterior two-thirds of the posterior limb of internal capsule. After descending through the midbrain and pons, they reach the medulla. In the medulla, these fibers form a bundle and produce a prominent elevation on the ventral side called "pyramids". Near the lower part of the medulla, about 90% of these fibers cross to the opposite side. This crossing-over is called the motor decussation. The crossed fibers descend through the lateral funiculus of the spinal cord and are called the lateral corticospinal tract. The uncrossed fibers descend through the anterior funiculus and are called the anterior corticospinal tract. Both these group of fibers finally terminate in the anterior horn cells of spinal cord, either directly or indirectly through the inter neurons. The

Applied Physiology

1. *Lesions of the pyramidal system:*
   The symptoms and signs may vary according to the site of lesion, i.e. whether the UMN or LMN is affected. The main differences between UMN and LMN lesions are given in Table 33.1.
2. *Hemiplegia:* It means paralysis of one half of the body. It is seen in UMN lesions of the pyramidal tract. Commonest site of lesion is the internal capsule, due to thrombosis or embolism of middle cerebral artery. As the motor fibers are very compactly arranged here, even a small lesion can cause wide involvement of the body. Usually the opposite half of the body is affected and is associated with the signs and symptoms of UMN lesions.
3. *Paraplegia:* It means paralysis of both lower limbs, due to transection of the spinal cord.
4. *Monoplegia:* It means paralysis of one limb, due to lesions in the motor cortex representing that limb.

Basal Ganglia

Basal ganglia are large masses of gray matter situated deep in the cerebral hemisphere. The components of basal ganglia include:

- Corpus striatum (made up of caudate nucleus and putamen)
- Globus pallidus
- Subthalamic nucleus or claustrum
- Substantia nigra

Table 33.1: Differences between UMN and LMN lesions

	UMN Lesion	LMN Lesion
Cause	Due to lesions of UMN which terminate on the anterior horn cells of spinal cord	Due to lesions of LMN which include the anterior horn cells and the fibers arising from them
Tone of muscles	Increased	Decreased or lost (flaccid muscles)
Power of muscles	Lost	Lost
Tendon jerks	Exaggerated	Lost
Atrophy of muscles	Absent	Present
Abnormal muscle contractions like clonus	Present	Absent
*Babinski's sign	Positive	Absent

*Babinski's sign or the extensor plantar response: When you stimulate the sole of the foot with a pin, if there is dorsiflexion of the great toe and fanning out of the small toes, the Babinski's sign is positive.

Globus pallidus and putamen together form the lentiform nucleus. For a detailed structure and relations of basal ganglia, refer Anatomy part of this book.

Direct close-circuit fibers viz, 'cortico-striato-pallido-thalamo-cortical tract' is the main pathway by which basal ganglia influence motor control. Also there are interconnections between different nuclei of basal ganglia. GABA, dopamine and acetyl choline are the neurotransmitters found in these neuronal terminals.

Functions

1. Along with prefrontal cortex, cerebellum and thalamus, basal ganglia play an important role in planning and programming of motor activities. It also has feedback control over all voluntary activities.
2. Corpus striatum regulates the muscle tone and makes the voluntary movements smooth.
3. Basal ganglia forms the integral part of extrapyramidal system through which it controls posture and equilibrium.
4. Control of associated automatic movements like swinging of hands while walking, facial expressions and emotions.
5. Through its connections to limbic system, basal ganglia is involved in cognition (initiation to do something) and affection.

■ APPLIED PHYSIOLOGY

Parkinsonism

This clinical condition is due to degenerative changes in the basal ganglia, which results in an imbalance between concentration of dopamine and acetyl choline. There is destruction of substantia nigra which secretes dopamine. The condition is characterized by marked rigidity, mask-like face, stooped posture, resting tremor, difficulty in speech and slow shuffling gait or short stepping gait.

Wilson's Disease

This is due to abnormal deposition of copper in the lenticular nucleus and also liver, which leads to hepato-lenticular degeneration.

Cerebellum

Cerebellum is the largest part of hind brain which lies in the posterior cranial fossa, behind the pons and the medulla. It is connected to the brain stem by three peduncles – superior cerebellar peduncle (connects with midbrain), middle cerebellar peduncle (connects with pons) and inferior cerebellar peduncle (connects with medulla). Through these peduncles, afferent and efferent fibers connect cerebellum with the extracerebellar regions.

Functional Anatomy

Cerebellum is divided by two transverse fissures into three parts – anterior lobe, middle (posterior) lobe, and flocculonodular lobe (Fig. 33.13). Middle portion of cerebellum is called vermis and the lateral portions are called cerebellar hemispheres (Fig. 33.13). The portion of cerebellar hemisphere lying adjacent to either sides of vermis are called the intermediate zones and the lateral most parts on either side form lateral cerebellum (Fig. 33.13).

Phylogenetically, cerebellum can be divided into three portions:
1. Archicerebellum (vestibulocerebellum): It consists of flocculonodular lobe. It is phylogenetically the oldest part. It is concerned with maintenance of equilibrium and eye movements.
2. Paleocerebellum (spinocerebellum): It consists of vermis and adjacent medial portions of cerebellar hemispheres. This region receives proprioceptive inputs from motor cortex and is concerned with coordinated control of axial and proximal limb muscles, whereas intermediate zones are concerned with control of distal limb muscles.

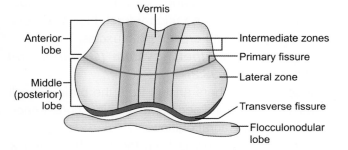

Fig. 33.13: Diagrammatic representation of functional lobes of cerebellum

3. Neocerebellum (cerebrocerebellum): This is phylogenetically the newest part. It consists of lateral portions of cerebellar hemispheres. Through its connections to motor and premotor cortex, it is involved in planning and programming of motor activities.

Functions

1. Cerebellum controls all voluntary motor activities. Along with thalamus, prefrontal cortex and basal ganglia, cerebellum plays an important role in planning and programming of motor activities.
2. It is concerned with error detection and correction of motor output from cortex.
3. Cerebellum maintains the muscle tone.
4. Timing of motor activity is controlled by cerebellum.
5. All skilled motor activities are coordinated by cerebellum.
6. Cerebellum is concerned with maintenance of body posture, equilibrium and movements of eyeballs.

Applied Physiology

Cerebellar syndrome: It consists of symptoms and signs of cerebellar damage due to any tumor, injury or thrombosis of the feeding artery.
Signs and symptoms are:
1. Ataxia or incoordination of movements
2. Asthenia or weakness of muscles, due to decreased tone
3. Atonia or hypotonia of muscles
4. Dysmetria or past-pointing, as the patient fails to assess the degree of contraction of the muscles
5. Intention tremor, i.e. the tremor which appears when the patient attempts to do something.
6. Dysarthria or defective speech or scanning speech. Patient breaks up every word and separately articulates them.
7. Nystagmus or rhythmic oscillation of eyes due to ataxia of ocular muscles.
8. Pendular tendon jerks.
9. Staggering gait or drunken gait.
10. Dysdiadochokinesia or inability to perform rapid fine movements.

Hypothalamus

Hypothalamus is a part of diencephalon, lying in the floor and lateral wall of third ventricle. It is considered to be the highest center for autonomic nervous system. Hypothalamus is composed of a number of nuclear masses, which can be broadly grouped into three –

Part 2

Physiology

anterior group, middle group and posterior group. Hypothalamic nuclei receive as well as send a number of projections from and to different parts of brain.

Functions

1. *Autonomic functions:* Hypothalamus is considered as the head ganglion of autonomic nervous system. Posterior group of nuclei control sympathetic nervous system while parasympathetic nervous system is under the control of middle and anterior group of nuclei.

2. *Endocrine functions:* The posterior pituitary hormones, viz. vasopressin and oxytocin are synthesized in the hypothalamus and are transported through the hypothalamo-hypophyseal tract to the posterior pituitary. Various hypophysiotropic hormones are released from hypothalamus that controls the secretion of anterior pituitary.

3. Regulation of body temperature (explained in the chapter 'Temperature Regulation'). Hypothalamus functions as a 'thermostat' which maintains the normal body temperature.

4. Control of circadian rhythm or day-night cycle.

5. Role in hunger and satiety: Lateral nucleus of hypothalamus is the feeding center and ventromedian nucleus is the satiety center. Normally, feeding center is chronically active and is transiently inhibited by the satiety center, after food intake.

6. Control of thirst mechanism via osmoreceptors present in the anterior hypothalamic nuclei. An increase in osmolality of plasma stimulates the osmoreceptors, which in turn cause secretion of ADH.

7. Control of emotional feelings and sexual behavior.

■ THE AUTONOMIC NERVOUS SYSTEM

The autonomic nervous system (ANS) or visceral nervous system is that portion of nervous system, which controls the visceral functions of body like heart rate, blood pressure, GI motility, secretions, sweating, etc. The ANS is made up of efferent and afferent neurons. Efferent autonomic neurons innervate:

a. Smooth muscles of blood vessels, respiratory tract, GIT, excretory tract, reproductive tract, pilomotor system
b. Cardiac muscle
c. Exocrine glands

Afferent autonomic neurons carry sensations from the viscera to the CNS and they mainly run through three nerves – the vagus, splanchnic and pelvic nerves. They synapse with the efferent autonomic neurons as well as with the somatic neurons, in the spinal cord. Therefore, the sensations originating from viscera can produce response in visceral as well as somatic structures.

ANS can be broadly classified into the sympathetic and the parasympathetic nervous systems.

1. *The sympathetic nervous system (Thoracolumbar outflow)*

   The pre-ganglionic fibers arise from the lateral gray horns of T_1-L_3 spinal segments; hence called thoracolumbar outflow.

   There are two sympathetic trunks, one on either side of the vertebral column – which consist of 22 ganglia, connected to each other. Generally, the sympathetic trunk bears one ganglion for each spinal nerve, but the total number is less due to fusion of some of the ganglia.

   – In the cervical region, there are 3 ganglia— superior, middle and inferior cervical ganglia.
   – In the thoracic region, there are 11 ganglia. The first thoracic ganglion fuses with the inferior cervical ganglion to form the stellate (cervico-thoracic) ganglion.
   – In the lumbar region, there are 4 ganglia and in the sacral region, there are 4 to 5 ganglia.

   The sympathetic nervous system supplies the eye, the lacrimal gland, salivary glands, lungs, heart, entire GIT, the urinary tract, blood vessels, sweat glands and erector pili muscles.

2. *The parasympathetic system (Cranio-sacral outflow) (Table 33.2)*

   It consists of fibers arising from the brain as well as from the sacral segments of spinal cord, hence called craniosacral outflow.

   The cranial outflow is from the III, VII, IX and X cranial nerves.

Table 33.2: Cranial outflow (parasympathetic)		
Parasympathetic fibers	*Origin*	*Supply*
Of III Cranial nerve (Oculomotor N)	Edinger Westphal nucleus in midbrain	Intrinsic muscles of eye (sphincter pupillae and ciliaris)
Of VII cranial nerve (Facial N)	Superior salivary nucleus in pons	Lacrimal gland, submandibular and sublingual salivary glands
Of IX cranial nerve (Glossopharyngeal nerve)	Inferior salivary nucleus in medulla	Parotid gland
Of X cranial nerve (Vagus N)	Dorsal nucleus of vagus in medulla	All thoracic and most of the abdominal viscera, except the area of supply from sacral outflow

Table 33.3: Comparison of sympathetic and parasympathetic stimulation		
Organ	*Sympathetic stimulation*	*Parasympathetic stimulation*
Heart	Rate ↑, contractility ↑, cardiac output ↑	Rate ↓, contractility ↓ Cardiac output ↓
Blood vessels	VD of coronary and skeletal vessels— VC of cutaneous and sympathetic vessels	No major action
Bronchial muscle	Relaxation	Constriction
GIT	↓ GI movements ↓ GI secretions	↑ GI movements ↑ GI secretions
Genitourinary system	Relaxation of detrusor muscle, retention of urine, ejaculation of sperm	Contraction of detrusor muscle. Helps in urination. Helps in erection
Eye	Pupillary dilatation	Pupillary constriction
Metabolism	Hyperglycemia, ↑ glycogenolysis	Hypoglycemia

↑ —Increase	VC—Vasoconstriction	GI—Gastrointestinal
↓ —Decrease	VD—Vasodilatation	

The sacral outflow is **from S₂,₃,₄ spinal segments**. They supply left 1/3rd of transverse colon, descending colon, rectum, urinary bladder, lower part of ureters and genitalia.

Neurotransmitters in ANS

Acetylcholine and nor-adrenaline are the main neurotransmitters in ANS. Acetylcholine is the neurotransmitter in:
a. Both preganglionic and postganglionic para-sympathetic neurons.
b. Postganglionic sympathetic neurons.
c. Preganglionic sympathetic neurons supplying sweat glands, blood vessels and erector pili.
    Noradrenaline is the neurotransmitter in all preganglionic sympathetic neurons except those supplying sweat gland, blood vessels and erector pili.

Actions of Autonomic Neurons (Table 33.3)

Sympathetic and parasympathetic neurons can have excitatory effects in some organs while inhibitory effect on others. Also, when sympathetic stimulation excites a particular organ, parasympathetic stimulation inhibits it, i.e. they act reciprocally. The important actions of both sympathetic and parasympathetic neurons are compared in Table 33.3.

Higher Control of ANS

Medulla is the chief higher center, which controls the ANS. Hypothalamus also helps in its function as it is the integrating center for somatic, autonomic and endocrine functions. Cerebral cortex has no control on the ANS and so this system is also called as involuntary nervous system.

Part 2

Physiology

34 The Special Senses

Special senses include cutaneous sensations, vision, audition, gustation and olfaction. Of these, gustation and olfaction form the chemical senses. Cutaneous sensations are already discussed in "Sensory System".

VISION

Functional Anatomy of Eye

Eyeballs are protected by bony sockets called orbits which are deficient anteriorly. Anteriorly, they are guarded by the eyelids. The two eyelids meet at the outer and inner canthi. Palpebral fissure is the opening between the two eyelids. Eyelids have tiny sensitive hairs over their margins called cilia, which protect and keep the cornea moist. In the upper and outer part of the orbit, lacrimal gland is situated (Fig. 34.1). The secretion of lacrimal gland is called tears. Tears are

secreted via 10-12 lacrimal ducts that open on the surface of conjunctiva. Tears can enter into the nasal cavity through the nasolacrimal duct. Tears contain many proteins and antibacterial substances like

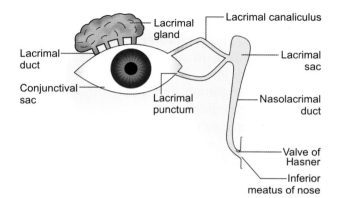

Fig. 34.1: Lacrimal apparatus

lyrozyme, lactoferrin, etc. They moisten and lubricate the cornea as well as wash away any irritant substances. As cornea is avascular, it uses the oxygen that is dissolved in tears.

Eyeball (Fig. 34.2)

The human eyeball is roughly spherical in shape, consisting of several layers and structures. The coverings of eyeball can be broadly grouped into three years:
1. Outer fibrous layer, which consists of sclera and cornea
2. Middle vascular layer or uveal tract
3. Inner neural layer or retina

Outer layer

Sclera is the tough, opaque, fibrous tissue that forms the outermost layer of eyeball. In the anterior one-sixth of the eyeball, sclera modifies in to the transparent 'cornea'. Sclera provides shape and protection to the eyeball. As cornea is transparent, light rays can pass through it. Cornea is an avascular structure (hence can be transplanted easily). Cornea receives nutrition from aqueous humor and dissolved oxygen from tears. Cornea is a highly sensitive structure, supplied by ophthalmic division of trigeminal nerve. The corneal epithelium continues as conjunctiva. It covers the anterior part of sclera and lines the inner surface of eyelid.

Middle layer

Uveal tract is bluish in color and consists of a highly vascular structure called choroid, ciliary body and iris (Fig.34.2). Choroid contains pigments that help to absorb any excess light rays, thus preventing scattering of light. Ciliary body is attached to the suspensory ligament of lens. It consists of longitudinal and circular smooth muscle fibers (ciliary muscles) as well as ciliary processes that secrete aqueous humor. Ciliary muscles receive innervations from parasympathetic fibers of III cranial nerve (oculomotor nerve). Situated anterior to the lens is the iris, which is surrounded by aqueous humor on both sides. It has a small central aperture called "pupil" through which light enters the eye. The color of iris can vary from blue to dark brown and black, according to the pigment

melanin in it. Iris contains circularly and radially placed smooth muscle fibers (Fig. 34.2). Contraction of circular smooth muscles causes constriction of pupil (miosis) and that of radial smooth muscles causes dilatation of pupil (mydriasis). Circular smooth muscles are supplied by parasympathetic nerve fibers, while radial smooth muscles receive sympathetic nerve fibers.

There is a space enclosed behind the cornea and in front of iris called anterior chamber (Fig. 34.2). Posterior chamber is the space situated in front of the lens and behind the iris (Fig. 34.2). Both anterior and posterior chambers are filled with aqueous humor.

Inner layer of retina

Retina is the neural layer that contains the photoreceptors – rods and cones. The retina is histologically and embryologically a part of the central nervous system which consists of 10 layers (Fig. 34.6). Nerve fibers arising from the layer of ganglion cells unite and form the optic nerve. Optic nerve leaves the eye as well as the retinal blood vessel enter it at a point about 3 mm medial to and above the posterior pole of eye. This region is known as 'optic disc', where there are no rods or cones. Hence, this region is also known as 'the blind spot' (Fig. 34.2). Near to the optic disc, there is a small yellowish area called macula lutea, which is specialized for very sharp color vision. At the center of macula is the fovea centralis – a depressed region where only cones are present.

■ INTRAOCULAR FLUIDS

Both anterior and posterior chambers are filled with a thin clear fluid called "aqueous humor", similar in composition to cerebrospinal fluid. It is continuously secreted from the ciliary processes of ciliary body.

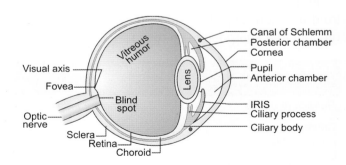

Fig. 34.2: Structure of human eyeball

As the fluid accumulates, it is drained through the canal of Schlemm (Fig. 34.2), situated at the iridocorneal angle, into the ciliary veins. The continuous outflow of aqueous humor results in the formation of intraocular pressure in the eyeball. The normal intraocular pressure is between 10-20 mm of Hg, which can be measured using a tonometer. The vitreous humor or vitreous body is a transparent gelatinous substance that fills the space between lens and retina (Fig. 34.2). It gives support to the retina.

■ APPLIED PHYSIOLOGY

Glaucoma

Glaucoma is the condition where intraocular pressure rises due to accumulation of aqueous humor inside the anterior chamber of eye. It is one of the common causes for blindness in our country. More often, the drainage of aqueous humor is affected. Because of increased intraocular pressure, the lens is pushed backwards resulting in compression of the optic disc and nerve fibers. This leads to progressive loss of vision.

■ TRANSMISSION OF LIGHT RAYS THROUGH EYE

Human eye has a complex optical system that allows light to pass through and focus the images onto the retina, where photoreceptors are present. Human eye can detect light rays between 400 nm and 750 nm wave length, which is known as the visible spectrum.

Light rays are reflected or bent when they pass from one medium to another of different intensity. The different refractory media through which light has to pass are given in the Table 34.1.

In a normal eye ("Emmetropic eye"), parallel rays from a distant object (more than 6 m distance) are brought to a sharp focus on retina.

Table 34.1: Refractory indices of different structures of eye	
Refractory media	Refractory index
Cornea	1.38
Aqueous humor	1.33
Lens	1.42
Vitreous body	1.34

Human eye possesses a crystalline, transparent and elastic lens which is biconvex. It is covered by an elastic capsule. The biconvex lens is related to iris anteriorly and vitreous humor posteriorly (Fig. 34.2). The lens is attached to ciliary body by suspensory ligaments. The anterior curvature of lens is more than the posterior curvature. The lens is an avascular structure with a refractory index of 1.42.

■ APPLIED PHYSIOLOGY

Cataract

Degenerative changes occur in the lens as the age advances. There is loss of water content due to which lens becomes opaque. This is known as cataract. Cataract first starts in the deeper parts of peripheral regions of lens ("immature cataract") and then spreads to the entire lens ("mature cataract").

Optical Defects in Eye (Errors of Refraction)

Myopia (Short sightedness)

In this condition, the parallel light rays coming from a distant object are focused at a point in front of the retina and not over it. So the person cannot see the distant objects clearly. This defect is due to increased anteroposterior diameter of eye than normal (Fig. 34.3). The condition can be corrected by using a concave lens which diverges the light rays and focuses them onto the retina (Fig. 34.3).

Hypermetropia (Far sightedness)

When the length of eyeball is too short, the parallel rays of light are brought to focus behind the retina. This condition is called hypermetropia (Fig. 34.4). The defect is corrected by using convex lens, which converges the rays and focuses them onto the retina.

Astigmatism

This defect is due to a difference in the horizontal and vertical axes or curvatures of cornea. As a result, images from all the portions of object cannot be clearly seen. For example, if he is given a graph-paper to see, he will fail to see both vertical and horizontal lines simultaneously. This defect is corrected by using cylindrical lenses.

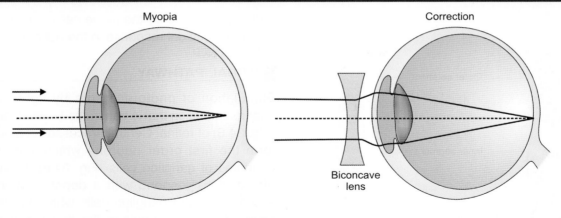

Fig. 34.3: Correction of myopia

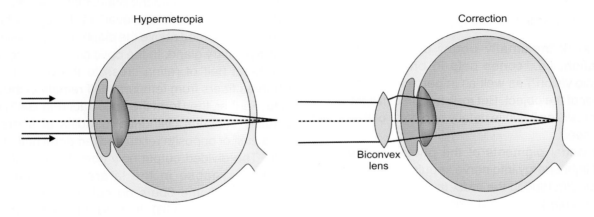

Fig. 34.4: Correction of hypermetropia

Presbyopia

Presbyopia is a defect of accommodation due to loss of elasticity of lens with advanced age. The lens becomes rigid and its power decreases. So the person cannot see small words when held at a usual distance for reading. But, when held at a greater distance, he may be able to read it. Such eyes are corrected using "reading glasses" – i.e., the convex lenses.

■ PHOTORECEPTORS

Rods and cones are the photoreceptors situated in the retina. Each eye contains about 125 million rods and 5.5 million cones. Rods are responsible for scotopic vision (night vision or 'black and white' vision) and cones are responsible for photopic or color vision. Their functions are basically similar even though they differ structurally and biochemically. A rod cell is long, slender and larger than a cone cell (Fig. 34.5). Its outer segment contains many "discs" that are filled with the photopigment rhodopsin. Cone cell has 'cone-shaped' outer segment (Fig. 34.5). Instead of discs, there are many saccules in the outer segment which contain three types of photopigments. The pigments differ in the wavelength of light that excites them. The inner segments of both rods and cones enlarge to form their corresponding nuclei (Fig. 34.5). According to the "duplicity theory of vision", there are two kinds of inputs

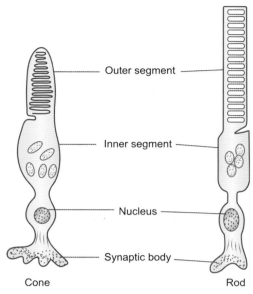

Fig. 34.5: Structure of photoreceptors

which work maximally under different types of illumination, i.e., cones help in bright light vision ("photopic vision") as well as in perceiving the details and color of the object. But rods help in dim light vision ("scotopic vision"), while details and color of the object are not seen.

The visual pigments of the photoreceptor cells convert light energy into nerve impulses. This process is best understood in the rods. In dim light, rhodopsin ("visual purple") of the rod exists as a complex of the protein – 'Scotopsin' and the aldehyde of vitamin A – "11-cis retinine"). When exposed to bright light, visual purple is bleached, i.e., the "11-cis" form of retinine is converted into "all trans" form. During this process, a number of intermediate products are formed, among which metarhodopsin II provides the critical link between this reaction series and the electrical response. Metarhodopsin II activates transducin (a "G-protein) which in turn activates the enzyme phosphodiesterase. Phosphodiesterase decreases the cGMP levels in the rod cells, causing hyperpolarization.

In the dark, the c-GMP levels are high which keeps the Na$^+$ channels open and the rod cell is relatively depolarized. Then there is a tonic release of neurotransmitters from the synaptic body of rod. But in bright light, as a result of light-induced reactions, Na$^+$ channels close and the cell becomes hyperpolarized. This reduces the release of neurotransmitters. This change is the signal which is further processed by the nerve cells of retina to form the final electrical response in the optic nerve.

VISUAL PATHWAY

Stimulation of photoreceptors by light produces hyperpolarization that is transmitted to the bipolar cells (Fig. 34.6) of retina. Central processes of bipolar cells form the first order neurons which synapse with dendrites of ganglion cells (Fig. 34.6). Some of the bipolar cells respond with a depolarization that is excitatory to the ganglion cells whereas other cells respond with a hyperpolarization that is inhibitory. The axons of ganglion cells form the optic nerve, which are the second order neurons. The optic nerves, each carrying about one million fibers from each retina, leave the orbit and enter into the brain. Optic nerves of both sides converge or "cross over" at the optic chiasma (Fig. 34.7) situated above the diaphragmatic sella, near to the pituitary gland. In the optic chiasma, fibers from the nasal half of retina cross to the opposite side, whereas fibers from temporal half remain in the same side. Further, they form the optic tract (Fig. 34.7) which actually contains cylindrical bundles of fibers from nasal half of opposite retina and temporal half of same side. The divided output now goes through the optic tract to synapse in the lateral geniculate body (Fig. 34.7). Some fibers do not relay in the lateral geniculate body (LGB). Instead, they leave the optic tract and end in the superior colliculus and pretectal region of midbrain (they are concerned with pupillary reflexes). Third order neurons arise from LGB and continue as the optic radiation (Fig. 34.7), which end

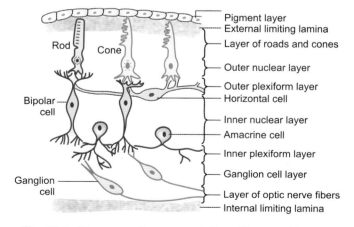

Fig. 34.6: Diagrammatic representation of layers of human retina

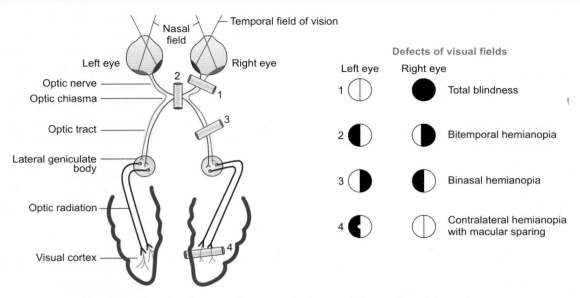

Fig. 34.7: Visual pathway and common lesions at differen sites of the pathway

in the area-17 of occipital cortex. Area-17 is also known as the primary visual area. Specific portions of each retina are mapped in specific areas of visual cortex. Greatest representations are for the fibers coming from foveal and macular regions, while the peripheral areas of retina have the least representation. Hence lesions of visual cortex largely affect peripheral fields of vision while macula is not much affected ("macular sparing"). Clusters of small cells, rich in cytochrome oxidase enzyme, called "blobs" are present in the primary visual cortex which are concerned with color vision.

Effects of Lesions in Visual Pathway

Defects in visual fields due to lesions in the visual pathway vary according to the area affected and are clearly summarized in (Fig. 34.7).

Loss of vision in one half of visual field is known as hemianopia. If vision is lost in the same half of visual fields of both eyes, it is of homonymous type. If vision is lost in opposite half of visual fields of both eyes, it is of heteronymous type. If vision is lost in one quadrant of visual field, it is called quadrantinopia.

■ DARK ADAPTATION

When a person shifts from a brightly illuminated environment to a dark room, at first his vision becomes

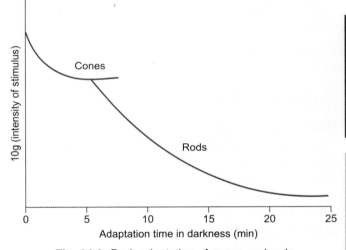

Fig. 34.8: Dark adaptation of cones and rods

poor. Gradually his vision in dim light improves after some time. This phenomenon is called dark adaptation. Normally about 20 minutes are taken for dark adaptation. The reason for this phenomenon can be explained as follows. When bright light falls in the retina, it bleaches the visual purple of rods, i.e., formation of all trans retinine. When the person suddenly moves to a dark room, re-synthesis of rhodopsin takes place. But it takes some time so that the person cannot initially see clearly in the dark room. Adaptation for rods takes more time than for cones (Fig. 34.8).

PUPILLARY LIGHT REFLEX

When light is flashed into one eye, the pupil of that eye constricts. This is called direct pupillary light reflex. If you observe the pupil of opposite eye, you can see that it also constricts. This is called consensual or indirect pupillary light reflex. The pathway of these reflexes is shown in (Fig. 34.9). Receptors are rods and cones. The afferent fibers travel along with the visual fibers via the optic nerve, cross in the optic chiasma and travel in the optic tract. But instead of reaching LGB, these fibers synapse in the pretectal nucleus and superior colliculus of midbrain. Fibers arising from this region reach the Edinger Westphal nucleus (parasympathetic nucleus of oculomotor nerve) of same side and opposite side. Then they synapse in the ciliary ganglion. Postganglionic fibers from ciliary ganglion innervate the circular muscles of iris via the short ciliary nerves, which on contraction cause constriction of pupil.

The crossing of some fibers to the opposite sided Edinger Westphal nucleus in midbrain explains the cause for consensual reflex.

Accommodation Reflex

When a person observing a distant object suddenly changes his focus onto a very near object, certain adjustments take place in his eyes so that the near object is clearly seen. This is termed as accommodation. There are three possible ways by which accommodation occurs in the eyes. They are : pupillary constriction, convergence of eyeballs and increase in anterior curvature of the lens. Pathway for accommodation is as follows. Receptors are rods and cones. The afferent fibers for accommodation travel in the visual pathway and reach the primary visual cortex in area 17. Some association fibers from area 17 project to the frontal eye field (area 8). Fibres from here pass down via the pyramidal tracts to reach the oculomotor nerve nucleus in the midbrain. From here, the fibers pass through the ciliary ganglion and innervate the circular muscles of iris. Some fibers from the third nerve nucleus pass to the medial rectus muscle of eyeball, which is concerned with convergence of eyeball.

COLOR VISION OR CHROMATIC VISION

There are three types of cones in the retina with three different pigments in the outer segment. They contain pigments for the three primary colors (red, green and blue). The pigments can be grouped as erythrolabe (red sensitive pigment), chlorolabe (green sensitive pigment) and cyanolabe (blue sensitive pigment).

According to Young-Helmholtz theory of trichromatic vision, the three types of cones, each containing a different photosensitive pigment, show maximum sensitivity to one type of primary color. e.g: blue light stimulates those cones with blue-sensitive pigment maximally and other cones feebly. Sensation of white is perceived when all the three types of cones are equally stimulated. This is evidenced by demonstrating the different absorption spectra of three types of cones (Fig. 34.10).

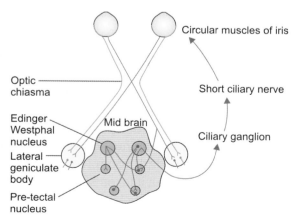

Fig. 34.9: Pupillary reflex pathway

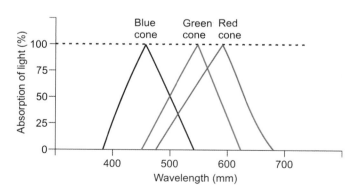

Fig. 34.10: Absorption spectra of the three-color receptive cones of human retina

APPLIED PHYSIOLOGY

Color Blindness

The genes for red and green sensitive pigments of cones are located on the X-chromosome and the abnormalities of these genes are inherited as sex-linked recessive. About 8% of male and 0.4% of female populations are color blind. The gene for blue sensitive cone is located on chromosome-7 (not on the X-chromosome) and so color blindness to blue is rare. Colour blindness may vary from total color blindness to weakness of any one of the primary colors.

Color blindness can be broadly classified into three:

Monochromats: Here there is total color blindness. The vision is said to be "black and white".

Dichromats: Blindness is present for a particular color. E.g: Protanopia (red color blindness), deuteranopia (green color blindness).

Trichromats: Here, there is only weakness in color perception, e.g. pratanomaly (red color weakness), deuteranomaly (green color weakness).

Color blindness can be tested by showing certain color charts called "Ishiharas Chart" to the patient.

AUDITION (HEARING)

Ear is the sense organ for hearing and equilibrium. Ear converts the sound waves into action potentials that are carried by the auditory nerve to the cortex.

Sound waves are mechanical disturbances produced by movement of molecules through an elastic medium, usually air or water. Sound waves travel at a speed of about 344m/sec at the seal level. The properties of sound can be explained by means of loudness, pitch and timbre of sound. The "loudness" of sound depends on amplitude of sound waves where-as "pitch" depends on its frequency. "Quality" or "timbre" of sound is the property by which different musical sounds having equal frequency and amplitude can be identified. Human ear can differentiate different sound frequencies between 20-20,000Hz. The average pitch of male voice is 120 Hz and that of female voice is 250 Hz. Measurement of sound is done using the decibel scale (dB). Zero decibel on this scale means just the threshold for hearing and not "absence of sound". Decibel scale for different common sounds is given in Table 34.2.

Table 34.2: Decibel scale for some common sounds

Decibel (dB)	Type of sound
160	Jet plane
120	Discomfort
80	Heavy traffic
60	Normal conversation
30	Whisper
0	Threshold for hearing

FUNCTIONAL ANATOMY OF HUMAN EAR

An overall picture of human ear is shown in Fig. 34.11. Human ear consists of outer ear, middle ear and inner ear. Only the functionally significant structural features are discussed here.

The Outer Ear

The outer ear consists of:
1. A cartilaginous "Pinna" that helps to collect and localize the sound waves;
2. An "external auditory canal" that transmits sound waves to the middle ear. Wax secreting ceruminous glands and sebaceous glands line the canal, and its inner end is sealed by the "tympanic membrane" or "ear drum". Tympanic membrane vibrates when sound waves strike it.

The Middle Ear

The middle ear is an air filled cavity in the mastoid region of temporal bone. It is connected to the pharynx by the Eustachian tube. The Eustachian tube is usually closed. During swallowing, chewing and yawning it opens, allowing equalization of pressures on either side of the tympanic membrane. Between the tympanic membrane and the inner ear is a chain of three small bones called ossicles (Fig. 34.11). They are malleus, incus and stapes. Back of the tympanic membrane is connected to the handle of malleus ("hammer"), so that the movement of tympanic membrane causes a rocking movement of malleus also. The incus ("anvil") connects the head of malleus to the stapes ("stirrup"). The foot plate of stapes is attached to the walls of oval window of the inner ear, by an annular ligament. Thus these ossicles function as a "lever system" to transmit the vibrations of tympanic membrane to the oval window. They increase the force of movement of sound waves about 1.3 times. The surface area of tympanic

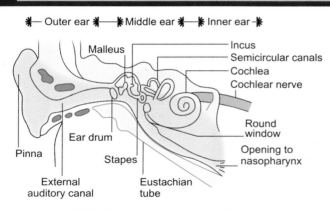

Fig. 34.11: Functional anatomy of human ear

membrane is about 55 mm² when compared to that of stapes, which is about 3.2 mm². This means that the waves of vibrations coming from the tympanic membrane are acting on a smaller surface area which will increase their total force by about 17 times. Both the above described phenomena help to increase the total force exerted by sound waves on the fluid of cochlea in the inner ear by about 22 times ! Sound transmission through the middle ear is also affected by the actions of two small muscles – tensor tympani and stapedius. These muscles contract in response to loud sound. Tensor tympani is attached to the head of malleus. When contracted, it pulls the malleus medially which will decrease the vibrations of tympanic membrane. Stapedius is attached to the stapes. When contracted, it pulls stapes out of the oval window. Thus contractions of both these muscle protect the inner ear from loud sounds ("tympanic reflex").

The Inner Ear

The sensory receptors for sound are located in the inner ear. The inner ear consists of a bony labyrinth and a membranous labyrinth. The bony labyrinth is formed by the petrous part of temporal bone, which encloses the membranous labyrinth. The membranous labyrinth consists of:
1. The vestibule, with utricle and saccule
2. The semicircular canals
3. The cochlea

Utricle, saccule and the semicircular canals are concerned with equilibrium. The auditory receptors or the hair cells are located in the cochlea, which is described below (Fig. 34.12).

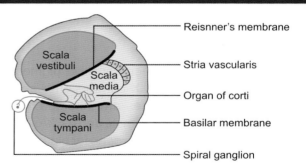

Fig. 34.12: Cross-section of cochlea

The cochlea (means "snail shell") is a fluid filled coiled tube which is 35 mm long and makes 2¾ turns. Throughout its length, it is partitioned into three compartments, by two membranes – Reisnner's membrane and basilar membrane (Fig. 34.12). The three compartments are:
1. Scala vestibule (upper compartment into which the oval window opens)
2. Scala media (middle compartment in which the organ of corti is located)
3. Scala tympani (lower compartment, which ends at the round window).

Scala vestibuli and scala tympani comminicate with each other at the apex of cochlea and are filled with perilymph.

Scala media is isolated from other compartments and consists of the fluid – endolymph. The endolymph is secreted by the "stria vascularis", located on the walls of scala media. Endolymph has a high concentration of K^+ and a low concentration of Na^+, whereas perilymph is high in Na^+ and low in K^+.

■ ORGAN OF CORTI (FIG. 34.13)

Organ of corti is located on the upper surface of the basilar membrane and extends from apex to the base of cochlea. It contains the hair cells, which are the primary receptors for audition. There is a single row about 3500 inner hair cells and three or four rows of outer hair cells (about 20,000). The inner and outer hair cells are separated by the Rods of corti or the pillar cells, that are arranged like an arch with a tunnel in between them called the tunnel of corti (Fig. 34.13). Tunnel of corti contains perilymph. Both sets of hairs are supported and anchored to the basilar membrane. The inner hair cells are supported by phalangeal cells

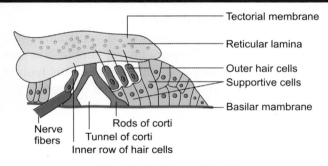

Fig. 34.13: Organ of corti

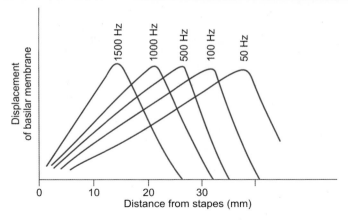

Fig. 34.14: Traveling waves generated by vibration of stapes at different frequencies

and the outer hair cells are supported by Deiter's cells. The tips of the hairs of the outer hair cells, and not the inner hair cells, are embedded in a thin elastic membrane called tectorial membrane that covers them (Fig. 34.13). The hair cells have tiny cilia on their apex called stereocilia. They are so arranged that the height of stereocilia increases progressively towards the outer edge (i.e. towards stria vascularis), giving a "sloping" appearance. Longest of these cilia is termed as kinocilium. Certain fine processes called "tip links" bind the tip of each stereocilium to the side of its higher neighbor. They are actually mechanically sensitive cation channels. The tips of hair cells are bathed in the endolymph of scala media. The endolymph cannot reach the base of the hair cells due to tight junctions between the hair cells and adjacent supporting cells. But the basilar membrane is permeable to perilymph so that it enters into the tunnel of corti and surrounds the base of the hair cells. In short, the processes of hair cells are embedded in endolymph and the bases of hair cells are embedded in perilymph. The fibers of auditory division of VIII cranial nerve or the cochlear nerve have their cell bodies in the spiral ganglion (Fig. 34.13). Majority of the nerve fibers synapse with the inner hair cells. The axons of auditory division of the VIII cranial nerve exit the inner ear via the internal auditory meatus and terminate on dorsal and ventral cochlear nerve nuclei in the medulla.

Integrated Function of Organ of Corti

When a sound wave is transmitted to the oval window by the ossicular chain, a pressure wave travels up the scala vestibule and down the scala tympani (Fig. 34.14). This sets up movements in the fluids of inner ear, producing an undulating distortion in the basilar membrane. The membrane deformation takes the form of a "traveling wave", which has its maximal amplitude at a position along the membrane corresponding to the particular frequency of the sound wave. Vibrations of the basilar membrane excite the cilia of outer hair cells, which are now subjected to a lateral shearing force. This mechanically opens the cation channels ("tip links") in the stereocilia, leading to either depolarization or hyperpolarization. If the stereocilia are pushed towards the kinocilium, the cell is depolarized. If the stereocilia are bent in the opposite direction, the cell is hyperpolarized. In the former case, the opening up of cation channels with influx of K^+ and Ca^{++} from endolymph into the cell, leads to depolarization. K^+ that enters the hair cells are recycled. K^+ passes through the supporting cells, reaches the stria vascularis, from where it is secreted back to the endolymph. Cells in the stria vascularis are rich in Na^+K ATPase pump which accounts for the positive electrical potential of scala media (about ^+80mV) when compared to that of scala vestibule and scala tympani (^-70mV). This positive potential is called endocochlear potential.

◼ AUDITORY PATHWAY (FIG. 34.15)

Receptors are the hair cells of organ of corti, which are depolarized on receival of sound waves. In response to this, an action potential is generated in the nerve fibers arising from the spiral ganglion (first order neurons). The impulses carried via the cochlear nerve enter the medulla and end in dorsal and ventral cochlear nuclei. Second order neurons arise from here,

Part 2

Physiology

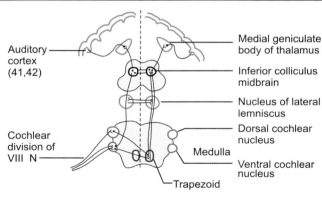

Fig. 34.15: Auditory pathway

ascend upwards and relay in the medial geniculate body (MGB) of thalamus. The third order neurons from MGB ascend and terminate in the primary auditory area (area 41, 42), situated in the superior temporal gyrus in cerebral cortex. In all the above relay stations, extensive crossing of fibers from both sides occur. Also both ears have equal representations on either sides of auditory cortex. Therefore, you will not get a unilateral hearing loss in lesions of auditory cortex of one side. Fibres also go to the auditory association area (area-22) for final interpretation of auditory sensation. Area 22 is also known as Wernicke's area or sensory speech area. Auditory pathway is also interconnected with the visual pathway.

PITCH DISCRIMINATION

Discrimination of sounds with different frequencies or pitches can be explained by the 'Place theory' or the 'travelling wave theory'. According to this theory, the movement of basilar membrane is like that of a travelling wave. Low frequency sounds cause maximal displacement of the membrane near the apex of cochlea, whereas high frequency sounds produce their maximal effect near the base of cochlea (Fig. 34.14).

Intensity Discrimination

The frequency of firing of action potentials in the auditory nerve fibers is directly proportional to the intensity of sound waves.

APPLIED PHYSIOLOGY

Deafness or Hearing Loss

There are generally two types of deafness – conduction deafness and sensory neural deafness. Conduction deafness is due to defective transmission of sound, e.g: damage of tympanic membrane, wax, damage to ossicular chain, etc. Sensory neural deafness is due to damage of cochlea or auditory nerve. This type of deafness is seen in soldiers, industrial workers, etc.

Tests for Hearing

Rinne's Test

A vibrating tuning fork is placed on the mastoid process of the patient. Instruct him to tell the clinician when he stops hearing. Immediately bring the tuning fork close to his ear and ask whether he can hear the vibrations now. Normally the person can. This is because air conduction of sound is better than bone conduction. This is called Rinne's positive (which is normal). In conduction deafness, bone conduction is better than air conduction (Rinne's negative). In neural deafness, both air and bone conductions are decreased, but still air conduction is better than bone conduction (reduced Rinne's positive).

Weber's Test

A vibrating tuning fork is kept over the vertex or forehead of the patient. A normal person can hear the vibrations in both ears equally. In conduction deafness, it is better heard in the affected ear or "Weber's is lateralized to the affected ear". In neural deafness, it is better heard in the normal ear or "Weber's is lateralized to the normal ear".

Absolute Bone Conduction (ABC) Test

A vibrating tuning fork is kept over the mastoid process of the patient. Instruct him to tell when he stops hearing. Immediately transfer the tuning fork onto the physician's mastoid process and check whether he can receive any sound. Here, it is assumed that the hearing power of the physician is normal.

GUSTATION (TASTE)

The receptors for taste sensation are located in the taste buds. There are about 10,000 taste buds. Each taste bud has a minor pore, opening into the oral cavity, through which saliva comes in contact with it. Numerous nerve fibers arise from the taste buds. Each taste bud has many microvilli on its apical end, which

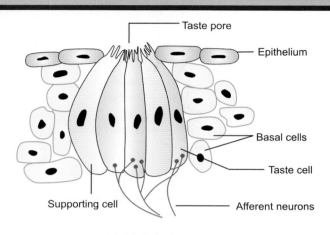

Fig. 34.16: Taste bud

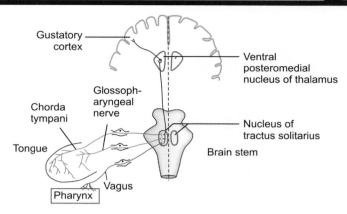

Fig. 34.17: Taste pathway

greatly increase their surface area (Fig. 34.16). At their basal ends, the nerve fibers form synapses with facial and glossopharyngeal nerves.

Location of Taste Buds

Taste buds are present in the following regions:
1. Mucous membrane of tongue
2. Epiglottis
3. Pharynx
4. Palate

Taste buds are present in the walls of circumvallate and fungiform papillae of tongue. The number of taste buds decreases as the age advances.

Types of Taste

In humans, there are about five taste modalities – sweet, sour, bitter, salt and umami. The fifth sense, "umami" was recently added to the four classic tastes. This taste is triggered by monosodium glutamate ("ajinomoto"), used extensively in Chinese cooking.

Previously, it was thought that there were special areas on the surface of tongue for the perception of each taste. But it is now clear that all are sensed from all parts of tongue and adjacent structures.

Taste Pathway (Fig. 34.17)

Receptors are the taste buds. Taste sensation can be evoked only after a substance has been dissolved in the saliva. Taste buds respond to it by production of an action potential that travels along the taste pathway to reach the somatosensory cortex.

The afferent fibers from the taste buds of anterior two-thirds of the tongue travel initially in the lingual nerve (branch of trigeminal nerve), but soon leave it and enter the chorda tympani (branch of facial nerve). Taste from posterior one-thirds of the tongue is carried via the glossopharyngeal nerve, and those from epiglottis, palate and pharynx, via the vagus nerve.

Ultimately, these three groups of nerve fibers unite in the gustatory portion of the nucleus of Tractus solitarius (NTS) in the medulla (Fig. 34.17). The second order neurons arising from here ascend in the ipsilateral medial lemniscus to the specific sensory nucleus of thalamus. From here, the third order neurons are projected to the face area of the somatosensory cortex, in the ipsilateral post central gyrus, and to the insula.

Applied Physiology

1. *Ageusia:* It is the absence of taste sensation, due to damage of lingual nerve or glossopharyngeal nerve. The damage can be triggered by chronic use of tobacco and certain drugs.
2. *Dysgeusia:* It is the unpleasant perception of taste.

▮ OLFACTION (SMELL)

In humans, this sense is poorly developed when compared to the animals. Also it is largely a subjective phenomenon.

The receptors for olfaction are the olfactory sensory neurons, located in a specialized portion of nasal mucosa called olfactory epithelium (Fig. 34.18). The olfactory epithelium in humans contains about 10-20 million bipolar olfactory sensory neurons, interspersed with sustentacular cells (supporting cells).

Part 2

Physiology

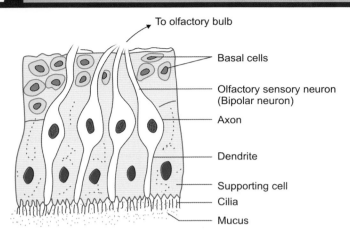

Fig. 34.18: Olfactory epithelium

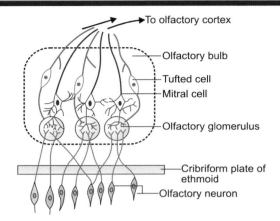

Fig. 34.19: Olfactory bulb and the neutonal circuits

Olfactory Pathway (Fig. 34.19)

Olfactory receptor cells are the bipolar olfactory neurons. The deep axons of these neurons pierce the cribriform plate of ethmoid bone on either side to reach the olfactory bulb (Fig. 34.19).

In the olfactory bulb, there are four types of cells – mitral cells, tufted cells, periglomerular cells and granule cells. The axons of olfactory neurons terminate by forming synapses with mitral as well as tufted cells to constitute the olfactory glomeruli. Periglomerular cells are inhibitory neurons that connect one olfactory

glomerulus to another. Granule cells make reciprocal synapses with dendrites of mitral and tufted cells. Periglomerular and granule cells control the transmission of impulses in mitral and tufted cells.

The second order neurons arising from mitral and tufted cells run in the olfactory tract. Posteriorly, the tract divides into medial and lateral olfactory stria. They reach specialized regions of cortex (olfactory cortex) without passing through the thalamus. The olfactory cortex forms a part of the limbic system. Hence olfactory stimuli can evoke emotional effects also.

Part 2

Physiology

35 The Urinary System

The kidneys maintain a stable internal environment by regulating the volume and composition of body fluids as well as by excreting the waste products and excess water. For maintaining homeostasis, kidneys do multiple functions as follows:

1. Excretion of metabolic waste products and chemicals like urea, uric acid, creatinine and many drugs.
2. Regulation of body fluid volume and osmolality by excreting either dilute or concentrated urine.
3. Regulation of concentration of electrolytes and various ions.
4. Regulation of acid-base balance by excreting either excess acid or base.
5. Regulation of arterial blood pressure by adjusting Na^+ and water excretion.

6. Secretion and production of some hormones like erythropoietin, 1,25- dihydroxy cholecalciferol, renin, prostaglandins, etc.
7. Metabolism of various hormones like insulin, glucagon, parathyroid hormone, etc.

FUNCTIONAL ANATOMY OF KIDNEYS

Kidneys are paired organs situated on either side of vertebral column, near the posterior wall of abdomen, behind the peritoneum. Each weighs about 150 gms. Each kidney is bean shaped with an indentation on the medial side called-hilum, through which the renal artery, renal vein, nerves, lymphatics and renal pelvis pass. A longitudinal section of kidney is shown in (Fig. 35.1). Two distinct regions are seen – an outer

Part 2

Physiology

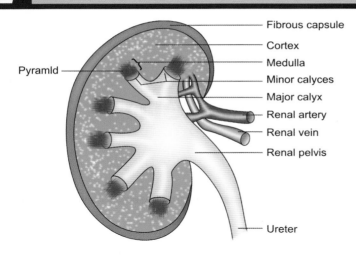

Fig. 35.1: Longitudinal section of human kidney

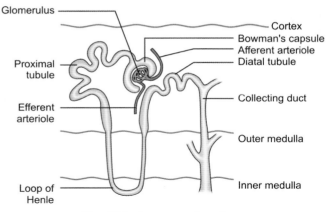

Fig. 35.2: Structure of nephron

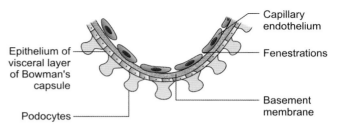

Fig. 35.3: Glomerular filtration barrier

dark colored cortex and an inner light colored medulla. The medulla of human kidney consists of multiple cone shaped masses called pyramids, whose base is directed towards cortex (Fig. 35.1). The apical region of each pyramid forms a renal papilla which fits into a structure called minor calyx. Minor calyces collect urine from each papilla. Several minor calyces unite to form a major calyx. 4 to 5 major calyces unite to form the renal pelvis, which continues downwards as the ureter. The urine formed from the nephron is propelled through the calyces, pelvis and ureters to the urinary bladder by the contraction of smooth muscles present on their walls.

Nephron

Nephron is the structural and functional unit of kidney. There are about 1.2 to 1.5 million nephrons in each kidney, the total length of which comes around 40 miles! Ultra structure of a nephron is shown in (Fig. 35.2). Nephron consists of a renal corpuscle and a renal tubule.

■ RENAL OR MALPIGHIAN CORPUSCLE

The renal corpuscle consists of a tuft of capillaries called glomerulus, surrounded by the Bowman's capsule (Fig. 35.2). Both are present in the cortex. Bowman's capsule consists of an outer parietal and an inner visceral layer, lined by squamous epithelial cells. The parietal layer is continuous with the proximal tubule. The space between visceral and parietal layers

is called Bowman's space. The networks of glomerular capillaries arise from the afferent arteriole and drain into the efferent arteriole (Fig. 35.2). The glomerular capillaries are highly permeable and have a high hydrostatic pressure, which favors filtration of plasma.

The endothelial cells of glomerular capillaries are covered by a basement membrane, which is surrounded by the epithelial cells of visceral layer of Bowman's capsule called podocytes. The endothelium, basement membrane and the foot processes of podocytes form the glomerular filtration membrane or filtration barrier (Fig. 35.3). The endothelium is fenestrated with many pores, through which small solutes and water can easily pass. Also the pores are lined by sialoproteins which provide a negative charge to them. The basement membrane is also porous. It is formed by type-IV collagen, proteoglycan and other proteins, which are negatively charged. The podocytes of visceral layer of Bowman's capsule have long foot – like processes that encircle the outer surface of the capillaries. In between these foot processes, there are filtration sites that are lined by negatively charged glycoproteins. Because of the negative charges lining the endothelium, basement membrane and filtration slits, the plasma proteins are repelled and their filtration is prevented.

All solutes up to 4 nm size can freely pass through the filtering membrane. Solutes with size more than 8 nm are not filtered. Among the charged ions, cations are more easily filtered than neutral ions, than anions. The cellular elements in the blood are not filtered.

APPLIED PHYSIOLOGY

Glomerular Nephritis

This is an inflammatory condition of kidney characterized by increased permeability of glomerular filtration barrier and dissipation of the negative charges. This causes filtration of proteins (especially albumin) and RBCs, resulting in proteinuria and hematuria respectively.

RENAL TUBULE

The renal tubule consists of several segments – the proximal tubule, the loop of Henle, the distal tubule and the collecting ducts (Fig. 35.2).

The proximal tubule has an initial coiled region called proximal convoluted tubule (PCT) and a straight portion called proximal straight tubule, that descends towards the medulla.

The loop of Henle consists of a thin descending limb and an ascending limb which has a thin and thick part (Fig. 35.2).

The distal tubule extends from the macula densa (initial portion of distal tubule) to the cortical part of collecting duct. The cortical collecting duct extends into the medulla and becomes the medullary collecting duct. The functions of each part of the nephron differ and are discussed later.

CORTICAL AND JUXTAMEDULLARY NEPHRONS

Not all the nephrons are alike. According to the location and length of the loop of Henle, nephrons can be grouped into cortical and juxtamedullary or medullary nephrons. The cortical nephrons, which constitute about 70% of total nephrons, possess short loops of Henle that are mainly present in the cortex. Only a

small portion extends into the medullary region (Fig. 35.4). 30% of the total nephrons are juxtamedullary nephrons that have long loops of Henle which extend deeply into the medullary region (Fig. 35.4). Another difference between the two groups is regarding the capillary network present around them. Cortical nephrons are surrounded by a rich network of peritubular capillaries (Fig. 35.4). The capillaries surrounding juxtamedullary nephrons are called "vasa recta" that also possess long loops which lie close to the loop of Henle (Fig. 35.4).

BLOOD SUPPLY OF KIDNEY

About 25% of the total cardiac output flows through kidneys, i.e., 1200 ml/minute at resting condition. However, kidney is the organ in our body possessing highest blood supply, i.e. about 400 ml/min/100 gms. A comparison of blood supply per 100 gms of some important organs is given in the Table 35.1.

As shown in the Figure 35.5, the renal artery (branch of abdominal aorta) branches progressively to form the interlobar arteries, the arcuate arteries, the cortical arteries or the interlobular arteries and the afferent arterioles. Each afferent arteriole breaks up into the glomerular capillary network through which

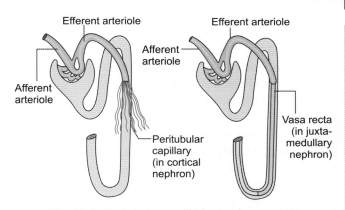

Fig. 35.4: Peritubular capillaries and vasa recta

Table 35.1: Blood Flow through different organs	
Organ	*Blood supply (per 100 gm of the tissue)*
Liver	20 ml / 100 gm/min.
Heart	80 ml/ 100 gm/min.
Skeletal muscle	27 ml/ 100 gm/min.
Kidney	400 ml / 100 gm/min

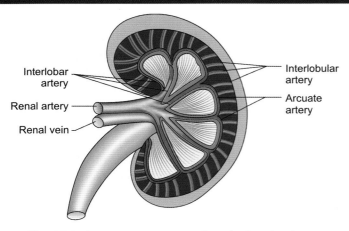

Fig. 35.5: Schematic representation of microcirculation

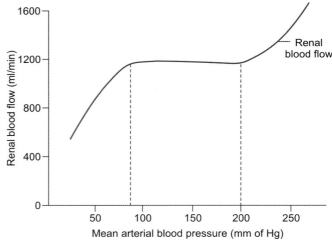

Fig. 35.6: Autoregulation of renal blood flow

20% of the plasma is filtered. The glomerular capillaries join together to form the efferent arteriole which again divides into a second capillary network. The second capillary networks around the cortical nephrons are called peritubular capillaries and those surrounding juxtamedullary nephrons are called vasa recta. Both peritubular capillaries and vasa recta drain into the vessels of venous system running parallel to the arterial vessels and branch progressively to form the interlobular or cortical vein, the arcuate vein, the interlobar vein and finally the renal vein. The renal circulation possesses the feature of double capillary network, which is also known as portal circulation. There is a regional difference in the blood flow. Majority (about 90%) of the blood flows to the cortical region and only about 10% of the blood flows to the medulla. The increased blood flow in the cortex favors adequate filtration of plasma. When compared with the systemic capillaries, the pressure in the glomerular capillaries is very high (about 45 mm of Hg). This high hydrostatic pressure causes rapid fluid filtration. Another important feature is that kidney can regulate its own blood flow in spite of changes in the systemic arterial blood pressure. This property is known as auto-regulation, which is observed between mean blood pressure of 80 to 200 mm of Hg (Fig. 35.6). This effect does not depend on the renal innervations since this can be seen even after cutting the renal nerves and in an isolated preparation of kidney perfused with isotonic saline. The myogenic theory put forward by Bayliss explains the reason for this phenomenon. It may be due to a direct contractile response of the smooth muscle of afferent arteriole to stretch. Experimentally

Bayliss proved that administration of drugs like procaine, which paralyzes the vascular smooth muscles, abolished this response. Also tubulo-glomerular feedback mechanism is thought to be responsible for auto-regulation of renal blood flow (explained later).

Innervation of Kidney

Kidneys have a rich innervations from the sympathetic nerve fibers ($T_{10,11,12}$ and L_1). They regulate the renal blood flow, glomerular filtration rate and reabsorption of salt and water. The renin secreting juxtaglomerular cells are also innervated by sympathetic nerves. Thus sympathetic stimulation increases the renin secretion also.

■ JUXTAGLOMERULAR APPARATUS (JGA)

Juxtaglomerular apparatus is a specialized structure composed of vascular and tubular components. The initial part of distal tubule up on approaching its own glomerulus, comes in contact with the afferent and efferent arterioles for a short distance. The JGA is located at this region of contact (Fig. 35.7).

The structures of JGA include:
1. Renin secreting granular cells or JG cells of the afferent arteriole.
2. Macula densa of the thick ascending limb of loop of Henle.
3. Extra glomerular mesangial cells or Lacis cells.

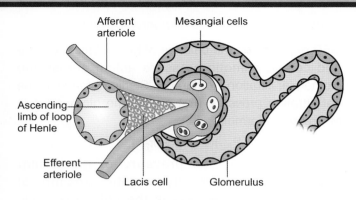

Fig. 35.7: Juxtaglomerular apparatus (JGA)

Granular Cells or Juxtaglomerular Cells (JG cells)

The smooth muscle cells of afferent arteriole at this region modify to become the granular cells that store and secrete granules containing a hormone called renin. Renin is involved in the formation of angiotensin II and ultimately in the secretion of aldosterone as follows:

Renin converts an inactive plasma globulin called angiotensinogen into angiotensin I. The angiotensin converting enzyme (ACE) present in the lungs converts angiotensin I into angiotensin II.

$$\text{Angiotensinogen} \xrightarrow{\text{Renin}} \text{Angiotensin I}$$
$$\xrightarrow{\text{ACE}} \text{Angiotensin II}$$

Angiotensin II is a potent vasoconstrictor which constricts the efferent arteriole thereby increasing the glomerular hydrostatic pressure and hence the glomerular filtration rate. Whenever there is a reduction in GFR, renin secretion will be stimulated and thus it helps in the regulation of GFR. Angiotensin II has other actions also. It stimulates adrenal cortex to secrete aldosterone. Aldosterone increases Na^+ and water reabsorption from the distal tubules. This is known as renin-angiotensin-aldosterone mechanism. This mechanism has an important role in the regulation of blood pressure (explained in "cardiovascular system"). Also, angiotensin II stimulates secretion of antidiuretic hormone (ADH) of posterior pituitary, which helps in the reabsorption of water from the collecting duct of kidney.

Macula Densa

Macula densa are the modified epithelial cells of the initial portion of distal tubule that lie in close contact with afferent and efferent arterioles (Fig. 35.7). These cells monitor the composition of the fluid especially Na^+, in the tubular lumen at this point and are involved in the "tubuloglomerular feedback" mechanism. Tubuloglomerular feedback mechanism is for the regulation of renal blood flow and GFR. The mechanism works as follows:

Whenever the glomerular filtration rate (GFR) or the renal blood flow goes high, the Na^+ load coming to the macula densa also increases. In response to the high Na^+ content, macula densa cells stimulate constriction of afferent arteriole. This will gradually bring down the GFR and renal blood flow back to the normal level.

Extraglomerular Mesangial Cells or Lacis Cells

They are modified smooth muscle cells seen in between the afferent and efferent arterioles (Fig. 35.7). By their contraction and relaxation, they can regulate the glomerular filtration.

▮ FORMATION OF URINE

Formation of urine involves three basic processes
  Glomerular filtration
  Tubular reabsorption
  Tubular secretion

About 1800 L/day (1.2 L/min) of blood flows through both kidneys. Of this, about 180 L/day is filtered through the glomeruli. But only less than 1% of the filtered water and variable amounts of many of the solutes are excreted in the urine. This is made possible by the processes of tubular reabsorption and secretion. As a result, the renal tubules precisely control the volume, composition and pH of the body fluids.

The urinary excretion of any solute = The GFR of the solute − Tubular reabsorption rate + Tubular secretion rate.

For a freely filtered substance which is neither reabsorbed nor secreted, the excretory rate will be same as that of GFR (Fig. 35.8), e.g.Inulin.

For a solute that is freely filtered and partially reabsorbed, excretory rate will be less than the GFR (Fig. 35.8), e.g. Electrolytes.

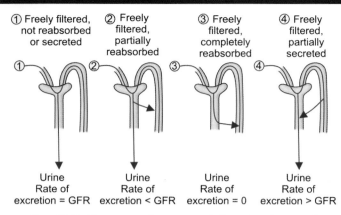

① Freely filtered, not reabsorbed or secreted
② Freely filtered, partially reabsorbed
③ Freely filtered, completely reabsorbed
④ Freely filtered, partially secreted

Urine Rate of excretion = GFR
Urine Rate of excretion < GFR
Urine Rate of excretion = 0
Urine Rate of excretion > GFR

Fig. 35.8: Diagrammatic representation showing renal excretion of different solutes

For a solute that is freely filtered and completely reabsorbed, the excretory rate will be zero, i.e. the substance is not usually excreted in urine (Fig. 35.8), e.g. Glucose, amino acids, lactate, etc.

For a solute that is freely filtered, but also secreted from peritubular capillaries into the tubules, the excretory rate will be more than the GFR (Fig. 35.8), e.g. organic acids and bases. In the following portions, these processes of urine formation are discussed in detail.

Glomerular Filtration

Glomerular filtration means the ultrafiltration of plasma by the glomerulus. The ultrafiltrate contains no blood cells or plasma proteins. The total quantity of ultrafiltrate formed in all the nephrons of both kidneys per minute is termed as glomerular filtration rate or GFR. It is the sum of the filtration rate of individual nephrons and hence the value gives an index of kidney function. Normal GFR is about 125 ml/min or 180 L/day. The number of functioning nephrons decreases as the age advances and hence the GFR decreases in old age.

Factors Affecting GFR

GFR is determined by the same factors that affect fluid movement across capillaries in general. In other words, ultrafiltration is driven by the 'Starling's forces' across the filtration barrier of glomerulus. Also the size as well as the permeability of glomerular filtration membrane and the changes in renal blood flow affects the GFR.

Starling's Forces Across Glomerular Capillaries

GFR depends on the balance of hydrostatic pressure and colloid osmotic pressure acting across the glomerular filtration membrane ('Starling's forces'). The net filtration pressure (P_F) across the glomerulus is given by the formula:

$$P_F = (P_{GC} - P_{BC}) - (\pi_{GC} - \pi_{BC})$$

P_{GC} is hydrostatic pressure of glomerular capillaries, P_{BC} is the hydrostatic pressure of Bowmann's capsule, π_{GC} is the colloid osmotic pressure in the glomerular capillaries and p_{BC} is the colloid osmotic pressure in the Bowmann's capsule. By distributing the normal values,

$$P_F = (45-10) - (25-0)$$
$$= 10 \text{ mm of Hg}$$

GFR = K_f x Net filtration pressure; where K_f is the glomerular ultrafiltration coefficient and is equal to 12.5

$$= 12.5 \times 10 = 125 \text{ ml/min.}$$

Hydrostatic pressure of glomerular capillaries (P_{GC}) varies according to changes in the vascular resistance of afferent and efferent arterioles. Constriction of afferent arteriole decreases the hydrostatic pressure, which in turn decreases the GFR and vice-versa. Angiotensin II causes efferent arteriolar constriction, thereby increasing the hydrostatic pressure and hence the GFR (explained before). Alterations of systemic arterial blood pressure above and below the values of auto-regulation (80 to 200 mm of Hg) and can increase or decrease the GFR respectively.

Hydrostatic pressure in the Bowman's capsule opposes glomerular filtration. In case of obstruction of urinary tract by stones and in inflammatory diseases of kidney, the Bowman's capsular hydrostatic pressure is increased, causing a marked decline in GFR.

The colloid osmotic pressure in the glomerular capillaries depends on the concentration of plasma proteins, and hence the plasma colloid osmotic pressure. In conditions causing dehydration like vomiting, and diarrhea, there is a relative increase in the plasma colloid osmotic pressure, leading to reduction in the GFR. On the other hand, hypoproteinemia can decrease the glomerular capillary osmotic pressure, causing an increase in the GFR.

As normally no plasma proteins are filtered, the Bowman's capsular colloid osmotic pressure is zero and hence neglected.

The glomerular capillaries are highly permeable and the factors affecting permeability of different solutes across the filtration barrier have been discussed earlier.

The auto-regulatory mechanism of renal blood flow works between a change in the mean blood pressure between 80-200 mm of Hg; above and below which there is a sharp decrease and increase in the renal blood flow and GFR respectively.

Filtration Fraction

Filtration fraction is the ratio of GFR to the renal plasma flow. Out of the 1.2 L of blood flowing through kidneys per minute, 700 ml is constituted by the plasma.

$$\text{Filtration fraction} = \frac{\text{GFR}}{\text{RPF}} = \frac{125}{700}$$

Normal range is 0.16 to 0.2. In other words, nearly 16 to 20% of plasma coming through afferent arteriole is filtered in the glomerulus. The remaining 80 to 84% of plasma flows to the efferent arteriole and returns back into the systemic circulation.

▌RENAL CLEARANCE AND MEASUREMENT OF GFR

To measure the excretory function of kidneys, several tests based on the renal clearance concept are used. These tests measure the GFR, tubular reabsorption as well as secretion of various substances and the renal blood flow. All these tests are based on the "Fick's principle". The renal clearance of a substance is defined as the volume of plasma from which that substance is completely removed in unit time.

The clearance, $C_x = U_x V$,
where 'Cx' is the clearance of the substance 'X'
'U_x' is the urinary concentration of 'X'
'P_x' is the plasma concentration of 'X'
'V' is the volume of urine

If a substance is freely filtered by the glomerulus and neither secreted nor reabsorbed by the tubules, the renal clearance value will be same as the GFR. The ideal substance to measure GFR is inulin, which is a fructose polymer. Inulin is preferred because of the following reasons:

1. It is freely filtered by the glomerulus
2. It is not reabsorbed or secreted by renal tubules
3. It is not synthesized, metabolized or stored in the kidney
4. It is a nontoxic substance, soluble in plasma
5. Its concentration in plasma and urine can be easily measured.

As inulin is freely filtered by glomerulus, the amount of inulin excreted in urine per minute will be equal to the amount filtered at glomerulus.
i.e., $U_I V = GFR \times P_I$, where U_I = Urine concentration of inulin
P_I = Plasma concentration of inulin

$$GFR = \frac{U_I V}{P_I} = \text{Clearance of inulin}$$

A loading dose of inulin is administered intravenously followed by a sustained infusion to keep the arterial plasma concentration constant. After inulin has equilibrated with the body fluids, an accurately timed sample of urine as well as plasma is obtained. From this, the clearance of inulin and thus the GFR is calculated. As inulin has to be administered intravenously, its clinical use is limited. Instead, the endogenous creatinine clearance is used clinically to estimate GFR. Creatinine is an end product of muscle metabolism formed in our body and so there is no need for intravenous infusion. Creatinine is completely cleared from body fluids and is easy to measure. But a small amount of creatinine is secreted by the renal tubules and therefore the amount of creatinine in urine will be more than that filtered. The GFR value measured by creatinine clearance will be slightly more than normal, i.e. about 140 ml/min.

▌REABSORPTION AND SECRETION OF WATER AND SOLUTES

Reabsorption and secretion of water and solutes by the renal tubules takes place mainly via two routes—transcellular and paracellular routes. Transcellular route means transport of substances across the cell (Fig. 35.9) e.g: water reabsorption. Paracellular route means transport of substances between the cells through lateral intercellular spaces (Fig. 35.9). Water reabsorption across the tubules occurs passively. But

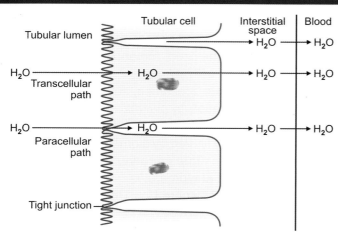

Fig. 35.9: Reabsorption of filtered water by transcellular and paracellular routes

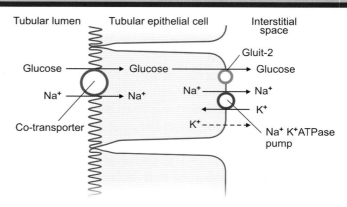

Fig. 35.10: Sodium-glucose symport

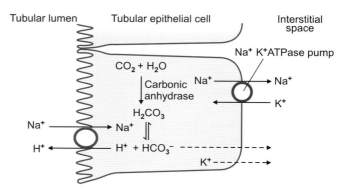

Fig. 35.11: Sodium-hydrogen antiport

transport of solutes can be either passive or active using energy.

TRANSPORT MAXIMUM

The maximum rate at which a solute can be transported across the tubules by active transport mechanism is known as transport maximum. This is because the carrier proteins and enzymes involved in the active processes become depleted when the tubular load of solute exceeds certain limits. In the following sections, reabsorption and secretion of water and solutes along the different parts of nephron are dicussed. Of these, mechanisms of Na^+ reabsorption, K^+ secretion, glucose rebasorption, H^+ secretion and water reabsorption are discussed in detail.

Transport in the Proximal Tubule

Various substances are reabsorbed and secreted in the proximal tubule, which are as follows:
1. Reabsorption of approximately 70% of the filtered water, Na^+, K^+ and Cl^-.
2. 100% reabsorption of glucose, amino acids and peptides.
3. Secretion of organic acids and bases
4. Reabsorption and secretion of urea and uric acid.

Reabsorption of Na^+ in Proximal Tubule

Reabsorption of Na^+ in the PCT takes place by three methods:

Na^+ - Solute symport
Na^+ - H^+ antiport, and
Cl^- driven Na^+ transport.

Na^+ - Solute Symport (Fig. 35.10)

Na^+ uptake into the tubular cell across the apical membrane is coupled with a solute like glucose, amino acids, phosphate and lactate. A common carrier protein is involved in the transport of Na^+ along with other solutes. This transport takes place down the concentration gradient. From the tubular cell, Na^+ is pumped into the circulation by $Na^+K^+ATPase$. This process is an example for secondary active transport as there is indirect involvement of energy.

Na^+ - H^+ Antiport (Fig. 35.11)

Carbonic anhydrase present in the tubular epithelial cells of PCT catalyzes the formation of carbonic acid from CO_2 and water. Carbonic acid then dissociates to form H^+ and HCO_3^-

$$CO_2 + H_2O \xrightarrow{\text{Carbonic anhydrase}}$$
$$H_2CO_3 \rightleftharpoons H^+ + HCO_3^-$$

The H^+ thus formed is secreted from the cell into the luminal fluid. In exchange for this, Na^+ enters into the cell. Thus Na^+ entry is coupled with pumping of H^+ out of the cell via a common carrier protein. This is also an example for secondary active transport.

HCO_3^- formed inside the tubular cell is transported into the circulation, along with Na^+.

Cl- driven Na$^+$ Transport (Fig. 35.12)

Na^+ is primarily reabsorbed along with chloride ion via both transcellular and paracellular pathways.

Glomerulotubular Feedback (Load dependent Na$^+$ Reabsorption)

The reabsorption of Na^+ in the PCT is load dependent. In other words, PCT tends to reabsorb a constant fraction (about 70%) of the filtered Na^+ rather than a constant amount. This is known as glomerulotubular feedback. For example, if there is an increase in GFR, reabsorption of solutes is enhanced and hence that of water also is increased. The net percentage of solute reabsorbed is held constant.

■ REABSORPTION OF GLUCOSE IN THE PCT

Under normal conditions, kidneys freely filter glucose at the glomerulus and then reabsorb it completely from the PCT itself so that the renal clearance for glucose is zero. Glucose reabsorption from tubular lumen is coupled with that of Na^+ via a common carrier protein – 'SGLT'(Sodium dependent glucose transporter). From the tubular epithelial cell, glucose is transported into the circulation via another carrier protein called 'Glucose transporter 2' or 'GLUT-2' (Fig. 35.10). The amount of glucose reabsorbed is proportional to the amount filtered at the glomerulus up to the transport maximum. When the amount of filtered glucose exceeds the transport maximum, it starts appearing in urine. Transport maximum for glucose or T_{MG} is about 375 mg/min in males and 300 mg/min in females. Glucose excretion in urine occurs only when the plasma concentration exceeds a certain level called renal threshold.

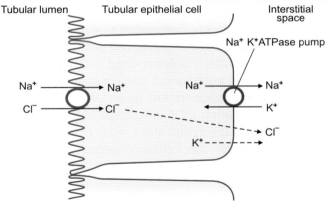

Fig. 35.12: Chloride driven sodium transport

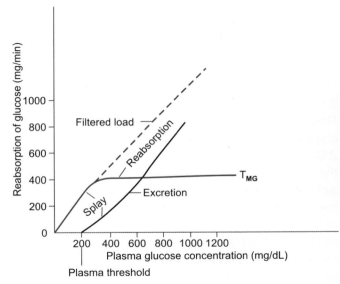

Fig. 35.13: Relation between plasma level and excretion of glucose

$$\text{Renal threshold of glucose} = \frac{T_{MG}}{GFR} = \frac{375}{125} = 300 \ \mu g/dL$$

But as the plasma glucose level exceeds 200 mg/dL, glucose starts appearing in urine. This anomaly between the predicted value and the practical value is called "splay" (Fig. 35.13). This occurs because all the nephrons may not have the same T_{MG} and some glucose may escape reabsorption.

Reabsorption of Water in the PCT

About 70% of the filtered water gets reabsorbed from PCT passively. The driving force that favors water reabsorption is the osmotic gradient created by reabsorption of Na$^+$, glucose, amino acids, K$^+$, Cl- and lactate. PCT is highly permeable to water. So water moves passively into the tubular cell along the osmotic gradient of solutes ("solvent drag"). Certain water channels called aquaporin-1 are present in the PCT through which reabsorption takes place. Thus the tubular fluid bearing PCT is isotonic to plasma. Any change in the reabsorption of solutes in PCT, especially Na$^+$ and glucose, can influence water reabsorption also.

■ TRANSPORT ACROSS THE LOOP OF HENLE

The descending limb of loop of Henle is highly permeable to water but impermeable to solutes. About 12 to 15% reabsorption of filtered water takes place here. The ascending limb is impermeable to water. But solutes like Na$^+$, Cl$^-$, K$^+$, etc. are actively transported across the tubular membrane. The reabsorption of Na$^+$ and K$^+$ takes place via Na$^+$ K$^+$ 2 Cl$^-$ co-transport or symport mechanism (Fig. 35.14). Reabsorption of Na$^+$ is coupled with that of K$^+$ by a common carrier protein. To maintain electrical neutrality 2 Cl$^-$ are also transported along with Na$^+$ and K$^+$. From the tubular cell, Na$^+$ is transported into the circulation actively by Na$^+$K$^+$ATPase pump. Because the ascending limb is impermeable to water, reabsorption of solutes reduces the osmolality of tubular fluid, i.e. the tubular fluid leaving the ascending limb of loop of Henle is hypo-osmotic. On the other hand, reabsorption of solutes increases the osmolality of medullary interstitium around the juxtamedullary nephrons. This helps in reabsorption of water via countercurrent mechanism (explained later).

Transport Across the Distal Tubule

Small amounts of Na$^+$, Cl$^-$ and Ca^{++} reabsorption take place in the distal tubules. Na$^+$ reabsorption occurs via Na$^+$–Cl$^-$ symport mechanism. Comparatively water reabsorption is less in this segment and therefore the tubular fluid remains hyposmotic.

Transport Across the Collecting Duct

The collecting duct consists of two special types of cells – 'P' cells and 'I' cells. 'P' cells (Principal cells) are concerned with reabsorption of Na$^+$ and water as well as secretion of K$^+$ (Fig. 35.15). Na$^+$ reabsorption is stimulated by the hormone aldosterone from adrenal cortex. Na$^+$ enters the tubular cells of collecting duct via epithelial Na$^+$ channels (ENaCs). Aldosterone increases the number of ENaCs, thereby increasing Na$^+$ reabsorption. From the tubular cell, Na$^+$ is pumped actively into the circulation by Na$^+$K$^+$ ATPase pump. Na$^+$ reabsorption is followed by water reabsorption as collecting duct is permeable to water. K$^+$ secretion occurs passively. Water reabsorption in the collecting duct also occurs under the influence of the posterior pituitary hormone – ADH (anti-diuretic hormone). ADH stimulates the synthesis and insertion of certain water channels called aquaporin-2 in the 'P' cells of collecting duct, thereby increasing their permeability to water.

'I' cells (intercalated cells) are involved in the acid-base balance. H$^+$- K$^+$ATPase pumps and H$^+$ATPase pumps are present in the luminal cell membrane of I-cells. Both these pumps become active in acidosis so that more H$^+$ will be secreted into the luminal fluid.

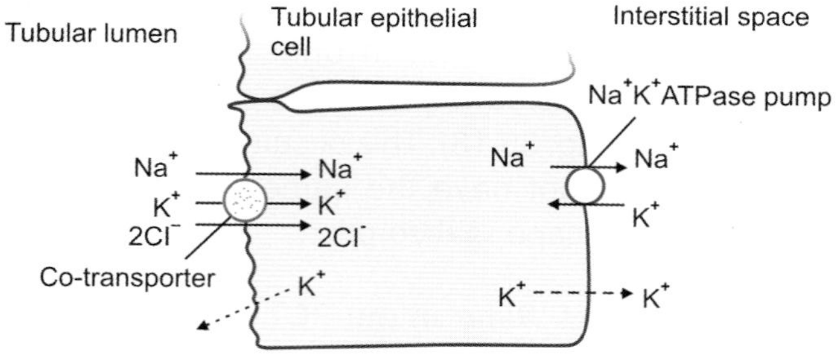

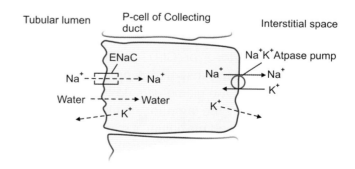

Fig. 35.14: Ion transport in the ascending limb of loop of Henle

Fig. 35.15: Reabsorption of sodium in the collecting duct, stimulated by aldosterone

URINE CONCENTRATING MECHANISM OF KIDNEY

We know that about 180 L/day of plasma is filtered by kidneys. But the daily urine output in an adult is about 1 to 1.5 L, with 290 mOsm/kg water osmolality. Kidneys are able to vary the urine osmolality from a minimum of 30 mOsm/kg water to a maximum of 1200 mOsm/kg water. The corresponding urine volume excreted varies from 23 L/day to about 0.5 L/day respectively. In the following section, let us discuss about the mechanisms involved in the production of concentrated and diluted urine.

There is a gradient of osmolality in renal medulla with the highest osmolality present at the tips of the renal papillae. The medullary hyperosmolality gradient plays an important role in the reabsorption of water and thus, in the concentration of urine. A 'countercurrent mechanism' is working in the kidney which is responsible for producing as well as maintaining the medullary hyperosmolality gradient. The term 'counter current' indicates a system in which the inflow runs parallel to and opposite to the outflow in closely placed structures. Two countercurrent processes occur in the renal medulla – countercurrent multiplication and countercurrent exchange.

COUNTERCURRENT MULTIPLICATION

The long loops of Henle of juxtamedullary nephrons are the countercurrent multipliers which create or produce the medullary osmotic gradient. The filtered fluid flows in opposite direction in the ascending and descending limbs of loop of Henle. Active movements of solutes out of the 'water-impermeable' ascending limb into the medullary interstitium, with passive reabsorption of water from the descending limb contribute to build up the gradient of osmolality. The processes of active reabsorption of solutes in ascending limb work till an osmotic gradient of 200 mOsm/kg water is created between the tubular lumen and medullary interstitium.

Let us imagine a loop of Henle that is filled with a fluid of osmolality about 300 mOsm/kg water, throughout the two limbs of loop of Henle and the medullary interstitium (Fig. 35.16). Now, the active $Na^+K^+2Cl^-$ pumps of thick ascending loop start working, which reabsorb solutes into the medullary interstitium till a gradient of 200 mOsm/kg water is created (Fig. 35.16). Note that the descending limb is impermeable to water. The medullary interstitial osmolality now changes from 300 mOsm/kg water to 400 mOsm/kg water. Since the descending limb is

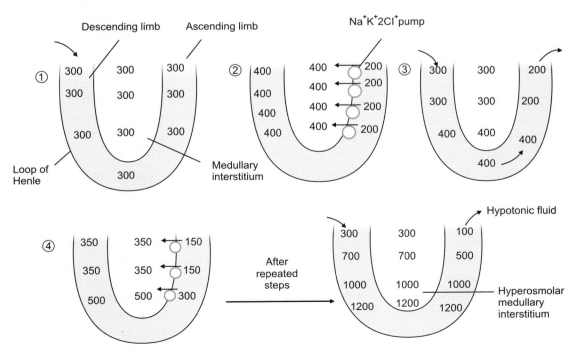

Fig. 35.16: Loop of Henle acting as countercurrent multiplier

freely permeable to water, it diffuses out till an osmotic equilibrium is reached between the descending limb fluid and medullary interstitium, i.e. 400 mOsm/kg water (Fig. 35.16). By the continuous transport of ions out of the ascending loop, medullary osmolality is maintained at 400 mOsm/kg water. Thus the fluid in the descending limb becomes hyperosmotic when compared to that in the ascending limb. This fluid starts flowing out of the loop of Henle and an additional flow of fluid comes from the PCT.

The hyperosmotic fluid (400 mOsm/kg water) in the descending limb now passes to the ascending limb (Fig. 35.16). This disturbs the osmotic gradient between ascending limb and medullary interstitium. Additional solutes are actively reabsorbed into medullary interstitium. In response to the increased interstitial osmolality, once again water reabsorption occurs in the descending limb, till it attains the state of equilibrium (Fig. 35.16). These steps are repeated continuously with the net effect of more and more reabsorption of solutes into the medullary interstitium in excess of water. This gradually multiplies the osmotic gradient till a maximum of 1200 mOsm/kg water is reached, at the tip of the renal papilla (Fig. 35.16).

The extent to which the countercurrent multiplication can establish a large gradient along the loop of Henle depends on several factors, including the length of the loop of Henle and the rate of flow of fluid. The greater the length of loop of Henle, the greater is the osmolality gradient created. If the rate of fluid flow in the loop of Henle is too high, enough time is not available for these processes and consequently the gradient is reduced.

Along with other solutes, urea also plays an important role in increasing the medullary osmotic gradient. Reabsorption of urea is a passive process. In the presence of ADH, the permeability of medullary part of collecting duct to urea increases and urea gets reabsorbed into the medullary interstitium. Along with this, some amount of water is also reabsorbed. A small amount of urea from medullary interstitium now diffuses into the descending limb of loop of Henle (Fig. 35.17). This reaches the collecting duct, where it is again reabsorbed along with water. This is termed as recirculation of urea. Urea recirculates several times before it is excreted. Even though urea is a waste

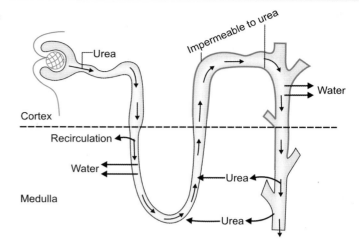

Fig. 35.17: Role of urea in countercurrent multiplication, by recirculation

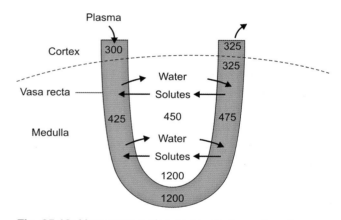

Fig. 35.18: Vasa recta acting as countercurrent exchanger

product, it contributes much in the creation of hyperosmolality in the renal medulla.

Countercurrent Exchange

The vasa recta around the loop of Henle of juxtamedullary nephrons function as the countercurrent exchangers. These long loops of capillaries minimize the wash out of medullary hypertonicity.

Vasa recta lie parallel to the loop of Henle (Fig. 35.18). As blood flows into the medullary region towards renal papilla, solutes from the medullary interstitium are transported into the blood. Also osmotic diffusion of water from vasa recta into the medullary interstitium takes palce. As a result, the blood reaching tips of vasa recta acquires an osmolality of

1200 mOsm/kg water – same as that of medullary interstitium (Fig. 35.18). But as the blood flow ascends back towards the cortex, there is progressive transport of solutes back into the medullary interstitium from blood. Simultaneously, water also moves back into the vasa recta (Fig. 35.18). In other words, the net loss of solutes from medullary interstitium is very minimal and therefore they tend to be trapped in the medulla. Countercurrent exchange is a purely passive process. The sluggish blood flow pattern in the medulla also helps in this process.

ROLE OF ADH IN THE CONCENTRATION OF URINE

ADH from posterior pituitary makes the tubular cells of collecting duct permeable to water (explained before). Whenever the water content in our body increases or the osmolality of ECF decreases, ADH is inhibited and water is not reabsorbed from the collecting duct. This leads to excretion of large amounts of diluted urine.

Applied Physiology

1. *Diabetes insipidus or Water Diuresis:* Already discussed in "Endocrinology"
2. *Osmotic Diuretics:* The presence of large quantities of unabsorbed solutes in the renal tubules, especially in PCT, decreases water reabsorption, leading to increased urine volume. Unabsorbed solutes in PCT exert an osmotic effect so that water remains in the tubular lumen itself. Such a condition is commonly seen in diabetes mellitus. Here, there is excessive filtration of glucose which remains unabsorbed in the PCT. Along with this, Na^+ reabsorption will also be affected as both of them have a common transport protein. The net result is that the loop of Henle receives a fluid with increased volume and osmolality. Reabsorption of water and Na^+ from the loop of Henle is also decreased and therefore medullary hypertonicity decreases. Finally, there is loss of increased volume of urine with Na^+ and other electrolytes.
3. *Diuretics:* Diuretics are agents or drugs which enhance urine output, by increasing the excretion of both solutes and water. Table 35.2 shows some important diuretic agents and their mode of action.

Table 35.2: Diuretic Agents

	Diuretic	Action
1.	Water, Ethanol	Inhibit ADH secretion, causing water diuresis
2.	Mannitol, Glucose	Cause osmotic diuresis
3.	Loop diuretics, e.g:Frusemide	Inhibit $Na^+K^+2Cl^-$ co-transport in ascending loop of Henle
4.	Thiazide diuretics e.g.Chlorthiazide	Inhibit $Na^+ - Cl^-$ symport in distal tubules
5.	Aldosterone antagonists e.g.Spironolactone	Inhibit aldosterone action in collecting ducts

ACIDIFICATION OF URINE

Along with respiratory system and the buffers of body fluids, kidney plays an important role in controlling the acid-base balance. H^+ secretion and HCO_3^- reabsorption occurs in almost all parts of nephron except descending limb of loop of Henle. The mechanisms of H^+ secretion are already discussed. If the ECF H^+ concentration rises above normal, kidneys excrete an acidic urine and vice-versa. The excess H^+ present in the tubular fluid are buffered by the following systems:

HCO_3^- Buffer System

Most of the H^+ secreted in the PCT reacts with HCO_3^- to form H_2CO_3 (carbonic acid)

$$H^+ + HCO_3^- \rightleftharpoons H_2CO_3 \xrightarrow{\text{Carbonic anhydase}} H_2O + CO_2$$

Carbonic acid dissociates into CO_2 and water in the presence of carbonic anhydrase enzyme in the cells.

Dibasic Phosphate Buffer System

The H^+ secreted in the DCT and collecting ducts react with dibasic phosphate (HPO_4^{2-}) present in the tubular fluid to form monobasic phosphate ($H_2PO_4^{2-}$).

$$HPO_4^{2-} + H^+ \longrightarrow H_2PO_4^{2-} + Na^+$$
$$\downarrow$$
$$NaH_2PO_4$$

NaH_2PO_4 is a weak acid, causing only a slight increase in pH.

Ammonia Buffer

In the renal tubular cells, ammonia is formed by deamination of the amino acid-glutamine.

$$\text{Glutamine} \xrightarrow{\text{Glutaminase}} \text{Glutamate} + NH_3$$

Since ammonia is lipid soluble, it can freely diffuse through all cell membranes. Hence it is secreted into the tubular fluid. There, it combines with H^+ to form ammonium (NH_4^+), which is not lipid soluble. Therefore NH_4^+ stays in the tubular fluid itself. This is called non-ionic diffusion.

$$NH_3 + H^+ \longrightarrow NH_4^+$$

Constituents of Urine

Normal urine output in adult is about 1 to 1.5 L/day. Urine is yellowish in color due to the presence of urochrome. It is acidic in nature, but the pH can vary widely from 4 to 8 according to the metabolic activity of body. Vegetarian diet makes the urine alkaline and a high protein diet makes it acidic. The osmolality of urine also varies widely between 30 mOsm/kg to 1200 mOsm/kg.

Micturition

Micturition is the process by which the urinary bladder empties when it becomes filled with urine. The urinary bladder fills progressively until the pressure inside it (called intravesical pressure) rises above a particular threshold level. Then it initiates the micturition reflex as follows: Several stretch receptors are present in the bladder wall which get stimulated when it is filled with urine. They send signals to the "micturition center" in the spinal cord via pelvic nerves and micturition contractions are initiated in the bladder. Initial contractions will further stimulate the stretch receptors and so micturition contractions are said to be self-regenerative. This lasts for about a few seconds to one minute. As the bladder becomes more and more filled, micturition reflexes occur more frequently and more powerfully, and urge to urination (micturition) occurs.

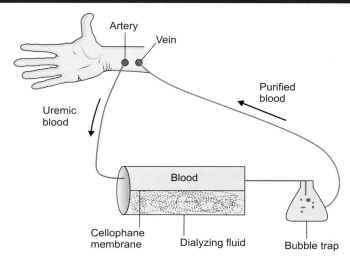

Fig. 35.19: Principle of dialysis

The micturition reflex is an autonomic spinal cord reflex, but it can be suppressed or facilitated voluntarily by several centers in brain, including cerebral cortex, posterior hypothalamus and midbrain. They keep the micturition reflex partially inhibited except when it is desired. Micturition can also be initiated voluntarily by contraction of abdominal muscles. After micturition, female urethra empties by gravity and male urethra, by contraction of bulbocavernosus muscle.

Dialysis (Artificial Kidney)

Dialysis is the process by which the waste products of metabolism accumulated in the blood are removed using an external device. This is usually done in severe renal failure, when kidneys cannot function adequately.

Blood from an artery is passed through a dialyzer unit. It consists of minute channels bounded by a thin semi-permeable membrane. On the other side of the membrane is a dialyzing fluid. The concentration of glucose, bicarbonate and calcium are high in the dialyzing fluid, whereas concentration of Na^+, K^+ and Cl^- are kept less. When blood passes through the dialyzer, the unwanted metabolic waste products and other solutes diffuse into the dialyzing fluid. The purified blood is then returned back through a vein (Fig. 35.19).

36 Temperature Regulation

INTRODUCTION

Man as well as other mammals are said to be homeo-thermic, i.e. their body temperature will not vary according to the environment. The thermoregulatory mechanism helps to keep the body temperature within a narrow range.

NORMAL TEMPERATURE

In healthy adults, normal oral temperature is 37°C (98.6°F). Rectal temperature is usually 0.5°C more than that of oral temperature. It is nearest to the core temperature (internal temperature).

Body temperature shows a circadian rhythm. It is usually highest in the afternoon and evening, and least at 5 am. Physiological variation in temperature occurs after exercise, in cold seasons, in females during the luteal phase of menstruation and in infants.

SOURCES OF BODY HEAT

Heat gain by the body occurs during rest as well as during physical activities. At rest, heat is produced by various metabolic reactions occurring in our body, including digestion. The internal muscular activities like pumping of heart, respiration, peristalsis, etc. can

generate heat. Also, heat is gained from the environment by radiation and conduction.

Brown fat is present in infants, which is an important source of heat production. Brown fat has very high metabolic rate and is located between and around the scapulae and neck, behind the sternum and around the kidneys.

CHANNELS OF HEAT LOSS

Heat is lost from the body through skin by conduction, convection, radiation and evaporation. Heat is also lost through urine, feces and the respiratory tract.

CONTROL MECHANISMS OF THERMOREGULATION

It is mainly by the hypothalamic thermostat and by the cutaneous response. There are hormonal as well as behavioral mechanisms for thermoregulation. All these mechanisms involve cutaneous responses.

Hypothalamic Thermostat

Hypothalamus plays an important role in keeping the body temperature near to the constant level. Normally the thermostat is set at 37°C. Various experimental studies have proved that the anterior hypothalamus integrates the mechanisms responding to increase in body temperature or exposure to hot environment. Posterior hypothalamus integrates body responses to cold. Both these responses consist of autonomic, somatic, hormonal and behavioral mechanisms. Cold and warm receptors present in the skin, abdominal visceral organs, internal organs and in the spinal cord sense the temperature changes and activate the hypothalamus.

MECHANISMS ACTIVATED BY HEAT

Heat gain by the body activates the anterior hypothalamus which causes the following:

Cutaneous Vasodilatation

By this process, the core heat is brought to the skin surface so that it can be lost to the surroundings by conduction, convection and radiation.

Sweating

This is the only mechanism of heat loss when a person is exposed to the environmental temperature greater than the body temperature.

Anorexia

Anorexia or loss of appetite will results in decreased food intake and decreased metabolic reactions, thus causing decreased heat production.

Behavioral Responses

This includes moving to a shade or cooler place and decreased muscular activities. All these decrease the body temperature.

Panting

This is seen in certain animals like dogs and not in humans. Panting means rapid shallow breathing. This can cause increased evaporation through the mouth and respiratory passages.

MECHANISMS ACTIVATED BY COLD

Exposure to cold can reduce the body temperature, which activates the following mechanisms to occur.

I. Mechanisms for heat conservation.
II. Mechanisms for heat production.

Mechanisms for Heat Conservation

Cutaneous Vasoconstriction

Activation of posterior hypothalamus increases the sympathetic discharge, when the body is exposed to a cold atmosphere. This can cause marked vasoconstriction of the cutaneous blood vessels throughout the body, thus decreasing the cutaneous blood flow. As a result, heat loss from the core of the body to the cold environment is prevented.

Piloerection

Cold can induce contraction of erector pili (which is a cutaneous muscle). The body hairs stand up forming an additional insulating layer to conserve heat ("Goose Flesh").

Measures for Increasing Heat Production

Shivering

Shivering is due to increased discharge from the anterior horn cells of spinal cord, causing increased muscle tone throughout the body. This occurs when the cutaneous vasodilatation becomes insufficient to conserve heat. Shivering can increase the heat production by 4 to 5 folds.

Increased Catecholamine Secretion

Cold stimulates secretion of catecholamines (Epinephrine and Norepinephrine) , which may induce chemical thermogenesis, by increasing the rate of cellular metabolism. This can increase the heat production by 10 to 15 percent. The heat production is doubled by this mechanism in infants.

Increased Secretion of Thyroxine

Continuous exposure to severe cold can increase the secretion of thyroid stimulating hormone (TSH), which can cause hyperplasia of thyroid gland. In infants, thyroxine production increases even for a short exposure of cold.

Heat Production by the Brown Fat

This occurs in infants in whom shivering does not occur. Increased metabolism of brown fat in infants causes increased heat production and this is called non-shivering thermogenesis.

Applied Physiology

1. *Fever:* Fever means, increase in body temperature above 99°F. Causes of fever are bacterial or viral infections, protozoal infections, tissue destruction, certain lesions of the central nervous system or due to certain drugs. Fever is due to increased production of heat by shivering and raised metabolism and decreased heat loss by cutaneous vasoconstriction.

2. *Mechanism*: The toxins liberated by the invading micro-organisms (endogenous pyrogen) reset the normal set point of hypothalamus (of 37°C) to a higher level. So, initially the patient feels chills and rigor. This can elevate the core temperature to the newly set point of hypothalamus. Now the skin becomes warm and flushed. Fever often subsides with sweating which causes heat loss. Fever is beneficial to the body as it can inhibit bacterial growth, promote antibody formation and increase the release of leukocytes from bone marrow.

3. *Heat Exhaustion:* Heat exhaustion is due to exposure to heat in a humid atmosphere. This causes increased sweating with salt and water loss. Symptoms like headache, dizziness, cramps, etc. occur. Skin becomes cold and clammy. This can even cause circulatory collapse.

4. *Heat Stroke:* Heat stroke occurs when heavy work is done along with prolonged exposure to hot and humid atmosphere. There is a sudden rise in temperature to above 41°C. This is a medical emergency and death occurs, if untreated. The person should be cooled immediately by moving him to a cooler place or by keeping ice packs.

5. *Hypothermia:* Hypothermia occurs due to prolonged exposure to cold, and the body temperature falls below 35°C. When the temperature falls < 25°C, death occurs. Induced hypothermia is used for a short duration in cardiac and brain surgery.

MODEL QUESTION PAPERS
(Physiology)

<u>MODEL QUESTION PAPER – I</u>

Time : 2 Hours   *Maximum : 75 Marks*

SECTION A

I. Define erythropoiesis. Discuss the factors influencing it. Add a note on reticulocyte. Give a schematic representation of different stages of erythropoiesis.   (2 + 5 + 3 + 3 = 13 marks)

II. Write short notes on:
   a. Lung volumes and capacities
   b. Second phase of deglutition
   c. Arterial pulse
   d. Juxtaglomerular apparatus (5 × 4 = 20 marks)

III. Write normal values of (including units)
   a. Glomerular filtration rate
   b. Duration of a cardiac cycle
   c. RMP of a nerve
   e. WBC count in an adult male (1 x 4 = 4 marks)

SECTION-B

IV. Name the hormones secreted from adrenal cortex. Describe the Physiological and Pharmacological actions of glucocorticoids.   (3 + 7 = 10 marks)

V. Write short notes on:
   a. Functions of thalamus
   b. Organ of corti
   c. Ovarian changes in menstrual cycle
   d. Visual pathway and lesions
                                    (5 × 4 = 20 marks)

VI. Name the following:
   a. Lens used in the correction of astigmatism
   b. Hormone responsible for milk ejection reflex
   c. Vitamin playing significant role in blood coagulation
   d. Abnormality due to increased growth hormone secretion after epiphyseal closure
   e. Volume of air present in the lungs even after a forceful expiration
   f. Macrophages present in liver
   g. Hormone responsible for water reabsorption from the collecting ducts of kidney
   h. A digestive enzyme present in saliva
                                    (1 x 8 = 8 marks)

MODEL QUESTION PAPER-II

Time: 2 hours   ***Maximum Marks: 75***

(Draw diagrams wherever necessary)

SECTION-A

I. Define blood pressure. Give its normal value. What is pulse pressure? Describe the mechanisms of regulation of blood pressure.

(2 + 1 + 2 + 7 = 12 marks)

II. Write short notes on:
   a. Factors affecting GFR
   b. Digestion of carbohydrates
   c. Immunity by T-lymphocytes
   d. Active transport mechanisms

(5 × 4 = 20 marks)

III. Write normal values of (including units)
   a. Platelet count
   b. Cardiac output in a young adult male
   c. Vital capacity in a young adult male
   d. PR interval   (1 × 4 = 4 marks)

SECTION – B

IV. Name the sensations carried by posterior column fibers. With the help of a neat diagram, trace the pathway   (4 + 7 = 11 marks)

V. Write short notes on:
   a. Indicators of ovulation
   b. Actions of insulin
   c. Functions of cerebellum
   d. Hypothyroidism

(5 × 4 = 20 marks)

VI. Name the following:
   a. Neurotransmitter secreted by postganglionic sympathetic nerve fibers
   b. Hormone responsible for milk secretion
   c. Tract which transmits impulses for pain sensation
   d. Protein present in the thick filament of muscle fiber
   e. Ion responsible for pacemaker potential in cardiac muscle
   f. Lens used in the correction of hypermetropia
   g. Cells producing antibodies
   h. A hormone secreted from kidney

(1 × 8 = 8 marks)

MODEL QUESTION PAPER – III

Total Marks : 50

1. Describe the mechanism of blood coagulation by extrinsic and intrinsic pathways. Add a note on hemophilia.   (17 marks)
2. Explain the formation, circulation and functions of CSF. Add a note on blood-brain barrier.

(15 marks)

3. Write short notes *on any three* of the following:
   a. Define
      1. Receptors
      2. Reflex
   b. Name four gastrointestinal hormones
   c. Functions of sertoli cells
   d. Functions of pancreatic juice   (3 x 6 = 18)

MODEL QUESTION PAPER – IV

Total Marks : 50

1. Name the ascending tracts of spinal cord. Trace the pathway for pain, with the help of a neat, labelled diagram. (15 marks)
2. Define cardiac output. Give its normal value. What are the factors affecting it ? (17 marks)

3. Write short notes on :
   a. Functions of growth hormone
   b. Morphological classification of leukocytes and their functions
   c. Uterine changes during menstrual cycle.
   (3 × 6 = 18)

MODEL QUESTION PAPER – V

Total Marks : 50

1. Name the hormones secreted from anterior pituitary. Describe the functions of thyroid hormone. Add a note on its abnormalities. (17 marks)
2. What is the normal respiratory rate? Describe the chemical regulation of respiration. (13 marks)

3. Write short notes on :
   a. Spermatogenesis
   b. Tests for hearing
   c. Taste pathway
   d. Factors affecting glomerular filtration rate.
   (4 × 5 = 20)

MODEL QUESTION PAPER – VI

Total Marks : 50

1. Define blood pressure. Give its normal value. Describe the factors influencing it. Add a note on hypertension. (1 + 1 + 8 + 2 = 12)
2. Name the descending tracts of spinal cord. Describe the pyramidal pathway and its function. Differentiate lower motor neuron lesions from upper motor neuron lesions. (2 + 8 + 3 = 13)
3. Draw a labelled diagram of :
   a. Visual pathway
   b. Important cortical areas of brain   (2 × 5 = 10)

4. Write the normal values of:
   a. Renal blood flow
   b. Vital capacity
   c. Sperm count
   d. Stroke volume
   e. Leukocyte count (5 × 1 = 5)
5. Write short notes on :
   a. Ovulation
   b. Rh incompatibility (2 × 5 = 10)

MODEL QUESTION PAPER – VII

Total Marks : 50

1. Describe the different phases of menstrual cycle and their hormonal background. (17 marks)
2. Draw a labelled diagram of the nephron. Give an outline of the process by which urine is formed. (15 marks)

3. Write short notes on :
   a. Normal spirogram
   b. Milk ejection reflex
   c. Functions of cerebellum
   d. Gastric juice
   e. Actions of testosterone
   f. Oral contraceptive pills (6 × 3 = 18 marks)

MODEL QUESTION PAPER – VIII

Total Marks : 25

1. Describe the process of erythropoiesis. Enumerate the factors affecting it.   (7 marks)
2. Illustrate lung volumes and capacities. Discuss lung function tests.   (3 + 4 = 7 marks)
3. Define blood pressure. Describe the factors affecting it.   (5 marks)
4. Write short notes on any two.   (3 × 2 = 6 marks)

   a. Placenta
   b. Ovarian cycle
   c. Intestinal motility
   d. Functions of liver
   e. Jaundice

MODEL QUESTION PAPER – IX

Total Marks : 25

1. Describe the intrinsic mechanism of blood coagulation. Mention the effect of heparin on it.   (3 + 4 = 7 marks)
2. Discuss the physiological basis of artificial respiration. Add a note on acute oxygen toxicity.   (3 + 4 = 7 marks)
3. Describe the movements of small intestine and discuss their neural control.   (5 marks)

4. Write short notes on any two: (3 × 2 = 6 marks)
   a. Blood volume
   b. Micturition reflex
   c. Function of skin
   d. Milieu Interior
   e. Antidiuretic hormone

MODEL QUESTION PAPER – X

Total Marks : 25

1. Enumerate the chemicals regulating respiration. Describe the role of one of them.   (3 + 4 = 7 marks)
2. Draw diagrams to illustrate
   a. Structure of eye
   b. Pathway for pain and touch   (3 + 4 = 7 marks)
3. Discuss the factors affecting cardiac output   (5 marks)

4. Write short notes on any two.   (3 + 2 = 6 marks)
   a. Motor unit
   b. Function of hypothalamus
   c. Pain
   d. Middle ear
   e. Juxtaglomerular apparatus

Papers

Physiology

Annexure

NORMAL VALUES (IN A HEALTHY YOUNG ADULT)

Blood, Formed Elements

Parameters	Normal values
pH of blood	7.4
Specific gravity	1.055-1.065
Osmotic pressure	290 milliosmoles/L
Viscosity	3.4-5.4
Plasma proteins	6.8 gm/100 ml
ESR	3-5 mm at the end of 1hour (in males)
	5-12 mm at the end of 1 hour (in females)
PCV	45% (males)
	42% (females)
Diameter of RBC	7.5 µm
RBC count	4.4-5.5 million/mm^3 of blood (in males)
	4-5 million/mm^3 of blood (in females)
Hb	14-18 gm% (in males)
	12-16 gm% (in females)
Total WBC count	4000-11,000 /mm^3 of blood
Differential count	Neutrophils: 50-70%,
	Eosinophils : 1-4%
	Basophils : 0-1%; Lymphocytes : 20-40%
	Monocytes : 2-8%
Platelet count	1.5-4 lakh/mm^3 of blood

Respiratory System

Parameters	Normal values
Tidal volume	0.5 L (males), 0. 5 L (females)
Inspiratory reserve volume	3.3 L (males), 1.9 (females)
Expiratory reserve volume	1 L (males); 0.7 (females)
Reserve volume	1.2 L (males); 1.1 (females)
Total lung capacity	6 L (males), 4.2 L (females)
O$_2$ content in 100% saturated Hb	1.34 ml

Cardiovascular System

Parameters	Normal values
PR interval	0.12 to 0.2 sec
QRS duration	0.08 to 0.1 sec
QT interval	0.4 sec
ST segment	0.32 sec
Duration of cardiac cycle	0.8 sec
Cardiac output	5 L/min
Stroke volume	70 ml
End diastolic volume	130 ml
Blood pressure	120/80 mmHg
Heart rate	72/min

Urinary System

Parameters	Normal values
Renal blood flow	1.2 L/min
GFR	125 ml/min or 180 L/day
Renal threshold for glucose	180 mg/dL
Tubular maximum for glucose	375 mg/min
Urine output	1.5 L/day

Index

Page numbers followed by *f* refer to figures and *t* refer to tables

Index

Index

Lubrication 445
Lumbar
    plexus 101, 305-307f
    puncture 278
        needle 352
    subarachnoid block 292
    sympathectomy 310
    triangle 107
    vertebrae 85
Lumbrical muscles 111, 112
Lung
    capacities 464
    volumes 464
Lungs 190, 191f
Luteinizing hormone releasing hormone
        314
Lymph nodes 33
Lymphatic drainage 39, 204, 213, 254,
        259, 269, 316
    of lungs 196
    of stomach 213f
Lymphocytes 16, 431
    of thymus 40
Lymphoid tissue 33
Lyon hypothesis 340
Lysosomes 388

M

Mackenrodt's ligaments 256
Macrocytosis 419
Macroglia 27
Macrophages 16
Macula densa 557
Maintaining blood pressure 490
Major
    endocrine glands 312
    salivary glands 207
Male
    pelvis 89f
    reproductive system 265, 516
    sterilization 526
    urethra 248, 249, 273
Malignant
    disease of skin 45
    melanoma 45
Malpighian corpuscle 242, 554
Mammary gland 44, 263
Mammography 265
Management of shock 496
Mandible 74
Mandibular nerve 73, 295
    block 296
Manubrium 80
    sterni 189
Marfan's syndrome 349
Marginal mandibular 298

Mass peristalsis 457
Mast cells 16
Mastication 454
Masticatory muscles 93
Mastoid
    angle of parietal bone 69
    part of temporal bone 69
Maternal serum screening 351
Mature
    cells 18, 21
    ovarian follicle 261, 261f
Maxilla 75, 76f
Maxillary
    artery 154
    nerve 295
    sinus 183
McBurney's point 222f
Mean
    arterial pressure 489
    corpuscular
        hemoglobin 427
        volume 427
Measurement of
    cardiac output 488
    pulmonary ventilation 463
Mechanics of pulmonary ventilation 461
Mechanism of
    action of hormones 498
    coagulation 433
    secretion of HCl 446
Meckel's diverticulum 239
Medial
    and lateral
        menisci 136
        plantar arteries 165
        pterygoid 94
    cuneiform 63
    meniscus 136
    plantar artery 165
    pterygoid 94
    rectus 330
    rotation of hip 116
    walls of orbit 77
Median
    cubital vein 171
    nerve 112, 131
        in carpal tunnel 304f
Mediastinal
    pleura 197
    surface 192
Mediastinum 187
Medium sized muscular artery 151
Medulla 34, 261, 319
    oblongata 276, 288
Medullary respiratory neurons 468f
Megaloblastic anemia 426
Membrane carbohydrates 386

Membranous labyrinth 327
Meninges 276
    and dural folds 276f
Menopause 522
Menstrual cycle 520
Merkel cell endings 45
Mesentery 228
    of jejunum and ileum 219
Mesoderm 9
Metacarpal bones 57
Metaphyseal arteries 23
Metatarsal bones 62, 64
Methods of measurement of BP 489
Microcytosis 419
Microglia 27
Microscopic structure of
    body of stomach 214f
    compact bone 21
    palatine tonsil 39
    spleen 37
    thyroid 316f
    uterine tube 259f
    uterus 257
Micturition 566
Mid inguinal point 359
Midbrain 276, 287
Middle
    cardiac vein 149
    cerebral artery 282
    colic artery 162
    ear 324, 547
    layer 541
    meatus 182
    mediastinum 187
    meningeal artery 155
    point of inguinal ligament 359
    radioulnar joint 129
    septum 187
    zonafasciculata 319
Mid-trimester termination of pregnancy
        352
Millard Gubler syndrome 288
Minor salivary glands 207
Miscarriage 352
Mismatched blood transfusion 437
Mitochondria 388
Modified
    sebaceous glands 44
    sweat glands 44
Monocytes 430
Mons pubis 252, 253
Morison's pouch 229
Morris' parallelogram 241f
Motor
    nerve supply 204
    speech area of broca 281
    system 534
    unit 534

Index

DATE DUE

Best Selling Dictionary for Nurses

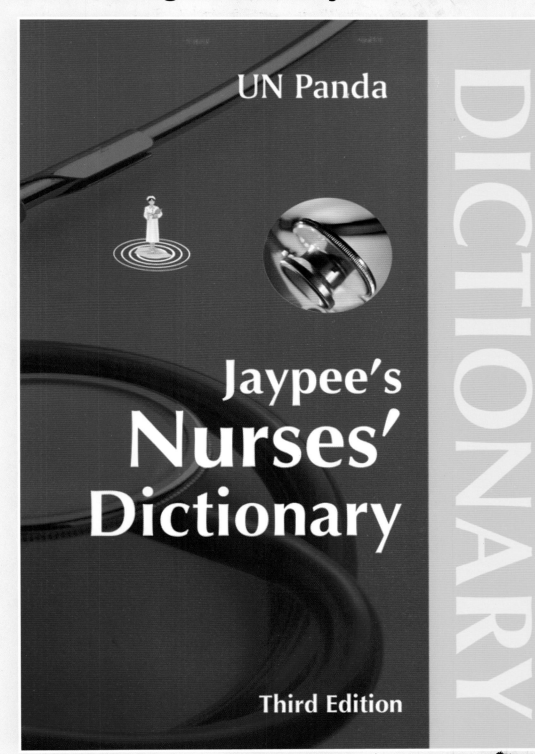

UN Panda

Jaypee's
Nurses'
Dictionary

DICTIONARY

Third Edition

Jaypee's Nurses' Dictionary

UN Panda

Third Edition

ISBN 978-81-8448-601-8

Textbook of
Anatomy & Physiology
for Nurses

Highlights of Third Edition

- Includes five new chapters: The Cell, Transport Across the Cell Membrane and the Resting Membrane Potential, The Nerve Physiology, The Muscle Physiology and Immunity
- 122 new color photographs on bones and histology slides
- 600 neatly labeled diagrams to make the study easy and interesting
- Notes on Embryology
- More Model Question Papers for self-assessment.

PR Ashalatha graduated from Calicut Medical College, Kerala, India, in the year 1988. She did her Masters in Anatomy in 1999 and is, at present, working as Associate Professor at Calicut Medical College. Her area of interest, besides teaching, is Cytogenetics. She has completed WHO In-country Fellowship in Genetics at AIIMS, New Delhi, India. She has also trained in FISH from CMC Vellore, India.

G Deepa graduated in Medicine from TD Medical College, Alappuzha, Kerala, India, and did her Masters in Physiology from Calicut Medical College. After Postgraduation in 2006, she joined Sree Gokulam Medical College and Research Foundation, Thiruvananthapuram, Kerala, where she worked as Assistant Professor for nearly five years. In 2010, she entered State Government service as Assistant Professor in Physiology at Medical College, Thiruvananthapuram, Kerala, India.

Available at all medical book stores
or buy directly from Jaypee Brothers through online shopping
at www.jaypeebrothers.com

or call + 91-11-32558559

JAYPEE BROTHERS
Medical Publishers (P) Ltd.
www.jaypeebrothers.com

ISBN 978-935025-423-3